Mathematics

ITS POWER AND UTILITY

SIXTH EDITION

MATHEMATICS

ITS POWER AND UTILITY

Karl J. Smith

Santa Rosa Junior College

Australia • Canada • Mexico • Singapore • Spain • United Kingdom • United States

Publisher: *Robert W. Pirtle*
Marketing Team: *Leah Thompson/Samantha Cabaluna*
Editorial Assistant: *Erin Wickersham*
Production Editor: *Tessa McGlasson Avila*
Production Service: *Susan L. Reiland*
Manuscript Editor: *Christine Levesque*
Permissions Editor: *Lillian Campobasso*
Interior Design: *Carolyn Deacy*
Cover Design: *Laurie Albrecht*
Cover Photo: *Steven Hunt/Image Bank*
Interior Illustration: *Lori Heckelman; Kathi Townes, TECHarts*
Cartoons: *Ryan Cooper*
Photo Researcher: *Terry Powell*
Print Buyer: *John Cronin*
Typesetting: *York Graphic Services, Inc.*
Printing and Binding: *R. R. Donnelley/Crawfordsville*

For more information, contact:
BROOKS/COLE
511 Forest Lodge Road
Pacific Grove, CA 93950 USA
www.brookscole.com

Printed in the United States of America

10 9 8 7 6 5 4 3 2 1

Library of Congress Cataloging-in-Publication Data

Smith, Karl J.
Mathematics: its power and utility / Karl J. Smith—6th ed.
Includes index.
ISBN 0-534-36455-1
1. Mathematics. I. Title.
QA39.2.S598 2000 99-34743
510—dc21 CIP

I dedicate this book
with love to my son, Shannon,
and his wife, Jane.

PREFACE

When we learn to drive a car we are able to "go places" easily and pleasantly instead of walking to them with a great deal of effort. And so you will see that the more mathematics we know the easier life becomes, for it is a tool with which we can accomplish things that we could not do at all with our bare hands.

LILLIAN R. LIEBER, *THE EDUCATION OF T. C. MITTS*

Situation

"What are you taking this semester?" "Well, I'm taking English and history, I made the swimming team, and I finally get to take band. Oh, I forgot, I also ***have*** *to take math."*

In this book we look at the power *of mathematics, not the drudgery of number crunching. We look at it in historical perspective, and we see how mathematics has revolutionized the world. We also look at the* utility *of mathematics, and you are NOT told that you should do this just because it might be useful* someday. *The first part of the book, The Power of Mathematics, develops some ideas in arithmetic, algebra, and geometry, and puts these ideas into a HISTORICAL PERSPECTIVE. The second part of the book, the Utility of Mathematics, develops the ideas around some SITUATION with which I hope you, the student, can identify. Each section begins with a SITUATION, develops some related concepts, and then concludes (in the problem set) by asking you to answer a question introduced in the SITUATION and by providing suggestions for writing a related paper.*

As a teacher and author, I am faced with transferring some knowledge of the **power,** the beauty, and the **utility** of mathematics to my students. But I can't instruct until my students have overcome their initial fears and attitudes about mathematics and are *ready* to listen. This week, I received a phone call from a student in Colorado complaining about the cost of this book. He asked if I could lower the price, and he told me that as an art student he would use this book for only 12 weeks and then the book would be worthless (or at best he could sell it back for pennies on the dollar). He also went on to tell me that if he were not required to spend the money for this book, he could use the money to buy some art supplies, which would allow him to do something that was really important to him. This is precisely the student that

I had in mind while I was writing this book. I know books are expensive. I know education is expensive. The goal of this book is to help the caller, and students like him, realize that there is worth in learning mathematics. My message to this student is my firm belief that if he were to take the content of this book seriously, his investment would be returned many times over in a very real financial way. A student using this book, regardless of academic major, should learn *something* that will make life easier, less costly, or more efficient. This book is not about "doing homework." It is about learning essential skills for living.

Education is much more than books, classrooms, teachers, and requirements—it is about knowledge and being able to use that knowledge. In 1958, when I was just starting to study high school mathematics, E. G. Begle, the leader of the School Mathematics Study Group (SMSG), wrote something that, at the time, meant little to me: "The world of today demands more mathematical knowledge on the part of more people than the world of yesterday, and the world of tomorrow will demand even more." Well, today is the world of tomorrow, and that prediction is absolutely true. It is my belief that it is even more true for the next generation as we enter the 21st century.

I have often wondered why, when I tell people that I'm a math teacher, more often than not I hear about their unpleasant experiences in mathematics. Why is it that we *get* to participate in sports, or art, or music, but ***have*** to take mathematics? For most people, mathematics is presented as the ultimate lesson in delayed gratification. Year after year, the students are told, "You must learn this thoroughly so you can do your math tomorrow, next week, next month, or next year." The implication is that in twenty or so years, we'll let you see the benefit of what you have been learning.

As I began to write this book, I asked myself, What do my students *need*? Why should there be a new mathematics book in the world? What can I hope to present that is not already available? I believe too much lip service is paid to "getting past the hurdles," and to asking, How little can I do and still finish this course? This book was written to give the student who has previously not been successful with mathematics a fresh and innovative approach to arithmetic, beginning algebra, and geometry to satisfy basic competencies in mathematics in today's world. I have long believed that, if students have avoided or have been unsuccessful with a particular aspect of mathematics in the past, simply to present it to them again in the same way will generally not meet with a great deal more success. Nevertheless, because of the tremendous importance of mathematics in the world today, a student must have some degree of mathematical competency to be successful in almost any discipline. Therefore, even though I have presented the essential ideas of arithmetic, beginning algebra, and geometry, I have presented them in settings different from the usual or traditional ones.

My students need to be able to relate to the real world. You will find this book filled with practical information rather than abstract (and meaningless) made-up word problems. For example, when we are looking at graphs (page 455), we consider graphs showing the blood alcohol level and its effects on driving when drinking or political graphs taken from a newspaper.

My students need to be able to solve problems ***outside*** *the classroom.* Take a look at the word problems we consider in Section 3.6. You will *not* find the usual mixture, distance, and age problems. Instead, we develop a *procedure* and work problems dealing with miles per gallon, price comparisons, and money exchange. We seek to develop higher-order reasoning in which the student must use, explain, and exploit newly learned knowledge. For example (page 403), we motivate probability by considering a Monopoly game.

My students need to be able to estimate. I believe this is one of the most avoided topics in mathematics textbooks. I have spent a great deal of effort discussing this topic, and you will notice that many of the problem sets have multiple-choice questions that ask the students to estimate, not calculate. My students must develop a *number sense,* which is different from "getting right answers." Answers in the back of the book, are, of course, desirable and necessary, but in the book of life and problem solving, there are no "answers in the back."

My students need money sense. Why do we learn to solve equations, but do not learn how to handle credit cards or purchase a car or home? (See Chapter 7.) I believe both skills are important in today's world.

My students need to develop communication skills and learn to think critically. In many of the sections of this book, I have assigned problems that solicit a written paper or a book report. Problems designated IN YOUR OWN WORDS encourage students to communicate mathematical ideas by using their own words. Students must be able to *communicate* in mathematics and should not stop after they can "find a few right answers." Throughout the book, problems encourage a higher-order reasoning. For example, we analyze the lyrics of "By the Time I Get to Phoenix," or look for the errors in an Anacin or a Saab advertisement. Finally, throughout the book, I have included problems labeled WHAT IS WRONG?, which include many common mistakes that must be avoided to understand the material. In Appendix E, I have included *all* the answers for this type of problem.

I believe we (teachers and authors) have done a disservice to most of our past students. We have taught mathematics as if we were preparing our students for a career in mathematics, when instead we should have been teaching them an appreciation of mathematics. It is not necessary to learn all the technical details to use and appreciate math. For example, most people like music of some kind; almost anyone can find a radio station playing something they like to hear. Suppose we were to learn about music in the same way we learn about mathematics. Suppose we said that before you listen to the radio, you had to take three years of sand blocks and recorder, followed by five years of practice piano (graded by your proficiency at scales and rhythm), and finally after some high school courses on theory of music, you would be allowed to hear your first composed piece. On the contrary, composers, musicians, conductors, and music teachers see themselves as engaged in a common enterprise of bringing music to the world, and they believe that whatever techniques they are trying to teach, it should always be in the context of the wonderful sound of music. We should expect no less of mathematics.

ORGANIZATION OF THIS BOOK

The main theme throughout the book is *problem solving.* In Part I, The Power of Mathematics, we begin by discussing math anxiety and how to formulate the problem. The most difficult first step for many students is to determine exactly the nature of the problem. All too often we try to solve a problem before we are even sure what we are trying to solve. Techniques from arithmetic, algebra, and geometry are all applied to problem solving.

These techniques of problem solving are then used in the second part of the book, The Utility of Mathematics. Each topic in this part of the book was selected because of its usefulness to the student. The topics include managing money using the ideas of interest, installment buying, credit cards, inflation, buying a car or home, sets, probability, contests, statistics, and surveys, as well as the influence of these topics on our lives.

The material of this book can be adapted to almost any course arrangement. Chapter 1 on calculators and arithmetic is required for the rest of the book, but it may be treated lightly or skipped by those familiar with its contents. Topics from beginning algebra are presented in Chapters 2, 3, and 4 and are then used in developing much of the material that follows. For example, percents in Chapter 4 are described with proportions and simple equations. Chapter 5 introduces the ideas of geometry and measurement in a practical and down-to-earth manner. Chapters 7–11 give the students a chance to use mathematics in a variety of ways including interest, consumer applications, sets, logic, probability, statistics, and graphs. I have written this book with the idea that different classes will pick different topics as interest and competency requirements dictate. These chapters are independent and can be covered in any order.

WHAT IS NEW TO THIS EDITION

This book has been very successful, and most users and reviewers offered only minor suggestions for this edition. I have added new sections on subsets, operations with sets, Venn diagrams, survey problems using sets, odds and conditional probability, measures of central tendency, measures of position, and measures of dispersion.

ACKNOWLEDGMENTS

I am most grateful to those who have assisted me in the development of this material: to my students who made many valuable suggestions about the material, and especially those students who were bold enough to share their fears and anxieties with me; to my colleagues who shared ideas and teaching suggestions; and to the reviewers who offered many valuable suggestions. In particular, I would like to thank Sharon Butler, Pikes Peak Community College; Rebecca Farrow, John Tyler Community College; Ann Loving, J. Sargeant Reynolds Community College; Carl Miller, TESST Technology Institute; Charles Peselnick, DeVry Institute; and Richard Watkins, Tidewater Community College.

I would like to thank the reviewers of previous editions: Carol Achs, Mesa Community College; Mari Jo Baker, Illinois Central College; John Beris, John Tyler Community College; Eugene Brown, Northern Virginia Community College; Elton

Beougher, Fort Hays State University; James W. Brown, Northern Essex Community College; Mary Clarke, Cerritos College; Gary Donica, Florence-Darlington Technical College; William Durand, Henderson State College; Carolyn Ehr, Fort Hays State University; Rebecca Farrow, John Tyler Community College; Margaret Finster, Erie Community College; Ben V. Flora, Jr., Moorhead State University; Elton Fors, Northern State College, South Dakota; Roy E. Garland, Minnersville State College; Armando Gingras, Metropolitan State College, Denver; Donald K. Hostetler, Mesa Community College; Dorothy Johnson, Cleveland State University; Glenn E. Johnston, Moorhead State University; Keith S. Joseph, Metropolitan State College, Denver; Virginia Keen, Western Michigan University; Roxanne King, Prince George's Community College; Genevieve M. Knight, Hampton Institute; Harvey Lambert, University of Nevada, Reno; Lara Langdon, J. Sargeant Reynolds Community College; Robert Levine, Community College of Allegheny County; Ann Loving, J. Sargeant Reynolds Community College; Chicha Lynch, Capuchino High School; Tucker Maney, Northern Virginia Community College; Sandra Manigault, Northern Virginia Community College, Annendale Campus; DeAnne Miller, Western New Mexico University; Glen E. Mills, Pensacola Junior College; Carol B. Olmstead, University of Nevada, Reno; Maryanne C. Petruska, Pensacola Junior College; Diana Pors, Eastside Union High School; Jane Rood, Eastern Illinois University; George C. Sethares, Bridgewater State College; Clifford Tremblay, Pembroke State College; Richard Watkins, Tidewater Community College; Erma Williams, Hampton University; and Susan Wood, J. Sargeant Reynolds Community College.

The production staff at Brooks/Cole also deserves special credit, especially Tessa McGlasson Avila. And, as always, it was a pleasure to work with Susan Reiland, and I give her my most heartfelt thanks for the exceptional job she did on this book.

Karl J. Smith
Sebastopol, California

smithkjs@yahoo.com

CONTENTS

Part I Foundations: The Power of Mathematics 1

CHAPTER 3 INTRODUCTION TO ALGEBRA 121

CHAPTER 4 PERCENTS AND PROBLEM SOLVING 168

CHAPTER 5 INTRODUCTION TO GEOMETRY 197

CHAPTER 6 MEASUREMENT AND PROBLEM SOLVING 232

*Optional section.

Part II Applications: The Utility of Mathematics 295

CHAPTER 7 APPLICATIONS OF PERCENT 297

CHAPTER 8 SETS AND LOGIC 358

CHAPTER 9 PROBABILITY 403

CHAPTER 10 STATISTICS 444

PART ONE

FOUNDATIONS: THE POWER OF MATHEMATICS

Historically, the prime value of mathematics has been that it enables us to answer basic questions about our physical world, to comprehend the complicated operations of nature, and to dissipate much of the mystery that envelops life. The simplest arithmetic, algebra, and geometry suffice to determine the circumference of the earth, the distances to the moon and the planets, the speeds of sound and light, and the reasons for eclipses of the sun and moon. But the supreme value of mathematics, insofar as understanding the world about us is concerned, is that it reveals order and law where mere observation shows chaos. . . .

MORRIS KLINE, *MATHEMATICS: AN INTRODUCTION TO ITS SPIRIT AND USE*, SAN FRANCISCO: W. H. FREEMAN, 1979, P. 1.

The first part of this book, The Power of Mathematics, attempts to develop an appreciation for mathematics by displaying the intrinsic **power** of the subject. We will begin by looking at some of the causes and effects of *math anxiety.* We take natural steps—small at first, and then a little larger as you gain confidence—to review and learn about calculators, fractions, percents, algebra, equations, metrics, and geometry.

Most people view mathematics as a series of techniques useful only to the scientist, the engineer, or the specialist. In fact, the majority of our population could be classified as math-avoiders, who consider the assertion that mathematics can be creative, beautiful, and significant not only as an "impossible dream," but also as something they don't even want to discuss.

At each turn of the page, I hope you will find something new and interesting. I want you to participate and become involved with the material. I want you to experience what I mean when I speak of the *beauty* of mathematics. I hope you are now ready to begin your study of a new course; I wish you success.

CHAPTER ONE

ARITHMETIC, CALCULATORS, AND PROBLEM SOLVING

1.1 MATH ANXIETY

Historical Perspective

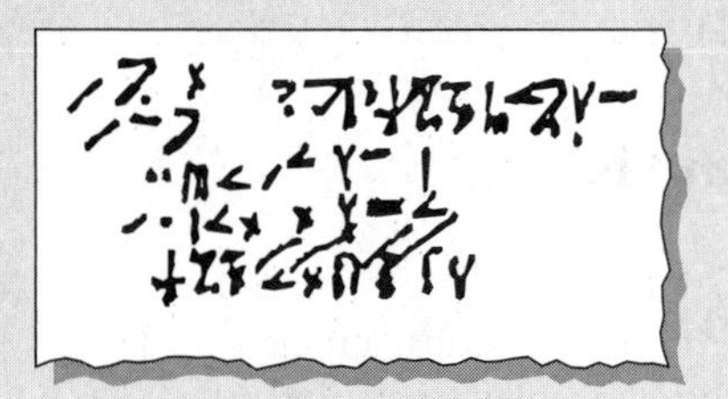

Rhind Papyrus (ca. 1575 B.C.): Ancient mathematical problems

The history of mathematics can be divided into six periods:

Chinese/Egyptian/Babylonian Period: 3000 B.C. to 601 B.C.

Greek Period: 600 B.C. to 499 A.D.

Hindu and Arabian Period: 500 to 1199

Transition Period: 1200 to 1599

Century of Enlightenment: 1600 to 1699

Modern Period: 1700 to present

We will profile each of these periods in Part I (Chapters 1–6), and then in Part II (Chapters 7–11) we will focus on situations in which you can apply the power of mathematics to your everyday life.

There are many reasons for reading a book, but the best reason is because you want to read it. Although you are probably reading this first page because you were requested to do so by your instructor, it is my hope that in a short while you will be reading this book because you want to read it.

Do you think that you are reasonably successful in other subjects but are unable to do math? Do you make career choices based on avoidance of mathematics courses? If so, you have *math anxiety.* If you reexamine your negative feelings toward mathematics, you can overcome them. In this book, I'll constantly try to help you overcome these feelings.

This book was written for people who are math-avoiders, people who think they can't work math problems, and people who think they are never going to use math. Do you see yourself making any of these statements?

Sheila Tobias, an educator, feminist, and founder of an organization called Overcoming Math Anxiety, has become one of our nation's leading spokespersons on math anxiety. She is not a mathematician, and in fact she describes herself as a math-avoider. She has written a book titled *Overcoming Math Anxiety* (New York: W. W. Norton & Company, 1978; available in paperback). I recommend this book to anyone who has ever said "I'm no good at numbers." In this book, she describes a situation that characterizes anxiety (p. 45):

> *Paranoia comes quickly on the heels of the anxiety attack. "Everyone knows," the victim believes, "that I don't understand this. The teacher knows. Friends know. I'd better not make it worse by asking questions. Then everyone will find out how dumb I really am." This paranoid reaction is particularly disabling because fear of exposure keeps us from constructive action. We feel guilty and ashamed, not only because our minds seem to have deserted us but because we believe that our failure to comprehend this one new idea is proof that we have been "faking math" for years.*

The reaction described in this paragraph sets up a vicious cycle. The more we avoid math, the less able we feel; and the less able we feel, the more we avoid it. The cycle can also work in the other direction. What do you like to do? Chances are, if you like it, you do it. The more you do something, the better you become at it. In fact, you've probably thought, "I like to do it, but I don't get to do it as often as I'd like to." This is the normal reaction toward something you like to do. In this book, I attempt to break the negative cycle concerning math and replace it with a positive cycle. However, I will need your help and willingness to try.

The central theme in this book is problem solving. Through problem solving, I'll try to dispel your feelings of panic. Once you find that you are capable of doing mathematics, we'll look at some of its foundations and uses. There are no prerequisites for this book; and as we progress through the book, I'll include a review of the math you never quite learned in school—from fractions, decimals, percents, and metrics to algebra and geometry. I hope to answer the questions that, perhaps, you were embarrassed to ask.

I hope you will enjoy reading this book; but if you feel an anxiety attack coming, STOP and put this book aside for a while. Talk to your instructor, or call me. My telephone number is

(707) 829-0606

I care about your progress with the course, and I'd like to hear your reactions to this book. You can write to me at the following address:

Karl Smith or smithkjs@yahoo.com
Mathematics Department
Santa Rosa Junior College
Petaluma Center
680 Sonoma Mountain Parkway
Petaluma, CA 94954

Throughout the book, I will take off my author hat and put on my teacher hat to write you notes about the material in this book. These notes will explain steps or give you hints on what to look for as you are reading the book. These notes are printed in this font.

At the end of each section in this book is a problem set. This first problem set is built around 12 math myths. These myths are in another book on math anxiety, *Mind Over Math,* by Stanley Kogelman and Joseph Warren (New York: Dial Press, 1978), which I highly recommend. These commonly believed myths have resulted in false impressions about how math is done, and they need to be dispelled.

Math Anxiety Bill of Rights*
by Sandra L. Davis

1. I have the right to learn at my own pace and not feel put down or stupid if I'm slower than someone else.
2. I have the right to ask whatever questions I have.
3. I have the right to need extra help.
4. I have the right to ask a teacher or TA for help.
5. I have the right to say I don't understand.
6. I have the right not to understand.
7. I have the right to feel good about myself regardless of my abilities in math.
8. I have the right not to base my self-worth on my math skills.
9. I have the right to view myself as capable of learning math.
10. I have the right to evaluate my math instructors and how they teach.
11. I have the right to relax.
12. I have the right to be treated as a competent adult.
13. I have the right to dislike math.
14. I have the right to define success in my own terms.

*From *Overcoming Math Anxiety,* by Sheila Tobias, pp. 236–237.

HINTS FOR SUCCESS

Mathematics is different from other subjects. One topic builds on another, and you need to make sure that you understand *each* topic before progressing to the next one.

You must make a commitment to attend each class. Obviously, unforeseen circumstances can come up, but you must plan to attend class regularly. Pay attention to what your teacher says and does, and take notes. If you must miss class, write an outline of the text corresponding to the missed material, including working out each text example on your notebook paper.

You must make a commitment to daily work. Do not expect to save up and do your mathematics work once or twice a week. It will take a daily commitment on your part, and you will find mathematics difficult if you try to "get it done" in spurts. You could not expect to become proficient in tennis, soccer, or playing the piano by practicing once a week, and the same is true of mathematics. Try to schedule a regular time to study mathematics each day.

Read the text carefully. Many students expect to get through a mathematics course by beginning with the homework problems, then reading some examples, and reading the text only as a desperate attempt to find an answer. This procedure is backward; do your homework only *after* reading the text.

I have included road signs throughout this book to help you successfully get through it.

When you see the stop sign, you should stop for a few moments and study the material by the stop sign. Memorize this material.

When you see the caution sign, you should make a special note of some mentioned fact because it will be used throughout the rest of the book.

When you see the yield sign, it means that you need to remember only the stated result and that the derivation is optional.

When you see the bump sign, some unexpected or difficult material follows, and you will need to slow down to understand the discussion.

Direct Your Focus

Read the following story. No questions are asked, but try to imagine yourself sitting in a living room with several others who share your feelings about math. Your job is to read the story and make up a problem you know how to solve from any part of the story. You should have a pencil and paper, and you can have as much time as you want; nobody will look at what you are doing, but I want you to keep track of your feelings as you read the story and follow the directions.

On the way to the market, which is 12 miles from home, I stopped at the drugstore to pick up a get-well card. I selected a series of cards with puzzles on them. The first one said, "A bottle and a cork cost $1.10 and the bottle is a dollar more than the cork. How much is the bottle and how much is the cork?" I thought that would be a good card for Joe, so I purchased it for $1.75, along with a six-pack of cola for $2.79. The total bill was $4.81, which included 6% sales tax. My next stop was the market, which was exactly 3.4 miles from the drugstore. I bought $15.65 worth of groceries and paid with a $20 bill. I deposited the change in a charity bank on the counter and left the store. On the way home I bought 8.5 gallons of gas for $13.60. Because I had gone 238 miles since my last fill-up, I was happy with the mileage on my new car. I returned home and made myself a ham and cheese sandwich.

Have you spent enough time on the story? Take time to reread it (spend at least 10 minutes with this exercise). Now, write down a math question, based on this study, that you could answer without difficulty. Can you summarize your feelings? If my experiences in doing this exercise with my students apply to you, I would guess that you encountered some difficulty, some discomfort, perhaps despair or anger, or even indifference. Most students tend to focus on the more difficult questions (perhaps a miles-per-gallon problem) instead of following the directions to formulate a problem that will give you no difficulty.

How about the question: What is the round-trip distance from home to market and back? *Answer*:

$$2 \times 12 \text{ miles} = 24 \text{ miles}$$

You say, "What does this have do with mathematics?" The point is that you need to learn to focus on what you know, rather than to focus on what you do not know. You may surprise yourself with the amount that you do know, and what *you* can bring to the problem-solving process.

Math anxiety builds on focusing on what you can't do rather than on what you can do. This leads to anxiety and frustration. Do you know what is the most feared thing in our society? It is the fear of speaking in public. And the fear of letting others know you are having trouble with this problem is related to that fear of speaking in public.

If you focus on a problem that is too difficult, you will be facing a blank wall. This applies to all hobbies or subjects. If you play tennis or golf, has your game improved since you started? If you don't play these games, how do you think you would feel trying to learn in front of all your friends? Do you think you would feel foolish?

Mathematicians don't start with complicated problems. If a mathematician runs into a problem that she can't solve, she will probably rephrase the problem as a simpler related problem that she can't solve. This problem is, in turn, rephrased as yet a simpler problem, and the process continues until the problem is manageable, and she has a problem that she *can* solve.

WRITING MATHEMATICS

The fundamental objective of education always has been to prepare students for life. A measure of your success with this book is a measure of its usefulness to you in your life. What are the basics for your knowledge "in life"? In this information age with access to a world of knowledge on the Internet, we still would respond by saying that the basics remain "reading, 'riting, and 'rithmetic." As you progress through the material in this book, we will give you opportunities to read mathematics and to consider some of the great ideas in the history of civilization, to develop your problem-solving skills ('rithmetic), and to communicate mathematical ideas to others ('riting). Perhaps you think of mathematics as "working problems" and "getting answers," but it is so much more. Mathematics is a way of thought that includes all three Rs, and to strengthen your skills you will be asked to communicate your knowledge in written form.

Journals

To begin building your skills in writing mathematics, you might keep a journal summarizing each day's work. Keep a record of your feelings and perceptions about what happened in class. How long did the homework take? What time of the day or night did I spend working and studying mathematics? What is the most important idea from the day's lesson? To help you with your journals, you will find problems in this text designated **IN YOUR OWN WORDS.** (For example, look at Problems 13–16 in the problem set at the end of this section.) There are no right answers or wrong answers to this type of problem, but you are encouraged to look at these for ideas of what you might write in your journal.

Journal ideas

Write in your journal every day.

Include important ideas.

Include new words, ideas, formulas, or concepts.

Include questions that you want to ask later.

If possible, carry your journal with you so you can write in it anytime you get an idea.

Reasons for keeping a journal

It will record ideas you might otherwise forget.

It will keep a record of your progress.

If you have trouble later, it may help you diagnose areas for change or improvement.

It will build your writing skills.

PROBLEM SET 1.1

LEVEL 1

In Your Own Words *In Problems* 1–12, *comment on each math myth. There are no right or wrong answers, but you will gain insight into your own attitudes as well as begin to dispel some false notions you might have about the subject.*

1. Myth 1: Men are better than women in math.
2. Myth 2: Math requires logic, not intuition.
3. Myth 3: You must always know how you got the answer.
4. Myth 4: Math is not creative.
5. Myth 5: There is a best way to do a math problem.
6. Myth 6: It's always important to get the answer exactly right.
7. Myth 7: It's bad to count on your fingers.
8. Myth 8: Mathematicians do problems quickly, in their heads.
9. Myth 9: Math requires a good memory.
10. Myth 10: Math is done by working intensely until the problem is solved.
11. Myth 11: Some people have a "math mind" and some don't.
12. Myth 12: There is a magic key to doing math.
13. **In Your Own Words** Summarize your math experiences in elementary school.
14. **In Your Own Words** Summarize your math experiences in high school.
15. **In Your Own Words** Summarize your feelings today about this course.
16. **In Your Own Words** At the beginning of this section, three hints for success were listed. Discuss each of these from your perspective. Are there any other hints that you might add to this list?

ESSENTIAL IDEAS

In Problems 17–20, *describe the meaning of each of the symbols, which you will find throughout the book.*

17. 

18. 

19. 

20. 

LEVEL 2

Symbols similar to the ones used in this book are common on the highways, and now we are seeing "universal" nonverbal symbols for other things. Problems 21–25 *show universal traffic signs. See if you can guess what each symbol means.*

21. 

22. 

23. 

24. 

25. 

In Problems 26–31, *symbols used by the fabric care industry are shown. Match each symbol with one of the choices below.*

A. Dry clean only
B. Nonchlorine bleach
C. Hand wash
D. Drip dry
E. Machine wash delicate
F. Tumble-dry medium

26. 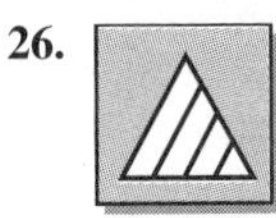

27. 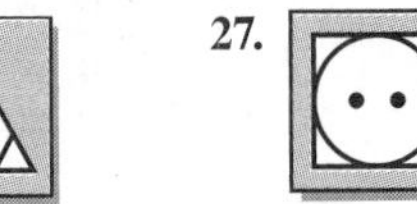

28. 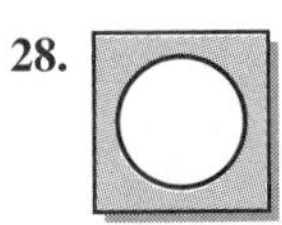

29. 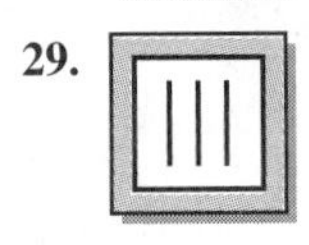

30. 

31. 

32. Read the following story, and make up a problem you can solve from some part of the story.

    Yesterday, I purchased five calves at the auction for $95 each, so my herd now consists of nineteen cows, one bull, twenty-six steers, and thirteen calves. The auction yard charged me $35 to deliver the calves to my ranch, but I figured it was a pretty good deal since I live 42 miles from the auction yard. Today, twelve tons of hay were delivered, and I paid $780 for it plus $10 a ton delivery charge. Yes, sir, if my crops do well, this will be a very good year.

33. Read the following story and make up a problem you can solve from some part of the story.

    Tickets for the concert go on sale for $65 each tomorrow; each person is allowed to purchase no more than four tickets. I really don't mind paying for the tickets, but why do they need to add a $3-per-ticket service charge? I am *going to pick up Jane at 5:00 A.M. so we can get in line early; the only problem is that Jane lives 23 miles from me and the ticket office is only 4 miles from my house. That means that we will not be able to line up before 6:00 A.M. Do you think that is early enough?*

34. In Your Own Words Describe some of your feelings as you worked Problem 32.

35. In Your Own Words Describe some of your feelings as you worked Problem 32.

LEVEL 3—Individual Project

36. Begin a journal for this course. You should write in your journal after each class and after you spend some time reading the book or doing homework problems. Make notes about how much time you spent reading the text, and how much time you spent working on homework problems.

As you progress through this book, keep the cartoon at the left in mind. When you think you are following the directions perfectly but things do not quite work out as they should, go back and reexamine your first few steps.

"... Now, after you've separated the four eggs, add..."
Reprinted by permission of Jerry Marcus.

LEVEL 3—Team Project

37. Working with others can be beneficial not only on the job, but also in the classroom. For this team project, introduce yourself to three or four classmates, and work with them on this problem. Spend at least 30 minutes getting to know one another, specifically focusing on these statements about your previous mathematics experiences:

"Everybody knows what to do, except me!"

"I got the right answer, but I don't know how!"

"I'm sure I learned it, but I can't remember what to do!"

"This may be a stupid question, but . . . ."

"I'm no good at numbers!"

"Math is unrelated to my life!"

"Math is my worst subject!"

"I don't have a math mind!"

Write a paper summarizing your discussion, and submit one paper for your team.

1.2 FORMULATING THE PROBLEM

Historical Perspective

George Pólya (1888–1985)

George Pólya is one of the most beloved mathematicians of the 20th century. His research and winning personality earned him a place of honor not only among mathematicians but among students and teachers as well. His discoveries are vast, but he is most remembered for his little book *How to Solve It,* which set the stage for a process of problem solving. He was born in Hungary, attended the universities of Budapest, Vienna, Göttingen, and Paris. He was a professor of mathematics at Stanford University. Here is his explanation of why he was a mathematician: "It is a little shortened but not quite wrong to say: I thought I am not good enough for physics and I am too good for philosophy. Mathematics is in between."

In mathematics, we generally focus our attention on some particular sets of numbers. The simplest of these sets is the set we use to count objects and is the first set that a child learns. It is called the set of **counting numbers** or **natural numbers.***

Natural Numbers and Whole Numbers

The set of numbers {1, 2, 3, 4, . . .} is called the set of **counting numbers** or the set of **natural numbers.** If the number zero is included, then the set {0, 1, 2, 3, 4, . . .} is called the set of **whole numbers.**

Addition, subtraction, multiplication, and division are called the **elementary operations** for the whole numbers, and it is assumed that you understand these operations.

However, certain agreements in dealing with these operations are necessary. Consider this arithmetic example:

Find: $2 + 3 \times 4$

There are two possible approaches to solve this problem:

Left to right: $\mathbf{2 + 3} \times 4 = \mathbf{5} \times 4 = \mathbf{20}$

Multiplication first: $2 + \mathbf{3 \times 4} = 2 + \mathbf{12} = \mathbf{14}$

Since there are different answers to this arithmetic example, it is necessary for us all to agree on one method or the other. At this point, you might be thinking, "Why on earth would I start at the right and do the multiplication first?" Consider the following example.

EXAMPLE 1 **Order of Operations**

Suppose that you sold a $2 benefit ticket on Monday and three $4 tickets on Tuesday. What is the total amount collected?

Solution

(SALES ON MONDAY) + (SALES ON TUESDAY) = (TOTAL SALES)

$2 + 3 × $4 = (TOTAL SALES)

CORRECT	*NOT CORRECT*
Multiplication first:	*Left to right:*
$2 + 3 \times 4 = 2 + 12 = \mathbf{14}$	$2 + 3 \times 4 = 5 \times 4 = \mathbf{20}$

■

*In mathematics, we use the word *set* as an undefined term. Although sets are discussed in Chapter 8, we assume that you have an intuitive idea of the word *set.* It is used to mean a collection of objects or numbers. Braces are used to enclose the elements of a set, and three dots (called ellipses) are used to indicate that some elements are not listed. We use three dots only if the elements not listed are clear from the given numbers.

Do you see why the correct solution to Example 1 requires multiplication before addition?

Many of you may use a calculator to help you work problems in this book. If you do use a calculator, it is important that you use a calculator that carries out the correct order of operations, as we illustrate with the next example.

EXAMPLE 2

Order of Operations Using a Calculator

In Example 1, we agreed that the correct simplification of the arithmetic problem

$$2 + 3 \times 4$$

is 14. Illustrate the correct buttons to press on a calculator, and check with the calculator you will use in this class to make sure you obtain the correct result.

Solution

You want the calculator you use in this class to accept the input of arithmetic problems in the same fashion as you would write it on your paper. Thus, the correct sequence of buttons to press is

[2] [+] [3] [×] [4] [=]

If you are using a calculator with this book, be sure you actually check out this answer.

Some calculators use a button labeled [ENTER] instead of the equal sign. The correct answer is 14, but some calculators will display the incorrect answer of 20. What is going on with a calculator that gives the answer 20? Such a calculator works from left to right without regard to the order of operations. For this book, you want to have a calculator that recognizes the order of operations correctly.

Notice that every time you push [+], a subtotal is shown in the display. After completing Example 2, you can either continue with the same problem or start a new problem. If the next button pressed is an operation button, the result 14 will be carried over to the new problem. If the next button pressed is a numeral, the 14 will be lost and a new problem started. For this reason, it is not necessary to press [C] to clear between problems. The button [CE] is called the *clear entry* key and is used if you make a mistake keying in a number and don't want to start over with the problem. For example, if you want 2 + 3 and accidentally push

[2] [+] [4]

you can then push

[CE] [3] [=]

to obtain the correct answer. This is especially helpful if you are in the middle of a long problem. ■

Consider another example.

EXAMPLE 3

Order of Operations: Multiplication First

Suppose you purchase a chair for \$100. If the sales tax is 6% (write this as 0.06), what is the total, including tax?

Solution

$$\text{(PRICE OF CHAIR)} + \text{(TAX)} = \text{(TOTAL PRICE)}$$
$$\$100 + 0.06 \times \$100 = \text{(TOTAL PRICE)}$$

CORRECT	*NOT CORRECT*
Multiplication first:	*Left to right:*
$100 + 0.06 \times 100 = 100 + 6$	$100 + 0.06 \times 100 = 100.06 \times 100$
$= \mathbf{106}$	$= \mathbf{10{,}006}$

Once again, we see that the correct answer is found when we do multiplication first. ■

If the operations are mixed, we agree to do multiplication and division *first* (from left to right) and *then* addition and subtraction from left to right.

EXAMPLE 4

Order of Operations

Perform the indicated operations.

a. $7 + 2 \times 6$ **b.** $3 \times 5 + 2 \times 5$ **c.** $1 + 3 \times 2 + 4 - 3 + 6 \times 3$

Solution

a. $7 + 2 \times 6 = 7 + \mathbf{12}$ Multiplication first
$= \mathbf{19}$ Addition next

b. $3 \times 5 + 2 \times 5 = \mathbf{15} + \mathbf{10}$ Multiplication first
$= \mathbf{25}$ Addition next

c. $1 + 3 \times 2 + 4 - 3 + 6 \times 3 = 1 + \mathbf{6} + 4 - 3 + \mathbf{18}$
$= \mathbf{7} + 4 - 3 + 18$
$= \mathbf{11} - 3 + 18$
$= \mathbf{8} + 18$
$= \mathbf{26}$ ■

If the order of operations is to be changed from this agreement, then parentheses are used to indicate this change.

EXAMPLE 5

Order of Operations with Parentheses

Perform the indicated operations.

a. $10 + 6 + 2$ **b.** $10 + (6 + 2)$ **c.** $10 - 6 - 2$ **d.** $10 - (6 - 2)$
e. Show the buttons you would press to work part **d** with a calculator.

Solution

a. $10 + 6 + 2 = (10 + 6) + 2$ Work from left to right; parentheses are understood (not necessary for this example).
$= \mathbf{16} + 2$
$= \mathbf{18}$

b. $10 + (6 + 2) = 10 + \mathbf{8}$ Parentheses change the order of operations.
$= \mathbf{18}$

Notice that parts **a** and **b** illustrate different problems that have the same answer. This is not the case with parts **c** and **d.**

c. $10 - 6 - 2 = (10 - 6) - 2$ *Parentheses understood.*
$= \mathbf{4} - 2$
$= \mathbf{2}$

d. $10 - (6 - 2) = 10 - \mathbf{4}$ *Parentheses given.*
$= \mathbf{6}$

e. Most calculators have parentheses keys. The one labeled $\boxed{(}$ is the open parenthesis (this one comes first) and the one labeled $\boxed{)}$ is the close parenthesis (this one comes last). Locate these keys on your calculator. Also note that the operations keys are usually found at the right side. These are labeled

$\boxed{\div}$ $\boxed{\times}$ $\boxed{-}$ $\boxed{\times}$

Do not confuse subtraction $\boxed{-}$ with other keys labeled $\boxed{(-)}$ or $\boxed{+/-}$. The correct sequence of buttons to press for this example is

$\boxed{10}$ $\boxed{-}$ $\boxed{(}$ $\boxed{6}$ $\boxed{-}$ $\boxed{2}$ $\boxed{)}$ $\boxed{=}$

Check the output of your calculator to make sure it gives the result shown in part **d,** namely, 6. ■

We can now summarize the correct **order of operations,** including the use of parentheses.

Order of Operations

1. Parentheses first
2. Multiplication and division, reading from left to right
3. Addition and subtraction, reading from left to right

Take a few minutes to memorize this order of operations agreement.

You can remember this by using the mnemonic "**P**lease **M**ind **D**ear **A**unt **S**ally." The first letters will remind you of "**P**arentheses, **M**ultiplication and **D**ivision, **A**ddition and **S**ubtraction." However, if you use this mnemonic, remember that the order is *left to right* for pairs of operations: multiplication and division (as they occur) and *then* addition and subtraction (as they occur).

EXAMPLE 6 **Mixed Operations**

Perform the indicated operations.

a. $12 + 9 \div 3$ **b.** $(12 + 9) \div 3$ **c.** $4 \times (3 + 2)$ **d.** $4 \times 3 + 4 \times 2$
e. $2 \times 15 + 9 \div 3 - 7 \times 2$ **f.** $2 \times (15 + 9) \div 3 - 7 \times 2$

Solution

a. $12 + 9 \div 3 = 12 + \mathbf{3}$
$= \mathbf{15}$

b. $(12 + 9) \div 3 = \mathbf{21} \div 3$
$= \mathbf{7}$

c. $4 \times (3 + 2) = 4 \times \mathbf{5}$
$= \mathbf{20}$

d. $4 \times 3 + 4 \times 2 = \mathbf{12} + \mathbf{8}$
$= \mathbf{20}$

e. $2 \times 15 + 9 \div 3 - 7 \times 2 = \mathbf{30} + \mathbf{3} - \mathbf{14}$
$= \mathbf{33} - 14$
$= \mathbf{19}$

f. $2 \times (15 + 9) \div 3 - 7 \times 2 = 2 \times \mathbf{24} \div 3 - 7 \times 2$
$= \mathbf{48} \div 3 - 7 \times 2$
$= \mathbf{16} - 7 \times 2$
$= 16 - \mathbf{14}$
$= \mathbf{2}$ ■

Look at parts **c** and **d** of Example 6. They illustrate a combined property of addition and multiplication called the **distributive law for multiplication over addition:**

$$\mathbf{4} \times (3 + 2) = \mathbf{4} \times 3 + \mathbf{4} \times 2$$

↑ Number outside parentheses

↑ ↑ Number outside parentheses is **distributed** to **each** number inside parentheses.

This property holds for all whole numbers.

EXAMPLE 7

Distributive Property

Write out each of the given expressions without parentheses by using the distributive property.

a. $8 \times (7 + 10)$ **b.** $3 \times (400 + 20 + 5)$

Solution

a. $8 \times (7 + 10) = \mathbf{8 \times 7 + 8 \times 10}$
b. $3 \times (400 + 20 + 5) = \mathbf{3 \times 400 + 3 \times 20 + 3 \times 5}$ ■

The ability to recognize the difference between reasonable answers and unreasonable ones is important not only in mathematics, but whenever you are doing problem solving. This ability is even more important when you use a calculator, because pressing the incorrect key can often cause very unreasonable answers. Whenever you try to find an answer, you should ask yourself whether the answer is reasonable. How do you decide whether an answer is reasonable? One way is to **estimate** an answer. Webster's *New World Dictionary* tells us that as a verb, to *estimate* means "to form an opinion or a judgment about" or to calculate "approximately." In the *1986 Yearbook* of the National Council of Teachers of Mathematics, we find:

> *The broad* mathematical context *for an estimate is usually one of the following types:*
> A. *An exact value is known but for some reason an estimate is used.*
> B. *An exact value is possible but is not known and an estimate is used.*
> C. *An exact value is impossible.*

We will work on building your estimation skills throughout this book.

EXAMPLE 8

Estimation

If your salary is $9.75 per hour, your annual salary is approximately
A. $5,000 B. $10,000 C. $15,000 D. $20,000 E. $25,000

Solution Problem solving often requires some assumptions about the problem. For this problem, we are not told how many hours per week you work, or how many weeks per year you are paid. We assume a 40-hour week, and we also assume that you are paid for 52 weeks per year.

Estimate: Your hourly salary is about $10 per hour.
A 40-hour week gives us $40 \times \$10 = \$400$ per week.

For the estimate, we calculate the wages for 50 weeks instead of 52:

50 weeks yields $50 \times \$400 = \$20{,}000$.

The answer is **D**. ■

There are two important reasons for estimation: (1) to form a reasonable opinion (as in Example 8), or (2) to check the reasonableness of an answer. If reason (1) is our motive, we should not think it necessary to follow an estimation like the one in Example 8 by direct calculation. To do so would defeat the purpose of the estimation. On the other hand, if we are using the estimate for reason (2)—to see whether an answer is reasonable—we might perform the estimate as a check on the calculated answer for Example 8, which is

[9.75] [×] [40] [×] [52] [=] *Display:* 20280

The actual annual salary is $20,280. In this case, our estimate confirms that our precise calculation does not contain a widely erroneous keying mistake.

One of the key steps in formulating the problem is developing skill in **translating** from English into math symbolism. The term **sum** is used to indicate the result obtained from addition, **difference** for the result from subtraction, **product** for the result of a multiplication, and **quotient** for the result of a division. When a problem involves mixed operations, it is classified as a sum, difference, product, or quotient according to the *last* operation performed, when using the order of operations agreement.

EXAMPLE 9

Translating and Classifying Arithmetic Operations

Write each verbal description in math symbols, and then use your calculator to simplify.

a. The sum of the first five natural numbers
b. The product of the first five natural numbers
c. The sum of 5 and twice the number 53
d. The product of 5 and twice the number 4
e. The product of 5 and the sum of 2 and 4
f. The sum of 5 and the product of 2 and 4
g. The difference of 7 and 2
h. The difference of 2 from 7

Solution

a. $1 + 2 + 3 + 4 + 5 = \mathbf{15}$
b. $1 \times 2 \times 3 \times 4 \times 5 = \mathbf{120}$
c. $5 + (2 \times 53) = \mathbf{111}$

[5] [+] [(] [2] [×] [53] [)] [=]

Note: Once we have shown you how to use a calculator for a certain type of problem, we will generally not repeat the calculator steps except for emphasis.

d. $5 \times (2 \times 4)$ or $5 \times 2 \times 4 = \mathbf{40}$
Note that the translation implies parentheses, but because of the order of operations, none are required.

e. $5 \times (2 + 4)$ or $5(2 + 4)$
Some calculators will recognize a number next to parentheses to mean multiplication. You should find out whether your calculator is one of those types, because in this book we will generally write such a calculation as $5(2 + 4)$, called **juxtaposition,** with the multiplication between the 5 and the parentheses understood.

CAUTION

You must remember whether your calculator needs a times sign when used next to a parenthesis.

Times sign required: [5] [×] [(] [2] [+] [4] [)] [=]
No times sign required: [5] [(] [2] [+] [4] [)] [=]

The correct answer is 30. Try the version without the times sign to see whether this works with your calculator.

f. $5 + (2 \times 4) = \mathbf{13}$

[5] [+] [(] [2] [×] [4] [)] [=]

The times sign *is* required for this problem, because without it, the answer would be $5 + 24 = 29$.

g. $7 - 2 = \mathbf{5}$
h. $7 - 2 = \mathbf{5}$

Pay attention to the wording on subtraction problems. Parts **g** and **h** are not the same as $2 - 7$. ■

EXAMPLE 10

Calculator Arithmetic with Mixed Operations

Show the calculator steps and display for each part, and classify as a sum, difference, product, or quotient.

a. $4 + 3 \times 7 - 5$ **b.** $4 + 3(7 - 5)$ **c.** $(4 + 3)(7 - 5)$

Solution

a. [4] [+] [3] [×] [7] [−] [5] [=] The result of this *difference* is 20.
If you have an algebraic-logic calculator, your machine will perform the correct order of operations. If it is an arithmetic-logic calculator, it will give the incorrect answer 45 unless you input the numbers using the agreement regarding order of operations.

b. [4] [+] [3] [(] [7] [−] [5] [)] [=] The result of this *sum* is 10.

c. [(] [4] [+] [3] [)] [(] [7] [−] [5] [)] [=] The result of this *product* is 14. ■

PROBLEM SET 1.2

LEVEL 1

WHAT IS WRONG, *if anything, with each of the statements in Problems 1–6? Explain your reasoning.*

1. $4 + 6 \times 8 = 80$
2. $2 + 8 \times 10 = 82$
3. $2 \times (3 + 4) = 14$ is an example of the distributive property.

4. The correct order of operations (for an expression with no parentheses) when simplifying is first multiply, then divide, then add, then subtract.
5. The number 0 is a natural number.
6. All natural numbers are also whole numbers.

ESSENTIAL IDEAS

7. State the order of operations.
8. Explain the distributive property in your own words.

LEVEL 1—Drill

Perform the indicated operations in Problems 9–36, and classify each as a sum, difference, product, or quotient.

9. $5 + 6 \times 2$ **10.** $8 + 2 \times 3$

11. $20 - 4 \times 2$ **12.** $10 - 5 \times 2$

13. $12 \div 6 + 3$ **14.** $100 \div 10 \div 2$

15. $12 + 6 \div 3$ **16.** $100 \div (10 \times 2)$

17. $15 + 6 \div 3$ **18.** $16 - 6 \div 3$

19. $(15 + 6) \div 3$ **20.** $(15 - 6) \div 3$

21. $4 \times 3 + 4 \times 5$ **22.** $8 \times 2 + 8 \times 5$

23. $4 \times (3 + 5)$ **24.** $8 \times (2 + 5)$

25. $2 + 15 \div 3 \times 5$ **26.** $5 + 12 \div 3 \times 2$

27. $2 + 15 \times 3 \div 5$ **28.** $5 + 12 \times 3 \div 2$

29. $(20 - 8) \div 4 \times 2 + 3$ **30.** $20 - 8 \div 4 \times 2 + 3$

31. $2 + 3 \times 4 - 12 \div 2$ **32.** $15 \div 5 \times 2 + 6 \div 3$

33. $2 \times 18 + 9 \div 3 - 5 \times 2$

34. $2 \times (18 + 9) \div 3 - 5 \times 2$

35. $4 \times (12 - 8) \div 2$ **36.** $4 \times 12 - 8 \div 2$

LEVEL 2—Drill

Write out the expressions in Problems 37–46, without parentheses, by using the distributive property.

37. $3 \times (4 + 8)$ **38.** $7 \times (9 + 4)$

39. $12 \times (4 + 6)$ **40.** $7 \times (70 + 3)$

41. $8 \times (50 + 5)$ **42.** $6 \times (90 + 7)$

43. $4 \times (300 + 20 + 7)$

44. $6 \times (500 + 30 + 3)$

45. $5 \times (800 + 60 + 4)$

46. $4 \times (700 + 10 + 5)$

In Problems 47–54, translate each of the word statements to numerical statements, and then simplify.

47. The sum of three and the product of two and four

48. The product of three and the sum of two and four

49. Ten times the sum of five and six

50. Ten times the product of five and six

51. Eight times five plus ten

52. Eight times the difference of seven and five

53. Eight times the difference of nine from eleven

54. The product of the sum of three and four with the sum of five and six

In Problems 55–70, perform the indicated operations on your calculator, and classify each as a sum, difference, product, or quotient.

55. $716 - 5 \times 91$

56. $143 + 12 \times 14$

57. $8 \times 14 + 8 \times 86$

58. $15 \times 27 + 15 \times 73$

59. $12 \times 63 + 12 \times 27$

60. $19 \times 250 + 19 \times 750$

61. $(18 + 2)(82 - 2)$

62. $(34 - 4)(16 + 4)$

63. $5 + 3 \times 7 + 65 - 8 \times 4$

64. $12 + 6 \times 9 - 5 \times 2 + 5 \times 14$

65. $27 \times 550 - 27 \times 450$

66. $23 \times 237 + 23 \times 763$

67. $1{,}214 - 18 \times 14 + 35 \times 8{,}121$

68. $862 + 328 \times 142 - 168$

69. $62 \times (48 - 12) + 13 \times (12 - 5)$

70. $12 \times (125 - 72) - 3 \times (18 - 3 \times 5)$

Level 2—Applications

In Problems 71–80, first estimate your answer and then calculate the exact answer.

71. How many hours are there in 360 days?

72. How many pages are necessary to make 1,850 copies of a manuscript that is 487 pages long? (Print on one side only.)

73. In 1998, the Internal Revenue Service allowed a $2,650 deduction for each dependent. If a family has four dependents, what is the allowed deduction?

74. If your payroll deductions are $255.83 per week and your weekly gross wages are $1,025.66, what is your net pay?

75. If your monthly salary is $1,543, what is your annual salary?

76. If you are paid $7.25 per hour, what is your annual salary?

77. If you are paid $18.00 per hour, what is your annual salary?

78. If you are paid $31,200 per year, what is your hourly salary?

79. If your car gets 23 miles per gallon, how far can you go on 15 gallons of gas?

80. If your car travels 492 miles on 12 gallons of gas, what is the number of miles per gallon?

Level 3—Mind Bogglers

Mind Bogglers are not required for an understanding of the material. They are challenging, fun, or especially interesting problems that you should attempt if your interest is aroused.

81. Children often learn to count using their fingers, and various methods of finger counting have been developed over the years. One of the first documented methods during the Middle Ages uses the finger positions for numbers less than 10,000, as shown in Figure 1.1.

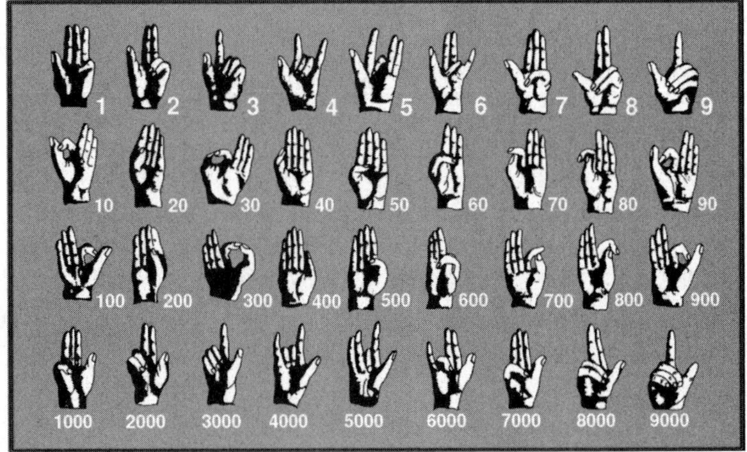

Figure 1.1 Finger calculation was important in the Middle Ages.

A method of finger calculation called *Chisanbop* has created a great deal of interest. It is described by one of its developers, Edwin Liberthal, and the director of the Psychological Research Laboratory, William Lamon, in "Chisanbop Finger Calculation" in *California Mathematics Journal* (Vol. 3, No. 2, October 1978, pp. 2–10). Write a report on Chisanbop or make a class demonstration on this method of finger calculation.

82. Read the article "Finger Multiplication" by Fred Balin in *The Arithmetic Teacher* (March 1979, pp. 34–37) and write a report on the article.

1.3 FRACTIONS AND DECIMALS

Historical Perspective

Stonehenge (1700 B.C.)

We have records of Chinese arithmetic and astronomy under the Huang-ti dynasty dated as early as 3000 B.C. About the time of Stonehenge, several Egyptian and Babylonian tablets were written. The Moscow papyrus presents 25 mathematical problems, and the Plimpton 322 introduces us to some additional mathematical problems. While the Phoenicians were inventing the modern alphabet around 1000 B.C., a major Chinese text on mathematics (called the *Chou-pei*) was being written. The history of the numeration system we use today is given at the beginning of the next section.

In the previous section, the symbol ÷ was used for division (for example, 10 ÷ 5). However, more often a division bar is used, as in $\frac{10}{5}$.

EXAMPLE 1 **Division Notation and Division by 0**

Perform the indicated operations.

a. $\dfrac{30}{6}$ **b.** $\dfrac{3,965}{305}$ **c.** $\dfrac{0}{5}$ **d.** $\dfrac{5}{0}$ **e.** $\dfrac{10+8}{7+2}$ **f.** $\dfrac{10}{3}$

Solution

a. $\dfrac{30}{6} = \mathbf{5}$ Check by multiplication: $6 \times 5 = 30$

b. If the division is lengthy, as in the case of $\dfrac{3,965}{305}$, you may need to do long division:

```
       13
305)3965
    305
     915
     915
       0
```
Check: $305 \times 13 = 3,965$

Thus, $\dfrac{3,965}{305} = \mathbf{13}$.

c. $\dfrac{0}{5} = \mathbf{0}$ Check by multiplication: $5 \times 0 = 0$

Remember, division by 0 is impossible.

d. $\dfrac{5}{0}$ To do this division, you would need to find a number such that, when it is multiplied by zero, the result is 5. There is no such number. For this reason, we say **division by zero is impossible.**

e. The division bar also acts like a grouping symbol. This division problem implies that we are to find the sum of 10 and 8 and then divide that answer by the sum of 7 and 2:

$$\frac{10+8}{7+2} = \frac{18}{9} = \mathbf{2}$$

You might wish to simplify this expression by using a calculator. When entering expressions such as this, if the number on top (or on the bottom) is more than a single number, then you must insert parentheses:

By calculator, think of this as $\dfrac{(10 + 8)}{(7 + 2)}$:

[( ] [10] [+] [8] [ )] [÷] [( ] [7] [+] [2] [ )] [=]

f. $\dfrac{10}{3}$ There is no answer to this division in the set of whole numbers. ■

The reason there is no answer to $\frac{10}{3}$ in the set of whole numbers in part **e** of Example 1 is that there is no whole number that can be multiplied by 3 to give 10. If you do long division for this problem, there will be a **remainder** that is not zero:

$$\begin{array}{r} 3 \\ 3\overline{)10} \\ \underline{9} \\ 1 \end{array} \leftarrow \text{Remainder}$$

What does this remainder mean? In this example, the remainder 1 is still to be divided by 3, so it can be written as $\frac{1}{3}$. Such an expression is called a **fraction** or a **rational number.*** The word *fraction* comes from a Latin word meaning "to break." A fraction involves two numbers: one "upstairs," called the **numerator,** and one "downstairs," called the **denominator.** The denominator tells us into how many parts the whole has been divided, and the numerator tells us how many of those parts we have.

Fraction

A **fraction** is a number that is the quotient of a whole number divided by a counting number. A fraction is usually written as a

$$\frac{\text{NUMERATOR}}{\text{DENOMINATOR}}$$

NUMERATOR ← *A whole number*
——— ← *Divided by*
DENOMINATOR ← *A counting number (so that division by zero is excluded)*

A fraction is called

1. A **proper fraction** if the numerator is less than the denominator.
2. An **improper fraction** if the numerator is greater than the denominator.
3. A **whole number** if the denominator divides into the numerator with no remainder.

*Rational numbers are discussed in Section 2.6.

EXAMPLE 2

Classifying Fractions

Classify each example as a proper fraction, an improper fraction, or a whole number.

a. $\frac{6}{7}$ **b.** $\frac{6}{8}$ **c.** $\frac{7}{6}$ **d.** $\frac{0}{4}$ **e.** $\frac{6}{3}$ **f.** $\frac{4}{0}$

Solution

a. $\frac{6}{7}$ **Proper** **b.** $\frac{6}{8}$ **Proper** **c.** $\frac{7}{6}$ **Improper**

d. $\frac{0}{4}$ **Whole number** **e.** $\frac{6}{3}$ **Whole number** **f.** $\frac{4}{0}$ **None of these** (Don't forget, you can't divide by zero.) ■

Improper fractions that are not whole numbers can also be written in a form called **mixed numbers** by carrying out the division and leaving the remainder as a fraction. A mixed number thus has two parts: a counting number part and a proper fraction part.

EXAMPLE 3

Writing an Improper Fraction as a Mixed Number

Write $\frac{23}{5}$ as a mixed number.

Solution

$$\begin{array}{r} 4 \\ 5\overline{)23} \\ \underline{20} \\ 3 \end{array}$$

4 ← Counting number part

3 ← Remainder means 3 to be divided by 5, or $\frac{3}{5}$.

$\frac{23}{5} = 4 + \frac{3}{5}$ or $\mathbf{4\frac{3}{5}}$. ■

EXAMPLE 4

Writing a Mixed Number as an Improper Fraction

Write $4\frac{3}{5}$ as an improper fraction.

Solution Reverse the procedure in Example 3. The remainder is 3, the quotient is 4, and the divisor is 5. Thus,

$$4 \times 5 + 3 = 23$$

Whole number ↓ (4) Numerator ↓ (3)

↑ (5) This is the divisor (the denominator of the fraction).

$$4\frac{3}{5} = \frac{\mathbf{23}}{\mathbf{5}}$$

← $23 = 4 \times 5 + 3$

↑ This is the divisor (or the denominator of the fraction). ■

We find certain fractions to be of special interest in our study of arithmetic. These are fractions with denominators that are powers of 10 (10, 100, 1,000, 10,000, . . .).

Tenths: $\frac{1}{10}, \frac{2}{10}, \frac{3}{10}, \frac{4}{10}, \frac{5}{10}, \frac{6}{10}, \frac{7}{10}, \frac{8}{10}, \frac{9}{10}, \frac{10}{10}, \frac{11}{10}, \ldots$

Hundredths: $\frac{1}{100}, \frac{2}{100}, \frac{3}{100}, \frac{4}{100}, \frac{5}{100}, \frac{6}{100}, \ldots$

Thousandths: $\frac{1}{1{,}000}, \frac{2}{1{,}000}, \frac{3}{1{,}000}, \frac{4}{1{,}000}, \frac{5}{1{,}000}, \ldots$

Each of these fractions can be written in **decimal form:**

$$\frac{1}{10} = 0.1 \qquad \frac{1}{100} = 0.01 \qquad \frac{1}{1{,}000} = 0.001$$

The word *fraction* is used with both forms. If we want to be specific, fractions written in the form

$$\frac{1}{10} \qquad \frac{1}{3} \qquad \frac{2}{5}$$

are called **common fractions,** and fractions written in the form

0.1 0.5 0.6

are called **decimal fractions.** When we simply use the word *fraction,* we are referring to a common fraction.

We will use division to show the relationship between fractions written as common fractions or as decimal fractions, but first let's look again at the positional notation of our number system:

If you do not already know these **place-value names,** you should learn them now.

Trillions	1,
Hundred billions	2
Ten billions	3
Billions	4,
Hundred millions	5
Ten millions	6
Millions	7,
Hundred thousands	8
Ten thousands	9
Thousands	1,
Hundreds	2
Tens	3
Units	4
← Decimal point	.
Tenths, $\frac{1}{10}$	8
Hundredths, $\frac{1}{100}$	9
Thousandths, $\frac{1}{1{,}000}$	0
Ten-thousandths, $\frac{1}{10{,}000}$	1
Hundred-thousandths, $\frac{1}{100{,}000}$	2
Millionths, $\frac{1}{1{,}000{,}000}$	3

Every whole number can be written in decimal form:

$$0 = 0. = 0.0 = 0.00 = 0.000 = 0.0000 = 0.00000 = \ldots$$
$$1 = 1. = 1.0 = 1.00 = 1.000 = 1.0000 = \ldots$$
$$2 = 2. = 2.0 = 2.00 = 2.000 = \ldots$$
$$3 = 3. = 3.0 = 3.00 = \ldots$$
$$4 = 4. = 4.0 = \ldots$$
$$5 = 5. = \ldots$$
$$\vdots$$

Sometimes zeros are placed after the decimal point or after the last digit to the right of the decimal point, as in

$$3.21 = 3.210 = 3.2100 = 3.21000 = 3.21\underbrace{0000\ldots}$$

These are called **trailing zeros.**

Now let's carry out division, bringing the decimal point straight up from the dividend to the quotient:

$$\begin{array}{r} 0.1 \\ 10\overline{)1.0} \\ \underline{1\,0} \\ 0 \end{array}$$

↑ ← Decimal point is carried straight up from dividend to quotient.

EXAMPLE 5

Changing Tenths, Hundredths, and Thousandths into Decimal Fractions by Division

Write the given fractions as decimals by performing long division.

a. $\frac{3}{10}$ **b.** $\frac{7}{100}$ **c.** $\frac{137}{1,000}$

Solution

a. $\frac{3}{10}$; $\begin{array}{r} \mathbf{0.3} \\ 10\overline{)3.0} \\ \underline{3\,0} \\ 0 \end{array}$

b. $\frac{7}{100}$; $\begin{array}{r} \mathbf{0.07} \\ 100\overline{)7.00} \\ \underline{7\,00} \\ 0 \end{array}$

c. $\frac{137}{1,000}$; $\begin{array}{r} \mathbf{0.137} \\ 1,000\overline{)137.000} \\ \underline{100\,0} \\ 37\,00 \\ \underline{30\,00} \\ 7\,000 \\ \underline{7\,000} \\ 0 \end{array}$ ← Place as many trailing zeros here as are necessary to complete the division.

■

Do you see a pattern in Example 5? Look at the next example.

EXAMPLE 6 **Changing Tenths, Hundredths, and Thousandths into Decimal Fractions by Inspection (without division)**

Write the given fractions in decimal form without doing any calculations.

a. $\frac{6}{10}$ **b.** $\frac{6}{100}$ **c.** $\frac{243}{1,000}$ **d.** $\frac{47}{10}$ **e.** $\frac{47}{100}$ **f.** $\frac{47}{1,000}$

Solution

a. $\frac{6}{10} = \mathbf{0.6}$ Tenth means one decimal place.

b. $\frac{6}{100} = \mathbf{0.06}$ Hundredth means two decimal places.

c. $\frac{243}{1,000} = \mathbf{0.243}$ Thousandth means three decimal places.

d. $\frac{47}{10} = \mathbf{4.7}$ **e.** $\frac{47}{100} = \mathbf{0.47}$ **f.** $\frac{47}{1,000} = \mathbf{0.047}$ ■

You can see that fractions with denominators of 10, 100, 1,000, and so on are what we've called **decimal fractions** or simply **decimals.** On the other hand, *any fraction,* even one whose denominator is not a power of 10, can be written in decimal form by dividing.

EXAMPLE 7 **Changing Common Fractions into Decimals by Using Division**

Write the given fractions in decimal form.

a. $\frac{3}{8}$ **b.** $\frac{7}{5}$ **c.** $2\frac{3}{4}$

Solution

a. $\frac{3}{8}$;

$$\begin{array}{r} 0.375 \\ 8\overline{)3.000} \\ \underline{2\,4} \\ 60 \\ \underline{56} \\ 40 \\ \underline{40} \\ 0 \end{array}$$

Remember, the decimal point is moved straight up from the dividend to the quotient. Otherwise, carry out the division in the usual fashion.

You may keep adding trailing zeros here as long as you wish.

Thus, $\frac{3}{8} = \mathbf{0.375}$

b. $\frac{7}{5}$;

$$\begin{array}{r} 1.4 \\ 5\overline{)7.0} \\ \underline{5} \\ 2.0 \\ \underline{2.0} \\ 0 \end{array}$$

Thus, $\frac{7}{5} = \mathbf{1.4}$

c. $2\frac{3}{4}$; Since $2\frac{3}{4} = \dfrac{11}{4}$, we divide 4 into 11:

$$\begin{array}{r} 2.75 \\ 4\overline{)11.00} \\ \underline{8} \\ 3\,0 \\ \underline{2\,8} \\ 20 \\ \underline{20} \\ 0 \end{array}$$

An alternative method is to notice that

$$2\tfrac{3}{4} = 2 + \frac{3}{4}$$

Divide 4 into 3:

$$\begin{array}{r} 0.75 \\ 4\overline{)3.00} \end{array}$$

and then add 0.75 to 2.

Thus, $2\frac{3}{4} = \mathbf{2.75}$ ■

We noted in part **a** of Example 7 that you can append as many trailing zeros after the 3.0 as you wish. For some fractions, you could continue to append zeros forever and never complete the division. Such fractions are called **repeating decimals.**

EXAMPLE 8

Changing Common Fractions to Decimal Fractions (repeating decimals)

Write $\dfrac{2}{3}$ in decimal form.

Solution

$$\begin{array}{r} 0.666 \\ 3\overline{)2.000} \\ \underline{1\,8} \\ 20 \\ \underline{18} \\ 20 \\ \underline{18} \\ 2 \end{array}$$

← The same remainder keeps coming up, so the process is never finished.

This is a *repeating decimal* and is indicated by three trailing dots or by a bar over the repeating digits:

$$\frac{2}{3} = \mathbf{0.666\ldots} \quad \text{or} \quad \frac{2}{3} = \mathbf{0.\overline{6}}$$ ■

When you write repeating decimals, be careful to include three dots or the overbar.

You cannot correctly write

$$\frac{2}{3} = 0.6666 \quad \text{or} \quad \frac{2}{3} = 0.6667$$

CAUTION These are wrong!

because $3 \times 0.6666 = 1.9998$ and $3 \times 0.6667 = 2.0001$. You can correctly write

$$\frac{2}{3} \approx 0.6666 \quad \text{or} \quad \frac{2}{3} = 0.\overline{6}$$

The symbol ≈ means "approximately equal to."

All fractions have a decimal form that either terminates (as in Example 7) or repeats (as in Example 8). Any number of digits may repeat; consider Example 9.

Calculators do not write repeating decimals, but finding $\frac{2}{3}$ on your calculator can be instructive.

Press: [2] [÷] [3] [=]

The display may vary: If you see 0.666666667, then your calculator *rounds* the last decimal place, whereas if you see 0.666666666, then your calculator *truncates* (cuts off) at the last decimal place. You should remember whether your calculator rounds or truncates. You should also count the number of decimal places shown in your calculator's display. The number of places shown by this example indicates the accuracy of your calculator. The one shown here is accurate to eight decimal places. (Note that the ninth decimal place is not accurate because of possible rounding.)

EXAMPLE 9

Changing a Mixed Fraction to a Decimal Fraction (with a repeating decimal)

Write $3\frac{5}{11}$ in decimal form.

Solution Write $\frac{5}{11}$ as a decimal:

```
     0.45 . . .
 11)5.00
    4 4
     60
     55
      5 ← Repeats
```

By calculator: [3] [+] [5] [÷] [11] [=]

Display: 3.4545454545

Remember, a calculator does not show beyond its display; for this one, you must notice that it is a repeating decimal when you write $3.\overline{45}$.

Thus, $3\frac{5}{11} = 3 + \frac{5}{11} = 3 + 0.\overline{45} = \mathbf{3.\overline{45}}$ ■

EXAMPLE 10

Estimating Parts Using Common Fractions

Estimate the size of each shaded portion as a common fraction.

a.

b.

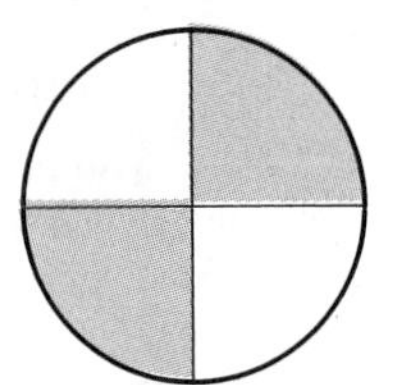

Solution

a. It looks as if the square is divided into 3 parts, and 2 of the 3 parts are shaded, so we **estimate that $\frac{2}{3}$ of the square is shaded.**

b. It looks as if the circle is divided into 4 parts, and 2 of the 4 parts are shaded, so we **estimate that $\frac{2}{4}$ or $\frac{1}{2}$ of the circle is shaded.** ■

EXAMPLE 11

Estimating Parts Using Decimal Fractions

Estimate the size of each shaded portion as a decimal.

a.

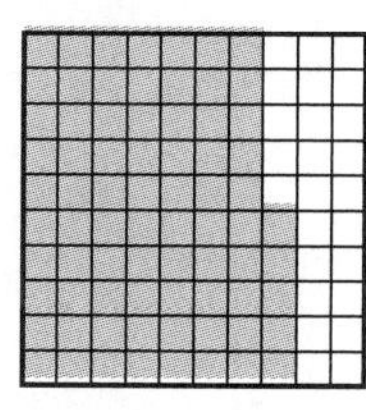

b.

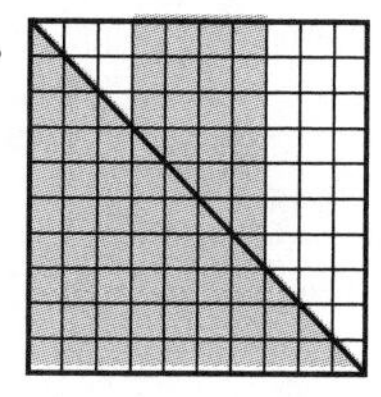

Solution

a. Divide the square in half:

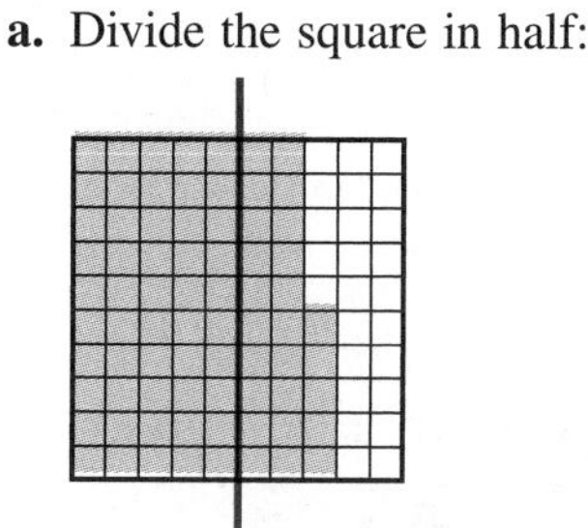

It looks as if half the remaining half is shaded, so we estimate that 75 of the 100 squares are shaded. The shaded portion is 0.75.

b. Divide into thirds:

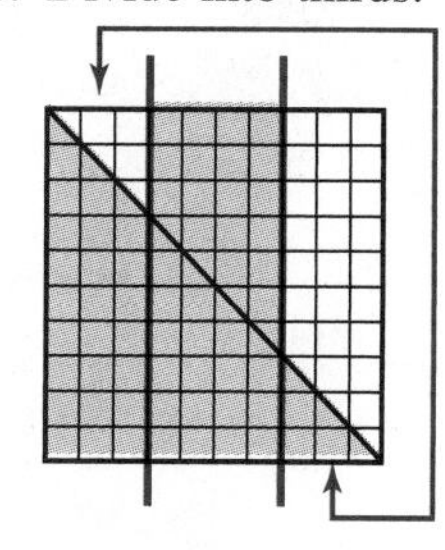

It appears that the number of squares not shaded in the first third is about the same as the number of small squares shaded in the last third, so we estimate the number of shaded squares to be $\frac{2}{3}$. ■

If you resort to counting squares, you are not estimating.

EXAMPLE 12

Application to a Stock Purchase

Suppose you purchase 400 shares of Disney stock selling at $49\frac{5}{8}$ per share. What is the total price you paid for this stock?

Solution

First, convert the stock price to dollars and cents by changing the mixed number to dollars and cents; then multiply by the number of shares:

$$\begin{aligned} 49\tfrac{5}{8} &= 49 + \tfrac{5}{8} \\ &= 49 + 0.625 \\ &= 49.625 \end{aligned}$$

You might wish to do this by calculator: $5 \div 8 = 0.625$

Thus, the total price is $400 \times \$49.625 =$ **$19,850.** ■

Do not round until you are ready to state your answer.

PROBLEM SET 1.3

Essential Ideas

1. **In Your Own Words** What is a fraction?

2. **In Your Own Words** Distinguish between a common fraction and a decimal fraction.

3. **In Your Own Words** Explain why division by zero is not permitted.

4. **In Your Own Words** Give the names of the positional columns from trillions to millionths.

Level 1—Drill

WHAT IS WRONG, *if anything, with each of the statements in Problems 5–10? Explain your reasoning.*

5. $\frac{0}{2} = 0$
6. $\frac{10}{0} = 0$
7. $0.05 = 0.050$
8. 0.75 is a fraction
9. $\frac{1}{3} = 0.333333333$
10. $0.\overline{6} = 0.666666666$

Show the sequence of buttons to be pressed in order to enter each of the fractions in Problems 11–14 into a calculator.

11. a. $1\frac{1}{2}$ b. $6\frac{5}{6}$
12. a. $\frac{8}{13}$ b. $\frac{17}{21}$
13. a. $14\frac{3}{4}$ b. $3\frac{2}{7}$
14. a. $4\frac{3}{7}$ b. $2\frac{1}{2}$

Write each of the improper fractions in Problems 15–26 as a mixed number or a whole number.

15. a. $\frac{3}{2}$ b. $\frac{4}{3}$
16. a. $\frac{5}{4}$ b. $\frac{19}{2}$
17. a. $\frac{16}{3}$ b. $\frac{25}{4}$
18. a. $\frac{141}{10}$ b. $\frac{163}{10}$
19. a. $\frac{1,681}{10}$ b. $\frac{1,493}{10}$
20. a. $\frac{833}{100}$ b. $\frac{1,457}{100}$
21. a. $\frac{27}{16}$ b. $\frac{177}{10}$
22. a. $\frac{33}{16}$ b. $\frac{27}{7}$
23. a. $\frac{41}{12}$ b. $\frac{61}{5}$
24. a. $\frac{118}{15}$ b. $\frac{89}{21}$
25. a. $\frac{83}{5}$ b. $\frac{42}{3}$
26. a. $\frac{18}{3}$ b. $\frac{125}{4}$

Write each of the mixed numbers in Problems 27–38 as an improper fraction.

27. a. $2\frac{1}{2}$ b. $1\frac{2}{3}$
28. a. $1\frac{3}{4}$ b. $3\frac{3}{5}$
29. a. $4\frac{3}{8}$ b. $5\frac{1}{4}$
30. a. $3\frac{2}{3}$ b. $5\frac{1}{5}$
31. a. $6\frac{1}{2}$ b. $3\frac{2}{5}$
32. a. $1\frac{3}{10}$ b. $4\frac{2}{5}$
33. a. $2\frac{2}{5}$ b. $3\frac{7}{10}$
34. a. $17\frac{2}{3}$ b. $12\frac{4}{5}$
35. a. $11\frac{3}{8}$ b. $2\frac{9}{10}$
36. a. $1\frac{14}{15}$ b. $2\frac{13}{17}$
37. a. $19\frac{3}{5}$ b. $17\frac{1}{8}$
38. a. $18\frac{7}{8}$ b. $3\frac{11}{12}$

Change the fractions and mixed numbers in Problems 39–50 into decimal fractions.

39. a. $\frac{1}{8}$ b. $\frac{3}{8}$
40. a. $\frac{5}{6}$ b. $\frac{7}{6}$
41. a. $\frac{3}{5}$ b. $\frac{7}{8}$
42. a. $2\frac{1}{2}$ b. $5\frac{1}{3}$
43. a. $\frac{3}{7}$ b. $3\frac{1}{6}$
44. a. $6\frac{1}{12}$ b. $6\frac{2}{3}$
45. a. $\frac{5}{9}$ b. $\frac{7}{9}$
46. a. $7\frac{5}{6}$ b. $4\frac{1}{15}$
47. a. $3\frac{1}{30}$ b. $\frac{2}{7}$
48. a. $2\frac{2}{3}$ b. $4\frac{1}{6}$
49. a. $\frac{7}{15}$ b. $\frac{8}{9}$
50. a. $4\frac{3}{8}$ b. $3\frac{1}{9}$

Level 2—Drill

Estimate as a common fraction the size of the shaded portion of the regions in Problems 51–52.

51. a.
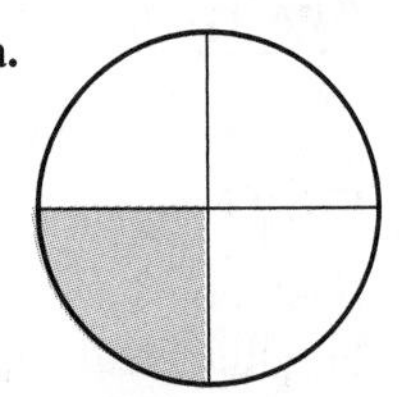
b.
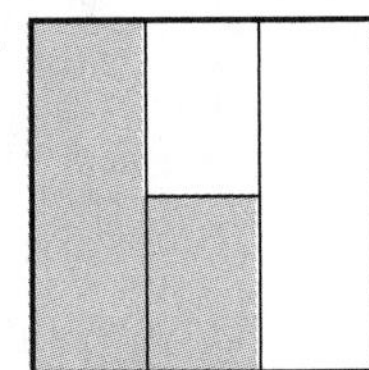

52. a.
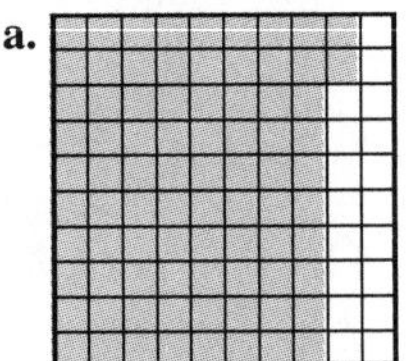
b.
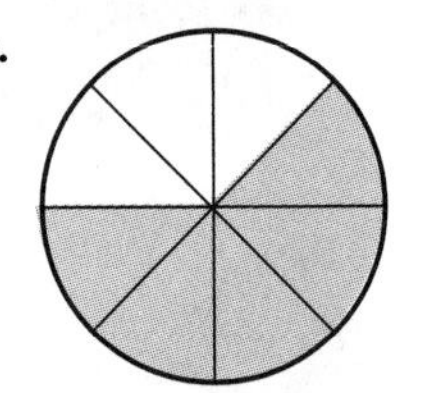

Estimate as a decimal fraction the shaded portions for each of the regions in Problems 53–54.

53. a. 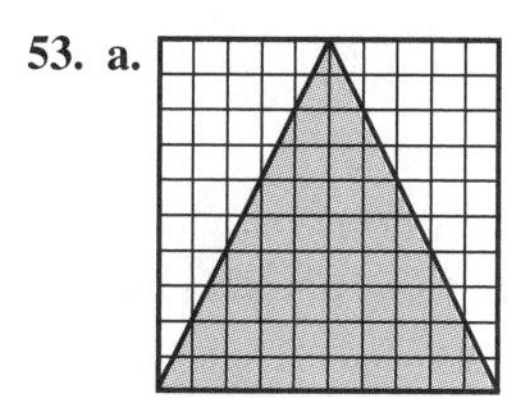 **b.** 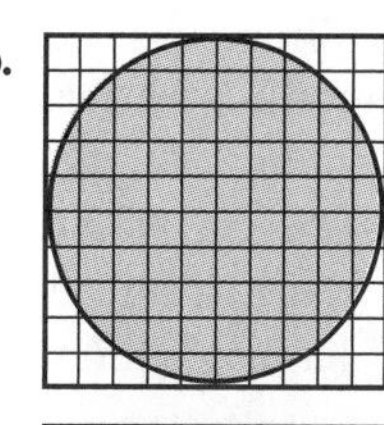

54. a. 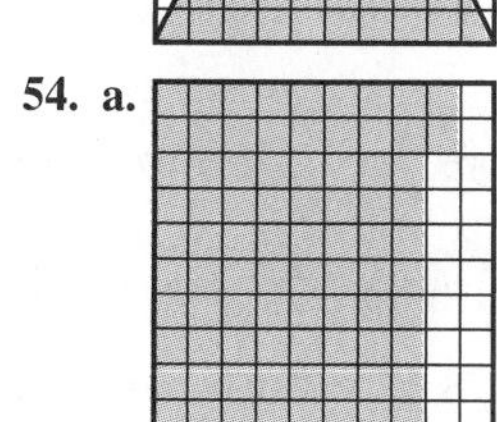 **b.** 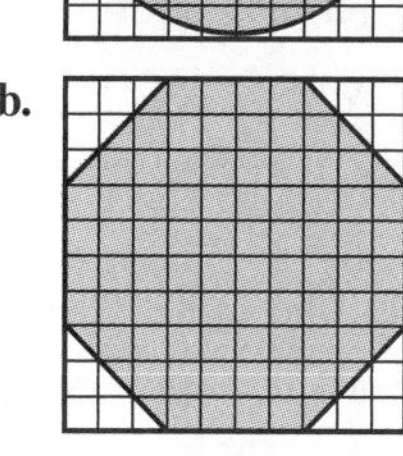

In Problems 55–57, use a calculator to write each fraction in decimal form. Give the exact decimal value, if possible. If you can't give the exact value, indicate that the answer you are giving is approximate.

55. a. $\frac{1}{11}$ **b.** $\frac{1}{12}$

56. a. $\frac{7}{15}$ **b.** $\frac{7}{22}$

57. a. $\frac{5}{19}$ **b.** $\frac{3}{17}$

Level 2—Applications

Stock prices are usually stated using eighths of a dollar. Find the total price of each transaction in Problems 58–63.

58. 100 shares of Citicorp at $52\frac{1}{2}$

59. 100 shares of Liz Claiborne at $18\frac{3}{4}$

60. 200 shares of Apple Computer at $42\frac{1}{4}$

61. 300 shares of National Semi at $126\frac{1}{2}$

62. 100 shares of Delta at $63\frac{1}{8}$

63. 100 shares of Genentech at $60\frac{5}{8}$

Level 3—Mind Boggler

64. Insert appropriate operation signs (+, −, ×, or ÷) between each digit so that the following becomes a true statement:

$$1 \quad 2 \quad 3 \quad 4 \quad 5 \quad 6 \quad 7 \quad 8 \quad 9 = 100$$

1.4 ROUNDING AND ESTIMATION

Historical Perspective

The numeration system we use every day, the one being presented in this chapter, is called the Hindu–Arabic numeration system because it had its origins in India and was brought to Baghdad and translated into Arabic (by the year 750 A.D.). It is also sometimes called the decimal numeration system because it is based on 10 symbols or, as they are sometimes called, digits. A brief discussion of the Hindu–Arabic numeration system is found in Appendix A.

A study of the Native Americans of California yields a wide variety of number bases different from 10. Several of these bases are discussed by Barnabas Hughes in an article entitled, "California Indian Arithmetic" in *The Bulletin of the California Mathematics Council* (Winter 1971/1972). According to Professor Hughes, the Yukis, who live north of Willits, have both the quaternary (base 4) and the octonary (base 8) system. Instead of counting on their fingers, these Native Americans enumerated the spaces between their fingers.

Very large or very small numbers are difficult to comprehend. Most of us are accustomed to hearing about millions, billions, and even trillions, but do we really understand the magnitude of these numbers?

If you were to count one number per second, nonstop, it would take about 278 hours, or approximately $11\frac{1}{2}$ days, to count to a million. Not a million days have elapsed since the birth of Christ (a million days is about 2,700 years). A large book of about 700 pages contains about a million letters. A million bottle caps placed in a single line would stretch about 17 miles.

But the age in which we live has been called the age of billions. How large is a billion? How long would it take you to count to a billion? Go ahead—make a guess. To get some idea about how large a billion is, let's compare it to some familiar units:

If you gave away $1,000 *per day,* it would take you more than 2,700 *years* to give away a billion dollars.

A stack of a billion $1 bills would be more than 59 miles high.

At 8% interest, a billion dollars would earn you $219,178.08 interest *per day*!

A billion seconds ago, John F. Kennedy was assassinated.

A billion minutes ago, was the second century A.D.

A billion hours ago, people had not yet appeared on earth.

To carry out the calculations necessary to make the preceding statements, we need two ideas. The first is rounding (discussed in this section), and the second is scientific notation (discussed in the next section). Rounding, estimation, and scientific notation can help increase the clarity of our understanding of numbers. For example, a budget of $252,892,988.18 for 99,200 students can be approximated as a $250 million budget for 100,000 students. A measurement of $19\frac{15}{16}$ inches might be recorded as 20 inches. You are often required to round decimals (for example, when you work with money). Estimates for the purpose of clarity are almost always found by rounding.

What do we mean by **rounding**? The following procedure should help explain.

Procedure for Rounding a Decimal Number

Step 1 **Locate the rounding place digit.** This is identified by the column name. (These are listed on page 22.)

Step 2 **Determine the rounding place digit.**

a. It stays the same if the first digit to its right is a 0, 1, 2, 3, or 4.

b. It increases by 1 if the digit to the right is a 5, 6, 7, 8, or 9. (If the rounding place digit is a 9 and 1 is added, there will be a carry in the usual fashion.)

Step 3 **Change digits.**

a. All digits to the left of the rounding digit remain the same (unless there is a carry).

b. All digits to the right of the rounding digit are changed to zeros.

(continued)

Step 4 **Drop zeros.**

a. If the rounding place digit is to the left of the decimal point, drop all trailing zeros.

b. If the rounding place digit is to the right of the decimal point, drop all trailing zeros to the right of the rounding place digit.

EXAMPLE 1

Rounding

Round 46.8217 to the nearest hundredth.

Solution

Step 1. The rounding place digit is in the hundredths column.

Step 2. Rounding place digit

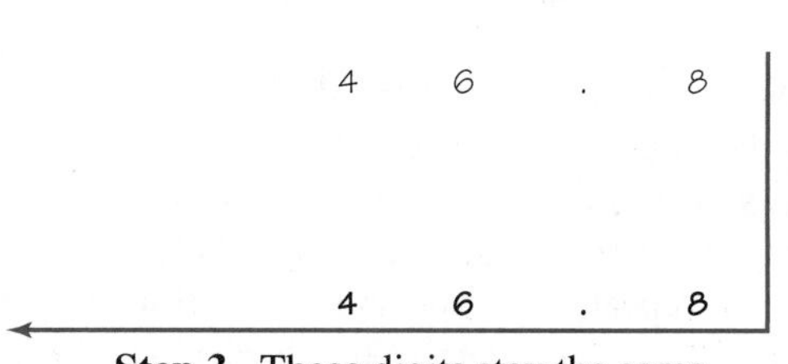

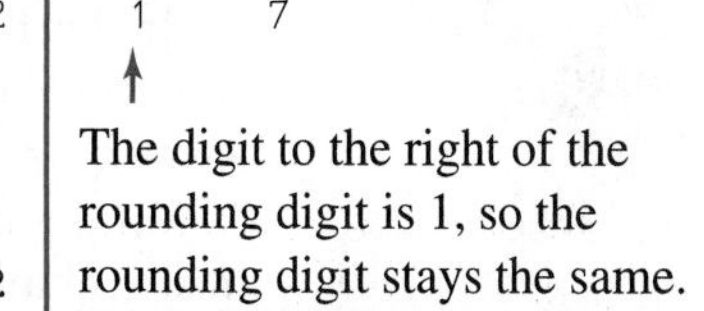

Step 3. These digits stay the same.

Step 4. These digits are changed into zeros; drop them because they are trailing zeros.

The rounded number is **46.82.**

EXAMPLE 2

Rounding with a Carry

Round 13.6992 to the nearest hundredth.

Solution

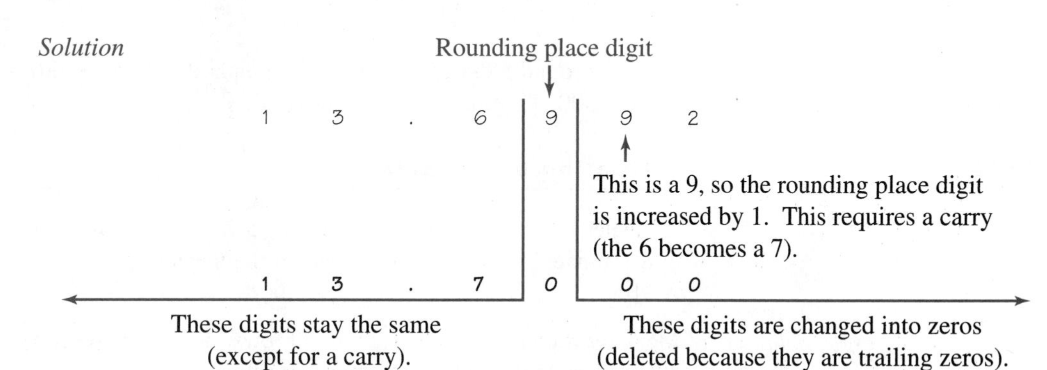

The rounded number is **13.70.** Notice that the zero that appears as the rounding place digit is *not* deleted. ■

EXAMPLE 3 **Rounding; Drop Trailing Zeros**

Round 72,416.921 to the nearest hundred.

Solution

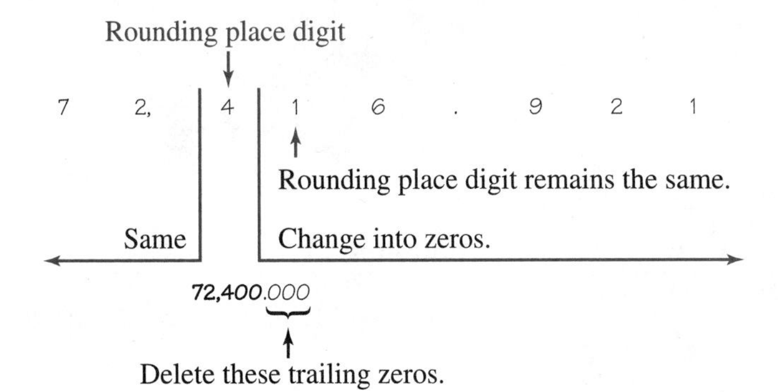

The rounded number is **72,400.** ■

Note the agreement about rounding used in this book.

In this book, we will round answers dealing with money to the nearest cent, and you should do the same in your work. Otherwise, you should not round any of your answers unless you are instructed to do so. This means, for example, that if you are converting fractions to decimals, you must include all decimal places until the decimal terminates or repeats (and you show this with an overbar). If you are using a calculator, you will need to interpret the display as representing either a terminating or a repeating decimal.

Rounding is often used as a means of estimation. Throughout this book, we will provide multiple-choice estimation questions so that you can practice and strengthen your estimating skills.

EXAMPLE 4 **Estimation by Rounding**

Divide $858.25 into three shares. Which of the following is the best estimate of one share?

A. $90 B. $300 C. $900

Solution Mentally round $858.25 to the nearest hundred dollars ($900); then divide by 3 to obtain $300. Thus, the best estimate is **B.** ■

EXAMPLE 5 **Estimation by Rounding**

A theater has 53 rows with 39 seats in each row. Which of the following is the best estimate of the number of seats in the theater?

A. 100 B. 1,000 C. 2,000

Solution Mentally round to the nearest ten: 50 rows of 40 seats each would give $50 \times 40 = 2{,}000$ seats. Thus, the best estimate is **C.** ■

PROBLEM SET 1.4

ESSENTIAL IDEAS

1. **In Your Own Words** What is a rounding place digit?
2. **In Your Own Words** Describe the process of rounding.

LEVEL 1

3. **In Your Own Words** Estimate the distance that would be covered by 1,000,000 textbooks, laid end to end.
4. **In your Own Words** Describe the size of a billion.

WHAT IS WRONG, *if anything, with each of the statements in Problems* 5–10? *Explain your reasoning.*

5. 30.05 rounded to the nearest tenth is 30.
6. 625.97555 rounded to the nearest hundredth is 600.
7. 3,684,999 rounded to the nearest ten thousand is 3,690,000.
8. 12,456.9099 rounded to the nearest tenth is 12,456.9000.
9. If a jar contains 23 jellybeans, then 12 similar jars will contain about 250 jelly beans.
10. If a box contains 144 pencils, then obtaining 1,000 pencils will require about 7 boxes.

LEVEL 1—Drill

Round the numbers in Problems 11–32 *to the given degree of accuracy.*

11. 2.312; nearest tenth
12. 14.836; nearest tenth
13. 6,287.4513; nearest hundredth
14. 342.355; nearest hundredth
15. 5.291; one decimal place
16. 5.291; two decimal places
17. 6,287.4513; nearest hundred
18. 6,287.4513; nearest thousand
19. 12.8197; two decimal places
20. 813.055; two decimal places
21. 4.81792; nearest thousandth
22. 1.396; nearest unit
23. 4.8199; nearest whole number
24. 48.5003; nearest whole number
25. \$12.993; nearest cent
26. \$6.4312; nearest cent
27. \$14.998; nearest cent
28. \$6.9741; nearest cent
29. 694.3814; nearest ten
30. 861.43; nearest hundred
31. \$86,125; nearest thousand dollars
32. \$125,500; nearest thousand dollars

Round the calculator display shown in Problems 33–38.

```
2/3
          .6666666667
2/17
          .1176470588
7/51
           .137254902
```

```
1/3
          .3333333333
5/11
          .4545454545
19/53
           .358490566
```

33. 2/3; 3-place accuracy
34. 1/3; 4-place accuracy
35. 2/17; 3-place accuracy
36. 5/11; 4-place accuracy
37. 7/51; 3-place accuracy
38. 19/53; 4-place accuracy

LEVEL 2—Drill

Estimate answers for Problems 39–44 *by choosing the most reasonable answer. You should not do any pencil-and-paper or calculator arithmetic for these problems.*

39. The distance to school is 4.82 miles, and you must make the trip to school and back five days a week. Estimate how far you drive each week.
    A. 9.64 miles B. 50 miles C. 482 miles
40. The length of a lot is 279 ft, and you must order fencing material that requires six times the length of the lot. Estimate the amount of material you should order.
    A. 1,600 ft B. 50 ft C. 10,000 ft

41. If a person's annual salary is $35,000 estimate the monthly salary.
A. $500 B. $1,000 C. $3,000

42. If 500 shares of a stock cost $19,087.50, estimate the value of each share of stock.
A. $40 B. $400 C. $4,000

43. If you must pay back $1,000 in 12 monthly installments, estimate the amount of each payment.
A. $25 B. $100 C. $12,000

44. If an estate of $22,000 is to be divided equally among three children, estimate each child's share.
A. $7,000 B. $733.33 C. $65

LEVEL 2 —Applications

A baseball player's batting average is found by dividing the number of hits by the number of times at bat. This number is then rounded to the nearest thousandth. Find each player's batting average in Problems 45–50.

45. Jack Foley was at bat 6 times with 2 hits.

46. Niels Sovndal was at bat 20 times with 7 hits.

47. Ben Becker had 5 hits in 12 times at bat.

48. Jeff Tredway had 98 hits in 306 times at bat.

49. Hal Morris had 152 hits in 478 times at bat.

50. Tony Gwynn had 168 hits in 530 times at bat.

51. If a person's annual salary is $15,000, what is the monthly salary?

52. If a person's annual tax is $512, how much is that tax per month?

53. If the sale of 150 shares of PERTEC stock grossed $1,818.75, how much was each share worth?

54. If an estate of $22,000 is to be divided equally among three children, what is each child's share?

55. If you must pay back $850 in 12 monthly payments, what is the amount of each payment?

56. If you must pay back $1,000 in 12 monthly payments, what is the amount of each payment?

57. A businesswoman bought a copy machine for her office. If it cost $674 and the useful life is six years, what is the cost per year?

58. A businessman bought a fax for his office. If it cost $890 and the useful life is seven years, what is the cost per year?

LEVEL 3 —Mind Boggler

59. In the B.C. cartoon, Peter has a mental block against 4s. See whether you can handle 4s by writing the numbers from 1 to 10 using four 4s for each.

Here are the first three completed for you:

$$\frac{4}{4} + 4 - 4 = 1 \qquad \frac{4}{4} + \frac{4}{4} = 2 \qquad \frac{4 + 4 + 4}{4} = 3$$

More than one answer is possible. For example,

$$\frac{4}{4 + 4 - 4} = 1 \quad \text{and} \quad 4 - \frac{4 + 4}{4} = 2$$

B.C. By permission of Johnny Hart and Creators Syndicate, Inc.

1.5 EXPONENTS AND PRIME FACTORIZATION

Historical Perspective

Karl Friedrich Gauss (1777–1855)

Karl Gauss is considered to be one of the greatest mathematicians in the history of mathematics. Gauss joked that he knew how to reckon before he could talk: It seems that when he was only 3 years old, he corrected his father's computations on a payroll report. Gauss graduated from college at the age of 15 and proved what was to become known as the Fundamental Theorem of Algebra for his doctoral thesis at the age of 22. One of the results of this theorem is used in this section when we assume that every number is either prime or can be factored as the product of primes. Gauss published only a small portion of the ideas that seemed to storm his mind because he felt that each published result had to be complete, concise, polished, and convincing. His motto was "Few, but ripe."

We often encounter numbers that are made by repeated multiplication of the same numbers. For example,

$$10 \times 10 \times 10 \qquad 6 \times 6 \times 6 \times 6 \times 6$$

$$15 \times 15 \times 15 \times 15 \times 15 \times 15 \times 15 \times 15 \times 15 \times 15 \times 15 \times 15 \times 15$$

These numbers can be written more simply by inventing a new notation:

$$10^3 = \underbrace{10 \times 10 \times 10}_{\text{3 factors}}$$

$$6^5 = \underbrace{6 \times 6 \times 6 \times 6 \times 6}_{\text{5 factors}}$$

$$15^{13} = \underbrace{15 \times 15 \times \cdots \times 15}_{\text{13 factors}}$$

We call this **power** or **exponential notation.** The number that is multiplied as a repeated factor is called the **base,** and the number of times the base is used as a factor is called the **exponent.** To use your calculator for exponents, locate the $\boxed{y^x}$, $\boxed{x^y}$, or $\boxed{\wedge}$ keys. For example, to find the value of 6^5,

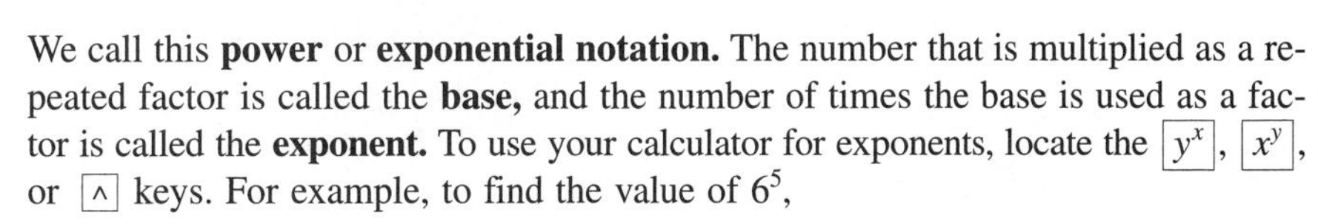

$\boxed{6}$ $\boxed{\wedge}$ $\boxed{5}$ $\boxed{=}$

You should see 7,776 displayed.

EXAMPLE 1 **Definition of Exponent**

Tell what each of the given expressions means, and then expand (multiply out). Use your calculator where appropriate.

a. 6^2 **b.** 10^3 **c.** 7^5 **d.** 2^{15}

Solution

a. $6^2 = 6 \times 6$ or **36**
The base is 6; the exponent is 2; this is pronounced "six squared."

b. $10^3 = 10 \times 10 \times 10 =$ **1,000**
The base is 10; the exponent is 3; this is pronounced "ten cubed."

c. $7^5 = 7 \times 7 \times 7 \times 7 \times 7 =$ **16,807** (by calculator)
The base is 7; the exponent is 5; this is pronounced "seven to the fifth power."

d. $2^{15} = \underbrace{2 \times 2 \times 2 \times \cdots \times 2 \times 2}_{15 \text{ factors}} =$ **32,768** ■

Historical Perspective

For historical, geometrical reasons, special terminology is used only with exponents of 2 or 3. The area of a square with side s is s^2, so the exponent 2 is pronounced "**squared.**" In the volume of a cube, the exponent 3 is pronounced "**cubed.**"

We now use this notation to observe a pattern for **powers of 10:**

$$10^1 = 10$$
$$10^2 = 10 \times 10 = 100$$
$$10^3 = 10 \times 10 \times 10 = 1{,}000$$
$$10^4 = 10 \times 10 \times 10 \times 10 = 10{,}000$$

Do you see a relationship between the exponent and the number?

$$10^5 = 1\underbrace{00{,}000}_{5 \text{ zeros}}$$

↑ Exponent is 5.

Notice that the exponent and the number of zeros are the same.

Could you write 10^{12} without actually multiplying?*

There is a similar pattern for multiplication of any number by a power of 10. Consider the following examples, and notice what happens to the decimal point:

$$9.42 \times 10^1 = 94.2$$
$$9.42 \times 10^2 = 942.$$
$$9.42 \times 10^3 = 9{,}420.$$
$$9.42 \times 10^4 = 94{,}200.$$

We find these answers by direct multiplication.

Do you see a pattern? If we multiply 9.42×10^5, how many places to the right will the decimal point be moved?

$$9.42 \times 10^5 = 9\ 42{,}000.$$

↓ ↑
5 places →

This answer is found by observing the pattern, not by directly multiplying.

Using this pattern, can you multiply the following *without direct calculation?*†

$$9.42 \times 10^{12}$$

We will investigate one final pattern of 10s, this time looking at smaller values in this pattern.

$$\vdots$$
$$100{,}000 = 10^5$$
$$10{,}000 = 10^4$$
$$1{,}000 = 10^3$$
$$100 = 10^2$$
$$10 = 10^1$$

*Answer: 1,000,000,000,000
†Answer: 9,420,000,000,000

Continuing with the same pattern, we have

$1 = 10^0$ We interpret the zero exponent as "decimal point moves 0 places."

$0.1 = 10^{-1}$ We will define negative numbers in Chapter 2.

$0.01 = 10^{-2}$ For now, we use the symbols −1, −2, −3, . . . as exponents to show the position of the decimal point when written as a decimal fraction.

$$0.001 = 10^{-3}$$
$$0.0001 = 10^{-4}$$
$$0.00001 = 10^{-5}$$
$$\vdots$$

When we multiply a power of 10 by some number, a pattern emerges:

$9.42 \times 10^2 = 942.$

$9.42 \times 10^1 = 94.2$

$9.42 \times 10^0 = 9.42$ Decimal moves 0 places.

$9.42 \times 10^{-1} = 0.942$ These answers are found by direct multiplication: $9.42 \times 0.1 = 0.942$.

$9.42 \times 10^{-2} = 0.0942$

$9.42 \times 10^{-3} = 0.00942$

Do you see that the same pattern holds for multiplying by a negative exponent? Can you multiply 9.42×10^{-6} *without direct calculation*? The solution is as follows:

$$9.42 \times 10^{-6} = 0.00000\,9\,42$$

← Moved 6 places to the left.

These patterns lead to a useful way of writing large and small numbers, called **scientific notation.**

Scientific Notation

The **scientific notation** of a number is that number written as a power of 10 or as a decimal number between 1 and 10 times a power of 10.

EXAMPLE 2 **Writing Numbers in Scientific Notation**

Write the given numbers in scientific notation.

a. 123,600 **b.** 0.00003 5 **c.** 48,300 **d.** 0.0821
e. 1,000,000,000,000 **f.** 7.35

Solution

a. $123{,}600 = 1.236 \times 10^?$

Step 1 Fix the decimal point after the first nonzero digit.

Step 2 From this number, count the number of decimal places to restore the number to its given form:

$$123{,}600 = 1.23600 \times 10^?$$

5 places to the right →

Step 3 The exponent is the same as the number of decimal places needed to restore scientific notation to the original given number:

$123{,}600 = \mathbf{1.236 \times 10^5}$

b. $0.000035 = 3.5 \times 10^?$

Step 1 ↑

Step 2 $0.000035 = 0\ 00003.5 \times 10^?$

← 5 places to the left; this is −5.

Step 3 $0.000035 = \mathbf{3.5 \times 10^{-5}}$

c. $48{,}300 = \mathbf{4.83 \times 10^4}$

d. $0.0821 = \mathbf{8.21 \times 10^{-2}}$

e. $1{,}000{,}000{,}000{,}000 = \mathbf{10^{12}}$

f. $7.35 = 7.35 \times 10^0$ or just **7.35** ■

When working with very large or very small numbers, it is customary to use scientific notation. For example, suppose we wish to expand 2^{63}. A calculator can help us with this calculation:

[2] [^] [63] [=]

The result is larger than can be handled with a calculator display, so your calculator will automatically output the answer in scientific notation, using one of the following formats:

9.223372037E18 9.223372037 18 $9.223372037 \times 10^{18}$

CAUTION

Do not confuse the exponent key [y^x] with scientific notation [EE]; they are NOT the same.

If you wish to enter a very large (or small) number into a calculator, you can enter these numbers using the scientific notation button on most calculators. Use the key labeled [EE], [EXP], or [SCI]. Whenever we show the [EE] key, we mean press the scientific notation key on your brand of calculator.

NUMBER	*SCIENTIFIC NOTATION*	*CALCULATOR INPUT*	*CALCULATOR DISPLAY*
468,000	4.68×10^5	[4.68] [EE] [5]	4.68 05 or 4.68 E5
93,000,000,000	9.3×10^{10}	[9.3] [EE] [10]	9.3 10 or 9.3 E10

EXAMPLE 3

Operations with Scientific Notation

If the federal budget is $1.5 trillion (1,500,000,000,000), how much does it cost each individual, on average, if there are 240,000,000 people?

Solution We must divide the budget by the number of people. We will use scientific notation and a calculator:

$$1{,}500{,}000{,}000{,}000 = 1.5 \times 10^{12} \text{ and } 240{,}000{,}000 = 2.4 \times 10^{8}$$

The desired calculation is $\dfrac{1.5 \times 10^{12}}{2.4 \times 10^{8}}$.

We carry out this calculation using the appropriate scientific notation keys:

[1.5] [EE] [12] [÷] [2.4] [EE] [8] [=]

The answer is **$6,250 per person.** ■

When you multiply numbers, the numbers being multiplied are called **factors.** The process of taking a given number and writing it as the product of two or more other numbers is called **factoring,** with the result called a **factorization** of the given number.

EXAMPLE 4

Finding Factors

Find the factors of the given numbers.

a. 1 **b.** 2 **c.** 3 **d.** 4 **e.** 5 **f.** 6 **g.** 7 **h.** 8

Solution

a. $1 = 1 \times 1$ **The factor is 1.**
b. $2 = 2 \times 1$ **Factors: 1, 2**
c. $3 = 3 \times 1$ **Factors: 1, 3**
d. $4 = 4 \times 1$ or 2×2 **Factors: 1, 2, 4**
e. $5 = 5 \times 1$ **Factors: 1, 5**
f. $6 = 6 \times 1$ or 2×3 **Factors: 1, 2, 3, 6**
g. $7 = 7 \times 1$ **Factors: 1, 7**
h. $8 = 8 \times 1$ or 4×2 or $2 \times 2 \times 2$ **Factors: 1, 2, 4, 8** ■

Let's categorize the numbers listed in Example 4:

	PRIMES	*COMPOSITES*
Fewer than two factors:	*Exactly two factors:*	*More than two factors:*
1	2	4
	3	6
	5	8
	7	

A **prime** number is a natural number with exactly two distinct factors, and a **composite** number is a number with more than two factors. Will any number besides 1 have fewer than two factors? The primes smaller than 100 are listed in the box at the top of page 40.

Primes Smaller Than 100

2, 3, 5, 7, 11, 13, 17, 19, 23, 29, 31, 37, 41, 43, 47, 53, 59, 61, 67, 71, 73, 79, 83, 89, 97

A **prime factorization** of a number is a factorization that consists exclusively of prime numbers.

EXAMPLE 5

Prime Factorization of a Number

Find the prime factorization of 36.

Solution

Step 1 From your knowledge of the basic multiplication facts, write any two numbers whose product is the given number. Circle any prime factor.

Step 2 Repeat the process for uncircled numbers.

Step 3 When all the factors are circled, their product is the *prime factorization.*

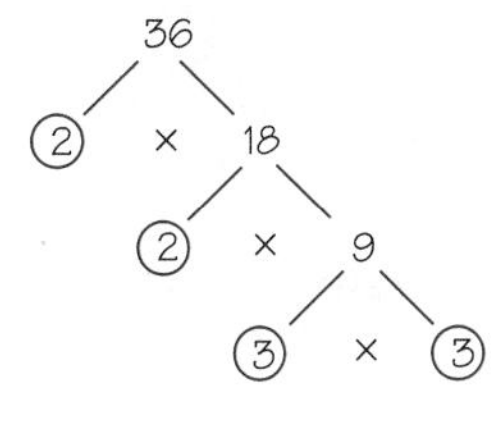

$36 = 2 \times 2 \times 3 \times 3 = \mathbf{2^2 \times 3^2}$ ■

EXAMPLE 6

Prime Factorization of a Number

Find the prime factorization of 48.

Solution

$48 = 2 \times 2 \times 2 \times 2 \times 3 = \mathbf{2^4 \times 3}$ ■

If you cannot readily find the prime factors of a number, you should look at the list of prime factors from the smallest to the larger numbers, as illustrated in Example 7.

EXAMPLE 7

Prime Factorization of a Large Number

Find the prime factorization of 34,153.

Solution First, try 2; you might notice that 34,153 is not an even number, so 2 is not a factor.

Next, try 3 (the next prime after 2); 3 does not divide evenly into 34,153, so 3 is not a factor. Next, try 5; it is also not a factor.

34,153
(7) × 4,879
(7) × 697
(17) × (41)

The next prime to try is 7, by calculator or long division:

Press: [34153] [÷] [7] [=] *Display:* 4879

Now focus on 4,879. Previously tried numbers cannot be factors, but since 7 *was* a factor, it might be again: $4{,}879 = 7 \times 697$.
Continue the process: 697 is not divisible (evenly) by 7, so try 11 (doesn't divide evenly), 13 (doesn't divide evenly), and 17 (does). The remaining number, 41, is a prime, so the process is complete:

$$34{,}153 = \mathbf{7^2 \times 17 \times 41}$$ ■

PROBLEM SET 1.5

Essential Ideas

1. **In Your Own Words** What is an exponent?

2. **In Your Own Words** What is scientific notation?

3. **In Your Own words** Describe a process for finding the prime factorization of a number.

4. **In Your Own Words** Describe the difference between the [y^x] and [EE] calculator keys.

Level 1

WHAT IS WRONG, *if anything, with each of the statements in Problems* 5–8? *Explain your reasoning.*

5. $5^2 = 10$

6. 2^3 means $2 + 2 + 2$

7. 4^3 means multiply 4 by itself 3 times

8. A number is in scientific notation when it is written as a number between 1 and 10 times a power of 10.

9. Consider the number 10^6.
   a. What is the common name for this number?
   b. What is the base?
   c. What is the exponent?
   d. According to the definition of exponential notation, what does the number mean?

10. Consider the number 10^3.
   a. What is the common name for this number?
   b. What is the base?
   c. What is the exponent?
   d. According to the definition of exponential notation, what does the number mean?

11. Consider the number 10^{-1}.
   a. What is the common name for this number?
   b. What is the base?
   c. What is the exponent?
   d. According to the definition of exponential notation, what does the number mean?

12. Consider the number 10^{-2}.
   a. What is the common name for this number?
   b. What is the base?
   c. What is the exponent?
   d. According to the definition of exponential notation, what does the number mean?

Level 1—Drill

Write each of the numbers in Problems 13–25 *in scientific notation.*

13. a. 3,200 **b.** 25,000

14. a. 18,000,000 **b.** 640

15. a. 0.004 **b.** 0.02

16. a. 0.0035 **b.** 0.00000 045

17. a. 5,624 **b.** 15,824

18. a. 23.79 **b.** 0.00081 7

19. a. 35,000,000,000 **b.** 63,000,000

20. a. 0.00001 **b.** 0.00000 00000 00000 00003 5

21. a. 0.00008 61 **b.** 249,000,000

22. a. 100 **b.** $11\frac{1}{2}$

23. a. 6.34E9 **b.** 5.2019E11

24. a. 4.093745 08 **b.** 8.291029292 12

25. a. 2.029283 −03 **b.** 5.209E−05

Write each of the numbers in Problems 26–40 without using exponents.

26. a. 7.2×10^{10} **b.** 4.5×10^3

27. a. 3.1×10^2 **b.** 6.8×10^8

28. a. 2.1×10^{-3} **b.** 4.6×10^{-7}

29. a. 2.05×10^{-1} **b.** 3.013×10^{-2}

30. a. 3.2×10^0 **b.** 8.03×10^{-4}

31. a. 5.06×10^3 **b.** 6.81×10^0

32. a. 7^2 **b.** 5^2

33. a. 2^6 **b.** 6^3

34. a. 2^8 **b.** 8^2

35. a. 4^3 **b.** 2^5

36. a. 10^4 **b.** 3^4

37. a. 4^5 **b.** 9^3

38. a. 2.18928271 10 **b.** 0.0000329 07

39. a. 0.00029214E12 **b.** 4.29436732478E19

40. a. 3.56 − 10 **b.** 3.8928E−14

Level 2—Drill

Find the prime factorization (written with exponents) for the numbers in Problems 41–49.

41. a. 12 **b.** 20

42. a. 120 **b.** 24

43. a. 256 **b.** 18

44. a. 150 **b.** 105

45. a. 400 **b.** 1,000

46. a. 10,000 **b.** 720

47. a. 4,459 **b.** 229,333

48. a. 2,098,987 **b.** 803,257

49. a. 45,733 **b.** 29,791

Level 2

Estimate answers for Problems 50–55 by choosing the most reasonable answer. You should not do any pencil-and-paper or calculator arithmetic for these problems.

50. The number of marbles that could be placed into a bathtub is about:
A. 10^3 B. 10^4 C. 10^9

51. The number of seconds since you have been born is about:
A. 10^9 B. 10^{70} C. 10^{700}

52. The cost of a pack (15 sticks) of gum (in dollars) is closest to:
A. 10^{-10} B. 10^{-1} C. 10^0

53. In a restaurant, the cost of a cup of coffee (in dollars) is closest to:
A. 10^{-10} B. 10^{-1} C. 10^0

54. The distance from the earth to the sun (in miles) is closest to:
A. 10^3 B. 10^7 C. 10^{18}

55. The number of grains of sand on the earth is closest to:
A. 10^{50} B. 10^{-50} C. 10^{100}

Level 2—Applications

56. A light year is the distance that light travels in 1 year; this is about 5,869,713,600 miles. Write this distance in scientific notation.

57. The world's largest library, the Library of Congress, has approximately 59,000,000 items. Write this number in scientific notation.

58. A thermochemical calorie is about 41,840,000 ergs. Write this number in scientific notation.

59. The estimated age of the earth is about 5×10^9 years. Write this number without using an exponent.

60. Saturn is about 8.86×10^8 miles from the sun. Write this distance without using an exponent.

61. The mass of the sun is about 3.33×10^5 times the mass of the earth. Write this without using an exponent.

62. The sun develops about 5×10^{23} horsepower per second. Write this without using an exponent.

63. A **googol** is a very large number that is defined in the cartoon. Write a googol in scientific notation.

64. Estimate the total price of the following grocery items: 1 tube toothpaste, \$1.89; 1 lb peaches, \$0.79; 1 lb elbow macaroni, \$1.39; 3 yogurts, \$1.61; lettuce, \$0.59; sliced turkey, \$2.50; chips, \$1.89; paper towels, \$0.99; half gallon of orange juice, \$1.97; 1 lb cheddar cheese, \$1.99; 3 cucumbers, \$0.99; 1 lb frozen corn, \$1.09; 2 bars soap, \$1.79

65. Estimate the total cost of the following plumbing items: 120 $1\frac{1}{4}$-in. PVC pipe, \$76.80; $1\frac{1}{4}$-in. ball valve, \$17.89; $1\frac{1}{4}$-in PVC fittings, \$24.28; 1-in. PVC fitting, \$14.16; $\frac{3}{4}$-in. PVC fittings, \$6.48; 1-in. ball valve, \$12.84.

Level 3—Mind Bogglers

66. The national debt is about \$5,200,000,000,000. Write this number in scientific notation.

Suppose that there are 222,000,000 people in the United States. Write this number in scientific notation.

If the debt is divided equally among the people, how much (rounded to the nearest hundred dollars) is each person's share?

67. If it took 1 second to write down each digit, how long would it take to write all the numbers from 1 to 1,000,000?

68. Imagine that you have written the numbers from 1 to 1,000,000. What is the total number of zeros you have recorded?

1.6 COMMON FRACTIONS

Historical Perspective

In everyday life, we use different names or titles for ourselves, depending on the context or situation. For example, in some situations I'm called Karl, in others Dr. Smith or Professor Smith, and in other places I'm Shannon's dad. We learn to deal with these different names quite effectively in everyday situations, and we need to transfer this concept to numbers.

PEANUTS Reprinted by permission of UFS, Inc.

A common fraction may have different names or representations, each of which is better than others in a certain context. For example, $\frac{1}{2}$, 0.5, $\frac{50}{100}$, and $\frac{8}{16}$ all name the same common fraction. We call $\frac{1}{2}$ the **reduced form,** but any of these may be the preferred form, depending on what we want to do. In this section, when we refer to a fraction, we are referring to a common fraction. We say that a fraction is **reduced** if no counting number other than 1 divides evenly into both the numerator and the denominator. The process of reducing a fraction relies on the **fundamental property of fractions** and on the process of factoring introduced in Section 1.5.

Fundamental Property of Fractions

If you multiply or divide both the numerator and the denominator by the same nonzero number, the resulting fraction will be the same.

Reducing fractions is essential to working with fractions; study this procedure.

We will use this fundamental property first to reduce fractions, and then to multiply fractions. Let's begin by stating the procedure for **reducing fractions.**

Procedure for Reducing Fractions

Step 1 Find all common factors (other than 1).

Step 2 Divide out the common factors.

EXAMPLE 1

Reduce a Common Fraction

Reduce $\frac{36}{48}$.

Solution

Step 1 Completely factor both numerator and denominator. We did this in the previous section.

Step 2 Write the numerator and denominator in factored form:

$$\frac{36}{48} = \frac{2 \times 2 \times 3 \times 3}{2 \times 2 \times 2 \times 2 \times 3}$$

Step 3 Use the fundamental property of fractions to eliminate the common factors. This process is sometimes called **canceling.***

$$\frac{36}{48} = \frac{\not{2} \times \not{2} \times 3 \times \not{3}}{\not{2} \times \not{2} \times 2 \times 2 \times \not{3}}$$

Step 4 Multiply the remaining factors. Treat the canceled factors as 1s.

$$\frac{36}{48} = \frac{\overset{1 \times 1 \times 3 \times 1 = 3}{\not{2} \times \not{2} \times 3 \times \not{3}}}{\underset{1 \times 1 \times 2 \times 2 \times 1 = 4}{\not{2} \times \not{2} \times 2 \times 2 \times \not{3}}} = \mathbf{\frac{3}{4}}$$

Notice how slashes are used as 1s.

This process is rather lengthy and can sometimes be shortened by noticing common factors that are larger than prime factors. For example, you might have noticed that 12 is a common factor in Example 1, so that

$$\frac{36}{48} = \frac{3 \times \not{12}}{4 \times \not{12}} = \frac{3}{4}$$

↑ Remember, this is 1.

■

EXAMPLE 2

Reducing Fractions

Reduce the fractions.

a. $\frac{4}{8}$ **b.** $\frac{75}{100}$ **c.** $\frac{35}{55}$ **d.** $\frac{160}{180}$ **e.** $\frac{1{,}200}{9{,}000}$

Solution

a. $\frac{4}{8} = \frac{1 \times 4}{2 \times 4} = \mathbf{\frac{1}{2}}$

b. $\frac{75}{100} = \frac{3 \times 25}{4 \times 25} = \mathbf{\frac{3}{4}}$

c. $\frac{35}{55} = \frac{7 \times 5}{11 \times 5} = \mathbf{\frac{7}{11}}$

*Notice that *cancel* does not mean "cross out or delete" factors. It means "use the fundamental property to eliminate common factors."

If the fractions are complicated, you may reduce them in several steps.

d. $\frac{160}{180} = \frac{10 \times 16}{10 \times 18} = \frac{2 \times 8}{2 \times 9} = \mathbf{\frac{8}{9}}$

e. $\frac{1,200}{9,000} = \frac{12 \times 100}{90 \times 100} = \frac{6 \times 2}{6 \times 15} = \mathbf{\frac{2}{15}}$ ■

We say that a fraction is **completely reduced** when there are no common factors of both the numerator and the denominator. CAUTION **In this book, all fractional answers should be reduced unless you are otherwise directed.**

We now turn to multiplying fractions.

Procedure for Multiplying Fractions

To multiply fractions, multiply numerators and multiply denominators.

EXAMPLE 3 **Multiplying Fractions**

Multiply the following.

a. $\frac{1}{3} \times \frac{2}{5}$ **b.** $\frac{2}{3} \times \frac{4}{7}$ **c.** $5 \times \frac{2}{3}$ **d.** $3\frac{1}{2} \times 2\frac{3}{5}$

Solution

a. $\frac{1}{3} \times \frac{2}{5} = \boxed{\frac{1 \times 2}{3 \times 5}} = \mathbf{\frac{2}{15}}$ ← This step can often be done in your head.

b. $\frac{2}{3} \times \frac{4}{7} = \mathbf{\frac{8}{21}}$

c. When multiplying a whole number and a fraction, as with $5 \times \frac{2}{3}$, write the whole number as a fraction and then multiply:

$$\frac{5}{1} \times \frac{2}{3} = \mathbf{\frac{10}{3}} \text{ or } 3\tfrac{1}{3}$$

d. When multiplying mixed numbers, as with $3\frac{1}{2} \times 2\frac{3}{5}$, write the mixed numbers as improper fractions and then multiply:

$$\frac{7}{2} \times \frac{13}{5} = \mathbf{\frac{91}{10}} \text{ or } 9\tfrac{1}{10}$$ ■

Both $\frac{10}{3}$ and $\frac{91}{10}$ are reduced fractions. Recall that a fraction is reduced if there is no number (other than 1) that divides into both the numerator and denominator evenly.

Don't fall into the trap of saying, "Why do I need to do this? I can find $\frac{4}{5}$ of $\frac{2}{3}$ by

$$\frac{4}{5} \times \frac{2}{3} = \frac{8}{15}$$

and that's all there is to it!" Figure 1.2 shows **why** multiplication of fractions works the way it does.

Multiplication of fractions can be visualized by considering the meaning of fractions as shown in Figure 1.2.

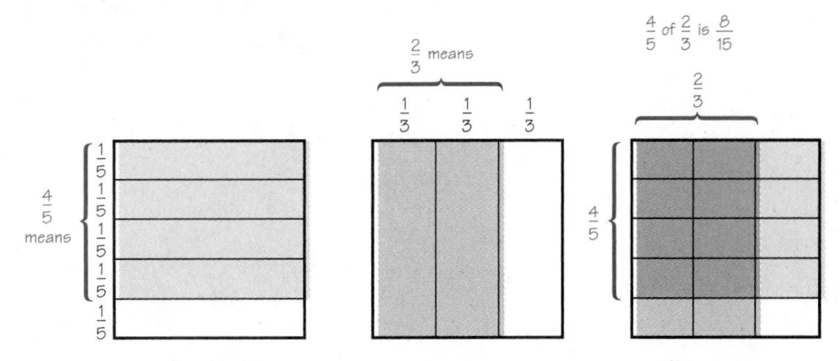

Figure 1.2 Geometrical justification of fractional multiplication

EXAMPLE 4

Illustrating the Meaning of Fractional Multiplication

Show that $\frac{3}{4}$ of $\frac{2}{3}$ is $\frac{6}{12}$.

Solution

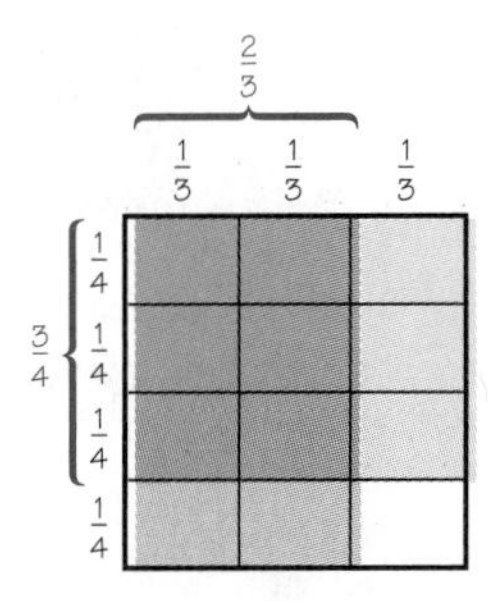

The result is $\mathbf{\frac{6}{12}}$ (6 shaded parts out of 12 parts). ■

After the multiplication has been done, the product is often a fraction (such as the answer in Example 4) that can and should be reduced. Remember, a fraction is in reduced form if the only counting number that can divide evenly into both the numerator and the denominator is 1.

EXAMPLE 5

Multiplying Fractions

Multiply the following.

a. $\frac{3}{4} \times \frac{2}{3}$ **b.** $\frac{2}{5} \times \frac{3}{4}$ **c.** $\frac{3}{5} \times \frac{1}{3}$ **d.** $3\frac{2}{3} \times 2\frac{2}{5}$ **e.** $\frac{4}{5} \times \frac{3}{8} \times \frac{2}{5}$

Solution **a.** $\frac{3}{4} \times \frac{2}{3} = \frac{6}{12}$ but $\frac{6}{12} = \frac{\cancel{3} \times \overset{1}{\cancel{2}}}{\underset{2}{\cancel{4}} \times \cancel{3}} = \mathbf{\frac{1}{2}}$

Notice that the actual multiplication was a wasted step because the answer was simply factored again to reduce it! Therefore, the proper procedure is to cancel common factors before you do the multiplication. Also, it is *not* necessary to write out prime factors. To reduce $\frac{6}{12}$, notice that 6 is a common factor.

These little numbers mean that → 6 divides into 6 one time, and 6 divides into 12 twice. →

$$\frac{\overset{1}{\cancel{6}}}{\underset{2}{\cancel{12}}} = \frac{1}{2}$$

6 is the common factor.

b. $\frac{2}{5} \times \frac{3}{4} = \underbrace{\frac{\overset{1}{\cancel{2}} \times 3}{5 \times \underset{2}{\cancel{4}}}} = \mathbf{\frac{3}{10}}$

Notice that this is very similar to the original multiplication as stated to the left of the equal sign. You can save a step by canceling with the original product, as shown in the next part.

c. $\frac{\overset{1}{\cancel{3}}}{5} \times \frac{1}{\underset{1}{\cancel{3}}} = \mathbf{\frac{1}{5}}$

d. $3\frac{2}{3} \times 2\frac{2}{5} = \frac{11}{\underset{1}{\cancel{3}}} \times \frac{\overset{4}{\cancel{12}}}{5} = \mathbf{\frac{44}{5}}$ or $8\frac{4}{5}$

e. With a more lengthy problem, such as $\frac{4}{5} \times \frac{3}{8} \times \frac{2}{5}$, you may do your canceling in several steps. We recopy the problem at each step for the sake of clarity, but in your work the result would look like the last step only.

Step 1 $\frac{\overset{1}{\cancel{4}}}{5} \times \frac{3}{\underset{2}{\cancel{8}}} \times \frac{2}{5}$

Step 2 $\frac{\cancel{4}}{5} \times \frac{3}{\underset{\underset{1}{\cancel{2}}}{\cancel{8}}} \times \frac{\overset{1}{\cancel{2}}}{5}$

Step 3 $\frac{\overset{1}{\cancel{4}}}{5} \times \frac{3}{\underset{\underset{1}{\cancel{2}}}{\cancel{8}}} \times \frac{\overset{1}{\cancel{2}}}{5} = \mathbf{\frac{3}{25}}$ ■

When you want to find the fractional part of any number, you can do so by multiplication. That is, the word *of* is often translated into multiplication:

$$\frac{4}{5} \text{ of } \frac{2}{3} \text{ is } \frac{4}{5} \times \frac{2}{3} = \frac{8}{15} \quad \textit{and} \quad \frac{3}{4} \text{ of } \frac{2}{3} \text{ is } \frac{3}{4} \times \frac{2}{3} = \frac{6}{12} = \frac{1}{2}$$

The following example shows how you can use this idea with an applied problem.

EXAMPLE 6 **Finding a Sale Price**

If a store is having a $\frac{1}{3}$-OFF sale, you must pay $\frac{2}{3}$ of the original price. What is the sale price of a suit costing \$355?

Solution We want to know "What is $\frac{2}{3}$ of \$355?"

$$\frac{2}{3} \text{ of } 355 = \frac{2}{3} \times 355 = \frac{710}{3}$$

This mixed operation can easily be done by calculator:

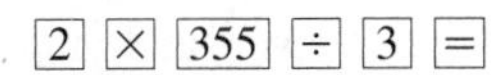

Since this is a problem involving a money answer, we want to write the answer in decimal form, rounded to the nearest cent. To do this, we divide 3 into 710 (and round) to obtain **\$236.67.** ■

If the product of two numbers is 1, then those numbers are called **reciprocals.** To find the reciprocal of a given number, write the number in fractional form and then **invert,** as shown in Example 7.

EXAMPLE 7 **Finding Reciprocals**

Find the reciprocal of each number, and then prove it is the correct reciprocal by multiplication.

Using a calculator, find the key labeled [1/x] or [x^{-1}] to find the reciprocal of a number in the display. For Example 7b, the reciprocal of 3 can be found: [3] [1/x]. Calculator reciprocals are given as decimal fractions.

a. $\frac{5}{11}$ Invert for reciprocal: $\mathbf{\frac{11}{5}}$ **Check:** $\frac{5}{11} \times \frac{11}{5} = 1$

b. 3 Write as a fraction: $\frac{3}{1}$

Invert for reciprocal: $\mathbf{\frac{1}{3}}$ **Check:** $3 \times \frac{1}{3} = 1$

c. $2\frac{3}{4}$ Write as a fraction: $\frac{11}{4}$

Invert for reciprocal: $\mathbf{\frac{4}{11}}$ **Check:** $2\frac{3}{4} \times \frac{4}{11} = 1$

d. 0.2 Write as a fraction: $\frac{2}{10} = \frac{1}{5}$

Invert for reciprocal: $\frac{5}{1} = \mathbf{5}$ **Check:** $0.2 \times 5 = 1$

e. Zero is the only whole number that does not have a reciprocal because any number multiplied by zero is zero (and not 1). ■

The process of division is very similar to the process of multiplication. To divide fractions, you must understand these three ideas:

1. How to multiply fractions
2. Which term is called the *divisor*
3. How to find the reciprocal of a number

Procedure for Dividing Fractions

To divide fractions, multiply by the reciprocal of the divisor. This is sometimes phrased as "invert and multiply."

EXAMPLE 8 **Dividing Fractions**

Divide the following.

a. $\frac{2}{3} \div \frac{4}{5}$ **b.** $\frac{3}{5} \div \frac{9}{20}$

c. $\frac{5}{8} \div 3$ **d.** $0 \div \frac{6}{7}$

Solution

a. $\frac{2}{3} \div \frac{4}{5} = \frac{2}{3} \times \frac{5}{4}$

$= \frac{\overset{1}{\cancel{2}}}{3} \times \frac{5}{\underset{2}{\cancel{4}}}$

$= \mathbf{\frac{5}{6}}$

b. $\frac{3}{5} \div \frac{9}{20} = \frac{\overset{1}{\cancel{3}}}{\underset{1}{\cancel{5}}} \times \frac{\overset{4}{\cancel{20}}}{\underset{3}{\cancel{9}}}$

$= \mathbf{\frac{4}{3}}$ or $\mathbf{1\frac{1}{3}}$

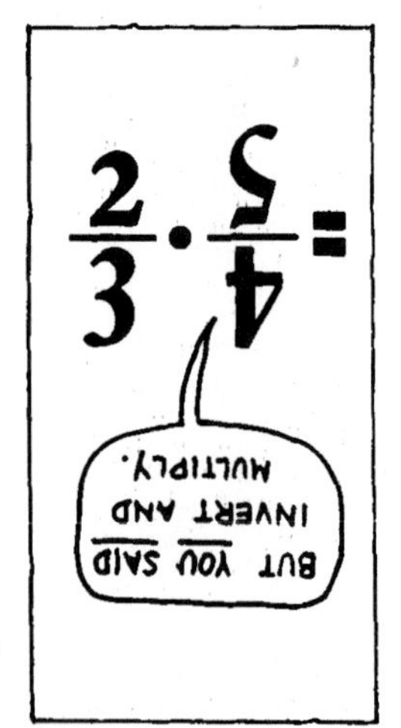

c. $\frac{5}{8} \div 3 = \frac{5}{8} \times \frac{1}{3}$

$= \mathbf{\frac{5}{24}}$

d. $0 \div \frac{6}{7} = 0 \times \frac{7}{6}$

$= \mathbf{0}$ ■

We have discussed how to change a fraction to a decimal by division. The reverse procedure—changing a terminating decimal to a fraction—can be viewed as a multiplication problem involving fractions. This process is summarized in the following box.

Procedure for Changing a Terminating Decimal into a Fraction

1. Multiply the given number without its decimal point by the decimal name of the last digit.
2. By the decimal name of the last digit, we mean:

   One place is tenth, or $\frac{1}{10}$.

   Two places is hundredth, or $\frac{1}{100}$.

   Three places is thousandth, or $\frac{1}{1,000}$.

   ⋮

EXAMPLE 9 **Changing a Terminating Decimal to a Fraction**

Change the given decimal fractions to common fractions.

a. 0.4 **b.** 0.75 **c.** 0.0014 **d.** $0.12\frac{1}{2}$ **e.** $0.3\frac{1}{3}$

Solution

a. 4 tenths; decimal position is tenth.

$$0.4 = 4 \times \frac{1}{10} = \frac{\overset{2}{\cancel{4}}}{1} \times \frac{1}{\underset{5}{\cancel{10}}} = \mathbf{\frac{2}{5}}$$

b. 75 hundredths; decimal position is hundredth.

$$0.75 = \overset{3}{\cancel{75}} \times \frac{1}{\underset{4}{\cancel{100}}} = \mathbf{\frac{3}{4}}$$

c. 14 ten-thousandths; decimal position is ten-thousandth.

$$0.0014 = \overset{7}{\cancel{14}} \times \frac{1}{\underset{5,000}{\cancel{10,000}}} = \mathbf{\frac{7}{5,000}}$$

d. $12\frac{1}{2}$ hundredths; decimal position is hundredth.

$$0.12\tfrac{1}{2} = 12\tfrac{1}{2} \times \frac{1}{100} = \frac{\overset{1}{\cancel{25}}}{2} \times \frac{1}{\underset{4}{\cancel{100}}} = \mathbf{\frac{1}{8}}$$

e. $3\frac{1}{3}$ tenths; decimal position is tenths.

$$0.3\tfrac{1}{3} = 3\tfrac{1}{3} \times \frac{1}{10} = \frac{\overset{1}{\cancel{10}}}{3} \times \frac{1}{\underset{1}{\cancel{10}}} = \mathbf{\frac{1}{3}}$$

■

Numbers in which decimal and fractional forms are mixed, as in parts **d** and **e** of Example 9, are called **complex decimals.**

PROBLEM SET 1.6

Essential Ideas

1. **In Your Own Words** State the fundamental property of fractions, and explain why you think it is so "fundamental."

2. **In Your Own Words** Explain a procedure for reducing fractions.

3. **In Your Own Words** How do you know when a fraction is reduced?

4. **In Your Own Words** Explain a procedure for multiplying fractions.

5. **In Your Own Words** Explain a procedure for dividing fractions.

6. **In Your Own Words** Explain a procedure for changing a terminating decimal to a fraction.

Level 1

WHAT IS WRONG, *if anything, with each of the statements in Problems 7–12? Explain your reasoning.*

7. $\frac{9}{2}$ is a reduced fraction.

8. If I want to find $\frac{1}{2}$ of 16, the correct operation is to divide 16 by $\frac{1}{2}$.

9. $8 \times \frac{3}{2} = \frac{8 \times 3}{8 \times 2} = \frac{24}{16} = \frac{3}{2}$

10. To find the reciprocal of any number, write $\frac{1}{\text{the number}}$.

11. $\frac{2}{3} \div \frac{4}{5} = \frac{4}{5} \div \frac{2}{3}$

12. "Invert and multiply" means that a division problem can be carried out by changing the division to a multiplication, and then multiplying by the reciprocal of the divisor.

Level 1—Drill

Completely reduce the fractions in Problems 13–19.

13. **a.** $\frac{2}{4}$ **b.** $\frac{3}{9}$ **c.** $\frac{4}{16}$ **d.** $\frac{2}{10}$

14. **a.** $\frac{3}{9}$ **b.** $\frac{6}{9}$ **c.** $\frac{2}{10}$ **d.** $\frac{4}{10}$

15. **a.** $\frac{3}{12}$ **b.** $\frac{4}{12}$ **c.** $\frac{6}{12}$ **d.** $\frac{8}{12}$

16. **a.** $\frac{14}{7}$ **b.** $\frac{38}{19}$ **c.** $\frac{92}{2}$ **d.** $\frac{160}{8}$

17. **a.** $\frac{72}{15}$ **b.** $\frac{42}{14}$ **c.** $\frac{16}{24}$ **d.** $\frac{128}{256}$

18. **a.** $\frac{18}{30}$ **b.** $\frac{70}{105}$ **c.** $\frac{50}{400}$ **d.** $\frac{35}{21}$

19. **a.** $\frac{140}{420}$ **b.** $\frac{150}{1{,}000}$ **c.** $\frac{2{,}500}{10{,}000}$ **d.** $\frac{105}{120}$

Illustrate the given parts of a whole in Problems 20–25.

20. $\frac{2}{5}$ of $\frac{3}{4}$

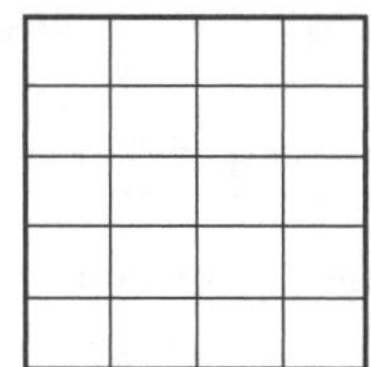

21. $\frac{3}{5}$ of $\frac{5}{6}$

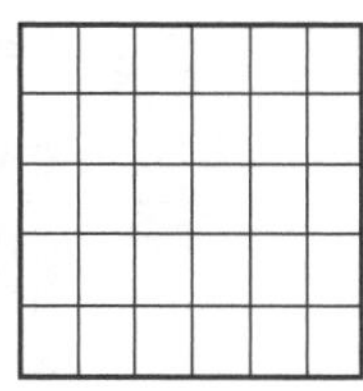

22. $\frac{4}{5}$ of $\frac{1}{3}$

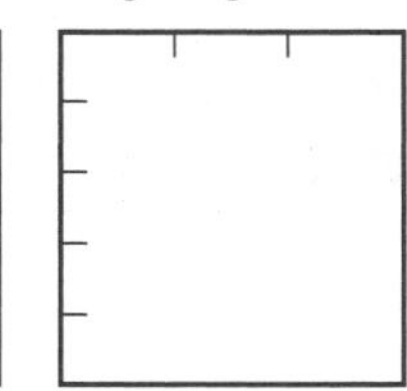

23. $\frac{3}{5}$ of $\frac{3}{4}$

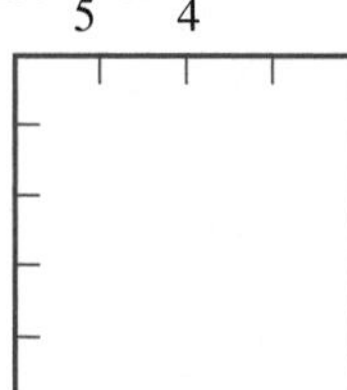

24. $\frac{1}{6}$ of $\frac{2}{3}$

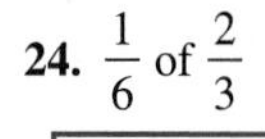

25. $\frac{3}{5}$ of $\frac{1}{2}$

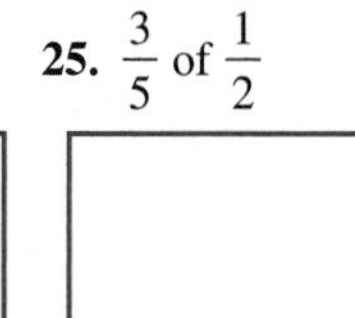

26. Name the divisor for each of the following problems.

a. $15 \div 5$ **b.** $\frac{3}{4} \div \frac{7}{3}$

c. $\frac{5}{8} \div 3$ **d.** $5 \div 0.2$

27. Name the divisor for each of the following problems.

a. $\frac{5}{6} \div \frac{2}{3}$ **b.** $4\frac{1}{5} \div 2\frac{3}{4}$

c. $0 \div \frac{6}{7}$ **d.** $\frac{3}{8} \div 0$

In Problems 28–31, first find the answer by direct division, and then find the answer by multiplying by the reciprocal. Both answers should be the same.

28. a. $15 \div 5$ **b.** $6 \div 3$

29. a. $9 \div 3$ **b.** $4 \div 2$

30. a. $20 \div 4$ **b.** $256 \div 16$

31. a. $5 \div 0.2$ **b.** $2.4 \div 0.6$

Level 2—Drill

Perform the indicated operations in Problems 32–48. Give your answers in reduced form.

32. a. $\frac{1}{4} \times \frac{1}{6}$ **b.** $\frac{2}{3} \times \frac{4}{5}$ **c.** $\frac{3}{4} \times \frac{1}{6}$ **d.** $\frac{7}{20} \times \frac{100}{14}$

33. a. $\frac{5}{9} \times \frac{18}{25}$ **b.** $\frac{2}{3} \times \frac{3}{8}$ **c.** $\frac{4}{5} \times \frac{13}{16}$ **d.** $\frac{18}{25} \times \frac{5}{36}$

34. a. $\frac{4}{5} \times \frac{3}{8}$ **b.** $\frac{4}{7} \times \frac{14}{9}$ **c.** $\frac{5}{12} \times \frac{4}{15}$ **d.** $\frac{9}{16} \times \frac{4}{27}$

35. a. $\frac{1}{2} \div \frac{1}{3}$ **b.** $\frac{1}{3} \div \frac{1}{2}$ **c.** $\frac{2}{3} \div \frac{1}{2}$ **d.** $\frac{3}{4} \div \frac{2}{3}$

36. a. $\frac{2}{5} \times \frac{15}{8}$ **b.** $\frac{5}{3} \times \frac{9}{15}$ **c.** $\frac{3}{5} \times \frac{20}{27}$ **d.** $\frac{3}{8} \div \frac{15}{16}$

37. a. $\frac{2}{3} \div \frac{5}{6}$ **b.** $\frac{4}{5} \div \frac{3}{10}$ **c.** $\frac{4}{7} \div \frac{4}{5}$ **d.** $\frac{5}{6} \div \frac{1}{3}$

38. a. $\frac{4}{5} \div \frac{4}{5}$ **b.** $\frac{7}{9} \div \frac{7}{9}$ **c.** $\frac{8}{3} \div \frac{8}{3}$ **d.** $\frac{4}{5} \times 5$

39. a. $\frac{3}{5} \div \frac{3}{7}$ **b.** $\frac{6}{7} \div \frac{2}{3}$ **c.** $\frac{4}{9} \div \frac{3}{4}$ **d.** $\frac{2}{3} \times 3$

40. a. $\frac{5}{6} \times 18$ **b.** $\frac{3}{8} \times 24$ **c.** $\frac{5}{8} \times 8$ **d.** $\frac{6}{7} \times 7$

41. a. $\frac{2}{5} \div 3$ **b.** $\frac{3}{8} \div 3$ **c.** $\frac{3}{5} \div 5$ **d.** $3 \div \frac{1}{6}$

42. a. $2\frac{1}{2} \div 3$ **b.** $6\frac{1}{2} \div 3$ **c.** $3\frac{4}{5} \div 0$ **d.** $7 \times \frac{9}{14}$

43. a. $5 \div 1\frac{1}{2}$ **b.** $4 \div 2\frac{2}{3}$ **c.** $6 \div 1\frac{5}{6}$ **d.** $52 \times \frac{5}{13}$

44. a. $2\frac{2}{5} \times 1\frac{4}{5}$ **b.** $5\frac{1}{2} \times 3\frac{2}{3}$ **c.** $4\frac{1}{6} \times 3\frac{3}{8}$ **d.** $1\frac{1}{6} \times 2\frac{1}{3}$

45. a. $6\frac{1}{2} \times \frac{5}{6}$ **b.** $3\frac{4}{5} \times \frac{1}{2}$ **c.** $2\frac{2}{5} \div 1\frac{2}{3}$ **d.** $5\frac{1}{2} \div 1\frac{4}{5}$

46. a. $2\frac{2}{3} \div 1\frac{1}{3}$ **b.** $5 \times \frac{3}{5}$ **c.** $2\frac{1}{2} \times \frac{3}{4}$ **d.** $4\frac{1}{2} \div 2\frac{3}{8}$

47. a. $\frac{1}{2} \times \frac{8}{9} \times \frac{3}{16}$ **b.** $\frac{2}{3} \times \frac{5}{8} \times \frac{16}{100}$
c. $\frac{2}{3} \times \frac{4}{5} \times \frac{15}{16}$ **d.** $2\frac{1}{2} \times 3\frac{1}{6} \times 1\frac{1}{5}$

48. a. $\left(\frac{1}{2} \div \frac{1}{2}\right) \div \frac{1}{4}$ **b.** $\frac{1}{2} \div \left(\frac{1}{3} \div \frac{1}{4}\right)$
c. $\frac{2}{3} \div \left(\frac{1}{2} \div \frac{3}{4}\right)$ **d.** $\left(\frac{2}{3} \div \frac{1}{2}\right) \div \frac{3}{4}$

Write the decimals in Problems 49–56 in fractional form.

49. a. 0.7 **b.** 0.9 **c.** 0.8

50. a. 0.25 **b.** 0.87 **c.** 0.375

51. a. 0.18 **b.** 0.48 **c.** 0.54

52. a. 0.78 **b.** 0.85 **c.** 0.246

53. a. 0.505 **b.** 0.015 **c.** 0.005

54. a. $0.66\frac{2}{3}$ **b.** $0.87\frac{1}{2}$ **c.** $0.16\frac{2}{3}$

55. a. $0.37\frac{1}{2}$ **b.** $0.8\frac{8}{9}$ **c.** $0.000\frac{1}{3}$

56. a. $0.1\frac{1}{9}$ **b.** $0.5\frac{5}{9}$ **c.** $0.08\frac{1}{3}$

Level 2—Applications

Some experts tell us that the amount of savings we need is $\frac{1}{10}$ times your age times your current salary. Calculate the amount of savings necessary in Problems 57–60.

57. age 20, salary \$24,000 **58.** age 30, salary \$46,000

59. age 25, salary \$25,000 **60.** age 50, salary \$125,000

61. If Karl owns $\frac{3}{16}$ of a mutual water system and a new well is installed at a cost of \$12,512, how much does Karl owe for his share?

62. A recipe calls for $\frac{5}{8}$ cup of sugar, one egg, and $\frac{7}{8}$ cup of flour. How much of each ingredient is needed to double this mixture?

63. Stock prices are quoted in eighths of dollars. If Sears stock is selling for $20\frac{1}{8}$ per share, how much would 100 shares cost?

64. What would 100 shares of Xerox stock cost when it is selling for $56\frac{5}{8}$ per share?

65. If you received \$277.50 for selling 20 shares of Brunswick stock, what is the price per share (stated as a mixed fraction)?

66. If $4\frac{2}{5}$ acres sell for \$44,000, what is the price per acre?

67. If Shannon drives $5\frac{3}{4}$ miles to work and he drives the same distance home five days each week, how many miles does he drive to and from work each week?

68. If two-thirds of a person's body weight is water, what is the weight of water in a person who weighs 180 pounds?

69. WHAT IS WRONG, if anything, with the following advertisement, which recently appeared in a Kansas newspaper? "Divide your age by one-half and that is the percent discount you will receive for this sale!"

1.7 ADDING AND SUBTRACTING FRACTIONS

Historical Perspective

Shirley Frye

Shirley Frye has actively taught mathematics to students and teachers of all levels during five decades. She is past president of both the National Council of Supervisors of Mathematics (NCSM) and the National Council of Teachers of Mathematics (NCTM). She served as a key leader in the development and introduction of the NCTM's *Standards* documents, which provided direction for the dramatic reform in mathematics education in the 1990s. "I appreciate the structure and beauty of mathematics that are continually being discovered and revealed in the world around us. I am excited about helping students and teachers learn school mathematics and developing a 'friendly feeling' towards numbers!" Shirley exclaimed with excitement in her voice. Sexual stereotypes are hard to break. One evening, Shirley was standing at the door of her classroom meeting parents who were attending an open house. As she greeted the parents of students in her calculus class a father said, "You can't be the calculus teacher, you are not a gray-haired old man!" Shirley added, "I then determined that changing the image of mathematics teachers was a necessity."

If the fractions you are adding or subtracting are similar (all halves, thirds, fourths, fifths, sixths, and so on), then the procedure is straightforward: Add or subtract the numerators. In this case, we say that the fractions have **common denominators.** If the fractions are not similar, then you *cannot* add or subtract them directly; you must change the form of the fractions so that they are similar. This process is called *finding common denominators.*

Procedure for Adding or Subtracting Fractions with Common Denominators

To add or subtract fractions with common denominators, add or subtract the numerators. The denominator of the sum or difference is the same as the common denominator.

EXAMPLE 1 **Adding and Subtracting Fractions with Common Denominators**

Perform the indicated operations.

a. $\dfrac{1}{5}+\dfrac{3}{5}$ **b.** $\dfrac{5}{9}-\dfrac{4}{9}$ **c.** $\dfrac{3}{2}+\dfrac{7}{2}$ **d.** $3\frac{2}{3}+1\frac{2}{3}+\frac{1}{3}+4\frac{2}{3}$

Solution **a.** $\dfrac{1}{5}+\dfrac{3}{5}=\mathbf{\dfrac{4}{5}}$

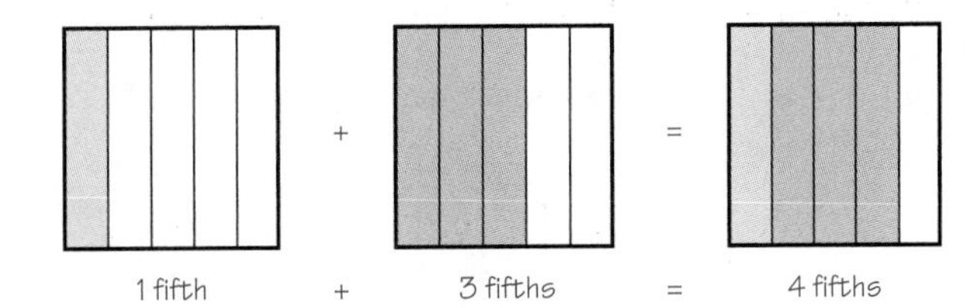

b. $\dfrac{5}{9}-\dfrac{4}{9}=\mathbf{\dfrac{1}{9}}$

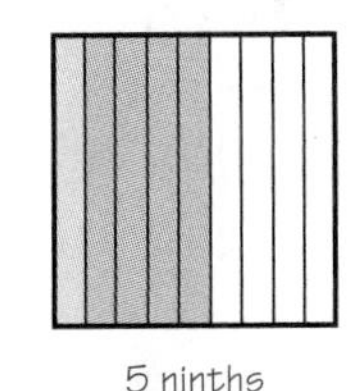

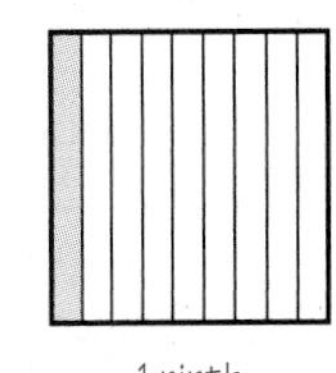

c. $\dfrac{3}{2}+\dfrac{7}{2}=\dfrac{10}{2}=\mathbf{5}$

This example could also be added in the form of mixed numbers:

Add fraction part first.

$$\begin{array}{r} 1\frac{1}{2} \\ +\ 3\frac{1}{2} \\ \hline 4\frac{2}{2} \end{array}$$

↑ Next, add whole number part.

Since $\dfrac{2}{2}=1$, it follows that $4\frac{2}{2}=4+\frac{2}{2}=4+1=\mathbf{5}$.

d.

$$\begin{array}{r} 3\frac{2}{3} \\ 1\frac{2}{3} \\ \frac{1}{3} \\ +\ 4\frac{2}{3} \\ \hline 8\frac{7}{3} \end{array} = \mathbf{10\tfrac{1}{3}}$$

Notice that the carry may be more than 1. In this example, $\frac{7}{3}=2\frac{1}{3}$. ■

Sometimes when doing subtraction, you must borrow from the units column to have enough fractional parts to carry out the subtraction. Remember,

$$1 = \frac{2}{2} = \frac{3}{3} = \frac{4}{4} = \frac{5}{5} = \frac{6}{6} = \frac{7}{7} = \frac{8}{8} = \frac{9}{9} = \frac{10}{10} = \cdots$$

EXAMPLE 2 **Subtraction with Common Denominators and Borrowing**

Borrow 1 from the units column and combine with the fractional part.

a. $3\frac{1}{3}$ **b.** $5\frac{3}{5}$ **c.** $3\frac{1}{2}$ **d.** $1\frac{1}{3}$ **e.** $14\frac{4}{5}$

Solution **a.** $3\frac{1}{3}$: You would do the following steps in your head and write only the answer:

Study this to understand rewriting numbers to carry out subtraction with mixed numbers and borrowing.

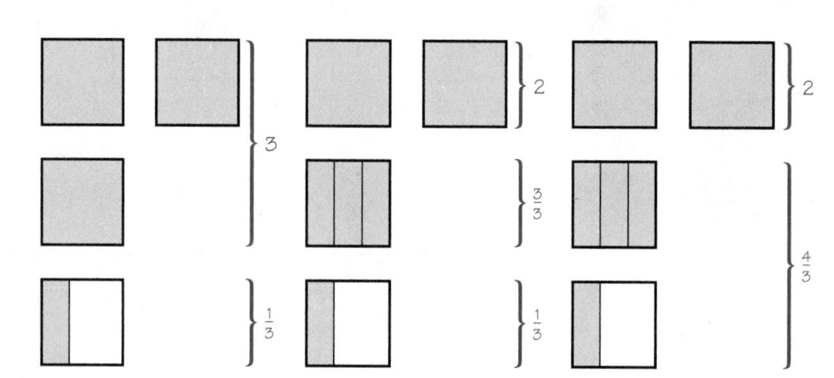

These all represent the same number: namely, $3 + \frac{1}{3}$, $2 + \frac{3}{3} + \frac{1}{3}$, $2 + \frac{4}{3}$.

Thus, $3\frac{1}{3} = \mathbf{2\frac{4}{3}}$.

b. $5\frac{3}{5} = \mathbf{4\frac{8}{5}}$ **c.** $3\frac{1}{2} = \mathbf{2\frac{3}{2}}$ **d.** $1\frac{1}{3} = \mathbf{\frac{4}{3}}$ **e.** $14\frac{4}{5} = \mathbf{13\frac{9}{5}}$ ■

Using the idea shown in Example 2, we can carry out some subtractions with mixed numbers.

EXAMPLE 3 **Subtraction of Mixed Numbers with Common Denominators**

Find the differences.

a. $3\frac{1}{3} - 1\frac{2}{3}$ **b.** $4\frac{3}{5} - 2\frac{4}{5}$ **c.** $1\frac{1}{3} - \frac{2}{3}$

Solution

a.
$$\begin{array}{rcr} 3\frac{1}{3} & = & 2\frac{4}{3} \\ -1\frac{2}{3} & = & -\ 1\frac{2}{3} \\ \hline & & \mathbf{1\frac{2}{3}} \end{array}$$

b.
$$\begin{array}{rcr} 4\frac{3}{5} & = & 3\frac{8}{5} \\ -2\frac{4}{5} & = & -2\frac{4}{5} \\ \hline & & \mathbf{1\frac{4}{5}} \end{array}$$

c.
$$\begin{array}{rcr} 1\frac{1}{3} & = & \frac{4}{3} \\ -\ \frac{2}{3} & = & -\ \frac{2}{3} \\ \hline & & \mathbf{\frac{2}{3}} \end{array}$$

■

If the fractions to be added or subtracted do not have common denominators, the sum or difference is not as easy to find. For example, consider the sum

$$\frac{1}{5} + \frac{3}{4}$$

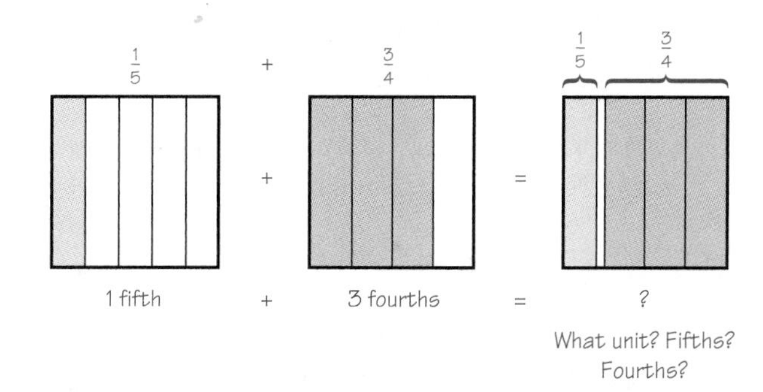

We need to use the fundamental property of fractions to change the form of one or more of the fractions so that they do have the same denominators. Use the following guidelines to find the best common denominator.

First: The common denominator should be a number that all the given denominators divide into evenly. This means that the product of the given denominators will always be a common denominator. Many students learn this and *always* find the common denominator by multiplication. Even though this works for all numbers, it is inefficient except for small numbers. Therefore, we have a second condition to find the best common denominator.

Second: The common denominator should be as small as possible. This number is called the **lowest common denominator,** denoted by **LCD.**

Procedure for Finding the Lowest Common Denominator (LCD)

STOP

Step 1 Factor all given denominators into prime factors; write this factorization using exponents.

Step 2 List each different prime factor you found in the prime factorization of the denominators.

Step 3 On each prime in the list from Step 2, place the largest exponent that appears on that prime factor anywhere in the factorization of the denominators.

Step 4 The LCD is the product of the prime factors with the exponents found in Step 3.

EXAMPLE 4 **Finding the Lowest Common Denominator**

Find the LCD for the given denominators.

a. 6 and 8 **b.** 8 and 12 **c.** 24 and 30 **d.** 8, 24, and 60
e. 300 and 144

Solution **a.** 6; 8

$6 = 2 \times \mathbf{3}$

$8 = \mathbf{2^3}$

↓ Largest exponent on prime factor 3 is 1 (remember $3 = 3^1$).

LCD: $2^3 \times 3 = 8 \times 3 = \mathbf{24}$

↑ Largest exponent on prime factor 2 is 3.

b. 8; 12

$8 = \mathbf{2^3}$

$12 = 2^2 \times \mathbf{3}$

LCD: $2^3 \times 3 = 8 \times 3 = \mathbf{24}$

c. 24; 30

$24 = \mathbf{2^3 \times 3}$

$30 = 2 \times 3 \times \mathbf{5}$

LCD: $2^3 \times 3 \times 5 = 8 \times 3 \times 5 = \mathbf{120}$

d. 8; 24; 60 The same procedure works no matter how many denominators are given.

$8 = \mathbf{2^3}$

$24 = 2^3 \times \mathbf{3}$

$60 = 2^2 \times 3 \times \mathbf{5}$

LCD: $2^3 \times 3 \times 5 = 8 \times 3 \times 5 = \mathbf{120}$

e. 300; 144 If the numbers are larger, you may need factor trees:

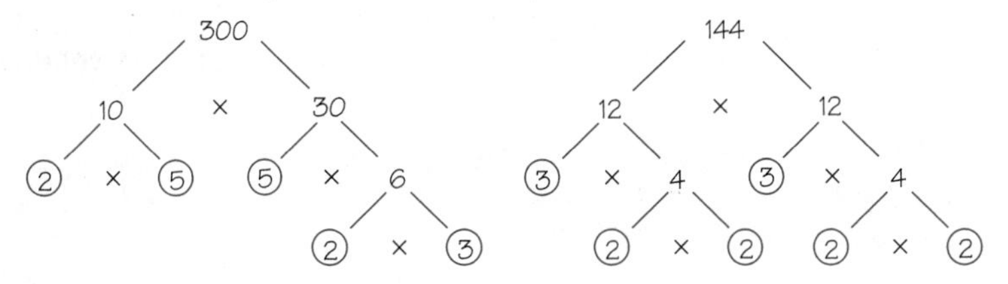

$300 = 2^2 \times 3 \times \mathbf{5^2}$

$144 = \mathbf{2^4 \times 3^2}$

LCD: $2^4 \times 3^2 \times 5^2 = 16 \times 9 \times 25 = \mathbf{3{,}600}$ ■

Notice the proper procedure for carrying out the order of operations in Example 4, which mixes exponents and multiplication. Since an exponent is an indicated multiplication, the proper procedure is first to simplify the exponent and then to carry out the multiplication. This leads to an **extended order-of-operations** agreement.

Extended Order-of-Operations Agreement

This is essential for almost all that follows. You must make its application second nature.

1. First, perform any operations enclosed in parentheses.
2. Next, perform any operations that involve raising to a power.
3. Perform multiplication and division, reading from left to right.
4. Do addition and subtraction, reading from left to right.

EXAMPLE 5

Using the Extended Order of Operations

Simplify $2^4 + 3(5 + 6)$.

Solution

$2^4 + 3(5 + 6) = 2^4 + 3(\mathbf{11})$	Parentheses first
$= \mathbf{16} + 3(11)$	Powers next
$= 16 + \mathbf{33}$	Multiplications/divisions
$= \mathbf{49}$	Additions/subtractions ■

We now turn to adding and subtracting fractions by finding the lowest common denominator.

Procedure for Adding or Subtracting Fractions that Do Not Have Common Denominators

Step 1 Find the LCD.

Step 2 Change the forms of the fractions to obtain forms with common denominators.

Step 3 Add or subtract the numerators of the fractions with common denominators.

EXAMPLE 6

Addition and Subtraction of Fractions

Perform the given operations.

a. $\frac{1}{6} + \frac{3}{4}$ **b.** $\frac{4}{9} - \frac{1}{4}$ **c.** $2\frac{1}{6} + 5\frac{3}{8}$ **d.** $12\frac{7}{24} - 5\frac{4}{30}$

e. $\frac{3}{8} + \frac{11}{12} + \frac{3}{20}$ **f.** $16\frac{1}{3} - 4\frac{1}{2}$

Solution

a. Write in column form: $\frac{1}{6} + \frac{3}{4}$ Find the LCD: $6 = 2 \times 3$, $4 = 2^2$

LCD: $2^2 \times 3 = 12$

Next, change form:

$$\frac{1}{6} = \frac{2}{12}$$
$$+\frac{3}{4} = \frac{9}{12}$$

Finally, add: $\mathbf{\frac{11}{12}}$

b. Combining the steps outlined in part **a,** you'll obtain

$$\frac{4}{9} = \frac{16}{36} \qquad 9 = 3^2$$
$$-\frac{1}{4} = -\frac{9}{36} \qquad 4 = 2^2$$
$$\mathbf{\frac{7}{36}} \qquad \text{LCD: } 2^2 \times 3^2 = 36$$

c.

$$\begin{array}{r} 2\frac{1}{6} = 2\frac{4}{24} \\ + 5\frac{3}{8} = 5\frac{9}{24} \\ \hline \mathbf{7\frac{13}{24}} \end{array}$$

From Example 4a, the LCD of 6 and 8 is 24; also, $\frac{1}{6} = \frac{4}{24}$ and $\frac{3}{8} = \frac{9}{24}$

d.

$$\begin{array}{r} 12\frac{7}{24} = \quad 12\frac{35}{120} \\ - 5\frac{4}{30} = - \quad 5\frac{16}{120} \\ \hline \mathbf{7\frac{19}{120}} \end{array}$$

From Example 4c, the LCD is 120;

120 ÷ 24 = 5 and 5 × 7 = 35: $\frac{7}{24} = \frac{35}{120}$

120 ÷ 30 = 4 and 4 × 4 = 16: $\frac{4}{30} = \frac{6}{120}$

e. Factoring the denominators, we obtain

$$\begin{aligned} 8 &= 2^3 \\ 12 &= 2^2 \times 3 \\ 20 &= 2^2 \quad\quad \times 5 \\ \text{LCD: } & 2^3 \times 3 \times 5 = 120 \end{aligned}$$

$\frac{3}{8} = \frac{45}{120}$ 120 ÷ 8 = 15 and 15 × 3 = 45

$\frac{11}{12} = \frac{110}{120}$ 120 ÷ 12 = 10 and 10 × 11 = 110

$+\frac{3}{20} = \frac{18}{120}$ 120 ÷ 20 = 6 and 6 × 3 = 18

$\frac{173}{120} = \mathbf{1\frac{53}{120}}$

f. Borrow so that you can complete the subtraction.

$$\begin{array}{r} 16\frac{1}{3} = \quad 16\frac{2}{6} = \quad 15\frac{8}{6} \\ - 4\frac{1}{2} = - \quad 4\frac{3}{6} = - \quad 4\frac{3}{6} \\ \hline \mathbf{11\frac{5}{6}} \end{array}$$

■

The order of operations is the same for fractions as it is for whole numbers.

EXAMPLE 7 **Mixed Operations with Fractions**

Perform the indicated operations.

a. $\frac{3}{4} \times \frac{1}{3} + \frac{3}{4} \times \frac{2}{3}$ **b.** $\frac{3}{4}\left(\frac{1}{3} + \frac{2}{3}\right)$

Solution Remember the correct order of operations.

a. $\frac{3}{4} \times \frac{1}{3} + \frac{3}{4} \times \frac{2}{3} = \frac{1}{4} + \frac{1}{2}$
$= \frac{1}{4} + \frac{2}{4}$
$= \mathbf{\frac{3}{4}}$

b. $\frac{3}{4}\left(\frac{1}{3} + \frac{2}{3}\right) = \frac{3}{4} \times 1$
$= \mathbf{\frac{3}{4}}$

Recall that multiplication is implied when there is no symbol between number and parentheses. This way of writing multiplication is called *juxtaposition.*

The distributive property tells us that the answers to parts **a** and **b** must be equal. ■

EXAMPLE 8 **Mixed Operations Using Juxtaposition for Multiplication**

Perform the indicated operations.

a. $\frac{5}{6}\left(\frac{3}{4} + \frac{1}{2}\right) + \frac{2}{3} \times \frac{9}{10}$ **b.** $\frac{1}{2} \div \frac{1}{3} \div \left(\frac{3}{4} + \frac{3}{5}\right)$

Solution

a. $\frac{5}{6}\left(\frac{3}{4} + \frac{1}{2}\right) + \frac{2}{3} \times \frac{9}{10} = \frac{5}{6}\left(\frac{3}{4} + \frac{2}{4}\right) + \frac{2}{3} \times \frac{9}{10}$ Common denominator, parentheses first

$= \frac{5}{6}\left(\frac{5}{4}\right) + \frac{\cancel{2}^{1}}{\cancel{3}_{1}} \times \frac{\cancel{9}^{3}}{\cancel{10}_{5}}$

$= \frac{25}{24} + \frac{3}{5}$

$= \frac{125}{120} + \frac{72}{120}$

$= \mathbf{\frac{197}{120}}$ or $\mathbf{1\frac{77}{120}}$

b. $\frac{1}{2} \div \frac{1}{3} \div \left(\frac{3}{4} + \frac{3}{5}\right) = \frac{1}{2} \div \frac{1}{3} \div \left(\frac{15}{20} + \frac{12}{20}\right)$

$= \frac{1}{2} \div \frac{1}{3} \div \frac{27}{20}$

$= \frac{1}{2} \times \frac{3}{1} \div \frac{27}{20}$ Multiplication and division from **left to right**

$= \frac{3}{2} \div \frac{27}{20}$

$= \frac{\cancel{3}^{1}}{\cancel{2}_{1}} \times \frac{\cancel{20}^{10}}{\cancel{27}_{9}}$

$= \mathbf{\frac{10}{9}}$ or $\mathbf{1\frac{1}{9}}$ ■

PROBLEM SET 1.7

ESSENTIAL IDEAS

1. **In Your Own Words** What is the extended order-of-operations agreement?

2. **In Your Own Words** State a procedure for finding the lowest common denominator for a set of fractions.

LEVEL 1

Estimate (do not calculate) the answers in Problems 3–10.

3. $\frac{19}{40}$ is about the same as

A. $\frac{1}{2}$ B. $\frac{1}{3}$ C. $\frac{5}{4}$ D. $\frac{1}{5}$

4. $\frac{3}{10}$ is about the same as

A. $\frac{1}{2}$ B. $\frac{1}{3}$ C. $\frac{5}{4}$ D. $\frac{1}{5}$

5. $\frac{19}{40} - \frac{3}{10}$ is about the same as

A. $\frac{1}{2}$ B. $\frac{1}{3}$ C. $\frac{5}{4}$ D. $\frac{1}{5}$

6. $\frac{1}{2} + \frac{1}{4} + \frac{1}{8} + \frac{1}{16} + \frac{1}{32}$ is about equal to

A. $\frac{1}{2}$ B. 1 C. $\frac{3}{2}$ D. 1,000

7. If I multiply a given nonzero number by $\frac{2}{3}$, the answer ________________ than that given number.
A. is larger B. is smaller
C. could be either larger or smaller

8. If I multiply a given nonzero number by $\frac{3}{2}$, the answer ________________ than the given number.
A. is larger B. is smaller
C. could be either larger or smaller

9. If I divide a given nonzero number by $\frac{2}{3}$, the answer ________________ than the given number.
A. is larger B. is smaller
C. could be either larger or smaller

10. If I divide a given nonzero number by $\frac{3}{2}$, the answer ________________ than the given number.
A. is larger B. is smaller
C. could be either larger or smaller

WHAT IS WRONG, *if anything, with each of the statements in Problems 11–16? Explain your reasoning.*

11. $\frac{3}{8} + \frac{4}{8} = \frac{3+4}{8+8} = \frac{7}{16}$

12. $3\frac{5}{8} - 2\frac{7}{8} = 1\frac{2}{8} = 1\frac{1}{4}$

13. $\frac{3}{8} \times \frac{5}{8} = \frac{3 \times 5}{8 \times 8} = \frac{15}{64}$

14. $\frac{3}{8} \div \frac{5}{8} = \frac{8}{3} \times \frac{5}{8} = \frac{5}{3}$

15. $\frac{3}{8} + \frac{2}{5} = \frac{3 \times 5 + 2 \times 8}{8 \times 5} = \frac{15 + 16}{40} = \frac{31}{40}$

16. $\frac{3}{5} + \frac{2}{5} \times \frac{1}{2} = 1 \times \frac{1}{2} = \frac{1}{2}$

LEVEL 1—Drill

Perform the indicated operations in Problems 17–22. Give your answers in reduced form.

17. **a.** $\frac{2}{5} + \frac{1}{5}$ **b.** $\frac{3}{7} + \frac{5}{7}$ **c.** $\frac{5}{11} + \frac{3}{11}$

18. **a.** $\frac{3}{2} + \frac{5}{2}$ **b.** $\frac{7}{3} - \frac{4}{3}$ **c.** $\frac{5}{9} + \frac{1}{9}$

19. **a.** $\frac{9}{13} - \frac{5}{13}$ **b.** $\frac{6}{23} - \frac{5}{23}$ **c.** $\frac{9}{7} - \frac{2}{7}$

20. **a.** $\frac{13}{15} + \frac{2}{15}$ **b.** $\frac{7}{12} - \frac{1}{12}$ **c.** $\frac{5}{8} - \frac{3}{8}$

21. **a.** $2\frac{2}{3} + 1\frac{1}{3}$ **b.** $6\frac{4}{5} + 2\frac{1}{5}$ **c.** $8\frac{3}{8} + 4\frac{5}{8}$

22. **a.** $5\frac{1}{3} - 3\frac{2}{3}$ **b.** $14\frac{1}{8} - 8\frac{5}{8}$ **c.** $6\frac{1}{4} - 5\frac{3}{4}$

Find the LCD for the denominators given in Problems 23–34.

23. **a.** 4; 8 **b.** 2; 6 **c.** 2; 5

24. **a.** 5; 12 **b.** 4; 12 **c.** 12; 90

25. **a.** 12; 336 **b.** 90; 210 **c.** 60; 72

26. **a.** 12; 20; 36 **b.** 6; 8; 10 **c.** 9; 12; 14

27. 60; 18 **28.** 450; 15

29. 735; 1,125 **30.** 315; 735

31. 420; 450 **32.** 600; 90; 30

33. 420; 315; 90 **34.** 210; 150; 245

Level 2—Drill

Perform the indicated operations in Problems 35–59. Give your answers in reduced form.

35. a. $\frac{1}{2} + \frac{2}{3}$ **b.** $\frac{1}{2} + \frac{3}{8}$ **c.** $\frac{1}{2} - \frac{1}{6}$

36. a. $\frac{1}{2} + \frac{2}{5}$ **b.** $\frac{1}{2} - \frac{2}{5}$ **c.** $\frac{5}{6} + \frac{2}{3}$

37. a. $\frac{5}{6} - \frac{1}{3}$ **b.** $\frac{5}{6} + \frac{5}{8}$ **c.** $\frac{5}{8} - \frac{1}{3}$

38. a. $\frac{3}{4} - \frac{5}{12}$ **b.** $\frac{3}{5} + \frac{1}{6}$ **c.** $\frac{3}{4} + \frac{1}{12}$

39. a. $\frac{2}{45} + \frac{1}{6}$ **b.** $\frac{41}{45} - \frac{5}{6}$ **c.** $\frac{4}{9} - \frac{5}{12}$

40. a. $\frac{3}{5} + \frac{1}{12}$ **b.** $\frac{5}{24} - \frac{2}{15}$ **c.** $\frac{5}{27} + \frac{1}{90}$

41. a. $2\frac{1}{2} + 4\frac{3}{4}$ **b.** $1\frac{2}{3} + 5\frac{1}{2}$ **c.** $3\frac{3}{8} + 5\frac{1}{2}$

42. a. $5\frac{1}{8} - 3\frac{3}{4}$ **b.** $17\frac{1}{2} - 6\frac{2}{3}$ **c.** $12\frac{1}{3} - 4\frac{1}{2}$

43. a. $6\frac{3}{8} - 5\frac{1}{6}$ **b.** $2\frac{1}{3} + 1\frac{5}{12}$ **c.** $6\frac{1}{16} - 5\frac{3}{8}$

44. a. $5 - \frac{3}{8}$ **b.** $7 - \frac{2}{3}$ **c.** $3\frac{3}{4} + 2\frac{1}{6}$

45. $\frac{1}{8} + 2\frac{2}{3} + \frac{1}{6}$ **46.** $\frac{4}{5} + \frac{3}{7} + \frac{3}{10}$

47. $6\frac{1}{8} + 3\frac{2}{5} + 1\frac{1}{4}$ **48.** $5\frac{1}{2} + 6\frac{3}{4} + 4\frac{1}{8}$

49. $7\frac{2}{3} + 5\frac{1}{2} + 12\frac{1}{6}$ **50.** $12\frac{3}{8} + 2\frac{1}{2} + 5\frac{1}{8}$

51. $\frac{1}{2} \times \frac{2}{3} + \frac{1}{5}$ **52.** $\frac{1}{2} + \frac{2}{3} \times \frac{1}{5}$

53. $\frac{1}{5} \div \frac{1}{3} \div \frac{1}{4}$ **54.** $\frac{1}{2} \div \left(\frac{1}{3} \div \frac{1}{4}\right)$

55. $\frac{3}{4}\left(\frac{9}{13} + \frac{4}{13}\right)$ **56.** $\frac{4}{5}\left(\frac{5}{16} + \frac{11}{16}\right)$

57. $\dfrac{3 \times 3 + 5 \times 2}{5 \times 3}$ **58.** $\dfrac{3 \times 5 + 7 \times 4}{7 \times 5}$

59. $\dfrac{1 \times 8 + 2 \times 5}{2 \times 8}$

Perform the indicated operations in Problems 60–65. You may use your calculator and leave answers in decimal form.

60. $\frac{3}{4} \times \frac{119}{200} + \frac{3}{4} \times \frac{81}{200}$ **61.** $\frac{4}{5} \times \frac{17}{95} + \frac{4}{5} \times \frac{78}{95}$

62. $\frac{7}{8} \times \frac{107}{147} + \frac{7}{8} \times \frac{40}{147}$ **63.** $\left(\frac{4}{5} + \frac{2}{3}\right) \div \frac{1}{5} + 2$

64. $\frac{19}{300} + \frac{55}{144} + \frac{25}{108}$ **65.** $\frac{15}{253} + \left(\frac{53}{104} - \frac{25}{208}\right)$

Level 2—Applications

Estimate the portion of each square occupied by the indicated letter in Problems 66–77.

66. A

67. B

68. C

69. A or B

70. A or C

71. B or C

72. R

73. K

74. G

75. Y

76. R or K or G

77. R or G or Y

78. A recipe calls for $\frac{2}{3}$ cup milk and $\frac{1}{2}$ cup water.

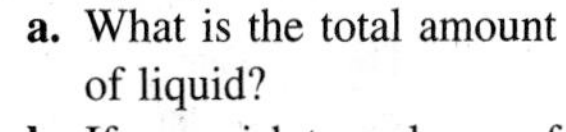

a. What is the total amount of liquid?

b. If you wish to make one-fourth of this recipe, how much of each ingredient is needed, and what is the total amount of liquid?

79. Suppose you have items to mail that weigh $1\frac{1}{4}$ lb, $2\frac{2}{3}$ lb, and $3\frac{1}{2}$ lb. What is the total weight of these packages?

80. Loretta received three boxes of candy for her birthday: a $1\frac{1}{2}$-lb box, a $\frac{3}{4}$-lb box, and a $2\frac{15}{16}$-lb box. What is the total weight of the candy she received?

81. Suppose you are installing molding around a table and you need pieces $5\frac{1}{4}$ in., $7\frac{1}{2}$ in., and $5\frac{3}{16}$ in. long. If each time a cut is made the saw chews up $\frac{1}{16}$ in. of material, what is the smallest single length of molding that can be used to do this job?

82. If the outside diameter of a piece of tubing is $\frac{15}{16}$ in. and the wall is $\frac{3}{16}$ in. thick, what is the inside diameter?

Level 3—Mind Bogglers

83. A rubber ball is known to rebound half the height it drops. If the ball is dropped from a height of 100 feet, how far will it have traveled by the time it hits the ground for:

a. The first time? **b.** The second time?
c. The third time? **d.** The fourth time?
e. The fifth time?

f. Look for a pattern, and decide whether there is a maximum distance the ball will travel if we assume that it will bounce indefinitely.

84. Look for a pattern in the following problem. Verify by division (show your work).

a. $\frac{1}{9} = 0.111\ldots$ **b.** $\frac{2}{9} = 0.222\ldots$ **c.** $\frac{3}{9} = 0.333\ldots$
d. $\frac{4}{9} = 0.444\ldots$ **e.** $\frac{8}{9} = 0.888\ldots$
f. What is $\frac{9}{9}$ according to the pattern?

1.8 SUMMARY AND REVIEW

Important Terms

(Numbers refer to sections of this chapter.)

Spending some time with the terms and objectives of this chapter will pay dividends in assuring your success.

Base [1.5]
Canceling [1.6]
Common denominator [1.7]
Common fraction [1.3, 1.6]
Complex decimal [1.6]
Composite [1.5]
Counting number [1.2]
Cubed [1.5]
Decimal [1.3]
Decimal fraction [1.3]
Denominator [1.3]
Difference [1.2]
Distributive law (for multiplication over addition) [1.2]
Division by zero [1.3]
Elementary operations [1.2]
Estimation [1.2]
Exponent [1.5]
Exponential notation [1.5]
Extended order of operations [1.7]
Factor [1.5]
Factoring [1.5]
Factorization [1.5]
Fraction [1.3]
Fundamental property of fractions [1.6]
Googol [1.5]
Improper fraction [1.3]
Invert [1.6]
Juxtaposition [1.2]
LCD [1.7]
Lowest common denominator [1.7]
Mixed number [1.3]
Natural number [1.2]
Numerator [1.3]
Order of operations [1.2; 1.7]
Place-value names [1.3]
Power [1.5]
Prime [1.5]
Prime factorization [1.5]
Product [1.2]
Proper fraction [1.3]
Quotient [1.2]
Rational number [1.3]
Reciprocal [1.6]
Reduced fraction [1.6]
Reducing fractions [1.6]
Remainder [1.3]
Repeating decimal [1.3]
Rounding numbers [1.4]
Scientific notation [1.5]
Squared [1.5]
Sum [1.2]
Trailing zeros [1.3]
Translation [1.2]
Whole numbers [1.2; 1.3]

Chapter Objectives

The material in this chapter is reviewed in the following list of objectives. A self-test (with answers and suggestions for additional study) is given. This self-test is constructed so that each problem number corresponds to a related objective. For example, Problem 7 is testing Objective 1.7. This self-test is followed by a practice test with the questions in mixed order.

[1.1]	*Objective 1.1*	Know some of the symptoms and possible cures for math anxiety.
[1.2]	*Objective 1.2*	Use the order-of-operations agreement to carry out calculations with mixed operations. Classify an expression as a sum, difference, product, or quotient.
[1.2]	*Objective 1.3*	Translate from words into math symbols.
[1.2]	*Objective 1.4*	Use the distributive property to eliminate parentheses.
[1.2–1.7]	*Objective 1.5*	Estimate answers to arithmetic problems.
[1.3]	*Objective 1.6*	Write an improper fraction as a mixed number or a whole number.
[1.3]	*Objective 1.7*	Write a mixed number as an improper fraction.
[1.3]	*Objective 1.8*	Change a common fraction to a decimal fraction.
[1.4]	*Objective 1.9*	Round a decimal fraction to a specified degree of accuracy.
[1.5]	*Objective 1.10*	Write a number in scientific notation.
[1.5]	*Objective 1.11*	Write a number without exponents.
[1.5]	*Objective 1.12*	Find the prime factorization of a given number.
[1.6]	*Objective 1.13*	Reduce a common fraction.
[1.6]	*Objective 1.14*	Multiply and divide common fractions.
[1.6]	*Objective 1.15*	Change a decimal fraction to a common fraction.
[1.7]	*Objective 1.16*	Add and subtract common fractions with common denominators.
[1.7]	*Objective 1.17*	Find the LCD for some given denominators.
[1.7]	*Objective 1.18*	Add and subtract common fractions.
[1.7]	*Objective 1.19*	Carry out mixed operations with fractions, including those using juxtaposition.
[1.2–1.7]	*Objective 1.20*	Work applied problems.

Self-Test

Each question of this self-test is related to the corresponding objective listed above.

1. Describe what is meant by math anxiety.

2. Simplify $40 + 20 \div 5 \times 3$, and classify as a sum, difference, product, or quotient.

3. Find (translate) and then simplify:
   a. the sum of the squares of three and eleven
   b. the square of the sum of three and eleven

4. Rewrite $8(500 + 60 + 7)$ using the distributive property.

5. If you spend $1.85 per day on tolls, estimate the amount you spend each year.

6. Write $\dfrac{85}{6}$ as a mixed number.

7. Write $7\frac{3}{8}$ as an improper fraction.

8. Change $6\frac{1}{3}$ to decimal form.

9. Round \$85.255 to the nearest cent.

10. Write 93,500,000 in scientific notation.

11. Write 8.92×10^{-9} without exponents.

12. Find the prime factorization of 1,330.

13. Reduce $\frac{1,330}{1,881}$.

14. Simplify: **a.** $5\frac{3}{8} \times 2\frac{2}{3}$ **b.** $\frac{12}{35} \div \frac{4}{7}$

15. Write $0.8\frac{1}{3}$ as a fraction.

16. Simplify $12\frac{2}{5} - 5\frac{4}{5}$.

17. Find the lowest common denominator for the numbers 120 and 700.

18. Simplify $\frac{3}{10} + \frac{4}{15} + \frac{5}{12}$.

19. Simplify $\frac{2}{3}\left(\frac{3}{8} - \frac{1}{8} \times 2\right)$.

20. If you join a book club and agree to buy six books at \$24.95 each plus \$3.50 postage and handling for each book, what is the total cost to fulfill this agreement?

STUDY HINTS *Compare your solutions and answers to the self-test. For each problem you missed, work some additional problems in the section listed in the margin. After you have worked these problems, you can test yourself with the practice test.*

Complete Solutions to the Self-Test

Additional Problems

[1.1] *Problems 1–12*

1. Answers vary: fear; inability to take constructive action; guilt; the more we avoid math, the less able we feel, and the less able we feel, the more we avoid it.

[1.2] *Problems 9–36; 55–70*

2. $40 + 20 \div 5 \times 3 = 40 + 4 \times 3$ Remember the order of operations.
$= 40 + 12$
$= 52$

This is a sum.

[1.2] *Problems 47–54*

3. a. $3^2 + 11^2 = 130$ **b.** $(3 + 11)^2 = 196$

[1.2] *Problems 37–46*

4. $8(500 + 60 + 7) = 8(500) + 8(60) + 8(7)$. Use the distributive property; do not carry out the arithmetic.

[1.2] *Problems 71–80*
[1.3] *Problems 51–54*
[1.4] *Problems 3–4; 39–44*
[1.5] *Problems 50–55*
[1.7] *Problems 3–10; 66–77*

5. $\$1.85 \times 365 \approx \$2 \times 350 = \$700$; you would spend about \$700 per year on tolls. Calculator check: $1.85 \times 365 = 675.25$.

[1.3] *Problems 15–26*

6. $\frac{85}{6} = 14\frac{1}{6}$. Find $85 \div 6 = 14$ with remainder 1.

[1.3] *Problems 27–38*

7. $7\frac{3}{8} = \frac{59}{8}$. Find $8 \times 7 + 3 = 59$.

[1.3] *Problems 39–50; 55–57*

8. $6.\overline{3}$. Do not round your answer; 6.3333333 is not correct; $6.33\frac{1}{3}$ is not a simplified form.

[1.4] *Problems 11–38*

9. \$85.26 The rounding place digit is the hundredth column: \$85.2[5]5

[1.5] *Problems 13–25*

10. 9.35×10^7 Count 7 decimal places to the right to restore the number to its given form.

[1.5] *Problems 26–40*

11. .00000 00089 2 Count 9 decimal places to the left.

[1.5] *Problems 41–49*

12. $2 \times 5 \times 7 \times 19$ Use a factor tree:

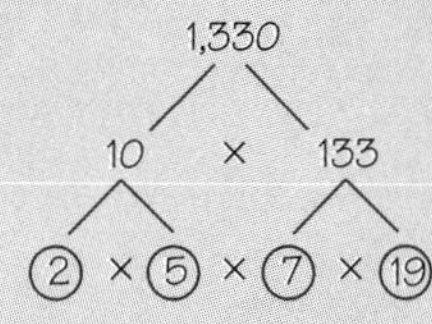

[1.6] *Problems 13–19*

13. $\frac{1{,}330}{1{,}881} = \frac{2 \times 5 \times 7 \times \cancel{19}}{3 \times 3 \times 11 \times \cancel{19}}$

$= \frac{2 \times 5 \times 7}{3 \times 3 \times 11}$

$= \frac{70}{99}$

The factor tree for 1,330 is shown in Problem 12. The factor tree for 1,881 is:

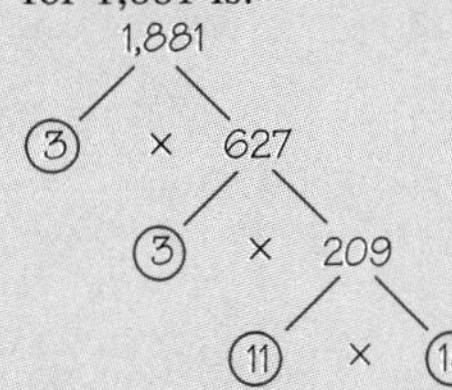

[1.6] *Problems 20–48*

14. a. $5\frac{3}{8} \times 2\frac{2}{3} = \frac{43}{8} \times \frac{8}{3} = \frac{43}{3}$ You can also write this answer as $14\frac{1}{3}$.

b. $\frac{12}{35} \div \frac{4}{7} = \frac{\overset{3}{\cancel{12}}}{\underset{5}{\cancel{35}}} \times \frac{\overset{1}{\cancel{7}}}{\underset{1}{\cancel{4}}} = \frac{3}{5}$

[1.6] *Problems 49–56*

15. $0.8\frac{1}{3} = 8\frac{1}{3} \times \frac{1}{10} = \frac{\overset{5}{\cancel{25}}}{3} \times \frac{1}{\underset{2}{\cancel{10}}} = \frac{5}{6}$

[1.7] *Problems 17–22*

16. $12\frac{2}{5} - 5\frac{4}{5} = 6\frac{3}{5}$ Write:

$$\begin{array}{rcr} 12\frac{2}{5} & = & 11\frac{7}{5} \\ -\ 5\frac{4}{5} & = & 5\frac{4}{5} \\ \hline & & 6\frac{3}{5} \end{array}$$

[1.7] *Problems 23–34*

17. $120 = \mathbf{2^3 \times 3} \times 5$

$700 = 2^2 \times \quad \mathbf{5^2 \times 7}$ LCD: $2^3 \times 3 \times 5^2 \times 7 = 4{,}200$

[1.7] *Problems 35–50*

18.

$$\begin{array}{rcl} \frac{3}{10} & = & \frac{18}{60} \\ \frac{4}{15} & = & \frac{16}{60} \\ +\ \frac{5}{12} & = & \frac{25}{60} \\ \hline & & \frac{59}{60} \end{array}$$

[1.7] *Problems 51–65*

19. $\frac{2}{3}\left(\frac{3}{8}-\frac{1}{8}\times 2\right)=\frac{2}{3}\left(\frac{3}{8}-\frac{2}{8}\right)$ Remember order of operations.

$=\frac{2}{3}\left(\frac{1}{8}\right)$

$=\frac{1}{12}$ Parentheses and juxtaposition (no operation symbol) mean multiplication.

[1.2] *Problems 71–80*
[1.3] *Problems 58–63*
[1.4] *Problems 45–58*
[1.5] *Problems 56–65*
[1.6] *Problems 57–68*
[1.7] *Problems 78–82*

20. Each book costs $24.95 + $3.50 = $28.45; thus six books cost 6($28.45) = $170.70.

Note: In studying for the exam, be sure you look at several different types of word problems.

Chapter 1 Practice Test

1. List a symptom of math anxiety that you have experienced at some time in your life.

2. Name a math myth that is difficult for you not to believe in, even if you know it is a myth.

3. Name a math myth that is easy for you to accept as a myth.

4. How do you think someone with a severe case of math anxiety could learn to deal with this anxiety?

5. A box of oranges contains approximately 96 oranges. If the U.S. production of oranges is 186,075,000 boxes, estimate the number of oranges produced each year in the United States. Leave your answer in scientific notation.

6. Your answer for Problem 5 is about
A. 2 million B. 20 million C. 2 trillion D. 2 billion E. 20 billion

7. Eliminate the parentheses, but do not carry out the arithmetic.
a. 5(8 + 2) **b.** 2(25 + 35)
c. 3(200 + 50 + 6) **d.** 5(400 + 50 + 9)

8. a. Write the sum of the cubes of 2 and 3.
b. Write the cube of the sum of 2 and 3.

9. Write each as a mixed number.
a. $\frac{114}{7}$ **b.** $\frac{25}{3}$ **c.** $\frac{167}{10}$ **d.** $\frac{153}{100}$

10. Write each as an improper fraction.
a. $4\frac{2}{3}$ **b.** $1\frac{5}{8}$ **c.** $3\frac{3}{4}$ **d.** $12\frac{5}{9}$

11. Write in decimal form.
a. $\frac{7}{8}$ **b.** $\frac{5}{6}$ **c.** $8\frac{2}{3}$ **d.** $2\frac{4}{5}$

12. **a.** Round 6.149 to the nearest tenth.
b. Round 45.5 to the nearest unit.
c. Round $45.31499 to the nearest dollar.
d. Round $104.996 to the nearest cent.

13. Write in scientific notation.
a. 0.0034 **b.** 4,000,300
c. 17,400 **d.** 5

14. Use a calculator to estimate 3 trillion dollars divided by the population of the United States, 215 million.

15. Indicate the calculator keys you would press to find the given expressions, and classify each as a sum, difference, product, or quotient.
a. $\frac{16.2 + 14.1 - 8.454}{2.13}$ **b.** $16.2 + 14.1 - \frac{8.454}{2.13}$

16. Write without exponents.
a. 4^3 **b.** 9^2 **c.** 5.79×10^{-4} **d.** 4.01×10^5

17. Write the prime factorization.
a. 86 **b.** 72 **c.** 486 **d.** 1,372

18. Reduce each fraction.
a. $\frac{15}{25}$ **b.** $\frac{32}{16}$ **c.** $\frac{192}{240}$ **d.** $\frac{128}{384}$

19. Write in fractional form.
a. 0.333 **b.** $0.2\frac{2}{9}$ **c.** 0.95 **d.** $0.00\frac{1}{2}$

20. Find the LCD.
a. 12; 15 **b.** 6; 10 **c.** 10; 15; 6 **d.** 24; 30; 18

21. Simplify and classify each as a sum, difference, product, or quotient.
a. $12 + 20 \div 2$ **b.** $(12 + 20) \div 2$
c. $8 + 3 \times 6 \div 2$ **d.** $(8 + 3) - (6 \div 2)$

Simplify the expressions in Problems 22–27.

22. **a.** $\frac{3}{5} \times \frac{25}{27}$ **b.** $\frac{4}{9} \times 27$ **c.** $4\frac{1}{6} \times 3\frac{2}{5}$ **d.** $2\frac{3}{4} \times \frac{4}{5}$

23. **a.** $\frac{5}{8} \div \frac{1}{2}$ **b.** $\frac{14}{25} \div \frac{7}{5}$ **c.** $1\frac{1}{2} \div \frac{3}{4}$ **d.** $6\frac{1}{2} \div 3\frac{3}{4}$

24. **a.** $\frac{5}{7} + \frac{3}{7}$ **b.** $\frac{6}{11} - \frac{2}{11}$ **c.** $12\frac{4}{5} + 6\frac{3}{5}$ **d.** $5\frac{1}{3} - 1\frac{2}{3}$

25. **a.** $\frac{3}{8} + \frac{1}{6}$ **b.** $\frac{7}{12} - \frac{2}{15}$ **c.** $7\frac{2}{15} - 3\frac{7}{12}$ **d.** $\frac{4}{10} + \frac{7}{15} - \frac{5}{6}$

26. **a.** $\frac{2}{3} + \frac{4}{5} \div \frac{1}{2}$ **b.** $\frac{4}{5}\left(\frac{12}{23} - \frac{2}{23}\right)$

27. **a.** $\frac{4}{5} \times \frac{12}{23} - \frac{4}{5} \times \frac{2}{23}$ **b.** $\frac{2 \times 8 + 3 \times 5}{3 \times 8}$

28. Rework Problem 27 using a calculator; leave your answer in decimal form.

29. Estimate the portion of the square that is shaded.

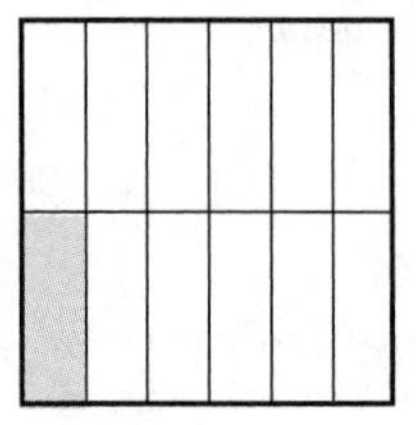

30. Enter your favorite number (a counting number from 1 to 9) into a calculator. Multiply by 259; then multiply this result by 429. What is your answer? Try it for three different choices.

CHAPTER TWO

SETS OF NUMBERS

2.1 SYMBOL SHOCK

Historical Perspective

Leonhard Euler
(1707–1783)

The development of the symbols of mathematics seems to have passed through three stages: a *rhetorical stage* in which problems and solutions were stated in words, a stage in which abbreviations for words were used, and a third *symbolic stage* in which the words were replaced by symbols. Letters to symbolize unknowns began to be used during the Middle Ages, but most people at that time still preferred abbreviations. One of the first persons to use letters to represent numbers was the French lawyer François Viète, who was ordered by Henry IV in 1591 to improve the symbolism of algebra. Descartes (see page 506) used a notation similar to our own in 1637. One of the greatest mathematicians of all time, Leonhard Euler (pronounced "Oiler") used the symbolism of Viète to advance mathematics to new levels. In 1735, the year after his wedding and when he was in Russia, Euler received a problem in celestial mechanics from the French Academy. Though other mathematicians had required several months to solve this problem, Euler, using improved methods of his own, solved it in 3 days. The strain, however, resulted in an illness in which he lost the sight in his right eye. He commented, "Now I will have less distraction." He went completely blind 31 years later, but his blindness did not slow down his mathematical work. Euler is known as the most prolific mathematician in the history of mathematics. It is said that he had almost total recall and could mentally calculate long and complicated problems. F. Argo described this ability by saying, "Euler calculated without any apparent effort, just as men breathe and as eagles sustain themselves in the air."

F. Argo in Howard Eves, *In Mathematical Circles,* p. 47.

Algebra is generous, she often gives more than is asked of her.
JEAN D'ALEMBERT

In arithmetic, you learned about numbers and operations with numbers. **Algebra** is defined as a generalization of arithmetic and has been described as one of the best labor-saving devices invented by the human mind. In algebra, extensive use of symbols is made, and that is where many of us go into "symbol shock." In the next chapter, we introduce some of the techniques of algebra, and in this chapter we introduce the building blocks of algebra, namely, the sets of integers, rationals, irrationals, and real numbers. To describe these sets of numbers, we need to use symbols. In this section, we look at some of the symbols used in algebra.

Our society is filled with symbols that we've learned to use intuitively. Books have been written to help us interpret the symbols of body language, we can learn to interpret our dreams, and the very language we use is a symbolic representation. For example, the letters FACE can stand for a variety of ideas, depending on the context in which they are used.

In algebra, we use letters of the alphabet to represent numbers with unknown values. A letter used in this way is called a **variable.**

EXAMPLE 1

Using Variables

Think of a counting number from one to ten. Add five. Multiply the result by two. Subtract six. Divide by two. Subtract the original number. Let me guess the result: It is two. How did I know the result, even though you were allowed to begin with *any* number? Why does this number trick work?

Solution

WITHOUT VARIABLES	*WITH VARIABLES*
Let □ represent the number you have chosen (its value is unknown to me). Let stars (∗ ∗ · · ·) represent the numbers stated in the question (known numbers).	Let n = UNKNOWN NUMBER.
Think of a number: □	n
Add five: □ ∗ ∗ ∗ ∗ ∗	$n + 5$
Multiply by two: □ ∗ ∗ ∗ ∗ ∗ □ ∗ ∗ ∗ ∗ ∗	$\mathbf{2}(n + 5) = 2n + 10$ Distributive property
Subtract six: □ ∗ ∗ ∗ ∗ ∗ − ∗ ∗ ∗ = □ ∗ ∗ □ ∗ ∗ ∗ ∗ ∗ − ∗ ∗ ∗ = □ ∗ ∗	$2n + 10 - \mathbf{6} = 2n + 4$ $= \mathbf{2(n + 2)}$ Distributive property

Divide by two:

$\dfrac{\square * *}{\boxed{\square} * *} = \square * *$	$\dfrac{2(n+2)}{2} = \mathbf{n + 2}$

Subtract the original number:

$\boxed{\square} * * = * *$	$n + 2 - \mathbf{n} = \mathbf{2}$

The final result is two! ■

Go slowly here! We are setting up some very important ideas about algebra.

As you look at Example 1, which solution seems easier: the one with or the one without variables? At this point, it may be that the one without variables is easier for you to understand. Our study of algebra is a study in becoming familiar enough with the variables to feel comfortable with manipulations like those shown at the right in Example 1.

A variable represents a number from a given set of numbers called the **domain** of the variable. To express the idea "sum of a number and two," you can write

$$n + 2$$

where n represents the unknown number. If an expression contains at least one variable and at least one defined operation, it is called a **variable expression.** The domain for this variable expression is the set of all numbers, because there is no reason to restrict the possible replacement of n (that is, n can be 6, 3, $\frac{1}{3}$, or any other number that comes to mind). On the other hand, if n represents the number of people on an elevator, then $n + 2$ represents the idea that two additional people board the elevator, and the domain for n is the set of whole numbers up to some maximum capacity of the elevator. If the domain is $D = \{0, 1, 2, 3, \ldots, 30\}$, then the possible number of people on the elevator after the two additional people board is $\{2, 3, 4, \ldots, 32\}$.

EXAMPLE 2

Variable Expression

Let x be the variable and let $D = \{0, 1, 5, 10\}$ be the domain. Then the variable expression

$$x + 7$$

represents

$\mathbf{0} + 7 = 7$ if $x = \mathbf{0}$ $\qquad$ $\mathbf{1} + 7 = 8$ if $x = \mathbf{1}$

$\mathbf{5} + 7 = 12$ if $x = \mathbf{5}$ $\qquad$ $\mathbf{10} + 7 = 17$ if $x = \mathbf{10}$

It *cannot* represent $\mathbf{9} + 7 = 16$ because 9 is not in the domain. ■

Some terms involving the operations with which you should be familiar are summarized in Table 2.1 on page 74.

Table 2.1 *Translating into Algebra*

Symbol	Verbal description
$=$	Is equal to; equals; are equal to; is the same as; is; was; becomes; will be; results in. Two expressions are **equal** if they are the same.
$+$	The numbers being added are called **terms,** and the result is called the **sum.** Plus; the sum of; added to; more than; greater than; increased by; taller; longer; larger; heavier.
$-$	The numbers being subtracted are called **terms,** and the result is called the **difference.** **Minus;** the difference of; the difference between; is subtracted from; less than; smaller than; decreased by; is diminished by; shorter; smaller; lighter.
$\times$	A times sign is used primarily in arithmetic.
$\cdot$	A raised dot is used to indicate multiplication in algebra, as in $11 \cdot 7$ or $x \cdot 1$.
( )	Parentheses are used to indicate multiplication in algebra, as in $3(4 + 10)$ or $6(x + 2)$ or $5(w)$.
no symbol	Juxtaposition (no symbol) is used to indicate multiplication in algebra, especially with variables, as in xy or $5x$, but not with numerals (for example, 24 still means twenty-four—not two times four).
CAUTION: Be careful to distinguish between the words **term** and **factor.**	The numbers being multiplied are called **factors,** and the result is called the **product.** Times; product; is multiplied by; of; twice; double.
$\div$	The division symbol is used primarily in arithmetic.
fractional bar	The fractional bar, as in $\frac{3}{4}$ (meaning 3 divided by 4) or $\frac{x}{y}$ (meaning x divided by y) or $\frac{x+2}{3}$ (meaning $x + 2$ divided by 3), is frequently used in algebra. In $\frac{x}{y}$, x is called the **dividend,** y is called the **divisor,** and the result is called the **quotient.** Divided by; quotient of; per.

EXAMPLE 3

Translating an English Expression into a Variable Expression

Choose a letter to represent the variable, and write a mathematical statement to express the idea.

a. The sum of a number and 13
b. The difference of a number subtracted from 10
c. The quotient of two numbers
d. The product of two consecutive numbers

Solution With an example such as this, we do not want to belabor an obvious answer. On the other hand, we realize the necessity of "building a procedure" for translating from English into algebra. In building this procedure and taking careful steps, we must also prevent "symbol shock" or undue avoidance of symbols.

a. "The sum of a number and 13"

Sum indicates addition, so the statement is

(NUMBER) + (THIRTEEN)

Now select some variable (your choice)—say, n = NUMBER. Then the symbolic statement is

$$\mathbf{n + 13}$$

b. "The difference of a number subtracted from 10"

(TEN) − (NUMBER)

Let x = NUMBER. Then the variable expression is

$$\mathbf{10 - x}$$

c. "The quotient of two numbers"

$$\frac{\text{NUMBER}}{\text{ANOTHER NUMBER}}$$

If there is more than one unknown in a problem, more than one variable may be needed. Let m = NUMBER and n = ANOTHER NUMBER; then

$$\frac{m}{n}$$

d. "The product of two consecutive numbers"

Do not use more variables than you need.

(NUMBER)(NEXT CONSECUTIVE NUMBER)

If there is more than one unknown in a problem, but a relationship between those unknowns is given, then *do not choose* more variables than you need for the problem. In this problem, a "consecutive number" means one more than the first number:

(NUMBER)(NUMBER + 1)

Let y = NUMBER. Then the variable expression is

$$\mathbf{y(y + 1)}$$ ■

In Chapter 1, you simplified numerical expressions. If $x = a$, then x and a name the same number; x may be replaced by a in any expression, and the value of the expression remains unchanged. When you replace variables by given numerical values and then simplify the resulting numerical expression, the process is called **evaluating an expression.**

EXAMPLE 4 **Evaluating an Expression**

Evaluate $a + cb$, where $a = 2$, $b = 11$, and $c = 3$.

Solution

$a + cb$ Remember, cb means c **times** b.

Step 1 Replace each variable with the corresponding numerical value. You may need additional parentheses to make sure you don't change the order of operations.

$a + c \cdot b$
↓ ↓ ↓↓ ↓
2 + 3(11) ← Parentheses are necessary so that the product cb is not changed to 311.

Step 2 Simplify:

$2 + 3(11) = 2 + 33$ Multiplication before addition
$= \mathbf{35}$

■

EXAMPLE 5

Evaluating an Expression

Evaluate the following, where $a = 3$ and $b = 4$.

a. $a^2 + b^2$ **b.** $(a + b)^2$

Solution

a. $a^2 + b^2 = 3^2 + 4^2$
$= 9 + 16$
$= \mathbf{25}$ Remember the order of operations; multiplication comes first, and $3^2 = 3 \cdot 3$, $4^2 = 4 \cdot 4$, which is multiplication.

Spend some time with this example. It is important because: (1) it illustrates the ideas of evaluation; and (2) you need to remember that $(a + b)^2 \neq a^2 + b^2$.

b. $(a + b)^2 = (3 + 4)^2$
$= 7^2$
$= \mathbf{49}$ Order of operations; parentheses first

Notice that $a^2 + b^2 \neq (a + b)^2$. ■

Remember that a particular variable is replaced by a single value when an expression is evaluated. You should also be careful to write capital letters differently from lowercase letters, because they often represent different values. This means that, just because $a = 3$ in Example 5, you should not assume that $A = 3$. On the other hand, it is possible that other variables *might* have the value 3. For example, just because $a = 3$, do not assume that another variable—say, t—cannot also have the value $t = 3$.

EXAMPLE 6

Number Puzzle

Let $a = 1$, $b = 3$, $c = 2$, and $d = 4$. Find the value of the given capital letters.

a. $G = bc - a$ **b.** $H = 3c + 2d$ **c.** $I = 3a + 2b$ **d.** $R = a^2 + b^2d$

e. $S = \dfrac{2(b + d)}{2c}$ **f.** $T = \dfrac{3a + bc + b}{c}$

Solution After you have found the value of a capital letter, write it in the box that corresponds to its numerical value. This exercise will help you check your work.

	37	9	5	14	6	

a. $G = bc - a$
$= 3(2) - 1$
$= 6 - 1$
$= \mathbf{5}$

b. $H = 3c + 2d$
$= 3(2) + 2(4)$
$= 6 + 8$
$= \mathbf{14}$

c. $I = 3a + 2b$
$= 3(1) + 2(3)$
$= 3 + 6$
$= \mathbf{9}$

d. $R = a^2 + b^2d$
$= 1^2 + 3^2(4)$
$= 1 + 9(4)$
$= \mathbf{37}$

e. $S = \dfrac{2(b + d)}{2c}$
$= \dfrac{2(3 + 4)}{2(2)}$
$= \dfrac{2(7)}{4}$
$= \dfrac{\mathbf{7}}{\mathbf{2}}$

f. $T = \dfrac{3a + bc + b}{c}$
$= \dfrac{3(1) + 3(2) + 3}{2}$
$= \dfrac{3 + 6 + 3}{2}$
$= \dfrac{12}{2}$
$= \mathbf{6}$

After you have filled in the appropriate boxes, the result is

	37	9	5	14	6	
	R	I	G	H	T	

■

PROBLEM SET 2.1

Essential Ideas

1. Numbers or variables that are added are called ________.

2. Numbers or variables that are multiplied are called ________.

3. **In Your Own Words** Give at least three translations for each of the following symbols:
a. + **b.** − **c.** × **d.** ÷ **e.** =

4. **In Your Own Words** What does it mean to "evaluate an expression"?

Level 1

5. **In Your Own Words** Write about your previous experiences with algebra, if any. If you have never had any previous experience with algebra, write about your feelings as you anticipate learning about algebra.

6. **In Your Own Words** Reread Example 1. Discuss which of the two solutions seems easier to you, and why.

Level 1—Drill

Let x be the variable and let D = {0, 1, 3, 7} be the domain. Find the values for the variable expressions in Problems 7–11.

7. $x + 5$ **8.** $x + 10$ **9.** $2x + 1$

10. x^2 **11.** $100 - 2x^2$

Let y be the variable and let $D = \{0, 2, 3, 10\}$ be the domain. Find the values for the variable expressions in Problems 12–17.

12. $y + 8$

13. $(y + 6)^2$

14. $3y + 2$

15. y^2

16. $10 + y^2$

17. $(10 + y)^2$

WHAT IS WRONG, *if anything, with each of the statements in Problems 18–23? Explain your reasoning.*

18. If $x = 10$ or 20, then $x + 9$ is 19 or 29. The domain is x.

19. xy is the sum of x and y.

20. In the expression, $10 \div 2$, the divisor is 2.

21. The quotient of 30 divided by 6 is the same as the quotient of 6 divided by 30.

22. Numbers that are added are called terms.

23. Numbers that are multiplied are called terms.

LEVEL 2—Drill

In Problems 24–47, let $w = 2$, $x = 1$, $y = 2$, and $z = 4$. Find the values of the given capital letters.

24. $A = x + z + 8$

25. $B = 5x + y - z$

26. $C = 10 - w$

27. $D = 3z$

28. $E = 25 - y^2$

29. $F = w(y - x + wz)$

30. $G = 5x + 3z + 2$

31. $H = 3x + 2w$

32. $I = 5y - 2z$

33. $J = 2w - z$

34. $K = wxy$

35. $L = x + y^2$

36. $M = (x + y)^2$

37. $N = x^2 + 2xy + y^2 + 1$

38. $P = y^2 + z^2$

39. $Q = w(x + y)$

40. $R = z^2 - y^2 - x^2$

41. $S = (x + y + z)^2$

42. $T = x^2 + y^2z$

43. $U = \dfrac{w + y}{z}$

44. $V = \dfrac{3wyz}{x}$

45. $W = \dfrac{3w + 6z}{xy}$

46. $X = (x^2z + x)^2z$

47. $Y = (wy)^2 + w^2y + 3x$

In Problems 48–72, choose a letter to represent the variable, and write a mathematical expression for the idea; you do not need to simplify.

48. A number plus five

49. Six plus a number

50. Five plus six plus a number

51. Two plus a number plus three

52. Twice a number

53. Three times a number

54. Five minus a number

55. Ten minus a number

56. A number minus five

57. A number minus two

58. A number x plus 3 divided by 2

59. The sum of $x + 3$ divided by 2

60. The difference of a number subtracted from 100

61. The difference of a number subtracted from 1

62. The difference of 1 subtracted from a number

63. The difference of 100 subtracted from a number

64. The product of seven and a number

65. The product of four and a number

66. The product of a number and five

67. The product of two, three, and a number

68. The product of three, five, and a number

69. The sum of five and a number

70. The sum of a number and seventeen

71. The quotient of five divided by a number

72. The quotient of a number divided by five

LEVEL 3—Mind Bogglers

73. This problem will help you check your work in Problems 24–47. Fill in the capital letters from Problems 24–47 to correspond with their numerical values (the letter O has been filled in for you). Some letters may not appear in the boxes. When you are finished, darken all the blank spaces to separate the words in the secret message. Notice that some of the blank spaces have also been filled in to help you.

13	5	19	21	3	11	13	14	2	49	22	50
17	7	21	23	19	11	21	13	17	21	49	17
5	13	3	0	11	■	49	13	48	2	10	19
12	21	48	2	8	21	26	21	48	21	11	22
2	10	48	21	10	17	21	12	.	■	■	■

74. Think of a counting number less than 20. Add six. Multiply by two. Subtract eight. Divide by two. Subtract your original number. The answer is 2. Explain why this trick works.

75. Think of a counting number less than 100. Add five. Multiply by two. Add ten. Subtract twenty. Divide by two. The answer is your original number. Explain why this trick works.

76. The pictures here describe a number trick. Describe it in words.

(1) □ (2) □ * * * (3) □ * * * / □ * * * (4) □ * * / □ * * (5) * * / * *

77. The algebraic steps below describe a number trick. Describe it in words.

(1) x
(2) $x + 7$
(3) $2(x + 7) = 2x + 14$
(4) $2x + 14 - 4 = 2x + 10 = 2(x + 5)$
(5) $\dfrac{2(x + 5)}{2} = x + 5$
(6) $x + 5 - x = 5$

78. My favorite number is 7. Make up a number trick in which you ask someone to think of a number and carry on some operations, with the final answer always 7.

79. Bill Leonard's favorite number is 23. Make up a number trick in which you ask someone to think of a number and carry on some operations, with the final answer always 23.

80. Translate the following symbolic message:

EZ4NE12CYU $\dfrac{\text{R}}{\text{WEIGHT}}$

81. Translate the following symbolic message:

U8N8N8N8NR $\dfrac{\text{2}}{\text{ACTIVE}}$

Level 3—Team Project

82. *"When will I ever use algebra?"* With a team of two or three other classmates, make a trip to your college library, where you each obtain a different college catalog. If possible, include one from a nearby liberal arts college, a state college, a university, and a world-class university. Compare the math requirements for entrance, and then outline the math requirements for at least 10 different baccalaureate degree programs.

2.2 ADDITION OF INTEGERS

Historical Perspective

It is not easy to understand the history of the positive and negative numbers. Everyone thinks they know, for example, what five means—until they try to define it or explain it. In 1884, the German logician and mathematician F. L. G. Frege (1848–1925) tried to define numbers precisely with what he called a cardinal number. He amplified his views in several volumes, but he was largely ignored during his life. As the historian Carl B. Boyer states, "History shows that novelty in ideas is more readily accepted if couched in relatively conventional form."

Funky Winkerbean reprinted by permission of Field Enterprises, Inc. © 1973.

You have probably seen the need for numbers to represent quantities less than zero many times. Some common examples are temperatures below zero, card scores that are "in the hole," business debts, and altitudes below sea level. To represent these ideas, we introduce **signed numbers.**

1. With each counting number, we associate a **positive sign:**

   $1 = +1 \qquad 2 = +2 \qquad 3 = +3 \ldots$

   Although the symbolism is the same as that of an addition sign, remember that addition requires two numbers, for example, $6 + 5$. Using positive signs, this addition problem would look like this: $(+6) + (+5)$.

2. For each counting number, we define a *new number,* called its *opposite,* by using a **negative sign.**

   The opposite of 1 is denoted by -1.

   The opposite of 2 is denoted by -2.

   The opposite of 3 is denoted by -3.

   And so on.

CAUTION

Do NOT confuse the opposite key [+/−] with the minus key [−].

The negative (or opposite) sign on a calculator is labeled [+/−] or [(−)] or [CHS].

The **positive numbers** and **negative numbers** are separated by a number called **zero.** The **opposite** of a given number is defined to be the number that, when added to the given number, gives a sum of zero:

Given positive number		Opposite	Given negative number		Opposite	
↓		↓	↓		↓	
$+1$	$+$	$(-1) = 0$	(-1)	$+$	1	$= 0$
$+2$	$+$	$(-2) = 0$	(-2)	$+$	2	$= 0$
$+3$	$+$	$(-3) = 0$	(-3)	$+$	3	$= 0$
	⋮	↑		⋮	↑	
		The opposite of a positive number is negative.			The opposite of a negative number is positive.	

Since $0 + 0 = 0$, the opposite of zero is zero.

3. The counting numbers, along with their opposites and the number 0, form a set called the **integers.**

Integers

The set of **integers** is the set

$$\{\ldots, -4, -3, -2, -1, 0, 1, 2, 3, 4, \ldots\}$$

Historical Perspective

Historically, the negative integers were developed quite late. There are indications that the Chinese had some knowledge of negative numbers as early as 200 B.C., and in the 7th century A.D., the Hindu Brahmagupta stated the rules for operations with positive and negative numbers. The Chinese represented negative integers by putting them in red (compare with the present-day accountant), and the Hindus represented them by putting a circle or a dot over the number. However, as late as the 16th century, some European scholars were calling numbers such as (-1) absurd. In 1545, Girolamo Cardano (1501–1576), an Italian scholar who presented the elementary properties of negative numbers, called the positive numbers "true" numbers and the negative numbers "fictitious" numbers. However, they did become universally accepted; and as a matter of fact, the word *integer* that we use to describe this set is derived from "numbers with integrity."

The easiest way to understand the operations and properties of integers is to represent them on a **number line,** as shown in Figure 2.1.

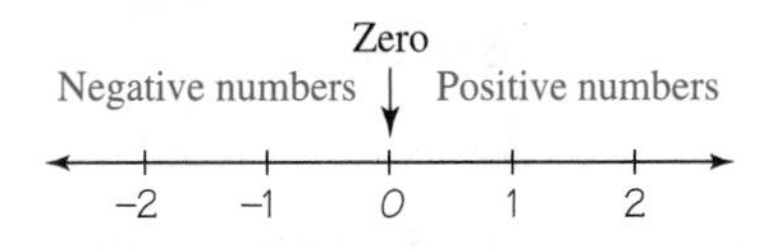

Figure 2.1 A number line

To draw a number line, locate any two convenient points; label the one on the left 0 (zero) and the one on the right $+1$ (positive one). The distance between these points is called a **unit scale** and can be used to mark off equal distances in both directions. These points correspond to the integers, as shown in Figure 2.1. A number line can be used to order numbers using three **order symbols,** $<$, $>$, and $=$.

Order Symbols

Less than (symbol $<$) means *to the left* on a number line.

Greater than (symbol $>$) means *to the right* on a number line.

Equal to (symbol $=$) means *the same point* on a number line.

EXAMPLE 1 **Comparing the Sizes of Numbers**

Write $<$, $>$, or $=$ in the blank.

a. 2 _____ 5 **b.** 6 _____ 3 **c.** 4 _____ 4 **d.** -2 _____ -3 **e.** 0 _____ -2

Solution

a. $2 < 5$, since 2 is to the left of 5 on a number line.
b. $6 > 3$
c. $4 = 4$
d. $-2 > -3$, since -2 is to the right of -3.
e. $0 > -2$ ■

With the introduction of negative values, we also need to introduce a symbol to represent distance, because distances are nonnegative.

Absolute Value

The **absolute value** of a number is the distance of that number from 0 on the number line.

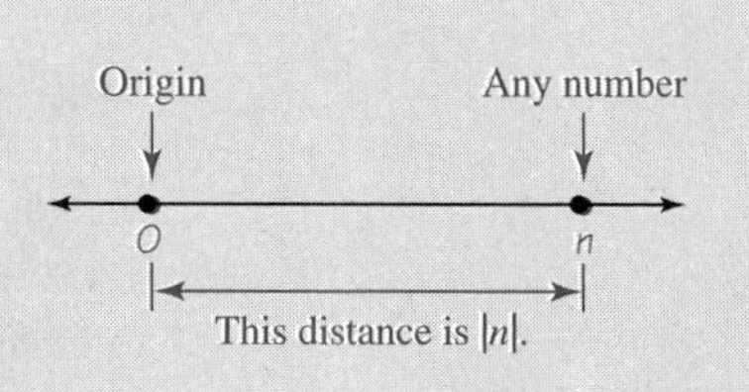

EXAMPLE 2 **Evaluating an Absolute Value**

Evaluate:

a. $|5|$ **b.** $|-5|$ **c.** $|-3|$ **d.** $|0|$ **e.** $|-349|$

Solution

a. The absolute value of 5, symbolized by $|5|$, is **5** because 5 is 5 units from 0.
b. The absolute value of -5, symbolized by $|-5|$, is **5** because -5 is 5 units from 0.
c. $|-3| = \mathbf{3}$ **d.** $|0| = \mathbf{0}$ **e.** $|-349| = \mathbf{349}$ ■

You will find an absolute value key on many calculators. Generally, you must enclose the number whose absolute value you want inside parentheses.

$|-3|$ is input as ABS (3 +/−)

$|0|$ is input as ABS (0)

$|-349|$ is input as ABS (349 +/−)

EXAMPLE 3

Finding the Larger Absolute Value

Tell which number is larger. Then show the larger absolute value.

a. 2; 5 **b.** −2; −5 **c.** −8; 10 **d.** 8; −10

Solution

	Larger number	*Larger absolute value*
a. 2; 5	**5**	**5**
b. −2; −5	**−2**	**5**
c. −8; 10	**10**	**10**
d. 8; −10	**8**	**10**

■

A number line is also used to illustrate the addition of integers. It is agreed that adding a positive number means moving to the right, and that adding a negative number means moving to the left. This process is illustrated in Example 4. Adding zero means no move at all.

EXAMPLE 4

Addition Using a Number Line

Add on a number line.

a. 2 + 6 **b.** 6 + (−4) **c.** (−10) + 6 **d.** (−2) + (−8)

Solution

a. Positive + Positive: 2 + 6

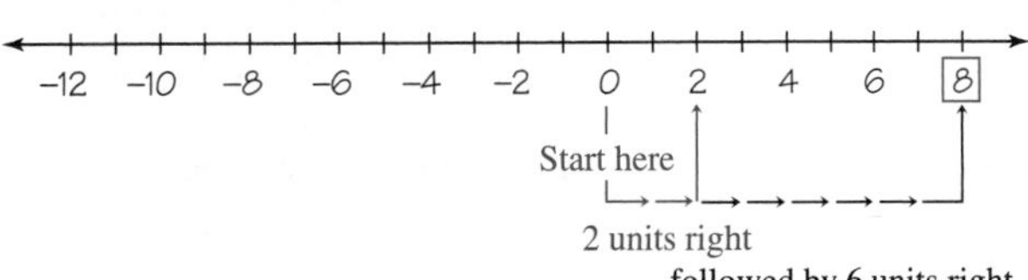

Ending point is 8.

b. Positive + Negative: 6 + (−4)

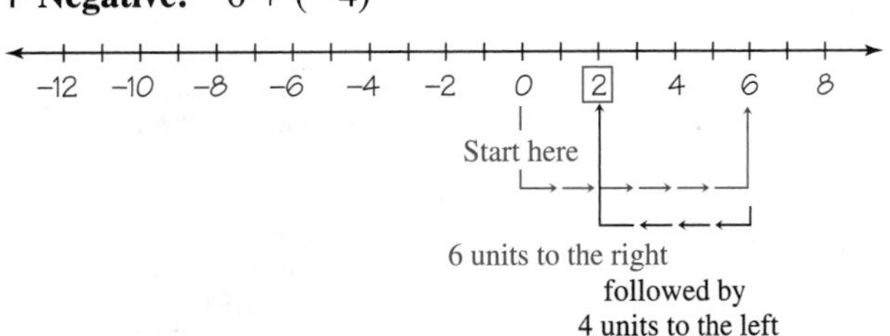

Ending point is 2.

CAUTION

To avoid confusion, negative numbers are often enclosed in parentheses when combined with other operations.

c. Negative + Positive: $(-10) + 6$

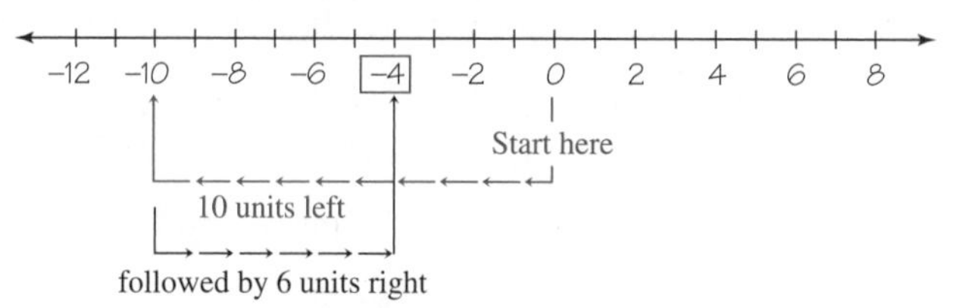

Ending point is -4.

d. Negative + Negative: $(-2) + (-8)$

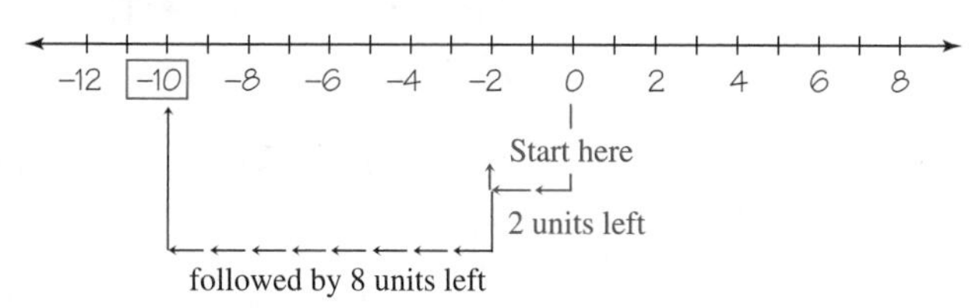

Ending point is -10. ■

The method of adding on a number line shown in Example 4 is not practical for continued use, so we use it to lead us to the following rules for the **addition of two nonzero integers.**

Procedure for Adding Integers

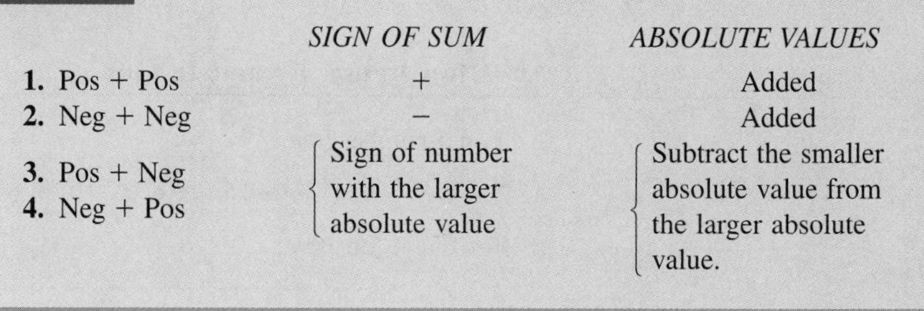

	SIGN OF SUM	*ABSOLUTE VALUES*
1. Pos + Pos	+	Added
2. Neg + Neg	−	Added
3. Pos + Neg **4.** Neg + Pos	Sign of number with the larger absolute value	Subtract the smaller absolute value from the larger absolute value.

This procedure can be summarized in what is called *flowchart form.* A flowchart is a step-by-step chart showing you precisely what to do, as in Figure 2.2.

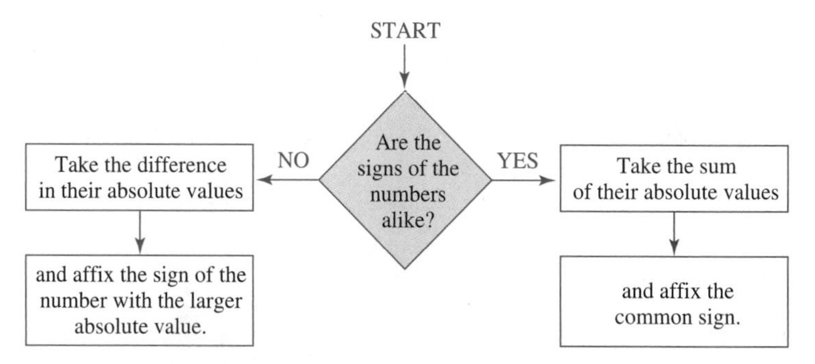

Figure 2.2 Flowchart for the addition of two integers

EXAMPLE 5 **Adding Integers**

Find the indicated sums. Note the use of parentheses.

a. $45 + 27$; Positive + Positive
b. $-18 + (-21)$; Negative + Negative
c. $13 + (-5)$; Positive + Negative
d. $-13 + 5$; Negative + Positive
e. $-12 + (-35)$ **f.** $-128 + (-19)$ **g.** $-8 + 12$
h. $-48 + 53$ **i.** $127 + (-127)$ **j.** $5 + (-3) + (-4)$
k. $-2 + [3 + (-5)]$

Solution

a. $45 + 27 = \mathbf{72}$ Positives are understood.

b. $-18 + (-21) = \mathbf{-39}$
↑ ↑
Neg + Neg is negative. Add absolute values: $18 + 21 = 39$

c. $13 + (-5) = \mathbf{+8}$
↑ ↑
Pos + Neg
Use the sign of the number with the larger absolute value. Subtract absolute values: $13 - 5 = 8$

d. $-13 + 5 = \mathbf{-8}$
↑
Sign of the number with the larger absolute value. Subtract absolute values: $13 - 5 = 8$

e. $-12 + (-35) = \mathbf{-47}$

f. $-128 + (-19) = \mathbf{-147}$

g. $-8 + 12 = \mathbf{+4}$
↑
Sign of the number with the larger absolute value. Subtract absolute values: $12 - 8 = 4$

h. $-48 + 53 = \mathbf{5}$
↑
Positive sign understood.

i. $127 + (-127) = \mathbf{0}$ Remember, adding opposites gives 0.

j. $5 + (-3) + (-4) = 2 + (-4)$ For more than two numbers,
$= \mathbf{-2}$ add two at a time.

k. $-2 + [3 + (-5)] = -2 + (-2)$
$= \mathbf{-4}$ ■

EXAMPLE 6 **Integer Operations on a Calculator**

Indicate the sequence of keys to enter $(-8) + (-5)$ into a calculator.

Solution

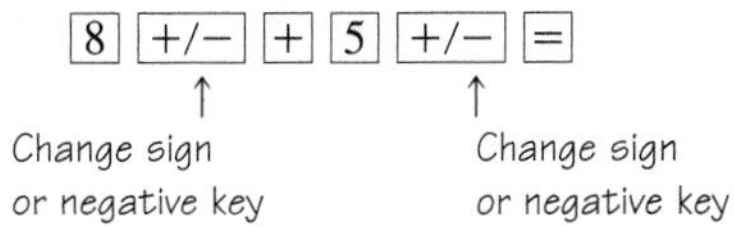

■

PROBLEM SET 2.2

Essential Ideas

1. **In Your Own Words** Describe how you add two positive integers.

2. **In Your Own Words** Describe how you add two negative integers.

3. **In Your Own Words** Describe how you add a positive and a negative integer.

Tell whether the "−" is a minus sign or an opposite sign in Problems 4–9.

4. $-(+4)$ **5.** $6 - 4$ **6.** $12 - 456$

7. $-(34)$ **8.** $-y$ **9.** $-x$

Level 1

Use a number line in Problems 10–15.

10. Start at 0 and move +5 units. Then move +2 units. Then move −10 units. What is the ending point?

11. Start at 0 and move −3 units. Then move +8 units. Then move −6 units. What is the ending point?

12. Start at 0, move +3 units, move −5 units, and then move −1 unit. What is the ending point?

13. Start at 0, move −5 units, move +12 units, and then move −6 units. What is the ending point?

14. Start at 0, move −3 units, move −2 units, move +8 units, and then move +2 units. What is the ending point?

15. Start at 0, move +6 units, move −8 units, move +5 units, and then move −9 units. What is the ending point?

WHAT IS WRONG, *if anything, with each of the statements in Problems* 16–21*? Explain your reasoning.*

16. The sum of a positive and a negative is negative.

17. The sum of two negatives is positive.

18. For $-6 + (-3)$, the larger number is -3.

19. The expression -10 symbolizes negative 10, but not minus 10.

20. The expression -5 symbolizes negative 5, but not minus 5.

21. The sum of opposites is positive.

Level 1—Drill

Find the sums in Problems 22–42.

22. a. $8 + 6$ **b.** $12 + 11$

23. a. $(-5) + 9$ **b.** $5 + (-9)$

24. a. $-5 + 9$ **b.** $-5 + (-9)$

25. a. $-7 + 6$ **b.** $7 + 6$

26. a. $-7 + (-6)$ **b.** $7 + (-6)$

27. a. $-7 + 6$ **b.** $-9 + 4$

28. a. $9 + 4$ **b.** $9 + (-4)$

29. a. $8 + (-6)$ **b.** $-8 + (-6)$

30. a. $-8 + 6$ **b.** $8 + 6$

31. a. $42 + (-60)$ **b.** $-43 + 80$

32. a. $-43 + (-80)$ **b.** $-76 + (-65)$

33. a. $-56 + 20$ **b.** $28 + (-76)$

34. a. $-72 + 56$ **b.** $-98 + (-84)$

35. a. $-62 + 79$ **b.** $71 + (-32)$

36. a. $-9 + 15$ **b.** $-82 + (-41)$

37. a. $162 + (-27)$ **b.** $-15 + 83$

38. a. $-14 + 27$ **b.** $42 + (-121)$

39. a. $62 + (-62)$ **b.** $-128 + 128$

40. a. $-64 + 64$ **b.** $247 + (-247)$

41. a. $-18 + (-4 + 3)$ **b.** $[-18 + (-4)] + 3$

42. a. $(-8 + 6) + (6 + 8)$ **b.** $-8 + (6 + 6) + 8$

Level 2—Drill

43. Suppose you are given the following number line:

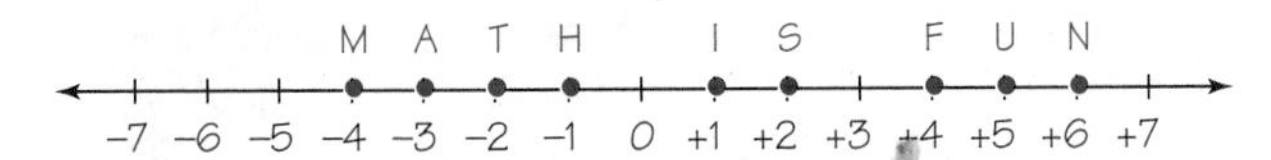

a. Explain why a move from H to N is described by $+7$.

b. Describe a move from T to F.

c. Describe a move from U to A.

Show the sums requested in Problems 44–49 *on a number line.*

44. $(+5) + (+3)$

45. $(-5) + (+3)$

46. $(+4) + (-7)$

47. $(+3) + (+5)$

48. $(-3) + (+5)$

49. $(-2) + (-4)$

Perform the indicated operations in Problems 50–59 *on your calculator.*

50. $561 + (-453)$

51. $-1{,}871 + 2{,}190$

52. $-671 + (-538)$

53. $-993 + (-482)$

54. $-37 + 92 + (-53)$

55. $-41 + 84 + (-45)$

56. $-50 + (-205)$

57. $-459 + (-340)$

58. $78 + 3{,}450 + (-583)$

59. $0 + (-689) + 482$

Level 2—Applications

60. In a game of rummy, a player's scores for five hands were 25, −120, 45, −10, and 60. What is the player's total score?

61. In a game of rummy, a player's scores for six hands were 45, 55, −30, −85, 35, and 50. What is the player's total score?

62. What is the final temperature in a freezer if at 9:00 A.M. it is −5°C and then it goes up by 10°, drops by 15°, drops by 6°, and finally goes up by 7°?

63. What is the final temperature if it is −8°C at 7:00 A.M., rises by 14°C, rises by 23°C, falls by 14°C, and then falls by 28°C?

64. The Rams took the kickoff on the 1-yard line, gained 22 yards, lost 4 yards, gained 5 yards, and then lost 8 yards. Where did the Rams finish this drive?

65. If a football drive begins on the 12-yard line, gains 15 yards, gains 3 yards, loses 8 yards, gains 4 yards, and finally gains 0 yards, where does this drive end?

2.3 SUBTRACTION OF INTEGERS

Historical Perspective

The second great period in the history of mathematics is called the Greek Period (600 B.C. to 499 A.D.). Before 600 B.C., the Mediterranean Period was marked by wars and violent social change. This upheaval was probably based on the invention of iron. The Greek Period saw Siddartha the Buddha in India, the birth of Confucius in China, and the birth of Christ in Bethlehem. During this period, we see the rise of the self-governing city states, and in 594 B.C. one of these city states set down a code of law that marked a change from rule of custom to rule of law. It was during this period that the new and important class of merchants emerged in Greek society. Mathematically, it was during this period that we find a gradual change of outlook from the practical point of view of the Egyptians and Babylonians to the logical viewpoint of the Greeks. There were many noted mathematicians during the time, including Thales, Pythagoras, Zeno, Euclid, and Hypatia of Alexandria, the first woman mentioned in the history of mathematics.

Greek Parthenon

Determining the amount of change a customer should receive from a purchase certainly uses subtraction. If the purchase is \$1.65 and you give the clerk \$2.00, the operation is $2.00 - 1.65 = 0.35$, or 35¢ change. But that is *not* the way a clerk determines your change. Usually, the clerk hands you a dime and a quarter and says, "That's \$1.65, \$1.75, and \$2.00." The necessary subtraction is done by adding: $1.65 + 0.10 + 0.25 = 2.00$.

Many children play "take away" games. "What is 6 take away 4?" Holding up six fingers, the child simply counts off "5, 4, 3, 2 is 6 take away 4." To take away, the child does not subtract but simply counts backward.

On the number line, subtraction involves going back in the opposite direction instead of going ahead as in addition. Let's look at the child's "take away" game using a number line. For $6 - 4$:

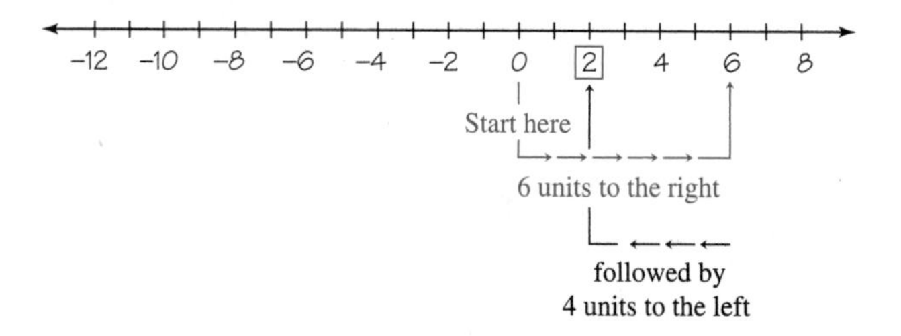

Notice that this looks the same as $6 + (-4)$; see Example 4b of the previous section. That is,

$$\left.\begin{array}{r} 6 - 4 = 2 \\ 6 + (-4) = 2 \end{array}\right\} \quad \text{Result is the same location.}$$

But what if we have negative numbers? Negative already indicates "going back" or to the left. Does subtracting a negative mean going to the right?

Look again at a sequence of problems and find the most logical solutions. Subtract each problem:

$$\begin{array}{r} 5 - 5 = 0 \\ 5 - 4 = 1 \\ 5 - 3 = 2 \\ 5 - 2 = 3 \\ 5 - 1 = 4 \\ 5 - 0 = 5 \\ 5 - (-1) = ? \\ 5 - (-2) = ? \\ 5 - (-3) = ? \end{array}$$

It does seem that subtracting successively smaller numbers from 5 should give successively larger differences. On the number line, $5 - (-3)$ means "going in the opposite direction from -3," but instead of trying to do subtraction on a number line, we look at a pattern:

$$\begin{array}{rl} 5 - 1 = 4 & \\ 5 - 0 = 5 & \\ 5 - (-1) = ? & \leftarrow \text{From the pattern, this is 6: } 5 - (-1) = 5 + 1 = 6 \\ 5 - (-2) = ? & \leftarrow \text{From the pattern, this is 7: } 5 - (-2) = 5 + 2 = 7 \\ 5 - (-3) = ? & \leftarrow \text{From the pattern, this is 8: } 5 - (-3) = 5 + 3 = 8 \end{array}$$

It does look as if $5 - (-3)$ is equal to 8. We note that this is the same as $5 + 3$.

Guided by these results, we define **subtraction of integers** as follows.

Procedure for Subtracting Integers

$$a - b = a + (-b)$$

To subtract, add the opposite of the number to be subtracted.

EXAMPLE 1 **Subtraction by Adding the Opposite**

$7 - 10 = 7 + \mathbf{(-10)}$

Add the opposite.

$= \mathbf{-3}$ ← Use the rules of **addition** to obtain this result. ■

EXAMPLE 2 **Subtraction of Integers**

Simplify the following.

a. $7 - (-10)$ **b.** $4 - (-8)$ **c.** $-2 - (-9)$ **d.** $2 - (-9)$
e. $-2 - 9$ **f.** $-345 - 527$ **g.** $8 - (-5) - 14 + (-3) - (-2)$

Solution

a. $7 - (-10) = 7 + (\mathbf{+10}) = \mathbf{17}$
b. $4 - (-8) = 4 + \mathbf{8} = \mathbf{12}$
c. $-2 - (-9) = -2 + \mathbf{9} = \mathbf{7}$
d. $2 - (-9) = 2 + \mathbf{9} = \mathbf{11}$
e. $-2 - 9 = -2 + (\mathbf{-9}) = \mathbf{-11}$
f. $-345 - 527 = -345 + (\mathbf{-527}) = \mathbf{-872}$
g. With more than two additions or subtractions, work from left to right. It is customary first to change all of the subtractions to additions by adding the opposites, and then to add the positive numbers as well as the negative numbers, and finally to carry out the addition.

$$\begin{aligned} 8 - (-5) - 14 + (-3) - (-2) &= 8 + 5 + (-14) + (-3) + 2 \\ &= [8 + 5 + 2] + [(-14) + (-3)] \\ &= 15 + (-17) \\ &= \mathbf{-2} \end{aligned}$$

■

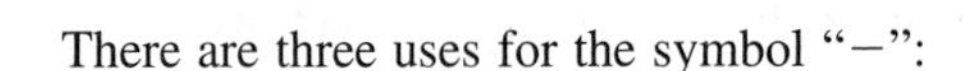

There are three uses for the symbol "−":

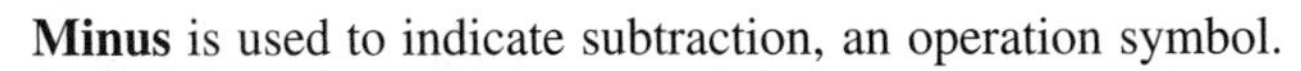

Minus is used to indicate subtraction, an operation symbol.

Much of your work will be simplified if your thinking on this point is clear.

Negative is used to indicate those numbers to the left of the origin on a number line.

Opposite is used to signify an equivalent distance from the origin, but in an opposite direction. This is a number that can be either positive or negative. For example, $5 - [-(-2)]$ means

5 minus the opposite of negative 2
↓ ↓ ↓
5 − [− (− 2)]

The next example illustrates the three different uses for the symbol "−".

EXAMPLE 3 **Practice with the "−" Symbol**

Indicate whether "minus," "negative," or "opposite" best describes the use of "−" in each expression.

a. $(-2) + (+9)$ **b.** $-(+6) + (+4)$ **c.** $(+4) - (+7)$ **d.** $-x$ **e.** $x - y$

Solution

a. $(-2) + (+9)$ **Negative**
b. $-(+6) + (+4)$ **Opposite**
c. $(+4) - (+7)$ **Minus**

d. $-x$ — **Opposite;** when the "$-$" symbol appears alone in front of a variable, as in this example, it always means opposite.

e. $x - y$ — **Minus;** when the "$-$" symbol appears between two variables, as in this example, it always means subtraction (minus). ■

Correct Use of $-x$ Symbol

The symbol $-x$ is read **"the opposite of x"** and should not be read "negative x" or "minus x." If you remember to do this, you will save yourself a lot of confusion.

When evaluating an expression involving negative numbers, you may sometimes need to use additional sets of parentheses to keep the meaning of the expression clear. This is illustrated in Example 4.

EXAMPLE 4 **Evaluation of Expression with Negative Numbers**

Evaluate the given expressions, where $a = -2$ and $b = -4$.

a. $a + b$ **b.** $a - b$

Solution

a. $a + b = -2 + (-4) = \mathbf{-6}$

b. $a - b = -2 - (-4) = -2 + 4 = \mathbf{2}$ ■

PROBLEM SET 2.3

Essential Ideas

1. **In Your Own Words** Explain the procedure for subtracting integers.
2. **In Your Own Words** Contrast the three uses for the "$-$" symbol, namely, minus, negative, and opposite.

Level 1

WHAT IS WRONG, *if anything, with each of the statements in Problems 3–8? Explain your reasoning.*

3. The opposite of -3 is 3.
4. $a - b$ means "a minus b" and $a + (-b)$ means "a plus the opposite of b."
5. "Add the opposite" applied to the expression $-5 - 3$ means $5 + 3$.
6. The symbol "$-a$" is read "minus a."
7. The symbol $-b$ means that b is negative.
8. $5 - (-15) = -10$

Write out each of the statements in Problems 9–16 in words.

9. $5 - (-3)$
10. $-2 - 5$

11. $-(-3)$

12. $-y$

13. $x - y$

14. $x + (-y)$

15. $6 - [-(-1)]$

16. $(-2) - [-(-3)]$

LEVEL 1—Drill

Find the differences in Problems 17–29. Show your solutions as two steps: First, rewrite the subtraction as addition; then carry out the addition to find the answer.

17. a. $8 - 5$ **b.** $12 - 7$

18. a. $15 - (-8)$ **b.** $9 - (-5)$

19. a. $-9 - 5$ **b.** $-9 - (-5)$

20. a. $17 - (-8)$ **b.** $-17 - 8$

21. a. $17 - 8$ **b.** $-17 - (-8)$

22. a. $-21 - 7$ **b.** $-21 - (-7)$

23. a. $21 - (-7)$ **b.** $21 - 7$

24. a. $-13 - (-6)$ **b.** $13 - 6$

25. a. $13 - (-6)$ **b.** $-13 - 6$

26. a. $8 - 23$ **b.** $-8 - (-23)$

27. a. $-8 - 23$ **b.** $8 - (-23)$

28. a. $46 - (-46)$ **b.** $-7 - (-18)$

29. a. $62 - (-112)$ **b.** $5 - (-416)$

Perform the indicated operations (simplify) in Problems 30–33.

30. a. $-4 - (-5) + 8$ **b.** $-8 + 9 - (-7)$ **c.** $8 - 6 - (-5)$

31. a. $-6 - 7 - (-9)$ **b.** $4 - 7 - 5$ **c.** $-7 - 3 - (-8)$

32. a. $-8 - 4 - 3$ **b.** $-8 - 4 - 5$ **c.** $11 - 14 - (-16)$

33. a. $-21 - 14 - (-52)$ **b.** $37 + (-15) - 21$ **c.** $-2 - 3 - 7 - 10$

LEVEL 2—Drill

In Problems 34–45, let $x = -2$, $y = -1$, and $z = -3$. Find the values of the given capital letters.

34. $A = x + y$ **35.** $B = x - z$ **36.** $C = z - y$

37. $D = y - z$ **38.** $E = x - y$ **39.** $F = -y$

40. $G = -x$ **41.** $H = -z$ **42.** $I = y + z - x$

43. $J = z - x - z$ **44.** $K = z - x - y$ **45.** $L = x - y - z$

Perform the indicated operations in Problems 46–53 on your calculator.

46. $487 - 843$

47. $-381 - (-843)$

48. $-1{,}439 - 816$

49. $-125{,}409 - (-34{,}817)$

50. $-4{,}567 + (-3{,}891) + 458$

51. $-982 - (-458) + (-402)$

52. $4{,}987 + (-4{,}583) - 478 + (-5{,}670)$

53. $-8{,}211 - 9 - 10{,}209 + 4{,}511 - (-4{,}529)$

LEVEL 2—Applications

54. In Death Valley, California, the temperature can vary from 134°F above zero to 25°F below zero. What is the difference between these temperature extremes?

55. What is the difference in elevation between the top of a mountain 8,520 ft above sea level and a point in a valley 253 ft below sea level?

56. What is the difference in elevation between a plane flying at 25,400 ft above sea level and a submarine traveling 450 ft below sea level?

57. IBM stock had the following changes during a certain week: $+1$, $+3$, -2, -1, and -3. What is the *net* change for the week?

58. If the Dow Jones Industrial Average of 30 stock prices for a particular week has the following changes, what is the *net* change for the week?

Monday, $+9$; Tuesday, -15; Wednesday, -29;
Thursday, $+7$; Friday, $+6$

The questions in Problems 59–61 are based on the following weather map.

59. What is the difference in temperature between Huntsville, AL, and Memphis, TN?

60. What is the difference in temperature between Salt Lake City, UT, and Grand Rapids, MI?

61. What is the difference in temperature between Colorado Springs, CO, and Huntsville, AL?

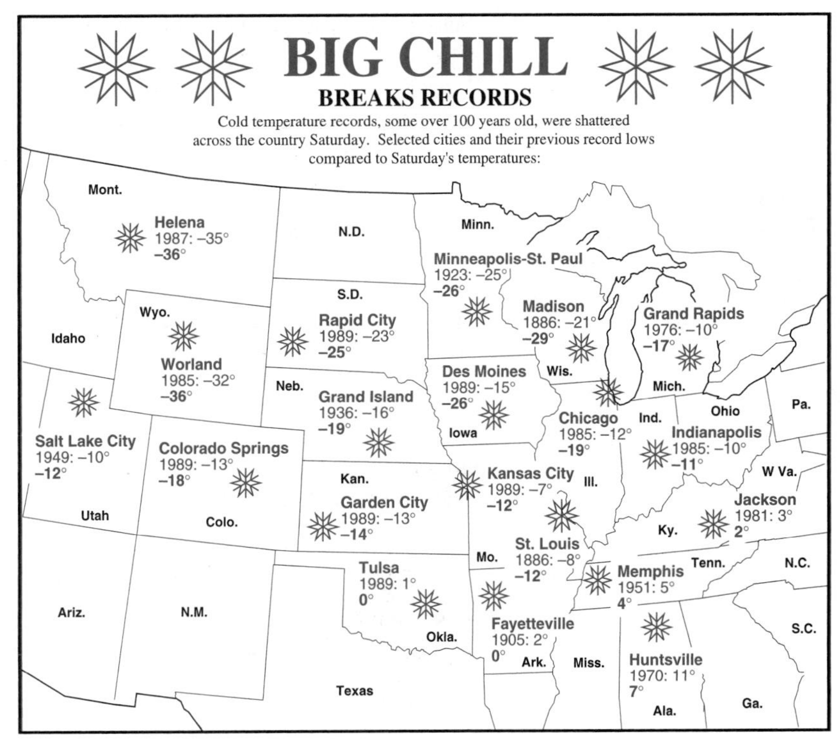

2.4 MULTIPLICATION OF INTEGERS

Historical Perspective

The first known woman in the history of mathematics is Hypatia. She was a professor of mathematics at the University of Alexandria. She wrote several major works, but her scientific rationalism caused her problems with the Church, and she was considered a heretic. At the age of 45, she was pulled from her carriage, tortured, and murdered for her scientific beliefs. The number of important women in the history of mathematics is small when compared to the number of men. One reason for this is that, for centuries, women were discouraged from studying mathematics and in most cases were unable to attend standard educational institutions.

Hypatia (370 – 415)

What is multiplication? We really should answer that question before considering *how* to multiply. What actually happens when we multiply? Well, what does "three times four" mean? "Twelve" you answer proudly. That *is* the product—the answer—

but how did you get that value? What did you do to multiply? You probably memorized that answer long ago, using a multiplication table, but what if you could no longer use that table or you forgot that particular entry? You might have to think about the meaning of multiplication.

For the whole numbers, multiplication can be defined as repeated addition, since we say that 4×3 means

$$\underbrace{3 + 3 + 3 + 3}_{4 \text{ addends}}$$

However, this *cannot* be done for all integers, since $(-4) \times 3$ or

$$\underbrace{3 + 3 + \cdots + 3}_{-4 \text{ addends}}$$

doesn't "make sense." Thus, it is necessary to generalize the definition of multiplication to include all integers. There are four cases to consider.

POSITIVE TIMES A POSITIVE

Positive integers are the same as the natural numbers, so the previous definition of repeated addition applies: **The product of two positive numbers is a positive number.**

POSITIVE TIMES A NEGATIVE

Now consider $(+3) \times (-4)$ by looking at a pattern:

$$(+3) \times (+4) = \mathbf{+12}$$
$$(+3) \times (+3) = \mathbf{+9}$$
$$(+3) \times (+2) = \mathbf{+6}$$

Would you know what to write next? Here it is:

$$(+3) \times (+1) = \mathbf{+3}$$
$$(+3) \times (0) = \mathbf{0}$$

What comes next? Do you see the pattern? *As the second factor decreases by* 1, *the product decreases by* 3:

$$(+3) \times (-1) = \mathbf{-3}$$
$$(+3) \times (-2) = \mathbf{-6}$$
$$(+3) \times (-3) = \mathbf{-9}$$
$$(+3) \times (-4) = \mathbf{-12}$$

Do you see how to continue? Try building a few more such patterns using different numbers. What do you see about the product of a positive and a negative number? **The product of a positive number and a negative number is a negative number.**

NEGATIVE TIMES A POSITIVE

The order in which two numbers are multiplied has no effect on the product, so

$$(-3) \times (+4) = (+4) \times (-3)$$
$$= \mathbf{-12}$$

The product of a negative number and a positive number is a negative number.

NEGATIVE TIMES A NEGATIVE

Consider the final example: the product of two negative integers, say, $(-3) \times (-4)$. Once again, we begin by looking for a pattern. Start with

$$(-3) \times (+4) = \mathbf{-12}$$
$$(-3) \times (+3) = \mathbf{-9}$$
$$(-3) \times (+2) = \mathbf{-6}$$

What would you write next? Here it is:

$$(-3) \times (+1) = \mathbf{-3}$$
$$(-3) \times (\ 0\ ) = \mathbf{0}$$

What comes next? You should notice that, *as the second factor decreases by* 1, *the product increases by* 3:

$$(-3) \times (-1) = \mathbf{+3}$$
$$(-3) \times (-2) = \mathbf{+6}$$
$$(-3) \times (-3) = \mathbf{+9}$$
$$(-3) \times (-4) = \mathbf{+12}$$

The product of two negative numbers is a positive number.

Finally, we note that if one (or both) of the integers is zero, then the product is zero. We now summarize these results for the **multiplication of nonzero integers** in the following box.

Procedure for Multiplying Integers

	SIGN OF PRODUCT	*ABSOLUTE VALUES*
1. Pos × Pos	+	Multiplied
2. Pos × Neg	−	Multiplied
3. Neg × Pos	−	Multiplied
4. Neg × Neg	+	Multiplied

To summarize what we have just established for multiplying two integers, we look at the signs of the numbers. If the signs are different, the product is negative. If the signs are alike, the product is positive. Using a flowchart, such as Figure 2.3 (page 96), we can break the process down into a few simple steps.

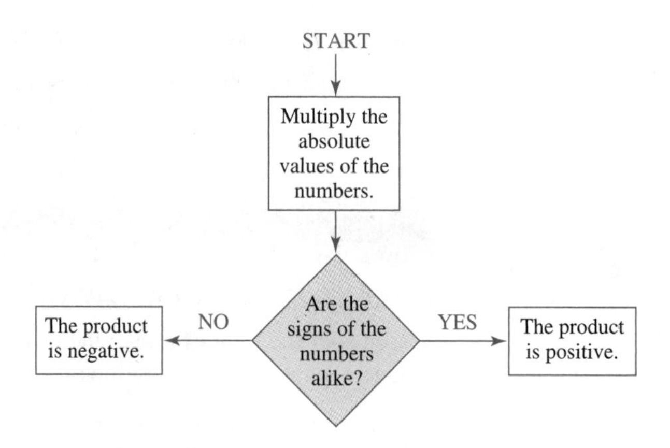

Figure 2.3 Flowchart for the multiplication of two nonzero integers

EXAMPLE 1

Multiplying Integers

Simplify the following.

a. 6×9 **b.** $6 \times (-2)$ **c.** $(-3)7$ **d.** $(-8)(-9)$
e. $-2(-3)(-4)$ **f.** $-2(-3)5$

Solution

a. $6 \times 9 = \mathbf{54}$ **b.** $6 \times (-2) = \mathbf{-12}$
↑ ↑
Positive understood

c. $(-3)7 = \mathbf{-21}$ Remember that numbers in parentheses written right next to each other mean multiplication.

d. $(-8)(-9) = \mathbf{72}$

e. $-2(-3)(-4) = 6(-4)$ Work from left to right.
$= \mathbf{-24}$

f. $-2(-3)5 = \mathbf{30}$
From left to right: $-2(-3)5 = (6)5 = 30$

■

Before leaving the topic of multiplication, we will consider one more example. Certainly $(-4) = (4)(-1) = (-1)(4)$, which can be shown on a number line:

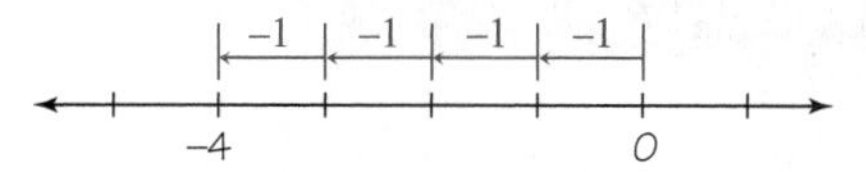

This example would seem to apply for any integer n.

Opposite of an Integer

$$(-1)n = -n$$

Negative one times a number is the **opposite** of that number.,

Examples 2 and 3 illustrate the process of evaluating expressions with integers.

EXAMPLE 2 **Evaluating Algebraic Expressions with Integers**

Evaluate $a^2 - 2ab + b^2$, where $a = -2$ and $b = -1$.

Solution

$$\begin{aligned} a^2 - 2ab + b^2 &= (-2)^2 - 2(-2)(-1) + (-1)^2 \\ &= 4 - 4 + 1 \\ &= \mathbf{1} \end{aligned}$$

■

EXAMPLE 3 **Evaluating Algebraic Expressions Dealing with Opposites**

Evaluate $-a$, a^2, $-a^2$ and $(-a)^2$ for $a = 2$ and $a = -3$.

Solution If $a = 2$ (the variable is a positive number):

$-a = \mathbf{-2}$ — The opposite of a is negative two.

$a^2 = 2^2 = 2 \cdot 2 = \mathbf{4}$ — The square of a is four.

$-a^2 = -2^2 = -(2 \cdot 2) = \mathbf{-4}$ — The opposite of a squared is negative four.

$(-a)^2 = (-2)^2 = (-2)(-2) = \mathbf{4}$ — The square of the opposite of a is four.

If $a = -3$ (the variable is a negative number):

$-a = -(-3) = \mathbf{3}$ — The opposite of a is three.

$a^2 = (-3)^2 = (-3)(-3) = \mathbf{9}$ — The square of a is nine.

$-a^2 = -(-3)^2 = -(-3)(-3) = \mathbf{-9}$ — The opposite of a squared is negative nine.

$(-a)^2 = (3)^2 = 3 \cdot 3 = \mathbf{9}$ — The square of the opposite of a is nine.

■

An important principle is illustrated in Example 3, and it involves an idea often missed by beginning algebra students—that is, the difference between

$$-2^2 \text{ and } (-2)^2 \qquad -3^2 \text{ and } (-3)^2 \qquad a^2 \text{ and } -a^2 \qquad (a \neq 0)$$

Students often use these pairs as if they were equal, but they are indeed opposites:

$-2^2 = -4$ $\quad -3^2 = -9$ $\quad -a^2$ is negative.

$(-2)^2 = 4$ $\quad (-3)^2 = 9$ $\quad a^2$ is positive.

PROBLEM SET 2.4

Essential Ideas

1. **In Your Own Words** What does m times n mean?
2. **In Your Own Words** Explain how to multiply two positive integers.
3. **In Your Own Words** Explain how to multiply two negative integers.
4. **In Your Own Words** Explain how to multiply a positive and a negative integer.

Level 1

WHAT IS WRONG, *if anything, with each of the statements in Problems 5–16? Explain your reasoning.*

5. The sum of two positives is positive.

6. The sum of a positive and a negative is negative.

7. The sum of two negatives is negative.

8. The product of two positives is positive.

9. The product of a positive and a negative is negative.

10. The product of two negatives is negative.

11. $-6^2 = 36$ **12.** $-10^2 = 100$ **13.** $-8^2 = 64$

14. $(-6)^2 = 36$ **15.** $(-10)^2 = 100$ **16.** $(-8)^2 = 64$

Level 1—Drill

Perform the indicated operations in Problems 17–38.

17. a. $6 \times (-9)$ **b.** $3 \times (-5)$

18. a. $(-2) \times 4$ **b.** $4 \times (-2)$

19. a. $4 \times (-5)$ **b.** $8 \times (-8)$

20. a. $-6 \times (-11)$ **b.** $-7 \times (-4)$

21. a. $6(-8)$ **b.** $(-4)(-7)$

22. a. $7(5)$ **b.** $3(9)$

23. a. $(-7)(4)$ **b.** $9(-9)$

24. a. $5(0)$ **b.** $(-8)4$

25. a. $(-6)(-9)$ **b.** $7(-6)$

26. a. $5(-2)$ **b.** $(-5)3$

27. a. $(-8)(-2)$ **b.** $-8(-9)$

28. a. $3(-6)$ **b.** $(-8)(-8)$

29. a. $(-1)(-5)(-6)$ **b.** $2(9)(-3)$

30. a. $(-4)(8)(-5)$ **b.** $(-4)(2)(-1)$

31. a. $5(-4)(-2)$ **b.** $(-2)(-1)3$

32. a. $(-8)(-2)(-1)$ **b.** $4(-7)3$

33. a. -4^2 **b.** $(-4)^2$

34. a. $(-5)^2$ **b.** -5^2

35. a. -6^2 **b.** $(-6)^2$

36. a. $(-7)^2$ **b.** -7^2

37. a. $(-3)^2 - (-4)^2$ **b.** $(-2)^2 - (-3)^2$

38. a. $(-2)^2 - (-1)^2$ **b.** $(-5)^2 - (-3)8$

Level 2—Drill

In Problems 39–53, let $x = -3$, $y = 2$, and $z = -1$. Find the values of the given capital letters.

39. $A = xy$ **40.** $B = xz$ **41.** $C = yz$

42. $D = xyz$ **43.** $E = x^2 + y^2$ **44.** $F = (x + y)^2$

45. $G = (x - y)^2$ **46.** $H = x^2 - y^2$

47. $I = x^2 - 2xy + y^2$ **48.** $J = x^2 + 2xy + y^2$

49. $K = -x^2$ **50.** $L = -y^2$

51. $M = z^2 - x^2$ **52.** $N = (z - x)^2$

53. $P = (x + y + z)^2$

Level 2—Applications

54. Walk to the right 15 steps three times in a row; then walk to the left 8 steps; repeat this eight-step move four more times. What is the net change from your starting position?

55. Walk to the left 12 steps; then walk to the right 20 steps; then walk to the left 7 steps; finally repeat this last seven-step move a total of six more times. What is the net change from your starting position?

56. Tim was not a very good card player and obtained a score of -23 four times in a row. What is Tim's final score?

57. Rebaldo obtained a score of 28, but then received three 25-point penalties. What is Rebaldo's final score?

58. Rose went down 5 floors three times in a row, and then turned around and reversed her steps. If 25 calories are used for each floor, what is the total number of calories Rose consumed?

59. Linda was climbing down a cliff and went down a 20-ft distance eight times, but then went up a 15-ft distance three times. What is the net change from her starting position?

60. What is the opposite of moving to the left 5 units a total of three times?

61. What is the opposite of moving down 20 units a total of 18 times?

62. What is the net result of the opposite of gaining $150 from each of three people?

63. What is the net result of the opposite of losing $150 to each of five people?

2.5 DIVISION OF INTEGERS

Historical Perspective

Herman Cain

"What can you do with mathematics? I don't want to be a mathematician, nor do I want to teach!" Herman Cain is the co-owner of Godfather's Pizza, Inc., and says, "I'm glad I majored in mathematics because it provided me a foundation for my current business career." He has received numerous awards and serves on several prestigious boards, one of which, the Federal Reserve Bank of Kansas City, he is vice-chairman. The story of his career is a story of work, determination, and a unique recipe for success: **focus.** He graduated from Morehouse College with a B.S. in mathematics, and then earned his Master's Degree in Computer Science from Purdue University. He attributes his success in business to a blending of people skills and analytical abilities that allows him to quickly diagnose a situation and plan and develop strategies for success. When I asked him what he enjoyed most about mathematics he said, "It is everywhere in daily life—in both my personal life and my business life. Mathematics has allowed me to make complex 'things' simple. I call it *focus,* which allows me to solve business problems of all types."

Since division can be thought of as multiplying by the reciprocal of a number, it follows that division can be written as a multiplication. For example,

$$(-10) \div 2 = (-10) \times \frac{1}{2}$$

$$= -5 \quad \text{Negative \textbf{times} positive is negative.}$$

This fact means that the rules for **division of nonzero integers** are identical to those for multiplication.

Procedure for Dividing Integers

	SIGN OF QUOTIENT	*ABSOLUTE VALUES*
1. Pos ÷ Pos	+	Divided
2. Pos ÷ Neg	−	Divided
3. Neg ÷ Pos	−	Divided
4. Neg ÷ Neg	+	Divided

EXAMPLE 1 Division of Integers

Simplify the following.

a. $10 \div 5$ **b.** $10 \div (-5)$ **c.** $(-10) \div 5$ **d.** $(-10) \div (-5)$
e. $90 \div (-9)$ **f.** $-282 \div (-6)$ **g.** $0 \div 10$ **h.** $10 \div 0$

Solution

a. $10 \div 5 = \mathbf{2}$ **b.** $10 \div (-5) = \mathbf{-2}$
c. $(-10) \div 5 = \mathbf{-2}$ **d.** $(-10) \div (-5) = \mathbf{2}$
e. $90 \div (-9) = \mathbf{-10}$ **f.** $-282 \div (-6) = \mathbf{47}$
g. $0 \div 10 = \mathbf{0}$ This checks, since $10 \times 0 = 0$.
h. $10 \div 0$ **Impossible;** you can't divide by zero because, if there were some number (say, x), then $\frac{10}{0} = x$ means $0 \cdot x = 10$. But $0 \times x = 0$ and is not 10 for *any* value of x. ■

You must remember that you can NEVER divide by zero.

Pay particular attention to parts **g** and **h** of Example 1. Students often confuse the two, so it is important that you specifically remember that **division by zero is impossible.**

In algebra, we rarely use the ÷ symbol; instead we use the fraction bar introduced in Section 1.3. The fraction bar is also used as a grouping symbol. For example,

$$\frac{2+3}{5} \quad \text{means} \quad (2+3) \div 5$$

whereas

$$2 + \frac{3}{5} \quad \text{means} \quad 2 + (3 \div 5)$$

This distinction is particularly important when you are evaluating expressions. Consider how you would use a calculator to evaluate these expressions.

For $\frac{2+3}{5}$, *press:* [(] [2] [+] 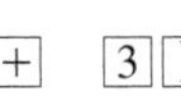  [5] [=] *Display:* 1

Note: You must press the equal key to group the numbers above the fractional bar before doing the division. Contrast this with the next sequence:

For $2 + \frac{3}{5}$, *press:* [2] [+] [3] [÷] [5] [=] *Display:* 2.6

EXAMPLE 2 Evaluating Expressions with a Fractional Grouping Bar

Evaluate the following expressions, where $x = -6$, $y = 3$, and $z = -3$.

a. $\dfrac{x - y}{x - z}$ **b.** $\dfrac{x^2 - y^2}{x + y}$

Solution

a. $\dfrac{x-y}{x-z} = \dfrac{-6-3}{-6-(-3)} = \dfrac{-9}{-6+3} = \dfrac{-9}{-3} = \mathbf{3}$

b. $\dfrac{x^2-y^2}{x+y} = \dfrac{(-6)^2-3^2}{-6+3} = \dfrac{36-9}{-3} = \dfrac{27}{-3} = \mathbf{-9}$ ■

Another common use of division of integers is in finding the **average** of a set of numbers. The average we are considering is sometimes called the **mean** and is defined as the *sum* of the quantities being averaged divided by the *number* of quantities being averaged.

EXAMPLE 3 Average Test Score

If a student's test scores are 72, 85, 79, and 92, what is this student's average test score?

Solution

$$\text{AVERAGE} = \frac{72+85+79+92}{4}$$

← Divide by 4 because there are 4 scores being averaged.

$$= \frac{328}{4}$$

$$= \mathbf{82}$$

Press: (72 + 85 + 79 + 92) ÷ 4 =

Don't forget to group the numbers on the top before doing the division. ■

Averages can also include negative numbers, as in finding the average low temperature for a week of very cold days.

EXAMPLE 4 Finding an Average

Find the average of 5, −7, 3, 0, 4, −15, and −4.

Solution

$$\text{AVERAGE} = \frac{5+(-7)+3+0+4+(-15)+(-4)}{7} = \frac{-14}{7} = \mathbf{-2}$$ ■

PROBLEM SET 2.5

Essential Ideas

1. **In Your Own Words** Describe a procedure for dividing integers.

2. **In Your Own Words** Discuss why we say division by 0 is impossible.

3. **In Your Own Words** Describe the difference between $2 - \dfrac{3}{4}$ and $\dfrac{2-3}{4}$.

4. **In Your Own Words** Describe the difference between 7 − 5/2 and (7 − 5)/2.

Level 1

WHAT IS WRONG, *if anything, with each of the statements in Problems 5–12? Explain your reasoning.*

5. In 5 ÷ 20, the divisor (the number we are dividing by) is 5.

6. 0 ÷ 2 is impossible.

7. 5 ÷ 0 = 0

8. A negative divided by a negative is negative.

9. An average must be positive.

10. $6 \div 3 + 3 = 1$

11. $2 + 6 \div 2 = 4$

12. $9 + 12 \div 3$ means $\dfrac{9 + 12}{3}$.

Find the quotient in Problems 13–24.

13. a. $42 \div 7$ **b.** $63 \div 9$

14. a. $144 \div 12$ **b.** $45 \div (-9)$

15. a. $48 \div (-12)$ **b.** $-32 \div 4$

16. a. $-51 \div 17$ **b.** $100 \div (-5)$

17. a. $-56 \div 8$ **b.** $-88 \div (-8)$

18. a. $(-8) \div 0$ **b.** $0 \div (-8)$

19. a. $\dfrac{+18}{-6}$ **b.** $\dfrac{-104}{4}$

20. a. $\dfrac{-63}{-9}$ **b.** $\dfrac{-15}{-3}$

21. a. $\dfrac{92}{-2}$ **b.** $\dfrac{-528}{-4}$

22. a. $\dfrac{0}{12}$ **b.** $\dfrac{0}{85}$

23. a. $\dfrac{0}{15}$ **b.** $\dfrac{0}{-5}$

24. a. $\dfrac{12}{0}$ **b.** $\dfrac{-6}{0}$

Find the value of each of the expressions in Problems 25–42.

25. $(-6)(-6) \div (+9)$

26. $2 \cdot (-14) \div (-4)$

27. $(-20)(+3) \div (-5)$

28. $5 \cdot 8 \div (-10)$

29. $(3 + 12) \div (-3)$

30. $3 + 12 \div (-3)$

31. a. $\dfrac{12 - 4}{2}$ **b.** $12 - \dfrac{4}{2}$

32. a. $18 - \dfrac{6}{3}$ **b.** $\dfrac{18 - 6}{3}$

33. $\dfrac{5 + 22}{-3}$

34. $\dfrac{2 + (-8)}{2}$

35. $\dfrac{-15 - 21}{6}$

36. $\dfrac{8 + (-3) - (-7)}{-4}$

37. $\dfrac{-7 - (-2) - 9}{-2}$

38. $\dfrac{-11 + 9 - 6 - 8}{(-2)(2)}$

39. $\dfrac{7 - 2 + 8 - 1}{(-1)(-3)}$

40. $\dfrac{(-7) - (-5)(-3)}{2 - (-9)}$

41. $\dfrac{(-5) - (-7)(-3)}{8 - (-5)}$

42. $\dfrac{12 - (-2)(-3)}{2}$

43. Find the average of the first 10 positive integers.

44. Find the average of 6, −10, 14, −25, and 0.

45. Find the average of 2, −5, 4, −7, and −4.

46. Find the average of −7, −18, 4, −9, and 5.

47. Find the average of −3, −6, 5, 8, 0, −2, and 5.

48. Find the average of 21, 0, −34, 45, and −12.

Given that $u = -1$, $v = 2$, $w = -3$, *and* $x = -5$, *evaluate each of the expressions in Problems* 49–62.

49. $\dfrac{w}{x + v}$

50. $w + vx$

51. $\dfrac{x + u}{w}$

52. $\dfrac{u - v}{v + x}$

53. $\dfrac{v^2w}{x + u}$

54. $\dfrac{x + w}{u - x}$

55. $\dfrac{u - w^2}{u - w}$

56. $\dfrac{x^2 - v^2}{v + x}$

57. $w + \dfrac{x}{u}$

58. $\dfrac{w + x}{u}$

59. $\dfrac{w}{u} + x$

60. $\dfrac{w + x}{v}$

61. $\dfrac{u + v + w + x}{4}$

62. $\dfrac{(u + v) - (w + x)}{4}$

2.6 RATIONAL AND IRRATIONAL NUMBERS

Historical Perspective

Pythagoras
(ca. 540 B.C.)

Very little is known about Pythagoras except that he was a leader of a secret society or school founded in the 6th century B.C. This group, known as the Pythagoreans, considered it impious for a member to claim any discovery for himself. Instead, each new idea was attributed to their founder, Pythagoras. This was more than a school; it was a philosophy, a religion, and a way of life. This group investigated music, astronomy, geometry, and number properties. Because of their strict secrecy, much of what we know about them is legend, and it is difficult to tell just what work can be attributed to the group or to its leader, Pythagoras himself. Every evening, each member of the Pythagorean Society had to reflect on three questions:

1. What good have I done?
2. At what have I failed?
3. What have I not done that I should have done?

The integers provide a set of numbers that includes all possible answers for addition, multiplication, and subtraction.* In the same way, we wish to consider a set of numbers that includes all possible answers for the operation of division. We start with the integers, and we also include the set of all possible fractions. With this larger set, we can add, subtract, multiply, and divide any two numbers (except division by zero) and get answers in the set.

This enlarged set is called the set of **rational numbers** and is denoted by $\mathbb{Q}$. (It is called $\mathbb{Q}$ because it is formed by considering all quotients, except those involving division by 0.) The set $\mathbb{Q}$ consists of integers and fractions—that is, numbers that can be written as $\frac{p}{q}$, where p is an integer and q is a nonzero integer. These sets are shown in Figure 2.4.

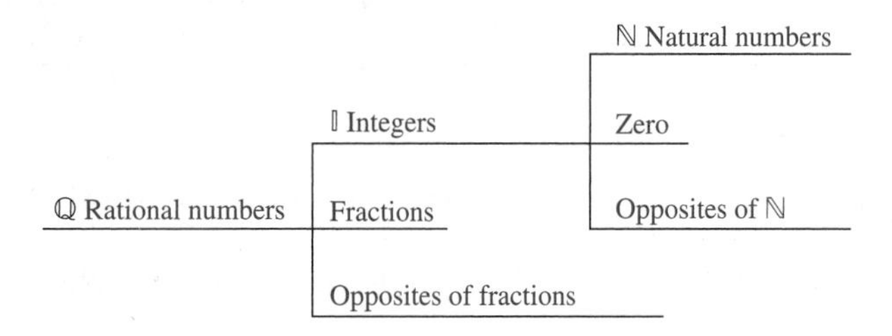

Figure 2.4 Relationships among sets of numbers

Consider any number in the set $\mathbb{N}$, say, 5. Then 5 is also contained in $\mathbb{I}$ and $\mathbb{Q}$. Consider some number in $\mathbb{Q}$, say, $\frac{1}{2}$. Since $\frac{1}{2}$ is in $\mathbb{Q}$ but not in $\mathbb{I}$, we say that $\frac{1}{2}$ is a rational number that is not an integer.

*In more advanced work, this idea is summarized by what is called *closure*. We say that the set of integers is closed for the operation of subtraction (for example) if any two elements in the set can be subtracted and the result is an integer.

Because the signs of the numerator and denominator of a fraction may be either positive or negative, there are several equivalent forms of any fraction. Suppose p and q are positive integers; then

$$\frac{p}{q} = \frac{-p}{-q} = -\frac{-p}{q} = -\frac{p}{-q}$$

Likewise, the fraction itself can be negative; then

$$-\frac{p}{q} = \frac{-p}{q} = \frac{p}{-q} = -\frac{-p}{-q}$$

Because of this variety of possible forms, we call $\frac{p}{q}$ and $\frac{-p}{q}$ the standard forms of a fraction, and we reduce all other forms to one of these.

Standard Forms of a Fraction

If p and q are positive integers, then $\frac{p}{q}$ and $\frac{-p}{q}$ are called the **standard forms of a fraction.** A **reduced fraction** is a fraction in standard form such that there is no number (other than 1 or -1) that divides into both p and q.

EXAMPLE 1 **Standard Forms of a Fraction**

Write each nonstandard form in standard form.

a. $\frac{-5}{-7}$ **b.** $\frac{5}{-7}$ **c.** $-\frac{-5}{7}$ **d.** $-\frac{5}{-7}$ **e.** $-\frac{5}{7}$ **f.** $-\frac{-5}{-7}$

Solution

a. $\frac{-5}{-7} = \mathbf{\frac{5}{7}}$ **b.** $\frac{5}{-7} = \mathbf{\frac{-5}{7}}$ **c.** $-\frac{-5}{7} = \mathbf{\frac{5}{7}}$

d. $-\frac{5}{-7} = \mathbf{\frac{5}{7}}$ **e.** $-\frac{5}{7} = \mathbf{\frac{-5}{7}}$ **f.** $-\frac{-5}{-7} = \mathbf{\frac{-5}{7}}$ ■

EXAMPLE 2 **Operations with Rational Numbers**

Perform the indicated operations, and leave your answers in lowest terms.

a. $\frac{-3}{2} \cdot \frac{1}{2}$ **b.** $\frac{4}{-9} \cdot \frac{-15}{8}$ **c.** $\frac{\frac{-2}{3}}{\frac{-5}{3}}$ **d.** $\frac{-2}{3} + \frac{5}{-3}$ **e.** $\frac{-3}{4} - \frac{1}{-5}$

f. $\frac{4}{5} + \frac{1}{5} \cdot \frac{-1}{2}$

Solution

a. $\frac{-3}{2} \cdot \frac{1}{2} = \mathbf{\frac{-3}{4}}$

b. $\dfrac{4}{-9} \cdot \dfrac{-15}{8} = \dfrac{\cancel{2} \cdot \cancel{2} \cdot \cancel{(-1)} \cdot \cancel{3} \cdot 5}{\cancel{(-1)} \cdot 3 \cdot \cancel{3} \cdot \cancel{2} \cdot \cancel{2} \cdot 2} = \dfrac{5}{3 \cdot 2} = \mathbf{\dfrac{5}{6}}$

c. $\dfrac{\frac{-2}{3}}{\frac{-5}{3}} = \dfrac{-2}{3} \div \dfrac{-5}{3} = \dfrac{-2}{3} \cdot \dfrac{3}{-5} = \mathbf{\dfrac{2}{5}}$

d. $\dfrac{-2}{3} + \dfrac{5}{-3} = \dfrac{-2}{3} + \dfrac{-5}{3} = \mathbf{\dfrac{-7}{3}}$

e. $\dfrac{-3}{4} - \dfrac{1}{-5} = \dfrac{-3}{4} - \dfrac{-1}{5}$

$= \dfrac{-3}{4} \cdot \dfrac{5}{5} - \dfrac{-1}{5} \cdot \dfrac{4}{4}$

$= \dfrac{-15}{20} - \dfrac{-4}{20}$

$= \dfrac{-15 - (-4)}{20}$

$= \dfrac{-15 + 4}{20}$

$= \mathbf{\dfrac{-11}{20}}$

f. $\dfrac{4}{5} + \dfrac{1}{5} \cdot \dfrac{-1}{2} = \dfrac{4}{5} + \dfrac{-1}{10}$

Don't forget the correct order of operations: multiplication first, then addition.

$= \dfrac{8}{10} + \dfrac{-1}{10}$

$= \mathbf{\dfrac{7}{10}}$ ■

For centuries it was thought that the set of rational numbers was complete in the sense that no other numbers existed or were needed, since all additions, subtractions, multiplications, and nonzero divisions result in rational numbers. The first indication that numbers other than rational numbers exist came from ancient Greece.

The secret Greek society called the Pythagoreans (see the Historical Perspective at the beginning of this section) is credited with discovering the famous property of square numbers that today bears Pythagoras' name (even though it was known to the Chinese long before). They found that, if they constructed any right triangle (a triangle with one right angle) and then constructed squares on each of the legs of the triangle, the area of the largest square was equal to sum of the areas of the smaller squares. We will consider this discovery when we consider areas in Chapter 6. Today, we state the Pythagorean theorem algebraically by saying that, if a and b are the lengths of the legs of a right triangle and c is the length of the **hypotenuse** (the longest side of the right triangle), then *the square of the length of the hypotenuse is equal to the sum of the squares of the lengths of the other two sides.*

There is a silly pun based on the Pythagorean theorem as shown at the right in italic: Three bunnies are proudly sitting side by side. The first, a 5-lb hare, sits on a buffalo skin. The second, a 7-lb hare, is on a deer skin. The third, papa hare, who weighs 12 lb, is on a hippopotamus skin. Therfore, *the hare on the hippopotamus is equal to the sum of the hares on the other two hides.*

Pythagorean Theorem

This is one of the most famous results in all of elementary mathematics. You should spend some time studying this theorem.

For any right triangle ABC, with sides of length a, b, and c,

$$a^2 + b^2 = c^2$$

where c is the length of the side opposite the right angle. Also, if

$$a^2 + b^2 = c^2$$

then the triangle ABC is a right triangle.

This property leads to a revolutionary idea in mathematics—one that caused the Pythagoreans many problems. It is the idea that there exist numbers that are not rational.

Consider a right triangle with each leg 1. Then the hypotenuse must be

$$1^2 + 1^2 = c^2$$
$$2 = c^2$$

If we denote the number whose square is 2 by $\sqrt{2}$, we have $\sqrt{2} = c$. The symbol $\sqrt{2}$ is read "**square root** of two," and the symbol $\sqrt{\ }$ is called a **square root symbol** or a **radical.** This means that $\sqrt{2}$ is that number such that, when multiplied by itself, as in

$$\sqrt{2} \times \sqrt{2}$$

the product is 2. You might notice your calculator has a button marked $\boxed{\sqrt{\ }}$. Some calculators require that you press the square root button first, and then the number, whereas others find the square root of the previously pressed number; you should try this on your calculator to see which way to do it:

$$\boxed{\sqrt{\ }}\,\boxed{2} \quad \text{or} \quad \boxed{2}\,\boxed{\sqrt{\ }}$$

Note that the display 1.414213562 is an approximation for the square root of 2, since if we actually square this number, we obtain a number close to 2, but not actually the number 2 as required for the square root of two:

$$1.414213562 \times 1.414213562 = 1.999999998944727844$$

We see that a calculator is no help in deciding whether $\sqrt{2}$ is rational or irrational.

The question for the Pythagoreans was whether $\sqrt{2}$ is rational. That is, does there exist a fractional or decimal representation for $\sqrt{2}$? We will not consider this question.

EXAMPLE 3 **Definition of Square Root**

Use the definition of square root to find the following numbers.

a. $\sqrt{3} \times \sqrt{3}$ **b.** $\sqrt{4} \times \sqrt{4}$ **c.** $\sqrt{5} \times \sqrt{5}$

d. $\sqrt{16} \times \sqrt{16}$ **e.** $\sqrt{155} \times \sqrt{155}$ **f.** $\sqrt{x} \times \sqrt{x}$ (x positive)

Solution **a.** $\sqrt{3} \times \sqrt{3} = \mathbf{3}$ **b.** $\sqrt{4} \times \sqrt{4} = \mathbf{4}$ **c.** $\sqrt{5} \times \sqrt{5} = \mathbf{5}$

d. $\sqrt{16} \times \sqrt{16} = \mathbf{16}$ **e.** $\sqrt{155} \times \sqrt{155} = \mathbf{155}$ **f.** $\sqrt{x} \times \sqrt{x} = \boldsymbol{x}$ ■

By considering the results of Example 3, we can show that some square roots are rational. From part **b** of Example 3, we have

$$\sqrt{4} \times \sqrt{4} = 4$$

and we know that $2 \times 2 = 4$, so it seems reasonable that $\sqrt{4} = 2$. But wait! We also know that $(-2) \times (-2) = 4$, so isn't it just as reasonable that $\sqrt{4} = -2$? Mathematicians have agreed that the *square root symbol be used to denote only positive numbers,* so that $\sqrt{4} = 2$ (*not* -2). If we want to indicate a negative root, we write

CAUTION

$$-\sqrt{4} = -2$$

From Example 3, you should notice that the square roots of numbers that are perfect squares will be rational. Look at this list of perfect squares:

$$1^2 = 1$$
$$2^2 = 4$$
$$3^2 = 9$$
$$4^2 = 16$$
$$5^2 = 25$$
$$\vdots$$

STOP

Note these perfect squares.

The **perfect squares** you should know for this book are 1, 4, 9, 16, 25, 36, 49, 64, 81, 100, 121, 144, 169, and 196.

EXAMPLE 4

Finding Square Roots

Find the positive square root of each of the given numbers.

a. 4 **b.** 49 **c.** $\frac{1}{4}$ **d.** -9 **e.** 324

Solution

a. $\sqrt{4} = \mathbf{2}$, since $2 \times 2 = 4$

b. $\sqrt{49} = \mathbf{7}$, since $7 \times 7 = 49$

c. $\sqrt{\frac{1}{4}} = \mathbf{\frac{1}{2}}$, since $\frac{1}{2} \times \frac{1}{2} = \frac{1}{4}$

d. $\sqrt{-9}$ **doesn't exist** because no number squared can be negative:* Pos × Pos = Pos
Neg × Neg = Pos

e. If the number is not one you recognize from the times table, such as $\sqrt{324}$, you can sometimes find it by trial-and-error multiplication:

$10^2 = 100$

$20^2 = 400$, so $\sqrt{324}$ is between 10 and 20

$15^2 = 225$, so $\sqrt{324}$ is between 15 and 20

$19^2 = 361$, so $\sqrt{324}$ is between 17 and 19

$18^2 = 324$

$\sqrt{324} = \mathbf{18}$ ■

*Actually, there are numbers whose squares are negative, but such numbers are outside the scope of this course.

Example 4 illustrates finding the square roots of perfect squares. What if the number is not a perfect square? To answer this question, we need to note that rational numbers can be written in a decimal form that either terminates or repeats, as shown by Example 5.

EXAMPLE 5

Writing a Rational Number as a Decimal

Show that each number is rational by representing it as a terminating or a repeating decimal.

a. 3 **b.** $\frac{1}{2}$ **c.** $\frac{1}{3}$ **d.** $\frac{1}{7}$ **e.** $-6\frac{1}{18}$ **f.** $\frac{2}{-35}$

Solution

a. $3 = \mathbf{3}$ terminating decimal

b. $\frac{1}{2} = \mathbf{0.5}$ terminating decimal

c. $\frac{1}{3} = \mathbf{0.\overline{3}}$ repeating decimal

d. $\frac{1}{7} = \mathbf{0.\overline{142857}}$ repeating decimal

e. $-6\frac{1}{18} = \mathbf{-6.0\overline{5}}$ repeating decimal

f. $\frac{2}{-35} = \mathbf{-0.0\overline{571428}}$ repeating decimal ■

Now consider the square root of a number that is not a perfect square—say, $\sqrt{2}$. You might try to represent this square root as a decimal, but you would not be successful:

$$1^2 = 1$$

$$2^2 = 4, \quad \text{so } \sqrt{2} \text{ is between 1 and 2}$$ Try 1.5 next; it is halfway between 1 and 2.

$$1.5^2 = 2.25, \quad \text{so } \sqrt{2} \text{ is between 1 and 1.5}$$ Try 1.4 next; it is between 1 and 1.5.

$$1.4^2 = 1.96, \quad \text{so } \sqrt{2} \text{ is between 1.4 and 1.5}$$

$$\vdots$$

You may even turn to a calculator:

[2] [$\sqrt{x}$] 1.414213562

but $1.414213562^2 \neq 2$, so even this number is not equal to the square root of 2. It is proved in more advanced courses that **$\sqrt{2}$ is not rational** and cannot be represented as a repeating or terminating decimal! Such a number is called an **irrational number** and is written as $\sqrt{2}$. Notice that at first we used the symbol $\sqrt{}$ to indicate a process (square root), but now we use it to indicate a *number.* Square roots of counting numbers that are not perfect squares are irrational numbers. Some irrational numbers are listed here:

$\sqrt{2},\ \sqrt{3},\ \sqrt{5},\ \sqrt{6},\ \sqrt{7},\ \sqrt{8},\ \sqrt{10},\ \sqrt{11},\ \sqrt{12},\ \sqrt{13},\ \sqrt{14},\ \sqrt{15},\ \sqrt{17}$

EXAMPLE 6

Classifying Radicals as Rational or Irrational

Classify each given number as rational or irrational. If it is irrational, place it between two integers.

a. $\sqrt{64}$ **b.** $\sqrt{65}$ **c.** $\sqrt{441}$ **d.** $\sqrt{965}$

Solution

a. $\sqrt{64}$ **Rational** because $8^2 = 64$.

b. $\sqrt{65}$ **Irrational** because 65 is not a perfect square; since $8^2 = 64$ and $9^2 = 81$, $\sqrt{65}$ is between 8 and 9.

Note: Your calculator shows $\sqrt{65} \approx 8.062257748$, but that does not answer the question asked because the *calculator* display 8.062257748 is a terminating decimal and terminating decimals are rational. The correct response, however, is that $\sqrt{65}$ is irrational because 65 is not a perfect square.

c. $\sqrt{441}$

$20^2 = 400$
$30^2 = 900$, so $\sqrt{441}$ is between 20 and 30
$25^2 = 625$, so $\sqrt{441}$ is between 20 and 25
$22^2 = 484$, so $\sqrt{441}$ is between 20 and 22
$21^2 = 441$

Rational because $\sqrt{441} = 21$.

d. $\sqrt{965}$

$30^2 = 900$
$40^2 = 1{,}600$, so $\sqrt{965}$ is between 30 and 40
$32^2 = 1{,}024$, so $\sqrt{965}$ is between 30 and 32
$31^2 = 961$, so $\sqrt{965}$ is between 31 and 32

Consecutive counting numbers, so 965 is not a perfect square.

Irrational; $\sqrt{965}$ is between 31 and 32 and cannot be represented as a repeating or terminating decimal. ■

Given a problem like Example 6, but one in which the number under the radical is a fraction, you can determine whether it is a rational or an irrational number by writing it as a common fraction and then handling the numerator and denominator separately, as shown in Example 7.

EXAMPLE 7

Classifying Radicals with Decimals as Rational or Irrational

Classify each number as rational or irrational.

a. $\sqrt{0.25}$ **b.** $\sqrt{1.5625}$

Solution

a. $\sqrt{0.25} = \sqrt{\dfrac{25}{100}}$

$= \dfrac{5}{10} = \dfrac{1}{2}$ **Rational**

b. $\sqrt{1.5625} = \sqrt{\dfrac{15{,}625}{10{,}000}}$ 15,625: $100^2 = 10{,}000$, $200^2 = 40{,}000$

$= \dfrac{125}{100} = \dfrac{5}{4}$ **Rational** $125^2 = 15{,}625$ ■

EXAMPLE 8

Using a Calculator to Determine Whether a Number Is Rational or Irrational

Classify each number as rational or irrational. You may use a calculator.

a. $\sqrt{20.4304}$ **b.** $\sqrt{3.525}$

Solution

a. $\sqrt{20.4304} = \sqrt{\dfrac{204{,}304}{10{,}000}}$ [204304] [$\sqrt{x}$] *Display:* 452

$= \dfrac{452}{100}$

$= 4.52$ **Rational**

b. $\sqrt{3.525} = \sqrt{\dfrac{3{,}525}{1{,}000}}$

$= \sqrt{\dfrac{141}{40}}$ Reduce [141] [$\sqrt{x}$] *Display:* 11.87434209

[40] [$\sqrt{x}$] *Display:* 6.32455532

Neither 141 nor 40 is a perfect square.

Irrational ■

After studying Example 8, you might wonder why you shouldn't simply press

[20.4304] [$\sqrt{x}$] 4.52

and

[3.525] [$\sqrt{x}$] 1.87749834

You must keep in mind the directions. You were not asked to *perform the operation* of square root, but rather to *classify the numbers* as rational or irrational. Calculators *always* make irrational numbers look like rational numbers because they represent them as terminating decimals. If the numerator and denominator of the reduced form of the number under the radical sign can be written as perfect squares, then the number is rational; if not, the number is irrational.

We will sometimes find it necessary to limit our work to the set of rational numbers; in these cases, we'll need to approximate an irrational number with a rational number. You can do this by successive approximations (trial-and-error multiplication) or by using your calculator.

EXAMPLE 9 **Approximating an Irrational Number as a Rational**

Approximate $\sqrt{18}$ with a rational number of two decimal places.

Solution

Method I Use a calculator: [18] [$\sqrt{x}$] *Display:* 4.242640687
Round to two decimal places: **4.24**

Method II Method II can be used if you have a calculator without a square root key.

$4^2 = 16$
$5^2 = 25$, so $\sqrt{18}$ is between 4 and 5

$4.1^2 = 16.81$ Try some numbers between 4 and 5.
$4.2^2 = 17.64$
$4.3^2 = 18.49$, so $\sqrt{18}$ is between 4.2 and 4.3

$4.21^2 = 17.7241$ Try some numbers between 4.2 and 4.3.
$4.22^2 = 17.8084$
$4.23^2 = 17.8929$
$4.24^2 = 17.9776$
$4.25^2 = 18.0625$, so $\sqrt{18}$ is between 4.24 and 4.25

$4.245^2 = 18.020025$, so $\sqrt{18}$ is closer to 4.24

Round to two decimal places: **4.24** ■

In algebra, we group together all the rational numbers and all the irrational numbers into a set called the **real numbers.** The real numbers may be classified in several ways. Any real number is:

Note these three different characterizations of real numbers.

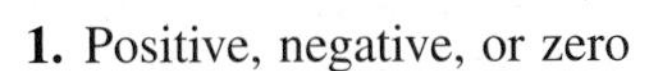

1. Positive, negative, or zero
2. A rational number or an irrational number
3. Expressible as a terminating, a repeating, or a nonterminating and nonrepeating decimal
   a. If it terminates, it is rational.
   b. If it eventually repeats, it is rational.
   c. If it does not terminate or eventually repeat, it is irrational.

When you first consider this third characterization of the real numbers, it is hard to imagine a decimal that does not terminate or repeat. All of the rational numbers we have considered have decimal representations that either terminate or repeat:

$$\frac{1}{2} = 0.5 \qquad \frac{5}{8} = 0.625 \qquad \frac{1}{3} = 0.333\ldots \qquad \frac{1}{6} = 0.1666\ldots$$

In fact, the Pythagoreans knew of numbers like $\sqrt{2}$ but they believed them to be rational (that is, a number whose decimal representation either terminates or repeats).

Legend tells us that one day while the Pythagoreans were at sea, one of their group came up with an argument that $\sqrt{2}$ could not be a rational number. This result shattered the Pythagorean philosophy, which said that all numbers were rational. This member of the group proved logically that $\sqrt{2}$ was not a rational number, so they were forced to change their philosophy or deny logic. Legend has it that they took the latter course: They set the man who discovered it to sea alone in a small boat and pledged themselves to secrecy.

Irrational numbers are needed when working with the **Pythagorean theorem.** Since the theorem asserts that $a^2 + b^2 = c^2$, and since c is the positive number whose square is $a^2 + b^2$, we see that

$$c = \sqrt{a^2 + b^2}$$

Also, if you want to find the length of one of the legs—say, a—when you know both b and hypotenuse c, you can use the formula

$$a = \sqrt{c^2 - b^2}$$

Examples 10 and 11 use this Pythagorean relationship.

EXAMPLE 10

Determining a Length Using the Pythagorean Relationship

If a 13-ft ladder is placed against a building so that the base of the ladder is 5 ft away from the building, how high up does the ladder reach?

Solution Consider Figure 2.5.

Since h is one of the legs, use the formula

$$h = \sqrt{13^2 - 5^2}$$

↑ Hypotenuse

↑ Subtract the square of the length of the known leg from the square of the known hypotenuse.

$$= \sqrt{169 - 25}$$
$$= \sqrt{144}$$
$$= \mathbf{12}$$

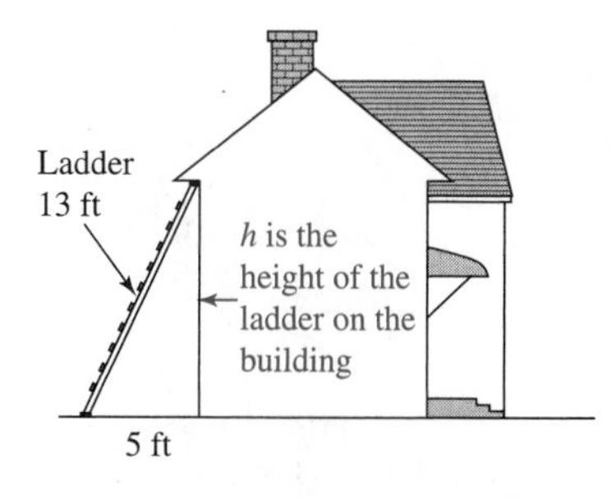

Figure 2.5

The ladder reaches 12 ft up the side of the building. ■

Your answer to Example 10 was rational. But suppose that the result is irrational. You can either leave your result in radical form or estimate your result, as shown in Example 11.

EXAMPLE 11

Estimating Distances Using the Pythagorean Theorem

Suppose that you need to attach several guy wires to your TV antenna, as shown in Figure 2.6. If one guy wire is attached 20 ft away from a 30-ft antenna, what is the exact length of that guy wire, and what is the length to the nearest foot?

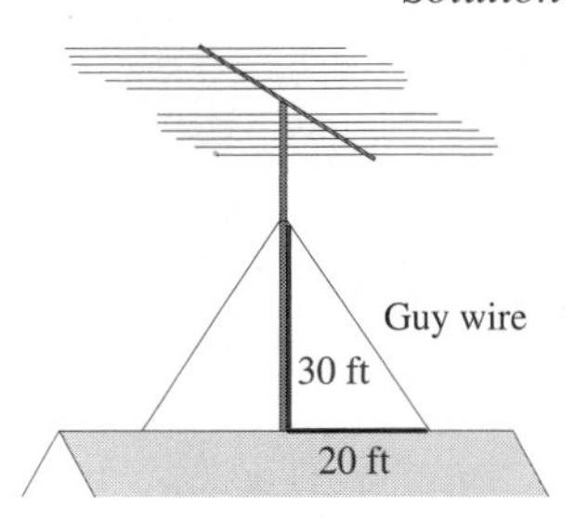

Figure 2.6

Solution The length of the guy wire is the length of the hypotenuse of a right triangle:

$$c = \sqrt{a^2 + b^2}$$
$$= \sqrt{20^2 + 30^2}$$
$$= \sqrt{1{,}300}$$

The exact length of the guy wire is $\sqrt{1{,}300}$; it is irrational, since 1,300 is not a perfect square. Estimate the length.

$30^2 = 900$ 1,300 is between 900 and 1,600 so $\sqrt{1{,}300}$ is between 30 and 40.

$40^2 = 1{,}600$

For a better approximation, you can use a calculator: $c \approx 36.05551275$. The guy wire is 36 ft long (to the nearest ft). *Note:* If the application will not allow any length less than $\sqrt{1{,}300}$ (as in real life), then instead of rounding, take the *next larger* foot—namely, 37 ft. ■

PROBLEM SET 2.6

Essential Ideas

1. **In Your Own Words** What is a reduced fraction? Include as part of your answer what is meant by the standard form of a fraction.
2. **In Your Own Words** State the Pythagorean theorem. Draw a picture of a right triangle, and describe in words what it means for a triangle to be a right triangle.
3. List the perfect squares less than 200.

Level 1

4. **In Your Own Words** What does $\sqrt{4}$ mean? *Hint:* Do not answer 2.
5. **In Your Own Words** What does $\sqrt{9}$ mean? *Hint:* Do not answer 3.
6. **In Your Own Words** What does $\sqrt{2}$ mean? *Hint:* Do not answer 1.4142 . . . .
7. **In Your Own Words** What does $\sqrt{3}$ mean? *Hint:* Do not answer 1.732 . . . .

WHAT IS WRONG, *if anything, with each of the statements in Problems 8–24? Explain your reasoning.*

8. $-\frac{1}{2}$ is a reduced form.
9. $\dfrac{-12}{5}$ is a reduced form.
10. $a^2 + b^2 = c^2$
11. $\sqrt{2} = 1.414213562$
12. $\sqrt{3} = 1.732050808$
13. $\dfrac{1}{3} = 0.3333333333$
14. $\dfrac{2}{3} = 0.6666666666$
15. $\sqrt{225}$ is an irrational number.
16. $\sqrt{100}$ is an irrational number.
17. 0.454545 . . . is an irrational number.
18. $\sqrt{9}$ is equal to both 3 and -3.
19. $\sqrt{16}$ is equal to both 4 and -4.
20. For a right triangle with sides a and b and hypotenuse c, $c^2 - a^2 = b^2$.
21. $\sqrt{8}$ is an irrational number between 2 and 3.
22. $\sqrt{10}$ is an irrational number between 2 and 3.
23. $\sqrt{280}$ is an irrational number between 10 and 20.
24. $\sqrt{441}$ is an irrational number between 20 and 30.

Level 1—Drill

Write each fraction in Problems 25–28 in standard form.

25. **a.** $-\frac{7}{8}$ **b.** $\frac{5}{-8}$ **c.** $-\frac{-7}{-9}$

26. **a.** $\frac{-6}{-11}$ **b.** $-\frac{1}{-2}$ **c.** $-\frac{2}{3}$

27. **a.** $\frac{-x}{-y}$ **b.** $\frac{-y}{-3}$ **c.** $\frac{1}{-x}$

28. **a.** $-\frac{5}{-z}$ **b.** $-\frac{-ab}{-cd}$ **c.** $-\frac{xy}{-2z}$

Simplify the expressions in Problems 29–44.

29. $\frac{-5}{7} \cdot \frac{14}{-10}$

30. $\frac{5}{-2} \div \frac{-2}{6}$

31. $\frac{-5}{14} + \frac{-3}{21}$

32. $\frac{-5}{8} - \frac{1}{12}$

33. $\frac{3}{-10} + \frac{-5}{14}$

34. $\frac{1}{9} - \frac{-2}{3}$

35. $\frac{-3}{5} - \frac{4}{-5}$

36. $\frac{-3}{-4} - \frac{3}{-10}$

37. $-6 + \frac{1}{6}$

38. $-5 - \frac{1}{5}$

39. $-3 - \frac{-1}{3}$

40. $2 - \frac{1}{-2}$

41. $\frac{2}{5} - \frac{1}{-25} + \frac{3}{10}$

42. $\frac{1}{-2} + \frac{-1}{4} - \frac{-3}{8}$

43. $\frac{3}{-10} + \frac{-1}{5} - \frac{-1}{3}$

44. $\frac{2}{-15} - \frac{1}{-3} + \frac{-1}{2}$

Find the positive square root of each of the numbers in Problems 45–50.

45. **a.** 9 **b.** 1 **c.** 0 **d.** -9

46. **a.** 36 **b.** 25 **c.** -16 **d.** 625

47. **a.** 81 **b.** -25 **c.** 169 **d.** 196

48. **a.** 225 **b.** $\frac{25}{36}$ **c.** $\frac{100}{144}$ **d.** $\frac{9}{49}$

49. **a.** 1,225 **b.** 2,025 **c.** 9,604 **d.** 6,084

50. **a.** 10,000 **b.** 1,089 **c.** 10,404 **d.** 3,364

Approximate each irrational number in Problems 51–54 with a rational number of two decimal places.

51. **a.** $\sqrt{15}$ **b.** $\sqrt{17}$ **c.** $\sqrt{20}$

52. **a.** $\sqrt{30}$ **b.** $\sqrt{40}$ **c.** $\sqrt{80}$

53. **a.** $\sqrt{190}$ **b.** $\sqrt{1{,}000}$ **c.** $\sqrt{2{,}000}$

54. **a.** $\sqrt{250}$ **b.** $\sqrt{875}$ **c.** $\sqrt{4{,}210}$

Level 2—Drill

Classify each of the numbers in Problems 55–60 as rational or irrational. If the number is rational, write it as a terminating or repeating decimal. If it is irrational, estimate it by placing it between two integers.

55. **a.** 5 **b.** $\sqrt{5}$ **c.** $\sqrt{25}$ **d.** $\frac{1}{4}$

56. **a.** $\sqrt{\frac{1}{4}}$ **b.** $\sqrt{\frac{1}{9}}$ **c.** $-2\frac{7}{10}$ **d.** $\sqrt{0.49}$

57. **a.** $\sqrt{10}$ **b.** $\sqrt{15}$ **c.** $\sqrt{16}$ **d.** $\sqrt{17}$

58. **a.** $\frac{1}{36}$ **b.** $\sqrt{784}$ **c.** $\sqrt{580}$ **d.** $\sqrt{18.49}$

59. **a.** $\sqrt{2{,}400}$ **b.** $\sqrt{2{,}401}$ **c.** $\sqrt{2{,}402}$ **d.** $\sqrt{\frac{1}{10}}$

60. **a.** $\sqrt{12.3904}$ **b.** $\sqrt{12.4}$ **c.** $\sqrt{1.2}$ **d.** $\sqrt{1.2321}$

Level 2—Applications

61. How far from the base of a building must a 26-ft ladder be placed so that it reaches 24 ft up the wall?

62. How far from the base of a building must a 10-ft ladder be placed so that it reaches 8 ft up the wall?

63. How high up on a wall does a 26-ft ladder reach if the bottom of the ladder is placed 10 ft from the building?

64. How high up on a wall does a 10-ft ladder reach if the bottom of the ladder is placed 6 ft from the building?

65. What is the exact length of the hypotenuse if the legs of a right triangle are 2 in. each?

66. What is the exact length of the hypotenuse if the legs of a right triangle are 3 ft each?

67. An empty rectangular lot is 400 ft by 300 ft. How many feet would you save by waking diagonally across the lot instead of walking the length and width?

68. An empty rectangular lot is 80 ft by 60 ft. How many feet would you save by walking diagonally across the lot instead of walking the length and the width?

69. A television antenna is to be erected and held by guy wires. If the guy wires are 20 ft from the base of the antenna and the antenna is 15 ft high, what is the exact length of each guy wire? If four guy wires are to be attached, how many feet of wire should be purchased if it can't be bought by a fraction of a foot?

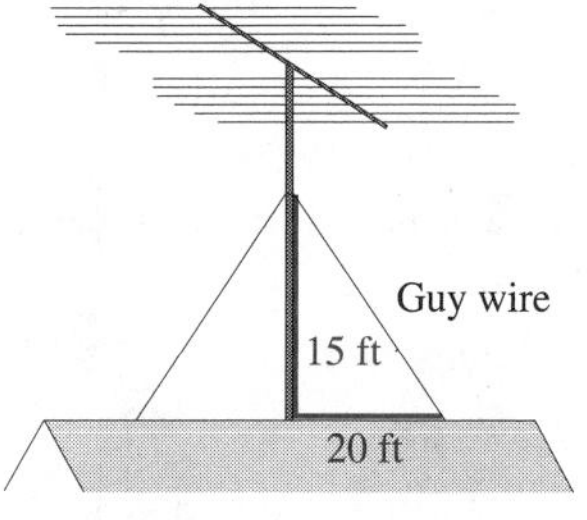

70. A television antenna is to be erected and held by guy wires. If the guy wires are 25 ft from the base of the antenna and the antenna is 15 ft high, what is the exact length of each guy wire? What is the length of each guy wire to the nearest foot? If four guy wires are to be attached, how many feet of wire should be purchased if it can't be bought by a fraction of a foot?

71. A television antenna is to be erected and held by guy wires. If the guy wires are 15 ft from the base of the antenna and the antenna is 10 ft high, what is the exact length of each guy wire? What is the length of each guy wire to the nearest foot? If four guy wires are to be attached, how many feet of wire should be purchased if it can't be bought by a fraction of a foot?

Level 3—Mind Bogglers

72. A person wants to board a plane with a 5-ft-long steel rod, but airline regulations say that the maximum length of any object or parcel permitted to be checked on board is 4 ft. Without bending or cutting the rod or altering it in any way, how did the person check it through without violating the rule?

73. How far from the base of a building (rounded to the nearest ft) must a 12-ft ladder be placed so that it reaches 10 ft up the wall?

2.7 SUMMARY AND REVIEW

Important Terms

Absolute value [2.2]
Addition of integers [2.2]
Algebra [2.1]
Average [2.5]
Difference [2.1]
Dividend [2.1]
Division by zero [2.5]
Division of integers [2.5]
Divisor [2.1]
Domain [2.1]
Equal to [2.1; 2.2]
Evaluating an expression [2.1]
Factor [2.1]
Greater than [2.2]
Hypotenuse [2.6]
Integer [2.2]
Irrational number [2.6]
Less than [2.2]
Mean [2.5]
Minus [2.1; 2.3]
Multiplication of integers [2.4]
Negative number [2.2]
Negative sign [2.2; 2.3]
Number line [2.2]
Opposite [2.2; 2.3; 2.4]
Order symbols [2.2]
Perfect square [2.6]
Positive number [2.2]
Positive sign [2.2]
Product [2.1]
Pythagorean theorem [2.6]
Quotient [2.1]
Radical [2.6]
Rational number [2.6]
Real number [2.6]
Reduced fraction [2.6]
Signed number [2.2]
Square root [2.6]
Square root symbol [2.6]
Standard form of a fraction [2.6]
Subtraction of integers [2.3]
Sum [2.1]
Term [2.1]
Unit scale [2.2]
Variable [2.1]
Variable expression [2.1]
Zero [2.2]

Chapter Objectives

The material in this chapter is reviewed in the following list of objectives. A self-test (with answers and suggestions for additional study) is given. This self-test is constructed so that each problem number corresponds to a related objective. The self-test is followed by a practice test with the questions in mixed order.

[2.1] *Objective 2.1* Given an algebraic expression and a domain, evaluate the expression.

[2.1] *Objective 2.2* Evaluate an algebraic expression using whole numbers.

[2.1] *Objective 2.3* Choose a letter to represent the variable, and write a mathematical statement to express an idea.

[2.2] *Objective 2.4* Be able to describe addition of integers, as well as to carry out the operation of addition of integers.

[2.2] *Objective 2.5* Be able to add integers using a number line.

[2.2; 2.3] *Objective 2.6* Contrast the three uses for the "−" sign.

[2.3] *Objective 2.7* Be able to describe subtraction of integers, as well as to carry out the operation of subtraction of integers.

[2.4] *Objective 2.8* Be able to describe multiplication of integers, as well as to carry out the operation of multiplication of integers.

[2.5] *Objective 2.9* Be able to describe division of integers, as well as to carry out the operation of division of integers.

[2.3–2.5] *Objective 2.10* Simplify expressions with mixed operations.

[2.5] *Objective 2.11* Find the average of a given set of numbers.

[2.3–2.5] *Objective 2.12* Evaluate an algebraic expression using integers.

[2.6] *Objective 2.13* Simplify fractional operations with positive and negative numerators and denominators. Leave answers in standard form.

[2.6] *Objective 2.14* Explain the meaning of square root.

[2.6] *Objective 2.15* Find the positive square root of a given number.

[2.6] *Objective 2.16* Approximate an irrational number with a rational number to a given number of decimal places.

[2.6] *Objective 2.17* Classify a number as rational or irrational.

[2.1–2.6] *Objective 2.18* Work applied problems.

Self-Test

Each question of this self-test is related to the corresponding objective listed above.

1. If $D = 0, 3, 9$, find the values for $100 - x^2$.

2. Evaluate $x^2 - 2xy + y^2$ for $x = 5$ and $y = 2$.

3. Write a mathematical statement for the product of two consecutive integers.

4. In your own words, describe how to add any two integers.

5. On a number line, start at 0 and move +6 units. Then move −4, followed by +2, and then −8. What is the ending point?

6. Write out the following statements in words.
a. -4 **b.** $-x$ **c.** $5 - 3$ **d.** $10 - [-(-3)]$

7. Find the following.
a. $-35 - (-58)$ **b.** $-48 - 56$

8. Find the following.
a. -5^2 **b.** $(-5)^2$ **c.** $(-1)(-2)(-5)$

9. Find the following.
a. $125 \div (-5)$ **b.** $\frac{-35}{0}$ **c.** $\frac{0}{-200}$

10. Find the following.
a. $10 - (-3)(-2)$ **b.** $\frac{6(-2) + 10}{-2}$

11. Find the average of 7, -4, -3, 0, and 10.

12. Evaluate $-x^2 + y^2$ for $x = -3$ and $y = -4$.

13. Find the following.
a. $\frac{-2}{3} + \frac{4}{-5}$ **b.** $\frac{-3}{12} - \frac{4}{-15}$

14. What does $\sqrt{20}$ mean?

15. Find the positive square root of 2,025.

16. Approximate $\sqrt{8.25}$ with a rational number to two decimal places.

17. Classify $\sqrt{1.69}$ as rational or irrational. If it is rational, write it as a terminating or repeating decimal. If it is irrational, place it between two consecutive integers.

18. A diagonal brace is to be placed in the wall of a room. The height of the wall is 8 ft, and the wall is 20 ft long. What is the exact length of the brace? What is the length of the brace to the nearest foot?

STUDY HINTS *Compare your solutions and answers to the self-test. For each problem you missed, work some additional problems in the section listed in the margin. After you have worked these problems, you can test yourself with the practice test.*

Complete Solutions to the Self-Test

Additional Problems

[2.1] *Problems 7–17*

1. If $x = 0$, then $100 - x^2 = 100 - 0 = 100$.
If $x = 3$, then $100 - x^2 = 100 - 9 = 91$.
If $x = 9$, then $100 - x^2 = 100 - 81 = 19$.

[2.1] *Problems 24–47*

2. $x^2 - 2xy + y^2 = 5^2 - 2(5)(2) + 2^2$
$= 25 - 20 + 4$
$= 9$

[2.1] *Problems 48–72*

3. Let x = AN INTEGER; then $x + 1$ = NEXT CONSECUTIVE INTEGER. $x(x + 1)$ represents the product of two consecutive integers.

[2.2] *Problems 1–3; 22–42; 50–59*

4. Answers vary; see page 84.

Section	Review	Answer
[2.2]	*Problems 10–15; 43*	**5.** -4; you should also draw a number line.
[2.2] [2.3]	*Problems 4–9* *Problems 2; 9–16*	**6. a.** negative four **b.** opposite of x **c.** five minus three **d.** ten minus the opposite of negative three
[2.3]	*Problems 1; 17–33; 46–49*	**7. a.** $-35-(-58)=-35+58=23$ **b.** $-48-56=-104$
[2.4]	*Problems 1–4; 17–36*	**8. a.** $-5^2=-25$ **b.** $(-5)^2=25$ **c.** $(-1)(-2)(-5)=-10$
[2.5]	*Problems 1–4; 13–24*	**9. a.** $125\div(-5)=-25$ **b.** Not defined **c.** 0
[2.3] [2.4] [2.5]	*Problems 50–53* *Problems 37–53* *Problems 25–42*	**10. a.** $10-(-3)(-2)=10-6=4$ **b.** $\frac{6(-2)+10}{-2}=\frac{-12+10}{-2}=\frac{-2}{-2}=1$
[2.5]	*Problems 43–48*	**11.** $\frac{7+(-4)+(-3)+0+10}{5}=2$
[2.3] [2.4] [2.5]	*Problems 34–45* *Problems 39–53* *Problems 49–62*	**12.** $-x^2+y^2=-(-3)^2+(-4)^2=-9+16=7$
[2.6]	*Problems 25–44*	**13. a.** $\frac{-2}{3}+\frac{4}{-5}=\frac{-2}{3}\cdot\frac{5}{5}+\frac{-4}{5}\cdot\frac{3}{3}=\frac{-10+(-12)}{15}=\frac{-22}{15}$ or $-1\frac{7}{15}$ **b.** $\frac{-3}{12}-\frac{4}{-15}=\frac{-1}{4}\cdot\frac{15}{15}-\frac{-4}{15}\cdot\frac{4}{4}=\frac{-15-(-16)}{60}=\frac{-15+16}{60}=\frac{1}{60}$
[2.6]	*Problems 4–7*	**14.** $\sqrt{20}$ is that number so that $\sqrt{20}\times\sqrt{20}=20$.
[2.6]	*Problems 45–50*	**15.** $40^2=1{,}600$ $50^2=2{,}500$ $45^2=2{,}025$ Thus, $\sqrt{2{,}025}=45$.
[2.6]	*Problems 51–54*	**16.** $\sqrt{8.25}\approx 2.872281323$ (by calculator); to two places, 2.87.
[2.6]	*Problems 55–60*	**17.** Rational; $\sqrt{1.69}=\sqrt{\frac{169}{100}}=\frac{13}{10}=1.3$
[2.2] [2.3] [2.4] [2.5] [2.6]	*Problems 60–65* *Problems 54–61* *Problems 54–63* *Problems 43–48* *Problems 61–71*	**18.** $\sqrt{8^2+20^2}=\sqrt{64+400}=\sqrt{464}$; this is the exact length. $\sqrt{464}$ is between 21 and 22, so to the nearest foot, it is 22 ft.

Chapter 2 Practice Test

1. **In Your Own Words** Describe how to add, subtract, multiply, and divide integers.

Simplify the expressions in Problems 2–7.

2. **a.** $5 + 123$ **b.** $-56 + 87$
   c. $93 + (-45)$ **d.** $-34 + (-27)$
   e. $3 + 19$ **f.** $-3 + 5$
   g. $41 + (-14)$ **h.** $-4 + (-6)$
   i. $-5 + 4$ **j.** $61 + 0$

3. **a.** $-43 + (-21)$ **b.** $73 + (-47)$
   c. $-61 + (-3)$ **d.** $81 + 23$
   e. $-19 + 41$ **f.** $77 + (-53)$
   g. $5 - 23$ **h.** $-5 - 9$
   i. $14 - (-10)$ **j.** $-3 - (-8)$

4. **a.** $-6 - 8$ **b.** $42 - (-27)$
   c. $0 - 47$ **d.** $-13 - (-63)$
   e. $-3 - (-15)$ **f.** $7 - (-32)$
   g. $15 - 23$ **h.** $-15 - 23$
   i. $14(-2)$ **j.** $6(5)$

5. **a.** $-3(-12)$ **b.** $-5(4)$
   c. $-4(-6)$ **d.** $0(-6)$
   e. $-10(60)$ **f.** $5(-30)$
   g. -4^2 **h.** $(-4)^2$
   i. $(-1)(-3)(-2)$ **j.** $(-1)(3)(6)$

6. **a.** $15 \div 3$ **b.** $125 \div (-5)$
   c. $-48 \div (-4)$ **d.** $-32 \div 8$
   e. $\frac{-14}{-7}$ **f.** $\frac{121}{-11}$
   g. $\frac{-19}{-19}$ **h.** $\frac{1{,}001}{13}$
   i. $\frac{-56}{8}$ **j.** $\frac{0}{-18}$

7. **a.** $(-4) + (-2)(-5)$ **b.** $[-4 + (-2)](-5)$
   c. $\frac{-24}{0}$ **d.** $\frac{0}{0}$
   e. $\frac{-24 + 14}{-2}$ **f.** $-24 - 14 \div (-2)$
   g. $\frac{-5}{12}\left(\frac{-2}{-25}\right)$ **h.** $\frac{-5}{12} + \frac{5}{-15}$
   i. $\frac{3}{-7} - \frac{-10}{21}$ **j.** $\frac{-4}{5} \div \frac{-1}{4}$

8. Simplify the expressions for $x = -3$, $y = 5$, and $z = 1$.
   a. $x - y - z$ **b.** $(xy - z)x$
   c. $\frac{y^2 - x^2}{x + y}$ **d.** $-x^2 + (-x)^2$

e. $(x + y)^2$ **f.** $x^2 + y^2$
g. $x + y - 5z$ **h.** $3x + y - 10z$
i. $-x$ **j.** $-y$

9. Find the positive square roots.
a. 100 **b.** -16 **c.** 0 **d.** 7,569

10. Classify the numbers as rational or irrational. If a number is rational, write it as a terminating or repeating decimal. If it is irrational, place it between two consecutive integers.
a. $\sqrt{529}$ **b.** $\sqrt{1,000}$ **c.** $\sqrt{\frac{4}{9}}$ **d.** $\sqrt{6.76}$

Write a mathematical statement to express the ideas in Problems 11–14.

11. The sum of a number and 20

12. The quotient of a number divided by the next consecutive integer

13. The product of a number and 3

14. The difference of a number subtracted from 50

15. Find the average of the sets of numbers.
a. 6, -5, -3, 2 **b.** 14, -5, -8, -11, 0
c. 12, 0, -40, 21, -9, -5, 0 **d.** 86, 90, 88, 72

16. If a 12-ft ladder is placed against a building so that the base of the ladder is 4 ft away from the building, how high up does the ladder reach (rounded to the nearest foot)?

17. A rectangular field measures 250 ft by 600 ft. If you walk across the field along the diagonal, how much shorter is the trip than if you walk along the edges?

18. What is the exact length of the hypotenuse if the legs of a right triangle are 4 in. each?

19. A diagonal brace is to be placed on a gate. If the height of the gate is 6 ft and the width is 4 ft, what is the exact length of the brace? What is the length of the brace to the nearest foot?

20. A balloon rises at a rate of 40 ft per minute when the wind is blowing horizontally at a rate of 30 ft per minute. After three minutes, how far away from the starting point, in a direct line, is the balloon?

CHAPTER THREE

INTRODUCTION TO ALGEBRA

3.1 POLYNOMIALS

Historical Perspective

Old Historical Manuscript

From Brahmagupta, ca. 630

A tree one hundred cubits high is distant from a well two hundred cubits; from this tree one monkey climbs down the tree and goes to the well, but the other leaps in the air and descends by the hypotenuse from the high point of the leap, and both pass over an equal space. Find the height of the leap.

The historical period known as the Hindu and Arabian Period is from 500 to 1199. The earliest civilization in India is recorded in the ruins of an ancient city at Mohenjo Daro on the Indus River, and around 2000 B.C. the Aryans established themselves and perfected the written and spoken Sanskrit language, as well as the caste system. In 326 B.C., Alexander the Great began his invasions of India, and a blending of the Persian and Greek cultures took place. After the decline of the Roman period, we entered the historical period that has become known as the Golden Age of India. India had become the center of learning. Around 500 A.D., the Hindu astronomer Aryabhata wrote a mathematical verse that gave rules for solving equations and for finding square and cube roots. The greatest Hindu mathematician was Brahmagupta, whose work in 628 A.D. became a book studied for centuries. One of the problems from this book is shown at the left.

The human mind has never invented a labor-saving machine equal to algebra.

THE NATIONS, VOL. 33, P. 327

In the first two chapters, we considered simplifying numerical expressions involving fractions and integers. We introduced the notion of a variable, and we discussed the nature of algebra. We will now turn our attention to simplifying variable expressions. Recall that a variable expression is an expression that contains at least one variable and at least one defined operation. Let's begin with multiplication. From arithmetic we know that the order in which we multiply numbers does not affect the answer we obtain. This means that, if we multiply numbers such as

$$2(3) \qquad \text{or} \qquad 3(2)$$

we obtain the same answer—namely, 6. The same is true if we multiply variables (remember that juxtaposition means multiplication):

$$xy \qquad \text{or} \qquad yx$$

The difference between the numerical expression and the variable expression is that 2(3) can be written more simply as 6, whereas xy is considered to be in simplified form. Also, even though

$$xy = yx$$

it is customary to write the letters in alphabetical order, as xy.

If you are multiplying a number and a variable, as in $3x$, the numerical part is usually written first and the juxtaposition tells us that it means 3 times x. In algebra we do not use a multiplication symbol, because to write

$$3 \times x$$

would cause confusion between the times sign and the variable x. For the product of a number and a variable, the form $3x$ is considered simplified form. In a product of two or more factors, any collection of factors is said to be the **coefficient** of the rest of the factors. Thus, in $5a$, 5 is the coefficient of a and a is the coefficient of 5. The constant factor is called the **numerical coefficient.** An expression may have many factors, such as $15x^2yz$. We say that 15 is the numerical coefficient, x^2 is the coefficient of $15yz$, $15y$ is the coefficient of x^2z, and so on. If no constant factor is shown, as in ab, we say that the numerical coefficient is 1, since $ab = 1 \cdot ab$.

Numerical Coefficient of 1

If no constant factor appears in an expression, the numerical coefficient is understood to be 1.

This seems like an easy idea, but one that is often forgotten.

To multiply variables, you will need to remember the meaning of an exponent. Consider Example 1.

EXAMPLE 1 **Multiplication with Variables**

Simplify:

a. x^2x^3 **b.** y^4y^5 **c.** $x^2y(xy^4)$

Solution

a. $x^2x^3 = xx \cdot xxx = \boldsymbol{x^5}$

b. $y^4y^5 = yyyy \cdot yyyyy = \boldsymbol{y^9}$

c. $x^2y(xy^4) = xxy \cdot xyyyy$
$= xxxyyyyy$
$= \boldsymbol{x^3y^5}$ Remember, you can rearrange the order of multiplication. ■

The product of two powers with the same base is a power with that base and with an exponent that is the sum of the exponents.

Addition Law of Exponents

The bases MUST be the same.

$$b^m \cdot b^n = b^{m+n}$$

To *multiply* two numbers with like bases, add the exponents.

EXAMPLE 2 **Using the Addition Law of Exponents**

Write the solutions in exponent form, applying the addition law of exponents whenever possible.

a. $2^2 \cdot 2^3$ **b.** x^3x^4 **c.** yy^2 **d.** x^2y^3

Solution

a. $2^2 \cdot 2^3 = 2^{2+3} = \mathbf{2^5}$ **b.** $x^3x^4 = x^{3+4} = \mathbf{x^7}$
c. $yy^2 = y^1y^2 = y^{1+2} = \mathbf{y^3}$ **d.** x^2y^3 cannot be simplified further. ■

$2^2 \cdot 2^3 \neq 4^5$

Not only can you rearrange the order of the factors in multiplication, but you can actually associate them to simplify an expression. When simplifying, you should associate the numerical coefficients as well as any variables that are the same. Multiply together the numerical coefficients and use the law of exponents on all base numbers that are the same, as illustrated in Example 3.

EXAMPLE 3 **Simplifying Algebraic Expressions**

Simplify:

a. $(-3x)(5x)$ **b.** $2a(3b)(4c)$ **c.** $(-2xy)(3x^2)(-y^3)$

Solution

a. $(-3x)(5x) = -3(5)xx$ Do this step mentally.
$= \mathbf{-15x^2}$

b. $2a(3b)(4c) = 2(3)(4)abc$
$= \mathbf{24abc}$

c. $(-2xy)(3x^2)(-y^3) = -2(3)(-1)xx^2yy^3$
$= \mathbf{6x^3y^4}$ ■

In the previous chapter, we defined the word *term* as a number that is being added or subtracted. In algebra, we need an expanded definition. Examples 1–3 illustrate the multiplication of numbers and variables. The answer in each case was still an indicated multiplication. Such expressions are called *terms.*

Term

This is an important word to know in algebra.

A **term** is a number, a variable, or the product of numbers and variables.

EXAMPLE 4 **Naming Terms**

Decide whether the following are terms.

a. 10 **b.** x **c.** $10x$ **d.** $x + 10$ **e.** $10abcd$

f. $\frac{10}{x}$ **g.** $\frac{10}{3}$ **h.** $\frac{x}{10}$

Solution

a. 10 is a **term** because it is a single number.
b. x is a **term** because it is a single variable.
c. $10x$ is a **term** because it is the product of a number and a variable.
d. $x + 10$ is **not a single term** because it is not a product; this is actually classified as a two-term expression.
e. $10abcd$ is a **term.**
f. $\frac{10}{x}$ is **not a term** because it is not a product. (It is a quotient of a number and a variable.)
g. $\frac{10}{3}$ is a **term** because it is a single number. Contrast this with part **f.** The number $\frac{10}{3}$ is a quotient, but it is also a single number.
h. $\frac{x}{10}$ is a **term** because it is the product of a number $\left(\frac{1}{10}\right)$ and a variable (x). ■

Terms are connected by addition or subtraction:

$$5x + 6 \qquad x^2 - x + 3 \qquad 2x + 7y - 3$$

We call expressions that consist of one or more terms *polynomials.* Think of subtractions as sums by adding the opposite.

Polynomial

A **polynomial** is a term or the sum of terms.

The polynomial is one of the fundamental building blocks in algebra.

A polynomial is classified by the number of terms it has:

A polynomial with one term is called a **monomial.**

A polynomial with two terms is called a **binomial.**

A polynomial with three terms is called a **trinomial.**

EXAMPLE 5 **Classifying Polynomials by Terms**

Classify each polynomial according to its number of terms.

a. $2x^2 + 3x - 5$ **b.** $15x^2y$ **c.** 6 **d.** $x^2 - y^2$

Solution

a. $2x^2 + 3x - 5$ is a **trinomial.**
b. $15x^2y$ is a **monomial.**
c. 6 is a **monomial.**
d. $x^2 - y^2$ is a **binomial.** ■

A polynomial is also classified according to the *degree* of its terms. Remember that terms are made up of numbers, variables, and products of numbers and variables. The number of *variable* factors is called the **degree** of the term.

EXAMPLE 6

Degree of a Term

Find the degree of each term.

a. $3x$ **b.** $5xy$ **c.** 10 **d.** $2x^2$ **e.** $9x^2y^3$

Solution

a. $3x$ is **degree 1** (one variable factor).
b. $5xy$ is **degree 2** (two variable factors).
c. 10 is **degree 0** (no variable factors).
d. $2x^2$ is **degree 2** (think what we mean by x^2; xx is two variable factors).
e. $9x^2y^3$ is **degree 5.** ■

A polynomial is the sum or difference of terms, and we also speak about the degree of a polynomial.

Degree

The **degree of a term** is the number of variable factors in that term.
The **degree of a polynomial** is the largest degree of any of its terms.

EXAMPLE 7

Classifying Polynomials by Degree

Classify each of the following polynomials by degree.

a. $2x^2 + 3x - 5$ **b.** $15x^2y$ **c.** 6 **d.** $x^2 - y^2$
e. $5x^3 - 4x^2 + 2x - 10$

Solution

a. $2x^2 + 3x - 5$ is **degree 2.**
b. $15x^2y$ is **degree 3.**
c. 6 is **degree 0.**
d. $x^2 - y^2$ is **degree 2.**
e. $5x^3 - 4x^2 + 2x - 10$ is **degree 3.** ■

CAUTION

Spend some time with this example.

When writing polynomials, it is customary to arrange the terms from the highest-degree term to the lowest-degree term, as shown in parts **a** and **e** of Example 7. If terms have the same degree, as in part **d** of Example 7, the terms are usually listed in alphabetical order.

PROBLEM SET 3.1

ESSENTIAL IDEAS

1. **In Your Own Words** What is a term?
2. **In Your Own Words** What is a polynomial?
3. **In Your Own Words** State the addition law of exponents.
4. **In Your Own Words** What is the degree of a term?

LEVEL 1—Drill

Give the degree and numerical coefficient of each term in Problems 5–8.

5. a. $3x^4$ **b.** $5x^2$ **6. a.** $8x^5$ **b.** $64x$

7. a. x **b.** x^3 **8. a.** x^2 **b.** $-5x^2$

Identify each polynomial in Problems 9–14 as a monomial, binomial, or trinomial. Also, give the degree of the polynomial.

9. a. $x^2 + 5$ **b.** $x^3 - 3$

10. a. $abcd$ **b.** $a + b + c$

11. a. 123,456 **b.** $4abc - d$

12. a. $5xy^3 + 5xy$ **b.** $a - b - c$

13. a. $2x^2 + 5x + 3$ **b.** $x^2 - 5x + 3$

14. a. $x - 5$ **b.** $x^2 + 5$

LEVEL 1

WHAT IS WRONG, *if anything, with each of the statements in Problems 15–22? Explain your reasoning.*

15. $x^4x^2 = x^8$ **16.** $y^3y^2 = y^6$

17. $x^2 + x^3 = x^5$ **18.** $y^3 + y^4 = y^7$

19. $2^3 \cdot 2^4 = 4^7$ **20.** $4^2 \cdot 4^3 = 16^5$

21. $2^2 \cdot 3^2 = 6^{2+2} = 6^4$ **22.** $3^2 \cdot 2^3 = 6^5$

Simplify the expressions in Problems 23–45 using the definition of exponent and the addition law of exponents.

23. a. x^3x^5 **b.** y^2y^4 **24. a.** $yyyyyy$ **b.** $x^3x^4y^2y^3$

25. a. $xxxxxxx$ **b.** $x^2x^3y^3y^3$ **26. a.** xxy^3y **b.** x^2xy^3y

27. a. $-6x^3y^2y^3z^2$ **b.** $49x^2xyy^3$

28. a. $25x^2x^3yz^2$ **b.** $a^2a^3a^5b^3b^4$

29. a. $7x(3x)$ **b.** $3x(7y)$ **30. a.** $(4x)(2y)$ **b.** $(9x)(-2y)$

31. a. $(2x)(3y)$ **b.** $(-9x)(3x)$

32. a. $5(2x)(3x)$ **b.** $-2(5x)(3y)$

33. a. $x(-3y)(3z)$ **b.** $(4a)(-2b)c$

34. a. $x(4y)(3z)$ **b.** $(-4a)(2y)(-c)$

35. a. $(2x)^2$ **b.** $(3y)^2$ **36. a.** $-(3y)^2$ **b.** $(2x)(5y^2)$

37. a. $-(5x)^2$ **b.** $(2x)(5y)^2$ **38. a.** $(2x)^2(5y)$ **b.** $(2x^2)(5y)$

39. a. $(4x)(-2x)^2$ **b.** $(2x)^2(-3x)$

40. a. $(4x)(-2x)^3$ **b.** $(2x^2)(-3x)$

41. a. $y^3(-5y^2)(-4y)$ **b.** $y(-5y^2)(-4y^3)$

42. a. $x^3(-x)^2(6x)$ **b.** $(-x^2)(-x^3)$

43. a. $(-x^4)(-x^3)$ **b.** $-x^2(x^3)(6x)$

44. a. $-(2x)^2(3x)$ **b.** $-(3x)^2(-2x)$

45. a. $-(4y)^2(-3y)$ **b.** $-(-5x)^2(x^4)$

LEVEL 2—Drill

Make up an example of a polynomial that uses the variable x and satisfies the description given in Problems 46–55.

46. A third-degree binomial **47.** A second-degree trinomial

48. A monomial of degree 0 **49.** A fifth-degree monomial

50. A second-degree monomial

51. A fourth-degree trinomial

52. A second-degree binomial

53. A first-degree monomial

54. A third-degree monomial

55. A first-degree binomial

Write each of the word phrases given in Problems 56–63 in symbols and simplify, if possible.

56. The sum of two squared and three squared

57. The square of the sum of two and three

58. The square of the sum of x and 5

59. The sum of x squared and 5

60. Three times the cube of a number

Spend some time on these word phrases. The problems are designed to help you "talk the talk of mathematics."

61. The square of a number plus four

62. The difference of squares

63. The difference of cubes

LEVEL 3—Team Project

64. Form teams in which team members have an interest in a similar type of business. With your team, make an appointment with a manager of the business and schedule an interview. Ask specific questions about the qualifications for a career in that business, and in particular find out the amount of mathematics or algebra required.

3.2 SIMILAR TERMS

Historical Perspective

Moslem manuscript, ca. 1258

The Byzantine Empire dates from 395 to 1453 A.D. and was formed by the division of the Roman Empire. The capital was Constantinople. It was formerly the Babylonian Empire, and although during this period the language changed from Greek and Latin to Arabic, the Greek cultural influence persisted. The most important mathematician of this period was Mohammed ibn Musa al-Khowârizmi, who lived around 825. He wrote an arithmetic book called (in Latin) *Liber algorismi de numero indorum,* and it was this book that introduced our numeration system to Western Europe. The word *algorismi* is the origin of our word *algorithm.* Al-Khowârizmi's treatise on algebra is called *Hisab al-jabr walmuqabals,* which means "the science of reduction and cancellation." As you might guess, our word *algebra* comes from this title. At the end of this historical period in 1110 A.D., the poet and mathematician Omar Khayyam solved cubic equations.

One of the most difficult tasks for students beginning to study algebra is to recognize when an expression is simplified and when it is not. In the previous section, we saw that numbers or variables can be multiplied to form a single new term. To simplify these products, you multiply together the numbers and use the law of exponents to simplify the variable parts of the terms. Now we turn our attention to addition and subtraction.

The key to simplification is the distributive property, which was introduced in Section 1.2. We will restate it in the next box for easy reference.

Distributive Property

This property is a virtual powerhouse property in algebra. You will use it almost every time you use algebra.

$$a(b + c) = ab + ac$$

Expressions such as

$$2x + 3y \qquad 2x^2 + 5x \qquad 5x^2 + 4x - 5$$

are simplified. However, if the variable parts of the individual terms are the same (both in variable and in degree), then the distributive property can be used to simplify:

$$\begin{aligned} 2x + 3x &= (2 + 3)x \quad \text{Distributive property} \\ &= 5x \end{aligned}$$

EXAMPLE 1 **Simplifying Algebraic Expressions**

Simplify:

a. $3a + 7a$ **b.** $2x^2 + 5x^2$ **c.** $5x + 4y + 4x$

Solution

a. $3a + 7a = (3 + 7)a$ You should do this in your head.
$= \mathbf{10a}$

b. $2x^2 + 5x^2 = \mathbf{7x^2}$

c. $5x + 4y + 4x = \mathbf{9x + 4y}$ Mentally: $5x + 4x = (5 + 4)x$
You cannot use the distributive property on $9x + 4y$. ■

We need some terminology to describe what we are doing in Example 1. We refer to terms that can be added together as **similar** or **like terms.**

Similar Terms

If two or more terms are exactly alike except for their numerical coefficients, we call them **similar terms.**

EXAMPLE 2 **Recognizing Similar Terms**

$5x$, $-3x$, and $9x$ are **similar** terms.

$2x^2$ and $2x$ are **not similar** terms.

abc, abc^2, and ab^2c are **not similar** terms.

$-3x^2y$, $-5x^2y$, and $14x^2y$ are **similar** terms. ■

Make SURE you study this paragraph until you really understand what is being said here.

To **simplify** expressions, you need to group together and combine the similar terms. You can usually do this in your head, but you need to be especially careful when the coefficients are integers rather than counting numbers. Note how we simplify the expressions in Example 3.

EXAMPLE 3 **Simplifying Algebraic Expressions**

Simplify:

a. $-12x - (-5x)$ **b.** $-3x - 6x$ **c.** $4x - 5x + 2x$
d. $(2x + 3y) + (5x - 2y)$ **e.** $(x - y) + (x - 5y)$

Solution

a. $-12x - (-5x) = \mathbf{-7x}$ Mentally, $-12 - (-5) = -12 + 5 = \mathbf{-7}$
b. $-3x - 6x = \mathbf{-9x}$ Mentally, $-3 - 6 = -3 + (-6) = \mathbf{-9}$
c. $4x - 5x + 2x = \mathbf{x}$ Mentally, $4 - 5 + 2 = 4 + (-5) + 2 = \mathbf{1}$
d. $(2x + 3y) + (5x - 2y) = \mathbf{7x + y}$ Mentally, $2x + 5x + 3y - 2y$
e. $(x - y) + (x - 5y) = \mathbf{2x - 6y}$ Mentally, $x + x - y - 5y$ ■

PROBLEM SET 3.2

Essential Ideas

1. **In Your Own Words** State the distributive property.
2. **In Your Own Words** What is meant by "similar terms?"

Level 1

3. Which of the following are similar terms?

   $3x$, $5xy$, $2x^2y$, $10x$, $3xy^2$, $7xy$, $4y$, $6x^2y$

4. Which of the following are similar terms?

   $4x$, $6x^2y$, $5xy$, $4y$, $12x$, $3xy^2$, $12x^2y$, $4xz$

WHAT IS WRONG, *if anything, with each of the statements in Problems 5–10? Explain your reasoning.*

5. $x + x + x = x^3$
6. $y + y + y + y = y^4$
7. $2x^2 + 3x^2 = 5x^4$
8. $6y + 2y = 8y^2$
9. $5x^2 + 3x + 1 = 9x^3$
10. $3x^2 - 2x + 4 = 5x^3$

Level 1—Drill

Combine similar terms in Problems 11–30.

11. $2x + 5x$
12. $3y + 5y$
13. $9z + 2z$
14. $7x - 3x$
15. $3y - 2y$
16. $8z - 2z$
17. $-3x + 5x$
18. $-5x + 2x$
19. $-3z + z$
20. $4x - 9x$
21. $3x - (-2)x$
22. $-5x - (-8x)$
23. $-6x - (8x)$
24. $-5x - (-2x)$
25. $-7x - (-5x)$
26. $5x - 5x$
27. $-x + x$
28. $-4x + 4x$
29. $2x + 3y$
30. $5 - 2x$

Level 2—Drill

Simplify the expressions in Problems 31–64.

31. $4x + 2x + 7x$
32. $2x + 3x + 8x$
33. $6x + 7x + (-4)x$
34. $9y + 2y + (-7)y$
35. $-2x + 3y + 2x$
36. $5z + 2z - 8z$

37. $5x + 4y + (-8)x$

38. $3a + (-5)a + (-2)a$

39. $2b + (-9)b + 6b$

40. $-2c + (-5)c + 9c$

41. $-4a + 7b + (-5)b$

42. $8x + 3y - 10x$

43. $5x + 15 - 15$

44. $x - 36 + 36$

45. $2y + 8 - 8$

46. $3x + 5 - 3x$

47. $2y + 8 - 2y$

48. $5x - 7 - 5x$

49. $5x + 3 - 4x + 5$

50. $6x + 3 - 5x + 7$

51. $10 - 3x - 10$

52. $12x - 9x + 5x - 8x$

53. $7y + 6y - 11y - 4y$

54. $7x + 3y - 5y + 2y - 8$

55. $8y - 2x + 3 - 5x + 15$

56. $(2x + 3y) + (5x - 6y)$

57. $(x - 2y) + (3x - 5y)$

58. $(x - y) + (2x - 3y)$

59. $(x + y) + (3x - 5y)$

60. $(2x - y) + (x - 3y) + 5y$

61. $(3x + 2y) + (x - 3y) + 5y$

62. $(2x^2 - 1) + (3x - 4) + (x^2 - 3x + 3)$

63. $(3x^2 + 4) + (2x - 5) + (x^2 - 5x)$

64. $(8x^2 - 3) + (5x + 3) + (x^2 - 8)$

3.3 SIMPLIFICATION

Historical Perspective

Sonja Kovalevsky (1850–1891)

Sonja Kovalevsky, earlier known as Sofia Korvin-Krukovsky, was a gifted mathematician who did work in calculus and in algebra. She is thought to be the first woman to receive a doctorate in mathematics. She was born in Moscow to Russian nobility. She left Russia in 1868 because universities were closed to women. She went to Germany since she wished to study with Karl Weierstrass in Berlin. The University in Berlin would not accept women, so she studied privately with Weierstrass and was awarded a doctoral degree in absentia from Göttingen in 1874, which excused her from oral examination on the basis of her outstanding thesis. When she won the *Prix Bordin* from the Institut de France in 1888, the prize was raised from 3,000 to 5,000 francs because of the exceptional merit of her paper. It was said that her early interest in mathematics was aroused by an odd wallpaper that covered her room in a summer house. Fascinated, she spent hours trying to make sense of it. The paper turned out to be lecture notes on higher mathematics purchased by her father during his student days.

In the previous section, we discussed the simplification of sums of polynomials by using the idea of similar terms. In this section, we simplify products of polynomials by using the distributive property.

First, consider a monomial times a polynomial.

EXAMPLE 1 **Using the Distributive Property to Simplify Algebraic Expressions**

Simplify:

a. $5(x + 3)$ **b.** $2x(x - y)$ **c.** $3x^2y(5xy^3 + x^2y)$
d. $(-1)(5x^2 - 2x + 3)$

Solution

a. $5(x + 3) = \mathbf{5x + 15}$
b. $2x(x - y) = \mathbf{2x^2 - 2xy}$

c. $3x^2y(5xy^3 + x^2y) = 3x^2y(5xy^3) + 3x^2y(x^2y)$
$= \mathbf{15x^3y^4 + 3x^4y^2}$

d. $(-1)(5x^2 - 2x + 3) = \mathbf{-5x^2 + 2x - 3}$ ■

You should pay particular attention to part **d** of Example 1. It is common to multiply a polynomial by -1, but this is often done in connection with a property of -1 that was stated in Chapter 2:

$$(-1)n = -n$$

This means that, if you are simplifying an expression such as

$$(6x^2 + 2x - 10) - (5x^2 - 2x + 3)$$

you need to remember that subtraction requires that you first add the opposite:

$$(6x^2 + 2x - 10) + (-1)(5x^2 - 2x + 3)$$

Next, simplify as shown in part **d** of Example 1:

$$6x^2 + 2x - 10 + (-5)x^2 + 2x + (-3)$$

Finally, add the similar terms:

$x^2 + 4x - 13$ Note: $6x^2 + (-5)x^2 = x^2$
$2x + 2x = 4x$
$(-10) + (-3) = -13$

EXAMPLE 2 **Adding Polynomials**

Simplify $(3x^2 + 3x - 5) + (2x^2 - 5x - 10)$.

Solution $(3x^2 + 3x - 5) + (2x^2 - 5x - 10) = \mathbf{5x^2 - 2x - 15}$

Mentally: $3x^2 + 2x^2 = 5x^2$
$3x - 5x = (3 - 5)x = -2x$
$-5 + (-10) = -15$ ■

EXAMPLE 3 **Subtracting Polynomials**

Simplify $(3x^2 + 3x - 5) - (2x^2 - 5x - 10)$.

Solution

$$(3x^2 + 3x - 5) - (2x^2 - 5x - 10) = 3x^2 + 3x - 5 - 2x^2 + 5x + 10$$
$$= \mathbf{x^2 + 8x + 5}$$

↑ ↑ ↑ Watch these signs.

Mentally eliminate parentheses by thinking:

$$3x^2 + 3x - 5 + (-1)(2x^2 - 5x - 10)$$

Distribute and write the result as shown above.

Finally, mentally associate the similar terms:

$3x^2 + (-2)x^2 = x^2$
$3x + 5x = 8x$
$-5 + 10 = 5$

■

Now we combine both multiplication and addition.

Be sure you remember the correct order of operations:

Parentheses first
Exponents next
Multiplication and division next
Addition and subtraction last

We say the polynomial is **simplified** when the similar terms are combined and the terms are written in descending degree.

EXAMPLE 4 **Simplifying Polynomials (Multiplication and Addition)**

Simplify:

a. $2(x - 3) + 4(x + 2)$ **b.** $2x(x - y) + 3y(x + 1)$
c. $3x(x + 2y) + 2y(y - 3)$

Solution

a. $2(x - 3) + 4(x + 2) = 2x - 6 + 4x + 8$ Parentheses first (distribute).
$= \mathbf{6x + 2}$ Similar terms next (add/subtract).

b. $2x(x - y) + 3y(x + 1) = 2x^2 - 2xy + 3xy + 3y$
$= \mathbf{2x^2 + xy + 3y}$

c. $3x(x + 2y) + 2y(y - 3) = \mathbf{3x^2 + 6xy + 2y^2 - 6y}$ ■

EXAMPLE 5 **Simplifying Polynomials (Multiplication and Subtraction)**

Simplify:

a. $2(4x - 5) - 4(3x - 7)$ **b.** $3(2x^2 - 4x + 9) - (x^2 + 5x - 6)$

Solution

a. $2(4x - 5) - 4(3x - 7) = 8x - 10 - 12x + 28$ Distributive property
↑ ↑ Watch signs.
$= \mathbf{-4x + 18}$ Combine similar terms.

b. $3(2x^2 - 4x + 9) - (x^2 + 5x - 6) = 6x^2 - 12x + 27 - x^2 - 5x + 6$
$= \mathbf{5x^2 - 17x + 33}$ ■

The distributive property is also used to multiply polynomials. In this course we will consider the product of two binomials, but the method illustrated here could be extended to any polynomials. Consider the product

$$(2x + 3)(3x + 5)$$

Recall the distributive property:

$$A(B + C) = AB + AC$$

Distribute the outside number $(2x + 3)$ to each number inside the parentheses:

$$(2x + 3)(3x + 5) = (2x + 3)3x + (2x + 3)5$$
$$= 2x(3x) + 3(3x) + (2x)(5) + 3(5)$$
Use the distributive property again.
$$= 6x^2 + 9x + 10x + 15$$
$$= \mathbf{6x^2 + 19x + 15}$$ ■

EXAMPLE 6 **Multiplying Binomials**

Simplify:

a. $(5x + 3)(2x + 3)$ **b.** $(x - 7)(x + 4)$ **c.** $(2x + 3)(x - 5)$
d. $(x - 5)(x - 3)$

Solution

a. $(5x + 3)(2x + 3) = (5x + 3)2x + (5x + 3)(3)$
$= 10x^2 + 6x + 15x + 9$
$= \mathbf{10x^2 + 21x + 9}$

b. $(x - 7)(x + 4) = (x - 7)x + (x - 7)4$
$= x^2 - 7x + 4x - 28$
$= \mathbf{x^2 - 3x - 28}$

c. $(2x + 3)(x - 5) = (2x + 3)x + (2x + 3)(-5)$
$= 2x^2 + 3x + (-10)x + (-15)$
$= \mathbf{2x^2 - 7x - 15}$

d. $(x - 5)(x - 3) = (x - 5)x + (x - 5)(-3)$
$= x^2 - 5x + (-3)x + 15$
$= \mathbf{x^2 - 8x + 15}$ ■

PROBLEM SET 3.3

ESSENTIAL IDEAS

1. **In Your Own Words** State the distributive property.
2. **In Your Own Words** What does it mean to simplify a polynomial?

LEVEL 1

WHAT IS WRONG, *if anything, with each of the statements in Problems 3–10? Explain your reasoning.*

3. $3(x + 5) = 3x + 5$
4. $-(x + 3) = -x + 3$
5. $-x = (-1)x$
6. $x - y = x + (-1)y$
7. $(x + 5) - (x + 4) = 9$
8. $(3y - 4) - (4 + 3y) = 0$
9. $(x + 5)(x + 4) = x + 5x + 20$
10. $(x - 3)(x - 2) = x^2 - 3x + 6$

LEVEL 1—Drill

Simplify the expressions in Problems 11–69.

11. $6(x + 1) + 3(x + 2)$
12. $3(x + 2) + 2(x + 3)$
13. $5(x + 3) + 2(x + 4)$
14. $6(x + 1) + 2(x + 4)$
15. $3(x - 5) + 2(x + 1)$
16. $2(x + 5) + (x - 4)$
17. $4(x - 2) + 6(x - 2)$
18. $(x - 6) + 2(x - 8)$
19. $(2x + 3y) + (5x - 6y)$
20. $(x - 2y) + (3x - 5y)$
21. $(x - y) + (2x - 3y)$
22. $(x + y) + (3x - 5y)$
23. $6(x + 2) - 3(x + 5)$
24. $3(x + 2) - 2(x + 1)$
25. $5(x + 3) - 2(x + 8)$
26. $6(x + 1) - 2(x + 6)$
27. $3(x - 5) - 2(x + 1)$
28. $2(x + 5) - (x - 4)$
29. $4(x - 2) - 6(x - 2)$
30. $(x - 6) - 2(x - 8)$
31. $(2x + 3y) - (5x - 6y)$
32. $(3x - 5y) - (x - 2y)$
33. $(x - y) - (2x - 3y)$
34. $(x + y) - (3x - 5y)$
35. $(x - 2y) - (3x - 5y)$
36. $(2x - y) - (x - 3y)$
37. $(3x + 2) - (2x + 1)$
38. $(5x + 3) - (3x - 5)$

39. $(x - 3) - (2x + 4)$

40. $(x - 4) - (3x - 5)$

41. $(3x - 5) - (2x + 3)$

42. $(5x + 3) - (3x + 5)$

Level 2—Drill

43. $(x + 1)(x + 2)$

44. $(x + 3)(x + 3)$

45. $(x - 5)(x + 7)$

46. $(x - 3)(x + 8)$

47. $(x + 3)(x - 2)$

48. $(x + 2)(x - 5)$

49. $(x - 4)(x - 1)$

50. $(x - 5)(x - 3)$

51. $(x + 2)(x - 2)$

52. $(x - 3)(x + 2)$

53. $(2x + 3)(3x - 1)$

54. $(3x - 2)(2x + 5)$

55. $(x + 5)(x + 5)$

56. $(x - 2)(x - 2)$

57. $(x - 3)^2$

58. $(x + 2)^2$

59. $(2x + 1)^2$

60. $(3x - 2)^2$

61. $(xy + z)(3xy - 2z)$

62. $(3x^2 - 2x + 4) - (x^2 + 2x + 1)$

63. $(3x^2 + 2x - 5) - (3x^2 - x + 3)$

64. $(x^2 + 4x - 1) - (3x^2 - 5x - 1)$

65. $(5x^2 - 3x - 4) - (7x^2 + 3x - 4)$

66. $2(x^2 + x - 1) + 4(x^2 - 2x + 3)$

67. $4x^2 (x^2 - x + 2) + 3x(x^2 + x - 1)$

68. $2(3x^3 + 2x - 1) + (5x^3 - 4x^2 + 3)$

69. $2(5x^3 - 4x^2 + 3x - 5) + 3(2x^2 - 5x - 6)$

3.4 EQUATIONS

Historical Perspective

Have you ever been asked to find the unknown?
CAN YOU TELL WHO'S WHO BY THEIR HAIR?

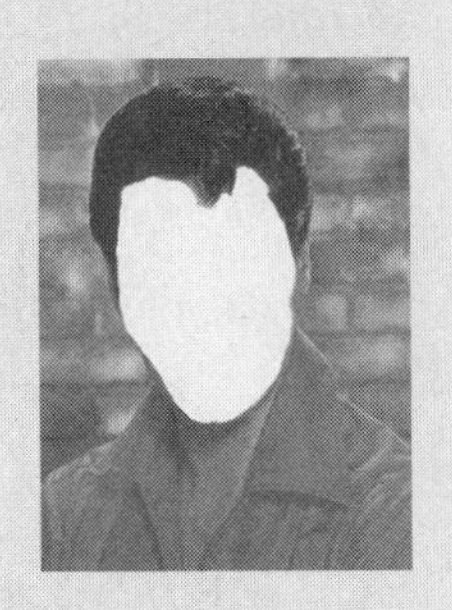

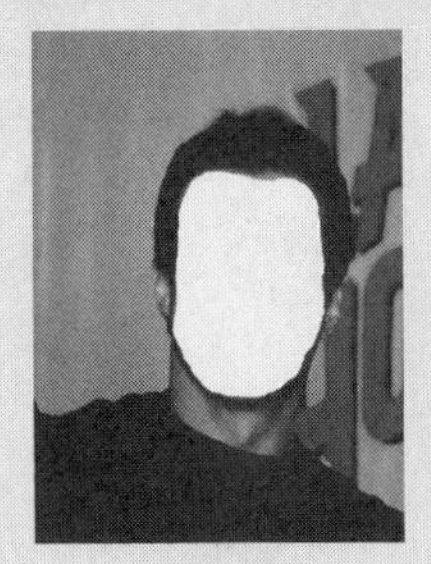

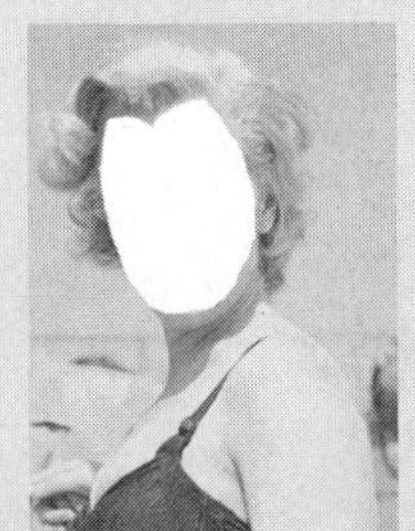

Just as you needed to look for visual clues to find the identities of the celebrities in these photographs, you must check the clues of a given equation to find the solution. Solving equations is as old as mathematics itself, but the use of the equal sign dates back to Robert Recorde in his book *The Whetstone of Witte,* written in 1557.

An **equation** is a statement of equality that can be true, false, or dependent on the values of the variable. If the equation depends on the values of the variable, it is called a **conditional equation.** The values that make the conditional equation true are said to **satisfy** the equation and are called the **solution** or **roots** of the equation. To **solve** an equation means to find the replacement for the unknown that makes the equation true. To do this, you must look for things to do to make the solution or roots more obvious. The key ideas are those of **opposites** (for addition and subtraction) and **reciprocals** (for multiplication and division).

Opposites and Reciprocals

Two numbers whose sum is 0 are called **opposites.**
Two numbers whose product is 1 are called **reciprocals.**

EXAMPLE 1 **Adding Opposites**

Use the idea of opposites to find the sums.

a. $5 + (-5)$ **b.** $-18 + 18$ **c.** $x + (-x)$
d. $x - x$ **e.** $175 + 7 + (-7)$ **f.** $14 + 23 - 23$
g. $x - 18 + 18$ **h.** $y + 10 - 10$ **i.** $x + y - x$

Solution

a. $5 + (-5) = \mathbf{0}$ **b.** $-18 + 18 = \mathbf{0}$
c. $x + (-x) = \mathbf{0}$ **d.** $x - x = x + (-x) = \mathbf{0}$
e. $175 + 7 + (-7) = \mathbf{175}$ **f.** $14 + 23 - 23 = \mathbf{14}$
g. $x - 18 + 18 = \boldsymbol{x}$ **h.** $y + 10 - 10 = \boldsymbol{y}$
i. $x + y - x = \boldsymbol{y}$ ■

The goal in solving an equation is to *isolate the variable* on one side of the equation. To begin our study, we state the first of four properties about equation solving.

Addition Property of Equations

The solution of an equation is unchanged by **adding the same number to both sides of the equation.**

You will use this property to add some number to both sides of an equation so that, after it is simplified, the only expression on one side of the equation is the variable.

EXAMPLE 2 **Using the Addition Property of Equations**

Solve the given equations.

a. $x - 7 = 12$ **b.** $y - 10 = 5$ **c.** $z - 12 = -25$ **d.** $-108 = k - 92$

Solution Remember that to *solve* an equation means to find the replacement of the variable that makes the equation true.

a. $\underbrace{x - 7}= 12$

The opposite of subtracting 7 is adding 7, so you **add 7 to both sides** of the equation.

$x - 7 \overbrace{+ 7} = 12 \overbrace{+ 7}$

$\mathbf{x = 19}$ Simplify.

This says that, **if** x is replaced by 19, the original equation is true. It does not say that x is always equal to 19. In fact, x is a variable, so it can be replaced by any number in the domain. For example, if x is replaced by 20, the equation is false. The form x = 19 is the desired form for the solution because in this form the solution is obvious.

Check: $19 - 7 = 12$

Replace x by 19 to check.

b. $y - 10 = 5$

$y - 10 + \mathbf{10} = 5 + \mathbf{10}$ Add 10 to both sides.

$\mathbf{y = 15}$ Simplify.

Can you check?

c. $z - 12 = -25$

$z - 12 + \mathbf{12} = -25 + \mathbf{12}$

$\mathbf{z = -13}$

Notice that, when solving equations, you write one equation under the other with the equal signs aligned.

d. $-108 = k - 92$

$-108 + \mathbf{92} = k - 92 + \mathbf{92}$

$\mathbf{-16 = k}$

It doesn't matter whether the variable ends up on the left or the right side. ■

The celebrities in the photographs on page 134 are Elvis Presley, Whoopi Goldberg, Sylvester Stallone, Bill Gates, and Marilyn Monroe.

The addition property is used whenever some number is being subtracted from the variable. **If a number is being added to the variable, then the opposite is subtraction,** and you can use the following property.

Subtraction Property of Equations

The solution of an equation is unchanged by **subtracting the same number from both sides of the equation.**

EXAMPLE 3 **Using the Subtraction Property of Equations**

Solve the given equations.

a. $x + 5 = 13$ **b.** $y + 8 = 5$ **c.** $17 = a + 12$ **d.** $19 = b + 48$

Solution **a.** $\underbrace{x + 5}$ $= 13$

↑ The opposite of adding 5 is subtracting 5, so you **subtract 5 from both sides** of the equation.

$x + 5 \mathbf{- 5} = 13 \mathbf{- 5}$

$\mathbf{x = 8}$ Simplify.

Check: $8 + 5 = 13$

If x is equal to 8, then the equation is true.

b. $y + 8 = 5$

$y + 8 \mathbf{- 8} = 5 \mathbf{- 8}$ Subtract 8 from both sides.

$\mathbf{y = -3}$ Simplify.

Can you check?

c. $17 = a + 12$

$17 \mathbf{- 12} = a + 12 \mathbf{- 12}$

$\mathbf{5 = a}$

d. $19 = b + 48$

$19 \mathbf{- 48} = b + 48 \mathbf{- 48}$

$\mathbf{-29 = b}$ ■

The idea of opposite operations extends to examples in which the variable is multiplied or divided by some number; that is, multiplication and division are opposite operations. Remember that when reciprocals are multiplied, the product is 1.

EXAMPLE 4 **Multiplying Reciprocals**

Simplify each expression.

a. $\frac{3x}{3}$ **b.** $\frac{5x}{5}$ **c.** $\frac{19p}{19}$ **d.** $\frac{-12q}{-12}$ **e.** $\left(\frac{x}{2}\right)(2)$ **f.** $\left(\frac{x}{7}\right)(7)$ **g.** $\left(\frac{t}{-3}\right)(-3)$

Solution The definition of a reciprocal tells us that the product of a number and its reciprocal is 1. This means, for example, that $3(\frac{1}{3}) = 1$ and $\frac{5}{5} = 1$.

a. $\frac{3x}{3}$: Think: "If some number x is multiplied by 3 and then the result is divided by 3, the result is the original number x." As you are thinking this, you write

$$\frac{3x}{3} = x$$

b. $\frac{5x}{5} = x$ **c.** $\frac{19p}{19} = p$ **d.** $\frac{-12q}{-12} = q$

e. $\left(\frac{x}{2}\right)(2)$ Think: "If some number x is divided by 2 and then the result is multiplied by 2, the result is the original number x." As you are thinking this, you write

$$\left(\frac{x}{2}\right)(2) = x$$

f. $\left(\frac{x}{7}\right)(7) = x$ **g.** $\left(\frac{t}{-3}\right)(-3) = t$ ■

To solve an equation in which some number is dividing the variable, use the following property.

Multiplication Property of Equations

The solution of an equation is unchanged by **multiplying both sides of the equation by the same nonzero number.**

EXAMPLE 5 **Using the Multiplication Property of Equations**

Solve each equation.

a. $\frac{x}{2} = 9$ **b.** $\frac{w}{8} = 7$ **c.** $\frac{r}{-5} = 12$ **d.** $-3 = \frac{m}{6}$ **e.** $0 = \frac{n}{4}$

Solution **a.** $\frac{x}{2} = 9$

↑ The opposite of dividing by 2 is multiplying by 2, so you **multiply both sides** of the equation **by 2.**

$$\left(\frac{x}{2}\right)(2) = 9(2)$$

$x = 18$ Simplify. **Check:** $\frac{18}{2} = 9$

b. $\frac{w}{8} = 7$

$$\left(\frac{w}{8}\right)(8) = 7(8)$$

$$\mathbf{w = 56}$$

c. $\frac{r}{-5} = 12$

$$\left(\frac{r}{-5}\right)(-5) = 12(-5)$$

$$\mathbf{r = -60}$$

d. $-3 = \frac{m}{6}$

$-3(\mathbf{6}) = \left(\frac{m}{6}\right)(\mathbf{6})$

$\mathbf{-18 = m}$

e. $0 = \frac{n}{4}$

$0(\mathbf{4}) = \left(\frac{n}{4}\right)(\mathbf{4})$

$\mathbf{0 = n}$

■

The last property of equations of this section is used if the variable is multiplied by some number.

Division Property of Equations

The solution of an equation is unchanged by **dividing both sides of the equation by the same nonzero number.**

EXAMPLE 6 **Using the Division Property of Equations**

Solve each equation.

a. $5a = 20$ **b.** $3b = 39$ **c.** $-7c = 42$ **d.** $-104 = 8d$ **e.** $-70 = -14e$

Solution **a.** $5a = 20$

↑ The opposite of multiplying by 5 is dividing by 5, so you **divide both sides** of the equation **by 5.**

$\frac{5a}{\mathbf{5}} = \frac{20}{\mathbf{5}}$

$\mathbf{a = 4}$ Simplify. **Check:** $5(4) = 20$

b. $3b = 39$

$\frac{3b}{\mathbf{3}} = \frac{39}{\mathbf{3}}$

$\mathbf{b = 13}$

c. $-7c = 42$

$\frac{-7c}{\mathbf{-7}} = \frac{42}{\mathbf{-7}}$

$\mathbf{c = -6}$

d. $-104 = 8d$

$\frac{-104}{\mathbf{8}} = \frac{8d}{\mathbf{8}}$

$\mathbf{-13 = d}$

e. $-70 = -14e$

$\frac{-70}{\mathbf{-14}} = \frac{-14e}{\mathbf{-14}}$

$\mathbf{5 = e}$

■

PROBLEM SET 3.4

ESSENTIAL IDEAS

1. **In Your Own Words** Explain what it means to "solve an equation."
2. **In Your Own Words** State the four properties of equations.
3. **In Your Own Words** What is a solution of an equation?
4. **In Your Own Words** What is the goal when solving an equation?

Level 1

WHAT IS WRONG, *if anything, with each of the statements in Problems 5–16? Explain your reasoning.*

5. $x + 2 = 8$
$x = 10$

6. $x - 5 = -6$
$x = -11$

7. $3x = 2$
$x = 6$

8. $\frac{x}{5} = -10$
$x = -2$

9. If you add opposites, the sum is 1.

10. If you multiply reciprocals, the product is 0.

11. The first step in solving $x + 8 = 5$ is to subtract 5 from both sides.

12. The first step in solving $x - 6 = 2$ is to subtract 2 from both sides.

13. The first step in solving $x - 5 = 11$ is to move the 5 to the other side.

14. The first step in solving $x + 3 = -14$ is to move the 3 to the other side.

15. To solve $5x = -20$, use the multiplication property of equations.

16. To solve $\frac{x}{-8} = -3$, use the division property of equations.

Level 1—Drill

Solve each equation in Problems 17–64. Show each step of the process until the variable is isolated. Be sure to keep your equal signs aligned.

17. $a - 5 = 10$
18. $b - 8 = 14$
19. $c - 2 = 0$
20. $d - 9 = -7$
21. $d - 7 = -10$
22. $e - 1 = -35$
23. $f - 3 = -12$
24. $6 = g - 2$
25. $-5 = h - 4$
26. $-40 = i - 92$
27. $-137 = j - 49$
28. $0 = k - 112$
29. $m + 2 = 7$
30. $n + 8 = 12$
31. $n + 19 = 20$
32. $p + 8 = 2$
33. $q + 7 = -15$
34. $r + 6 = -14$
35. $8 + s = 4$
36. $12 + t = 15$
37. $36 + u = 40$
38. $18 + v = 10$
39. $-10 + w = 14$
40. $-12 + x = -6$
41. $\frac{y}{4} = 8$
42. $\frac{z}{2} = 18$
43. $\frac{A}{3} = 7$
44. $\frac{B}{12} = 8$
45. $\frac{C}{5} = -12$
46. $\frac{D}{8} = -15$
47. $\frac{E}{-4} = 11$
48. $\frac{F}{-12} = 16$
49. $\frac{G}{-13} = -3$
50. $7 = \frac{H}{-8}$
51. $-4 = \frac{I}{-10}$
52. $-18 = \frac{J}{-4}$
53. $\frac{K}{19} = 0$
54. $4L = 12$
55. $3M = 33$
56. $4N = -48$
57. $-8P = -96$
58. $-Q = 5$
59. $-R = 14$
60. $-S = 19$
61. $-12T = 168$
62. $13U = -234$
63. $-5V = 225$
64. $13W = 0$

Level 2

65. In Your Own Words To solve $\frac{2}{3}x = 5$, do you think it would be easier to divide both sides by $\frac{2}{3}$ or to multiply both sides by $\frac{3}{2}$? Are both these steps permissible? Explain.

66. In Your Own Words To solve $5x = \frac{2}{3}$, do you think it would be easier to divide both sides by 5 or to multiply both sides by $\frac{1}{5}$? Are both these steps permissible? Explain.

67. WHAT IS WRONG, *if anything, with the following "proof"?*

i.	12 eggs = 1 dozen	
ii.	24 eggs = 2 dozen	Multiply both sides of step i by 2.
iii.	6 eggs = $\frac{1}{2}$ dozen	Divide both sides of step i by 2.
iv.	144 eggs = 1 dozen	Multiply step iii times step ii (equals times equals are equal).
v.	12 dozen = 1 dozen	Substitute, since 144 eggs = 12 dozen.

68. WHAT IS WRONG, *if anything, with the following reasoning?*

A hippopotamus and a professor want to play on a teeter-totter. The professor says that it's impossible, but the hippopotamus assures the professor that it will work out and that she will prove it, since she has had a little algebra. She presents the following argument:

Let H = WEIGHT OF HIPPOPOTAMUS

p = WEIGHT OF PROFESSOR

Now there must be some weight, w (probably very large), so that

$$H = p + w$$

Multiply both sides by $H - p$:

$$H(H - p) = (p + w)(H - p)$$

Using the distributive property, we have

$$H^2 - Hp = pH + wH - p^2 - wp$$

Subtract wH from both sides:

$$H^2 - Hp - wH = pH - p^2 - wp$$

Use the distributive property again:

$$H(H - p - w) = p(H - p - w)$$

Divide both sides by $H - p - w$:

$$H = p$$

Thus, the weight of the hippopotamus is the same as the weight of the professor. "Now," says the hippopotamus, "since our weights are the same, we'll have no problem on the teeter-totter."

"Wait!" hollers the professor. "Obviously this is false." But where is the error in the reasoning?

3.5 SOLVING EQUATIONS

Historical Perspective

The formal development of algebra is beyond of the scope of this book, but it depends on some properties that we have assumed without formal development. These properties are stated here:

COMMUTATIVE PROPERTY
$a + b = b + a$
$ab = ba$

The order is which you add or multiply numbers does not change the result of the answer. For example,

$$3 + 5 = 8 \quad \text{and} \quad 5 + 3 = 8$$

Also,

$$3 \times 5 = 15 \quad \text{and} \quad 5 \times 3 = 15$$

ASSOCIATIVE PROPERTY
$(a + b) + c = a + (b + c)$
$(ab)c = a(bc)$

When adding three numbers, you can add the third number to the sum of any pair of those numbers; for example,

$$(3 + 5) + 7 = 15 \quad \text{and} \quad 3 + (5 + 7) = 15$$

The same result holds for multiplication; namely,

$$(3 \times 5) \times 7 = 105 \quad \text{and} \quad 3 \times (5 \times 7) = 105$$

One of the most important tools in problem solving is solving equations. In this section, we'll take the fundamentals you learned in the previous section and apply them to more complicated equations. The first step is to learn to solve the equations from the previous section *mentally*.

EXAMPLE 1 **Basic Properties of Solving Equations**

Solve each equation.

a. $7h = -56$ **b.** $j + 7 = -56$ **c.** $\frac{r}{7} = -56$ **d.** $s - 7 = -56$

Solution

a. $7h = -56$
$\boldsymbol{h = -8}$
Mentally divide both sides by 7.

b. $j + 7 = -56$
$\boldsymbol{j = -63}$
Mentally subtract 7 from both sides.

c. $\frac{r}{7} = -56$
$\boldsymbol{r = -392}$
Mentally multiply both sides by 7.

d. $s - 7 = -56$
$\boldsymbol{s = -49}$
Mentally add 7 to both sides. ■

CAUTION

Don't forget that whatever is done to an equation must be done to **both** sides.

More difficult equations than those in Example 1 can now be considered. In these examples, you must simplify by adding similar terms.

EXAMPLE 2 **Solving Equations by First Simplifying**

Solve each equation.

a. $3x - 2 - 2x = 5 + 2x - 2x$ **b.** $4 + 5x - 5x = 7 + 6x - 5x$

c. $\frac{x}{2} + 4 - 4 = 5 - 4$ **d.** $4 - 10 = 10 - 3x - 10$

Solution

a. $3x - 2 - 2x = 5 + 2x - 2x$

$x - 2 = 5$ First simplify.

$\mathbf{x = 7}$ Add 2 to both sides (done mentally).

b. $4 + 5x - 5x = 7 + 6x - 5x$

$4 = 7 + x$ Simplify.

$\mathbf{-3 = x}$ Subtract 7 from both sides (mentally).

c. $\frac{x}{2} + 4 - 4 = 5 - 4$

$\frac{x}{2} = 1$ Simplify.

$\mathbf{x = 2}$ Multiply both sides by 2.

d. $4 - 10 = 10 - 3x - 10$

$-6 = -3x$ Simplify.

$\mathbf{2 = x}$ Divide both sides by −3. ■

Sometimes it is necessary to use more than one of the equation properties when solving a simple equation.

Procedure for Solving Simple Equations

1. Simplify the left and right sides.
2. Use equation properties to isolate the variable on one side.
   a. First, use the addition or subtraction property.
   b. Next, use the multiplication or division property.

EXAMPLE 3 **Solving Equations**

Solve each equation.

a. $4x + 3 = 7$ **b.** $2x - 7 = 13$ **c.** $41 = \frac{x}{2} - 7$ **d.** $-4 = \frac{x}{3} + 2$

Solution

a. $4x + 3 = 7$

$4x = 4$ Subtract 3 from both sides.

$\mathbf{x = 1}$ Divide both sides by 4.

b. $2x - 7 = 13$

$2x = 20$ Add 7 to both sides.

$\mathbf{x = 10}$ Divide both sides by 2.

c. $41 = \frac{x}{2} - 7$

$48 = \frac{x}{2}$ Add 7 to both sides.

$\mathbf{96 = x}$ Multiply both sides by 2.

d. $-4 = \frac{x}{3} + 2$

$-6 = \frac{x}{3}$ Subtract 2 from both sides.

$\mathbf{-18 = x}$ Multiply both sides by 3. ■

Sometimes you need to solve equations with the variable on both sides of the equal sign. In such cases, use equation properties to get the variable *on one side only*.

EXAMPLE 4 **Solving Equations with a Variable on Both Sides**

Solve each equation.

a. $4x + 3 = 3x + 8$ **b.** $2x - 4 = 5x + 2$

Solution **a.** $4x + 3 = 3x + 8$

$x + 3 = 8$ Subtract $3x$ from both sides.

$\mathbf{x = 5}$ Subtract 3 from both sides.

b. $2x - 4 = 5x + 2$

$-4 = 3x + 2$ Subtract $2x$ from both sides.

$-6 = 3x$ Subtract 2 from both sides.

$\mathbf{-2 = x}$ Divide both sides by 3. ■

PROBLEM SET 3.5

Level 1

WHAT IS WRONG, *if anything, with each of the statements in Problems 1–8? Explain your reasoning.*

1. The best first step in solving $4x + 3 = 15$ is to divide both sides by 4.
2. The best first step in solving $2x - 5 = -4$ is to divide both sides by 2.
3. The best first step in solving $\frac{x}{3} + 5 = 2$ is to multiply both sides by 3.
4. The best first step in solving $\frac{x}{5} - 3 = 10$ is to multiply both sides by 5.

Solve the equations in Problems 5–8.

5. $3x + 4 = x - 8$
   $4x + 4 = -8$
   $4x = -12$
   $x = -3$

6. $2x + 5 = x - 10$
$x + 5 = -10$
$x = -5$

7. $\frac{x}{3} + 5 = 6$
$x + 5 = 18$
$x = 13$

8. $1 - x = 8$
$-x = 7$

LEVEL 1—Drill

Simplify the expressions in Problems 9–16.

9. a. $3x + 2x$ **b.** $5y + 2y$

10. a. $9x - 5x$ **b.** $12y - 5y$

11. a. $y + 8 - 8$ **b.** $3x - 10 + 10$

12. a. $6 - x + x$ **b.** $2x - 10 - 2x$

13. a. $x - y - x$ **b.** $10x - 5 - 10x$

14. a. $10 - x + x$ **b.** $6 - x - 6$

15. a. $3x + 5 + x - 5$ **b.** $9x + 8 - 2x - 8$

16. a. $5x + 8 - 3x - 8$ **b.** $5 - 2x + 6 + 2x$

Solve the equations in Problems 17–28. Show all your work and be sure to keep the equal signs aligned.

17. a. $2x + 1 = 9$ **b.** $3x + 4 = 19$

18. a. $5x + 7 = 47$ **b.** $4y - 3 = 77$

19. a. $6y - 9 = 21$ **b.** $9y - 23 = 49$

20. a. $\frac{z}{5} + 3 = 4$ **b.** $\frac{z}{3} + 1 = 0$

21. a. $\frac{z}{9} + 12 = 6$ **b.** $\frac{w}{2} - 5 = 4$

22. a. $\frac{w}{-4} - 7 = 6$ **b.** $\frac{w}{3} - 19 = 47$

23. a. $6 + 2t = 4$ **b.** $4 + 7t = 53$

24. a. $9 + 3t = 45$ **b.** $-s + 2 = 9$

25. a. $-s - 5 = 14$ **b.** $-s - 7 = 10$

26. a. $4 + (-u) = 17$ **b.** $3 + (-s) = 15$

27. a. $17 + (-s) = 21$ **b.** $4 - s = 17$

28. a. $3 - s = 15$ **b.** $18 - s = 31$

Solve the equations in Problems 29–52. Show all your work and be sure to keep the equal signs aligned.

29. $3A + 7 = 49$ **30.** $2B + 5 = 65$

31. $3C - 8 = 115$ **32.** $2D - 10 = 52$

33. $\frac{E}{7} + 12 = 12$ **34.** $\frac{F}{5} - 12 = 53$

35. $9 = \frac{G}{2} + 5$ **36.** $-8 = \frac{H}{6} - 3$

37. $4 + 3I = 67$ **38.** $2 - 5J = 62$

39. $\frac{K}{7} - 10 = 0$ **40.** $\frac{L}{12} + 4 = -3$

41. $-M = 14$ **42.** $6 + \frac{N}{2} = -10$

43. $-P + 5 = -12$ **44.** $7 + (-Q) = -15$

45. $8 - R = 5$ **46.** $5 - S = 10$

47. $10 - T = -5$ **48.** $4 - 2U = 8$

49. $7 - 3V = 10$ **50.** $6 - 5W = 26$

51. $1 - 5X = 126$ **52.** $12 - 21Y = 75$

LEVEL 2

Solve the equations in Problems 53–66. Show all your work and be sure to keep the equal signs aligned.

53. $2x + 3 = x + 5$ **54.** $4x - 7 = 3x + 2$

55. $3x + 4 = 4x - 5$ **56.** $2x + 1 = 3x + 8$

57. $4x - 11 = 3x + 2$ **58.** $6x + 4 = 7x - 3$

59. $3x + 2 = x + 6$ **60.** $5x - 8 = 3x + 2$

61. $4x - 3 = 2x + 1$ **62.** $5x - 6 = 2x + 3$

63. $6x - 2 = 3x - 11$ **64.** $5x - 3 = 7x + 5$

65. $3x + 4 = 5x - 6$ **66.** $2x - 3 = 4x - 1$

67. In Your Own Words When solving $5x + 3 = 5$, as a first step, would you subtract 3 or 5 from both sides? Explain.

68. In Your Own Words When solving $\frac{x}{4} + 5 = 3$, as a first step, would you multiply both sides by 4 or subtract 5 from both sides? Explain.

69. In Your Own Words Explain the steps in solving $-x = -5$.

70. In Your Own Words Explain the steps in solving $3 = -x$.

71. In Your Own Words Explain the steps in solving $5x - 4 = 24$.

72. In Your Own Words Explain the steps in solving $\frac{x}{3} + 1 = 5$.

73. In Your Own Words Explain the steps in solving $7x - 3 = 2x + 2$.

74. In Your Own Words Explain the steps in solving $x + 5 = 3x - 1$.

Level 3—Mind Boggler

75. This problem will help you check your work in Problems 29–52. Fill in the capital letters from Problems 29–52 to correspond with their numerical values shown in the blanks. (The letter O has been filled in for you.) Some letters may not appear in the boxes. When you finish putting letters in the boxes, darken all the blank spaces to separate the words in the message.

21	–8	–30	O	17	0	1	–3	O	–2	4
14	3	0	4	325	21	–32	31	21	–32	8
35	15	–30	21	–5	5	0	14	–5	–3	–15
14	–32	31	7	14	3	0	24	2	12	12
–84	0	14	3	–32	21	–32	8	–15	2	2
–14	14	15	–30	0	–14	14	15	21	41	–5

3.6 PROBLEM SOLVING WITH ALGEBRA

Historical Perspective

The following quotation is from an address at a meeting of the National Academy of Sciences and the National Academy of Engineering:*

> *Good problem-solvers do not rush in to apply a formula or an equation. Instead, they try to understand the problem situation; they consider alternative representations and relations among variables. Only when satisfied that they understand the situation and all the variables in a qualitative way do they start to apply the quantification.*

*From an address by Lauren B. Resnick, National Convocation of Precollege Education in Mathematics, National Academy of Sciences and National Academy of Engineering, Washington, D.C., May 1982. (Quoted in *Science*, Vol. 220, No. 4596, April 26, 1983, p. 29.)

Our goal in applying algebra to problem solving is to find solutions to a wide variety of problems. In the last two sections, you've seen techniques for solving equations. In this section, we'll introduce a *technique for solving problems.* Eventually, you want to be able to solve applied problems to obtain *answers;* but when beginning, we must start with simple problems—ones for which the answers are obvious or trivial. You must, therefore, keep in mind that we are seeking not answers in this section, but a strategy for attacking word problems. You should also pay particular attention to the way we translate from the words of the problem to the symbols of algebra, which leads to the use of algebraic variables.

We begin by considering an equation with more than one variable. An equation with more than one variable is called a **literal equation.** When working with literal equations, we carry out the same steps that were explained in the last section, except that we treat one variable as the unknown and the other variables as known quantities.

EXAMPLE 1

Solving a Literal Equation

Solve the equation $P = S - C$ for C.

Solution

Since we are solving for C, we treat C as the unknown and the other variables as known quantities. This means that we wish to isolate C on one side.

$$\begin{aligned} P &= S - C && \text{Given equation} \\ P + C &= S - C + C && \text{Add } C \text{ to both sides.} \\ P + C &= S \\ \mathbf{C} &= \mathbf{S - P} && \text{Subtract } P \text{ from both sides.} \end{aligned}$$

■

Literal equations are often given as **formulas,** where each of the variables has a specific meaning. In Example 1, the given literal equation is sometimes remembered as the **profit formula,** where P represents the profit, S represents the selling price (or revenue), and C the cost (or overhead). Another example of a literal equation is the Pythagorean theorem, which we stated in Section 2.6: If $a^2 + b^2 = c^2$, then

$$a^2 = c^2 - b^2 \quad \text{Subtract } b^2 \text{ from both sides.}$$

and

$$b^2 = c^2 - a^2 \quad \text{Subtract } a^2 \text{ from both sides.}$$

EXAMPLE 2

Solving a Literal Equation

Solve the equation $2x + 3y - 6 = 0$ for y.

Solution

$$\begin{aligned} 2x + 3y - 6 &= 0 && \text{Given equation} \\ 2x + 3y &= 6 && \text{Add 6 to both sides.} \\ 3y &= -2x + 6 && \text{Subtract } 2x \text{ from both sides.} \\ \mathbf{y} &= \mathbf{\tfrac{-2}{3}x + 2} && \text{Divide both sides by 3.} \end{aligned}$$

■

> Stand firm in your refusal to remain conscious during algebra. In real life, I assure you, there is no such thing as algebra.
>
> Fran Lebowitz, "Tips for Teens," *Newsweek*, 1/1/79

When confronted with a word problem, many people begin by looking at the last sentence to see what is being asked for, declare that this quantity is the unknown, and then proceed to read the problem backward to come up with an equation. This procedure complicates the thinking process and can be frustrating. Instead, we begin with the problem as stated, using as many unknowns as we wish with our initial statement. We then let this relationship *evolve* until we have a statement with a single unknown. At *this* point a variable is chosen as a natural result of the thinking process, and the resulting equation can be solved.

The first type of problem involves number relationships.

EXAMPLE 3 **Number Problem**

If you add twelve to twice a number, the result is six. What is the number?

Solution

Step 1 Read the problem carefully. Make sure you know what is given and what is wanted.

Step 2 Write a verbal description of the problem using operations signs and an equal sign, but still using key words.

$$2(\text{A NUMBER}) + 12 = 6$$

Step 3 If there is a single unknown, choose a variable.

Let $n = \text{A NUMBER}$

Step 4 Replace the verbal phrase by the variable. This is called **substitution.**

$$2n + 12 = 6$$

Step 5 Solve the equation and check the solution in the original problem to see whether it makes sense.

$$2n + 12 = 6$$
$$2n = -6 \quad \text{Subtract 12 from both sides.}$$
$$n = -3 \quad \text{Divide both sides by 2.}$$

Check: Add 12 to twice -3 and the result is 6.

Step 6 State the solution to the word problem in words.

The number is −3. ■

Notice that we illustrated a *procedure* with this example. A summary of this procedure is now given, before we consider more examples.

Procedure for Problem Solving

Spend some time with this procedure.

1. **Read the problem.** Note what it is all about. Focus on processes rather than numbers. You can't work a problem you don't understand.
2. **Restate the problem.** Write a verbal description of the problem using operations signs and an equal sign. Look for equality. If you can't find equal quantities, you will never formulate an equation.
3. **Choose a variable.** If there is a single unknown, choose a variable.
4. **Substitute.** Replace the verbal phrases by known numbers and by the variable.

(continued)

5. **Solve the equation.** This is the easy step. Be sure your answer makes sense by checking it with the original question in the problem. Use estimation to eliminate unreasonable answers.
6. **State the answer.** There were no variables defined when you started, so $x = 3$ is not an answer. Pay attention to units of measure and other details of the problem. Remember to answer the question that was asked.

The second type of problem involves consecutive numbers. If n = A NUMBER, then

THE NEXT CONSECUTIVE NUMBER $= n + 1$

THE THIRD CONSECUTIVE NUMBER $= n + 2$

Also, if E = AN EVEN NUMBER, then

$E + 2$ = THE NEXT CONSECUTIVE EVEN NUMBER

or if F = AN ODD NUMBER, then

$F + 2$ = THE NEXT CONSECUTIVE ODD NUMBER

EXAMPLE 4

Consecutive Integer Problem

Find two consecutive even integers whose sum is 34.

Solution

Step 1 Read the problem.

Step 2 Write a verbal description of the problem, using operations signs and an equal sign. **Do not attempt to skip this step.**

EVEN INTEGER + NEXT EVEN INTEGER = SUM

Step 3 Choose a variable.

Let x = EVEN INTEGER

$x + 2$ = NEXT EVEN INTEGER

Step 4 Substitute the variable into the verbal equation; also substitute known or given numbers (such as SUM is 34).

$$x + (x + 2) = 34$$

Step 5 Solve the equation.

$$2x + 2 = 34 \quad \text{Simplify (combine terms).}$$
$$2x = 32 \quad \text{Subtract 2 from both sides.}$$
$$x = 16 \quad \text{Divide both sides by 2.}$$

Thus, $x + 2 = 18$.

Step 6 State the solution to the word problem in words.

The even integers are 16 and 18. ■

Examples 3 and 4 are good problems to help you learn the *procedure* for working with word problems, but the ultimate goal of problem solving is to work applied problems. One of the most common applications that we can consider at this point deals with the gasoline mileage of your car. The gasoline consumption of a car is measured in *miles per gallon* (MPG) and is related to the distance traveled by the following formula:

$$\begin{pmatrix}\text{MILES PER}\\ \text{GALLON}\end{pmatrix} \times \begin{pmatrix}\text{NUMBERS OF}\\ \text{GALLONS}\end{pmatrix} = \begin{pmatrix}\text{DISTANCE}\\ \text{TRAVELED}\end{pmatrix}$$

Examples 5 and 6 illustrate how you might use this formula.

EXAMPLE 5

Applied Problem; mpg

An advertisement for an Oldsmobile had the information shown. What is the size of the tank?

336	441
Miles per tankful Estimated city driving range EPA estimated MPG (16) CITY	Miles per tankful Estimated highway driving range EPA estimated 21 HIGHWAY

Solution

Example 5 uses the city mileage. You might show that you obtain the same answer if you use the highway mileage.

Step 1 Do you understand the question?
Can you rephrase it in your own words?

Step 2
$$\begin{pmatrix}\text{MILES PER}\\ \text{GALLON}\end{pmatrix} \times \begin{pmatrix}\text{NUMBERS OF}\\ \text{GALLONS}\end{pmatrix} = \begin{pmatrix}\text{DISTANCE}\\ \text{TRAVELED}\end{pmatrix}$$

Step 3 Let g = NUMBER OF GALLONS.

Step 4 Substitute:

$$\begin{pmatrix}\text{MILES PER}\\ \text{GALLON}\end{pmatrix} \times \begin{pmatrix}\text{NUMBERS OF}\\ \text{GALLONS}\end{pmatrix} = \begin{pmatrix}\text{DISTANCE}\\ \text{TRAVELED}\end{pmatrix}$$
$$\updownarrow \qquad\qquad\qquad\qquad \updownarrow$$
$$16 \times \begin{pmatrix}\text{NUMBER OF}\\ \text{GALLONS}\end{pmatrix} = 336$$

Step 5
$$16g = 336$$
$$g = 21 \quad \text{Divide both sides by 16.}$$

Step 6 **The car has a 21-gallon tank.** ■

EXAMPLE 6

Applied Problem; mpg

Suppose you take a trip in your car. At the beginning of the trip, the odometer reads 13,689.4, and at the end of the trip it reads 15,274.9. You used a total of 46.2 gallons of gasoline for the trip. What gas mileage (to the nearest $\frac{1}{10}$ gallon) did you get on this trip?

Solution Divide both sides of the given MPG formula to solve for *miles per gallon:*

$$\begin{pmatrix}\text{MILES PER}\\ \text{GALLON}\end{pmatrix} \times \begin{pmatrix}\text{NUMBERS OF}\\ \text{GALLONS}\end{pmatrix} = \begin{pmatrix}\text{DISTANCE}\\ \text{TRAVELED}\end{pmatrix}$$

Divide both sides by NUMBER OF GALLONS:

$$\begin{pmatrix}\text{MILES PER}\\ \text{GALLON}\end{pmatrix} = \frac{\text{DISTANCE TRAVELED}}{\text{NUMBER OF GALLONS}}$$

$$= \frac{1{,}585.5}{46.2} \qquad \leftarrow \text{This is found by subtraction: } \begin{array}{r} 15{,}274.9 \\ -\ 13{,}689.4 \\ \hline 1{,}585.5 \end{array}$$

$$\approx 34.318\ldots$$

To the nearest $\frac{1}{10}$, this number is 34.3. Thus, **the car averaged 34.3 MPG for the trip.** ■

Another application that requires problem-solving ability involves determining the best price for an item in a grocery store. Consider Example 7.

EXAMPLE 7

Applied Problem; Price Comparison

Three prices for dog food are mentioned in the cartoon. What are the prices per can for the three types of dog food (assuming that all the cans are the same size)? If the boy buys 10 cans, what does he save by buying the least expensive as compared to the most expensive?

Quincy © 1970 King Features Syndicate. Reprinted with special permission of King Features Syndicate, Inc.

Solution

$$\begin{pmatrix}\text{NUMBER OF}\\ \text{CANS}\end{pmatrix}\begin{pmatrix}\text{PRICE PER}\\ \text{CAN}\end{pmatrix} = \begin{pmatrix}\text{TOTAL}\\ \text{COST}\end{pmatrix}$$

Apply this general formula for each of the dog foods. Let *a, b,* and *c* be the prices per can for the three brands, respectively.

First brand	*Second brand*	*Third brand*
$3\begin{pmatrix}\text{PRICE PER}\\ \text{CAN}\end{pmatrix} = 78$	$2\begin{pmatrix}\text{PRICE PER}\\ \text{CAN}\end{pmatrix} = 68$	$10\begin{pmatrix}\text{PRICE PER}\\ \text{CAN}\end{pmatrix} = 254$
$\downarrow$	$\downarrow$	$\downarrow$
$3a = 78$	$2b = 68$	$10c = 254$
$a = 26$	$b = 34$	$c = 25.4$

Notice that we worked in pennies (this is often easier than working in dollars). The third dog food is the cheapest.

$$\text{SAVINGS} = \begin{pmatrix}\text{TOTAL COST}\\ \text{OF MOST}\\ \text{EXPENSIVE}\end{pmatrix} - \begin{pmatrix}\text{TOTAL COST}\\ \text{OF LEAST}\\ \text{EXPENSIVE}\end{pmatrix}$$

$$= \begin{pmatrix}\text{NO. OF CANS}\\ \text{OF MOST}\\ \text{EXPENSIVE}\end{pmatrix}\begin{pmatrix}\text{COST PER}\\ \text{CAN OF MOST}\\ \text{EXPENSIVE}\end{pmatrix} - \begin{pmatrix}\text{NO. OF CANS}\\ \text{OF LEAST}\\ \text{EXPENSIVE}\end{pmatrix}\begin{pmatrix}\text{COST PER}\\ \text{CAN OF LEAST}\\ \text{EXPENSIVE}\end{pmatrix}$$

$$\updownarrow \qquad \updownarrow \qquad \updownarrow \qquad \updownarrow$$

$$\begin{aligned} &= (10)\quad(34) - (10)\quad(25.4)\\ &= 340 - 254\\ &= 86 \end{aligned}$$

The savings is $.86. ■

The final example will test your problem-solving ability in a new setting to see whether you can use the techniques in a situation that is not exactly like one of the given types of problems. You will begin to be able to accomplish problem solving only when you begin to use it in a wide variety of contexts.

EXAMPLE 8

Applied Problem

A dispatcher must see that 75,000 condensers are delivered immediately. He has two sizes of trucks; one will carry 15,000 condensers, and the other will carry 12,000 condensers. How many smaller trucks are necessary if the dispatcher is required by union contract to use exactly two of the larger trucks for this order?

Solution

$$\begin{pmatrix}\text{AMOUNT DELIVERED}\\ \text{BY LARGER TRUCKS}\end{pmatrix} + \begin{pmatrix}\text{AMOUNT DELIVERED}\\ \text{BY SMALLER TRUCKS}\end{pmatrix} = \begin{pmatrix}\text{TOTAL}\\ \text{AMOUNT}\\ \text{DELIVERED}\end{pmatrix}$$

$$\begin{pmatrix}\text{NO. OF}\\ \text{LG. TRUCKS}\end{pmatrix}\begin{pmatrix}\text{CAPACITY OF}\\ \text{LG. TRUCKS}\end{pmatrix} + \begin{pmatrix}\text{NO. OF SM.}\\ \text{TRUCKS}\end{pmatrix}\begin{pmatrix}\text{CAPACITY OF}\\ \text{SM. TRUCKS}\end{pmatrix} = 75{,}000$$

$$\updownarrow \qquad \updownarrow \qquad \updownarrow \qquad \updownarrow$$

$$2 \quad (15{,}000) + \begin{pmatrix}\text{NO. OF SM.}\\ \text{TRUCKS}\end{pmatrix} (12{,}000) = 75{,}000$$

Let n = NUMBER OF SMALLER TRUCKS.

$$\begin{aligned} 2(15{,}000) + n(12{,}000) &= 75{,}000\\ 30{,}000 + 12{,}000n &= 75{,}000\\ 12{,}000n &= 45{,}000\\ n &= 3.75 \end{aligned}$$

Notice that the solution of the equation is not necessarily the answer to the problem. You must interpret this solution to answer the question that was asked. **The dispatcher will need to send 4 smaller trucks.** ■

"Great scott! You got the order backwards! We wanted twenty 75,000-microfared condensers!"

Example 8 illustrates a key ingredient in problem solving. Look at your answers to make sure of the following:

CAUTION

1. **They are possible.** In Example 8, it is impossible to send 3.75 trucks. If a question asks how many people can get onto an elevator and you obtain the answer $25\frac{1}{4}$, you know that answer is not possible, and must be rounded.
2. **They answer the question that is asked.**
3. **They are reasonable;** that is, make some estimation in your own mind about what the answer should be before you begin. For example, in Example 5, we asked for the size of the gas tank on an Oldsmobile. If the answer to your arithmetic had turned out to be 210 gallons or 2 gallons, you would know that these answers are not reasonable and you would begin looking for some error in your calculations. Problems 2–11 ask you to practice your estimation skills.

PROBLEM SET 3.6

ESSENTIAL IDEAS

1. **In Your Own Words** Explain the procedure for problem solving.

LEVEL 1

Estimate answers for Problems 2–11 by choosing the most reasonable answer. You should not do any pencil-and-paper or calculator arithmetic for these problems.

2. Estimate the size of the gas tank on the Ford Probe.
   A. 5 gal B. 30 gal C. 15 gal
3. Estimate the gas mileage you might obtain with a compact car.
   A. 28 mpg B. 89 mpg C. 12 mpg
4. Estimate the distance you might be able to drive on a full tank of gas.
   A. 50 mi B. 250 mi C. 850 mi
5. Estimate the price of a 12-oz can of corn.
   A. $.25 B. $.89 C. $1.49
6. Estimate the price of a six-pack of cola.
   A. $.75 B. $1.79 C. $4.50
7. Estimate the price of dinner for two at an elegant restaurant.
   A. $12.00 B. $80.00 C. $350.00
8. Estimate the average price of a new home.
   A. $25,000 B. $165,000 C. $650,000
9. Estimate the average monthly rent for an apartment.
   A. $225 B. $550 C. $4,250
10. Estimate the starting wage for a new college graduate.
    A. $6.00/hr B. $20.00/hr C. $80.00/hr
11. If you add three consecutive numbers and the sum is 600, estimate the size of the first of the three numbers you are adding.
    A. 20 B. 50 C. 200

LEVEL 2—Drill

In Problems 12–23, solve for the variable that is capitalized. Assume that none of the variables is 0.

12. $a + B = c$
13. $c = d + E$
14. $x - Y = 5$
15. $2x - Y = 7$
16. $i = Prt$
17. $x = 5Yz$
18. $p = 2\ell + 2W$
19. $p = 2L + 2w$
20. $3x + 2Y + 4 = 0$
21. $5x + 3Y - 9 = 0$
22. $3x - 2Y + 4 = 0$
23. $5x - 3Y - 9 = 0$

LEVEL 2—Applications

Solve Problems 24–38. Because you are practicing a procedure, you must show all your work.

24. If you add 7 to twice a number, the result is 17. What is the number?

25. If you subtract 12 from twice a number, the result is 6. What is the number?

26. If you add 15 to twice a number, the result is 7. What is the number?

27. If you subtract 9 from three times a number, the result is 0. What is the number?

28. If you multiply a number by 5 and then subtract -10, the difference is -30. What is the number?

29. If you multiply a number by 4 and then subtract -4, the result is 0. What is the number?

30. The sum of a number and 24 is equal to four times the number. What is the number?

31. The sum of 6 and twice a number is equal to four times the number. What is the number?

32. If 12 is subtracted from twice a number, the difference is four times the number. What is the number?

33. If 6 is subtracted from three times a number, the difference is twice the number. What is the number?

34. Find two consecutive numbers whose sum is 117.

35. Find two consecutive even numbers whose sum is 94.

36. The Toyota Corolla has an estimated MPG of 22, with a cruising range of 286 miles. What is the size of the tank?

37. The Rolls-Royce Silver Shadow I has an estimated MPG of 9, with a cruising range of 234 miles. What is the size of the tank?

38. The Nissan Sentra sedan has an estimated MPG of 35, with a cruising range of 455 miles. What is the size of the tank?

Find the miles per gallon in Problems 39–46 to the nearest tenth of a gallon. First, estimate an answer; then calculate.

	ODOMETER		
	START	*FINISH*	*GALLONS USED*
39.	02316	02480	13.4
40.	47341	47576	17.3
41.	19715	19932	9.2
42.	13719	14067	11.1
43.	21812	22174	8.7
44.	16975	17152	18.6
45.	23485	24433	48.4
46.	08271.6	10373.0	53.2

47. Del Monte asparagus spears sell for \$1.39 for 8 oz, and Green Giant asparagus spears cost \$1.99 for 12 oz. Which is the better buy, and by how much, rounded to the nearest $\frac{1}{10}$ cent?

48. Libby's green beans cost \$0.79 for 15 ounces and Diet Delight green beans cost \$0.49 for 5 ounces. Which is the better buy, and by how much, rounded to the nearest $\frac{1}{10}$ cent?

49. Chicken of the Sea tuna sells for \$1.39 for a 7-oz can, whereas Star Kist diet pack costs \$0.99 for a 5-oz can. Which is the better buy, and by how much, rounded to the nearest $\frac{1}{10}$ cent?

50. Which of the three sizes in the following ad is the least expensive per oz?

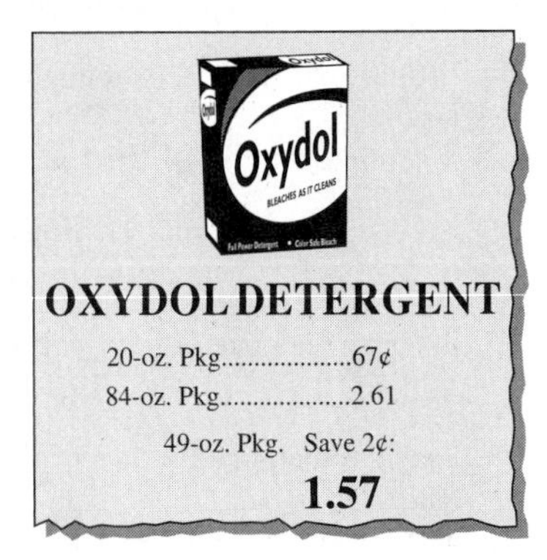

51. An eight-pack of AA Duracell batteries is \$8.09 and a four-pack is \$3.79. Which is the better buy, and by how much, rounded to the nearest cent?

52. Palmolive liquid detergent is $3.99 for a 64-oz bottle, and $1.59 for a 22-oz bottle. Which is the better buy, and by how much, rounded to the nearest cent?

53. A box of Total cereal is $5.09 for 1 lb 2 oz, and $3.49 for the 12-oz size. Which box is less expensive, and by how much, rounded to the nearest tenth of a cent?

54. Yuban coffee is $3.69 for a 12-oz size (reg) and $8.79 for a 24-oz size (decaf). What is the better buy, and by how much, rounded to the nearest cent?

55. A house and a lot together are appraised at $112,200. If the house is worth five times the value of the lot, how much is the lot worth?

56. A cabinet shop produces two types of custom-made cabinets for a customer. If one cabinet costs four times as much as the other, and if the total price for both cabinets is $2,075, how much does each cabinet cost?

Currency exchange rates fluctuate daily, but we will use Table 3.1 as the foreign currency exchange rate. To convert from a foreign currency into dollars, use the following equation:

$$\begin{pmatrix}\text{NUMBER OF}\\\text{DOLLARS}\end{pmatrix}\begin{pmatrix}\text{EXCHANGE}\\\text{RATE}\end{pmatrix}=\begin{pmatrix}\text{AMOUNT OF}\\\text{FOREIGN}\\\text{CURRENCY}\end{pmatrix}$$

Table 3.1
Foreign Currency Exchange Rates

Country	Currency	Rate
Australia	Dollar	1.3888
Britain	Pound	0.6309
Canada	Dollar	1.3730
France	Franc	5.0994
Germany	Mark	1.3884
Italy	Lira	1,569.86
Japan	Yen	84.30
Mexico	Peso	5.88
The Netherlands	Guilder	1.6303
Switzerland	Franc	1.1481

For example, if a ring costs 300 *marks in Germany, the equivalent amount in U.S. dollars is found as follows:*

From Table 3.1 (use local currency)

$$\begin{pmatrix}\text{NUMBER OF}\\\text{DOLLARS}\end{pmatrix}(1.3884) = 300$$

$$\begin{pmatrix}\text{NUMBER OF}\\\text{DOLLARS}\end{pmatrix} \approx 216.08$$

Divide both sides by 1.3884. (Rounded to the nearest cent)

Press: [300] [÷] [1.3884] [=]

Display: 216.0760588

The ring costs $216.08 *in U.S. dollars.*

Find the values of the items in Problems 57–65 *expressed in U.S. dollars, assuming the given prices are in local currency.*

57. A dinner in a Canadian restaurant is $15.00.

58. A dinner in a Japanese restaurant is 10,300 yen.

59. A sweater sells for $125 in Canada.

60. A hotel room in Mexico is 200 pesos.

61. A prize tulip bulb in Holland is selling for 8 guilder.

62. A German Black Forest clock is selling for 435 marks.

63. A set of bone china in England is 346 pounds.

64. A tray in an Italian store is marked 24,120 lira.

65. One admission to EuroDisney is 400 francs.

LEVEL 3—Mind Boggler

66. One day Perry White sent Lois Lane and Clark Kent out to the Coliseum to cover a big story. On the way, Clark, disguised as Superman, made a quick trip to Kmart to pick up a present for Lois. It took Lois two hours to reach the Coliseum from the *Daily Planet,* but it took Clark only 10 minutes, even though he traveled twice as far as Lois. If Lois drove at 50 mph, how fast did Clark (Superman) travel?

3.7 INEQUALITIES

Historical Perspective

Archimedes (287–212 B.C.)

The four most famous people in the history of mathematics include Karl Gauss (1772–1855), whose work has influenced almost every branch of mathematics (see Historical Perspective in Section 1.5; Isaac Newton (1642–1723), who entered the university when he was 18 and who invented the calculus when the university closed for a year because of the bubonic plague (see Historical Perspective in Section 5.2); Gottfried Leibniz (1646–1716), who is also credited with inventing the calculus and an early calculating machine, and at the age of 14 attempted to reform mathematics. The fourth person who is a part of this distinguished company is the greatest mathematician of antiquity, Archimedes (287–212 B.C.). He formulated many of our formulas for area and volume, calculated an approximation for π, found centers of gravity, invented engines and catapults, and formulated principles of levers. He was, in every sense of the word, the first universal mathematical genius.

The techniques of the previous sections can be applied to quantities that are not equal. If we are given any two numbers x and y, then

$$\text{either} \quad x = y \quad \text{or} \quad x \neq y$$

If $x \neq y$, then

$$\text{either} \quad x < y \quad \text{or} \quad x > y$$

Comparison Property*

For any two numbers x and y, exactly one of the following is true:

1. $x = y$; x is equal to (the same as) y

2. $x > y$; x is greater than (bigger than) y

3. $x < y$; x is less than (smaller than) y

This means that if two quantities are not exactly equal, we can relate them with a greater-than or a less-than symbol (called an **inequality symbol**). The solution of

$$x < 3$$

has more than one value, and it becomes very impractical to write "The answers are 2, 1, -110, 0, $2\frac{1}{2}$, 2.99, ..." Instead, we relate the answer to a number line, as shown in Figure 3.1. The fact that 3 is not included (3 is not less than 3) in the solution set is indicated by an open circle at the point 3.

*Sometimes this is called the trichotomy property.

Figure 3.1 Graph of $x < 3$

If we want to include the endpoint $x = 3$ with the inequality $x < 3$, we write $x \le 3$ and say "x is less than or equal to 3." We define two additional inequality symbols

$x \ge y$ means $x > y$ or $x = y$

$x \le y$ means $x < y$ or $x = y$

Our language translates inequalities in many ways, and the following list gives some of the more common translations:

SYMBOLS	*TRANSLATIONS*
$x > y$	x is greater than y
or	x is bigger than y
$y < x$	x is larger than y
These symbolic	x is more than y
statements have	y is smaller than x
the same meaning.	y is less than x
	x is to the right of y
	y is to the left of x
$x \ge y$	x is greater than or equal to y
or	x is bigger than or equal to y
$y \le x$	x is at least y
These symbolic	y is less than or equal to x
statements have	y is smaller than or equal to x
the same meaning.	y is no more than x
$x > 0$	x is greater than 0
	x is positive
	x is to the right of the origin
$x \ge 0$	x is greater than or equal to 0
	x is nonnegative
	x is not negative
$x < 0$	x is less than 0
	x is negative
	x is to the left of the origin
$x \le 0$	x is less than or equal to 0
	x is nonpositive
	x is not positive

EXAMPLE 1 **Graphing Inequalities**

Graph the solution set for each given inequality.

a. $x \leq 3$ **b.** $x > 5$ **c.** $-2 \leq x$

Solution

a. Notice that, in this example, the endpoint is included because with $x \leq 3$, it is possible that $x = 3$. This is shown as a solid dot on the number line:

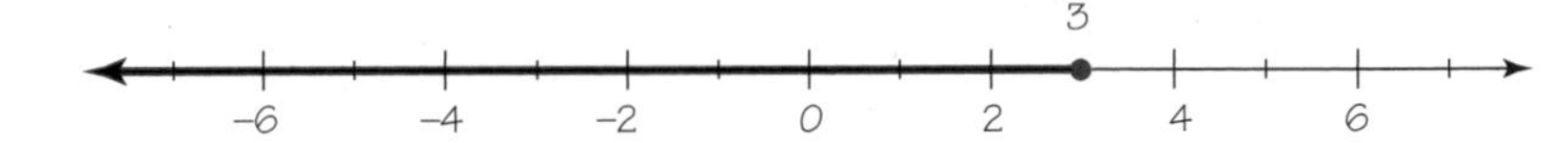

b. $x > 5$

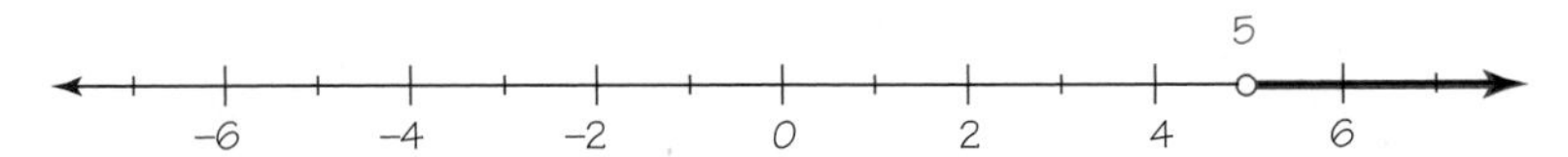

c. $-2 \leq x$ Although you can graph this directly, you will have less chance of making a mistake when working with inequalities if you rewrite them so that the variable is on the left; that is, reverse the inequality to read $x \geq -2$. Notice that the direction of the arrow has been changed. This is because the symbol requires that the arrow always point to the smaller number. When you change the direction of the arrow, we say that you have *changed the order of the inequality.* For example, if $-2 \leq x$, then $x \geq -2$, and we say the order has been changed. The graph of $x \geq -2$ is shown on the following number line:

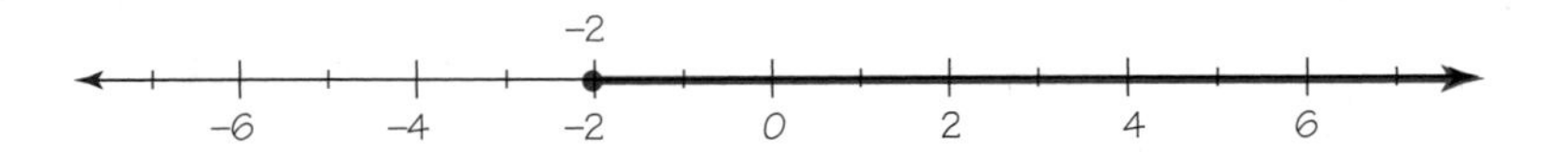

■

On a number line, if $x < y$, then x is to the left of y, Suppose the coordinates x and y are plotted as shown in Figure 3.2.

Figure 3.2 Number line showing two coordinates, x and y

If you add 2 to both x and y, you obtain $x + 2$ and $y + 2$. From Figure 3.3, you see that $x + 2 < y + 2$.

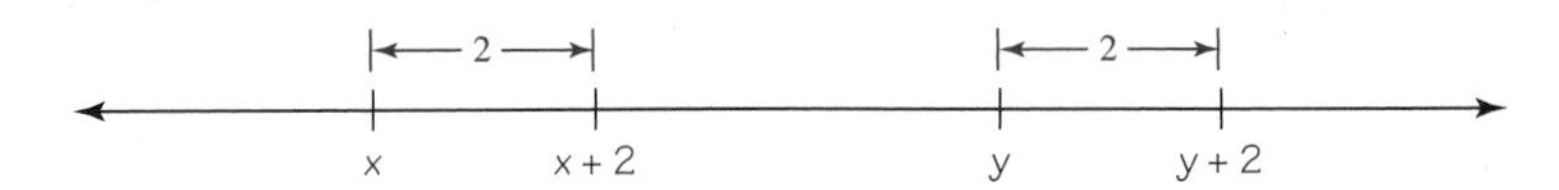

Figure 3.3 Number line with 2 added to both x and y

If you add some number c, there are two possibilities:

$c > 0$ ($c > 0$ is read "c is positive")

$c < 0$ ($c < 0$ is read "c is negative")

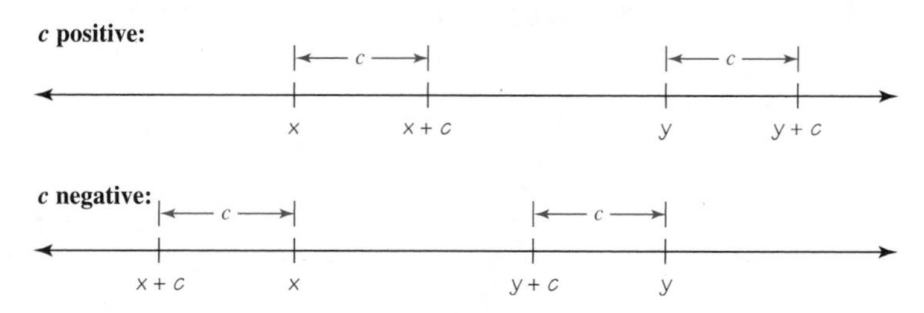

Figure 3.4 Adding positive and negative values to x and y

If $c > 0$, then $x + c$ is still to the left of $y + c$. If $c < 0$, then $x + c$ is still to the left of $y + c$, as shown in Figure 3.4. In both cases, $x < y$, which justifies the following property.

Addition Property of Inequality

If $x < y$, then

$$x + c < y + c$$

Also, if $x \leq y$, then $x + c \leq y + c$
if $x > y$, then $x + c > y + c$
if $x \geq y$, then $x + c \geq y + c$

Because this **addition property of inequality** is essentially the same as the addition property of equality, you might expect that there is also a multiplication property of inequality. We would hope that we could multiply both sides of an inequality by some number c without upsetting the inequality. Consider some examples. Let $x = 5$ and $y = 10$, so that $5 < 10$.

Let $c = 2$: $5 \cdot 2 < 10 \cdot 2$
$10 < 20$ True

Let $c = 0$: $5 \cdot 0 < 10 \cdot 0$
$0 < 0$ False

Let $c = -2$: $5(-2) < 10(-2)$
$-10 < -20$ False

You can see that you cannot multiply both sides of an inequality by a constant and be sure that the result is still true. However, if you restrict c to a positive value, then you can multiply both sides of an inequality by c. On the other hand, if c is a negative number, then the order of the inequality should be reversed. This is summarized by the **multiplication property of inequality.**

Multiplication Property of Inequality

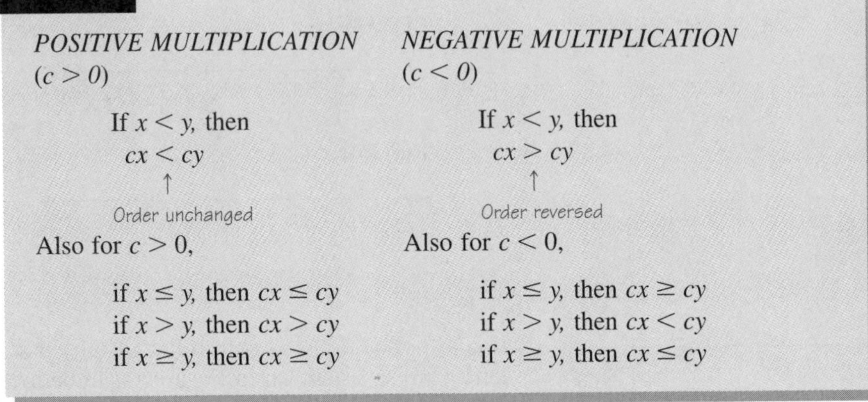

POSITIVE MULTIPLICATION ($c > 0$)	*NEGATIVE MULTIPLICATION* ($c < 0$)
If $x < y$, then $cx < cy$ ↑ Order unchanged	If $x < y$, then $cx > cy$ ↑ Order reversed
Also for $c > 0$,	Also for $c < 0$,
if $x \le y$, then $cx \le cy$	if $x \le y$, then $cx \ge cy$
if $x > y$, then $cx > cy$	if $x > y$, then $cx < cy$
if $x \ge y$, then $cx \ge cy$	if $x \ge y$, then $cx \le cy$

The same properties hold for positive and negative division. To **solve an inequality** is to find the replacement(s) for the variable that make the inequality true. We can summarize the procedure for solving an inequality with the following statement.

Solution of Inequalities

The procedure for solving inequalities is the same as the procedure for solving equations except that, if you multiply or divide by a negative number, you reverse the order of the inequality.

In summary, given $x < y$, $x \le y$, $x > y$, or $x \ge y$,

The **inequality symbols are the *same*** if we

1. Add the same number to both sides.
2. Subtract the same number from both sides.
3. Multiply both sides by a positive number.
4. Divide both sides by a positive number.

This works the same as equations.

The **inequality symbols are *reversed*** if we

1. Multiply both sides by a negative number.
2. Divide both sides by a negative number.
3. Interchange the x and the y.

This is where inequalities differ from equations.

EXAMPLE 2 **Solving Inequalities**

Solve:

a. $-x \geq 2$ **b.** $\frac{x}{-3} < 1$ **c.** $5x - 3 \geq 7$

Solution **a.** $-x \geq 2$

$\boldsymbol{x \leq -2}$ Multiply both sides by -1, and remember to reverse the order of the inequality.

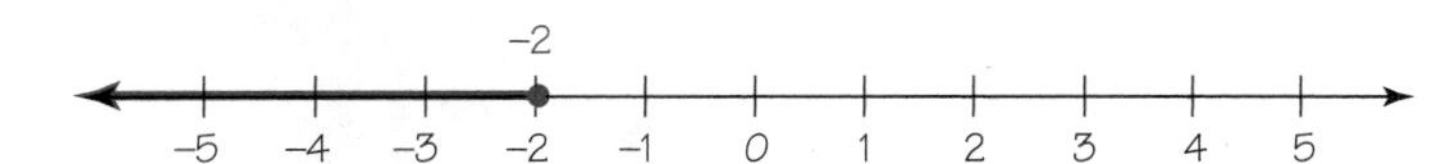

b. $\frac{x}{-3} < 1$

$\boldsymbol{x > -3}$ Multiply both sides by -3, and reverse the order of the inequality.

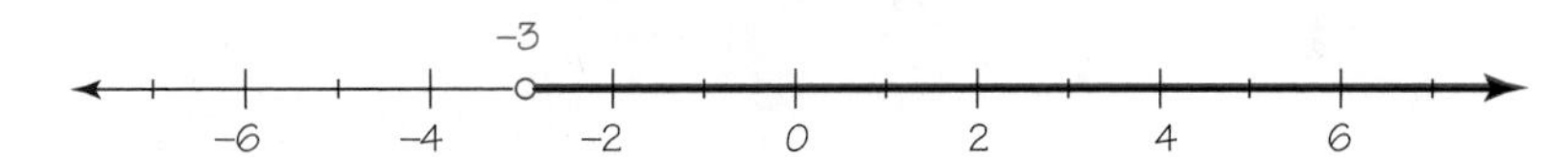

c. $5x - 3 \geq 7$

$5x - 3 + \mathbf{3} \geq 7 + \mathbf{3}$ Add 3 to both sides.

$5x \geq 10$ Simplify.

$\frac{5x}{\mathbf{5}} \geq \frac{10}{\mathbf{5}}$ Divide both sides by 5.

$\boldsymbol{x \geq 2}$ Simplify.

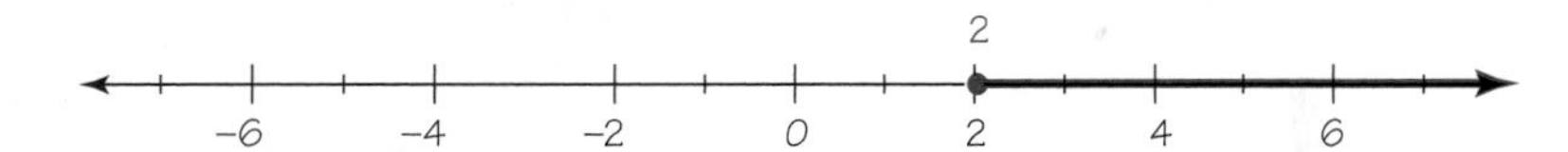

■

PROBLEM SET 3.7

ESSENTIAL IDEAS

Translate each symbolic statement in Problems 1–4 *into words.*

1. a. $19 > 4$ **b.** $x < 8$ **2. a.** $y \leq 20$ **b.** $p \geq 2$

3. a. $-5 \leq -2$ **b.** $-x > 2$ **c.** $k \geq -2$ **d.** $m < -4$

4. a. $x > 0$ **b.** $y \leq 0$ **c.** $w < 0$ **d.** $t \geq 0$

Translate each verbal statement in Problems 5–8 *into symbols.*

5. a. P is positive. **b.** N is negative.

6. a. Q is not zero. **b.** R is not negative.

7. a. M is no more than 4. **b.** Q is no less than 4.

8. a. x is a nonnegative. **b.** y is not positive.

LEVEL 1—Drill

Graph the solution sets in Problems 9–32.

9. $x < 5$ **10.** $x \geq 6$

11. $x \geq -3$ **12.** $x \leq -2$

13. $4 \geq x$

14. $-1 < x$

15. $-5 \leq x$

16. $1 > x$

17. $-3 > x$

18. $\frac{x}{2} > 3$

19. $4 < \frac{x}{2}$

20. $-2 > \frac{x}{-4}$

21. $x < 50$

22. $x \geq 100$

23. $x \geq -125$

24. $x \leq -75$

25. $45 \geq x$

26. $-40 < x$

27. $-50 \leq x$

28. $100 > x$

29. $-30 > x$

30. $\frac{x}{2} > 50$

31. $40 < \frac{x}{2}$

32. $-20 > \frac{x}{4}$

LEVEL 1

WHAT IS WRONG, *if anything, with each of the statements in Problems* 33–38*? Explain your reasoning.*

33. If $x \geq 5$, then the endpoint of the ray is an open dot.

34. If $x < -5$, then the endpoint of the ray is an open dot.

35. If $-x > 5$, then $x > -5$.

36. If $-x \leq 5$, then $x \leq -5$.

37.
$$\begin{aligned} 2 - 3x &\leq 8 \\ -3x &\leq 6 \\ x &\leq -2 \end{aligned}$$

38.
$$\begin{aligned} x + 3 &> 2x - 5 \\ 3 &> x - 5 \\ 8 &> x \\ x &> 8 \end{aligned}$$

LEVEL 2—Drill

Solve the inequalities in Problems 39–56.

39. $x + 7 \geq 3$

40. $x - 2 \leq 5$

41. $x - 2 \geq -4$

42. $10 < 5 + y$

43. $-4 < 2 + y$

44. $-3 < 5 + y$

45. $2 > -s$

46. $-t \leq -3$

47. $-m > -5$

48. $5 \leq 4 - y$

49. $3 > 2 - x$

50. $5 \geq 1 - w$

51. $2x + 6 \leq 8$

52. $3y - 6 \geq 9$

53. $3 > s + 9$

54. $2 < 2s + 8$

55. $4 \leq a + 2$

56. $3 > 2b - 13$

LEVEL 2—Applications

57. If the opposite of a number must be less than 5, what are the possible numbers satisfying this condition?

58. If the opposite of a number must be greater than twice the number, what are the possible numbers?

59. Suppose that three times a number is added to 12 and the result is negative. What are the possible numbers?

60. Suppose that seven times a number is added to 35 and the result is positive. What are the possible numbers?

61. Suppose that twice a number is subtracted from 8 and the result is positive. What are the possible numbers?

62. Suppose that five times a number is subtracted from 15 and the result is negative. What are the possible numbers?

63. If you solve an inequality and obtain $2 < 5$, what interpretation can you give to the solution?

64. If you solve an inequality and obtain $2 > 5$, what interpretation can you give to the solution?

Level 3—Mind Bogglers

65. If a number is four more than its opposite, what are the possible numbers?

66. If a number is six less than twice its opposite, what are the possible numbers?

67. If a number is less than four more than its opposite, what are the possible numbers?

68. If a number is less than six minus twice its opposite, what are the possible numbers?

69. Current postal regulations state that no package may be sent if its combined length, width, and height exceed 72 in. What are the possible dimensions of a box to be mailed with equal height and width if the length is four times the height?

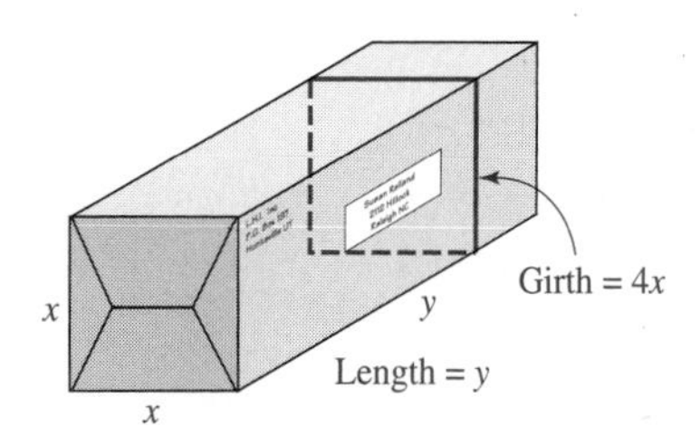

3.8 SUMMARY AND REVIEW

Important Terms

Addition law of exponents [3.1]
Addition property of equations [3.4]
Addition property of inequality [3.7]
Binomial [3.1]
Coefficient [3.1]
Comparison property [3.7]
Conditional equation [3.4]
Degree [3.1]
Distributive property [3.2]
Division property of equations [3.4]
Equation [3.4]
Formula [3.6]
Inequality symbols [3.7]
Like terms [3.2]
Literal equation [3.6]
Monomial [3.1]
Multiplication property of equations [3.4]
Multiplication property of inequality [3.7]
Numerical coefficient [3.1]
Opposite [3.4]
Polynomial [3.1]
Problem-solving procedure [3.6]
Profit formula [3.6]
Reciprocal [3.4]
Root of an equation [3.4]
Satisfy an equation [3.4]
Similar terms [3.2]
Simplify [3.2; 3.3]
Solution [3.4; 3.7]
Solve [3.4; 3.5]
Substitution [3.6]
Subtraction property of equations [3.4]
Term [3.1]
Trinomial [3.1]

Chapter Objectives

The material in this chapter is reviewed in the following list of objectives. A self-test (with answers and suggestions for additional study) is given. This self-test is constructed so that each problem number corresponds to a related objective. This self-test is followed by a practice test with the questions in mixed order.

[3.1] *Objective 3.1* Give the degree and numerical coefficient of a term.

[3.1] *Objective 3.2* Identify a polynomial as a monomial, binomial, or trinomial, and give the degree of the polynomial.

[3.1]	*Objective 3.3*	Use the definition of exponent and the addition law of exponents to simplify an expression.
[3.1; 3.7]	*Objective 3.4*	Write out a word phrase in symbols and simplify.
[3.2; 3.3; 3.5]	*Objective 3.5*	Simplify an expression by combining similar terms (both addition and subtraction).
[3.3]	*Objective 3.6*	Simplify expressions involving both multiplication and addition (or subtraction).
[3.3]	*Objective 3.7*	Multiply binomials.
[3.4]	*Objective 3.8*	Solve equations using the addition property.
[3.4]	*Objective 3.9*	Solve equations using the subtraction property.
[3.4]	*Objective 3.10*	Solve equations using the multiplication property.
[3.4]	*Objective 3.11*	Solve equations using the division property.
[3.5]	*Objective 3.12*	Solve equations with mixed operations.
[3.6]	*Objective 3.13*	Estimate answers for applied problems.
[3.6]	*Objective 3.14*	Solve literal equations.
[3.6; 3.7]	*Objective 3.15*	Solve applied problems with algebra.
[3.4]	*Objective 3.16*	Solve (and graph) linear inequalities.

Self-Test

Each question of this self-test is related to the corresponding objective listed above.

1. What are the degree and the numerical coefficient of $-2x^4$?

2. Identify each polynomial as a monomial, binomial, or trinomial, and give the degree.
a. $2x - 3y$ **b.** $x^2 + y^3$ **c.** -5^2 **d.** $x^2 - x^2y^2 + y^2$

3. Simplify $(5x)^2(2y)^3$.

4. a. Write out a difference of the squares of a and b.
b. Write out the square of the difference of a from b.

5. Simplify $5x - 3y + 5 - 8x + 3y - 12$.

6. Simplify $3(2x + 5) - (x + 3)$.

7. Simplify $(2x - y)(x + 3y)$.

8. Solve $-3 = x - 5$.

9. Solve $x + 8 = -2$.

10. Solve $\frac{x}{-5} = 13$.

11. Solve $8x = -32$.

12. Solve: **a.** $5x - 3 = 12$ **b.** $\frac{x}{2} + 8 = -2$

13. If you drive 243 miles and use 8.2 gallons of gasoline, estimate the miles per gallon.

14. Solve: **a.** $A = \ell w$ for w; $(\ell \neq 0)$ **b.** $6x - 2y - 4 = 0$ for y

15. Suppose that the exchange rate in Great Britian is 0.6154 pound per dollar. What is the cost (in dollars) for an item marked 250 pounds in Great Britain?

16. Solve $-x > 5$.

STUDY HINTS *Compare your solutions and answers to the self-test. For each problem you missed, work some additional problems in the section listed in the margin. After you have worked these problems, you can test yourself with the practice test.*

Complete Solutions to the Self-Test

Additional Problems	
[3.1] *Problems 5–8*	**1.** The degree is 4 and the numerical coefficient is −2.
[3.1] *Problems 9–14; 46–55*	**2. a.** first-degree binomial **b.** third-degree binomial **c.** zero-degree monomial **d.** fourth-degree trinomial
[3.1] *Problems 23–45*	**3.** $(5x)^2(2y)^3 = 5^2x^2(2^3y^3)$ $= 200\,x^2y^3$ $5^2 \cdot 2^3 = 25 \cdot 8 = 200$
[3.1] *Problems 56–63* [3.7] *Problems 5–8*	**4. a.** $a^2 - b^2$ or $b^2 - a^2$ **b.** $(b - a)^2$
[3.2] *Problems 11–64* [3.3] *Problems 62–69* [3.5] *Problems 9–16*	**5.** $5x - 3y + 5 - 8x + 3y - 12 = (5x - 8x) + (-3y + 3y) + (5 - 12)$ This step should be done mentally. $= -3x - 7$
[3.3] *Problems 11–42*	**6.** $3(2x + 5) - (x + 3) = 6x + 15 - x - 3$ Don't forget to distribute the number −1. $= 5x + 12$
[3.3] *Problems 43–61*	**7.** $(2x - y)(x + 3y) = (2x - y)x + (2x - y)3y$ $= 2x^2 - xy + 6xy - 3y^2$ $= 2x^2 + 5xy - 3y^2$
[3.4] *Problems 17–28*	**8.** $-3 = x - 5$ $2 = x$ Add 5 to both sides.
[3.4] *Problems 29–40*	**9.** $x + 8 = -2$ $x = -10$ Subtract 8 from both sides.
[3.4] *Problems 41–53*	**10.** $\frac{x}{-5} = 13$ $x = -65$ Multiply both sides by −5.
[3.4] *Problems 54–64*	**11.** $8x = -32$ $x = -4$ Divide both sides by 8.
[3.5] *Problems 17–66*	**12. a.** $5x - 3 = 12$; $5x = 15$; $x = 3$ **b.** $\frac{x}{2} + 8 = -2$; $\frac{x}{2} = -10$; $x = -20$
[3.6] *Problems 2–11; 39–46*	**13.** $\frac{243}{8.2} \approx \frac{240}{8} \approx 30$; estimate 30 mpg
[3.6] *Problems 12–23*	**14. a.** $A = \ell w$; $\frac{A}{\ell} = w$ **b.** $6x - 2y - 4 = 0$; $6x - 4 = 2y$; $3x - 2 = y$

[3.6] *Problems 24–65*
[3.7] *Problems 57–64*

15. (DOLLARS)(EXCHANGE RATE) = (AMOUNT OF FOREIGN CURRENCY)
(NUMBER OF DOLLARS)(0.6154) = 250
NUMBER OF DOLLARS = 406.239844 Divide both sides by 0.6154.
The cost (in dollars) is \$406.24.

[3.7] *Problems 9–32; 39–56*

16. $-x > 5$
$x < -5$ Multiply both sides by −1; don't forget to reverse the order of the inequality.

Chapter 3 Practice Test

1. Give the degree and the numerical coefficient.
a. $5x^3$ **b.** $2x$ **c.** x^4 **d.** $-3x^5$

2. Identify each polynomial as a monomial, binomial, or trinomial, and give the degree.
a. $6x^2 + xy - 5y^2$ **b.** $xy - 1$
c. 6^2 **d.** $x^2 - y^2$

Simplify the expressions in Problems 3–7.

3. a. $x^2x^7y^3y^8$ **b.** $(3x)^2(4y)$
c. $3x^2(4y)^2$ **d.** $(-x^2)(-xy)$

4. a. $6x + 5 - 5$ **b.** $3x - 2 - 3x$
c. $-6x - (-4)x$ **d.** $3y - 2x + 5 - 3y + 4x$

5. a. $2(x + 3) + 4(x - 5)$ **b.** $3(2x - 1) + 2(1 - 5x)$
c. $(5x - 3y) - 2(x + 3y)$ **d.** $(3x + 2) - (5x + 3)$

6. a. $(x - 5)(x + 4)$ **b.** $(2x + 1)(x - 3)$

7. a. $(3x + 4)(3x - 4)$ **b.** $(ab - c)(2ab + 3c)$

Solve the equations or inequalities in Problems 8–16.

8. a. $a - 10 = 40$ **b.** $b - 6 = -31$
c. $13 = c - 20$ **d.** $-5 = d + 14$

9. a. $A + 5 = 2$ **b.** $B + 23 = -8$
c. $9 + C = 30$ **d.** $-5 + D = -15$

10. a. $\frac{E}{6} = 5$ **b.** $\frac{F}{3} = 12$
c. $\frac{G}{-2} = 60$ **d.** $-12 = \frac{H}{-5}$

11. a. $2x = 46$ **b.** $-x = 5$

12. a. $-4x = 20$ **b.** $-25 = -5x$

13. a. $3x + 2 = 8$ **b.** $1 - 5y = 101$

14. a. $\frac{x}{2} - 5 = 4$ **b.** $\frac{x - 5}{2} = 4$

15. **a.** $x + 2 > -1$ **b.** $2 - x < -1$

16. **a.** $-x \leq 25$ **b.** $6 \geq 3 - x$

17. Solve $I = Prt$ for t.

18. Solve $3x + 2y + 8 = 0$ for y.

19. Solve $2x - y = 0$ for y.

20. Solve $2x - y = 0$ for x.

21. If you add three consecutive even numbers and the sum is 78, estimate the first of the three numbers.

22. If you subtract eight from twice a number and the result is equal to negative twelve, what is the number?

23. The Mercedes-Benz 560SL has an estimated MPG of 12 and a fuel tank capacity of 23.8 gallons. What is the estimated cruising range?

24. If a 32-oz bottle of ketchup costs $0.99 and a 44-oz bottle costs $1.49, which size is less expensive (per ounce), and by how much per ounce (rounded to the nearest tenth of a cent)?

25. Suppose that the odometer read 46312 at the start of your vacation and 48132 at the end. If your trip required 52 gallons of gas, what was the car's MPG for this trip?

CHAPTER FOUR

PERCENTS AND PROBLEM SOLVING

4.1 RATIO AND PROPORTION

Historical Perspective

Euclid (ca. 300 B.C.)

The first study of ratios and proportions dates back to Pythagoras. At first, proportions were regarded as geometrical magnitudes. As we will see in this section, a ratio is nothing more than a fraction, and whereas today we would write a/b or $\frac{a}{b}$, historically this was written as $a:b$. The Greeks made use of the idea that four quantities are in proportion, $a:b = c:d$, if the two ratios $a:b$ and $c:d$ are the same. One of the most important sources from this period is a 13-volume masterpiece of mathematical thinking attributed to Euclid, a professor of mathematics at the Museum of Alexandria. Even though we know very little about Euclid, we do have his work, *Elements,* and in Book V, he states the property of proportions we study in this section.

One of the most common uses of mathematics in our everyday lives involves percents. If you want to develop your problem-solving ability and apply it to everyday problems, you will first need to understand the ideas of *ratio, proportion,* and *percents.* We will discuss ratios and proportions in this section, solve proportions in the next section, and then spend the remainder of this chapter working with percents.

A **ratio** expresses a size relationship between two sets and is defined as a quotient of two numbers. It is written using the word *to,* a colon, or a fraction; that is, if the ratio of men to women is 5 **to** 4, this could also be written as $5:4$ or $\frac{5}{4}$. We will emphasize the idea that a ratio can be written as a fraction (or as a quotient of two numbers). Since a fraction can be reduced, a ratio also can be reduced.

EXAMPLE 1 **Writing a Ratio in Lowest Terms**

Reduce the given ratios to lowest terms.

a. 4 to 52 **b.** 15 to 3 **c.** $1\frac{1}{2}$ to 2 **d.** $1\frac{2}{3}$ to $3\frac{3}{4}$

Solution

a. A ratio of 4 to 52

$$\frac{4}{52} = \frac{1}{13}$$

A ratio of 1 to 13

b. A ratio of 15 to 3

$$\frac{15}{3} = 5$$

Write this as $\frac{5}{1}$ because a ratio compares two numbers.
A ratio of 5 to 1

c. A ratio of $1\frac{1}{2}$ to 2

$$\frac{1\frac{1}{2}}{2} = 1\frac{1}{2} \div 2 = \frac{3}{2} \times \frac{1}{2} = \frac{3}{4}$$

A ratio of 3 to 4

d. A ratio of $1\frac{2}{3}$ to $3\frac{3}{4}$

$$\frac{1\frac{2}{3}}{3\frac{3}{4}} = 1\frac{2}{3} \div 3\frac{3}{4} = \frac{5}{3} \div \frac{15}{4} = \frac{5}{3} \times \frac{4}{15} = \frac{4}{9}$$

A ratio of 4 to 9 ■

A **proportion** is a statement of equality between ratios. In symbols,

$$\frac{a}{b} = \frac{c}{d}$$

↑ "a is to b" ↑ "as" ↑ "c is to d"

The notation used in some books is $a:b::c:d$. Even though we won't use this notation, we will use words associated with this notation to name the terms:

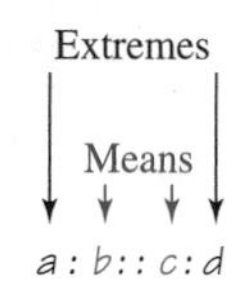

In the more common fractional notation, we have

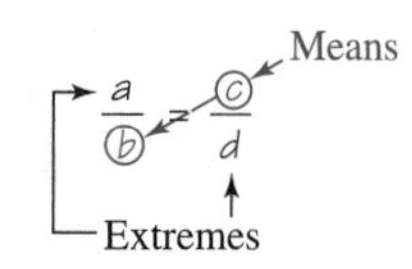

EXAMPLE 2 **Notation of Proportions**

Read each proportion, and name the means and the extremes.

a. $\frac{2}{3} = \frac{10}{15}$ **b.** $\frac{m}{5} = \frac{3}{8}$

Solution **a.** $\frac{2}{3} = \frac{10}{15}$ *Read:* Two is to three as ten is to fifteen.

Means: 3 and 10 *Extremes:* 2 and 15

b. $\frac{m}{5} = \frac{3}{8}$ *Read:* m is to five as three is to eight.

Means: 5 and 3 *Extremes:* m and 8 ■

The following **property of proportions** is fundamental to our study of proportions and percents.

Property of Proportions

If the product of the means equals the product of the extremes, the ratios form a proportion.

ALSO,

If the ratios form a proportion, then the product of the means equals the product of the extremes.

You will use this frequently in this chapter.

The property of proportions can be stated in symbols:

$$\frac{a}{b} = \frac{c}{d}$$

$$\underbrace{b \times c}_{\uparrow} = \underbrace{a \times d}_{\uparrow}$$

Product of means = *Product of extremes*

EXAMPLE 3 **Finding Ratios That Form a Proportion**

Tell whether each pair of ratios forms a proportion.

a. $\frac{3}{4}, \frac{36}{48}$ **b.** $\frac{5}{16}, \frac{7}{22}$

Solution

a. *Means:* 4×36 *Extremes:* 3×48

$144 = 144$

Thus,

$$\frac{3}{4} = \frac{36}{48}$$

They form a proportion.

b. *Means:* 16×7 *Extremes:* 5×22

$112 \neq 110$

Thus,

$$\frac{5}{16} \neq \frac{7}{22}$$

They do not form a proportion. ■

PROBLEM SET 4.1

ESSENTIAL IDEAS

1. **In Your Own Words** What is a ratio? What is a proportion?

2. **In Your Own Words** What is the property of proportions?

LEVEL 1—Drill

Write the ratios given in Problems 3–24 as reduced fractions.

3. The ratio of "yes" to "no" answers is 3 to 2.
4. The vote was affirmative by a 5 to 4 margin.
5. The ratio of cars to people is 1 to 3.
6. The ratio of taxis to cars is 1 to 20.
7. The ratio of cats to dogs is 4 to 7.
8. The ratio of dogs to cats is 7 to 4.
9. The ratio of gallons to miles is 20 to 4.
10. The ratio of miles to gallons is 25 to 1.
11. The ratio of dollars to people is 92 to 4.
12. The ratio of dollars to donuts is 5 to 6.
13. The ratio of wins to losses is 12 to 4.
14. The ratio of losses to wins is 6 to 15.
15. The ratio of apples to oranges is 6 to 2.
16. The ratio of peanuts to cashews is 10 to 2.
17. The ratio of sand to gravel is 2 to 5.
18. The ratio of children to adults is $2\frac{1}{5}$ to 2.
19. The ratio of cars to people is 1 to $2\frac{1}{2}$.
20. The ratio of adults to children is 500 to 150.
21. Find the ratio of $2\frac{3}{4}$ to $6\frac{1}{4}$.
22. Find the ratio of $6\frac{1}{4}$ to $2\frac{3}{4}$.
23. Find the ratio of $5\frac{2}{3}$ to $2\frac{1}{3}$.
24. Find the ratio of $6\frac{1}{3}$ to $4\frac{1}{3}$.

In Problems 25–32, read each proportion and name the means and extremes.

25. $\frac{5}{8}=\frac{35}{56}$
26. $\frac{94}{47}=\frac{2}{1}$
27. $\frac{5}{3}=\frac{2}{x}$
28. $\frac{5}{x}=\frac{1}{2}$
29. $\frac{a}{2}=\frac{c}{3}$
30. $\frac{8}{b}=\frac{c}{2}$
31. $\frac{a}{b}=\frac{c}{d}$
32. $\frac{w}{x}=\frac{y}{z}$

LEVEL 1

WHAT IS WRONG, *if anything, with each of the statements in Problems 33–38? Explain your reasoning.*

33. A simplified form of a ratio of 6 to 4 is 1.5.
34. A simplified form of a ratio of 6 to 1 is 6.
35. The symbol $\frac{2}{3}$ is used to represent a division, a fraction, and a ratio.
36. The expression

$$\frac{6}{9}=\frac{2}{3}$$

is a proportion.

37. A ratio of $2\frac{1}{2}$ to $6\frac{1}{3}$ is not permitted because a ratio cannot contain fractions.
38. A ratio of 12 to 4 is the same as the ratio of 3 to 1.

Tell whether each pair of ratios in Problems 39–54 forms a proportion.

39. $\frac{7}{1}, \frac{21}{3}$
40. $\frac{6}{8}, \frac{9}{12}$
41. $\frac{3}{6}, \frac{5}{10}$
42. $\frac{7}{8}, \frac{6}{7}$
43. $\frac{9}{2}, \frac{10}{3}$
44. $\frac{6}{5}, \frac{42}{35}$
45. $\frac{85}{18}, \frac{42}{9}$
46. $\frac{403}{341}, \frac{13}{11}$
47. $\frac{20}{70}, \frac{4}{14}$
48. $\frac{3}{4}, \frac{75}{100}$
49. $\frac{2}{3}, \frac{67}{100}$
50. $\frac{5}{3}, \frac{7\frac{1}{3}}{4}$
51. $\frac{3}{2}, \frac{5}{3\frac{1}{2}}$
52. $\frac{5\frac{1}{5}}{7}, \frac{4}{5}$
53. $\frac{1}{3}, \frac{33\frac{1}{3}}{100}$
54. $\frac{2}{3}, \frac{66\frac{2}{3}}{100}$

LEVEL 2—Applications

55. A cement mixture calls for 60 pounds of cement for 3 gallons of water. What is the ratio of cement to water?

56. What is the ratio of water to cement in Problem 55?

57. About 106 baby boys are born for every 100 baby girls. Write this as a simplified ratio of males to females.

58. Use Problem 57 to write a simplified ratio of females to males.

59. A certain school has 40 freshmen, 30 sophomores, 15 juniors, and 10 seniors.
a. What is the ratio of sophomores to freshmen?
b. What is the ratio of seniors to freshmen?
c. What is the ratio of juniors to seniors?
d. What is the ratio of juniors to those in school?

60. If Jesse can do a job in four hours and Mia can do the same job in three hours, what is the ratio of the rate of work of Jesse to that of Mia?

61. If you drive 119 miles on $8\frac{1}{2}$ gallons of gas, what is the simplified ratio of miles to gallons?

62. If you drive 180 miles on $7\frac{1}{2}$ gallons of gas, what is the simplified ratio of miles to gallons?

63. If you drive 279 miles on $15\frac{1}{2}$ gallons of gas, what is the simplified ratio of miles to gallons?

64. If you drive 151.7 miles on 8.2 gallons of gas, what is the simplified ratio of miles to gallons?

65. What is the gear ratio of a bicycle that has 30 teeth on the rear sprocket and 52 teeth on the front sprocket?

A baseball player's batting average is the ratio of hits to times at bat. For example, how many hits per times at bat does a player with a batting average of 0.275 *have?*

$$0.275 = \frac{275}{1{,}000} = \frac{11}{40}$$

Mark McGwire hitting record-setting home run

This is a ratio of 11 *to* 40*, which means the player has an average of* 11 *hits for every* 40 *times at bat. Find the average number of hits per times at bat for the batting champions named in Problems* 66–71.

66. Jake Daubert (Brooklyn, 1913); 0.350

67. Ty Cobb (Detroit, 1912); 0.390

68. Ernie Lombardi (Boston, 1942); 0.330

69. Pete Runnels (Boston, 1960); 0.320

70. Julio Franco (Texas, 1991); 0.342

71. Wade Boggs (Boston, 1988); 0.366

4.2 PROBLEM SOLVING WITH PROPORTIONS

Historical Perspective

In the previous section, we mentioned Euclid's *Elements.* This work is not only the earliest known major Greek mathematical book, but it is also the most influential textbook of all time. It was composed around 300 B.C. and first printed in 1482. Except for the Bible, no other book has been through so many printings. Here is what the property of proportions of the last section looks like in Euclid's book:

> *Magnitudes are said to be in the same ratio, the first to the second and the third to the fourth, when if any equimultiples whatever be taken of the first and the third, and any equimultiples whatever of the second and fourth, the former equimultiples alike exceed, are alike equal to, or alike less than, the latter equimultiples taken in corresponding order.*
>
> *The Thirteen Books of Euclid's Elements,* ed. by T. L. Heath (Cambridge, 1908, 3 vols.), II, 114.

Euclid's *Elements*, first printed edition, 1482

In Section 4.1, we tested pairs of ratios to see whether they formed a proportion. In the usual setting for a proportion, however, three of the terms of the proportion are known and one of the terms is unknown. It is always possible to find the missing term.

EXAMPLE 1 **Finding a Missing Term of a Proportion**

Find the missing term of each proportion.

a. $\frac{3}{4} = \frac{w}{20}$ **b.** $\frac{3}{4} = \frac{27}{y}$ **c.** $\frac{2}{x} = \frac{8}{9}$ **d.** $\frac{t}{15} = \frac{3}{5}$

Solution **a.** $\frac{3}{4} = \frac{w}{20}$:

$$\text{PRODUCT OF MEANS} = \text{PRODUCT OF EXTREMES}$$

$$4w = 3(20)$$

$$w = \frac{3(\overset{5}{\cancel{20}})}{\underset{1}{\cancel{4}}}$$

Solve the equation by dividing both sides by 4. Notice that 4 is the number opposite the unknown: $\frac{3}{4} = \frac{w}{20}$

$$w = \mathbf{15}$$

b. $\dfrac{3}{4} = \dfrac{27}{y}$:

PRODUCT OF MEANS = PRODUCT OF EXTREMES

$$4(27) = 3y$$

$$\frac{4(\overset{9}{\cancel{27}})}{\underset{1}{\cancel{3}}} = y$$

Divide both sides by 3; notice that 3 is the number opposite the unknown: $\dfrac{3}{4} = \dfrac{27}{y}$

$$\mathbf{36} = y$$

c. $\dfrac{2}{x} = \dfrac{8}{9}$:

PRODUCT OF MEANS = PRODUCT OF EXTREMES

$$8x = 2(9)$$

$$x = \frac{\overset{1}{\cancel{2}}(9)}{\underset{4}{\cancel{8}}}$$

Divide both sides by 8; notice that 8 is the number opposite the unknown: $\dfrac{2}{x} = \dfrac{8}{9}$

$$x = \mathbf{\frac{9}{4}}$$

d. $\dfrac{t}{15} = \dfrac{3}{5}$:

PRODUCT OF MEANS = PRODUCT OF EXTREMES

$$3(15) = 5t$$

$$\frac{3(\overset{3}{\cancel{15}})}{\underset{1}{\cancel{5}}} = t$$

Divide both sides by 5; notice that 5 is the number opposite the unknown: $\dfrac{t}{15} = \dfrac{3}{5}$

$$\mathbf{9} = t$$

■

Notice that the unknown term can be in any one of four positions, as illustrated by the four parts of Example 1. But even though you can find the missing term of a proportion (called **solving the proportion**) by the technique used in Example 1, it is easier to think in terms of **the cross-product divided by the number opposite the unknown.** This method is easier than actually solving the equation because it can be done quickly using a calculator, as shown in the following examples.

Procedure for Solving Proportions

This procedure is essential when working with percents.

Step 1 Find the product of the means or the product of the extremes, whichever does not contain the unknown term.

Step 2 Divide this product by the number that is opposite the unknown term.

EXAMPLE 2 **Solving a Proportion**

Solve the proportion for the unknown term.

a. $\frac{5}{6} = \frac{55}{y}$ **b.** $\frac{5}{b} = \frac{3}{4}$ **c.** $\frac{2\frac{1}{2}}{5} = \frac{a}{8}$

Solution **a.** $\frac{5}{6} = \frac{55}{y}$

$$y = \frac{6 \times 55}{5} \quad \begin{array}{l} \leftarrow \text{Product of the means} \\ \leftarrow \text{Number opposite the unknown} \end{array}$$

$$= \frac{6 \times \overset{11}{\cancel{55}}}{\underset{1}{\cancel{5}}} \quad \text{You can cancel to simplify many of these problems.}$$

$$= \mathbf{66}$$

The procedure for solving a proportion is easy to carry out when using a calculator: [6] [×] [55] [÷] [5] [=]

b. $\frac{5}{b} = \frac{3}{4}$

$$b = \frac{5 \times 4}{3} \quad \begin{array}{l} \leftarrow \text{Product of the extremes} \\ \leftarrow \text{Number opposite the unknown} \end{array}$$

$$= \mathbf{\frac{20}{3}} \text{ or } \mathbf{6\frac{2}{3}}$$

c. $\frac{2\frac{1}{2}}{5} = \frac{a}{8}$

$$a = \frac{2\frac{1}{2} \times 8}{5} \quad \begin{array}{l} \leftarrow \text{Product of the extremes} \\ \leftarrow \text{Number opposite the unknown} \end{array}$$

$$= \frac{\frac{5}{2} \times \frac{8}{1}}{5}$$

$$= \frac{20}{5}$$

$$= \mathbf{4}$$

■

Many applied problems can be solved using a proportion. Whenever you are working an applied problem, you should estimate an answer so that you will know whether the result you obtain is reasonable.

When setting up a proportion with units, be sure that like units occupy corresponding positions, as illustrated in Examples 3–5.

EXAMPLE 3 **Problem Solving Using Proportions**

If 4 cans of cola sell for 89¢, how much will 12 cans cost?

Solution

$$\frac{4}{89} = \frac{12}{x}$$

(cans → 4, 12; cents → 89, x)

$$x = \frac{89 \times \overset{3}{\cancel{12}}}{\underset{1}{\cancel{4}}} = 267$$

Press: 89 × 12 ÷ 4 =

12 cans will cost $2.67. ■

EXAMPLE 4 **Problem Solving Using Proportions**

If a 120-mile trip took $8\frac{1}{2}$ gallons of gas, how much gas is needed for a 240-mile trip?

Solution

$$\frac{120}{8\frac{1}{2}} = \frac{240}{x}$$

(miles → 120, 240; gallons → $8\frac{1}{2}$, x)

$$x = \frac{8\frac{1}{2} \times \overset{2}{\cancel{240}}}{\underset{1}{\cancel{120}}} = 17$$

Press: 8.5 × 240 ÷ 120 =

The trip will require 17 gallons. ■

EXAMPLE 5 **Problem Solving Using Proportions**

If the property tax on a $65,000 home is $416, what is the tax on an $85,000 home?

Solution

$$\frac{65{,}000}{416} = \frac{85{,}000}{x}$$

(value → 65,000, 85,000; tax → 416, x)

$$x = \frac{416 \times 85{,}000}{65{,}000}$$

You can do the arithmetic by canceling or by using a calculator:

$$x = \frac{416 \times \overset{\overset{17}{\cancel{85}}}{\cancel{85{,}000}}}{\underset{\underset{13}{\cancel{65}}}{\cancel{65{,}000}}}$$

Press: 416 × 85000 ÷ 65000 =

$$= 544$$

The tax is $544. ■

PROBLEM SET 4.2

LEVEL 1

WHAT IS WRONG, *if anything, with each of the statements in Problems 1–2? Explain your reasoning.*

1. To solve a proportion means to reduce the ratios.

2. The variable of a proportion is equal to the product of the means or the product of the extremes divided by the number opposite the variable.

LEVEL 1—Drill

Solve each proportion in Problems 3–26 for the item represented by a letter.

3. $\frac{5}{1} = \frac{A}{6}$ 4. $\frac{1}{9} = \frac{4}{B}$ 5. $\frac{C}{2} = \frac{5}{1}$ 6. $\frac{7}{D} = \frac{1}{8}$

7. $\frac{12}{18} = \frac{E}{12}$ 8. $\frac{12}{15} = \frac{20}{F}$ 9. $\frac{G}{24} = \frac{14}{16}$ 10. $\frac{4}{H} = \frac{3}{15}$

11. $\frac{2}{3} = \frac{I}{24}$ 12. $\frac{4}{5} = \frac{3}{J}$ 13. $\frac{3}{K} = \frac{2}{5}$ 14. $\frac{L}{18} = \frac{5}{6}$

LEVEL 2—Drill

15. $\frac{7\frac{1}{5}}{9} = \frac{M}{5}$ 16. $\frac{4}{2\frac{2}{3}} = \frac{3}{N}$ 17. $\frac{P}{4} = \frac{4\frac{1}{2}}{6}$

18. $\frac{5}{2} = \frac{Q}{12\frac{3}{5}}$ 19. $\frac{5}{R} = \frac{7}{12\frac{3}{5}}$ 20. $\frac{1\frac{1}{3}}{\frac{1}{9}} = \frac{S}{2\frac{2}{3}}$

21. $\frac{33}{2\frac{1}{5}} = \frac{3\frac{3}{4}}{T}$ 22. $\frac{U}{1\frac{1}{2}} = \frac{\frac{1}{2}}{\frac{3}{4}}$ 23. $\frac{\frac{1}{5}}{\frac{2}{3}} = \frac{\frac{3}{4}}{V}$

24. $\frac{\frac{3}{5}}{\frac{1}{2}} = \frac{X}{\frac{2}{3}}$ 25. $\frac{9}{Y} = \frac{1\frac{1}{2}}{3\frac{2}{3}}$ 26. $\frac{Z}{2\frac{1}{3}} = \frac{1\frac{1}{2}}{4\frac{1}{5}}$

LEVEL 2—Applications

27. If 4 melons sell for \$2.80, how much would 7 melons cost?

28. If you can read a 120-page book in 4 hours, how long will it take to read 150 pages?

29. If a 184-mile trip took $11\frac{1}{2}$ gallons of gas, how much gas is needed for a 160-mile trip?

30. If a 121-mile trip took $5\frac{1}{2}$ gallons of gas, how many miles can be driven with a full tank of 13 gallons?

31. If Roger can type at a rate of 65 words per minute, at this rate, how long (to the nearest minute) will it take him to type a 550-word letter?

32. If Ginger can type a 15,120-word report in 4 hours, how many words per minute does Ginger type (assume a constant rate)?

33. If a family uses $3\frac{1}{2}$ gallons of milk per week, how much milk will this family need for four days?

34. If 2 gallons of paint are needed for 75 ft of fence, how many gallons are needed for 900 ft of fence?

35. If Jack jogs 3 miles in 40 minutes, how long will it take him (to the nearest minute) to jog 2 miles at the same rate?

36. If Jill jogs 2 miles in 15 minutes, how long will it take her (to the nearest minute) to jog 5 miles at the same rate?

37. A moderately active 140-pound person will use 2,100 calories per day to maintain that body weight. How many calories per day are necessary to maintain a moderately active 165-pound person?

38. You've probably seen advertisements for posters that can be made from any photograph. If the finished poster will be 2 ft by 3 ft, it's likely that part of your original snapshot will be cut off. Suppose that you send in a photo that measures 3 in. by 5 in. If the shorter side of the enlargement will be 2 ft, what size should the longer side of the enlargement be so that the entire snapshot is shown in the poster?

39. Suppose you wish to make a scale drawing of your living room, which measures 18 ft by 25 ft. If the shorter side of the drawing is 6 in., how long is the longer side of the scale drawing?

40. If the property tax on a \$180,000 home is \$1,080, what is the tax on a \$130,000 home?

Level 3—Mind Bogglers

41. This problem will help you check your work in Problems 3–26. Fill in the capital letters from Problems 3–26 to correspond with their numerical values in the boxes. For example, if

$$\frac{W}{7} = \frac{10}{14}$$

then

$$W = \frac{7 \times 10}{14} = 5$$

Now find the box or boxes with number 5 in the corner, and fill in the letter W. This has already been done for you. (The letter O has also been filled in for you.) Some letters may not appear in the boxes. When you are finished filling in the letters, darken all the blank spaces to separate the words in the secret message. Notice that one of the blank spaces has also been filled in to help you.

1/4	20	16	32	■	3	9	O	36	15	8	4
1	2	56	O	1	36	1/4	8	56	15	22	7
5 W	16	15	15	1/3	30	32	32	16	32	1/4	12/5
22	O	1	1/2	16	2	18	25	16	2	56	-
16	2	21	11	8	9	9	O	9	32	!	!

42. At a certain hamburger stand, the owner sold soft drinks out of two 16-gallon barrels. At the end of the first day, she wished to increase her profit, so she filled the soft-drink barrels with water, thus diluting the drink served. She repeated the procedure at the end of the second and third days. At the end of the fourth day, she had 10 gallons remaining in the barrels, but they contained only 1 pint of pure soft drink. How much pure soft drink was served in the four days?

4.3 PERCENT

Historical Perspective

Although the use of percent in computing taxes, interest, and the like probably dates back at least to the Roman Empire, the symbol for percent is not that ancient. The percent symbol is the product of a time-saving shortcut used by scribes and clerks in the 15th century. At that time, there was a tremendous increase in the volume of trade all over Europe, especially in the Italian city-states. An increase in trade meant an increase in record keeping. And to lighten their work load, Italian scribes began to abbreviate the often-used phrase "per cento." Some abbreviated it as "P 100," some as "p cento," and some as "p c^0." Our symbol probably comes from "p c^0," which became "P $-^0$," then later "$\frac{0}{0}$," and finally "%."

31st Yearbook of the National Council of Teachers of Mathematics (1969), pp. 146–147.

Percent is a commonly used word, which you can find daily in any newspaper.

Percent

Percent is the ratio of a given number to 100. This means that a percent is the numerator of a fraction whose denominator is 100.

The symbol % is used to indicate percent. Consider some examples from one issue of a local newspaper.

EXAMPLE 1

Meaning of Percent

Illustrate the meaning of percent in the following quotes.

a. "The President recommended an 8 percent cost-of-living raise in Social Security payments."

b. "$33\frac{1}{3}$% OFF"

Solution

a. "8 percent" means "the ratio of 8 to 100."

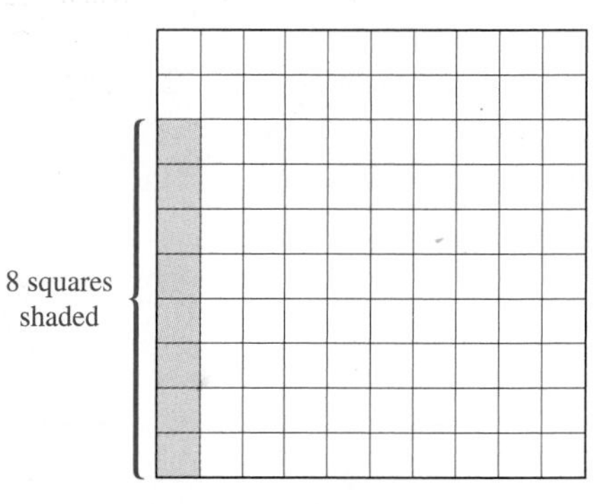

b. "$33\frac{1}{3}$%" means "the ratio of $33\frac{1}{3}$ to 100."

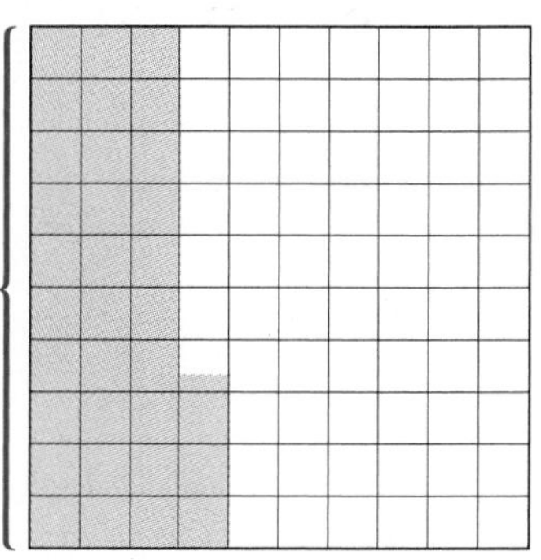

■

Since a percent is a ratio, percents can easily be written in fractional form.

EXAMPLE 2 **Writing Percents as Fractions**

Write the following percents as simplified fractions.

a. "Sale 75% OFF" **b.** "SALARIES UP 6.8%"

Solution **a.** 75% means a "ratio of 75 to 100": $\dfrac{75}{100} = \mathbf{\dfrac{3}{4}}$

b. 6.8% means a "ratio of 6.8 to 100":

$$\begin{aligned} 6.8 \div 100 &= 6\tfrac{8}{10} \div 100 \\ &= 6\tfrac{4}{5} \div 100 \\ &= \frac{\overset{17}{\cancel{34}}}{5} \times \frac{1}{\underset{50}{\cancel{100}}} \\ &= \mathbf{\frac{17}{250}} \end{aligned}$$

■

Percents can also be written as decimals. Since a percent is a ratio of a number to 100, we can divide by 100 by moving the decimal point, as we discussed in Chapter 1.

Procedure for Changing a Percent to a Decimal

To express a percent as a decimal, shift the decimal point two places to the *left* and delete the % symbol. If the percent involves a fraction, write the fraction as a decimal; *then* shift the decimal point.

EXAMPLE 3 **Writing Percents as Decimals**

Write each percent in decimal form.

a. 8 percent **b.** 6.8% **c.** $33\frac{1}{3}\%$ **d.** $\frac{1}{2}\%$

Solution **a.** 8%

If a decimal point is not shown, it is always understood to be at the right of the whole number.

0.08%

Shift the decimal point two places to the left; add zeros as placeholders, if necessary. Delete percent symbol.

Answer: 8% = **0.08**

b. 6.8%

Think: 6.8%

Shift two places, add placeholders as necessary, and delete percent symbol.

Answer: 6.8% = **0.068**

c. $33\frac{1}{3}\%$

Think: $33\frac{1}{3}\%$

↑ Decimal point is understood.

Answer: $33\frac{1}{3}\% = \mathbf{0.33\frac{1}{3}}$ or **0.333 . . .**

d. $\frac{1}{2}\%$

$\frac{1}{2} = 0.5$, so $\frac{1}{2}\% = 0.5\%$ *Think:* **00.5%**

Answer: $0.5\% = \mathbf{0.005}$ ■

As you can see from the examples, every number can be written in three forms: fraction, decimal, and percent. Even though we discussed changing from fraction to decimal form earlier in the text, we'll review the three forms in this section.

EXAMPLE 4

Changing Fractions to Percents

Write each fraction in percent form.

a. $\frac{5}{8}$ **b.** $\frac{5}{6}$

Solution

a. Look under the fraction heading in Table 4.1 (page 182), and follow the directions for changing a fraction to a percent.

Step 1 $\frac{5}{8} = 0.625$

$$8\overline{)5.000} = 0.625$$

48, 20, 16, 40, 40, 0

Step 2 $0.62\,5 = 62.5\%$

Two places; add zeros as placeholders as necessary; add percent symbol.

Answer: **62.5%**

b. Step 1 $\frac{5}{6} = 0.83\frac{1}{3}$

In Chapter 1, you would have written this as 0.833 . . . or $0.8\overline{3}$.

$$6\overline{)5.000} = 0.83$$

48, 20, 18, 2 ← Remainder

Carry out the division two places and save the remainder as a fraction: $\frac{2}{6} = \frac{1}{3}$

Step 2 $0.83\frac{1}{3} = 83\frac{1}{3}\%$

The decimal point is understood.

Answer: $\mathbf{83\frac{1}{3}\%}$ ■

Spend some time with this table. In fact, you may want to place a marker on this page for future reference.

Table 4.1 *Fraction/Decimal/Percent Conversion Chart*

To From	Fraction	Decimal	Percent
Fraction		Divide the numerator (top) by the denominator (bottom). Write as a terminating or as a repeating decimal (bar notation).	First change the fraction to a decimal by carrying out the division to two decimal places and writing the remainder as a fraction. *Then* move the decimal point two places to the right, and affix a percent symbol.
Terminating decimal	Write the decimal without the decimal point, and multiply by the decimal name of the last digit (rightmost digit).		Shift the decimal point two places to the *right,* and affix a percent symbol.
Percent	Write as the ratio to 100, and reduce the fraction. If the percent involves a decimal, first write the decimal in fractional form, and then multiply by $\frac{1}{100}$. If the percent involves a fraction, delete the percent symbol and multiply by $\frac{1}{100}$.	Shift the decimal point two places to the *left,* and delete the percent symbol. If the percent involves a fraction, first write the fraction as a decimal, and then shift the decimal point.	

EXAMPLE 5 **Changing from Decimal Form**

Write decimal forms as percents and fractions.

a. 0.85 **b.** 2.485

Solution **a. Step 1** 0.85 = **85%**

↑_↑ Two places ↑ Decimal understood

Step 2 85% means $\frac{85}{100} = \mathbf{\frac{17}{20}}$

b. Step 1 2.485 = **248.5%**

Step 2 248.5% means $248.5 \times \frac{1}{100} = 248\frac{1}{2} \times \frac{1}{100}$

$$= \frac{497}{2} \times \frac{1}{100}$$

$$= \mathbf{\frac{497}{200}} \text{ or } \mathbf{2\frac{97}{200}}$$

■

PROBLEM SET 4.3

Essential Ideas

1. **In Your Own Words** What is a percent?
2. **In Your Own Words** How do you convert a percent to a decimal?
3. **In Your Own Words** How do you convert a percent to a fraction?
4. **In Your Own Words** How do you convert a terminating decimal to a percent?
5. **In Your Own Words** How do you convert a terminating decimal to a fraction?
6. **In Your Own Words** How do you convert a decimal to a percent?
7. **In Your Own Words** How do you convert a decimal to a fraction?

Level 1

Write the percents in Problems 8–14 as simplified fractions.

8. SALE 50% OFF
9. In 1974, there were two persons out of work in Luxembourg, and by the end of the year, unemployment had spiraled to 11, making the unemployment figure jump by 550%.
10. There was a 13.4 percent increase in the use of electricity.
11. The weight of a certain model of automobile was decreased by 3.7 percent.
12. Ivory soap is $99\frac{44}{100}$% pure.
13. In 1995, the sales tax in Pennsylvania was 6%.
14. In 1998, the stock market dropped 15% during a one-month period.

Level 1—Drill

In Problems 15–62, change the given form into the two missing forms.

	FRACTION	DECIMAL	PERCENT
15.	$\frac{1}{2}$	____	____
16.	$\frac{1}{4}$	____	____
17.	____	0.75	____
18.	____	0.2	____
19.	____	____	40%
20.	____	____	80%
21.	____	____	100%
22.	____	____	200%
23.	____	0.85	____
24.	____	0.95	____
25.	____	____	60%
26.	$\frac{1}{8}$	____	____
27.	$\frac{3}{8}$	____	____
28.	____	____	65%
29.	____	____	45%
30.	____	____	120%
31.	____	____	250%
32.	____	0.7	____
33.	____	0.9	____
34.	$\frac{1}{3}$	____	____
35.	$\frac{1}{5}$	____	____

	FRACTION	*DECIMAL*	*PERCENT*
36.	$\frac{2}{5}$	____	____
37.	$\frac{2}{3}$	____	____
38.	____	0.05	____
39.	____	0.15	____
40.	____	0.35	____
41.	____	0.65	____
42.	____	0.025	____
43.	____	0.175	____
44.	____	____	4%
45.	____	____	8%
46.	____	____	$6\frac{1}{2}\%$
47.	____	____	$62\frac{1}{2}\%$
48.	____	0.375	____
49.	$\frac{3}{20}$	____	____
50.	$\frac{11}{20}$	____	____
51.	____	0.0375	____
52.	$\frac{1}{6}$	____	____
53.	$\frac{5}{6}$	____	____
54.	____	0.875	____
55.	____	0.0025	____
56.	$\frac{1}{12}$	____	____
57.	$\frac{5}{36}$	____	____
58.	____	____	$22\frac{2}{9}\%$
59.	____	____	$11\frac{1}{9}\%$
60.	$\frac{5}{9}$	____	____
61.	$\frac{7}{9}$	____	____
62.	$\frac{1}{7}$	____	____

LEVEL 2—Applications

Grades in a classroom are often given according to the percentage score obtained by the student. Suppose that a teacher grades according to the following scheme, where all scores are rounded to the nearest percent:

A	90%–100%
B	80%–89%
C	65%–79%
D	50–64%
F	0%–49%

To determine a percent grade, form a ratio of score received to possible score and then write this ratio as a percent. For example, if a student gets seven answers right out of a possible ten, this score is

$$\frac{7}{10} = 70\%$$

Calculate the percent (nearest percent) and determine the letter grade for each score given in Problems 63–66.

63. Possible 10 points

a. 7 **b.** 8 **c.** $7\frac{1}{2}$ **d.** $8\frac{1}{4}$ **e.** $9\frac{1}{2}$

64. Possible 100 points

a. 75 **b.** 92 **c.** 38 **d.** $66\frac{1}{2}$ **e.** $95\frac{1}{2}$

65. Possible 200 points

a. 160 **b.** 150 **c.** 155 **d.** 120 **e.** 195

66. Possible 1200 points

a. 900 **b.** 780 **c.** 816 **d.** 1000 **e.** 768

Level 3—Mind Bogglers

67. A man goes into a store and says to the salesperson, "Give me as much money as I have with me and I will spend \$10 here." It is done. The operation is repeated in a second and a third store, after which he has no money left. How much did he have originally?

68. This is an old problem, but it is still fascinating. One day three men went to a hotel and were charged \$300 for their room. The desk clerk then realized that he had overcharged them \$50, and he sent the refund up with the bellboy. Now the bellboy, being an amateur mathematician, realized that it would be difficult to split the \$50 three ways. Therefore, he kept a \$20 "tip" and gave the men only \$30. Each man had originally paid \$100 and was returned \$10. Thus it cost each man \$90 for the room. This means that they spent \$270 for the room plus the \$20 "tip," for a total of \$290. What happened to the other \$10?

4.4 PROBLEM SOLVING WITH PERCENTS

Historical Perspective

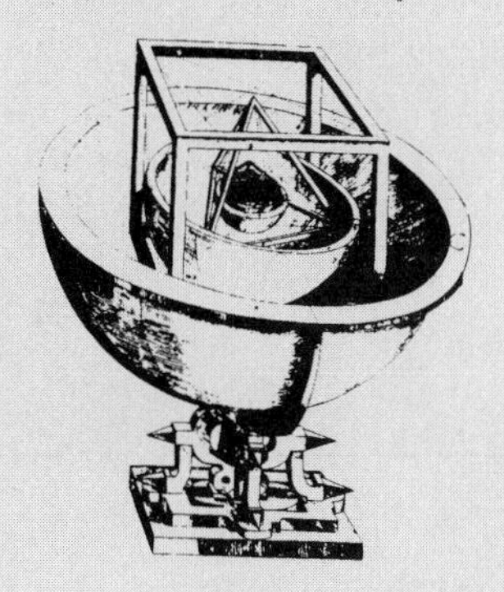

With the beginning of the Dark Ages around 950 A.D., we enter what is known as the Transition Period in the history of mathematics (1200–1599). The 14th and 15th centuries showed very few mathematical advances. In 1348–1350, bubonic plague killed a third of the European population, and from 1337 to 1453, England and France were engaged in the Hundred Years' War. In 1281, Li Yeh introduced a notation for negative numbers, Renaissance painters begin to use perspective to give depth to their paintings, Pellos used a decimal point, and Tartaglia and Cardano participated in mathematical contests to solve equations. It was during this period that the printing press was invented in 1438, Columbus came to America, Luther was excommunicated, and Copernicus' work challenged the heliocentric theory of the universe.

Unit Fraction Comparison

Percent	Fraction
10%	$\frac{1}{10}$
25%	$\frac{1}{4}$
$33\frac{1}{3}$%	$\frac{1}{3}$
50%	$\frac{1}{2}$

Percent problems are very common. We begin this section with a discussion of estimation, and then we conclude with some percent calculations.

The first estimation method is the *unit fraction-conversion method,* which can be used to estimate the common percents of 10%, 25%, $33\frac{1}{3}$%, and 50%. To estimate the size of a part of a whole quantity, which is sometimes called a **percentage,** rewrite the percent as a fraction and mentally multiply, as shown by the following example.

50% of 800: $\frac{1}{2} \times 800 = 400$; THINK: $800 \div 2 = 400$

25% of 1,200: $\frac{1}{4} \times 1{,}200 = 300$; THINK: $1{,}200 \div 4 = 300$

$33\frac{1}{3}$% of 600: $\frac{1}{3} \times 600 = 200$; THINK: $600 \div 3 = 200$

10% of 824: $\frac{1}{10} \times 824 = 82.4$; THINK: $824 \div 10 = 82.4$

(Move the decimal point one place to the left.)

If the numbers for which you are finding a percentage are not as "nice" as those given here, you can estimate by rounding the number, as shown in Example 1.

EXAMPLE 1

Estimating Percentages

Estimate the following percentages:

a. 25% of 312 **b.** 50% of 843 **c.** $33\frac{1}{3}$% of 1,856 **d.** 25% of 43,350

Solution

a. Estimate 25% of 312 by rounding 312 so that it is easily divisible by 4:

$\frac{1}{4} \times 320 = \mathbf{80}$ Find $320 \div 4 = 80$.

b. Estimate 50% of 843 by rounding 843 so that it is easily divisible by 2:

$\frac{1}{2} \times 840 = \mathbf{420}$ Find $840 \div 2 = 420$.

c. Estimate $33\frac{1}{3}$% of 1,856 by rounding 1,856 so that it is *easily* divisible by 3:

$\frac{1}{3} \times 1{,}800 = \mathbf{600}$ Find $1{,}800 \div 3 = 600$.

d. Estimate 25% of 43,350 by rounding 43,350 so that it is *easily* divisible by 4:

$\frac{1}{4} \times 44{,}000 = \mathbf{11{,}000}$ ■

The second estimation procedure uses a multiple of a unit fraction. For example,

Think of 75% as $\frac{3}{4}$, which is $3 \times \frac{1}{4}$.
Think of $66\frac{2}{3}$% as $\frac{2}{3}$, which is $2 \times \frac{1}{3}$.
Think of 60% as $\frac{6}{10}$, which is $6 \times \frac{1}{10}$. ■

EXAMPLE 2

Estimating Percents Using Multiples

Estimate the following percentages.

a. 75% of 943 **b.** $66\frac{2}{3}$% of 8,932 **c.** 60% of 954 **d.** 80% of 0.983

Solution

a. 75% of 943 $\approx \frac{3}{4} \times 1{,}000 = 3(\frac{1}{4} \times 1{,}000) = 3(250) = \mathbf{750}$
b. $66\frac{2}{3}$% of 8,932 $\approx \frac{2}{3} \times 9{,}000 = 2(\frac{1}{3} \times 9{,}000) = 2(3{,}000) = \mathbf{6{,}000}$
c. 60% of 954 $\approx \frac{6}{10} \times 1{,}000 = 6(\frac{1}{10} \times 1{,}000) = 6(100) = \mathbf{600}$
d. 80% of 0.983 $\approx \frac{8}{10} \times 1 = \mathbf{0.8}$ ■

Many percentage problems are more difficult than those thus far considered in this chapter. The following quotation was found in a recent publication: "An elected official is one who gets 51 percent of the vote cast by 40 percent of the 60 percent of voters who registered." Certainly, most of us will have trouble understanding the percent given in this quotation; but you can't pick up a newspaper without seeing dozens of examples of ideas that require some understanding of percents. A difficult job for most of us is knowing whether to multiply or divide by the given numbers. In this section, I will provide you with a sure-fire method for knowing what to do. The first step is to understand what is meant by **the percent problem.**

The Percent Problem

STOP

Study this percent problem. If you learn this, you get a written guarantee for correctly working percent problems.

A is $P\%$ of W

- A — This is the given amount.
- $P\%$ — The percent is written $\frac{P}{100}$
- W — This is the whole quantity. It always follows the word "of."

The percent problem won't always be stated in this form, but notice that three quantities are associated with it:

1. The *amount*—sometimes called the **percentage**
2. The *percent*—sometimes called the **rate**
3. The *whole quantity*—sometimes called the **base**

Now, regardless of the form in which you are given the percent problem, follow these steps to write a proportion:

Read these three steps SLOWLY!

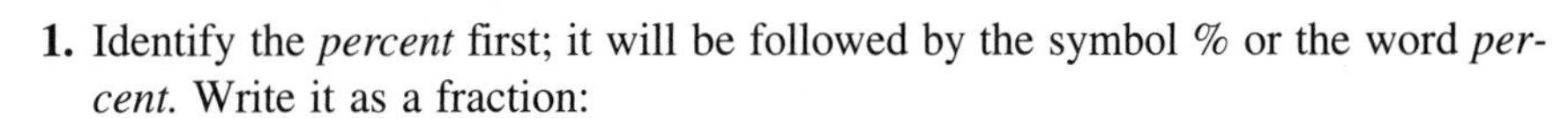

1. Identify the *percent* first; it will be followed by the symbol % or the word *percent.* Write it as a fraction:

$$\frac{P}{100}$$

2. Identify the *whole quantity* next; it is preceded by the word *of.* It is the denominator of the second fraction in the proportion:

$$\frac{P}{100} = \frac{}{W}$$ ← This is the quantity following the word "of."

3. The remaining number is the partial amount; it is the numerator of the second fraction in the proportion:

$$\frac{P}{100} = \frac{A}{W}$$ ← This is the last quantity to be inserted into the proportion.

EXAMPLE 3 **Identifying the Parts of a Percent Problem**

For each of the following cases, identify the percent, the whole quantity, and the amount (the percentage or part), and then write a proportion.

a. What number is 18% of 200? **b.** 18% of 200 is what number?
c. 150 is 12% of what number? **d.** 63 is what percent of 420?
e. 18% of what number is 72? **f.** 120 is what percent of 60?

Solution

	PERCENT P (%)	*WHOLE* W (*"of"*)	*AMOUNT* A (*part*)	*PROPORTION* $\frac{P}{100} = \frac{A}{W}$
a. What number is 18% of 200?	18	200	unknown	$\frac{18}{100} = \frac{A}{200}$
b. 18% of 200 is what number?	18	200	unknown	$\frac{18}{100} = \frac{A}{200}$
c. 150 is 12% of what number?	12	unknown	150	$\frac{12}{100} = \frac{150}{W}$
d. 63 is what percent of 420?	unknown	420	63	$\frac{P}{100} = \frac{63}{420}$
e. 18% of what number is 72?	18	unknown	72	$\frac{18}{100} = \frac{72}{W}$
f. 120 is what percent of 60?	unknown	60	120	$\frac{P}{100} = \frac{120}{60}$

↑ *Regardless of the arrangement of the question, identify P first.*

↑ *Second, identify the number following the word "of."*

↑ *This number is identified last.*

■

Since there are only three letters in the proportion

$$\frac{P}{100} = \frac{A}{W}$$

there are three types of percent problems. These possible types were illustrated in Example 3. To answer a question involving a percent, write a proportion and then solve the proportion. Try solving each proportion in Example 3. The answers are: **a.** $A = 36$; **b.** $A = 36$; **c.** $W = 1{,}250$; **d.** $P = 15$; **e.** $W = 400$; **f.** $P = 200$.

EXAMPLE 4

Problem Solving with Percents

In a certain class there are 500 points possible. The lowest C grade is 65% of the possible points. How many points are equal to the lowest C grade?

Solution What is 65% of 500 points?

$$\frac{65}{100} = \frac{A}{500} \quad \leftarrow \text{The whole amount follows the word "of."}$$

$$A = \frac{65 \times \overset{5}{\cancel{500}}}{\underset{1}{\cancel{100}}} = 325$$

Check by estimation: 65% of 500 $\approx 6(\frac{1}{10} \times 500) = 300$.

The lowest C grade is 325 points. ■

EXAMPLE 5 **Problem Solving with Percents**

If your monthly salary is $4,500 and 21% is withheld for taxes and Social Security, how much money will be withheld from your check on payday?

Solution How much is 21% of $4,500?

$$\frac{21}{100} = \frac{A}{4,500}$$

$$A = \frac{21 \times 4,500}{100} = 945$$

Check by estimation: 21% of $4,500 \approx 2(\frac{1}{10} \times 4,500) = 900$.

The withholding is $945. ■

EXAMPLE 6 **Problem Solving with Percents**

You make a $25 purchase, and the clerk adds $2.25 for sales tax. This doesn't seem right to you, so you want to know what percent tax has been charged.

Solution What percent of $25 is $2.25?

$$\frac{P}{100} = \frac{2.25}{25}$$

$$\frac{100 \times 2.25}{25} = P$$

$$9 = P$$

Check by estimation: 9% of $25 \approx \frac{1}{10} \times 25 = 2.50$.

The tax charged was 9%. ■

EXAMPLE 7 **Problem Solving with Percents**

Your neighbors tell you that they paid $4,437 in taxes last year, and this amounted to 29% of their total income. What was their total income?

Solution 29% of total income is $4,437.

$$\frac{29}{100} = \frac{4,437}{W}$$

$$\frac{100 \times 4,437}{29} = W$$

$$15,300 = W$$

Check by estimation: 29% of $15,300 \approx 3(\frac{1}{10} \times 15,000) = 3(1,500) = 4,500$.

Since $4,500 is an estimate for $4,437, we conclude the result is correct.
Their total income was $15,300. ■

WARNING! You must be careful not to add percents. For example, suppose you have \$100 and spend 50%. How much have you spent, and how much do you have left?

AMOUNT SPENT	*REMAINDER*
\$50	\$50

Now, suppose you spend 50% of the remainder? How much have you spent, and how much is left?

NEW SPENDING	*OLD SPENDING*	*REMAINDER*
\$25	\$50	\$25

This means you have spent \$75 or 75% of your original bankroll. A common ERROR is to say "50% spending + 50% spending = 100% spending." **Remember, if you add percents, you often obtain incorrect results.**

EXAMPLE 8

A Common Error from the Newspaper

A newspaper headline proclaimed

Teen drug use soars 105%

WASHINGTON – Teen drug use rose 105% between 1995 and 1997.
A national survey showed that between 1995 and 1996 youth drug use rose 30%, but between 1996 and 1997 usage soared to 75%.
Over the two year period, the rise of 105% was attributed to ...

What is wrong with headline?

Solution We are not given all the available numbers, but consider the following possibility:

Suppose there are 100 drug users, so a rise of

100 to 130 is a 30% increase
130 to 227 is a 75% increase

100 to 227 is a 127% increase NOT 105%

Remember, adding percents can given faulty results. ■

PROBLEM SET 4.4

Essential Ideas

1. **In Your Own Words** Describe the "percent problem."
2. **In Your Own Words** Explain what is meant by each of the following words: percentage, rate, base.

Level 1

WHAT IS WRONG, *if anything, with each of the statements in Problems 3–6? Explain your reasoning.*

3. The percent problem has three parts: the percent, the whole quantity, and the partial amount.
4. The first quantity to be identified when solving the percent problem is the percent. It is the quantity preceding a percent symbol or the word "percent."
5. The second quantity to be identified when solving the percent problem is the whole quantity. It is the quantity preceded by the word "of."
6. The newspaper clipping below was printed in a newspaper as a letter to the editor during the 1998 military build-up in the Middle East. What is wrong with John's complaint about "the new math"?

Military math

Editor: Let me see if I have this right. During the Gulf War, I kept hearing news reports that allied forces had destroyed 90 percent of Iraq's war-making capability. Then the news reports said that brave U.N. inspectors had forced Iraq to destroy another 90 percent several times. In the last few months news reports said that Iraq itself had destroyed 90 percent of its weapons at least a couple of times as a show of earnest good faith.

All told, then, some 450 percent of the Iraqi military was wiped out. and yet somehow they are massing forces against Iran. Now I hear news reports say that Saddam Hussein is still a threat and that there may be another October surprise.

Did I miss something in those news reports? Did the news people miss something? 450 percent? Must be the new math.

JOHN
Rohnert Park

Level 1—Drill

Estimate the percentage in Problems 7–20.

7. 50% of 2,000
8. 25% of 400
9. 10% of 95,000
10. 10% of 85.6
11. 50% of 9,800
12. $33\frac{1}{3}$% of 3,600
13. 25% of 819
14. 25% of 790
15. 75% of 1,058
16. 75% of 94
17. 40% of 93
18. 90% of 8,741
19. $66\frac{2}{3}$% of 8,600
20. $66\frac{2}{3}$% of 35

Level 1—Applications

The book ***100% American*** *by Daniel Evan Weiss is a book of "facts" about American opinions. The percentages in Problems 21–28 are taken from this book (published in 1989 by Poseidon Press). In estimating the numbers in Problems 21–28, assume that the questions refer to the 180 million adult Americans.*

21. 50% of American adults are men.
22. 50% of American men (see Problem 21) are less than 5 ft 9 in.
23. 25% of Americans never exercise at all.
24. 10% of Americans say the car is the greatest invention of all time.
25. 90% of Americans consider themselves happy people.
26. 40% of Americans do not think a college education is important to succeed in the business world.
27. 3% of Americans think Elvis Presley was history's most exciting figure.
28. 6% of Americans believe the single greatest element in happiness is great wealth.

LEVEL 1—Drill

Write each sentence in Problems 29–46 as a proportion, and then solve to answer the question.

29. What number is 15% of 64?

30. What number is 120% of 16?

31. 14% of what number is 21?

32. 40% of what number is 60?

33. 10 is what percent of 5?

34. What percent of $20 is $1.20?

35. 4 is what percent of 5?

36. 2 is what percent of 5?

37. What percent of 12 is 9?

38. What percent of 5 is 25?

39. 49 is 35% of what number?

40. 3 is 12% of what number?

41. 120% of what number is 16?

42. 21 is $66\frac{2}{3}\%$ of what number?

43. 12 is $33\frac{1}{3}\%$ of what number?

44. What is 8% of $2,425?

45. What is 6% of $8,150?

46. 400% of what number is 150?

LEVEL 2—Applications

47. If 11% of the 180 million adult Americans live in poverty, how many adult Americans live in poverty?

48. If 6.2% of the 180 million adult Americans are unemployed, how many adult Americans are unemployed?

49. If the sales tax is 6% and the purchase price is $181, what is the amount of tax?

50. If the sales tax is 5.5% and purchase price is $680, what is the amount of tax?

51. If you were charged $151 in taxes on a $3,020 purchase, what percent tax were you charged?

52. If a government worker will receive a pension of 80% of her present salary, what will the pension be if her monthly salary is $4,250?

53. Government regulations require that, for certain companies to receive federal grant money, 15% of the total number of employees must meet minority requirements. If a company employs 390 people, how many minority people should be employed to meet the minimum requirements?

54. If $14,300 has been contributed to the United Way fund drive and this amount represents 22% of the goal, what is the United Way goal?

55. If Brad's monthly salary is $8,200, and 32% is withheld for taxes and Social Security, how much money is withheld each month?

56. In a certain class 500 points are possible. The lowest B grade is 80%. How many points are needed to obtain the lowest B grade?

57. A certain test is worth 125 points. How many points (rounded to the nearest point) are needed to obtain a score of 75%?

58. If you correctly answer 8 out of 12 questions on a quiz, what is your percentage right?

59. If Carlos answered 18 out of 20 questions on a test correctly, what was his percentage right?

60. If Wendy answered 15 questions correctly and obtained 75%, how many questions were on the test?

61. Shannon Sovndal received an 8% raise, which amounted to $100 per month. What was his old wage, and what will his new wage be?

62. An advertisement for a steel-belted radial tire states that this tire delivers 15% better gas mileage. If the present gas mileage is 25.5 MPG, what mileage would you expect if you purchased these tires? Round your answer to the nearest tenth of a mile per gallon.

LEVEL 3—Mind Bogglers

63. A drop from 50 to 10 is a loss of 80%; what is the percent gain from 10 to 50?

64. A saleswoman complained to her friend that she had had a bad day. She had made only two sales, for $1,500 each. On the first sale she had made a profit of 30% on the cost price, but on the second one she had taken a 30% loss on the cost price. "That doesn't seem to be any loss at all," said the friend. "Your profit and loss balance each other." "On the contrary," said the saleswoman, "I lost almost $300, overall." Who was right, the saleswoman or the friend? Justify your answer.

65. What is the square root of 25%? Write your answer as a percent.

66. What is the product of 5% and 20%? Write your answer as a percent.

LEVEL 3 — Team Project

67. With a team of two or three others, keep track of each time any team member sees a fraction or percent being used outside the classroom. For example, a half-off sale, or a bank interest rate, or a recipe calling for $2\frac{1}{2}$ tablespoons of an ingredient.

Make up a master list for the team (eliminate duplications), and discuss why a fraction or percent is used, rather than a whole number. Also comment on why a fraction is used rather than a percent, or vice versa.

4.5 SUMMARY AND REVIEW

Important Terms

STOP

Base [4.4]	Percentage [4.4]	Rate [4.4]
Extremes [4.1]	Percent problem [4.4]	Ratio [4.1]
Means [4.1]	Property of proportions [4.1]	Solve a proportion [4.2]
Percent [4.3]	Proportion [4.1]	

Chapter Objectives

The material in this chapter is reviewed in the following list of objectives. A self-test (with answers and suggestions for additional study) is given. This self-test is constructed so that each problem number corresponds to a related objective. This self-test is followed by a practice test with the questions in mixed order.

[4.1] *Objective 4.1* Reduce a ratio to lowest terms.

[4.1] *Objective 4.2* Read a proportion, and name the means and extremes. Decide whether a given pair of ratios forms a proportion.

[4.1] *Objective 4.3* Solve applied ratio problems.

[4.2] *Objective 4.4* Solve a proportion.

[4.2] *Objective 4.5* Solve applied proportion problems.

[4.3] *Objective 4.6* Change fractions to decimals and percents.

[4.3] *Objective 4.7* Change decimals to fractions and percents.

[4.3] *Objective 4.8* Change percents to fractions and decimals.

[4.4] *Objective 4.9* Estimate percentages.

[4.4] *Objective 4.10* Solve percent problems.

[4.3; 4.4] *Objective 4.11* Solve applied percent problems.

Self-Test

Each question of this self-test is related to the corresponding objective listed above.

1. If the ratio of miles to gallons is 154 to 5.5, what is this as a reduced ratio?
2. Do $\frac{7}{2}$ and $\frac{21}{8}$ form a proportion?
3. The ratio of errors to correct answers is 4 per 100; write this as a simplified ratio.
4. Solve the proportion $\frac{25}{x} = \frac{575}{138}$.
5. If the ratio of wins to losses is 3 to 5, how many losses would you expect if there are 882 wins?
6. Write the fraction $\frac{5}{12}$ as a decimal and as a percent.
7. Write the decimal 0.005 as a fraction and as a percent.
8. Write the percent $8\frac{1}{2}\%$ as a fraction and as a decimal.
9. Estimate 25% of 412.
10. 85% of what number is 170?
11. If inflation is 0.9% and your salary is $42,500, what should your salary be next year to keep pace with inflation?

STUDY HINTS *Compare your solutions and answers to the self-test. For each problem you missed, work some additional problems in the section listed in the margin. After you have worked these problems, you can test yourself with the practice test.*

Complete Solutions to the Self-Test

Additional Problems

[4.1] *Problems 3–24*

1. $\frac{154}{5.5} = \frac{1{,}540}{55} = \frac{28}{1}$; the ratio of miles to gallons is 28 to 1.

[4.1] *Problems 25–32; 39–54*

2. *MEANS* *EXTREMES*
   2×21 7×8
   $42 \neq 56$
   These ratios do not form a proportion.

[4.1] *Problems 55–71*

3. $\frac{\text{ERRORS}}{\text{CORRECT ANSWERS}} = \frac{4}{100} = \frac{1}{25}$; the error rate is 1 to 25.

[4.2] *Problems 3–26*

4. $\frac{25}{x} = \frac{575}{138}$
   $$x = \frac{25 \times 138}{575} = 6$$
   The value of x is 6.

[4.2] *Problems 27–40*

5. Set up $\frac{\text{WINS}}{\text{LOSSES}}$; $\frac{3}{5} = \frac{882}{x}$
   $$x = \frac{5 \times 882}{3} = 1{,}470$$
   We would expect 1,470 losses.

[4.3] *Problems 15; 16; 26; 27; 34–37; 49; 50; 52; 53; 56; 57; 60–62*

6. Divide; $5 \div 12 = 0.41666\ldots$ or $0.41\overline{6}$

Divide; shift decimal two places to the right, and save the remainder: $0.41\frac{2}{3} = 41\frac{2}{3}\%$

[4.3] *Problems 17; 18; 23; 24; 32; 33; 38–43; 48; 51; 54; 55*

7. 0.005 as a fraction is $\frac{5}{1{,}000} = \frac{1}{200}$

As a percent, $0.005 = 0.5\%$ (move decimal point two places to the right)

[4.3] *Problems 8–14; 19–22; 25; 28–31; 44–47; 58; 59*

8. $8\frac{1}{2}\% = 8\frac{1}{2} \times \frac{1}{100} = \frac{17}{2} \times \frac{1}{100} = \frac{17}{200}$

$8\frac{1}{2}\% = 8.5\% = 0.085$ (move decimal point two places to the left)

[4.4] *Problems 7–28*

9. Think of 25% of 412 as $\frac{1}{4}$ of $412 \approx \frac{1}{4} \times 400 = 100$.

[4.4] *Problems 29–46*

10. $\frac{85}{100} = \frac{170}{x}$

$x = \frac{100 \times 170}{85}$ Carry out this calculation on your calculator: [100] [×] [170] [÷] [85] [=]

$x = 200$

The number is 200.

[4.3] *Problems 63–66*

[4.4] *Problems 47–62*

11. What is 0.9% of $42,500?

$\frac{0.9}{100} = \frac{x}{42{,}500}$

$x = \frac{0.9 \times 42{,}500}{100}$ Carry out this calculation on your calculator: [.9] [×] [42500] [÷] [100] [=]

$x = 382.5$

The raise due to inflation is $382.50, so the expected salary is $42,500 + $382.50 = $42,882.50.

Chapter 4 Practice Test

1. Express the ideas as reduced ratios.

a. The ratio of cement to water is 120 to 6.

b. The gear ratio is 34 teeth to 17 teeth.

c. The ratio of wins to losses is 155 to 75.

d. The ratio of miles to gallons is 117 to $6\frac{1}{2}$.

2. a. Read $\frac{5}{8} = \frac{x}{2}$ as a proportion.

b. What are the means in the proportion $\frac{4}{y} = \frac{19}{7}$?

c. What are the extremes in the proportion $\frac{x}{2} = \frac{1}{12}$?

d. Do $\frac{3}{2}$ and $\frac{9}{6}$ form a proportion?

3. Change the fractions to decimals and percents.

a. $\frac{3}{5}$ **b.** $\frac{3}{2}$ **c.** $\frac{4}{3}$ **d.** $\frac{1}{6}$

4. Change the decimals to fractions and percents.

a. 0.25 **b.** 0.8 **c.** 1.05 **d.** 0.125

5. Change the percents to fractions and decimals.

a. 35% **b.** 240% **c.** 6% **d.** $37\frac{1}{2}\%$

6. Solve each proportion.

a. $\frac{5}{8} = \frac{A}{2}$ **b.** $\frac{4}{5} = \frac{3}{B}$

c. $\frac{C}{100} = \frac{2}{3}$ **d.** $\frac{12}{D} = \frac{4}{5}$

7. a. 45% of 120 is what number?

b. 60 is what percent of 80?

c. What number is 25% of 300?

d. 90 is 120% of what number?

8. If a car went 351 miles on $13\frac{1}{2}$ gallons of gas, how many miles will it go on 5 gallons of gas?

9. If 1 gallon of paint covers 250 sq ft, how many gallons are needed for 1,250 sq ft?

10. If rolls are sold at 12 for $3.50, how much do 3 rolls cost?

11. If the ratio of cement to water is 120 lb to 6 gallons, how many gallons should be added to 150 lb of cement?

12. Suppose you get 16 correct out of 20 on a test. What is your score, expressed as a percent?

13. If you received 85% on a test consisting of 20 items, how many questions did you get correct?

14. If your gross taxable income is $35,240 and the tax rate is 32%, what is the amount of tax due?

15. If the sales tax rate on a $59 item is 6%, what is the total price, including tax?

CHAPTER FIVE

INTRODUCTION TO GEOMETRY

5.1 EUCLIDEAN GEOMETRY

Historical Perspective

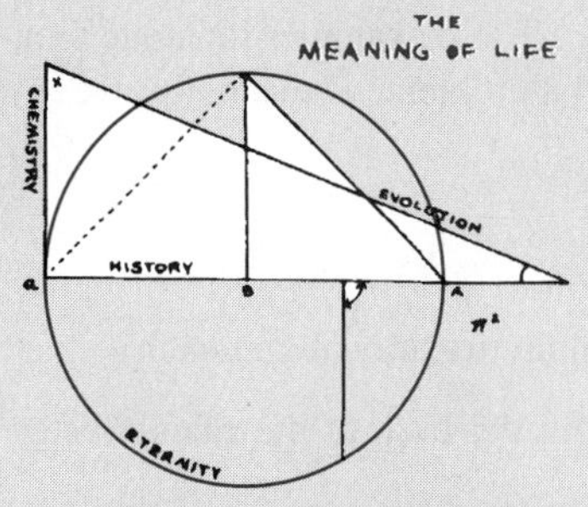

"I wish I could see the expression on the faces of my students who said there was no value in studying geometry!"

A book called Euclid's *Elements* (see Historical Perspective, p. 173) collected all the material about geometry known at that time and organized it into a logical deductive system. Geometry, or "earth measure," was one of the first branches of mathematics. Both the Egyptians and the Babylonians needed geometry for construction, land measurement, and commerce. They both discovered the Pythagorean theorem, although it was not proved until the Greeks developed geometry formally. This formal development utilizes deductive logic (which is discussed in Chapter 8), beginning with certain assumptions, called **axioms** or **postulates.** Historically, the first axioms that were accepted seemed to conform to the physical world.

Geometry involves **points** and sets of points called **lines, planes,** and **surfaces.** Certain concepts in geometry are called **undefined terms.** For example, what is a line? You might say, "I know what a line is!" But try to define a line. Is it a set of points? Any set of points? What is a point?

1. A point is something that has no length, width, or thickness.

2. A point is a location in space.

Certainly these are not satisfactory definitions because they involve other terms that are not defined. We will therefore take the terms *point, line,* and *plane* as undefined.

What do you see? A square? A cube? Figures can be ambiguous (not clear, or with hidden meaning).

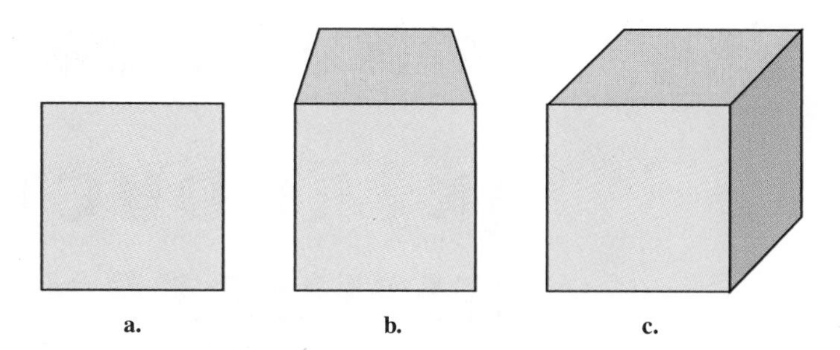

Figure 5.1 Three views of the same object

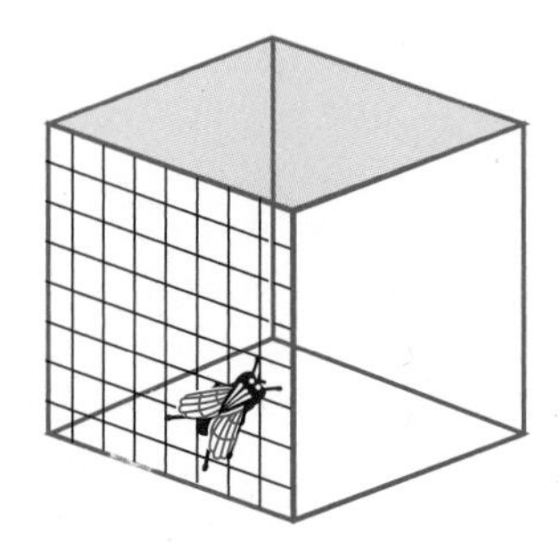

Figure 5.2 Is the fly on the cube or in the cube? On which face?

We often draw physical models or pictures to represent these concepts; however, we must be careful not to try to prove assertions by looking at pictures, since a picture may contain hidden assumptions or ambiguities. For example, consider Figure 5.1. If you look at Figure 5.1(a), you might call it a square. If you look at Figure 5.1(b), you might say it is something else. But what if we have in mind a cube, as shown in Figure 5.1(c)? Even if you view this object as a cube, do you see the same cube as everyone else?

Is the fly in Figure 5.2 inside or outside the cube? These examples illustrate that, although we may use a figure to help us understand a problem, we cannot prove results by this technique.

Geometry can be separated into two categories:

1. Traditional (which is the geometry of Euclid)
2. Transformational (which is more algebraic than the traditional approach)

In the remainder of this section, we will briefly consider each of the categories of geometry.

When Euclid was formalizing traditional geometry, he based it on five postulates, known today as **Euclid's postulates.** In mathematics, a result that is proved on the basis of some agreed-upon postulates is called a **theorem.**

Euclid's Postulates

1. A straight line can be drawn from any point to any other point.
2. A straight line extends infinitely far in either direction.
3. A circle can be described with any point as center and with a radius equal to any finite straight line drawn from the center.
4. All right angles are equal to each other.
5. Given a straight line and any point not on this line, there is one and only one line through that point that is parallel to the given line.*

*The fifth postulate stated here is the one usually found in high school geometry books. It is sometimes called Playfair's axiom and is equivalent to Euclid's original statement as translated from the original Greek by T. L. Heath: "If a straight line falling on two straight lines makes the interior angle on the same side less than two right angles, the two straight lines, if produced infinitely, meet on that side on which the angles are less than the two right angles."

The first four of these postulates were obvious and noncontroversial, but the fifth one was different. This fifth postulate looked more like a theorem than a postulate. It was much more difficult to understand than the other four postulates, and for more than 20 centuries mathematicians tried to derive it from the other postulates or to replace it by a more acceptable equivalent. Two straight lines in the same plane are said to be **parallel** if they do not intersect.

Today we can either accept Postulate 5 (without proof) or we can deny it. If it is accepted, then the geometry that results is consistent with our everyday experiences and is called **Euclidean geometry.** If it is denied, it turns out that no contradiction results; in fact, if it is not accepted, other *non-Euclidean geometries* result.

Let's look at each of Euclid's postulates. The first one says that a straight line can be drawn from any point to any other point.

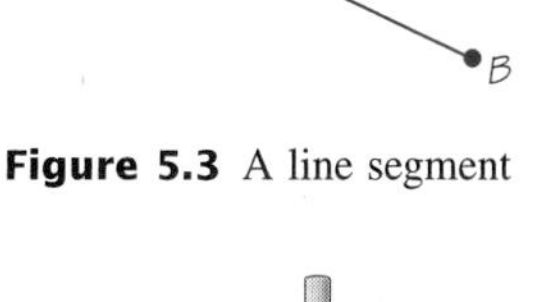

Figure 5.3 A line segment

To connect two points, you need a device called a **straightedge** (a device that we assume has no markings on it; you will use a ruler, but not to measure, when you are treating it as a straightedge). The portion of the line that connects points A and B in Figure 5.3 is called a **line segment.** The line segment from point A to point B is denoted by $\overline{AB}$.

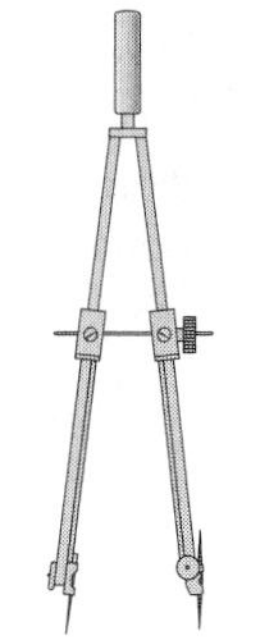

Pointer Pencil

Figure 5.4 A compass

The second postulate says that we can draw a straight line. This seems straightforward and obvious, but we should point out that we will indicate a line by putting arrows on each end. If the arrow points in only one direction, it is called a **ray.** To construct a line segment of length equal to that of a given line segment, we need a device called a **compass.** Figure 5.4 shows a compass, which is used to mark off and duplicate lengths, but not to measure them. If objects have exactly the same size and shape, they are called **congruent.**

To **construct** a figure means to use a straightedge and compass so that it meets certain requirements. To *construct a line segment congruent to a given line segment,* copy a segment $\overline{AB}$ on any line ℓ. First fix the compass so that the pointer is on point A and the pencil is on B, as shown in Figure 5.5(a). Then, on line ℓ, choose a point C. Next, without changing the compass setting, place the pointer on C and strike an arc at D, as shown in Figure 5.5(b).

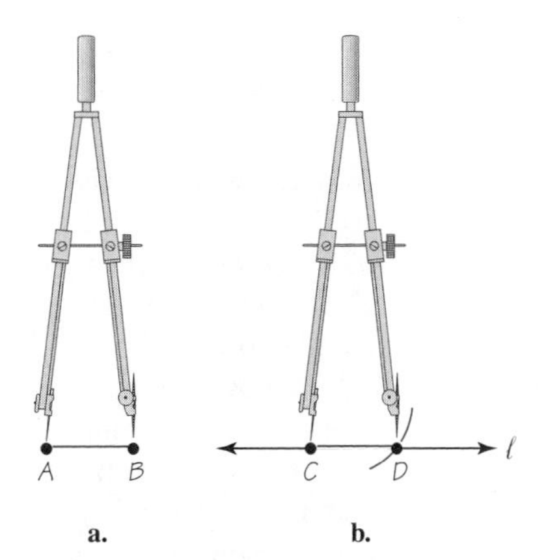

Figure 5.5 Constructing a line segment

Euclid's third postulate leads us to a second construction. The task is to construct a circle, given its center and radius. These steps are summarized in Figure 5.6.

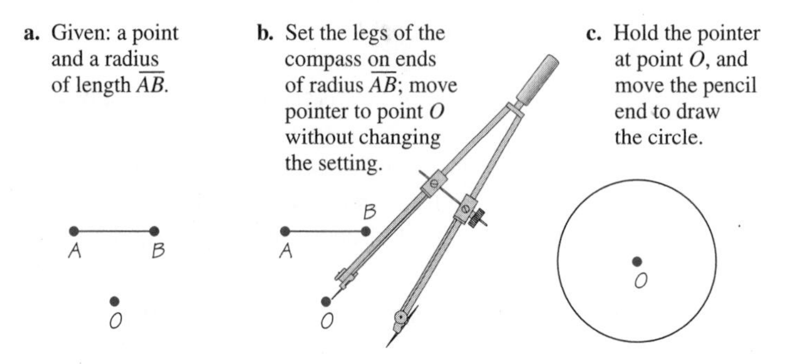

Figure 5.6 Construction of a circle

Figure 5.7 Given a point and a line

We will demonstrate the fourth postulate in the next section when we consider angles.

The final construction of this section will demonstrate the fifth postulate. The task is to construct a line through a point P parallel to a given line ℓ, as shown in Figure 5.7. First, draw any line through P that intersects ℓ at a point A, as shown in Figure 5.8(a).

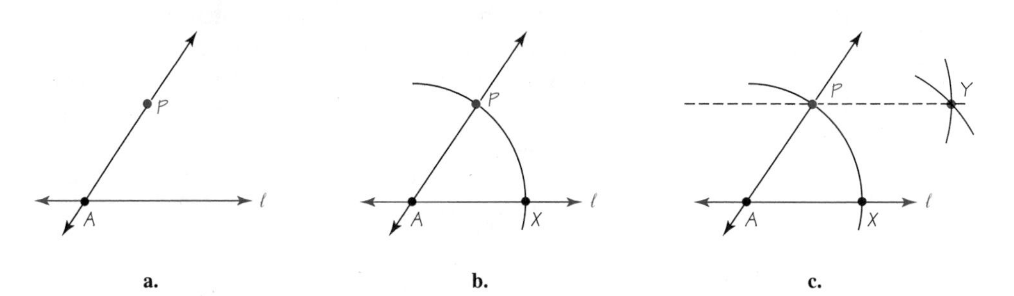

Figure 5.8 Construction of a line through a given point parallel to a given line

Now, draw an arc with the pointer at A and radius $\overline{AP}$, and label the point of intersection of the arc and the line X, as shown in Figure 5.8(b). With the same opening of the compass, draw an arc first with the pointer at P and then with the pointer at X. This will determine a point Y. Draw the line through both P and Y. This line is parallel to ℓ.

We now turn our attention to the second category of geometry. The transformational category is quite different from traditional geometry. It begins with the idea of a **transformation.** For example, one way to transform one geometric figure into another is by a *reflection.* Given a line ℓ and a point P, as shown in Figure 5.9, we call the point P' the **reflection** of P about the line L if $\overline{PP'}$ is perpendicular to L and is also bisected by L. Each point in the plane has exactly one reflection point

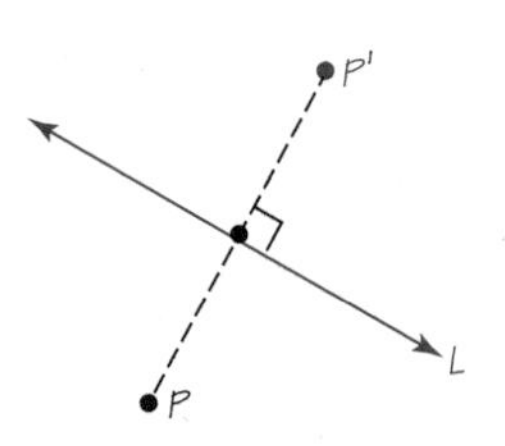

Figure 5.9 A reflection

corresponding to a given line ℓ. A reflection is called a *reflection transformation,* and the line of reflection is called the **line of symmetry.**

Other transformations include *translations, rotations, dilations,* and *contractions,* which are illustrated in Figure 5.10.

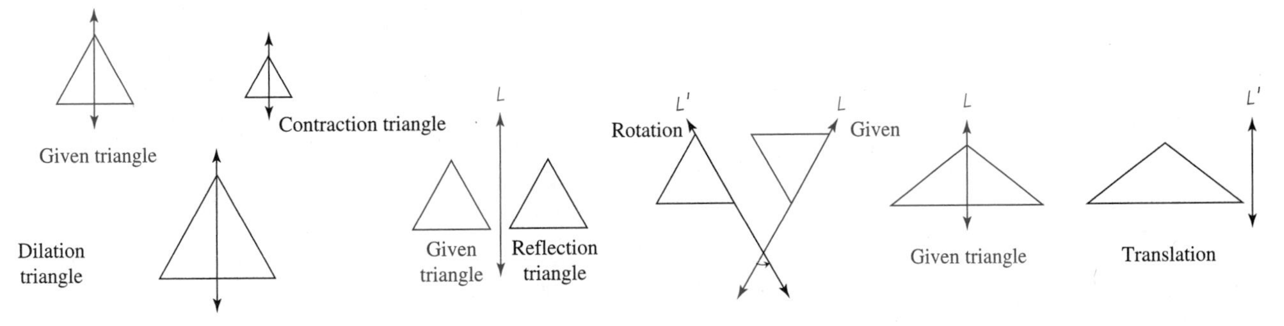

Figure 5.10 Transformations of a fixed geometric figure

PROBLEM SET 5.1

LEVEL 1

1. **In Your Own Words** What is the difference between an axiom and a theorem?

2. **In Your Own Words** Is the woman in the figure a young woman or an old woman?
3. **In Your Own Words** What are the two categories into which geometry is usually separated?

WHAT IS WRONG, *if anything, with each of the statements in Problems 4–6? Explain your reasoning.*

4. Every fact in mathematics can be proved, if only we are careful enough and take our time.
5. Two lines are parallel if they never intersect.
6. A construction allows only a straightedge and a compass.

LEVEL 1—Drill

Using only a straightedge and a compass, reproduce the figures in Problems 7–15.

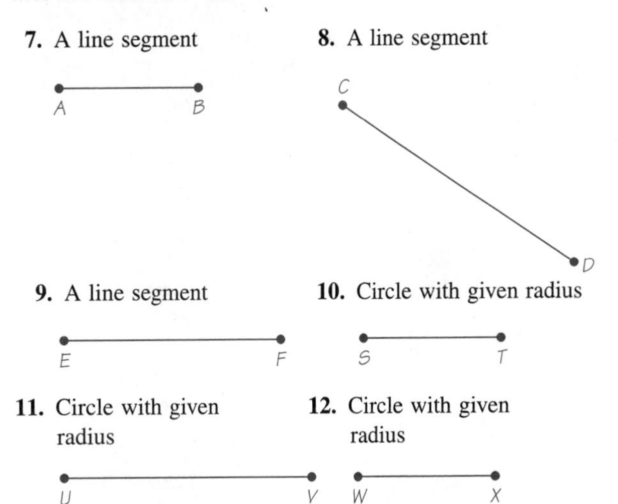

13. Line through P parallel to ℓ

14. Line through Q parallel to m

15. Line through R parallel to n

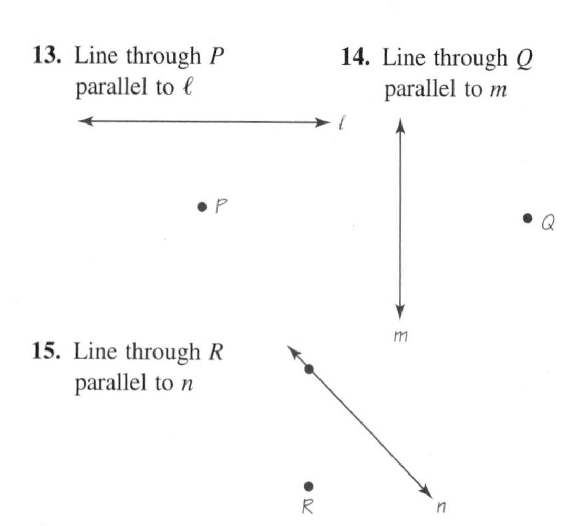

LEVEL 2

Which of the pictures in Problems 16–24 *illustrate a line symmetry?*

16. Chambered nautilus

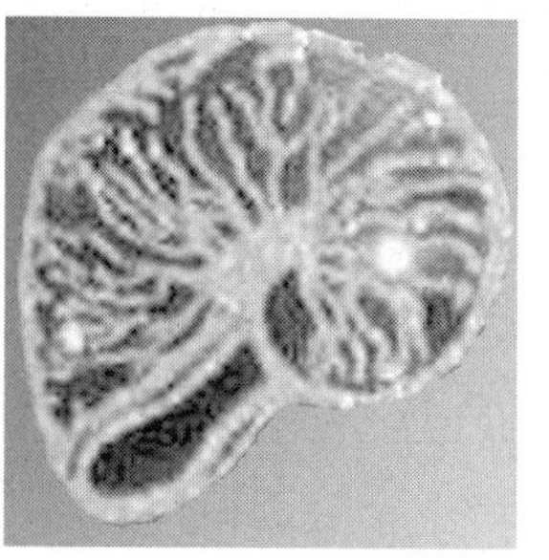

17. Butterfly

18. *Winged Lion,* China 7th–8th century A.D.

19. *Dodecahedron,* drawn by Leonardo da Vinci

20. Human brain

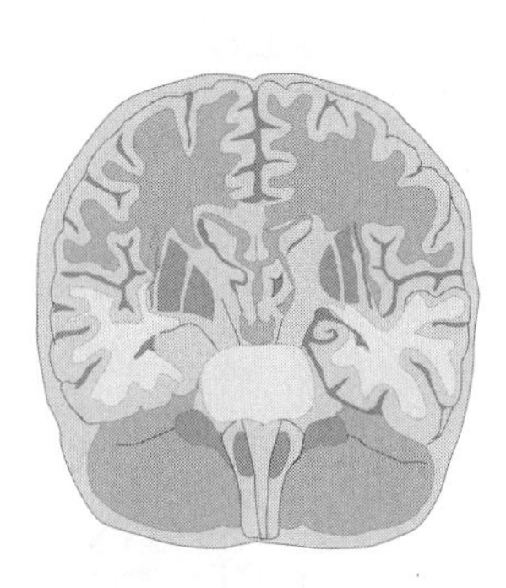

21. Human circulatory system

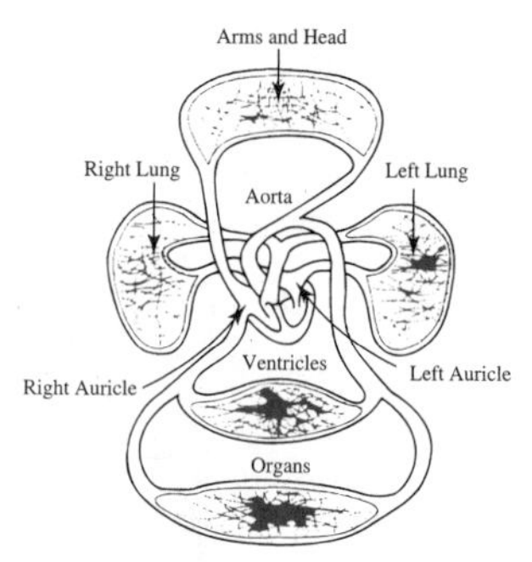

22. Restaurant at Los Angeles International Airport

23. Empire State Building, New York City

24. *Kaiser porcelain vase*

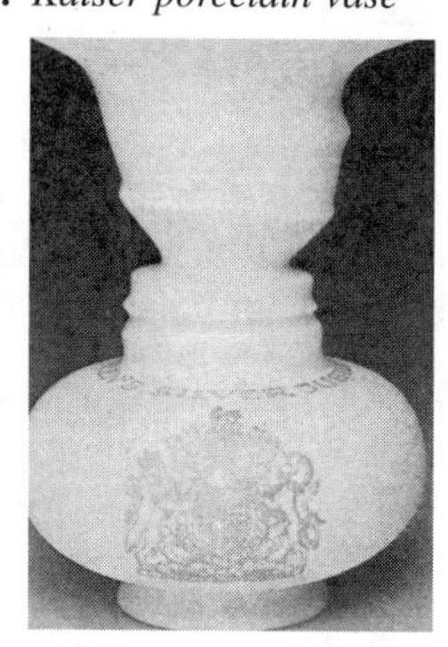

In Problems 25–26, *label each cartoon as illustrating a translation, rotation, dilation, or contraction.*

25.

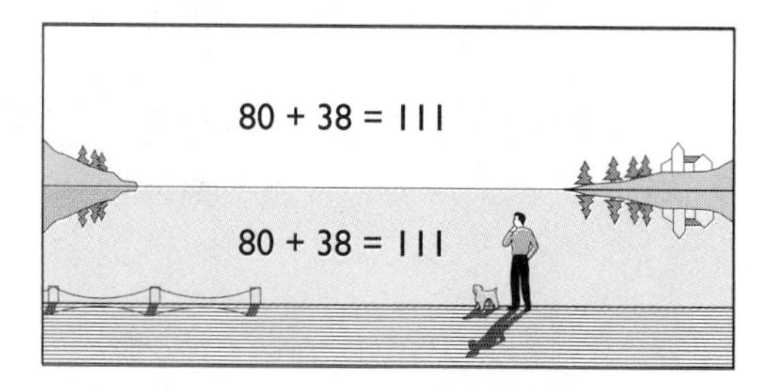

26.

LEVEL 2—Applications

27. Many curves can be illustrated by using only straight line segments. The basic design is drawn by starting with an angle, as shown in Figure 5.11.

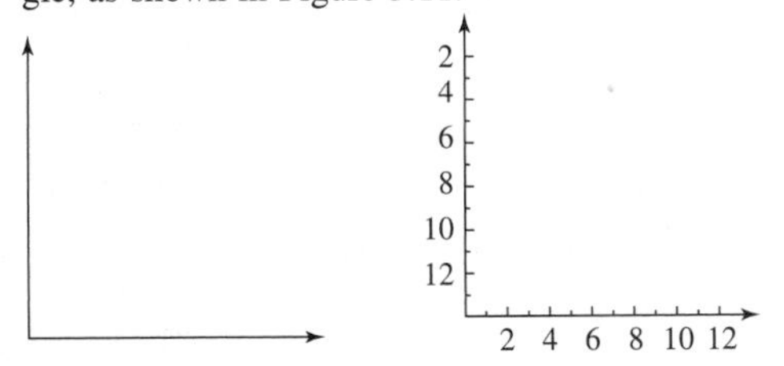

Step 1: Draw an angle with two sides of equal length.

Step 2: Mark off equally distant units on both rays, using a compass.

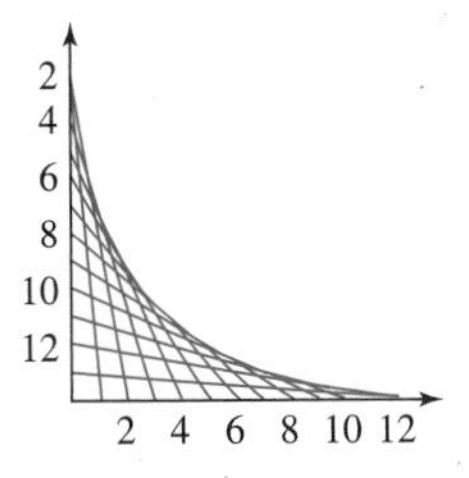

Step 3: Connect 1 to 1, connect 2's, connect 3's ... .

Figure 5.11 Constructing an angle design

The result is called *aestheometry* and is depicted in Figure 5.12. Make up your own angle design.

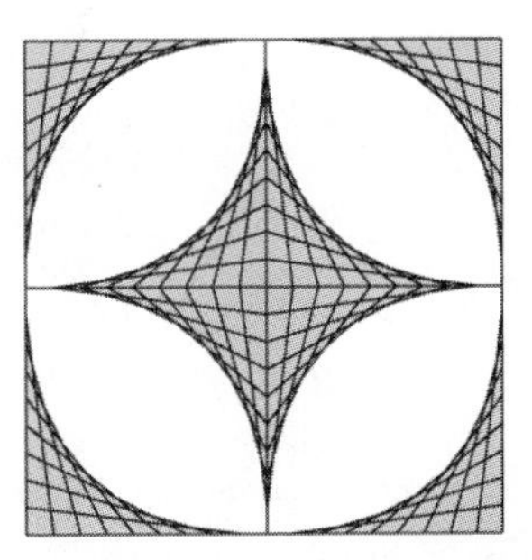

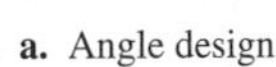

a. Angle design

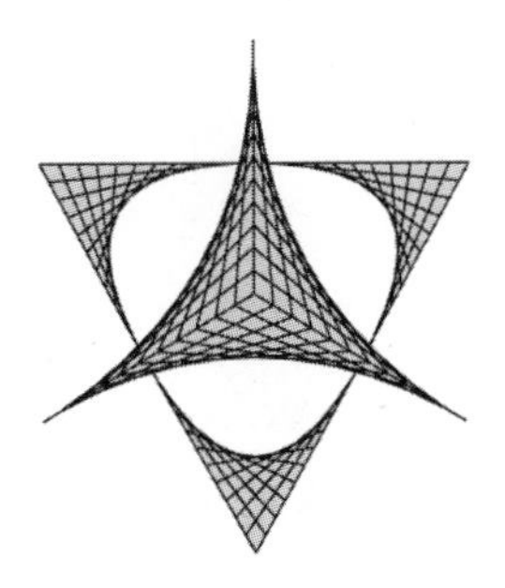

b. Angle design

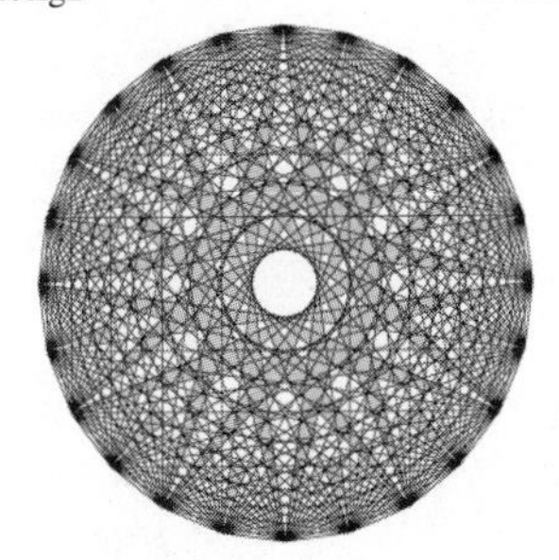

c. Circle design

Figure 5.12 Aestheometry designs

28. A second basic *aestheometric design* (see Problem 27) begins with a circle, as shown in Figure 5.13.

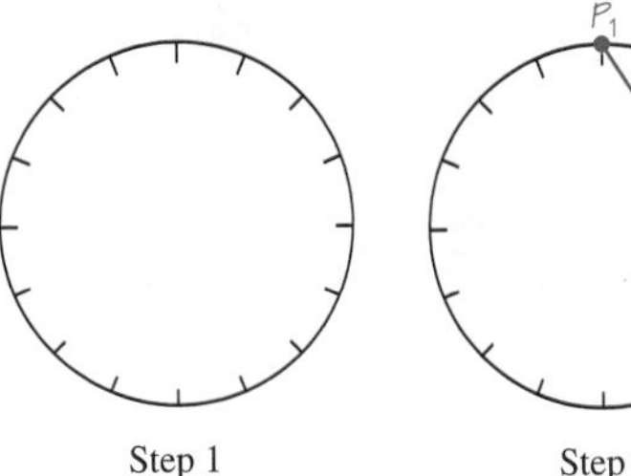

Step 1

a. Draw a circle and mark off equally spaced points.

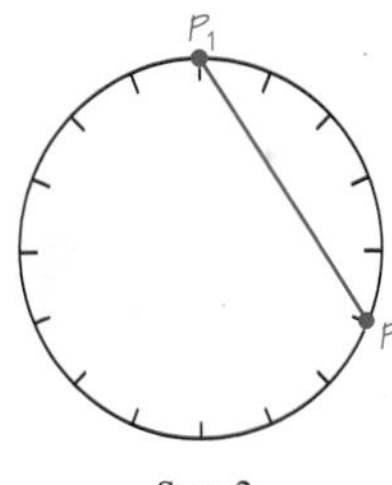

Step 2

b. Choose any two points and connect them.

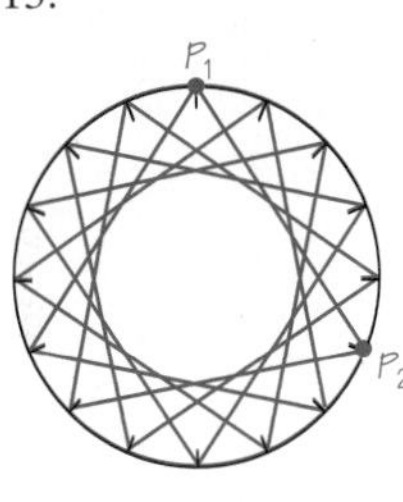

Step 3

c. Connect succeeding points around the circle.

Figure 5.13 Constructing a circle design

Make up your own design using circles or parts of a circle.

LEVEL 3—Mind Boggler

29. Connect the nine dots with four line segments, but do not lift your pencil from the paper.

•   •   •

•   •   •

•   •   •

5.2 POLYGONS AND ANGLES

Historical Perspective

Isaac Newton (1642–1727)

Gottfried Leibniz (1646–1716)

The Century of Enlightenment in the history of mathematics refers to the period from 1600 to 1699. This century begins with the invention of logarithms by Napier in 1614 and ends with the invention of calculus by Newton and Leibniz. In this century, the Pilgrims landed at Plymouth, Harvard College was founded, and there was a great deal of creativity in the mathematical world. New results were produced in a variety of subjects: algebra, analytic geometry, number theory, and the new fields of calculus and probability. Some of the greatest names in the history of mathematics are associated with this period: René Descartes (1596–1650) revolutionized the world with the development of a coordinate system (see Chapter 11) and analytic geometry; Pierre de Fermat (1601–1665) founded number theory, and, along with Blaise Pascal (1623–1662), invented the theory of probability (see Chapter 9). And finally, two of the greatest mathematicians of all time, Isaac Newton (1642–1727) and Gottfried Leibniz (1646–1716), independently invented the calculus.

This invention of calculus is one of the most profound events in the history of the world, but it was marked during their lives by a bitter priority dispute over who actually invented calculus first. This dispute expanded into a political and national rivalry, with followers of Newton in England and followers of Leibniz in Germany. In fact, it has been said that this dispute retarded mathematical progress in England for almost 100 years. This story illustrates the rivalry between Newton and Leibniz: In 1716, when Newton was 74 years old, Leibniz posed a challenge problem to all the mathematicians in Europe. (Challenge problems were common in those days.) Newton received the problem after a day's work at the mint (he was 74 years old at the time!) and solved the challenge problem that evening. His intellect was monumental. On the other hand, Leibniz' influence can also be described as monumental. At the age of 14, he attempted to reform logic. He called his logic the universal characteristic and wrote, in 1666, that he wanted to create a general method in which truths of reason would be reduced to a calculation so that errors of thought would appear as computational errors.

A **polygon** is a geometric figure that has three or more straight sides that all lie on a flat surface or plane so that the starting point and the ending point are the same. Polygons are classified according to their number of sides, as shown in Figure 5.14.

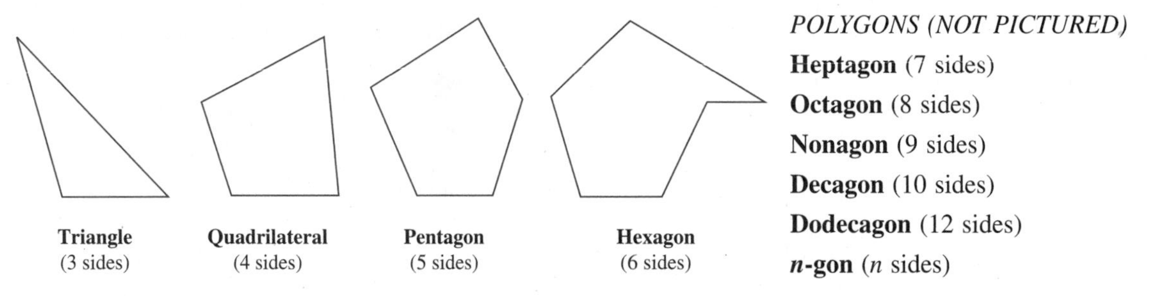

Figure 5.14 Polygons classified according to number of sides

A connecting point of two sides is called a **vertex** (plural **vertices**) and is usually designated by a capital letter. An **angle** is composed of two rays or segments with a common endpoint. The angles between the sides of a polygon are sometimes also denoted by a capital letter, but other ways of denoting angles are shown in Figure 5.15.

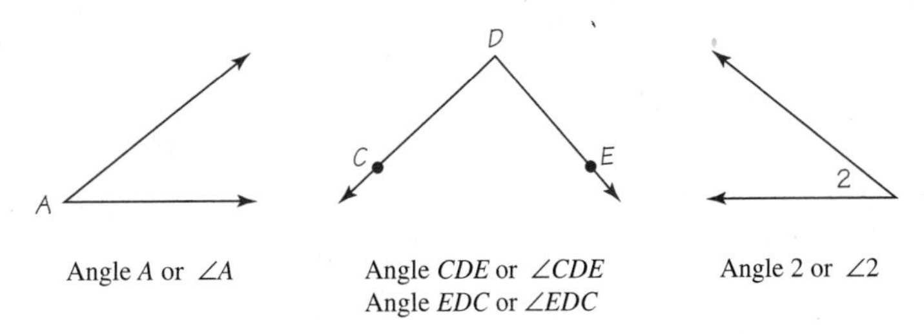

Figure 5.15 Ways of denoting angles

EXAMPLE 1

Locating Angles

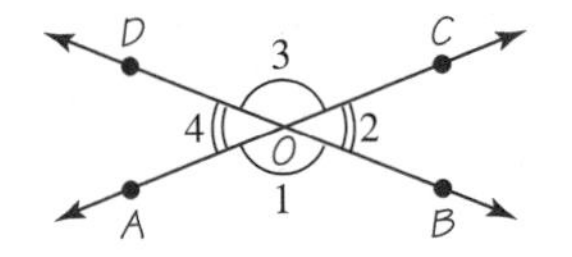

Figure 5.16

Locate each of the following angles in Figure 5.16.

a. $\angle AOB$ **b.** $\angle COB$ **c.** $\angle BOA$
d. $\angle DOC$ **e.** $\angle 3$ **f.** $\angle 4$ ■

To construct an angle congruent to a given angle B, first draw a ray from B', as shown in Figure 5.17(a). Next, mark off an arc with the pointer at the vertex of the given angle and label the points A and C. Without changing the compass, mark off a similar arc with the pointer at B', as shown in Figure 5.17(b). Label the point C' where this arc crosses the ray from B'. Place the pointer at C and set the compass to the distance from C to A. Without changing the compass, put the pointer at C' and strike an arc to make a point of intersection with the arc from C', as shown in Figure 5.17(c). Finally, draw a ray from B' through A'.

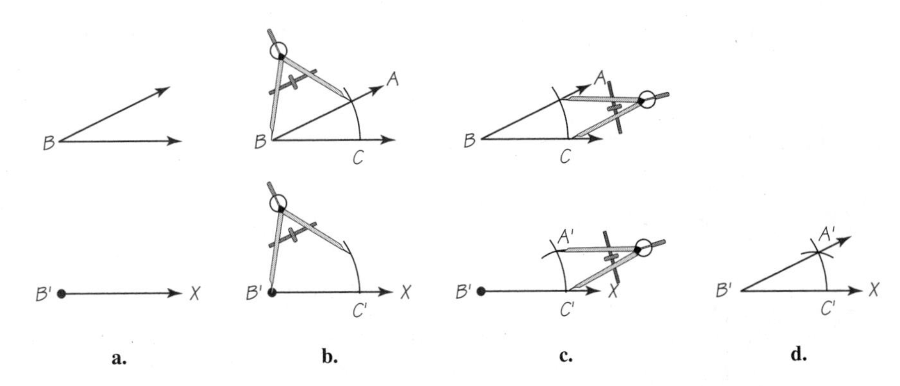

Figure 5.17 Construction of an angle congruent to a given angle

Two angles are said to be **equal** if they have the same measure. If we write m in front of an angle symbol, we mean the measure of the angle rather than the angle itself. Notice in Example 1 that parts **a** and **c** name the *same angle,* so $\angle AOB = \angle BOA$. Also notice the single and double arcs used to mark the angles in Example 1; these are used to denote *equal angles,* so $m\angle COB = m\angle AOD$ and $m\angle COD = m\angle AOB$. Denoting an angle by a single letter is preferred, except in the case (as shown by Example 1) where several angles share the same vertex.

Angles are usually measured using a unit called a **degree,** which is defined to be $\frac{1}{360}$ of a full revolution. The symbol ° is used to designate degrees. To measure an angle, you can use a **protractor,** but in this book the angle whose measures we need will be labeled as in Figure 5.18.

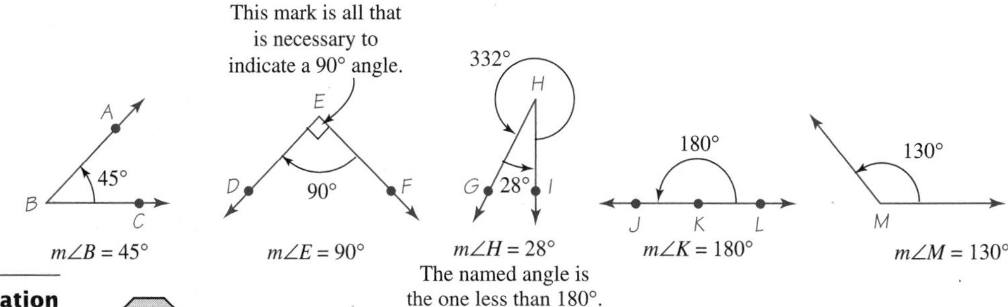

Figure 5.18 Labeling angles

STOP

Table 5.1
Types of Angles

Type	Classification
less than 90°	**Acute**
equal to 90°	**Right**
between 90° and 180°	**Obtuse**
equal to 180°	**Straight**

Angles are sometimes classified according to their measures, as shown in Table 5.1. It is also easy to see the plausibility of Euclid's fourth postulate that all right angles are equal to one another.

EXAMPLE 2 **Classification of Angles**

Label the angles B, E, H, K, and M in Figure 5.18 by classification.

Solution $\angle B$ is **acute;** $\angle E$ is **right;** $\angle H$ is **acute;** $\angle K$ is **straight;** $\angle M$ is **obtuse.** ■

PROBLEM SET 5.2

LEVEL 1

WHAT IS WRONG, *if anything, with each of the statements in Problems 1–3? Explain your reasoning.*

1. Two angles are said to be equal if they are the same measure.
2. A right angle is an angle whose measure is 180°.
3. If the rays of an angle point in opposite directions so that the rays form a straight line, we say the angle is a straight angle.

Level 1—Drill

Name the polygons in Problems 4–15 *according to the number of sides.*

4. 

5. 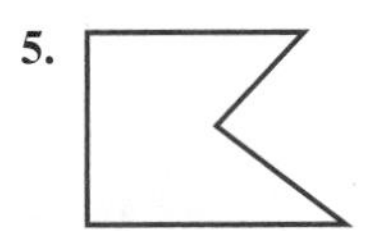

6. 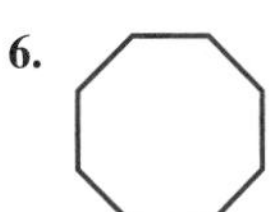

7. 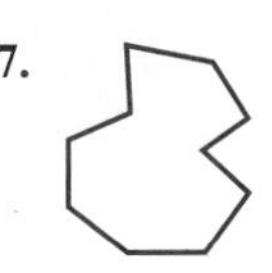

8. 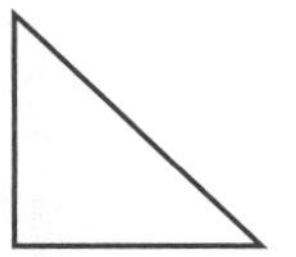

9. 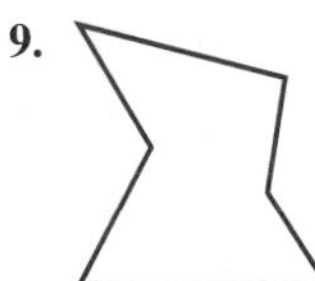

10.

11. 

12. 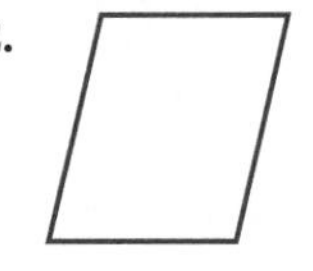

13. 

14. 

15. 

Level 2

Using only a straightedge and a compass, reproduce the angles given in Problems 16–21.

16. 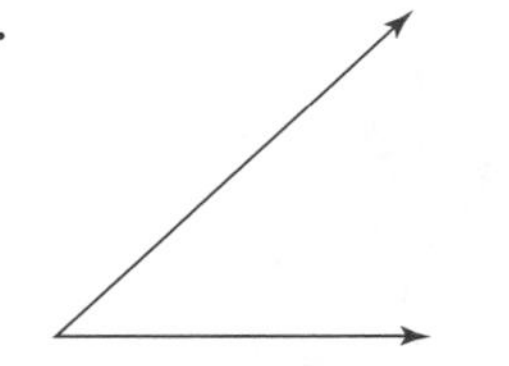

17. 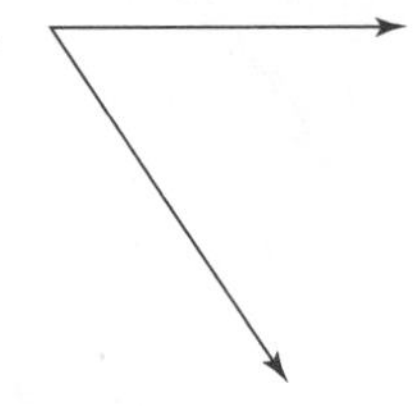

18. **19.** **20.** **21.**

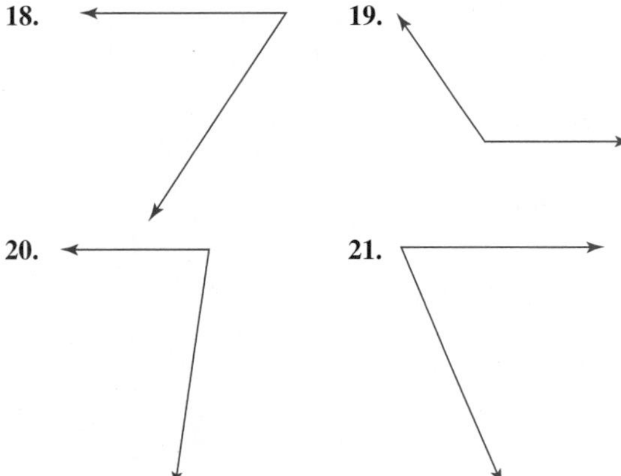

Consider Figure 5.19 *for Problems* 22–25.

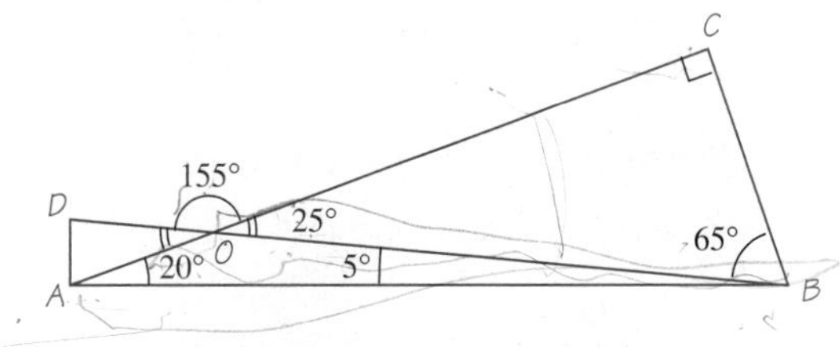

Figure 5.19 Angles in a triangle

22. Classify the angles as acute, right, straight, or obtuse.
a. $\angle DOC$ **b.** $\angle AOB$ **c.** $\angle DBC$ **d.** $\angle CAB$ **e.** $\angle DOB$

23. Classify the angles as acute, right, straight, or obtuse.
a. $\angle C$ **b.** $\angle COB$ **c.** $\angle AOC$ **d.** $\angle DOA$ **e.** $\angle OBA$

24. Name a pair of equal angles.

25. Name an angle equal to $\angle DOB$.

Consider Figure 5.20 *for Problems* 26–29.

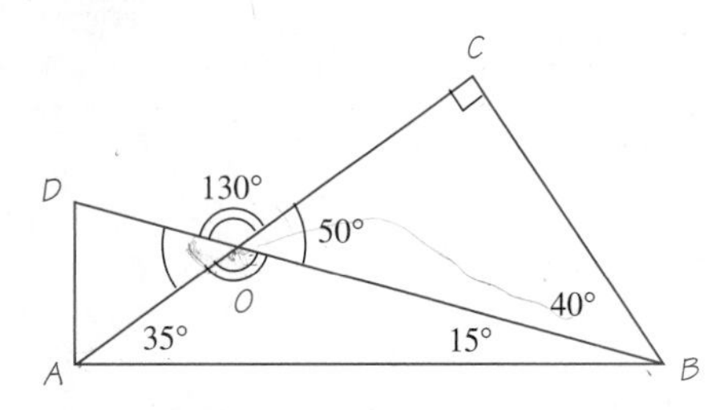

Figure 5.20 Angles in a triangle

26. Classify the angles as acute, right, straight, or obtuse.
a. $\angle DOC$ **b.** $\angle AOB$ **c.** $\angle DBC$ **d.** $\angle CAB$ **e.** $\angle DOB$

27. Classify the angles as acute, right, straight, or obtuse.
a. $\angle C$ **b.** $\angle COB$ **c.** $\angle AOC$ **d.** $\angle DOA$ **e.** $\angle OBA$

28. Name a pair of equal angles.

29. Name an angle equal to $\angle DOB$.

Consider Figure 5.21 *for Problems* 30–33.

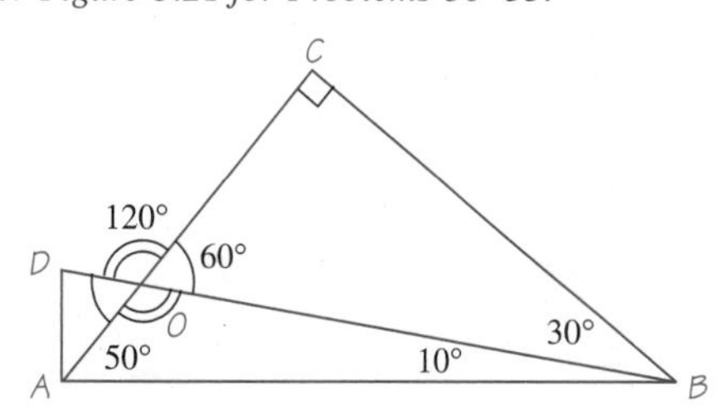

Figure 5.21 Angles in a triangle

30. Classify the angles as acute, right, straight, or obtuse.
a. $\angle DOC$ **b.** $\angle AOB$ **c.** $\angle DBC$ **d.** $\angle CAB$ **e.** $\angle DOB$

31. Classify the angles as acute, right, straight, or obtuse.
a. $\angle C$ **b.** $\angle COB$ **c.** $\angle AOC$ **d.** $\angle DOA$ **e.** $\angle OBA$

32. Name a pair of equal angles.

33. Name an angle equal to $\angle DOB$.

Level 3—Mind Bogglers

34. The first illustration in the accompanying figure shows a cube with the top cut off. Use solid lines and shading to depict seven other different views of a cube with one side cut off.

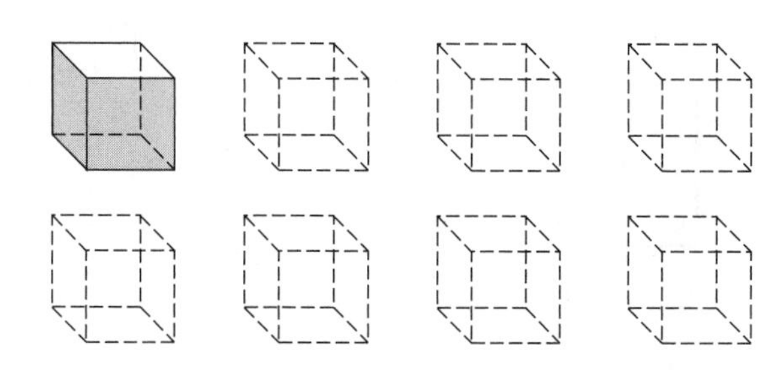

5.3 TRIANGLES

Historical Perspective

Benjamin Banneker (1731–1806)

Geometrical concepts are extremely important in surveying. One of the most perfectly planned cities is Washington, D.C. The center of the city is the Capitol, and the city is divided into four sections: Northwest, Northeast, Southwest, and Southeast. These are separated by North, South, East, and Capitol streets, which all converge at the Capitol. The initial surveying of this city was done by a group of mathematicians and surveyors, which included the first distinguished black mathematician, Benjamin Banneker. He was born in 1731 in Maryland and showed extreme talent in mathematics without the benefit of formal education. His story, as well as those of other prominent black mathematicians, can be found in a book by Virginia K. Newell, Joella H. Gipson, L. Waldo Rich, and Beauregard Stubblefield, titled *Black Mathematicians and Their Works* (Ardmore, Pennsylvania: Dorrance, 1980).

One of the most frequently encountered polygons is the triangle. Every triangle has six parts: three sides and three angles. We name the sides by naming the endpoints of the line segments, and we name the angles by identifying the vertex (see Figure 5.22).

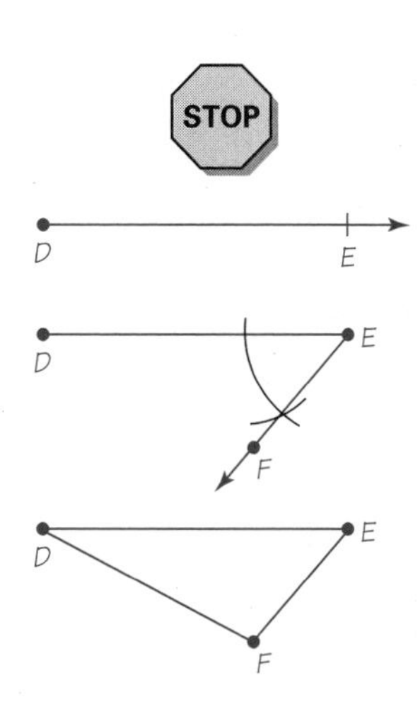

Figure 5.23 Constructing congruent triangles

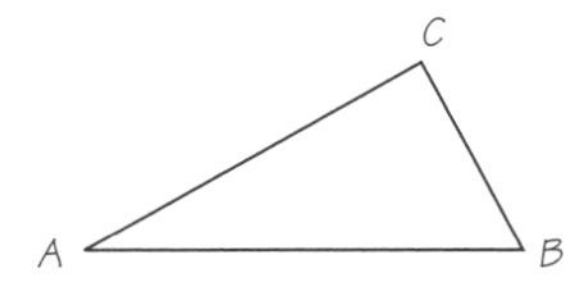

SIDES	*ANGLES*
$\overline{AB}$	$\angle A$
$\overline{CB}$	$\angle B$
$\overline{AC}$	$\angle C$

Figure 5.22 A standard triangle showing the six parts

We say that two triangles are **congruent** if they have the same size and shape. Suppose that we wish to construct a triangle congruent to ΔABC with vertices D, E, and F, as shown in Figure 5.23. We would proceed as follows:

1. Mark off segment $\overline{DE}$ so that it is congruent to $\overline{AB}$. We write this as $\overline{DE} \simeq \overline{AB}$.
2. Construct angle E so that it is congruent to angle B. We write this as $\angle E \simeq \angle B$.
3. Mark off segment $\overline{EF} \simeq \overline{BC}$.

You can now see that, if you connect points D and F with a straightedge, the resulting ΔDEF has the same size and shape as ΔABC. The procedure we used here is called SAS, meaning we constructed two sides and an *included* angle (an angle between two sides) congruent to two sides and an included angle of another triangle. We call these **corresponding parts.** There are other procedures for constructing congruent triangles. Some of these are discussed in Problem Set 5.3. For this example, we say $\Delta ABC \simeq \Delta DEF$. From this we conclude that all six corresponding parts are congruent.

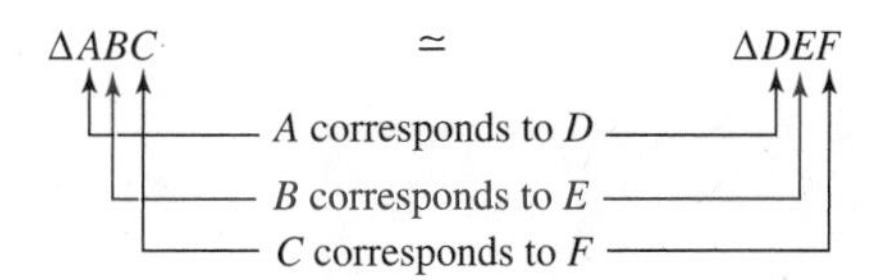

EXAMPLE 1

Finding Corresponding Parts of a Given Triangle

Name the corresponding parts of the given triangles.

a. $\Delta ABC \simeq \Delta A'B'C'$

b. $\Delta RST \simeq \Delta UST$

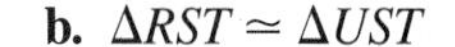

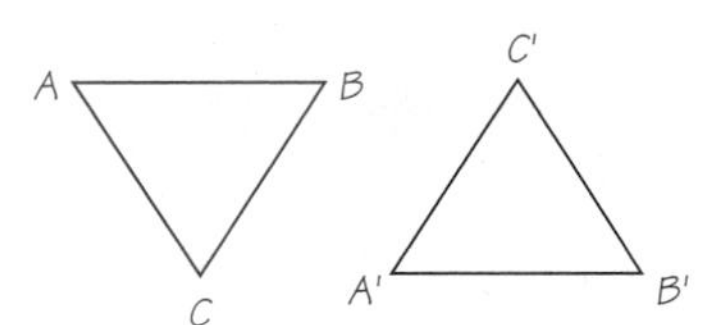

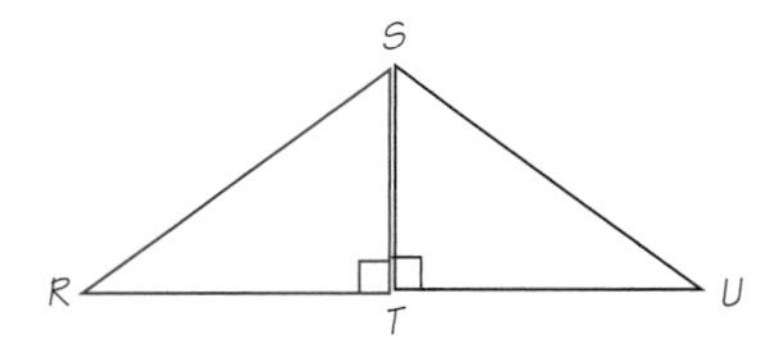

Solution

a. $\overline{AB}$ corresponds to $\overline{A'B'}$
$\overline{AC}$ corresponds to $\overline{A'C'}$
$\overline{BC}$ corresponds to $\overline{B'C'}$
$\angle A$ corresponds to $\angle A'$
$\angle B$ corresponds to $\angle B'$
$\angle C$ corresponds to $\angle C'$

b. $\overline{RS}$ corresponds to $\overline{US}$
$\overline{RT}$ corresponds to $\overline{UT}$
$\overline{ST}$ corresponds to $\overline{ST}$
$\angle R$ corresponds to $\angle U$
$\angle RTS$ corresponds to $\angle UTS$
$\angle RST$ corresponds to $\angle UST$ ■

One of the most basic properties of triangles involves the sum of the measures of the angles of a triangle. To discover this property for yourself, place a pencil with an eraser along one side of any triangle as shown in Figure 5.24(a).

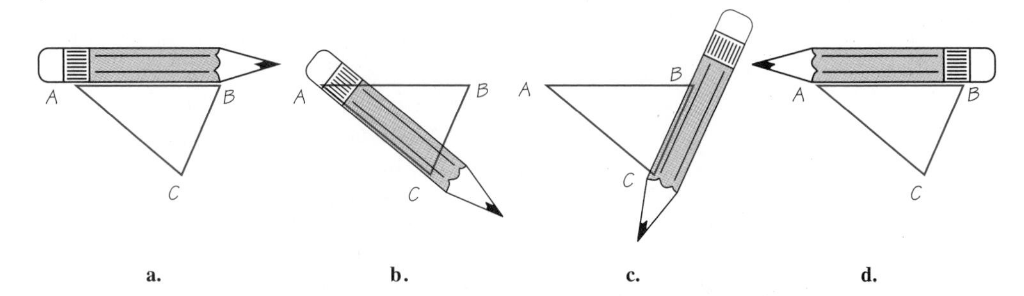

Figure 5.24 Demonstration that the sum of the measures of the angles in a triangle is 180°

You should remember the result shown in the box.

Now rotate the pencil to correspond to the size of $\angle A$, as shown in Figure 5.24(b). You see your pencil is along side $\overline{AC}$. Next, rotate the pencil through $\angle C$, as shown in Figure 5.24(c). Finally, rotate the pencil through $\angle B$. Notice that the pencil has been rotated the same amount as the sum of the angles of the triangle. Also notice that the orientation of the pencil is exactly reversed from the starting position. This leads us to the following important property.

Sum of the Measures of the Angles in a Triangle

The sum of the measures of the angles in any triangle is 180°.

EXAMPLE 2 **Finding a Missing Angle of a Triangle**

Find the missing angle measure in the triangle in Figure 5.25.

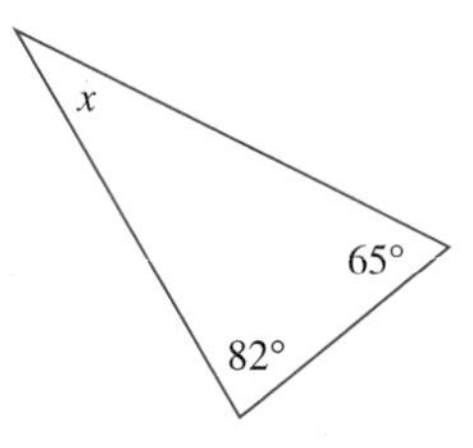

Figure 5.25 What is x?

Solution Let x represent the missing angle's measure.

$$65 + 82 + x = 180$$
$$147 + x = 180$$
$$x = 33$$

The missing angle's measure is 33°. ■

EXAMPLE 3 Using Algebra to Find the Angles of a Triangle

Find the measures of the angles of a triangle if it is known that the measures are x, $2x - 15$, and $3(x + 17)$ degrees.

Solution Using the theorem for the sum of the measures of the angles in a triangle, we have:

$$x + (2x - 15) + 3(x + 17) = 180$$
$$x + 2x - 15 + 3x + 51 = 180 \quad \text{Eliminate parentheses.}$$
$$6x + 36 = 180 \quad \text{Combine similar terms.}$$
$$6x = 144 \quad \text{Subtract 36 from both sides.}$$
$$x = 24 \quad \text{Divide both sides by 6.}$$

Now find the angle measures:

$$x = 24$$
$$2x - 15 = 2(24) - 15 = 33$$
$$3(x + 17) = 3(24 + 17) = 123$$

The angles have measures of 24°, 33°, and 123°. ■

An **exterior angle** of a triangle is the angle on the other side of an extension of one side of the triangle. An example is the angle whose measure is marked as x in Figure 5.26. Notice that the following relationships are true for any ΔABC with exterior angle x:

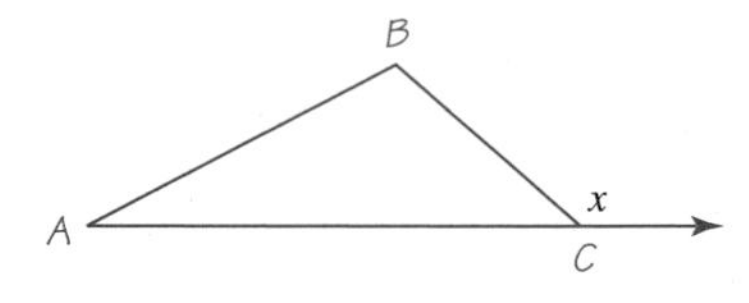

Figure 5.26 Exterior angle x

$$m\angle A + m\angle B + m\angle C = 180° \quad \text{and} \quad m\angle C + x = 180°$$

Thus,

$$m\angle A + m\angle B + m\angle C = m\angle C + x$$
$$m\angle A + m\angle B = x \quad \text{Subtract } m\angle C \text{ from both sides.}$$

Exterior Angles of a Triangle

The measure of the exterior angle of a triangle equals the sum of the measures of the two opposite interior angles.

EXAMPLE 4 Finding an Exterior Angle

Find the value of x in Figure 5.27.

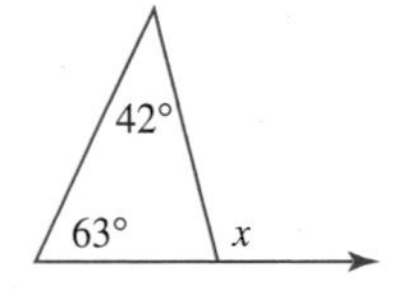

Figure 5.27 What is x?

Solution

Sum of interior angles = Exterior angle

$$\underbrace{63 + 42} = x$$
$$105 = x$$

The measure of the exterior angle is 105°. ■

PROBLEM SET 5.3

Essential Ideas

1. What is the sum of the angles of any triangle?
2. What is the sum of the measures of the acute angles of a right triangle?

Level 1—Drill

Name the corresponding parts of the triangles in Problems 3–6.

3. **4.**

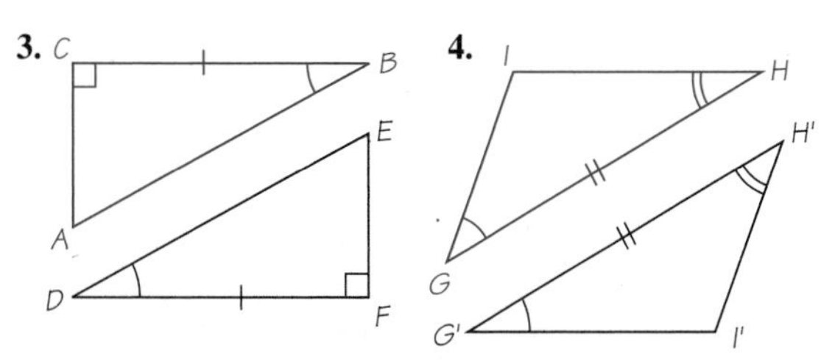

5.

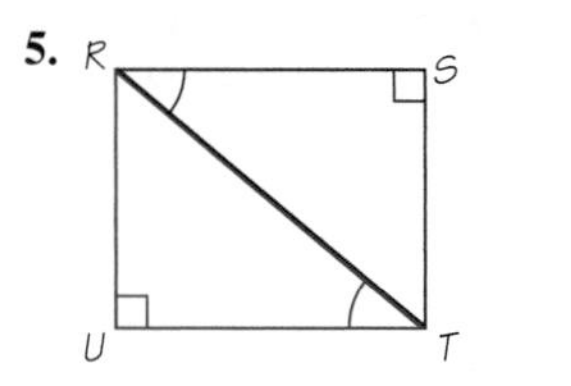

6.

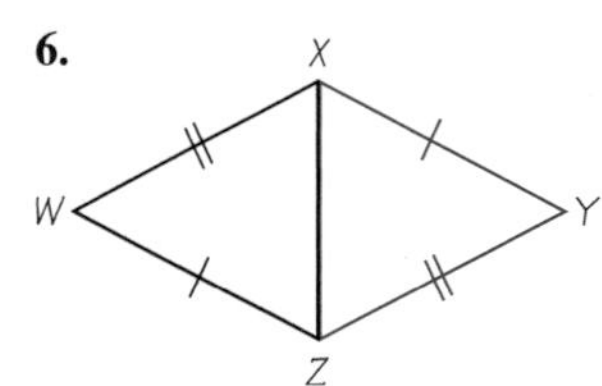

In Problems 7–12, find the measure of the third angle in each triangle.

7. **8.** **9.** **10.** **11.** **12.**

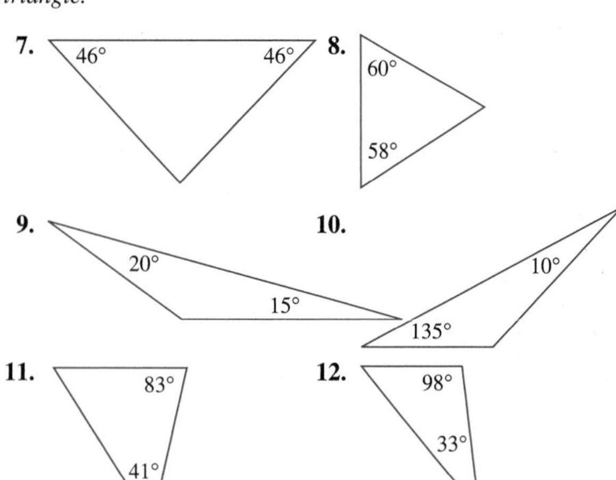

Find the measure of the indicated exterior angle in each of the triangles in Problems 13–18.

13.

35°
40°
x

14.

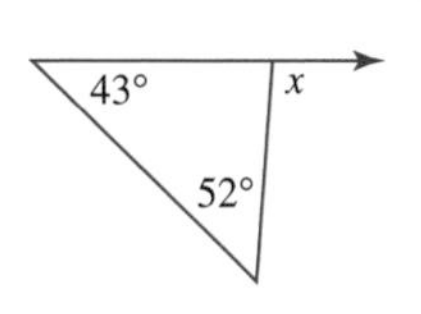

15.

x
100°
20°

16.

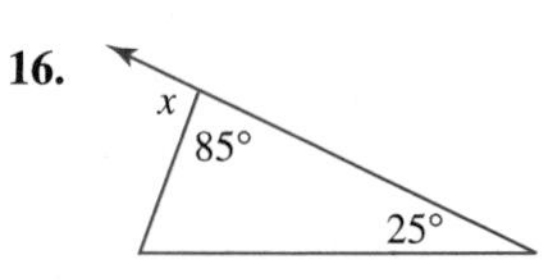

17.

x
80°
30°
70°

18.

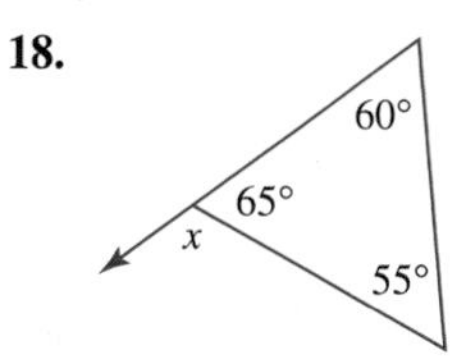

Level 2—Drill

Using only a straightedge and a compass, reproduce the triangles given in Problems 19–24.

19.

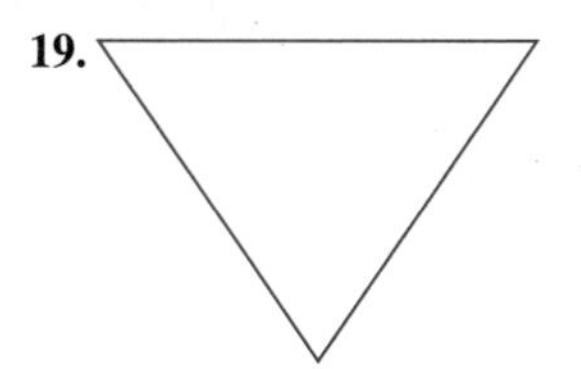

20.

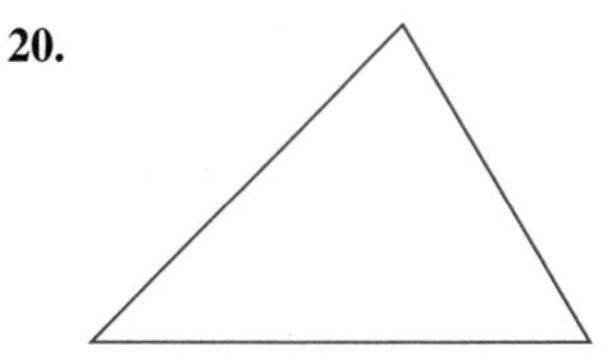

21.

22.

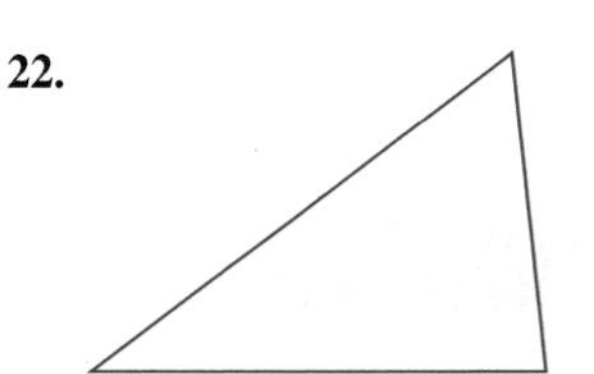

23.

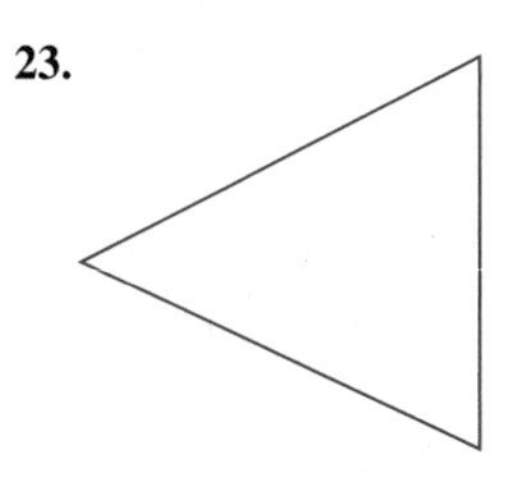

24.

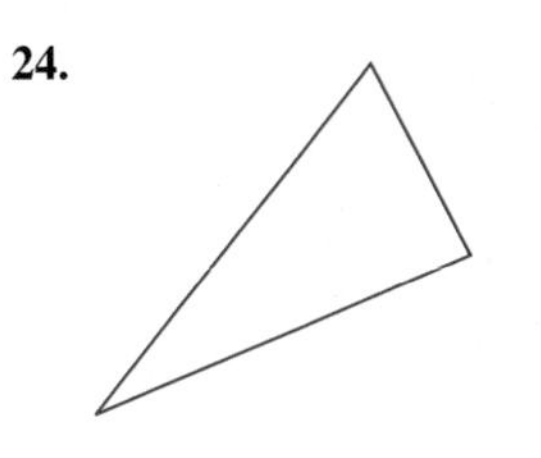

Level 2—Applications

In Problems 25–28, assume that the given angles have been measured, and calculate the angles (marked as x) that could not be found by direct measurement.

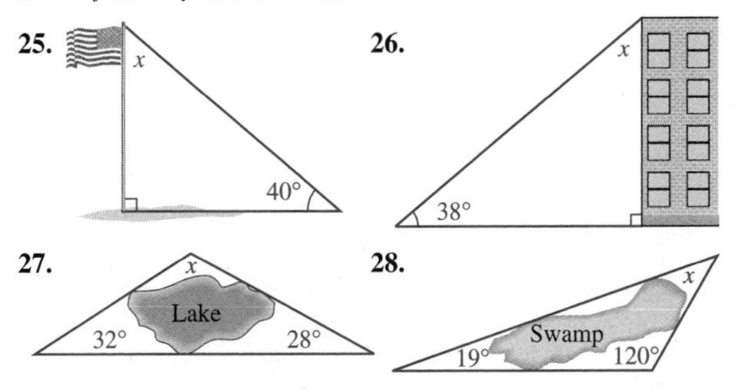

Level 2—Drill

Use algebra to find the value of x in each of the triangles in Problems 29–34. Notice that the measurement of the angle is not necessarily the same as the value of x.

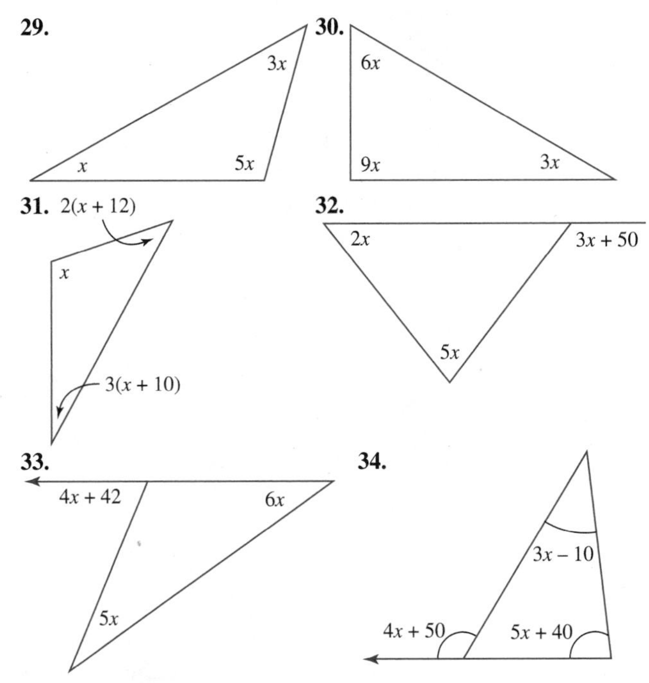

35. Find the measures of the angles of a triangle if it is known that the measures of the angles are x, $2x$, and $3x$.

36. Find the measures of the angles of a triangle if it is known that the measures of the angles are $2x + 30°$, $3x - 50°$, and $4x + 20°$.

37. Find the measures of the angles of a triangle if it is known that the measures of the angles are x, $14 + 3x$, and $3(x + 25)$.

38. Find the measures of the angles of a triangle if it is known that the measures of the angles are x, $3x - 10$, and $3(55 - x)$.

39. In the text we constructed congruent triangles by using SAS. Reproduce the triangle shown in Problem 21 by using SSS. This means constructing the triangle by using the lengths of the three sides.

40. In the text we constructed congruent triangles by using SAS. Reproduce the triangle shown in Problem 22 by using SSS. This means constructing the triangle by using the lengths of the three sides.

41. In the text we constructed congruent triangles by using SAS. Reproduce the triangle shown in Problem 23 by using ASA. This means constructing the triangle by using a side included between two angles.

42. In the text we constructed congruent triangles by using SAS. Reproduce the triangle shown in Problem 24 by using ASA. This means constructing the triangle by using a side included between two angles.

43. Show that *the sum of the measures of the interior angles of any triangle is* 180° by carrying out the following steps.

a. Draw three triangles: one with all acute angles, one with a right angle, and a third with an obtuse angle.

b. Tear apart the angles of each triangle you've drawn.

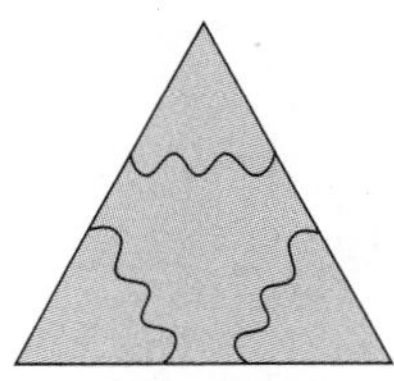

c. Place the pieces together to form a straight angle.

44. Show that *the sum of the measures of the interior angles of any quadrilateral is* 360° by carrying out the following steps.

a. Draw any quadrilateral, as illustrated (but draw your quadrilateral so it has a different shape from the one shown here).

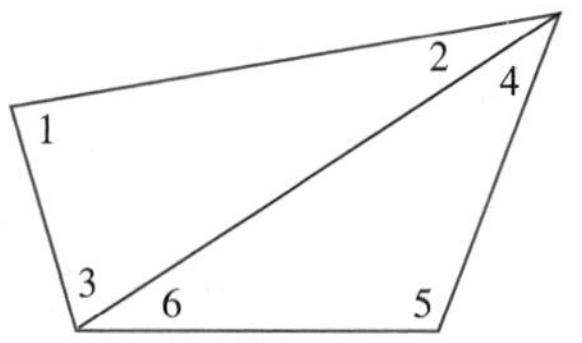

b. Divide the quadrilateral into two triangles by drawing a diagonal (a line segment connecting two nonadjacent vertices). Label the angles of your triangles as shown in the figure.

c. What is the sum $m\angle 1 + m\angle 2 + m\angle 3$?

d. What is the sum $m\angle 4 + m\angle 5 + m\angle 6$?

e. The sum of the measures of the angles of the quadrilateral is

$(m\angle 1 + m\angle 2 + m\angle 3) + (m\angle 4 + m\angle 5 + m\angle 6)$

What is this sum?

f. Do you think this argument will apply for *any* quadrilateral?

LEVEL 3—Mind Bogglers

45. Look at Problem 44. What is the sum of the measures of the interior angles of any pentagon?

46. Look at Problem 44. What is the sum of the measures of the interior angles of any octagon?

LEVEL 3—Team Project

47. With a team of two or three others, prepare a portfolio of photographs of buildings with interesting architecture. Find the architectural name for the building style, as well as the name of the mathematical solid that most closely approximates the shape of the building.

5.4 SIMILAR TRIANGLES

Historical Perspective

As Europe passed from the Middle Ages to the Renaissance, artists were at the forefront of the intellectual revolution. No longer satisfied with flat-looking scenes, they wanted to portray people and objects as they looked in real life. The artists' problem was one of dimension, and dimension is related to mathematics, so many of the Renaissance artists had to solve some original mathematical problems. The ideas of shapes, rectangles, triangles, and circles, and how these figures related to each other were of importance not only to mathematicians, but also to artists and architects.

Renaissance perspective

Congruent figures have exactly the same size and shape. However, it is possible for figures to have exactly the same shape without having the same size. Such figures are called **similar figures.** In this section, we will focus on **similar triangles.** If ΔABC is similar to ΔDEF, we write

$$\Delta ABC \sim \Delta DEF$$

Similar triangles are shown in Figure 5.28. Since these figures have the same shape, we talk about **corresponding angles** and **corresponding sides.** The corresponding angles of similar triangles are those angles that are equal. The corresponding sides are those sides that are opposite equal angles.

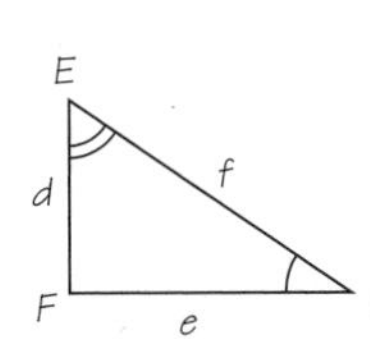

$m\angle A = m\angle D$, so these are corresponding angles.
$m\angle B = m\angle E$, so these are corresponding angles.
$m\angle C = m\angle F$, so these are corresponding angles.
Side $\overline{BC}$ is opposite $\angle A$ and side $\overline{EF}$ is opposite $\angle D$, so we say that $\overline{BC}$ corresponds to $\overline{EF}$.
$\overline{AC}$ corresponds to $\overline{DF}$.
$\overline{AB}$ corresponds to $\overline{DE}$.

Figure 5.28 Similar triangles

Even though corresponding angles are equal, corresponding sides do not need to have the same length. If they do have the same length, the triangles are congruent. However, when they are not the same length, we can say they are proportional. From Figure 5.28, we see that the lengths of the sides are labeled *a, b,* and *c* and *d, e,* and *f.* When we say the sides are proportional, we mean

PRIMARY RATIOS	*RECIPROCALS*
$\frac{a}{b}=\frac{d}{e} \quad \frac{a}{c}=\frac{d}{f} \quad \frac{b}{c}=\frac{e}{f}$	$\frac{b}{a}=\frac{e}{d} \quad \frac{c}{a}=\frac{f}{d} \quad \frac{c}{b}=\frac{f}{e}$

Similar Triangles

Two triangles are similar if two angles of one triangle are equal to two angles of the other triangle. If the triangles are similar, then their corresponding sides are proportional.

EXAMPLE 1 **Corresponding Parts of Similar Triangles**

Identify pairs of triangles that are similar in Figure 5.29.

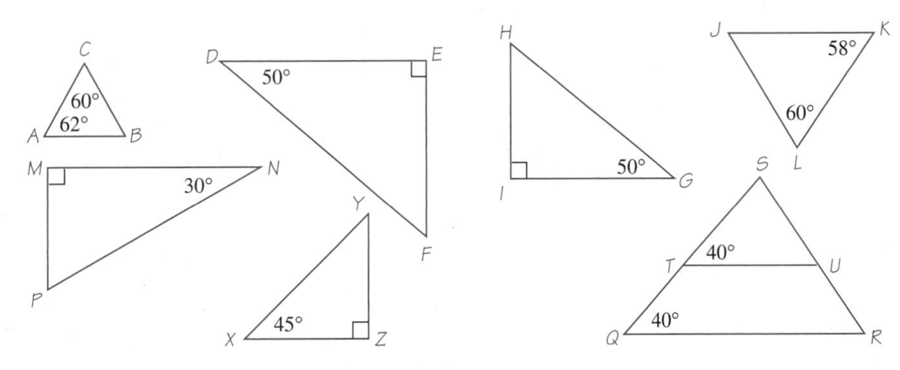

Figure 5.29 Which of these triangles are similar?

Solution $\Delta ABC \sim \Delta JKL$; $\Delta DEF \sim \Delta GIH$; $\Delta SQR \sim \Delta STU$ ■

EXAMPLE 2

Finding Lengths of Sides of Similar Triangles

Given the similar triangles in Figure 5.30, find the unknown lengths marked b' and c'.

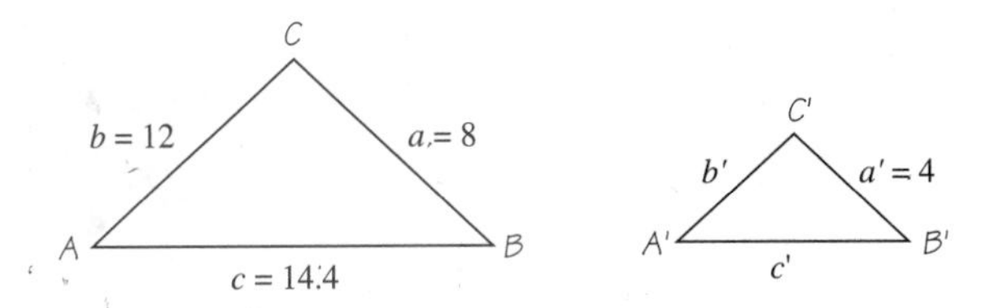

Figure 5.30 Given $\Delta ABC \sim \Delta A'B'C'$

Solution Since corresponding sides are proportional (other proportions are possible), we have

$$\frac{a'}{a} = \frac{b'}{b} \qquad \frac{a}{c} = \frac{a'}{c'}$$

$$\frac{4}{8} = \frac{b'}{12} \qquad \frac{8}{14.4} = \frac{4}{c'}$$

$$b' = \frac{4 \times 12}{8} \qquad c' = \frac{14.4 \times 4}{8}$$

$$b' = \mathbf{6} \qquad c' = \mathbf{7.2}$$

■

Finding similar triangles is simplified even further if we know that the triangles are right triangles, because then the triangles are similar if one of the acute angles of one triangle has the same measure as an acute angle of the other.

EXAMPLE 3

Problem Solving with Similar Triangles

Suppose that a tree and a yardstick are casting shadows as shown in Figure 5.31. If the shadow of the yardstick is 3 yards long and the shadow of the tree is 12 yards long, use similar triangles to estimate the height h of the tree if you know that $m\angle S = m\angle S'$.

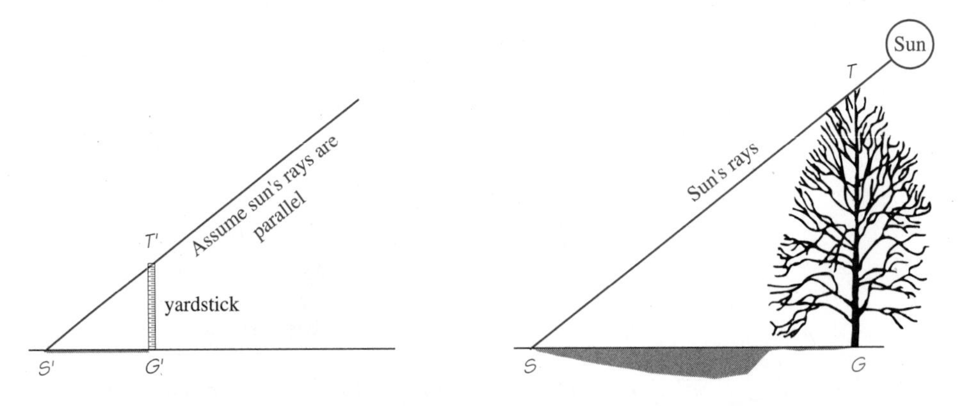

Figure 5.31 Finding the height of a tall object by using similar triangles

Solution Since $\angle G$ and $\angle G'$ are right angles, and since $m\angle S = m\angle S'$, we know that $\Delta SGT \sim \Delta S'G'T'$. Therefore corresponding sides are proportional.

$$\frac{\text{LENGTH OF YARDSTICK}}{\text{DISTANCE FROM } S' \text{ TO } G'} = \frac{\text{HEIGHT OF TREE}}{\text{DISTANCE FROM } S \text{ TO } G}$$

$$\frac{1}{3} = \frac{h}{12}$$ You solved proportions like this in Chapter 4.

$$h = \frac{1 \times 12}{3}$$

$$h = 4$$

The tree is 4 yards (or 12 feet) tall. ■

There is a relationship between the sizes of the angles of a right triangle and the ratios of the lengths of the sides. In a right triangle, the side opposite the right angle is called the **hypotenuse.** Each of the acute angles of a right triangle has one side that is the hypotenuse. The other side of that angle is called the **adjacent side.**

EXAMPLE 4 Sides of a Right Triangle

In ΔABC with right angle at C, as shown in Figure 5.32,

the hypotenuse is c,
the side adjacent to $\angle A$ is b,
the side adjacent to $\angle B$ is a.

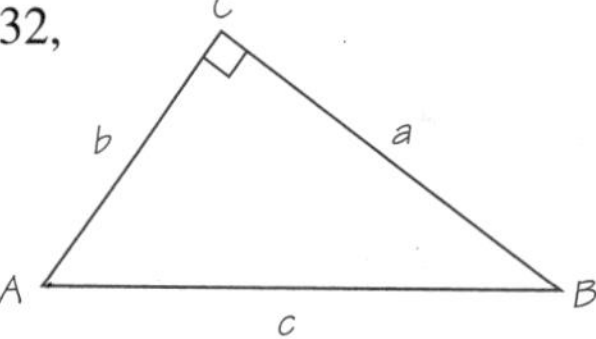

Figure 5.32 A right triangle

We also talk about an **opposite side.** The side opposite $\angle A$ is a, and the side opposite $\angle B$ is b. ■

You might recall the Pythagorean theorem, which was introduced in Chapter 2, which we now expand to help us decide if a given triangle is a right triangle.

Pythagorean Theorem

For any right triangle with sides a and b and hypotenuse c,

$$a^2 + b^2 = c^2$$

Furthermore, if $a^2 + b^2 = c^2$ for three sides of a triangle, then the triangle is a right triangle.

EXAMPLE 5 **Problem Solving Using the Pythagorean Theorem**

A carpenter wants to make sure that the corner of a room is square (a right angle). If she measures out sides of 3 feet and 4 feet, how long should she make the diagonal (hypotenuse)?

Solution The hypotenuse is the unknown, so use the Pythagorean theorem:

$$\begin{aligned} c &= \sqrt{a^2 + b^2} \\ &= \sqrt{3^2 + 4^2} && \text{The sides are 3 and 4.} \\ &= \sqrt{9 + 16} \\ &= \sqrt{25} \\ &= 5 \end{aligned}$$

If she makes the diagonal 5 feet long, then the triangle will be a right triangle and the corner of the room will be square. ■

The primary ratios in a right triangle are called **trigonometric ratios** and are defined in the following box.

Trigonometric Ratios

In a right triangle ABC with right angle at C,

sin A (pronounced "**sine** of A") is the ratio $\dfrac{\text{opposite side of } A}{\text{hypotenuse}}$

cos A (pronounced "**cosine** of A") is the ratio $\dfrac{\text{adjacent side of } A}{\text{hypotenuse}}$

tan A (pronounced "**tangent** of A") is the ratio $\dfrac{\text{opposite of side } A}{\text{adjacent side of } A}$

EXAMPLE 6 **Finding Sides of a Triangle Using Trigonometric Ratios**

Given a right triangle with sides 5 and 12, find the trigonometric ratios for the angles A and B. Show your answers in both common-fraction and decimal-fraction form, with decimals rounded to four places.

Solution First, use the Pythagorean theorem to find the length of the hypotenuse:

$$\begin{aligned} c &= \sqrt{5^2 + 12^2} \\ &= \sqrt{25 + 144} \\ &= \sqrt{169} \\ &= 13 \end{aligned}$$

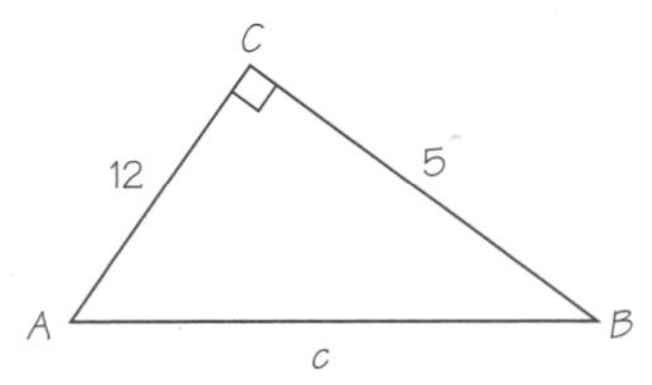

$$\sin A = \frac{5}{13} = \mathbf{0.3846}; \quad \cos A = \frac{12}{13} = \mathbf{0.9231}; \quad \tan A = \frac{5}{12} = \mathbf{0.4167}$$

$$\sin B = \frac{12}{13} = \mathbf{0.9231}; \quad \cos B = \frac{5}{13} = \mathbf{0.3846}; \quad \tan B = \frac{12}{5} = \mathbf{2.4}$$ ■

Tables of ratios for different angles are available (see, for example, Table I in Appendix D). Certain calculators have keys for the sine, cosine, and tangent ratios. If you have such a calculator, you should press the angle first and then the key for trigonometric ratio.*

EXAMPLE 7 **Finding Trigonometric Ratios**

Find the trigonometric ratios by using either Table I in Appendix D or a calculator. (Round calculator answers to four decimal places.)

a. sin 45° **b.** cos 32° **c.** tan 19°

Solution

a. sin 45° = **0.7071** from Table I; by calculator, *press:* [45] [sin]
b. cos 32° = **0.8480** from Table I; *press:* [32] [cos]
c. tan 19° = **0.3443** from Table I; *press:* [19] [tan] ■

Trigonometric ratios are useful in a variety of situations, as illustrated in Example 8.

EXAMPLE 8 **Problem Solving with Trigonometic Ratios**

The angle from the ground to the top of the Great Pyramid of Cheops is 52° if a point on the ground directly below the top is 351 ft away. What is the height of the pyramid (rounded to the nearest foot)?

Solution

From Figure 5.33, we see that for height h:

$$\tan 52^\circ = \frac{h}{351}$$

Solving for h by multiplying both sides by 351, we obtain

$$h = 351 \tan 52^\circ$$

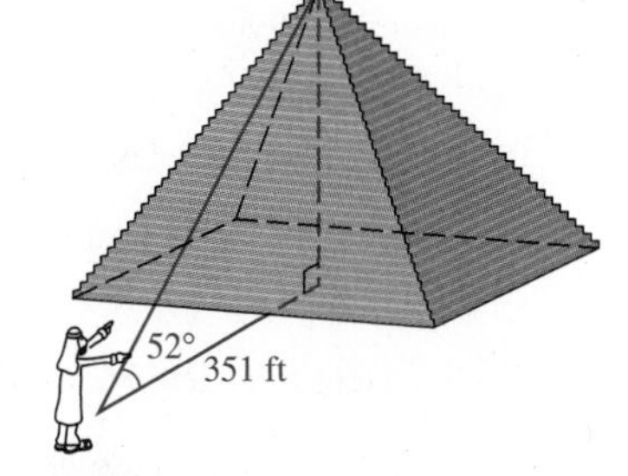

Figure 5.33 Calculating the height of the Great Pyramid of Cheops

Now we need to know the ratio for tan 52°. From Table I, tan 52° = 1.2799, so $h = 351(1.2799) = 449.2449$ or about **449 ft.**

By calculator, press: [351] [×] [52] [tan] [=]

Display: 449.2595129 or about 449 ft. ■

*Some brands of calculator require that you press the trigonometric ratio first and the angle second. For example, press [sin] [45] to obtain sin 45°.

Notice that the calculator and table answers in Example 8 are not identical. This is because both the table and the calculator give approximations of the exact value of tan 52°.

PROBLEM SET 5.4

ESSENTIAL IDEAS

1. **In Your Own Words** What are similar triangles?
2. **In Your Own Words** What is the important property regarding similar triangles?
3. **In Your Own Words** Explain the Pythagorean theorem.
4. What are the trigometric ratios?

LEVEL 1

WHAT IS WRONG, *if anything, with each of the statements in Problems 5–10? Explain your reasoning.*

5. If $\Delta ABC \sim \Delta A'B'C'$, then $\frac{a}{c} = \frac{a'}{b'}$.
6. If $\Delta ABC \sim \Delta A'B'C'$, then $\overline{BC}$ and $\overline{AB}$ are corresponding parts.
7. If two angles of one triangle are congruent to two angles of another triangle, then the triangles are similar.
8. In a right triangle ABC with right angle at C,
$$\sin B = \frac{\text{opposite side of } B}{\text{hypotenuse}}$$
9. In a right triangle ABC with right angle at C,
$$\cos B = \frac{\text{opposite side of } B}{\text{hypotenuse}}$$
10. In a right triangle ABC with right angle at C,
$$\tan B = \frac{\text{opposite side of } B}{\text{adjacent side of } B}$$

Use the right triangle in Figure 5.34 to answer the questions in Problems 11–21.

11. What is the side opposite $\angle A$?
12. What is the side opposite $\angle B$?
13. What is the side adjacent $\angle A$?
14. What is the side adjacent $\angle B$?

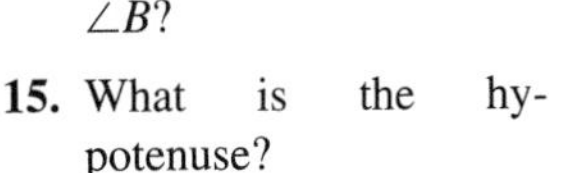

15. What is the hypotenuse?

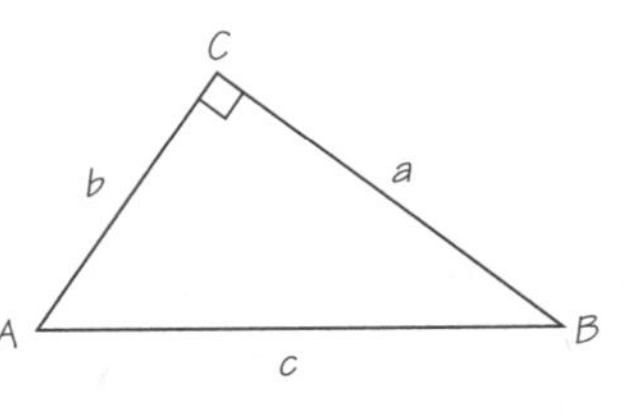

Figure 5.34 ΔABC

16. What is $\sin A$?
17. What is $\sin B$?
18. What is $\cos A$?
19. What is $\cos B$?
20. What is $\tan A$?
21. What is $\tan B$?

In Problems 22–27, tell whether it is possible to conclude that the pairs of triangles are similar. If they are similar, list the proportional parts.

22.

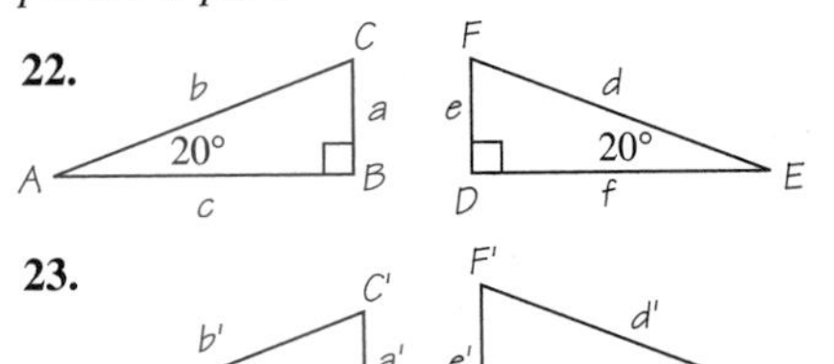

23.

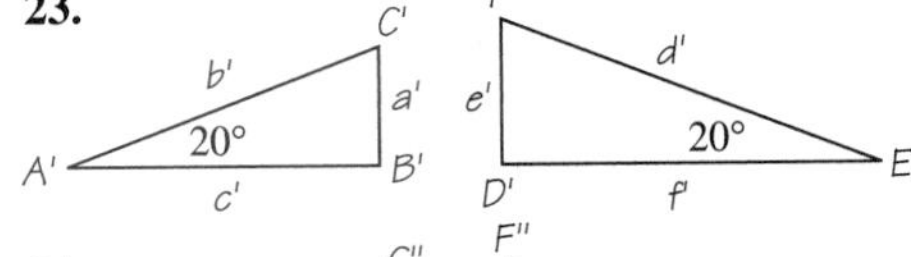

24.

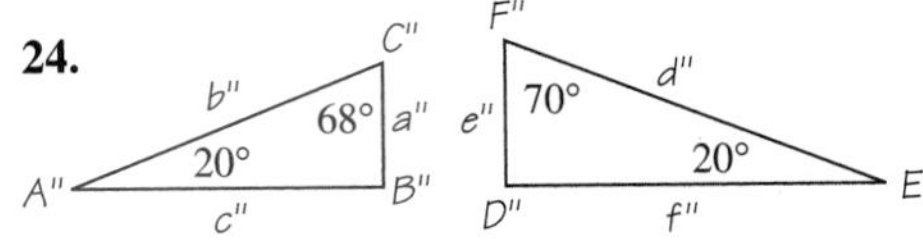

25.

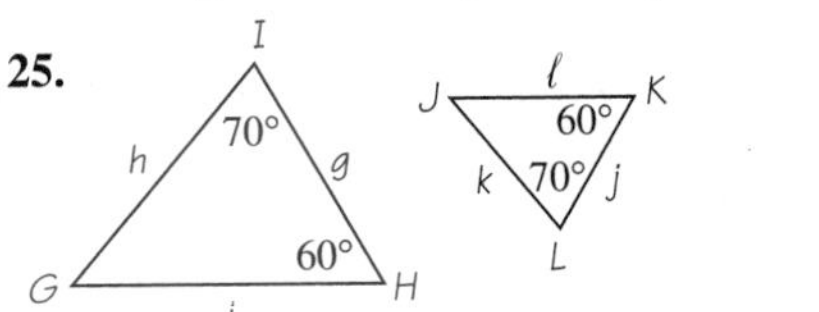

26.

27.

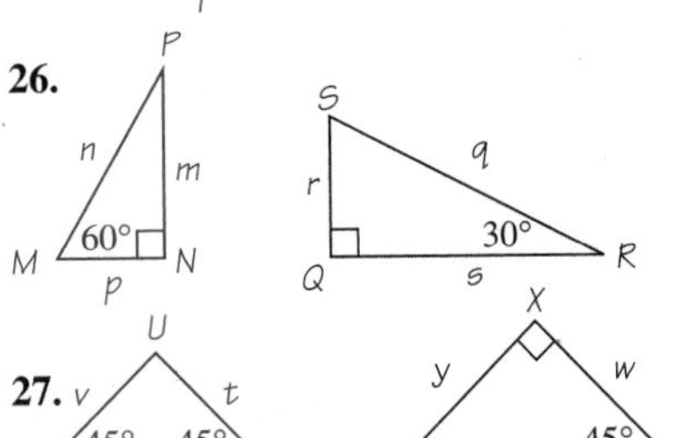

Level 2—Drill

Given two similar triangles, as shown in Figure 5.35, find the unknown lengths in Problems 28–35.

28. $a = 4$, $b = 8$; find c.

29. $a' = 7$, $b' = 3$; find c'.

Figure 5.35 Similar triangles

30. $a = 4$, $b = 8$, $a' = 2$; find b'.

31. $b = 5$, $c = 15$, $b' = 3$; find c'.

32. $c = 6$, $a = 4$, $c' = 8$; find a'.

33. $a' = 7$, $b' = 3$, $a = 5$; find b.

34. $b' = 8$, $c' = 12$, $c = 4$; find b.

35. $c' = 9$, $a' = 2$, $c = 5$; find a.

Find the trigonometric ratios in Problems 36–47 by using either Table I or a calculator. Round your answers to four decimal places.

36. $\sin 56°$ **37.** $\sin 15°$
38. $\sin 61°$ **39.** $\sin 18°$
40. $\cos 54°$ **41.** $\cos 8°$
42. $\cos 90°$ **43.** $\cos 34°$
44. $\tan 24°$ **45.** $\tan 52°$
46. $\tan 75°$ **47.** $\tan 89°$

In Problems 48–57, find the sine, cosine, and tangent for the angle A.

48.

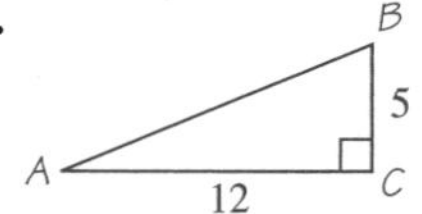

49.

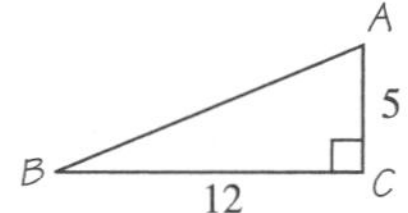

50.

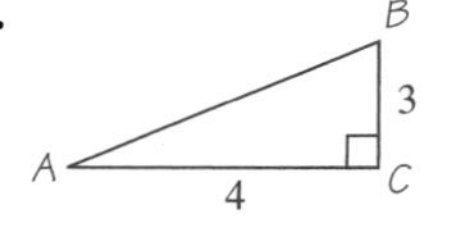

51.

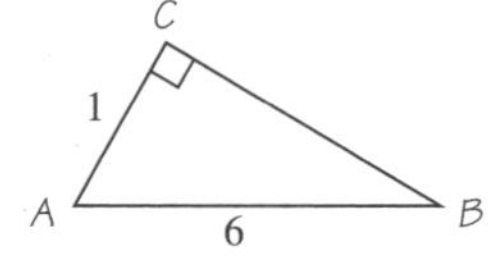

52.

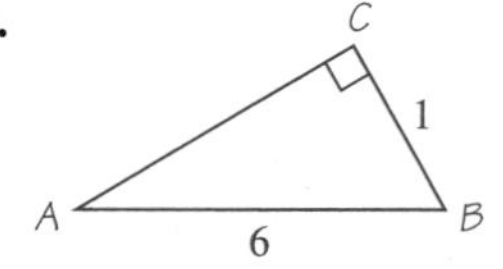

53.

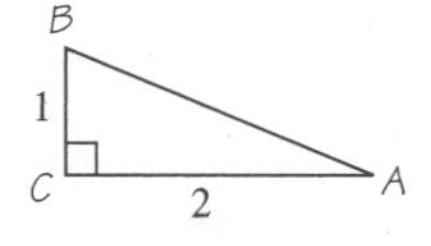

54.

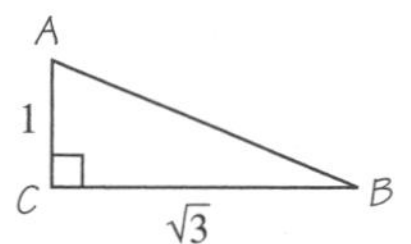

55.

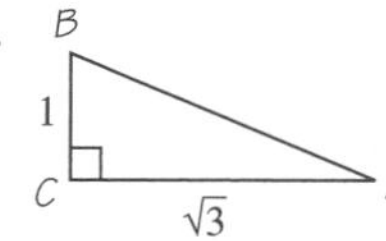

56.

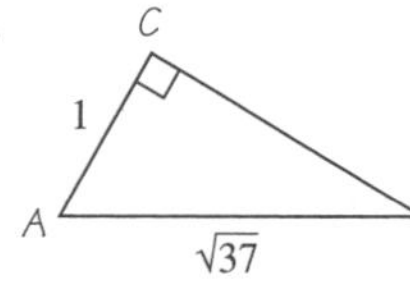

57.

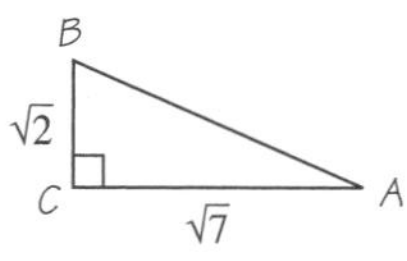

Level 2—Applications

58. Use similar triangles and a proportion to find the length of the lake shown in Figure 5.36.

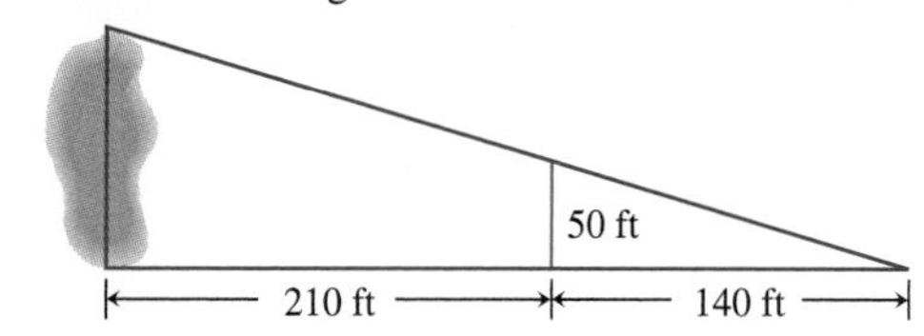

Figure 5.36 Determining the length of a lake

59. Use similar triangles and a proportion to find the length of the swamp shown in Figure 5.37.

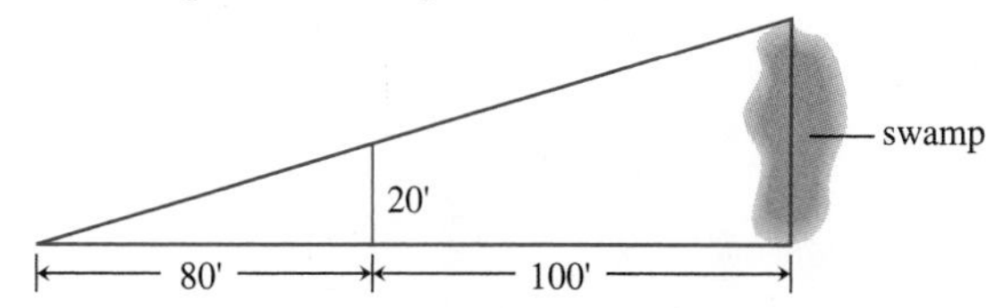

Figure 5.37 Determining the distance across a swamp

60. Use similar triangles and a proportion to find the height of the building shown in Figure 5.38.

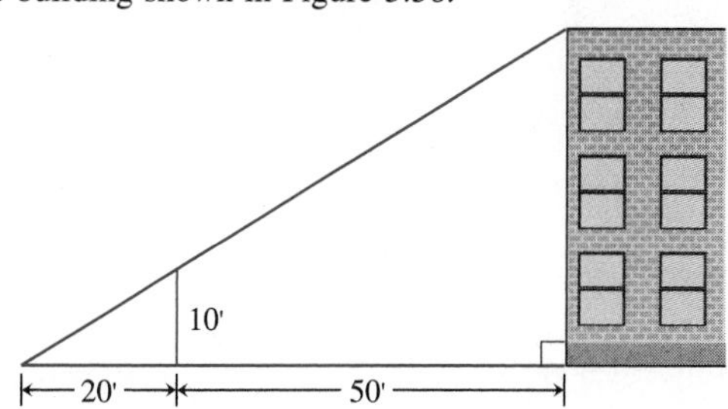

Figure 5.38 Determining the height of a building

61. Use similar triangles and a proportion to find the height of the house shown in Figure 5.39.

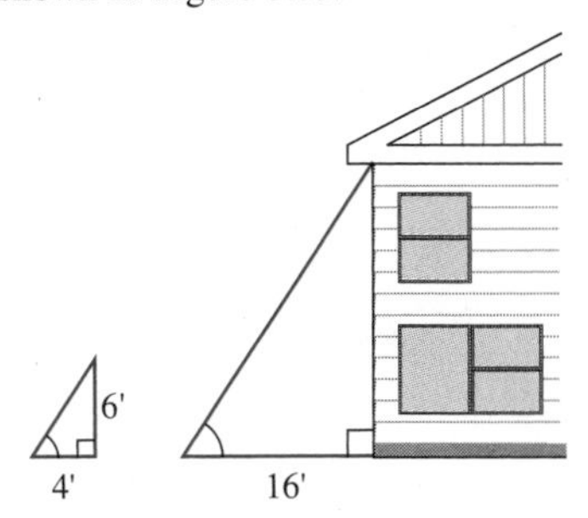

Figure 5.39 Determining the height of a house

62. Find the height of a flagpole, as shown in Figure 5.40, if it casts a shadow of 10 ft at the same time that the shadow of a 6-ft person is 5 ft.

Figure 5.40 Determining the height of a flagpole

63. If a carpenter wants to make sure that the corner of a room is square and measures 5 ft and 12 ft along the walls, how long should he make the diagonal?

64. If a carpenter wants to be sure that the corner of a building is square and measures 6 ft and 8 ft along the sides, how long should the diagonal be?

65. Find the height of a flagpole (rounded to the nearest foot) as shown in Figure 5.40 if it casts a shadow of 15 ft at the same time that the shadow of a 5-ft person is 4 ft.

66. If a tree casts a shadow of 12 ft at the same time that a 6-ft person casts a shadow of $2\frac{1}{2}$ ft, find the height of the tree (to the nearest foot).

67. If the angle from the horizontal to the top of a building is 38° and the horizontal distance from its base is 90 ft, what is the height of the building (to the nearest foot)?

68. If the angle from the horizontal to the top of a tower is 52° and the horizontal distance from its base is 85 ft, what is the height of the tower (to the nearest foot)?

Level 3—Mind Bogglers

69. If the distance from the earth to the sun is 92.9 million miles, and the angle formed between Venus, the earth, and the sun is 47° (as shown in the illustration), find the distance from the sun to Venus (to the nearest hundred thousand miles).

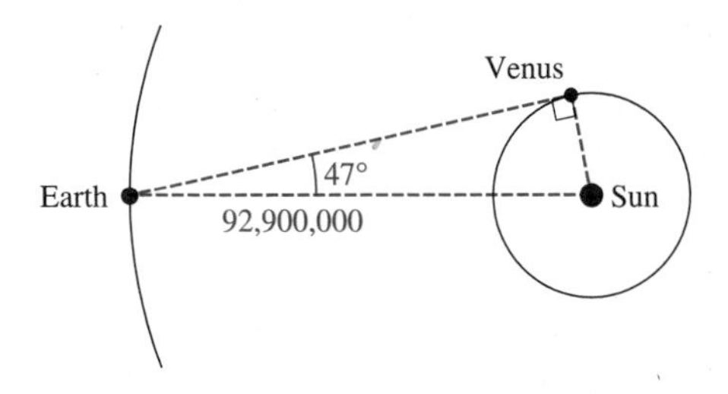

70. Use the information and illustration in Problem 69 to find the distance from the earth to Venus (to the nearest hundred thousand miles).

5.5 SUMMARY AND REVIEW

Important Terms

Acute angle [5.2]
Adjacent side [5.4]
Angle [5.2]
Axiom [5.1]
Compass [5.1]
Congruent [5.1]
Congruent triangles [5.3]
Construction [5.1]
Corresponding angles [5.4]
Corresponding parts [5.3]
Corresponding sides [5.4]
Cosine [5.4]

Decagon [5.2]
Degree [5.2]
Dodecagon [5.2]
Equal angles [5.2]
Euclidean geometry [5.1]
Euclid's postulates [5.1]
Exterior angle [5.3]
Heptagon [5.2]
Hexagon [5.2]
Hypotenuse [5.4]
Line [5.1]
Line of symmetry [5.1]
Line segment [5.1]
n-gon [5.2]
Nonagon [5.2]
Obtuse angle [5.2]
Octagon [5.2]
Opposite side [5.4]
Parallel lines [5.1]
Pentagon [5.2]
Plane [5.1]
Point [5.1]
Polygon [5.2]
Postulate [5.1]
Protractor [5.2]
Pythagorean theorem [5.4]
Quadrilateral [5.2]
Ray [5.1]
Reflection [5.1]
Right angle [5.2]
Similar figures [5.4]
Similar triangles [5.4]
Sine [5.4]
Straight angle [5.2]
Straightedge [5.1]
Surface [5.1]
Tangent [5.4]
Theorem [5.1]
Transformation [5.1]
Triangle [5.2, 5.3]
Trigonometric ratios [5.4]
Undefined terms [5.1]
Vertex [5.2]

Chapter Objectives

The material in this chapter is reviewed in the following list of objectives. A self-test (with answers and suggestions for additional study) is given. This self-test is constructed so that each problem number corresponds to a related objective. This self-test is followed by a practice test with the questions in mixed order.

[5.1] *Objective 5.1* Using a straightedge and a compass, construct a line segment congruent to a given segment and construct a circle with a given radius.

[5.1] *Objective 5.2* Construct a line through a given point parallel to a given line.

[5.1] *Objective 5.3* Identify line symmetry in a given picture or figure.

[5.1] *Objective 5.4* Recognize examples of translations, rotations, dilations, and contractions.

[5.2] *Objective 5.5* Classify polygons according to the number of sides.

[5.2] *Objective 5.6* Construct an angle congruent to a given angle.

[5.2] *Objective 5.7* Classify angles as acute, right, straight, or obtuse.

[5.3] *Objective 5.8* Name the corresponding parts of congruent triangles.

[5.3] *Objective 5.9* Find the measure of the third angle of a triangle.

[5.3] *Objective 5.10* Find the measure of the exterior angle of a triangle.

[5.3] *Objective 5.11* Construct a triangle congruent to a given triangle.

[5.3] *Objective 5.12* Use algebra to find the angles of a triangle.

[5.4] *Objective 5.13* Identify the parts of a correctly labeled triangle.

[5.4] *Objective 5.14* Determine whether two triangles are similar.

[5.4] *Objective 5.15* Use proportions and similar triangles to find the length of a side of a triangle.

[5.4] *Objective 5.16* Find a trigonometric ratio by using either Table I (Appendix D) or a calculator.

[5.4] *Objective 5.17* Use the sine, cosine, or tangent to find the length of a side in a right triangle.

[5.4] *Objective 5.18* Use similar triangles, the Pythagorean theorem, or trigonometric ratios to find an unknown distance.

Self-Test

Each question of this self-test is related to the corresponding objective listed previously.

1. **a.** Construct a segment congruent to the given segment. ————
   b. Draw a circle with radius equal to the segment given in part **a.**
2. Construct a line through the given point parallel to the given line.

3. Decide which of the given pictures illustrate line symmetry.

a.

b.

c.

4. Start with this figure:

Identify each part as a translation, rotation, dilation, or contraction.

a.

b.

c.

d.

e.

5. Identify the name of the polygon with the given number of sides.
   a. six **b.** four **c.** eight **d.** three **e.** five

6. Construct an angle congruent to $\angle POS$ in Figure 5.41.

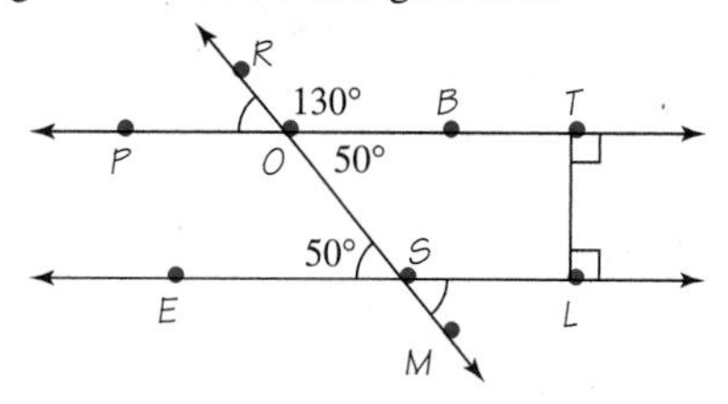

Figure 5.41

7. In Figure 5.41, classify the given angle as acute, right, straight, or obtuse.
   a. $\angle BOS$ **b.** $\angle ESO$ **c.** $\angle ROS$ **d.** $\angle BTL$ **e.** $\angle POS$

8. If $\Delta TRI \simeq \Delta ARI$ in Figure 5.42, name the corresponding parts.

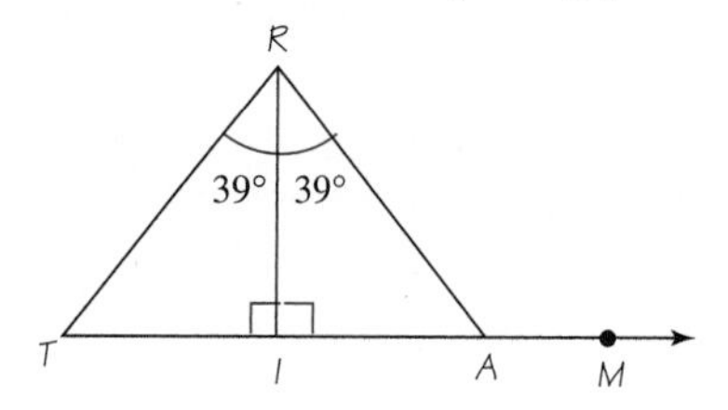

Figure 5.42

9. What is the measure of $\angle T$ in Figure 5.42?

10. What is the measure of $\angle RAM$ in Figure 5.42?

11. Construct a triangle congruent to ΔTAR in Figure 5.42.

12. Find the angles of a triangle if it is known that the measures of the angles are $2x + 10$, $5x - 60$, and $x - 10$.

13. Draw a correctly labeled right triangle.

14. Is $\Delta AIR \sim \Delta TIR$ in Figure 5.42? Tell why or why not.

15. If $\Delta ABC \sim \Delta A'B'C'$ and $a' = 5$, $b' = 12$, and $a = 7\frac{1}{2}$, what is b?

16. Find the value of the given trigonometric ratios by using Table I or a calculator. Round your answers to four decimal places.
    a. sin 59° **b.** tan 0° **c.** cos 18° **d.** tan 82°

17. If $\overline{RI}$ in Figure 5.42 is 5 in., what is the length of $\overline{AI}$ (to the nearest inch)?

18. **a.** Use similar triangles and a proportion to find the height of the cliff shown in Figure 5.43.

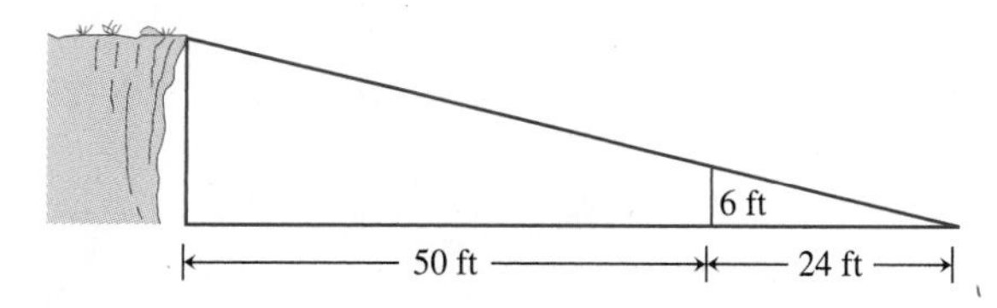

Figure 5.43 Determining the height of a cliff

b. The world's most powerful lighthouse is on the coast of Brittany, France, and is about 160 ft tall. Suppose you are in a boat just off the coast, as shown in Figure 5.44. Determine your distance (to the nearest foot) from the base of the lighthouse if $\angle B = 12°$.

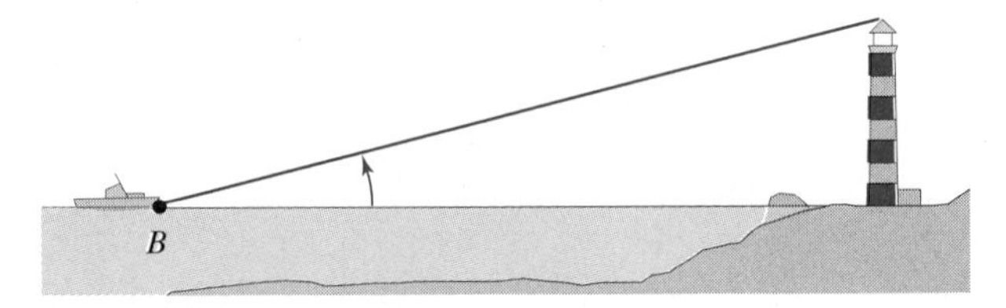

Figure 5.44 Distance from ship to shore

STUDY HINTS *Compare your solutions and answers to the self-test. For each problem you missed, work some additional problems in the section listed in the margin. After you have worked these problems, you can test yourself with the practice test.*

Complete Solutions to the Self-Test

Additional Problems

[5.1] *Problems 7–12*

1. a. **b.**

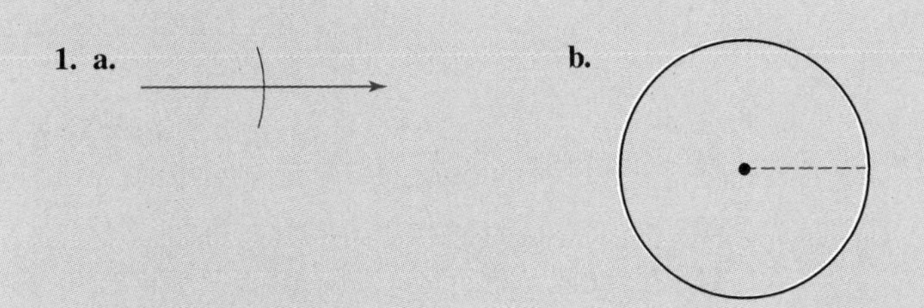

[5.1] *Problems 13–15*

2.

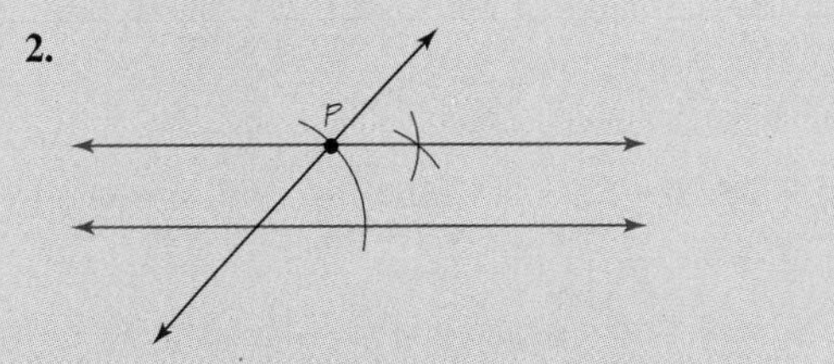

[5.1] *Problems 16–24*

3. a. not symmetric (look at the eagle's head) **b.** symmetric **c.** symmetric

[5.1] *Problems 25–26*

4. a. rotation **b.** translation **c.** dilation **d.** contraction **e.** rotation and contraction

[5.2] *Problems 4–15*

5. a. hexagon **b.** quadrilateral **c.** octagon **d.** triangle **e.** pentagon

[5.2] *Problems 16–21*

6.

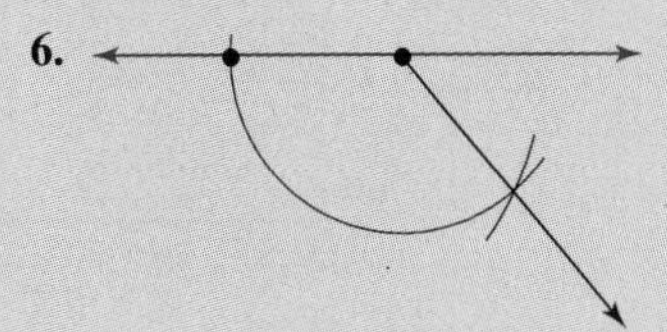

[5.2] *Problems 22–33*

7. **a.** acute **b.** acute **c.** straight **d.** right **e.** obtuse

[5.3] *Problems 3–6*

8. $\angle T \simeq \angle A$; $\angle TIR \simeq \angle AIR$; $\angle TRI \simeq \angle ARI$; $\overline{TI} \simeq \overline{AI}$; $\overline{TR} \simeq \overline{AR}$; $\overline{RI} \simeq \overline{RI}$

[5.3] *Problems 7–12; 25–28*

9. Since the sum of the measures of the angles is 180°,

$$\angle T = 180° - 90° - 39° = 51°$$

[5.3] *Problems 13–18*

10. $\angle RAM = 90° + 39° = 129°$

[5.3] *Problems 19–24; 39–42*

11.

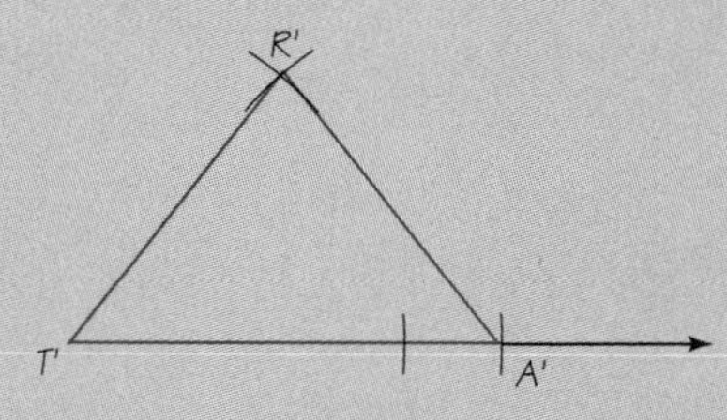

[5.3] *Problems 29–38*

12. Since the sum of the measures of the angles is 180°,

$$\begin{aligned}(2x + 10) + (5x - 60) + (x - 10) &= 180\\ 8x - 60 &= 180\\ 8x &= 240\\ x &= 30\end{aligned}$$

The angles are 70°, 90°, and 20°.

[5.4] *Problems 11–15*

13.

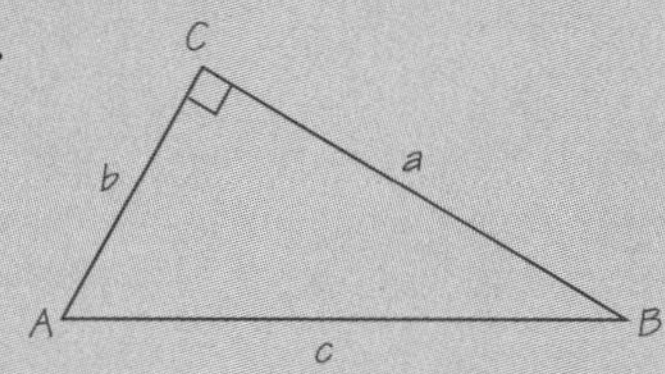

[5.4] *Problems 22–27*

14. They are similar because two angles of one triangle are equal to two angles of the other.

[5.4] *Problems 28–35*

15. Since the triangles are similar, corresponding sides are proportional.

$$\frac{a'}{b'} = \frac{a}{b}$$

$$\frac{5}{12} = \frac{7\frac{1}{2}}{b}$$

$$b = \frac{12 \times 7.5}{5}$$

$$b = 18$$

By calculator: 

[5.4] *Problems 16–21; 36–47*

16. **a.** $\sin 59° = 0.8572$ **b.** $\tan 0° = 0$
c. $\cos 18° = 0.9511$ **d.** $\tan 82° = 7.1154$

[5.4] *Problems 48–57*

17. $\tan 39° = \frac{\text{opposite side}}{\text{adjacent side}} = \frac{\overline{AI}}{} = \frac{\overline{AI}}{5}$

$5 \tan 39° = \overline{AI}$ Multiply both sides by 5.

$4.0489 = \overline{AI}$ By calculator: 5 × 39 tan =

This side is 4 in. Display: 4.0489202

By table: Use $\tan 39° = 0.8098$.

[5.4] *Problems 58–68*

18. a. Use proportional parts.

$\frac{6}{24} = \frac{h}{50 + 24}$ Let h be the height of the cliff.

$h = \frac{6 \times 74}{24}$ Note: $50 + 24 = 74$

$h = 18.5$ By calculator: 6 × 74 ÷ 24 =

The height of the cliff is 18.5 ft.

b. Use the definition of the tangent ratio.

$\tan 12° = \frac{160}{d}$ Let d be the distance to the lighthouse.

$d \tan 12° = 160$ Multiply both sides by d.

$d = \frac{160}{\tan 12°}$ By calculator: 160 ÷ 12 tan =

Display: 752.7408175

$d = 753$ By table: Use 0.2126 for $\tan 12°$.

The distance from the boat to the lighthouse is approximately 753 ft.

Chapter 5 Practice Test

1. Using only a straightedge and compass, carry out the requested constructions.

a. Construct a segment congruent to the given segment: ———

b. Construct a circle with radius congruent to the segment in part **a.**

c. Construct a line through the point Q parallel to the given line.

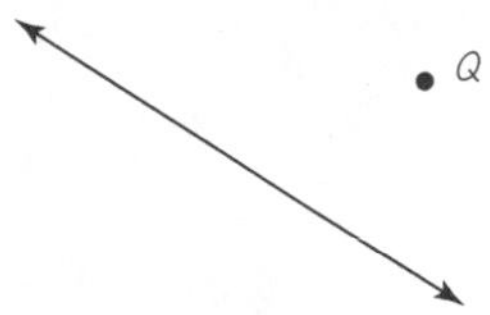

d. Construct a triangle congruent to ΔSTU.

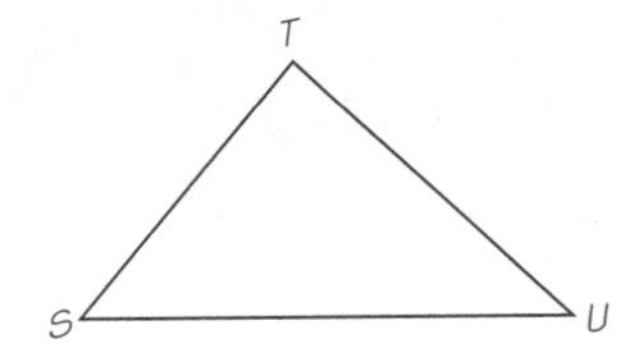

2. Give the name of each polygon.

a. **b.**

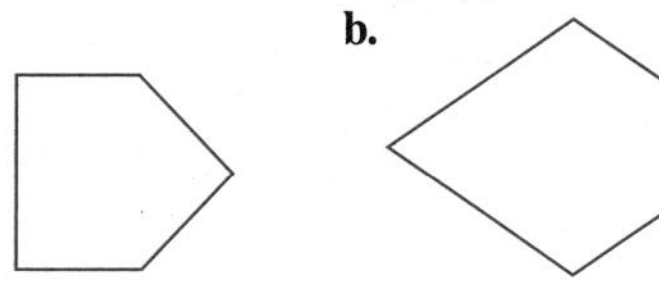

c.

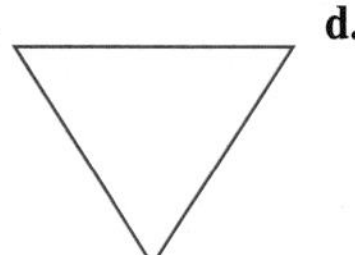

d.

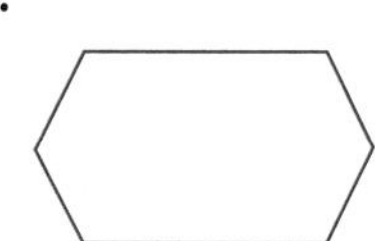

3. In ΔABC, what is the measure of angle A if $m\angle B = 58°$ and $m\angle C = 37°$?

4. What is the test to determine whether two triangles are similar?

5. Classify the angles as acute, right, straight, or obtuse, and construct an angle congruent to each. Refer to Figure 5.45.

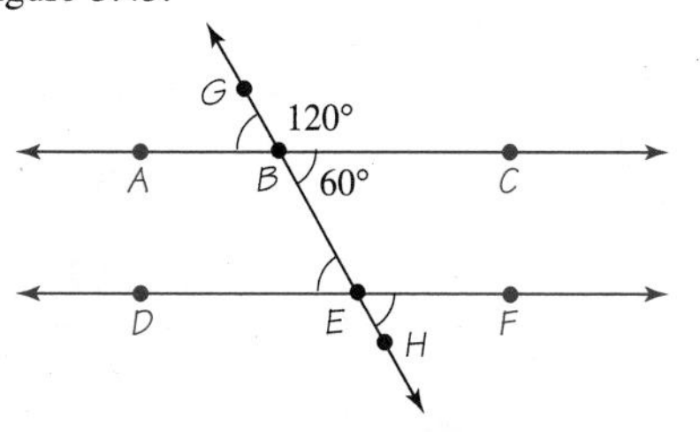

Figure 5.45

a. $\angle ABC$ **b.** $\angle GBC$ **c.** $\angle FEH$ **d.** $\angle BED$

6. Suppose $\Delta ABD \simeq \Delta CBD$, as shown in Figure 5.46.

a. What angle corresponds to $\angle A$?

b. What side corresponds to $\overline{AD}$?

c. What side corresponds to $\overline{BD}$?

d. What angle corresponds to $\angle BDC$?

e. If $m\angle C = 20°$ and $m\angle CDB = 30°$ in Figure 5.46, what is the measure of $\angle DBC$?

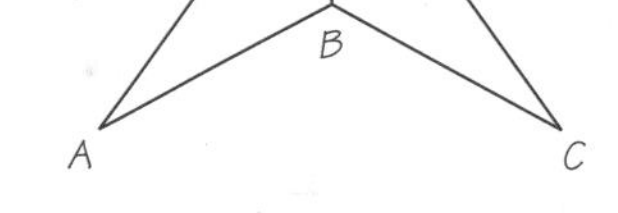

Figure 5.46 $\Delta ABD \simeq \Delta CBD$

7. What are the measures of the angles in Figure 5.47?

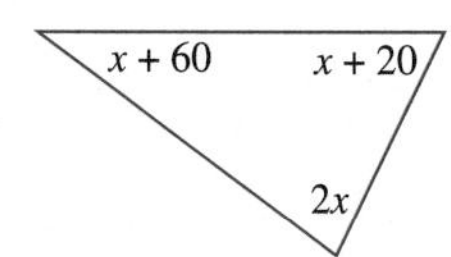

Figure 5.47

8. Find the value of x in each of the figures.

a.

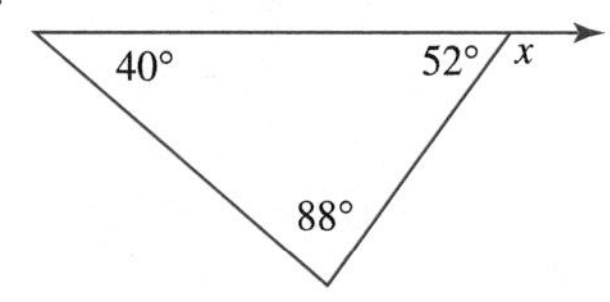

b.

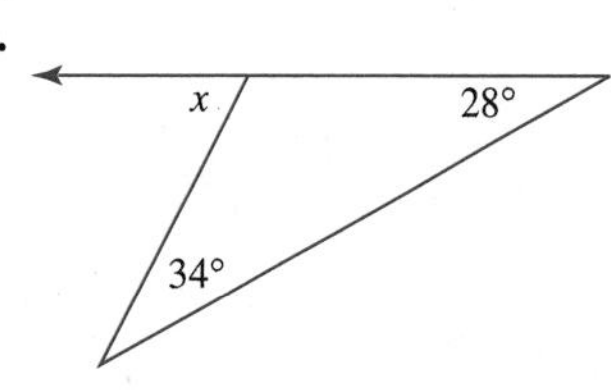

c.

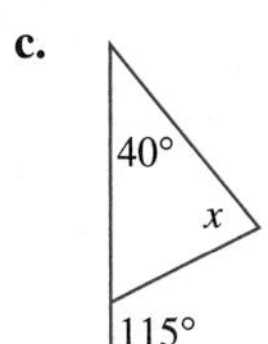

d.

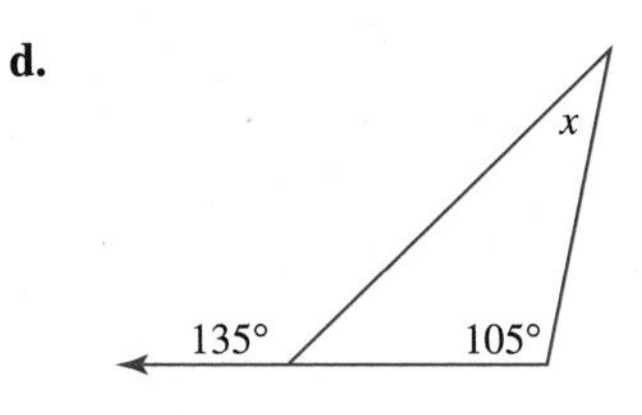

9. Is $\Delta DEF \sim \Delta HGF$?

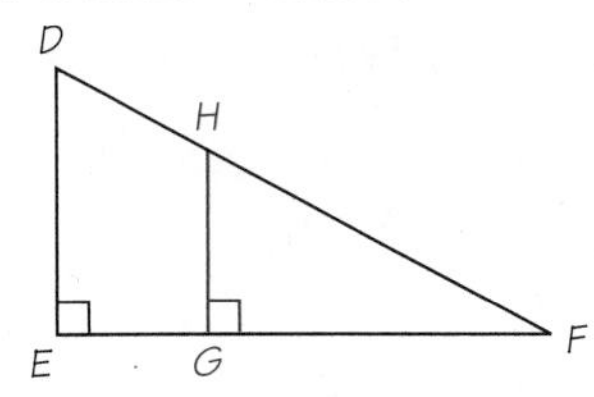

10. Find the unknown lengths (to the nearest tenth) as shown in Figure 5.48.

a. If $a = 4$, find b.

b. If $b = 4$, find a.

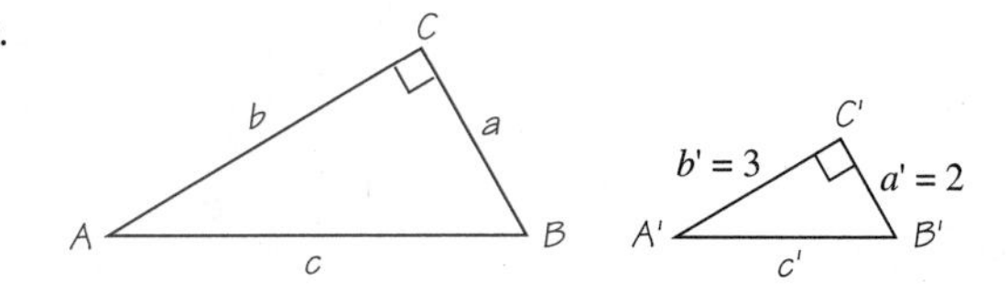

Figure 5.48 $\Delta ABC \simeq \Delta A'B'C'$

c. From the Pythagorean theorem, $c' = \sqrt{13}$. If $a = 4$, find c.

d. From the Pythagorean theorem, $c' \approx 3.6$. If $b = 4$, find c.

e. If $m\angle A = 35°$ and $a = 45$, find b.

f. If $m\angle B = 22°$ and $a = 38$, find b.

g. If $m\angle B = 17°$ and $b = 19$, find c.

h. If $m\angle A = 28°$ and $b = 128$, find c.

11. Find the values by using Table I or a calculator (rounded to four decimal places).

a. tan 77° **b.** cos 83° **c.** cos 18° **d.** sin 69°

12. If a tree casts a shadow of 18 ft at the same time that a yardstick (3 ft high) casts a shadow of 2 ft, find the height of the tree (to the nearest foot).

13. Use similar triangles to find the length of the sand pit in Figure 5.49.

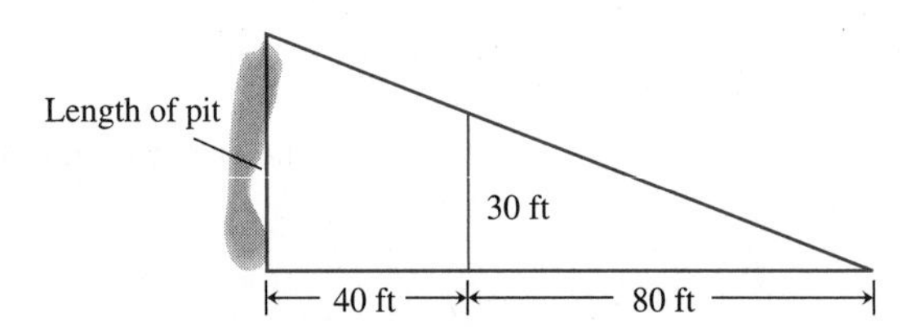

Figure 5.49 Finding the length of a sand pit

14. Find the distance $\overline{PA}$ (rounded to the nearest foot) across the river shown in Figure 5.50, if it is known that distance $\overline{DA}$ is 50 ft and that $\angle D$ is 38°.

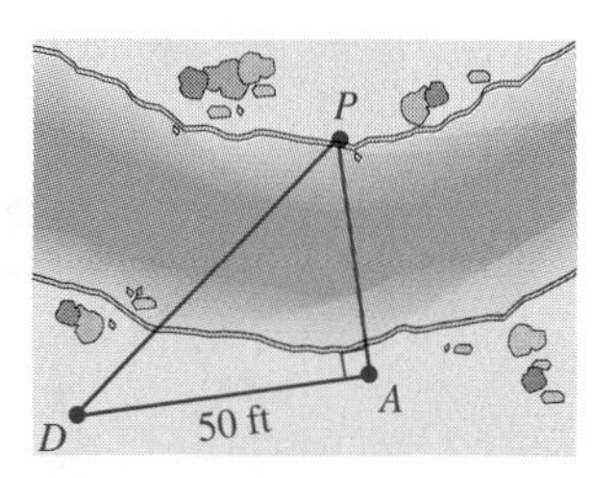

Figure 5.50 Distance across a river

15. Find the distance $\overline{PA}$ (rounded to the nearest foot) across the river shown in Figure 5.50, if it is known that the distance $\overline{DA}$ is 50 ft and that $\angle D$ is 42°.

CHAPTER SIX

MEASUREMENT AND PROBLEM SOLVING

6.1 PRECISION, ACCURACY, AND ESTIMATION

Historical Perspective

Early measurements were made in terms of the human body (digit, palm, cubit, span, and foot). Eventually, measurements were standardized in terms of the physical measurements of certain monarchs. King Henry I, for example, decreed that one yard was the distance from the tip of his nose to the end of his thumb. In 1790, the French Academy of Science was asked by the government to develop a logical system of measurement, and the original **metric system** came into being. By 1900, it was adopted by more than 35 major countries. In 1906, there was a major effort to convert to the metric system in the United States, but it was opposed by big business and the attempt failed. In 1960, the metric system was revised and simplified to what is now known as the **SI system** (an abbreviation of *Système International d'Unités*). In this book, the term *metric* will refer to the SI system, and the term **U.S. system** will refer to the current measurement system in the United States.

Numbers are used to count and to measure. In counting, the numbers are considered exact unless the result has been rounded. In the previous chapter, we considered some measures related to angles and triangles. In this chapter, we study the topic of measurement in more detail.

Measure of Length

To measure an object is to assign a number to its size. The number representing its linear dimension, as measured from end to end, is called its **measure** or **length.**

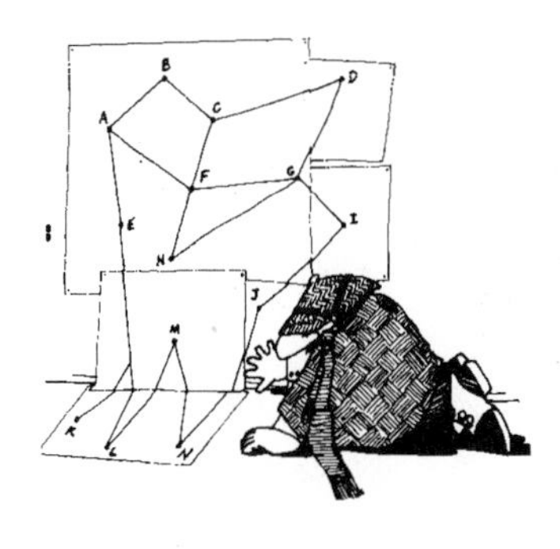

Measurement is never exact, and you therefore must decide how **precise** the measure should be. For example, the measurement might be to the nearest inch, nearest foot, or even nearest mile. The precision of a measurement depends not only on the instrument used but also on the purpose of your measurement. For example, if you are measuring the size of a room to lay carpet, the precision of your measurement might be different from when you are measuring the size of an airport hangar.

The **accuracy** refers to your answer. Suppose that you use an instrument that measures to the nearest tenth of a unit. You find one measurement to be 4.6 and another measurement to be 2.1. If, in the process of your work, you need to multiply these numbers, the result you obtain is

$$4.6 \times 2.1 = 9.66$$

This product is calculated to two decimal places, using the procedures for the multiplication of decimals discussed in Chapter 1. However, it does not seem quite right that you obtain an answer that is more accurate (two decimal places) than the instrument you are using to obtain your measurements (one decimal place). In this book, we will require that the accuracy of your answers not exceed the precision of the measurement. This means that after doing your calculations, the final answer should be rounded. The principle we will use is stated in the following box.

Accuracy of Answers in This Book

All measurements are as precise as given in the text. If you are asked to make a measurement, the precision will be specified. Carry out all calculations without rounding. After obtaining a *final answer,* round this answer to be as accurate as *the least precise measurement.*

This means that, to avoid round-off error, you should round only once (at the end). This is particularly important if you are using a calculator that displays 8, 10, 12, or even more decimal places.

You will also be asked to *estimate* the size of many objects in this chapter. As we introduce different units of measurement, you should remember some reference points so that you can make intelligent estimates. Many comparisons will be mentioned in the text, but you need to remember only those that are meaningful for you to estimate other sizes or distances. You will also need to choose appropriate units of measurement. For example, you would not measure your height in yards or miles, or the distance to New York City in inches.

There have been numerous attempts to make the metric system mandatory in the United States. Figure 6.1 (page 234) shows that the United States is the only major country not using the metric system. Today, big business is supporting the drive toward metric conversion, and it appears inevitable that the metric system will eventually come into use in the United States. In the meantime, it is important that you understand how to use both the United States and metric systems.

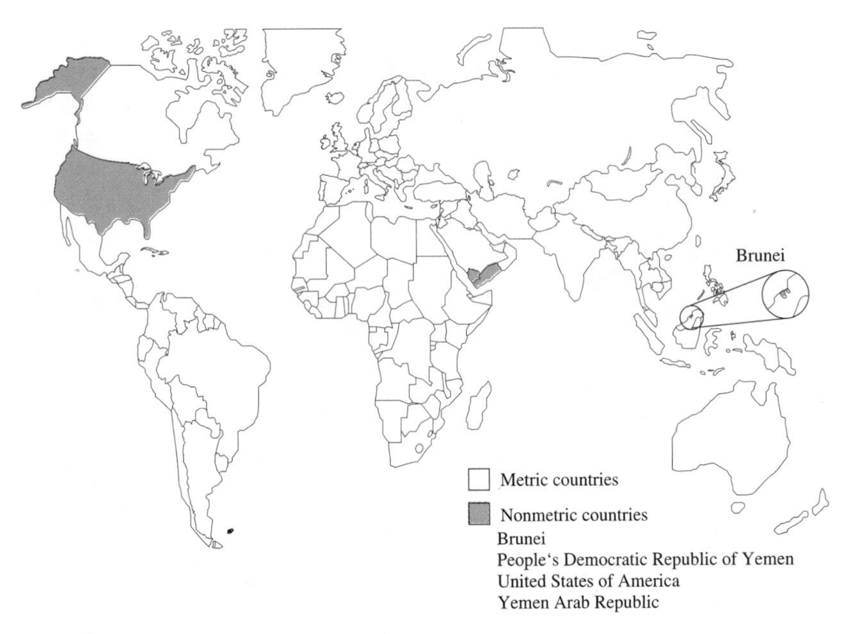

Figure 6.1 The metric world

The real advantage of using the metric system is the ease of conversion from one unit of measurement to another. Do you remember the difficulty you had in learning to change tablespoons to cups? Or pints to gallons?

In this book we will work with both the U.S. and the metric measurement systems. You should be familiar with both and be able to make estimates in both systems. The following box gives the standard units of length.

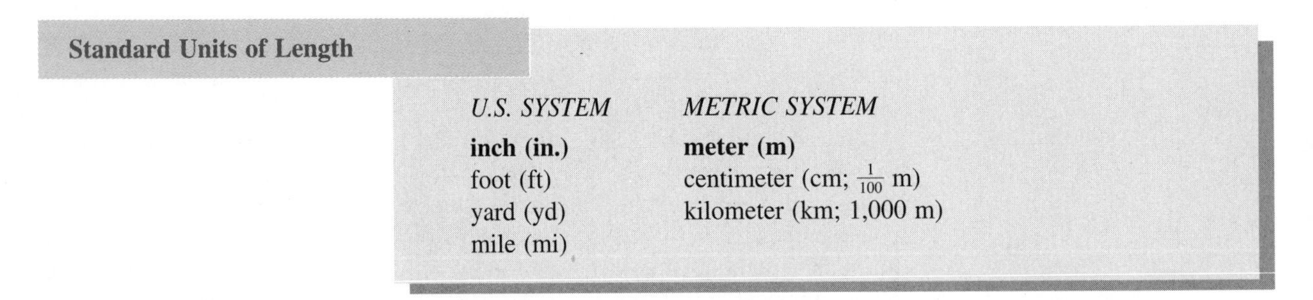

Standard Units of Length

U.S. SYSTEM	*METRIC SYSTEM*
inch (in.)	**meter (m)**
foot (ft)	centimeter (cm; $\frac{1}{100}$ m)
yard (yd)	kilometer (km; 1,000 m)
mile (mi)	

To understand the size of any measurement, you must see it, have experience with it, and take measurements using it as a standard unit. The basic unit of measurement for the U.S. system is the inch; it is shown in Figure 6.2. You can remember that an inch is about the distance from the joint of your thumb to the tip of your thumb. The basic unit of measurement for the metric system is the meter; it is also shown in Figure 6.2. You can remember that a meter is about the distance from your left ear to the tip of the fingers on the end of your outstretched right arm.

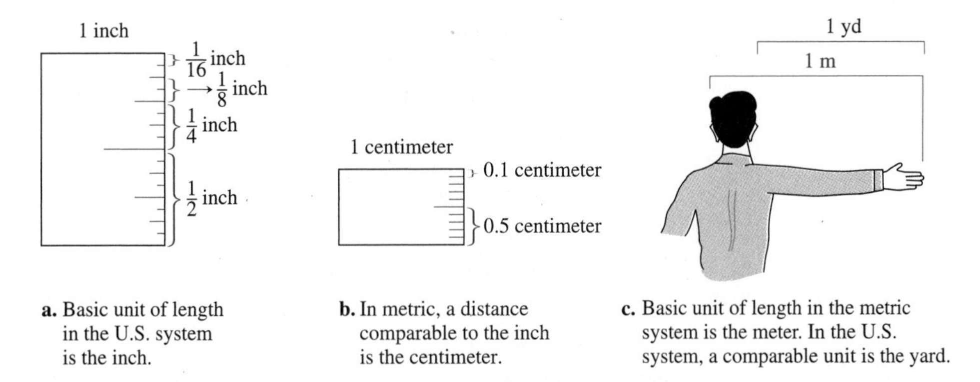

a. Basic unit of length in the U.S. system is the inch.

b. In metric, a distance comparable to the inch is the centimeter.

c. Basic unit of length in the metric system is the meter. In the U.S. system, a comparable unit is the yard.

Figure 6.2 Standard units of measurement for length

For the larger distances of a mile and a kilometer, you must look at maps or at the odometer of your car. However, you should have some idea of these distances.*

Later in the book, we will consider conversions of units within the U.S. system as well as conversions within the metric system, and we will use a variety of prefixes in the metric system. It might help to have a visual image of these prefixes as you progress through this chapter, even though we will not consider conversions until Section 6.6. The Greek prefixes **kilo-, hecto-,** and **deka-** are used for measurements larger than the basic metric unit, and the Latin prefixes **deci-, centi-,** and **milli-** are used for quantities smaller than the basic unit (see Figure 6.3). As you can see from Figure 6.3, a centimeter is $\frac{1}{100}$ of a meter. This means that 1 meter is equal to 100 centimeters.

You need to learn these prefixes for the metric system.

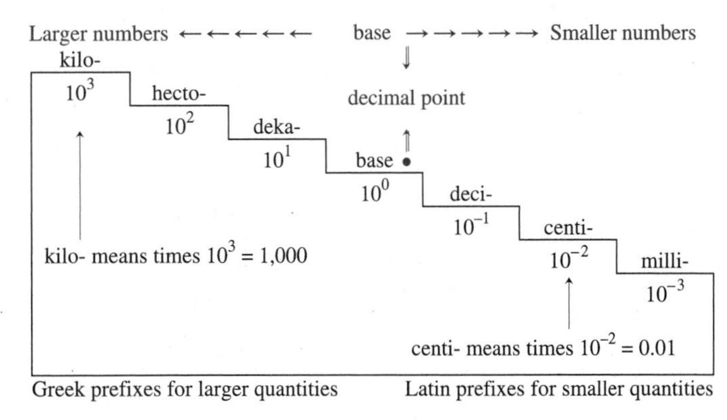

Figure 6.3 Metric prefixes

*We could tell you that a mile is 5,280 ft or that a kilometer is 1,000 m, but to do so does not give you any feeling for how far these distances really are. You need to get into a car and watch the odometer to see how far you travel in going 1 mile. Most cars in the United States do not have odometers set to kilometers, and until they do it is difficult to measure in kilometers. You might, however, be familiar with a 10-kilometer race. It takes a good runner about 30 minutes to run 10 kilometers and an average runner about 45 minutes. You can walk a kilometer in about 6 minutes.

Now we will measure given line segments with different levels of precision. We will consider two different rulers, one marked to the nearest centimeter and another marked to the nearest $\frac{1}{10}$ centimeter.

EXAMPLE 1 Measuring Segments

Measure the given segment

a. to the nearest centimeter.

b. to nearest $\frac{1}{10}$ centimeter.

A

Solution **a.** To measure the segment to the nearest centimeter, place a ruler showing centimeters next to the segment.

A End of *A* is nearest to 6.

1 2 3 4 5 6 7 8

b. To measure the segment to the nearest tenth of a centimeter, place a ruler showing tenths of a centimeter next to the segment. Notice that it looks as if the length of segment *A* is right on 6 cm. When measuring to the nearest tenth of a centimeter, we write 6.0 cm to indicate that this measurement is correct to the nearest tenth of a unit.

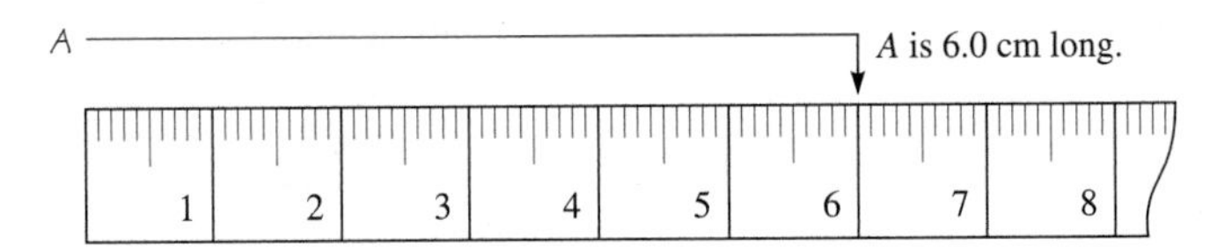

■

EXAMPLE 2 Measuring Length

Measure the given segment

a. to the nearest centimeter.

b. to the nearest $\frac{1}{10}$ centimeter.

B

Solution **a.**

B End of *B* is nearer to 5 than to 6.

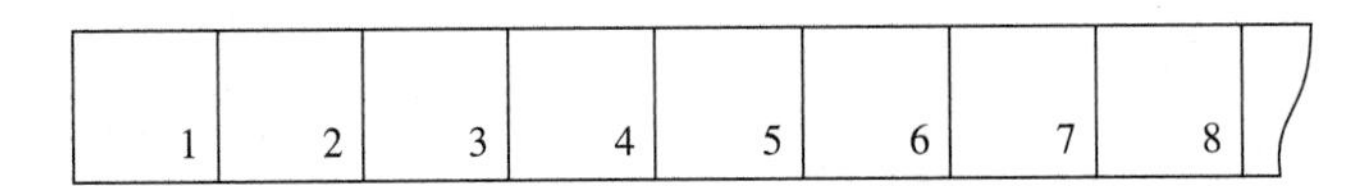

b.

B *B* is 5.3 cm long.

1 2 3 4 5 6 7 8

■

EXAMPLE 3 Measuring Length

Measure the given segment

a. to the nearest centimeter.

b. to the nearest $\frac{1}{10}$ centimeter.

C ————————————

Solution **a.**

C ———————— End of C is nearer to 5 than to 4.

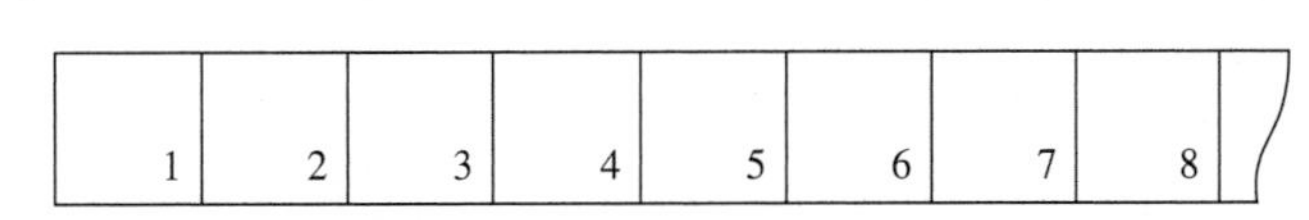

b.

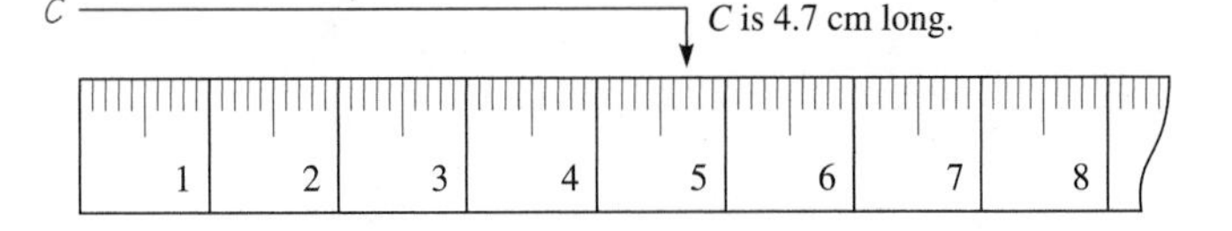

■

PROBLEM SET 6.1

Level 1

1. **In Your Own Words** Discuss the differences between *precision* and *accuracy.*
2. **In Your Own Words** Write a short essay discussing your opinion about using the metric measurement system.

Pick the best choices in Problems 3–17 by estimating. Do not measure. For metric measurements, do not attempt to convert to the U.S. system. The hardest part of the transition to the metric system is the transition to thinking in metrics.

3. The length of your math textbook is about
   A. 9 in. B. 9 cm C 2 ft
4. The length of a car is about
   A. 1 m B. 4 m C. 10 m
5. The length of a dollar bill is about
   A. 3 in. B. 6 in. C. 9 in.
6. The width of a dollar bill is about
   A. 1.9 cm B. 6.5 cm C. 0.65 m
7. The distance from your home to the nearest grocery store is most likely to be
   A. 1 cm B. 1 m C. 1 km
8. The length of a 100-yard football field is
   A. 100 m B. more than 100 m C. less than 100 m
9. The distance from San Francisco to New York is about 3,000 miles. This distance is
   A. less than 3,000 km B. more than 3,000 km
   C. about 3,000 km
10. Suppose someone could run the 100-meter dash in 10 seconds flat. At the same rate, this person should be able to run the 100-yard dash in
    A. less than 10 sec B. more than 10 sec C. 10 sec
11. Your height is closest to
    A. 5 ft B. 10 ft C. 25 in.
12. An adult's height is most likely to be about
    A. 6 m B. 50 cm C. 170 cm
13. The distance around your waist is closest to
    A. 10 in. B. 36 in. C. 30 cm
14. The distance from floor to ceiling in a typical home is about
    A. 2.5 m B. 0.5 m C. 4.5 m
15. The length of a new pencil is about
    A. 7 cm B. 17 cm C. 7 m
16. The prefix *centi-* means
    A. one thousand B. one-thousandth
    C. one-hundredth

17. The prefix *kilo-* means
A. one thousand B. one-thousandth
C. one-hundredth

LEVEL 1—Drill

From memory, and without using any measuring devices, estimate the length of the line segments indicated in Problems 18–25.

18. 1 in.

19. 2 in.

20. 3 in.

21. $\frac{1}{2}$ in.

22. 1 cm

23. 5 cm

24. 10 cm

25. 3 cm

Measure the segments given in Problems 26–36 *with the indicated precision.*

26. To the nearest centimeter: ______________

27. To the nearest centimeter: ______________

28. To the nearest $\frac{1}{10}$ centimeter: ______________

29. To the nearest $\frac{1}{10}$ centimeter: ______________

30. To the nearest $\frac{1}{10}$ centimeter: ______________

31. To the nearest inch: ______________

32. To the nearest inch: ______________

33. To the nearest inch: ______________

34. To the nearest eighth of an inch: ______________

35. To the nearest eighth of an inch: ______________

36. To the nearest eighth of an inch: ______________

37. In Your Own Words Find your own metric measurements.

	WOMEN	*MEN*
a.	Height	Height
b.	Bust	Chest
c.	Waist	Waist
d.	Hips	Seat
e.	Distance from waist to hemline	Neck

LEVEL 3—Team Project

38. Form a team of three or four others to discuss the advantages and disadvantages of the United States completely changing to the metric system. You might consider organizing a debate with another team to consider the following statement: *Resolved, the United States should convert to the metric system.*

6.2 PERIMETER

Historical Perspective

As we discuss the Modern Period in the history of mathematics, which dates from 1700 to the present, there are so many significant events that we must be selective about what we say. In fact, there is so much that we will discuss this historical period throughout this chapter. The best summary of this period is given by Vivian Groza:*

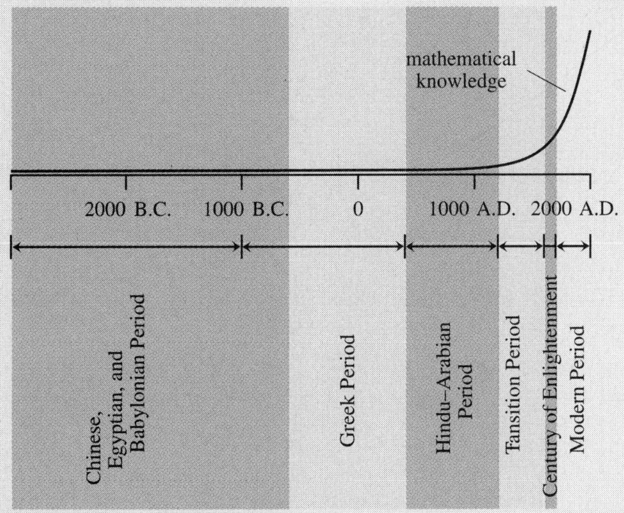

Periods in the history of mathematics

Stimulated by the French Revolution near the end of the eighteenth century, the new democratic way of life had its impact on both the social and the scientific worlds. As the Industrial Revolution spread from England to the continent of Europe, schools and universities were revised and modernized in response to the increased interest in scientific and technical education. The royal courts and academies declined as the centers of learning. Scientific Latin was gradually replaced by the national languages, and the universities and technical schools included in their responsibilities both research and instruction.

There was a pessimistic feeling at this time among some of the leading mathematicians that everything significant in mathematics had been discovered. Astronomy and mechanics had long been the motivating forces beginning with the ancient Babylonians and now apparently ending with the works of Euler and Laplace. How wrong this attitude was, soon became evident. As the old ways of thought were severely criticized and carefully scrutinized, a vast new world of mathematics was revealed.

*From *A Survey of Mathematics* by Vivian Shaw Groza, New York, NY: Holt, Rinehart and Winston, Inc., 1968, pp. 206 – 207.

One application of both measurement and geometry involves finding the distance around a polygon. This distance around is called the *perimeter* of the polygon.

Perimeter

The **perimeter** of a polygon is the sum of the lengths of the sides of that polygon.

EXAMPLE 1 **Finding a Perimeter**

Find the perimeter of the polygon in Figure 6.4 (page 240) by measuring each side to the nearest $\frac{1}{10}$ centimeter.

Solution

SIDE	*LENGTH*
$\overline{AB}$	3.2 cm
$\overline{BC}$	3.5 cm
$\overline{CD}$	6.7 cm
$\overline{DE}$	3.8 cm
$\overline{EA}$	3.0 cm
Total:	20.2 cm

Figure 6.4 Perimeter of a polygon

The perimeter is **20.2 cm.** ■

Following are some formulas for finding the perimeters of the most common polygons. You should remember these formulas.

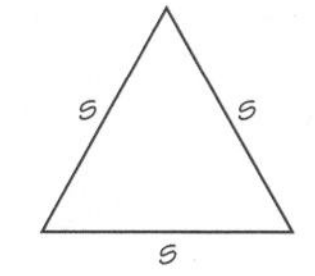

Equilateral triangle

An **equilateral triangle** is a triangle with sides that are equal.

$$\text{PERIMETER} = 3(\text{SIDE})$$
$$P = 3s$$

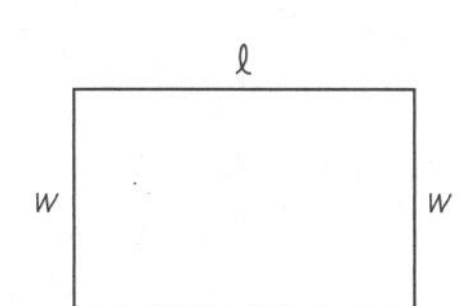

Rectangle

A **rectangle** is a quadrilateral with angles that are all right angles.

$$\text{PERIMETER} = 2(\text{LENGTH}) + 2(\text{WIDTH})$$
$$P = 2\ell + 2w$$

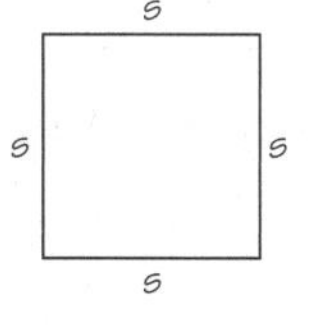

Square

A **square** is a rectangle with sides that are equal.

$$\text{PERIMETER} = 4(\text{SIDE})$$
$$P = 4s$$

EXAMPLE 2 **Finding Perimeters**

Find the perimeter of each polygon.

a.

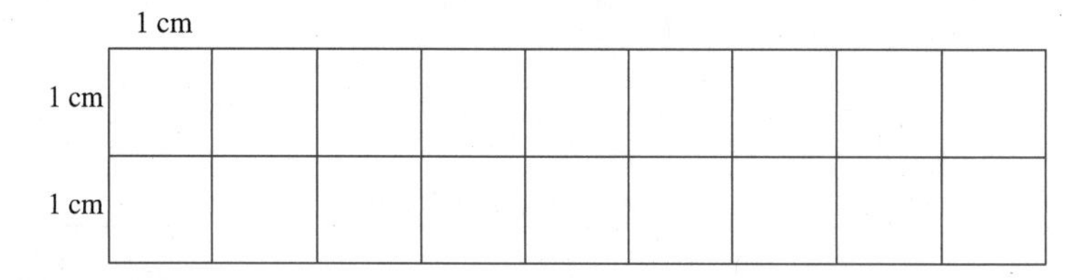

b.

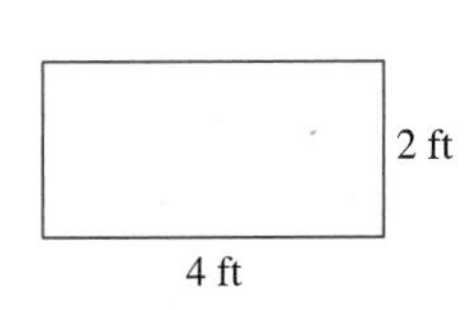

c.

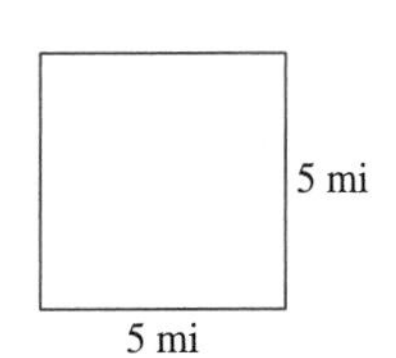

d.

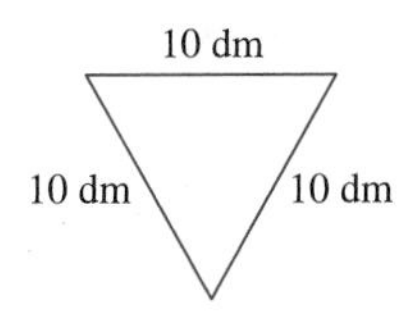

e.

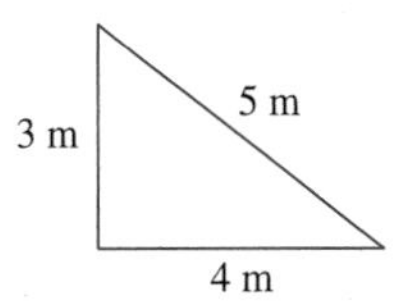

Solution

a. Rectangle is 2 cm by 9 cm, so

$P = 2\ell + 2w$
$= 2(\mathbf{9}) + 2(\mathbf{2})$
$= 18 + 4 = \mathbf{22\ cm}$

b. Rectangle is 2 ft by 4 ft, so

$P = 2\ell + 2w$
$= 2(\mathbf{4}) + 2(\mathbf{2}) = 8 + 4$
$= \mathbf{12\ ft}$

c. Square, so

$P = 4s = 4(5) = \mathbf{20\ mi}$

d. Equilateral triangle, so

$P = 3s = 3(10) = \mathbf{30\ dm}$

e. Triangle (add lengths of sides), so

$P = 3 + 4 + 5 = \mathbf{12\ m}$

■

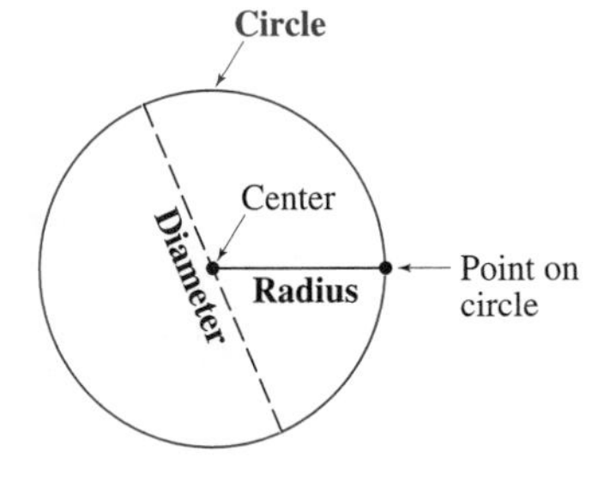

Figure 6.5 Circle

Although a **circle** (see Figure 6.5) is not a polygon, sometimes we need to find the distance around a circle. This distance is called the **circumference.** For *any circle,* if you divide the circumference by the diameter, you will get the *same number.* This number is given the name **pi** (pronounced "pie") and is symbolized by the Greek letter π. The number π is an irrational number and is about 3.14 or $\frac{22}{7}$. We need this number π to state a formula for the circumference C:

$C = d\pi$ or $C = 2\pi r$
d = DIAMETER r = RADIUS

In a circle, the radius is half the diameter. This means that, if you know the radius and want to find the diameter, you simply multiply by 2. If you know the diameter and want to find the radius, divide by 2.

EXAMPLE 3 **Finding the Circumference of a Circle**

Find the circumference of each circle (or the distance around each figure).

a.

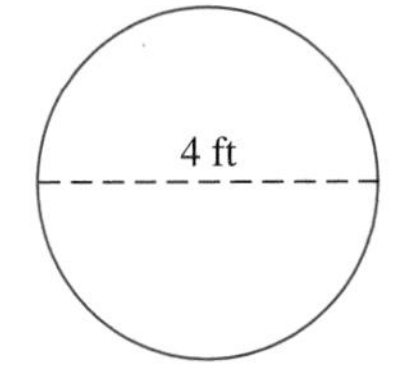

b.

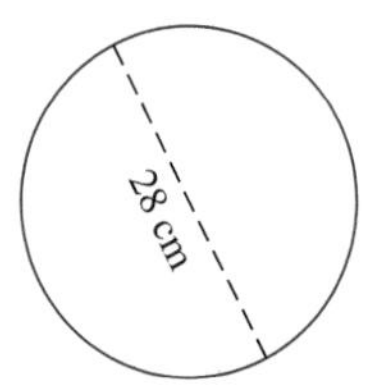

c.

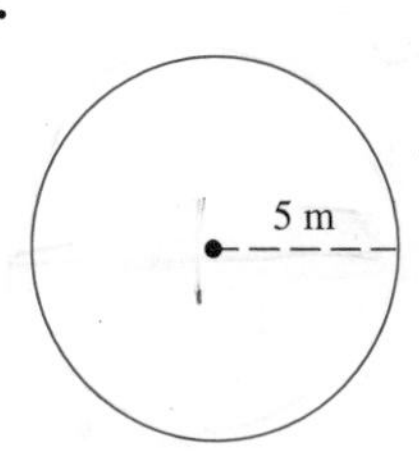

d.

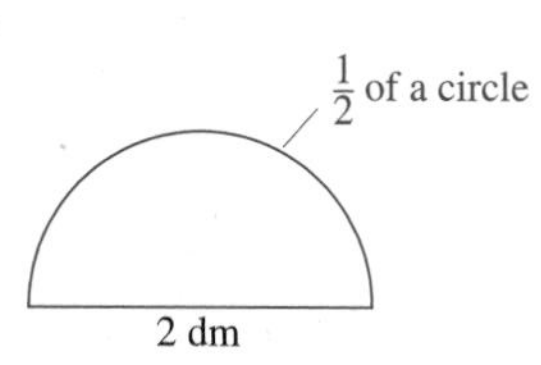

Answer to the nearest dm.

e.

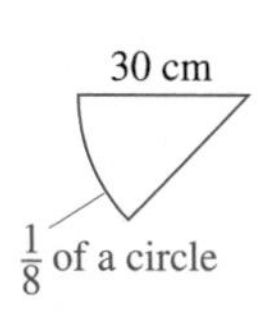

Answer to the nearest cm.

Solution

a. $C = \mathbf{4\pi}$

A decimal approximation (using the π key on a calculator) is

$C = 4\pi \approx 12.56637061$

The circumference is $\mathbf{4\pi}$ ft, which is about 13 ft. Note that the symbol "$\approx$" means *approximately equal to* and is used when π is approximated by a decimal.

b. $C = \mathbf{28\pi}$**;** this is about 88 cm.

c. $C = 2\pi(5) = \mathbf{10\pi}$**;** this is about 31.4 m.

d. This is half of a circle (called a **semicircle**); thus, the curved part is half of the circumference ($C = 2\pi$), or π, which is added to the diameter:

$\pi + 2 \approx 5.141592654$ *By calculator:* [π] [+] [2] [=]

The distance around the figure is about **5 dm.**

e. This is one-eighth of a circle. The curved part is one-eighth of the circumference [$C = 2\pi(30) \approx 188.5$], or about 23.6. Thus, the distance around the figure is

$2\pi(30) \div 8 + 30 + 30 \approx 83.5619449$ *By calculator:*
[2] [×] [π] [×] [30] [÷] [8] [+] [30] [+] [30] [=]

The distance around the figure is about **84 cm.** ■

The ideas involving perimeter and circumference are sometimes needed to solve certain types of problems.

EXAMPLE 4

Problem Solving Using a Perimeter Formula

Suppose you have enough material for 70 ft of fence and want to build a rectangular pen 14 ft wide. What is the length of this pen?

Solution

$$\text{PERIMETER} = 2(\text{LENGTH}) + 2(\text{WIDTH})$$ This is the formula for perimeter.

$$\mathbf{70} = 2(\text{LENGTH}) + 2(\mathbf{14})$$ Fill in the given information.

Let L = LENGTH OF PEN

$70 = 2L + 28$

$42 = 2L$

$21 = L$

The pen is 21 ft long. ■

PROBLEM SET 6.2

Essential Ideas

1. **In Your Own Words** Explain the meaning of the concepts of perimeter and circumference.

2. **In Your Own Words** What is π? (*Hint:* The correct answer is NOT 3.1416.) Explain why the calculator display for π, namely, 3.141592654, does not answer this question.

3. State the following formulas:
   a. perimeter of an equilateral triangle
   b. perimeter of a rectangle
   c. perimeter of a square
   d. circumference of a circle

Level 1

Pick the best choices in Problems 4–13 by estimating. Do not measure. For metric measurements, do not attempt to convert to the U.S. system.

4. The perimeter of a dollar bill is
   A. 18 in. B. 6 in. C. 46 in.

5. The perimeter of a five-dollar bill is
   A. 18 cm B. 6 cm C. 46 cm

6. The length of a new pencil is
   A. 4 in. B. 18 in. C. 7 in.

7. The circumference of a new pencil is
   A. 2 cm B. 2 in. C. 7 in.

8. The circumference of an automobile tire is
   A. 60 cm B. 60 in. C. 1 m

9. The perimeter of this textbook is
   A. 34 in. B. 34 cm C. 11 in.

10. The perimeter of a VISA credit card is
    A. 30 in. B. 30 cm C. 1 m

11. The perimeter of a sheet of notebook paper is
    A. 1 cm B. 10 cm C. 1 m

12. The perimeter of the screen of a console TV set is
    A. 30 in. B. 100 cm C. 100 in.

13. The perimeter of a classroom is
    A. 100 ft B. 100 m C. 100 yd

WHAT IS WRONG, *if anything, with each of the statements in Problems 14–19? Explain your reasoning.*

14. The number π is equal to 3.1416.

15. The number π is 3.141592654.

16. The circumference of a circle is the distance across.

17. The radius of a circle is twice the diameter.

18. Perimeter is a linear measure.

19. The distance around a semicircle of radius r is $\pi r + 2r$.

Level 1—Drill

Find the perimeter or circumference of the figures given in Problems 20–34 by using the appropriate formula. Round approximate answers to two decimal places.

20.

4 in.
5 in.

21.

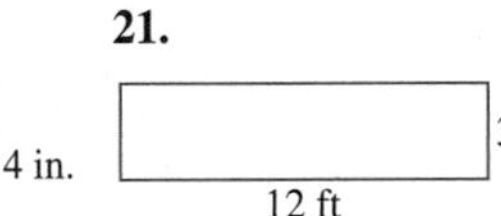

22.

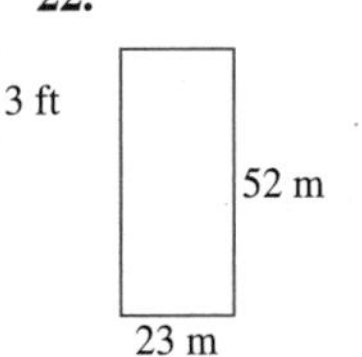

23. **24.**

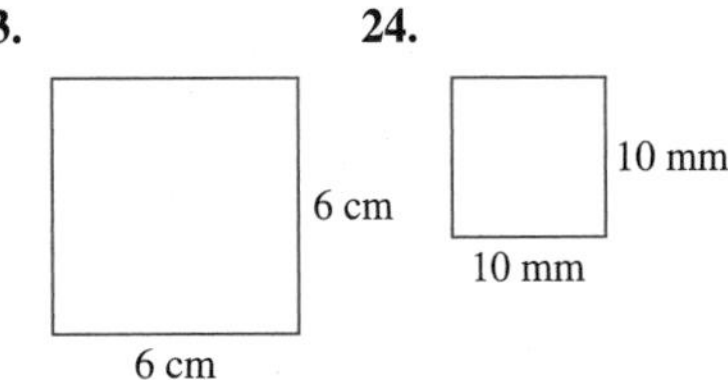

25.

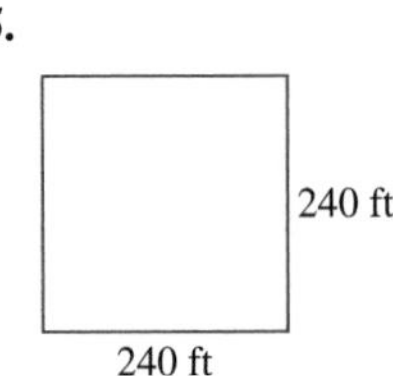

26.

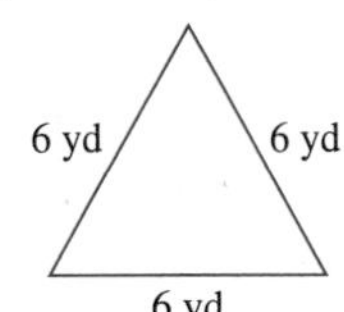

27.

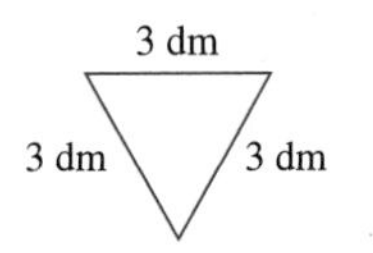

28.

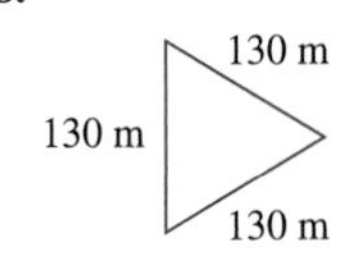

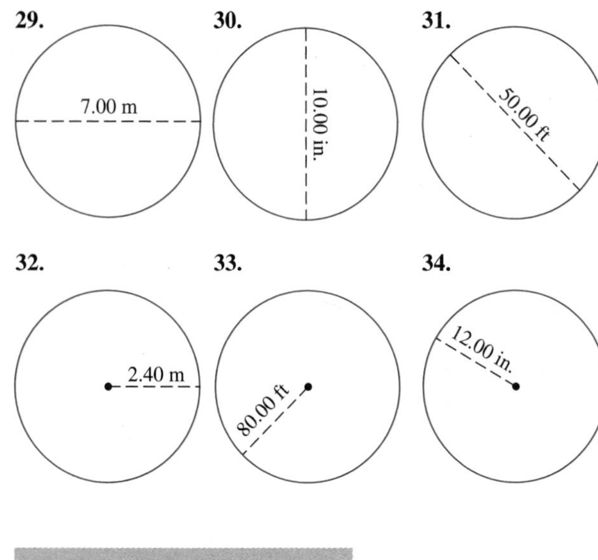

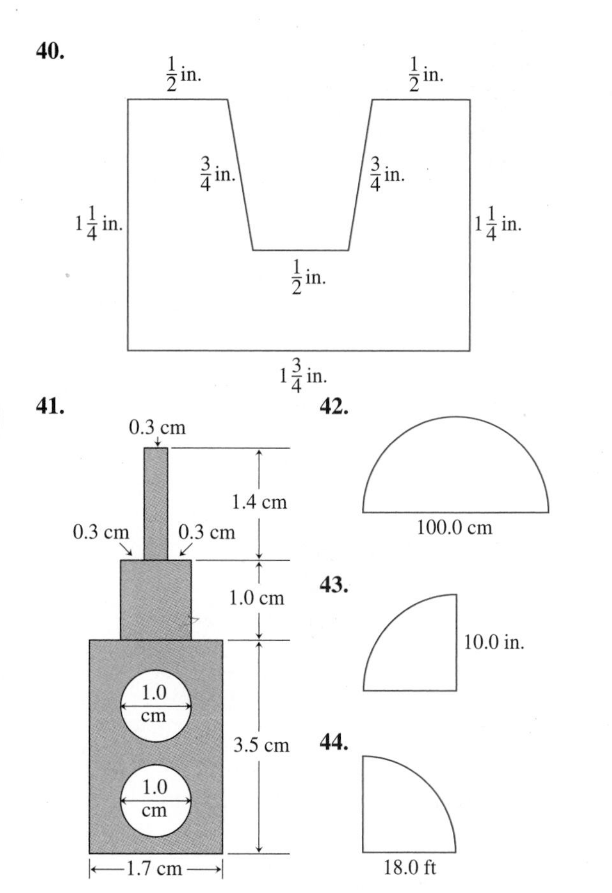

Level 2—Drill

Find the distance around the figures in Problems 35–46. Round approximate answers to one decimal place.

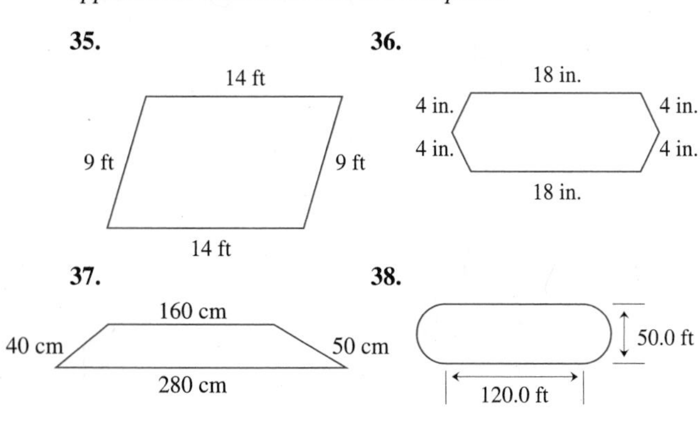

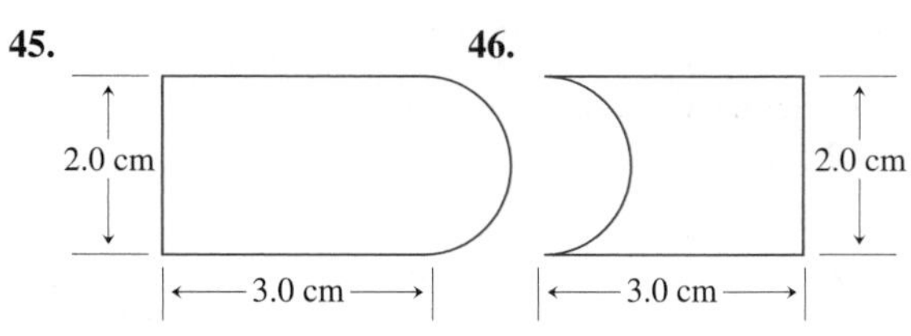

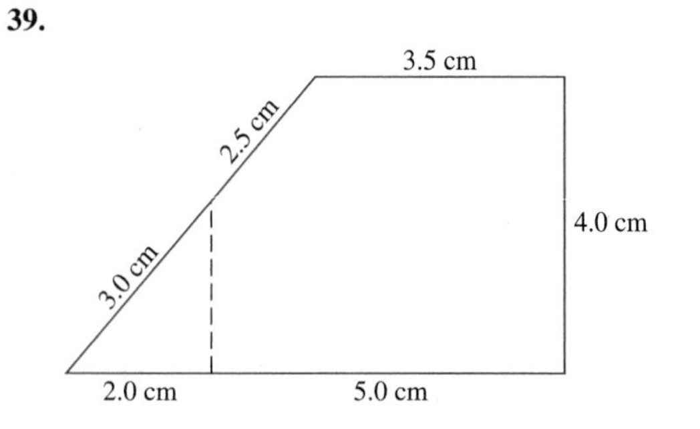

Level 2—Applications

47. What is the width of a rectangular lot that has a perimeter of 410 ft and a length of 140 ft?

48. What is the length of a rectangular lot that has a perimeter of 750 m and a width of 75 m?

49. The perimeter of ΔABC is 117 in. Find the lengths of the sides.

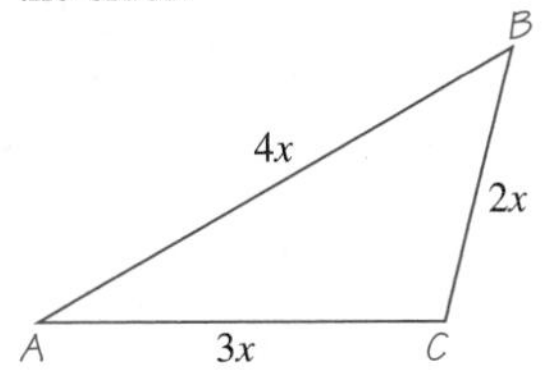

50. The perimeter of the pentagon is 280 cm. Find the lengths of the sides.

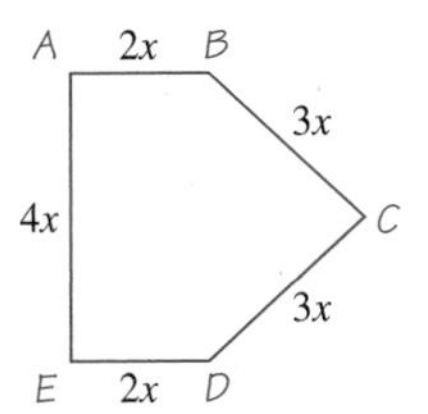

51. Find the dimensions of an equilateral triangle that has a perimeter of 198 dm.

52. Find the dimensions of a rectangle with a perimeter of 54 cm if the length is 5 less than three times the width.

Level 3—Mind Bogglers

53. **Santa Rosa street problem** On Saturday evenings, a favorite pastime of the high school students in Santa Rosa, California, is to cruise certain streets. The selected routes are shown on the map in Figure 6.6. Is it possible to choose a route so that all the permitted streets are traveled exactly once?

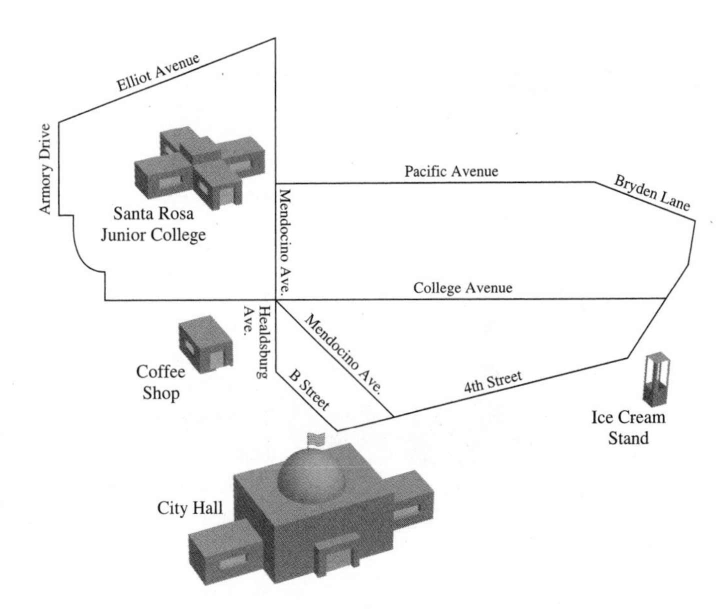

Figure 6.6 Santa Rosa street problem

54. Suppose that we fit a band tightly around the earth at the equator. We wish to raise the band so that it is uniformly supported 6 ft above the earth at the equator.
   a. Guess how much extra length would have to be added to the band (not the supports) to do this.
   b. Calculate the amount of extra material that would be needed.

6.3 AREA

Historical Perspective

Johann Bernoulli (1667–1748)

The first great mathematician of the Modern Period is not an individual, but rather a family. The remarkable Bernoulli family of Switzerland produced at least eight noted mathematicians over three generations. Two brothers, Jakob (1654–1705) and Johann, were bitter rivals. In spite of their disagreements, they maintained continuous communication with each other and with the great mathematician Leibniz (see Historical Perspective on page 204). The brothers were extremely influential in advocating the new calculus. Johann was the most prolific of the clan, but he was jealous and cantankerous; he tossed a son (Daniel) out of the house for winning an award he had expected to win himself. He was, however, the premier teacher of his time. Leonhard Euler (see Historical Perspective on page 71) was his most famous student. Curiously, Johann and Jakob's father opposed his sons' study of mathematics and tried to force Jakob into theology. Johann chose the motto *Invito patre sidera verso,* which means "I study the stars against my father's will."

Suppose that you want to carpet your living room. The price of carpet is quoted as a price per square yard. A square yard is a measure of **area.** To measure the area of a plane figure, you fill it with **square units.** (See Figure 6.7.)

Square units and area are important ideas not only in mathematics, but in real-world measurements.

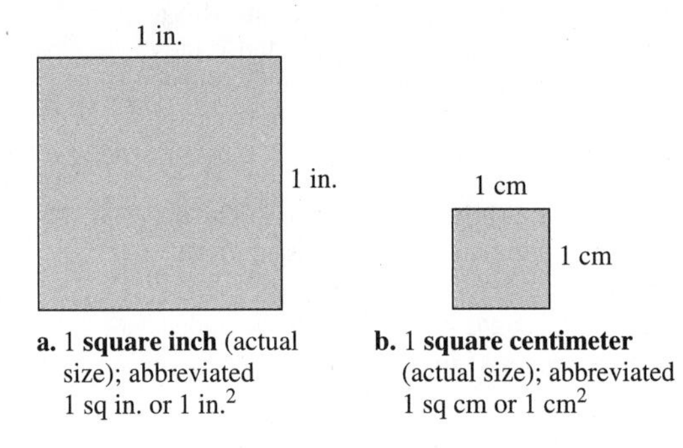

a. 1 **square inch** (actual size); abbreviated 1 sq in. or 1 in.^{2}

b. 1 **square centimeter** (actual size); abbreviated 1 sq cm or 1 cm^2

Figure 6.7 Common units of measurement for area

EXAMPLE 1 **Finding Areas by Counting Square Units**

What is the area of the shaded region?

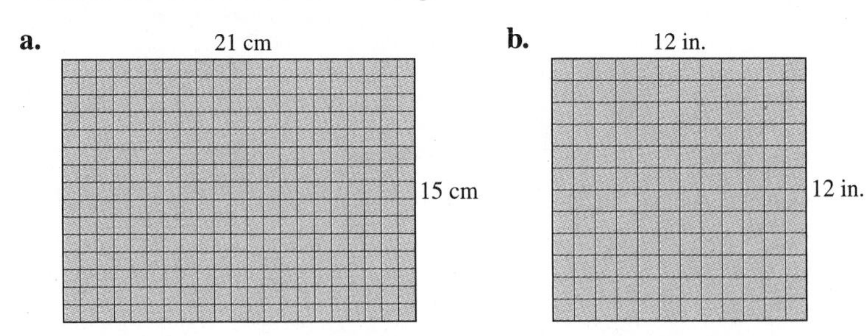

Solution **a.** You can count the number of square centimeters in the shaded region; there are 315 squares. Also notice:

Across Down
$21 \text{ cm} \times 15 \text{ cm} = 21 \times 15 \text{ cm} \times \text{cm}$
$= \mathbf{315\ cm^2}$

b. The shaded region is a **square foot.** You can count 144 square inches inside the region. Also notice:

Across Down
$12 \text{ in.} \times 12 \text{ in.} = \mathbf{144\ in.^2}$ ■

As you can see from Example 1, the area of a rectangular or square region is the product of the distance across (length) and the distance down (width).

Area of Rectangles and Squares

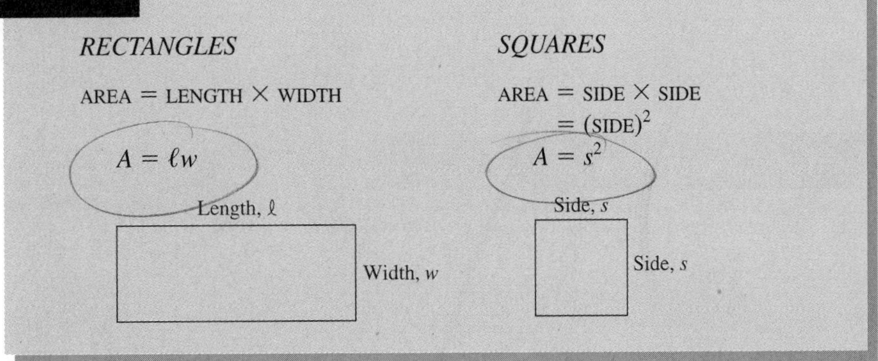

EXAMPLE 2 **Areas by Formula**

How many square feet are there in a square yard?

Solution Since 1 yd = 3 ft, we see from Figure 6.8 that

$$1 \text{ yd}^2 = (3 \text{ ft})^2$$
$$= \mathbf{9\ ft^2}$$

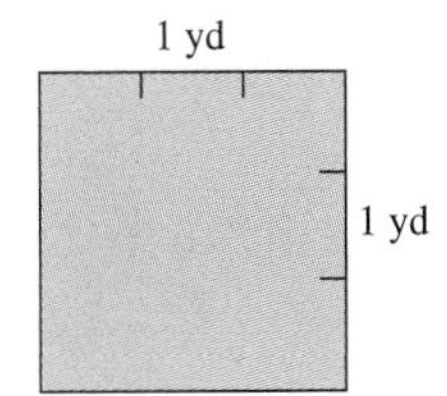

Figure 6.8 $1 \text{ yd}^2 = 9 \text{ ft}^2$ ■

Figure 6.9 Parallelograms

A **parallelogram** is a quadrilateral with two pairs of parallel sides, as shown in Figure 6.9. To find the area of a parallelogram, we can estimate the area by counting the number of square units inside the parallelogram (which may require estimation of partial square units), or we can show that the formula for the area of a parallelogram is the same as the formula for the area of a rectangle.

Geometric Justification of the Area Formula for a Parallelogram

Cut here.

Cut this off and move it to the other side.

Move this triangular piece to form a rectangle.

This is the rectangle that has been formed from the parallelogram. This means that the formula for the area of a parallelogram is the same as that for the rectangle.

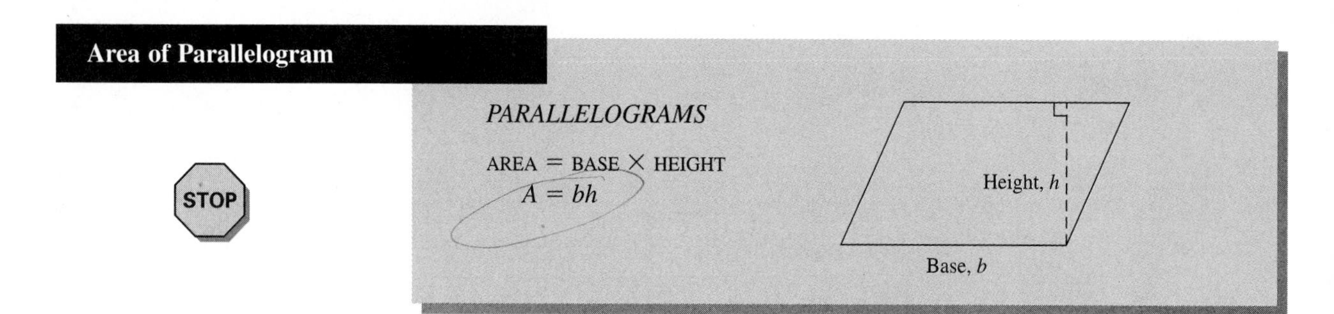

EXAMPLE 3 **Area of Parallelograms by Formula**

Find the area of each shaded region.

a.

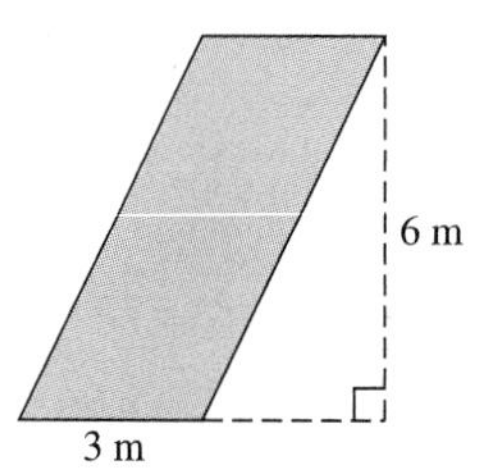

b.

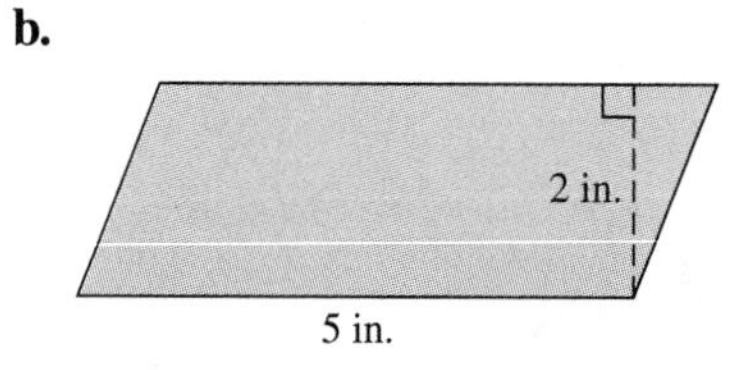

Solution

a. $A = 3 \text{ m} \times 6 \text{ m}$
$= \mathbf{18 \text{ m}^2}$

b. $A = 5 \text{ in.} \times 2 \text{ in.}$
$= \mathbf{10 \text{ in.}^2}$

■

You can find the area of a triangle by filling in and approximating the number of square units, by rearranging the parts, or by noticing that *every* triangle has an area that is exactly half that of a corresponding parallelogram.

Geometric Justification of the Area Formula for a Triangle

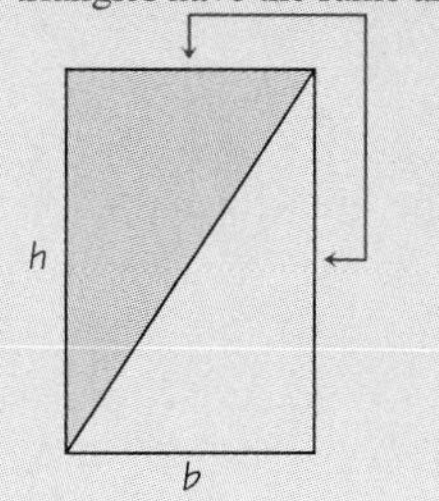

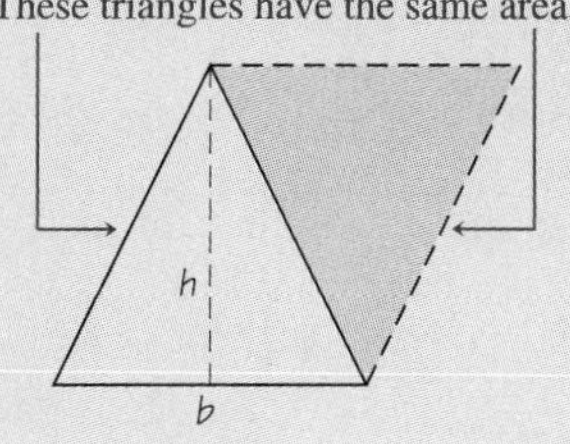

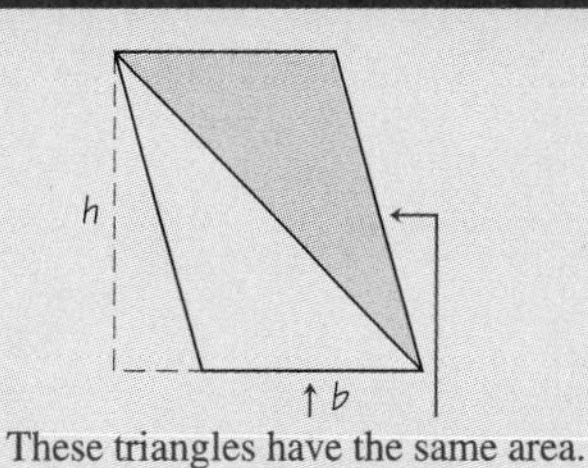

We can therefore state the following result.

Area of Triangles

TRIANGLES

AREA $= \frac{1}{2} \times$ BASE $\times$ HEIGHT

$A = \frac{1}{2}bh$

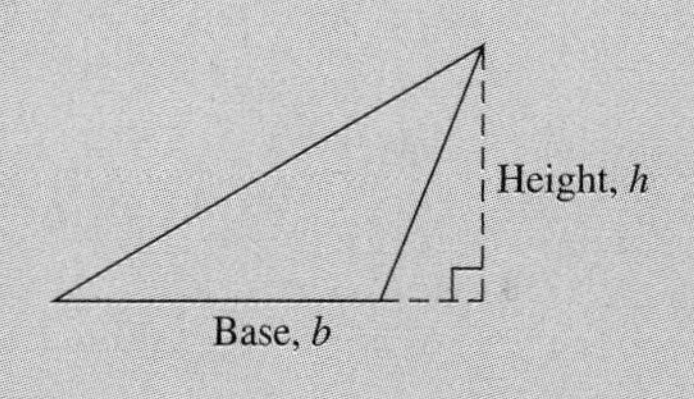

EXAMPLE 4 **Areas of Triangles by Formula**

Find the area of each shaded region.

a.

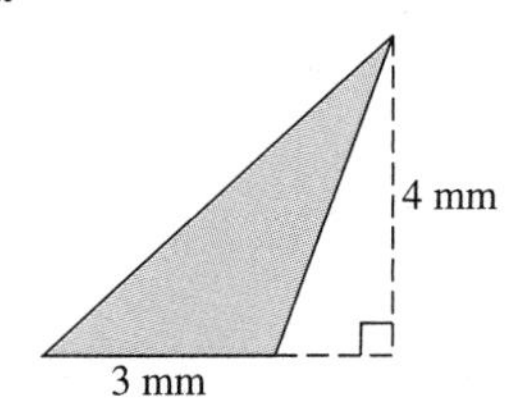

b.

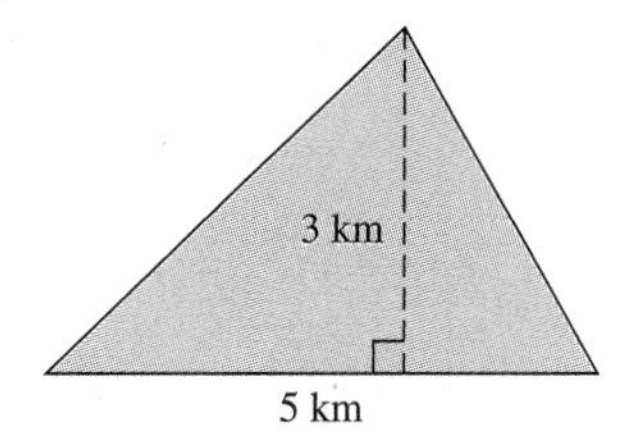

Solution **a.** $A = \frac{1}{2} \times 3 \text{ mm} \times 4 \text{ mm}$

$= \mathbf{6\ mm^2}$

b. $A = \frac{1}{2} \times 5 \text{ km} \times 3 \text{ km}$

$= \mathbf{\frac{15}{2}\ km^2}$ or $\mathbf{7\frac{1}{2}\ km^2}$ ■

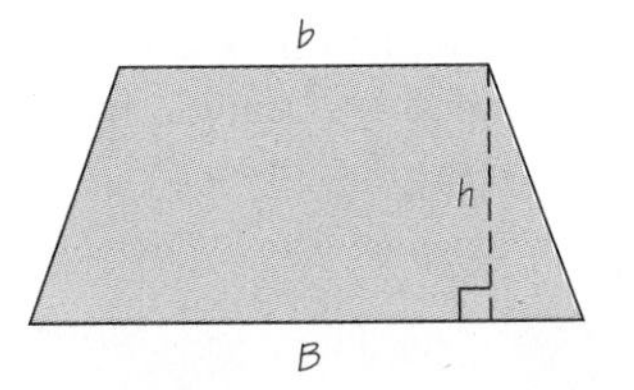

Figure 6.10 Trapezoid

We can also find the area of a trapezoid by finding the area of triangles. A **trapezoid** is a quadrilateral with two sides parallel. These sides are called the *bases,* and the perpendicular distance between the bases is the *height,* as shown in Figure 6.10. The area can be found as the sum of the areas of triangle I and triangle II:

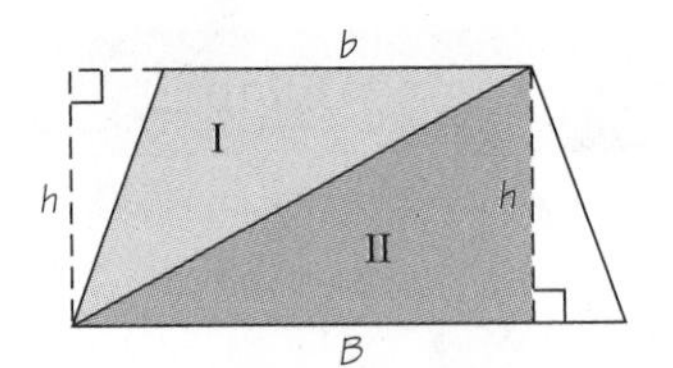

Area of triangle I: $\frac{1}{2}bh$
Area of triangle II: $\frac{1}{2}Bh$
Total area: $\frac{1}{2}bh + \frac{1}{2}Bh$

If we use the distributive property, we obtain the area formula for a trapezoid, as shown in the following box.

Area of Trapezoids

AREA $= \frac{1}{2}h(b + B)$

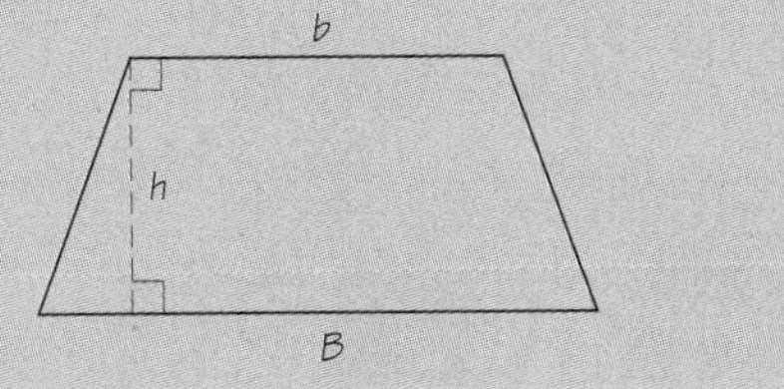

EXAMPLE 5 **Area of Trapezoids by Formula**

Find the area of each shaded region.

a.

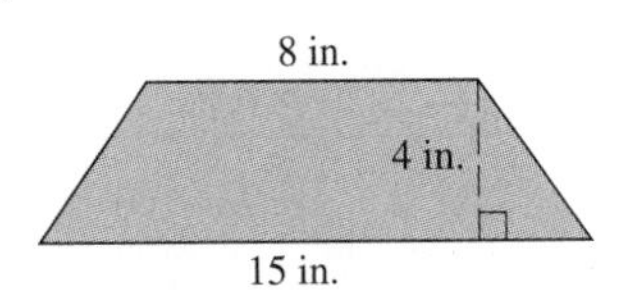

b.

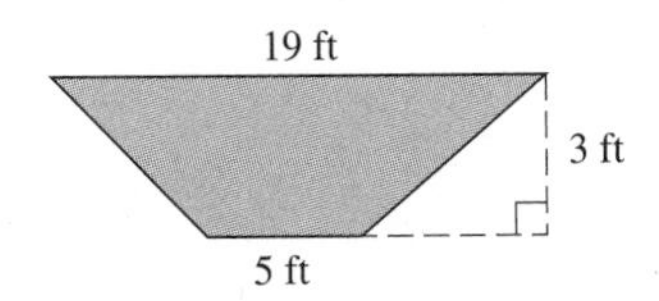

Solution

a. $h = 4;\ b = 8;\ B = 15$

$$A = \tfrac{1}{2}(4)(8 + 15) = 2(23) = 46$$

The area is 46 in.^{2}.

b. $h = 3;\ b = 19;\ B = 5$

$$A = \tfrac{1}{2}(3)(19 + 5) = \tfrac{3}{2}(24) = 36$$

The area is 36 ft^2. ■

The last of our area formulas is that of a circle.

Area of Circles

CIRCLES

$A = r \times r \times \pi$
$A = r^2 \times \pi$
$A = \pi r^2$

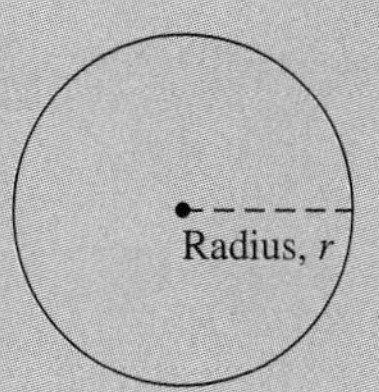

Even though it is beyond the scope of this course to derive a formula for the area of a circle, we can give a geometric justification that may appeal to your intuition.

Geometric Justification of the Area Formula for a Circle

Consider a circle with radius r.

Cut the circle in half.

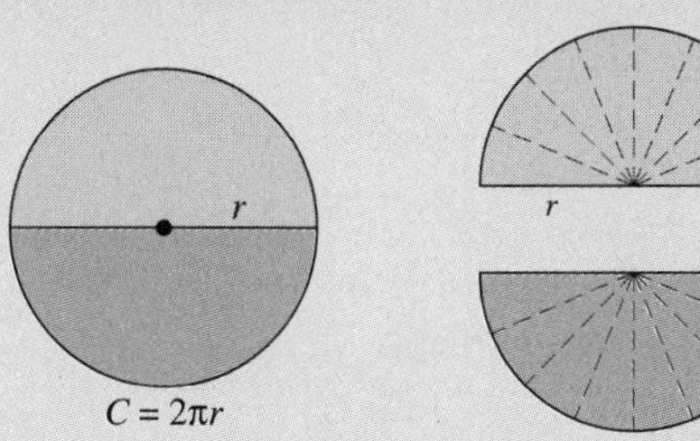

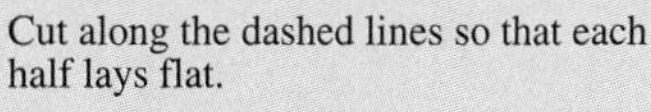

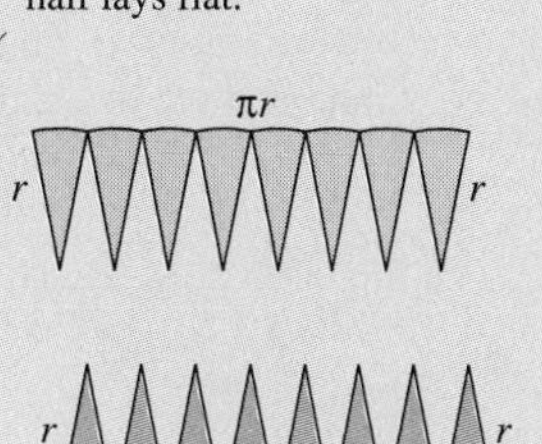

Fit these two pieces together:

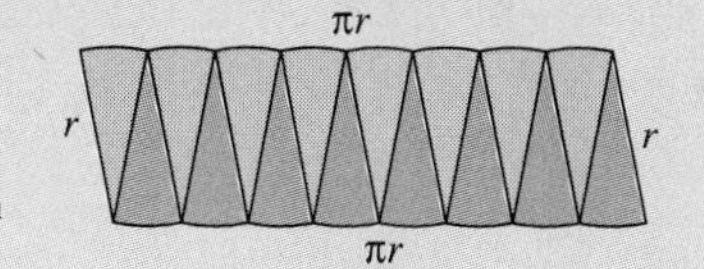

Therefore, it looks as if the area of a circle of radius r is about the same as the area of a rectangle of length πr and width r; that is, the area of a circle of radius r is πr^2.

EXAMPLE 6 **Area of Circles by Formula**

Find the area of each shaded region to the nearest tenth unit.

a.

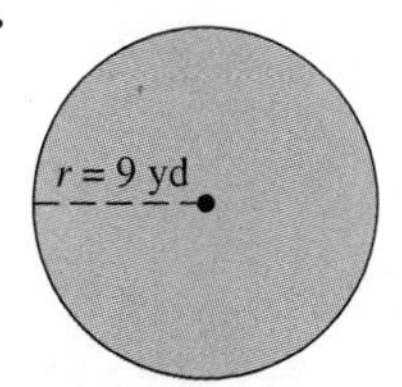

b.

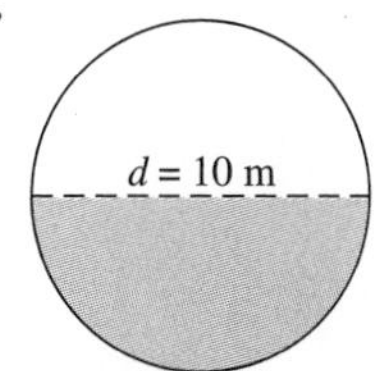

Solution

a. On a calculator:
Press: 81 × π =
Display: 254.4690049
To the nearest tenth, the area is **254.5 yd²**.

b. The shaded portion is half the area of the circle.
On a calculator:
Press: 25 × π ÷ 2 =
Display: 39.26990817 ↑ semicircle
To the nearest tenth, the area is **39.3 m²**. ■

Sometimes we obtain area using one unit of measurement and then we want to convert the result to another.

EXAMPLE 7 **Problem Solving with Areas**

Suppose your living room is 12 ft by 15 ft and you want to know how many square yards of carpet you need to cover this area.

Solution

Method I.

$$\begin{aligned} A &= 12 \text{ ft} \times 15 \text{ ft} \\ &= 180 \text{ ft}^2 \\ &= 180 \times 1 \text{ ft}^2 \\ &= 180 \times (\tfrac{1}{9} \text{ yd}^2) \\ &= \mathbf{20\ yd^2} \end{aligned}$$

Since 1 yd = 3 ft,
$1 \text{ yd}^2 = (1 \text{ yd}) \times (1 \text{ yd})$
$1 \text{ yd}^2 = (3 \text{ ft}) \times (3 \text{ ft})$
$1 \text{ yd}^2 = 9 \text{ ft}^2$
$\frac{1}{9} \text{ yd}^2 = 1 \text{ ft}^2$

Method II. Change feet to yards to begin the problem:

$$12 \text{ ft} = 4 \text{ yd} \quad \text{and} \quad 15 \text{ ft} = 5 \text{ yd}$$
$$\begin{aligned} A &= 4 \text{ yd} \times 5 \text{ yd} \\ &= \mathbf{20\ yd^2} \end{aligned}$$
■

If the area is large, as with property, a larger unit is needed. This is called an *acre*.

Acre

An **acre** is 43,560 ft².

To convert from ft² to acres, divide by 43,560.

Estimation Hint: Real estate brokers estimate an acre as 200 ft by 200 ft or 40,000 ft².

When working with acres, you usually need a calculator to convert square feet to acres, as shown in Example 8.

EXAMPLE 8 **Using Acre as an Area Measurement**

How many acres are there in a rectangular piece of property measuring 363 ft by 180 ft?

Solution

$$\text{AREA} = 363 \text{ ft} \times 180 \text{ ft}$$
$$= 65{,}340 \text{ ft}^2$$

To change ft^2 to acres, divide by 43,560:

$$\text{AREA} = 65{,}340 \text{ ft}^2$$
$$= (65{,}340 \div 43{,}560) \text{ acres}$$
$$= \mathbf{1.5 \text{ acres}}$$

Estimate: $65{,}340 \approx 65{,}000$ and $65{,}000 \div 40{,}000$
$= 65 \div 40 = 13 \div 8 \approx 12 \div 8$
$= 1.5$

■

EXAMPLE 9 **Problem Solving with Areas**

You want to paint 200 ft of a three-rail fence that is made up of three boards, each 6 inches wide. You want to know

a. the number of square feet on one side of the fence.
b. the number of square feet to be painted if the posts and edges of the boards comprise 100 ft^2.
c. the number of gallons of paint to purchase if each gallon covers 325 ft^2.

a. $A = (200 \text{ ft}) \times (6 \text{ in.}) \times 3$
$= (200 \text{ ft}) \times (\frac{1}{2} \text{ ft}) \times 3$
$= \mathbf{300 \text{ ft}^2}$

b. $\text{AMOUNT TO BE PAINTED} = 2(\text{AMOUNT ON ONE SIDE}) + \text{EDGES AND POSTS}$
$= 2(300 \text{ ft}^2) + 100 \text{ ft}^2$
$= \mathbf{700 \text{ ft}^2}$

c. $$\begin{pmatrix}\text{NUMBER OF SQUARE}\\ \text{FEET PAINTED}\end{pmatrix} = \begin{pmatrix}\text{NUMBER OF SQUARE}\\ \text{FEET PER GALLON}\end{pmatrix}\begin{pmatrix}\text{NUMBER OF}\\ \text{GALLONS}\end{pmatrix}$$

$$700 = 325\begin{pmatrix}\text{NUMBER OF}\\ \text{GALLONS}\end{pmatrix}$$

$$2.15 \approx \begin{pmatrix}\text{NUMBER OF}\\ \text{GALLONS}\end{pmatrix}$$ Divide both sides by 325.

If paint must be purchased by the gallon (as implied by the question), the amount to purchase is **3 gallons.** ■

PROBLEM SET 6.3

ESSENTIAL IDEAS

1. **In Your Own Words** Describe what is meant by area.

2. **In Your Own Words** From memory, estimate the following areas. **a.** 1 sq in. **b.** 1 sq cm

3. **In Your Own Words** Distinguish between perimeter and area.

4. **In Your Own Words** What is a trapezoid?

LEVEL 1

Pick the best choices in Problems 5–16 by estimating. Do not measure. For metric measurements, do not attempt to convert to the U.S. system.

5. The area of a dollar bill is
   A. 18 in. B. 6 in.^{2} C. 18 in.^{2}

6. The area of a five-dollar bill is
   A. 100 cm B. 10 cm^2 C. 100 cm^2

7. The area of the front cover of this textbook is
   A. 70 in.^{2} B. 70 cm^2 C. 70 in.

8. The area of a VISA credit card is
   A. 8 in. B. 8 in.^{2} C. 8 cm^2

9. The area of a sheet of notebook paper is
   A. 90 cm^2 B. 10 in.^{2} C. 600 cm^2

10. The area of a sheet of notebook paper is
    A. 90 in.^{2} B. 10 cm^2 C. 600 in.^{2}

11. The area of the screen of a console TV set is
    A. 4 ft^2 B. 19 in.^{2} C. 100 in.^{2}

12. The area of a classroom is
    A. 100 ft^2 B. 1,000 ft^2 C. 0.5 acre

13. The area of the floor space of the Superdome in New Orleans is
    A. 1 mi^2 B. 1,000 m^2 C. 10 acres

14. The state of California has an area of
    A. 5,000 acres B. 150,000 mi^2 C. 5,000 m^2

15. The area of the bottom of your feet is
    A. 1 m^2 B. 400 in.^{2} C. 400 cm^2

16. It is known that your body's surface area is about 100 times the area that you will find if you trace your hand on a sheet of paper. Using this estimate, your body's surface area is
    A. 300 in.^{2} B. 3,000 in.^{2} C. 3,000 cm^2

WHAT IS WRONG, *if anything with each of the statements in Problems 17–22? Explain your reasoning.*

17. The area of a square whose side is 3 ft is 9 ft.

18. The area of a circle with diameter 3 ft is 9π ft.

19. If one side of a rectangle is measured in km and the other side is measured in km, then the unit of measurement for area is km.

20. The area of a triangle with base ℓ and height h is ℓh.

21. The area of a parallelogram of length ℓ and height h is ℓh.

22. A trapezoid is a quadrilateral with no sides parallel.

LEVEL 1—Drill

Estimate the area of each figure in Problems 23–30 to the nearest square centimeter.

23.

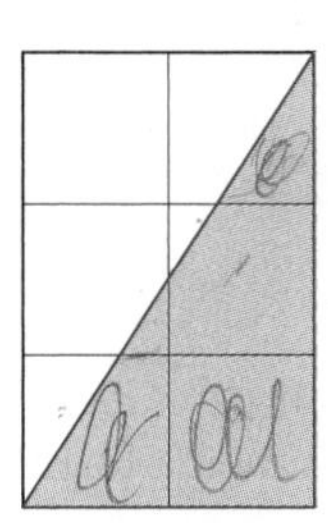

24.

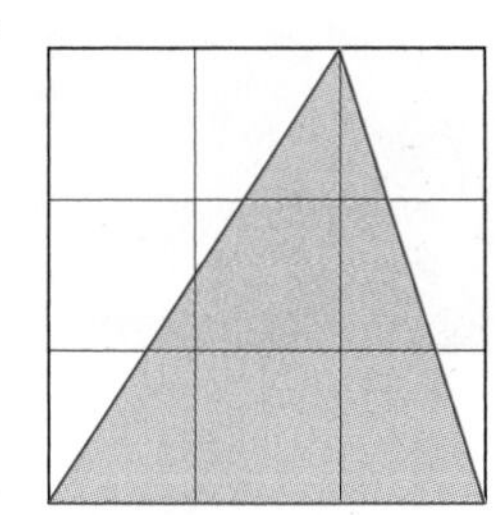

25.

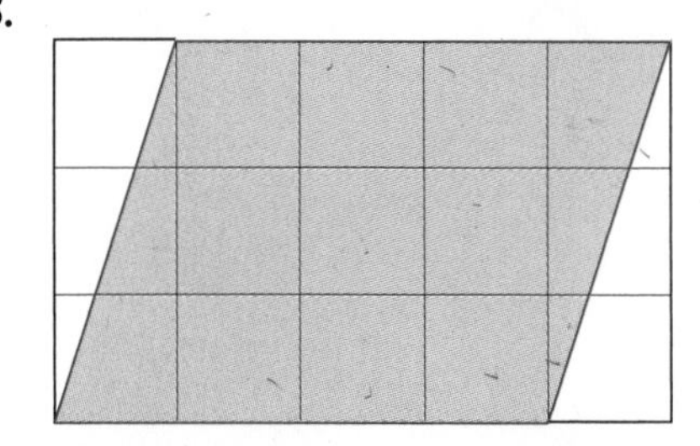

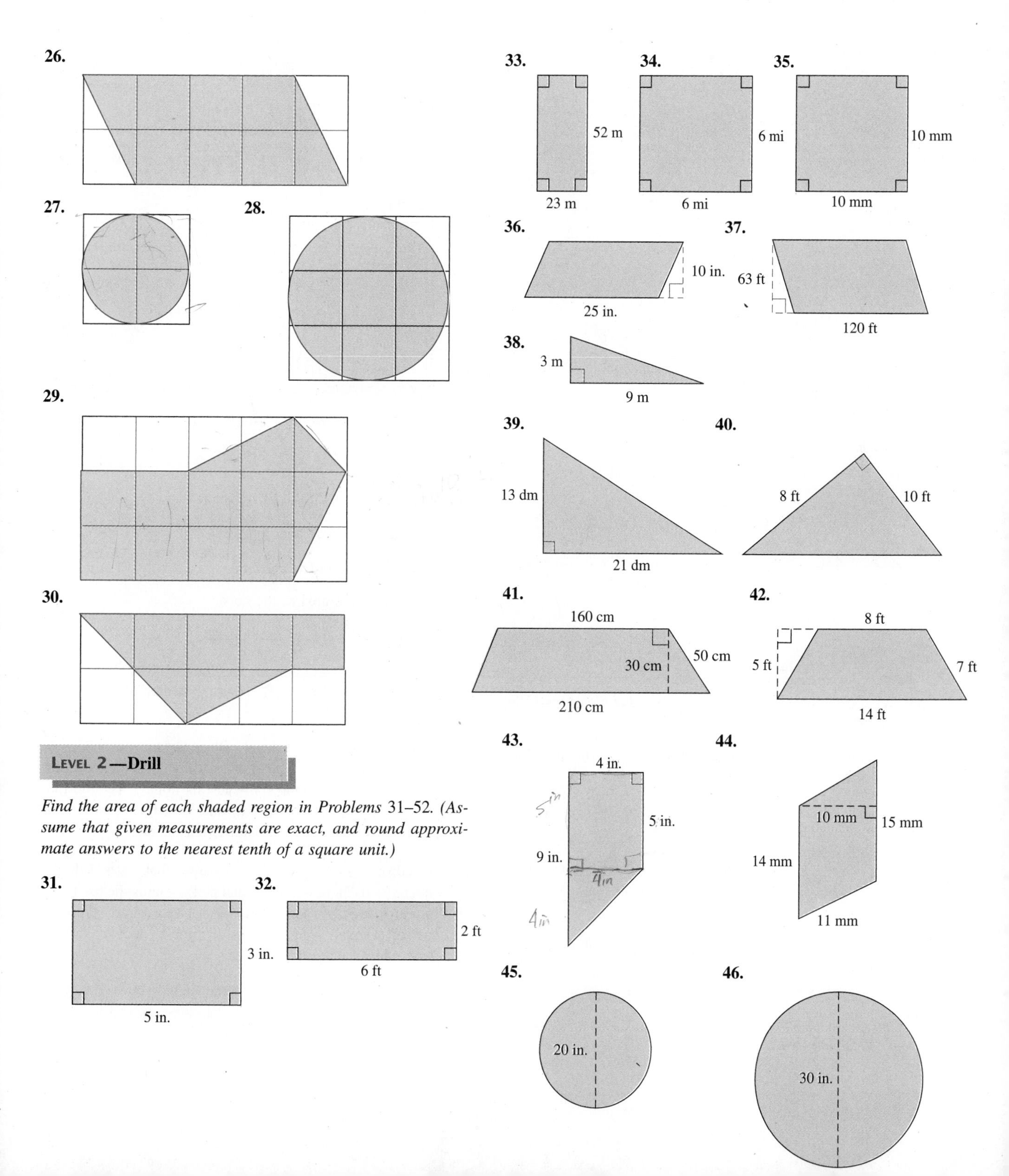

LEVEL 2—Drill

Find the area of each shaded region in Problems 31–52. (Assume that given measurements are exact, and round approximate answers to the nearest tenth of a square unit.)

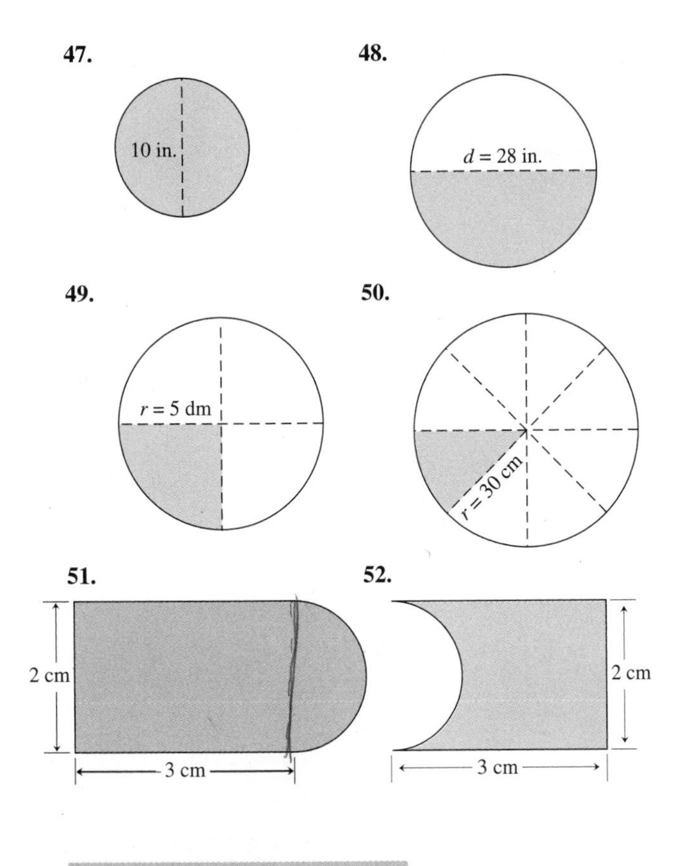

Level 2—Applications

53. What is the area of a television screen that measures 12 in. by 18 in.?

54. What is the area of a rectangular building lot that measures 185 ft by 75 ft?

55. What is the area of a piece of $8\frac{1}{2}$-in. by 11-in. typing paper?

56. If a certain type of material comes in a bolt 3 feet wide, how long a piece must be purchased to have 24 square feet?

57. Find the cost of pouring a square concrete slab of uniform thickness with sides of 25 ft if the cost (at that uniform thickness) is $5.75 per square foot.

Consider the house plan for Problems 58–64 as shown in Figure 6.11. Dimensions are in feet.

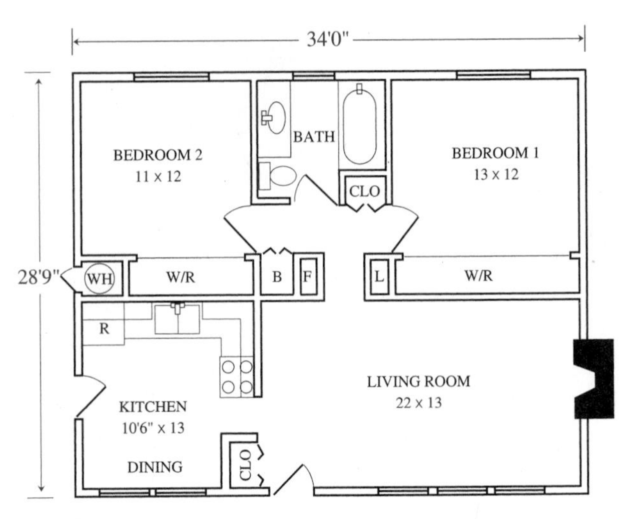

Figure 6.11 House floor plan

58. What is the area of the living room?

59. What is the area of bedroom 1?

60. What is the area of bedroom 2?

61. a. Estimate the area of the house.

b. What is the area of the home?

62. The kitchen and dining area is labeled $10'6'' \times 13$. This means 10 ft 6 in. by 13 ft. What is the area of the kitchen and dining area?

63. If the living room is carpeted with carpet that costs $52 per square yard (including labor and pad), what is the total cost for carpeting the living room in this home? Assume that you cannot purchase part of a square yard. Use estimation to decide whether your answer is reasonable.

64. If bedroom 1 is carpeted with carpet that costs $45 per square yard (including labor and pad), what is the total cost for carpeting this room? Assume that you cannot purchase part of a square yard. Use estimation to decide whether your answer is reasonable.

65. What is the cost of seeding a rectangular lawn 100 ft by 30 ft if 1 pound of seed costs $5.85 and covers 150 square feet? Use estimation to decide whether your answer is reasonable.

66. If a mini pizza has a 6-in. diameter, what is the number of square inches (to the nearest square inch)?

67. If a small pizza has a 10-in. diameter, what is the number of square inches (to the nearest square inch)?

68. If a medium pizza has a 12-in. diameter, what is the number of square inches (to the nearest square inch)?

69. If a large pizza has a 14-in. diameter, what is the number of square inches (to the nearest square inch)?

To find the price per square inch, divide the price by the number of square inches in the pizza. Use the price of the original style pizza in Problems 70–75.

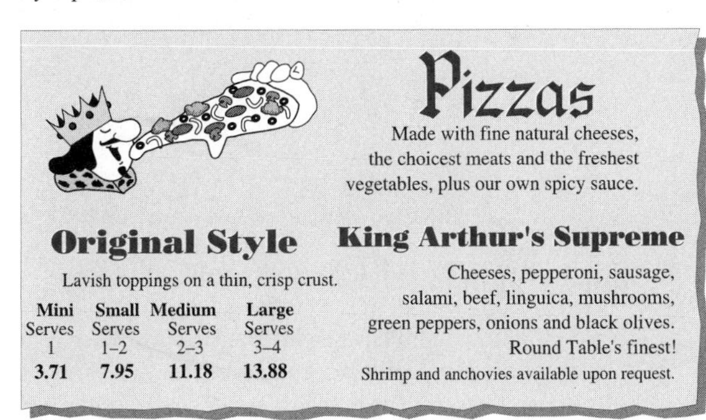

70. What is the price per square inch for a small pizza? (See Problem 67.)

71. What is the price per square inch for a medium pizza? (See Problem 68.)

72. What is the price per square inch for a large pizza? (See Problem 69.)

73. If you order a pizza, what size should you order if you want the best price per square inch? (See Problems 70–73).

74. Suppose you were the owner of a pizza restaurant and were going to offer a 16-in.-diameter pizza. What would you charge for the pizza if you wanted the price to be comparable with the prices of the other sizes?

75. Suppose you were the owner of a pizza restaurant and were going to offer an 8-in.-diameter pizza. What would you charge for the pizza if you wanted the price to be comparable with the prices of the other sizes?

76. Which property costs less per square foot?

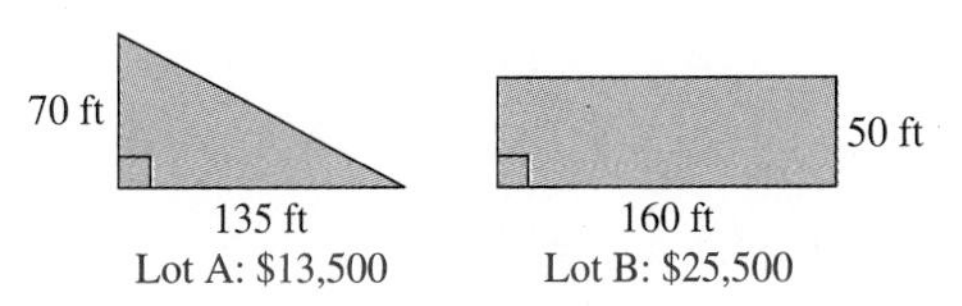

77. Which property costs less per square foot?

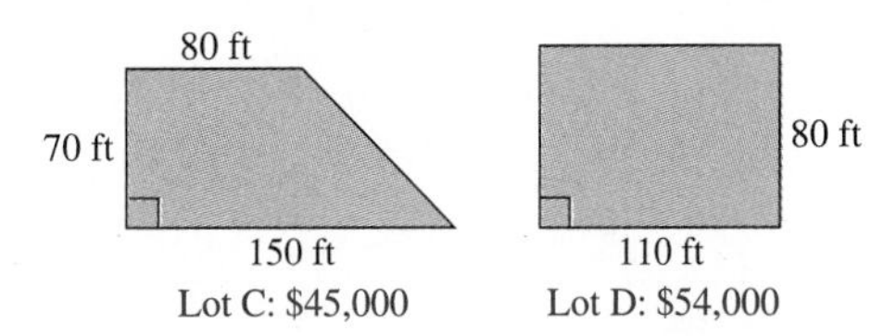

78. If a rectangular piece of property is 750 ft by 1,290 ft, what is the acreage? Round your answer to the nearest tenth acre.

79. How many square feet are there in $4\frac{1}{2}$ acres?

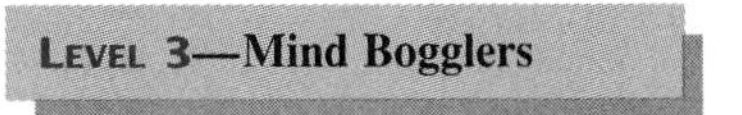

Level 3—Mind Bogglers

80. Find the area of the shaded portion in Figure 6.12 correct to the nearest square centimeter.

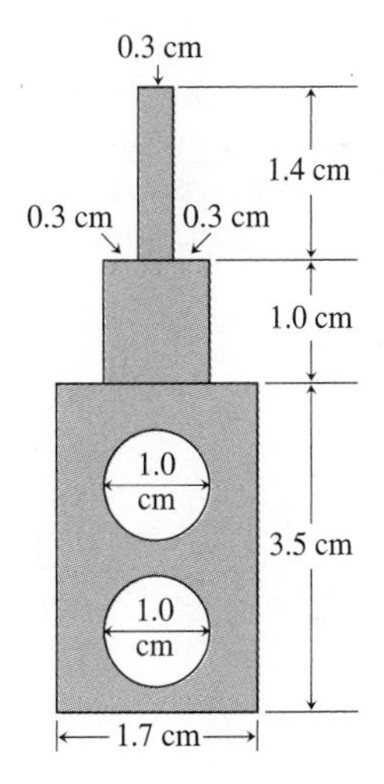

Figure 6.12

81. What would happen if the *entire world population* moved to California?

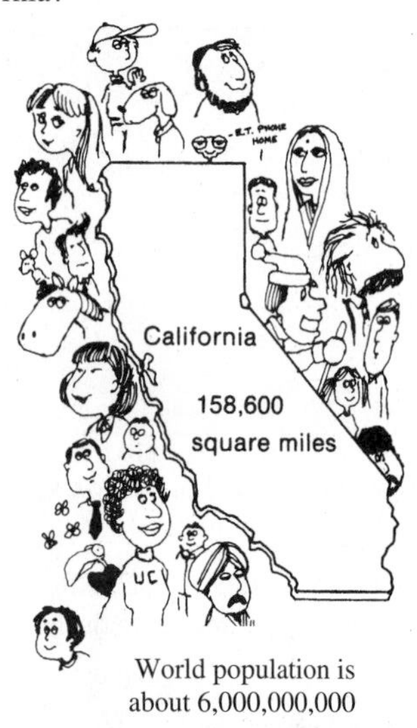

World population is about 6,000,000,000

a. How much space would each person have (make a guess)?
A. 9 in.^{2} B. 9 ft^2 C. 90 ft^2 D. 900 ft^2
E. 9,000 ft^2

b. Now, using the information in the figure, calculate the answer to the question asked in part **a.**

82. Figure 6.13 illustrates a strange and interesting relationship. The square in part **a** has an area of 64 cm^2 (8 cm by 8 cm). When *this same figure* is cut and rearranged as shown in part **b**, it appears to have an area of 65 cm^2. Where did this "extra" square centimeter come from?

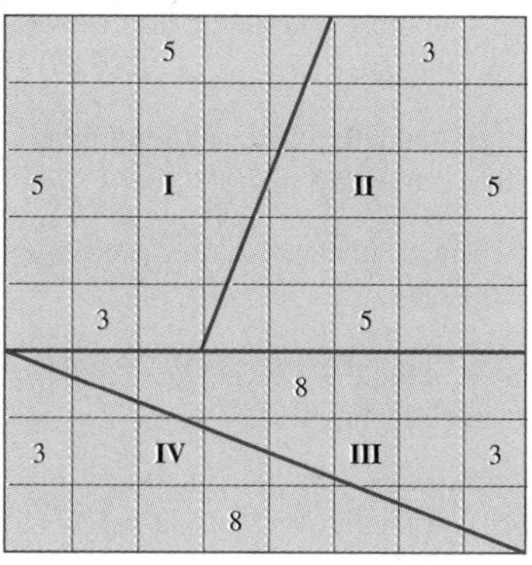

a. 8 cm x 8 cm = 64 cm^2

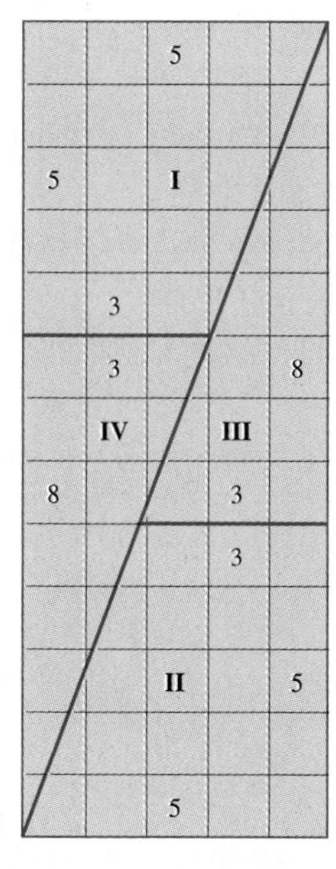

b. 13 cm x 5 cm = 65 cm^2

Figure 6.13 Extra square inch?

[*Hint:* Construct your own square 8 cm on a side, and then cut it into the four pieces as shown. Place the four pieces together as illustrated. Be sure to do your measuring and cutting very carefully. Satisfy yourself that this "extra" square centimeter has appeared. Can you explain this relationship?]

6.4 VOLUME AND CAPACITY

Historical Perspective

Bill Gates (1956–)

The range of discoveries during the Modern Period of mathematics is vast: the maturation of the calculus by Joseph Lagrange (1736–1813), Augustin-Louis Cauchy (1789–1857), Karl Weierstrass (1815–1864), Jules-Henri Poincaré (1854–1912), and David Hilbert (1862–1943); advances in geometry by Adrien-Marie Legendre (1752–1833), Janos Bolyai (1802–1860), and Nikolai Lobachevski (1793–1856); number theory by Sophie Germain (1776–1831) and Srinivasa Ramanujan (1887–1920); computers by Charles Babbage (1792–1871), John Atanasoff, and Stephen Jobs; and computer software by John Kemeny (1927–1992), Thomas Kurtz, and Bill Gates. Since 1995, Bill Gates, founder of Microsoft, has been named the world's wealthiest person. His company owns MS-DOS, Windows, Word, Excel, and is currently acquiring rights to banking and interactive television. He is even buying digital reproduction rights to the world's greatest works of art.

To measure area, we covered a region with square units and then found the area by using a mathematical formula. A similar procedure is used to find the amount of space inside a solid object, which is called its **volume.** We can imagine filling the space with **cubes.** A **cubic inch** and a **cubic centimeter** are shown in Figure 6.14.

Cubic units and volume are important ideas not only in mathematics, but in real-world measurements.

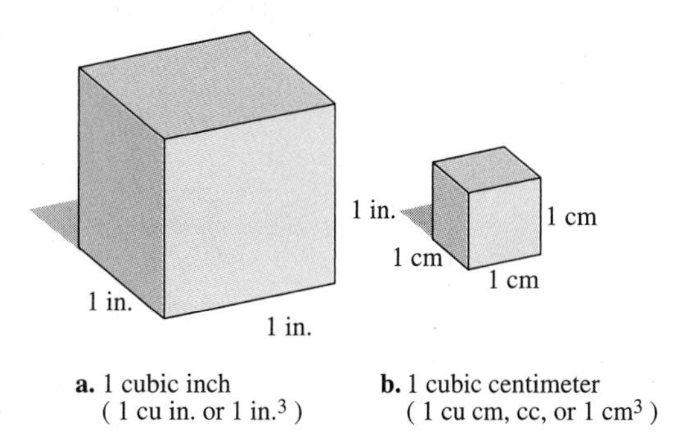

Figure 6.14 Cubic units used for measuring volume

Volume of a Cube

The volume V of a cube with edge s is found by

$$\text{VOLUME} = \text{EDGE} \times \text{EDGE} \times \text{EDGE}$$
$$= (\text{EDGE})^3$$

or

$$V = s^3$$

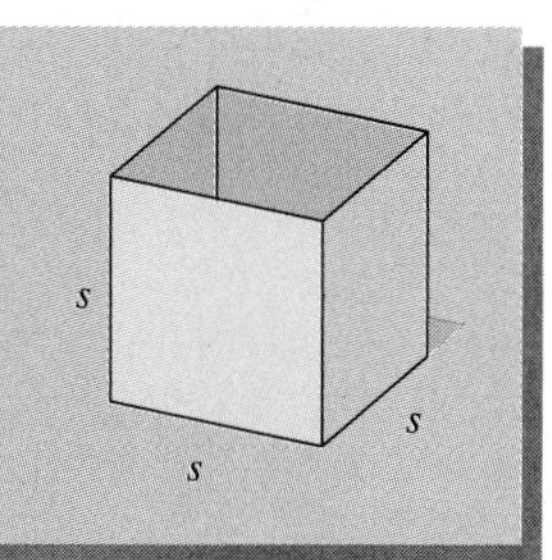

If the solid is not a cube but is a box with edges of different lengths (called a **rectangular parallelepiped**), the volume can be found similarly.

EXAMPLE 1 **Finding the Volume of a Box**

Find the volume of a box that measures 4 ft by 6 ft by 4 ft.

Solution There are 24 cubic feet on the bottom layer of cubes. Do you see how many layers of cubes will fill the solid? (See Figure 6.15.) Since there are four layers with 24 cubes in each, the total number of cubes is

$$4 \times 24 = 96$$

The volume is 96 ft^3. ■

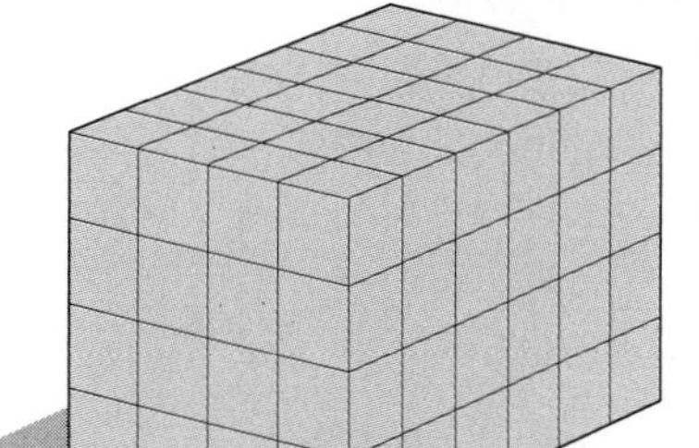

Figure 6.15 What is the volume?

Volume of a Box (Parallelepiped)

The volume V of a box (parallelepiped) with edges ℓ, w, and h is

VOLUME = LENGTH × WIDTH × HEIGHT

or

$$V = \ell \times w \times h$$
$$= \ell wh$$

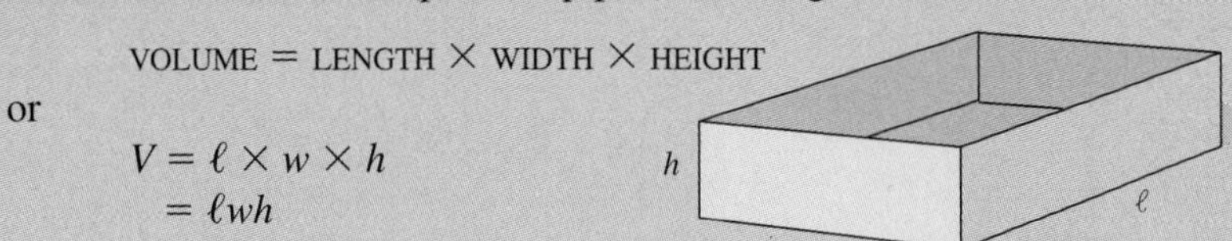

EXAMPLE 2 **Finding Volumes**

Find the volume of each solid.

a.

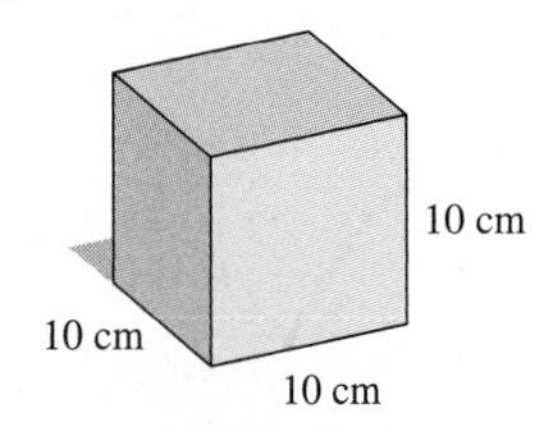

b.

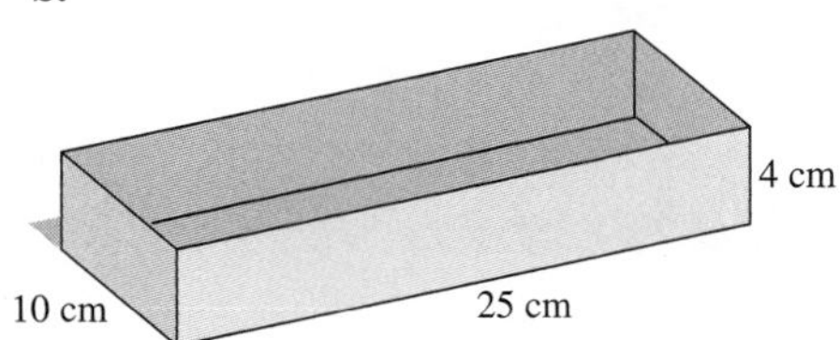

c.

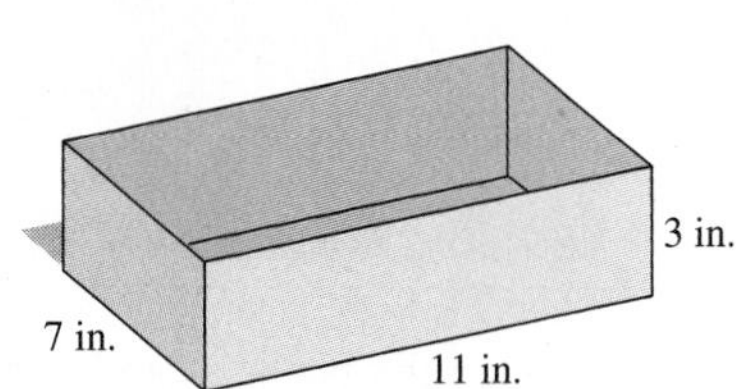

Solution

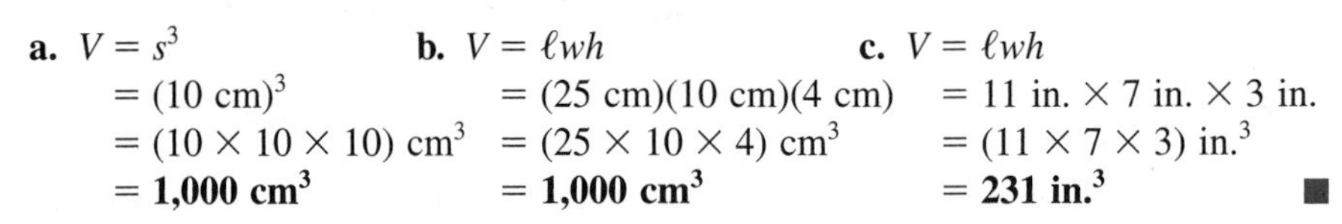

a. $V = s^3$
$= (10\text{ cm})^3$
$= (10 \times 10 \times 10)\text{ cm}^3$
$= \mathbf{1{,}000\ cm^3}$

b. $V = \ell wh$
$= (25\text{ cm})(10\text{ cm})(4\text{ cm})$
$= (25 \times 10 \times 4)\text{ cm}^3$
$= \mathbf{1{,}000\ cm^3}$

c. $V = \ell wh$
$= 11\text{ in.} \times 7\text{ in.} \times 3\text{ in.}$
$= (11 \times 7 \times 3)\text{ in.}^3$
$= \mathbf{231\ in.^3}$ ■

Sometimes the dimensions for the volume we are finding are not given in the same units. In such cases, you must convert all dimensions to the same units. The common conversions are as follows:

1 ft = 12 in.	To convert feet to inches, multiply by 12. To convert inches to feet, divide by 12.
1 yd = 3 ft	To convert yards to feet, multiply by 3. To convert feet to yards, divide by 3.
1 yd = 36 in.	To convert yards to inches, multiply by 36. To convert inches to yards, divide by 36.

EXAMPLE 3 **Problem Solving with Volumes**

Suppose you are pouring a rectangular driveway with dimensions of 24 ft by 65 ft. The depth of the driveway is 3 in., and concrete is ordered by the yard. By a "yard" of concrete, we mean a cubic yard. You cannot order part of a yard of concrete. How much concrete should you order?

> Do you know where the expression "the whole 9 yards" comes from? It's related to concrete. A standard-size "cement mixer" has a capacity of 9 cubic yards of concrete. Thus, a job requiring the mixer's full capacity demands "the whole 9 yards."

Solution There are three different units of measurement in this problem: inches, feet, and yards. Since we want the answer in cubic yards, we will convert all of the measurements to yards:

$65 \text{ ft} = (65 \div 3) \text{ yd} = \frac{65}{3} \text{ yd}$ This is the length, ℓ.

$24 \text{ ft} = (24 \div 3) \text{ yd} = 8 \text{ yd}$ This is the width, w.

$3 \text{ in.} = (3 \div 36) \text{ yd} = \frac{3}{36} \text{ yd} = \frac{1}{12} \text{ yd}$ This is the height (depth), h.

$$\begin{aligned} V &= \ell wh \\ &= \tfrac{65}{3}(8)(\tfrac{1}{12}) \text{ yd}^3 \quad \text{Think of 8 as } \tfrac{8}{1}. \\ &= \frac{65 \times 8 \times 1}{3 \times 12} \\ &= \frac{130}{9} \\ &= 14\tfrac{4}{9} \end{aligned}$$

You must order 15 yards of concrete. ■

One of the most common applications of volume involves measuring the amount of liquid a container holds, which we refer to as its **capacity.** For example, if a container is 2 ft by 2 ft by 12 ft, it is fairly easy to calculate the volume:

$$2 \times 2 \times 12 = 48 \text{ ft}^3$$

But this still doesn't tell us how much water the container holds. The capacities of a can of cola, a bottle of milk, an aquarium tank, the gas tank in your car, and a swimming pool can all be measured by the amount of fluid they can hold.

Standard Units of Capacity

U.S. SYSTEM	*METRIC SYSTEM*
gallon (gal)	**liter (L)**
quart (qt; $\frac{1}{4}$ gal)	kiloliter (kL; 1,000 L)
ounce (oz; $\frac{1}{128}$ gal)	milliliter (mL; $\frac{1}{1,000}$ L)
cup (c; 8 oz)	

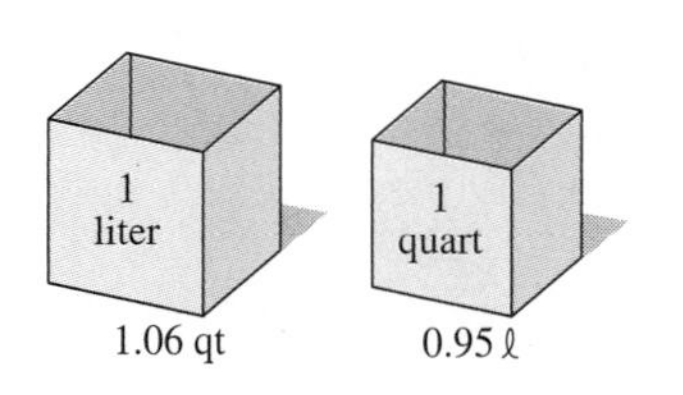

Figure 6.16 Standard capacities

Most containers of liquid that you buy have capacities stated in both milliliters and ounces, or quarts and liters (see Figure 6.16). Some of these size statements are listed in Table 6.1 (page 262). The U.S. Bureau of Alcohol, Tobacco, and Firearms

You should remember some of these references for purposes of estimation. For example, know that a can of Coke is 355 mL, and be able to recognize the size of a liter of milk. A cup of coffee is about 300 mL and a spoonful of medicine about 5 mL.

Table 6.1 *Capacities of Common Grocery Items, as Shown on Labels*

Item	U.S. capacity	Metric capacity
Milk	$\frac{1}{2}$ gal.	1.89 L
Milk	1.06 qt	1 L
Budweiser	12 oz	355 mL
Coke	67.6 oz	2 L
Hawaiian Punch	1 qt	0.95 L
Del Monte pickles	1 pt 6 oz	651 mL

has made metric bottle sizes mandatory for liquor, so the half-pint, fifth, and quart have been replaced by 200-mL, 750-mL, and 1 L sizes. A typical dose of cough medicine is 5 mL, and 1 kL is 1,000 L, or about the amount of water one person would use for all purposes in two or three days.

Since it is common practice to label capacities both in U.S. and in metric measuring units, it will generally not be necessary for you to make conversions from one system to another. But if you do, it is easy to remember that a liter is just a little larger than a quart, just as a meter is a little longer than a yard.

To measure capacity, you use a measuring cup.

EXAMPLE 4 **Measuring Capacity**

Measure the amount of liquid in the measuring cup in Figure 6.17, both in the U.S. system and in the metric system.

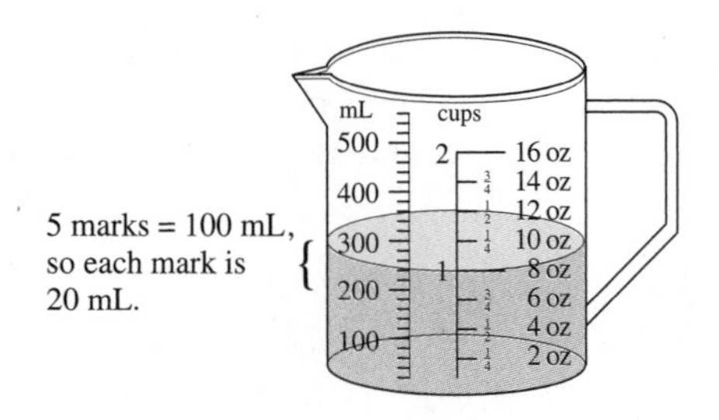

Figure 6.17 Standard measuring cup with both metric and U.S. measurements

Solution Metric: **240 mL** U.S.: **About 1 c or 8 oz** ■

Some common relationships among volume and capacity measurements in the U.S. system are shown in Figure 6.18.

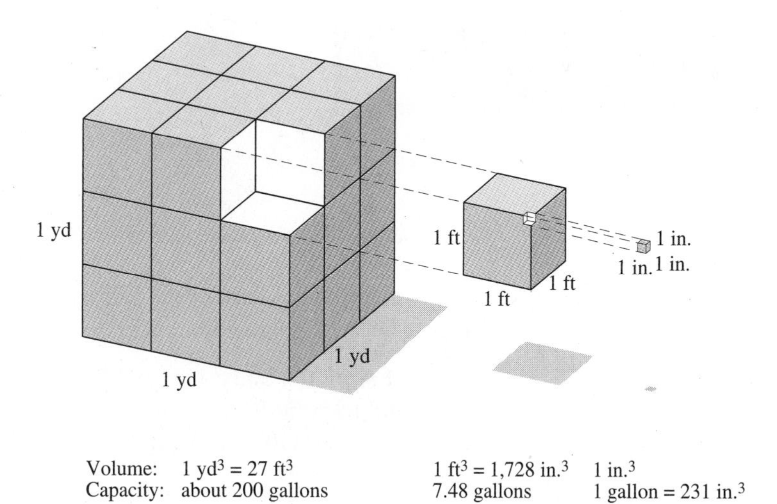

Figure 6.18 U.S. measurement relationship between volume and capacity

In the U.S. system of measurement, the relationship between volume and capacity is not particularly convenient. One gallon of capacity occupies 231 in.^{3}. This means that, since the box in part **c** of Example 2 has a volume of 231 in.^{3}, we know that it will hold exactly 1 gallon of water.

The relationship between volume and capacity in the metric system is easier to remember. One cubic centimeter is one-thousandth of a liter. Notice that this is the same as a milliliter. For this reason, you will sometimes see cc used to mean cm^3 or mL. These relationships are shown in Figure 6.19.

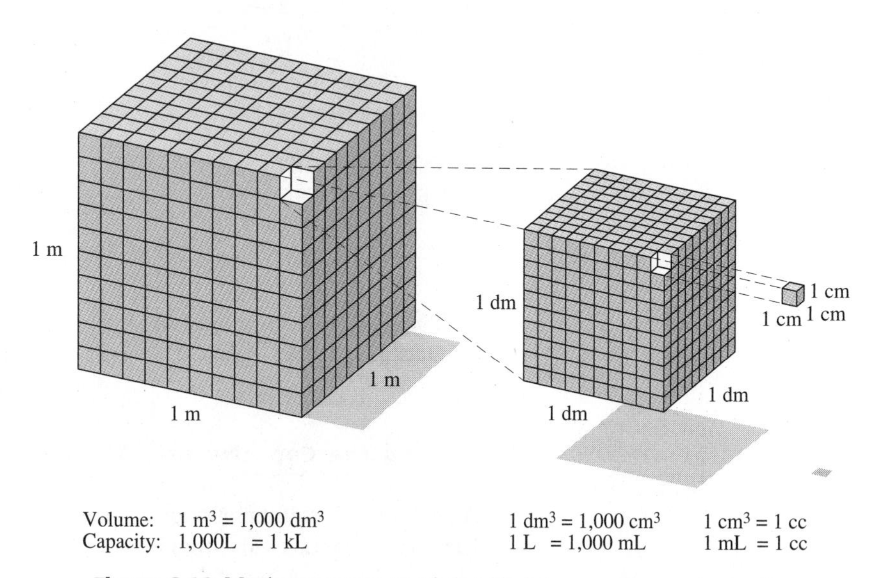

Figure 6.19 Metric measurement relationship between volume and capacity

Relationship Between Volume and Capacity

1 liter = 1,000 cm^3

1 gallon = 231 in.^{3}; 1 ft^3 $\approx$ 7.48 gal

To find the capacity of the 2-ft by 2-ft by 12-ft box mentioned previously, we must change 48 ft^3 to cubic inches:

Estimate: Since 1 ft^3 $\approx$ 7.5 gal
48 ft^3 $\approx$ 50 ft^3
$\approx$ (50 × 7.5) gal
= 375 gal

48 ft^3 = 48 × (1 ft) × (1 ft) × (1 ft)
= 48 × 12 in. × 12 in. × 12 in.
= 82,944 in.^{3} A calculator would help here.

Since 1 gallon is 231 in.^{3}, the final step is to divide 82,944 by 231 to obtain approximately 359 gallons.

EXAMPLE 5 **Calculating Capacities**

How much water would each of the following containers hold?

a.

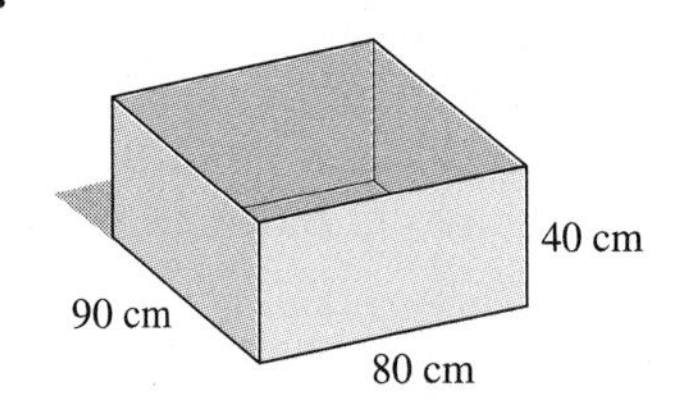

b.

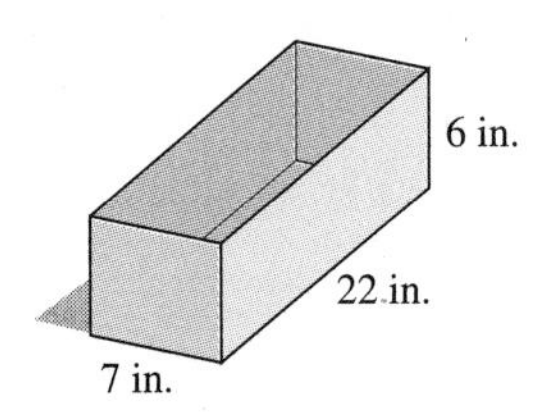

Solution

Estimate for part **b:**

Container is approximately $\frac{1}{2}$ ft × 2 ft × $\frac{1}{2}$ ft = $\frac{1}{2}$ ft^3
$\approx$ ($\frac{1}{2}$ × 7.5) gal = 3.75 gal

a. $V = 90 \times 80 \times 40$ cm^3
= 288,000 cm^3

Since each 1,000 cm^3 is 1 liter,

$$\frac{288{,}000}{1{,}000} = 288$$

This container would hold 288 liters.

b. $V = 7 \times 22 \times 6$ in.
= 924 in.^{3}

Since each 231 in.^{3} is 1 gallon,

$$\frac{924}{231} = 4$$

This container would hold 4 gallons. ■

EXAMPLE 6 **Finding the Capacity of a Swimming Pool**

An ecological swimming pool is advertised as being 20 ft × 25 ft × 5 ft. How many gallons will it hold?

Solution

$$V = 20 \text{ ft} \times 25 \text{ ft} \times 5 \text{ ft}$$
$$= 2{,}500 \text{ ft}^3$$

Since $1 \text{ ft}^3 \approx 7.48$ gal, the swimming pool contains

$$2{,}500 \times 7.48 \approx \mathbf{18{,}700 \text{ gallons}}$$

■

PROBLEM SET 6.4

ESSENTIAL IDEAS

1. **In Your Own Words** Contrast area and volume.

2. **In Your Own Words** Compare or contrast 1 liter and 1 cm^3.

LEVEL 1

3. **In Your Own Words** Write a report listing the volumes or capacities of several items on your kitchen shelf, stating the amounts in both the metric and U.S. measurement systems.

WHAT IS WRONG, *if anything, with each of the statements in Problems* 4–10*? Explain your reasoning.*

4. A cubic inch is larger than a cubic centimeter.

5. A quart is larger than a liter.

6. A meter is longer than a yard.

7. A mL is larger than a cubic centimeter.

8. Volume is a square measure.

9. $1{,}000 \text{ in.}^3 = 1 \text{ yd}^3$

10. Since $3 \text{ ft} = 1 \text{ yd}$, $3 \text{ ft}^2 = 1 \text{ yd}^2$, and $3 \text{ ft}^3 = 1 \text{ yd}^3$.

LEVEL 1—Drill

In Problems 11–13*, find the volume of each solid by counting the number of cubic centimeters in each box.*

11.

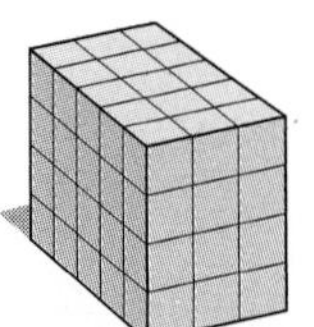

12.

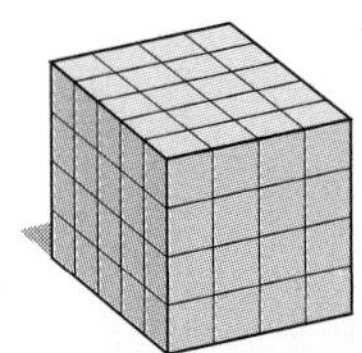

13.

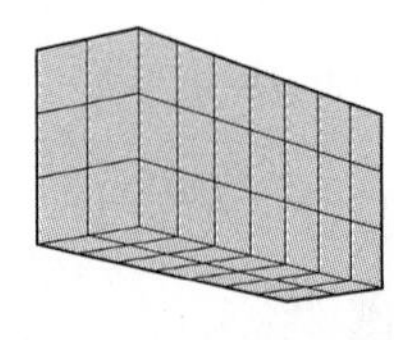

Find the volume of each solid in Problems 14–21.

14. **15.**

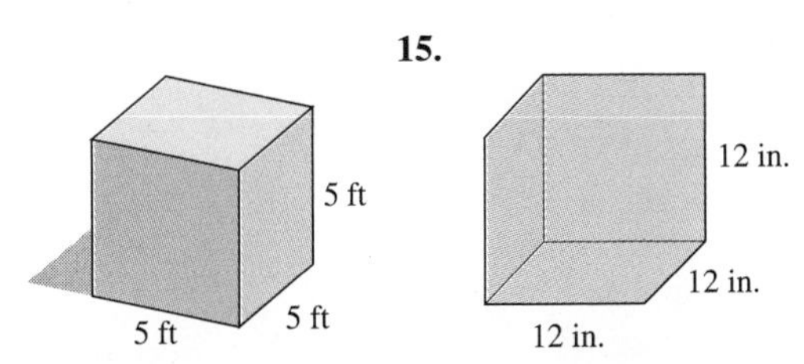

16. **17.**

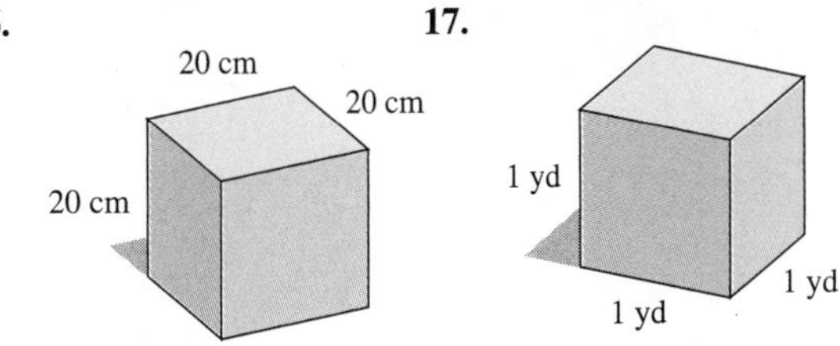

18. **19.**

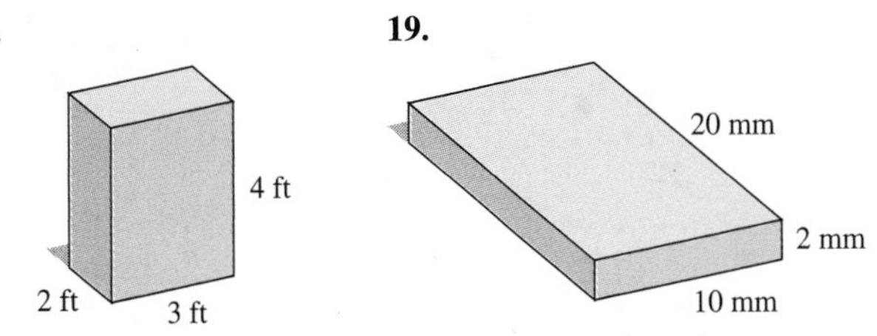

20. **21.**

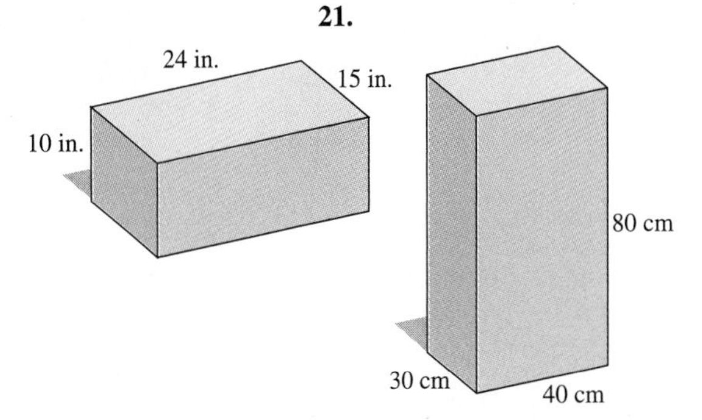

Measure each amount given in Problems 22–33.

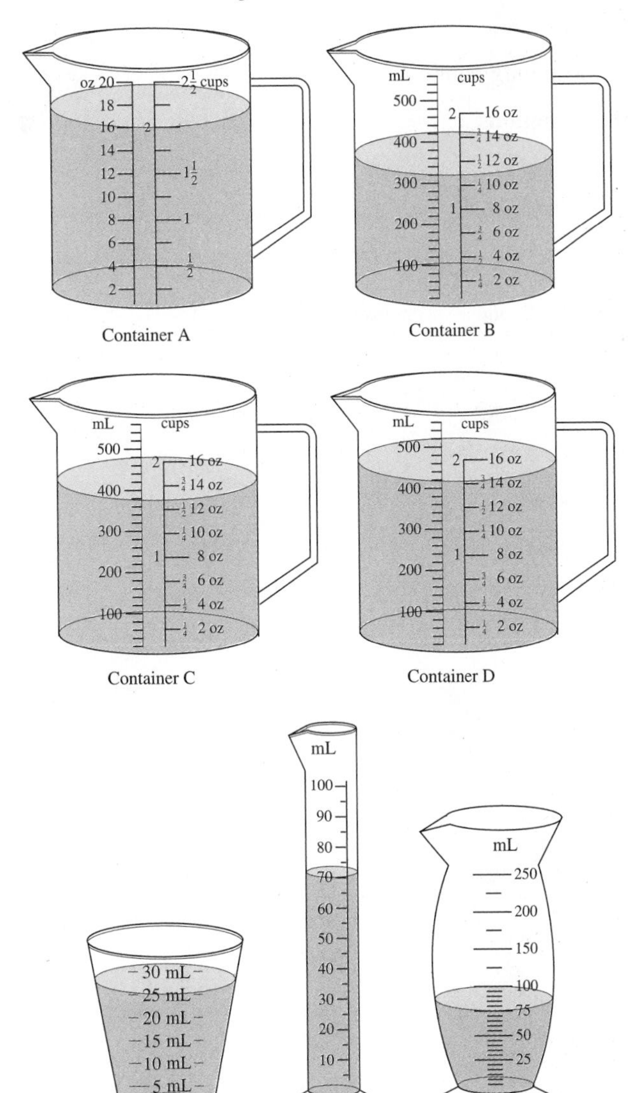

Container A Container B Container C Container D Container E Container F Container G

22. Container A in cups

23. Container A in ounces

24. Container B in ounces

25. Container C in ounces

26. Container B in milliliters

27. Container C in milliliters

28. Container D in cups

29. Container D in ounces

30. Container D in milliliters

31. Container E in milliliters

32. Container F in milliliters

33. Container G in milliliters

Level 2—Drill

The ability to estimate capacities is an important skill to develop. Without measuring, pick the best answer in Problems 34–46.

34. An average cup of coffee is about
A. 250 mL B. 750 mL C. 1 L

35. If you want to paint some small bookshelves, how much paint would you probably need?
A. 1 mL B. 100 mL C. 1 L

36. A six-pack of beer would contain about
A. 2 mL B. 200 mL C. 2 L

37. The dose of a strong cough medicine might be
A. 2 mL B. 200 mL C. 2 L

38. A glass of water served at a restaurant is about
A. 200 mL B. 2 mL C. 2 L

39. Enough water for a bath would be about
A. 300 mL B. 300 L C. 300 kL

40. Enough gas to fill your car's empty tank would be about
A. 15 L B. 200 mL C. 70 L

41. 50 kL of water would be about enough for
A. taking a bath B. taking a swim
C. supplying the drinking water for a large city

42. Which measurement would be appropriate for administering some medication?
A. mL B. L C. kL

43. You order some champagne for yourself and one companion. You would most likely order
A. 2 mL B. 700 mL C. 20 L

44. The prefix *centi-* means
A. one thousand B. one-thousandth
C. one-hundredth

45. The prefix *milli-* means
A. one thousand B. one-thousandth
C. one-hundredth

46. The prefix *kilo-* means
A. one thousand B. one-thousandth
C. one-hundredth

Level 2—Drill

What is the capacity for each of the containers in Problems 47–56*? (Give answers in the U.S. system to the nearest tenth of a gallon or in metric to the nearest tenth of a liter.)*

47.

10 in.
10 in.
10 in.

48.

25 cm
25 cm
25 cm

49.

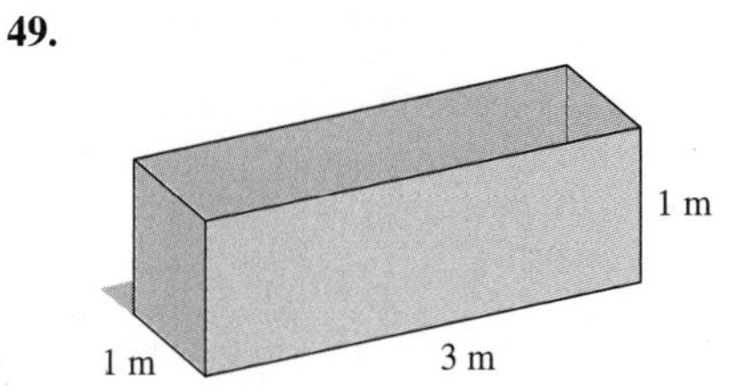

50. **51.**

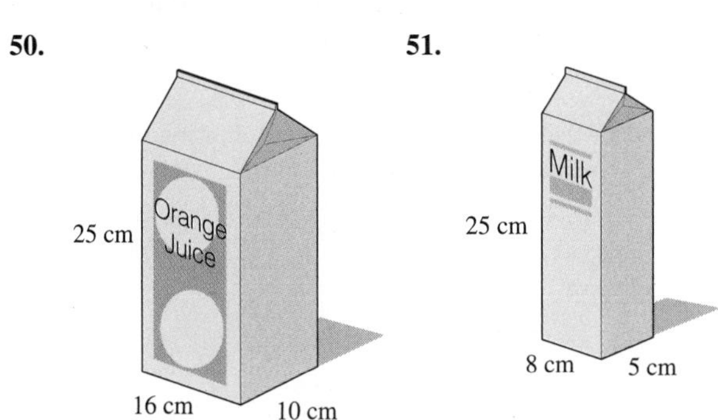

52.

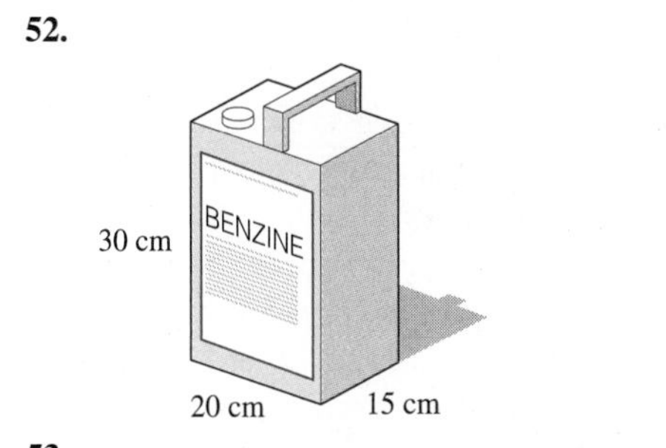

53.

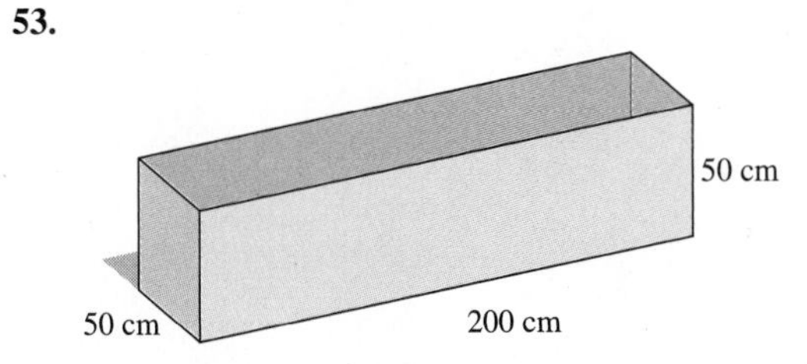

54. **55.** **56.**

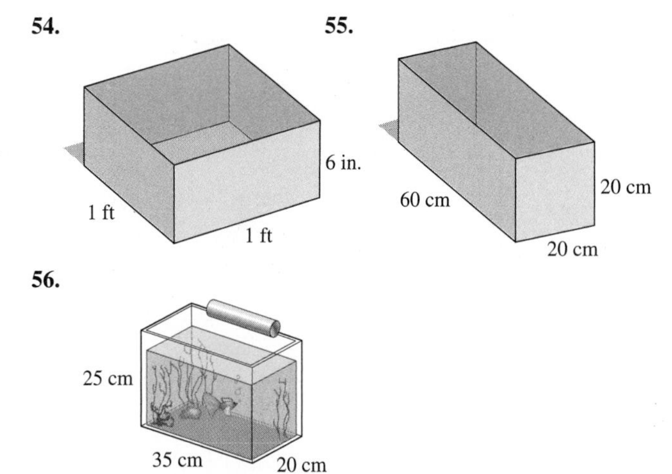

Level 2—Applications

57. The exterior dimensions of a refrigerator are shown in the figure.

33 in.
36 in.
66 in.

a. How many cubic feet are contained within the refrigerator?

b. If it is advertised as a 19-cu-ft refrigerator, how much space is taken up by the motor, insulation, and so on?

58. The exterior dimensions of a freezer are 48 inches by 36 inches by 24 inches, and it is advertised as being 27.0 cu ft. Is the advertised volume correctly stated?

59. Suppose that you must order concrete for a sidewalk 50 ft by 4 ft to a depth of 4 in. How much concrete is required? (Answer to the nearest $\frac{1}{2}$ cubic yard.)

Use the plot plan shown in Figure 6.20 (page 268) *to answer Problems* 60–64. *Answer to the nearest cubic yard.**

60. How many cubic yards of sawdust are needed for preparation of the lawn area if it is to be spread to a depth of 6 inches?

*In practice, you would not round to the nearest yard, but rather would round up to ensure that you had enough material. However, for consistency in this book, we will round according to the rules developed in the first chapter.

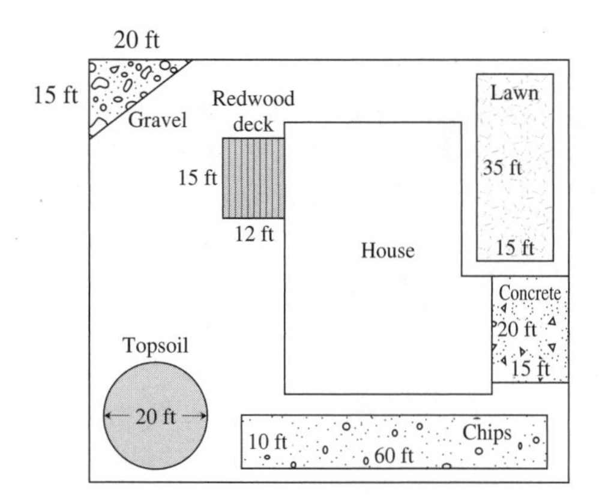

Figure 6.20 Plot plan

61. How many cubic yards of chips are necessary if they are to be placed to a depth of 3 inches?

62. How much topsoil must be hauled in if it is to be spread to a depth of 12 inches?

63. How much gravel is necessary if it is to be laid to a depth of 3 inches?

64. Suppose that you wish to pave the area labeled "Concrete." How much concrete is needed if it is to be poured to a depth of 4 inches?

65. How much water will a 7-m by 8-m by 2-m swimming pool contain (in kiloliters)?

66. How much water will a 21-ft by 24-ft by 4-ft swimming pool contain (rounded to the nearest gallon)?

Level 3—Mind Bogglers

67. The total human population of the earth is about 6.0×10^9.

a. If each person has the room of a prison cell (50 sq ft), and if a square mile is about 2.8×10^7 sq ft, how many people could fit into a square mile?

b. How many square miles would be required to accommodate the entire human population of the earth?

c. If the total land area of the earth is about 5.2×10^7 sq mi, and if all the land area were divided equally, how many acres of land would each person be allocated (1 sq mi = 640 acres)?

68. a. Guess what percentage of the world's population could be packed into a cubical box measuring $\frac{1}{2}$ mi on each side. [*Hint:* The volume of a typical person is about 2 cu ft.]

b. Now calculate the answer to part **a,** using the earth's population as given in Problem 67.

6.5 MISCELLANEOUS MEASUREMENTS

Historical Perspective

Amalie (Emmy) Noether (1882–1935)

Noether has been described as the greatest female mathematician. She chose a career in mathematics during a period of time when it was very unusual for a woman to seek a higher education. In fact, when she attended the University of Erlangen, the Academic Senate declared that women students would "overthrow all academic order" and a historian of the day wrote that talk of "surrendering our university to the invasion of women . . . is a shameful display of moral weakness." Emmy Noether is famous for her work in algebra, which laid a cornerstone for Einstein's work in general relativity. But in addition to her theoretical work, she was a teacher and a caring person.

In this chapter we have been discussing measurement.

LENGTH

If you are measuring length, you will use linear measures:

in.; ft; yd; mi   cm; m; km

AREA

For area, you will use square measure:

in.^{2}; ft^2; yd^2; mi^2 cm^2; m^2; km^2

VOLUME

For volumes, you will use cubic measures:

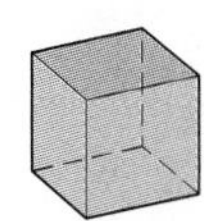

in.^{3}; ft^3 cm^3 (or cc); m^3; km^3

In the last section we found the volume and the capacity of boxes; now we can extend this to other solids. In Figure 6.21 we show some of the more common solids, along with the volume formulas. We will use B in each case to signify the area of the base and h for the height.

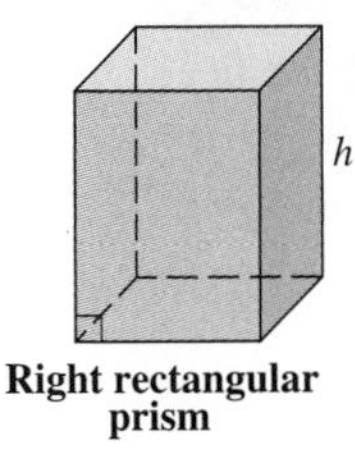

Right rectangular prism

$V = Bh$

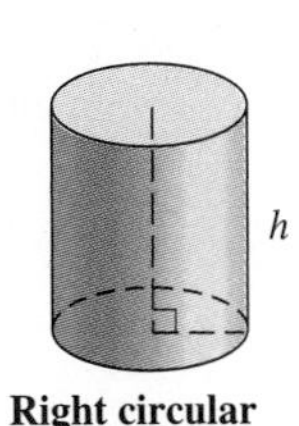

Right circular cylinder

$V = Bh$

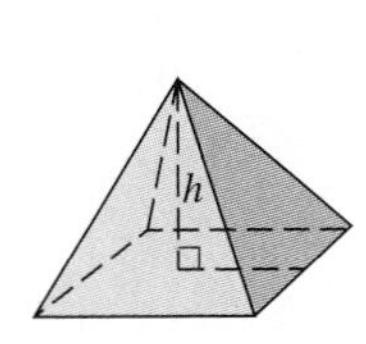

Pyramid

$V = \frac{1}{3}Bh$

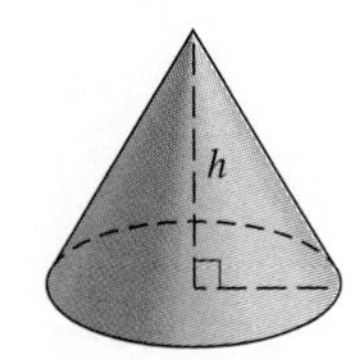

Right circular cone

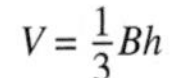

$V = \frac{1}{3}Bh$

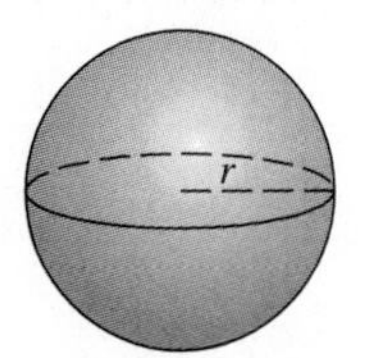

Sphere

$V = \frac{4}{3}\pi r^3$

Figure 6.21 Common solids, with accompanying volume formulas

EXAMPLE 1

Volume of a Right Circular Cylinder

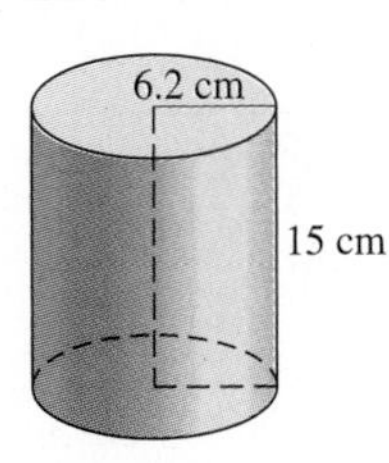

Find the volume of the accompanying solid to the nearest cubic unit.

Solution This is a right circular cylinder. Hence,

$$\begin{aligned} V &= Bh \\ &= \pi(6.2)^2(15) \\ &\approx 1811.4423 \end{aligned}$$

Notice that $B = \pi r^2$, where $r = 6.2$; the height of the cylinder is 15 ($h = 15$).

The volume is 1,811 cm³. ■

EXAMPLE 2

Volume of a Right Circular Cone

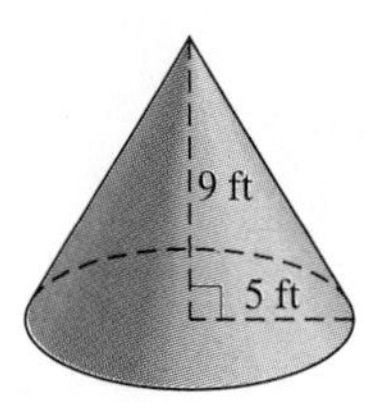

Find the volume of the accompanying solid to the nearest cubic unit.

Solution This is a right circular cone. Hence,

$$\begin{aligned} V &= \tfrac{1}{3}Bh \\ &= \tfrac{1}{3}\pi(5)^2 9 \\ &= 75\pi \\ &\approx 235.6 \end{aligned}$$

Notice that $B = \pi r^2$ and $r = 5$.

The volume is 236 ft³. ■

EXAMPLE 3

Volume of a Right Triangular Prism

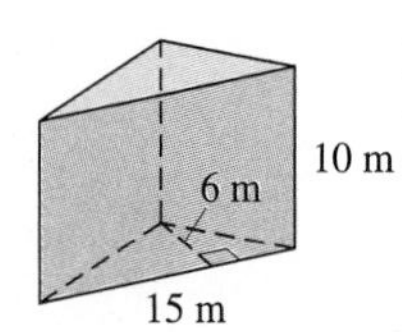

Find the volume of the accompanying solid to the nearest cubic unit.

Solution This is a right triangular prism. Hence,

$$\begin{aligned} V &= Bh \\ &= \tfrac{1}{2}(15)(6)(10) \\ &= 450 \end{aligned}$$

Notice that $B = \frac{1}{2}ba$, where $a = 6$ (height of triangle), $b = 15$ (base of triangle), and $h = 10$ (height of prism).

The volume is 450 m³. ■

EXAMPLE 4

Volume of a Sphere

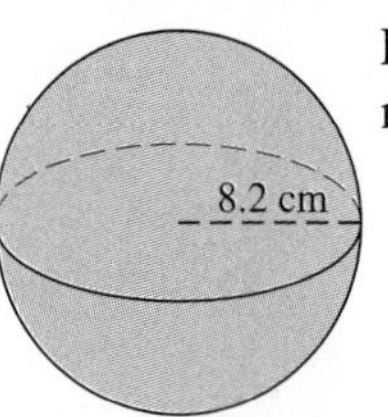

Find the volume of the accompanying solid to the nearest cubic unit.

Solution This is a sphere. Hence,

$$V = \tfrac{4}{3}\pi r^3$$
$$= \tfrac{4}{3}\pi(8.2)^3$$
$$\approx 2{,}309.564878$$

The volume is 2,310 cm³. ■

EXAMPLE 5 **Volume of a Pyramid**

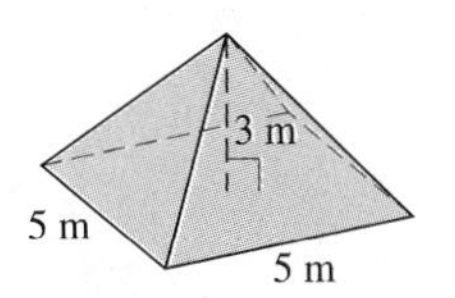

Find the volume of the accompanying solid to the nearest cubic unit.

Solution This is a pyramid. Hence,

$$V = \tfrac{1}{3}Bh$$
$$= \tfrac{1}{3}(5)^2(3) \quad \text{Notice that the base is a square, so } B = s^2 = 5^2.$$
$$= \tfrac{25}{3}(3)$$
$$= 25$$

The volume is 25 m³. ■

The measurements of length, area, volume, and capacity were discussed in the previous section. Two additional measurements to consider are mass and temperature.

MASS

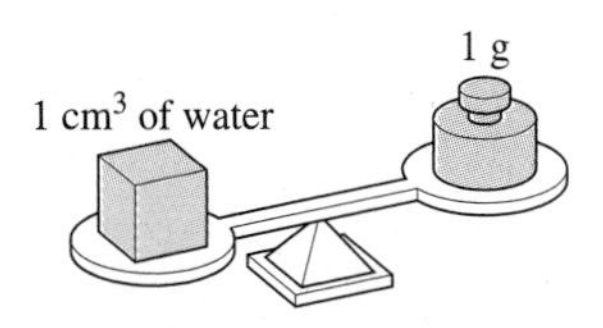

Figure 6.22 Mass of 1 g

The **mass** of an item is the amount of matter it comprises. The **weight** of an item is the heaviness of the matter.* The U.S. and metric units of measurement for mass or weight are given in the following box. Notice that, in the U.S. measurement system, ounces are used as a weight measurement; this is not the same as the capacity measurement for ounces that we used previously. The basic unit of measurement for mass in the metric system is the **gram,** which is defined as the mass of 1 cm^3 of water, as shown in Figure 6.22.

Standard Units of Weight

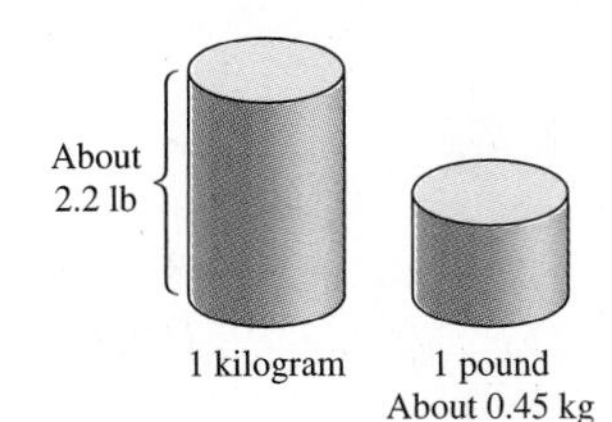

U.S. SYSTEM	*METRIC SYSTEM*
ounce (oz)	**gram (g)**
pound (lb; 16 oz)	milligram (mg; $\frac{1}{1{,}000}$ g)
ton (T; 2,000 lb)	kilogram (kg; 1,000 g)

*Technically, weight is the force of gravity acting on a body. The mass of an object is the same on the moon or on the earth, whereas the weight of that same object would be different on the moon and on the earth, since they have different forces of gravity. For our purposes, you can use either of the words *mass* or *weight,* since we are weighing things only on earth.

A paper clip weighs about 1 g, a cube of sugar about 3 g, and a nickel about 5 g. This book weighs about 1 kg, and an average-size person weighs from 50 to 100 kg. Some common weights of grocery items are listed in Table 6.2.

Pay attention to the weight of common everyday items.

Table 6.2 *Weights of Common Grocery Items, as Shown on Labels*

Item	U.S. weight	Metric weight
Kraft cheese spread	5 oz	142 g
Del Monte tomato source	8 oz	227 g
Campbell cream of chicken soup	$10\frac{3}{4}$ oz	305 g
Kraft marshmallow cream	11 oz	312 g
Bag of sugar	5 lb	2.3 kg
Bag of sugar	22 lb	10 kg

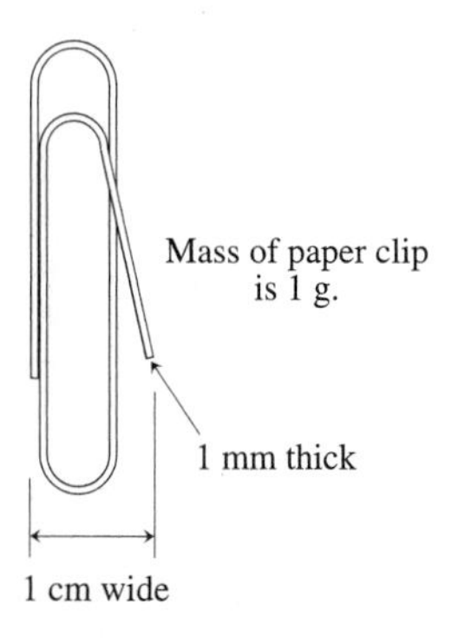

We use a scale to measure weight. To weigh items, or ourselves, in metric units, we need only replace our U.S. weight scales with metric weight scales. As with other measures, we must begin to think in terms of metric units and to estimate the weight of various items. The multiple-choice questions in Problem Set 6.5 are designed to help you do this.

TEMPERATURE

The final quantity of measure that we'll consider in this chapter is the degree of hotness or coldness, or **temperature.**

Standard Units of Temperature

THE U.S. SYSTEM HAS ONE COMMON TEMPERATURE UNIT:	*THE METRIC SYSTEM HAS ONE COMMON TEMPERATURE UNIT:*
Fahrenheit (°F)	**Celsius** (°C)

To work with temperatures, it is necessary to have some reference points.

U.S. TEMPERATURE		*METRIC TEMPERATURE*	
Water freezes:	32°F	Water freezes:	0°C
Water boils:	212°F	Water boils:	100°C

We are usually interested in measuring temperature in three areas: atmospheric temperature (usually given in weather reports), body temperature (used to determine illness), and oven temperature (used in cooking). The same scales are used, of course, for measuring all of these temperatures. But notice the difference in the ranges of temperatures we're considering. The comparisons for Fahrenheit and Celsius are shown in Figure 6.23.

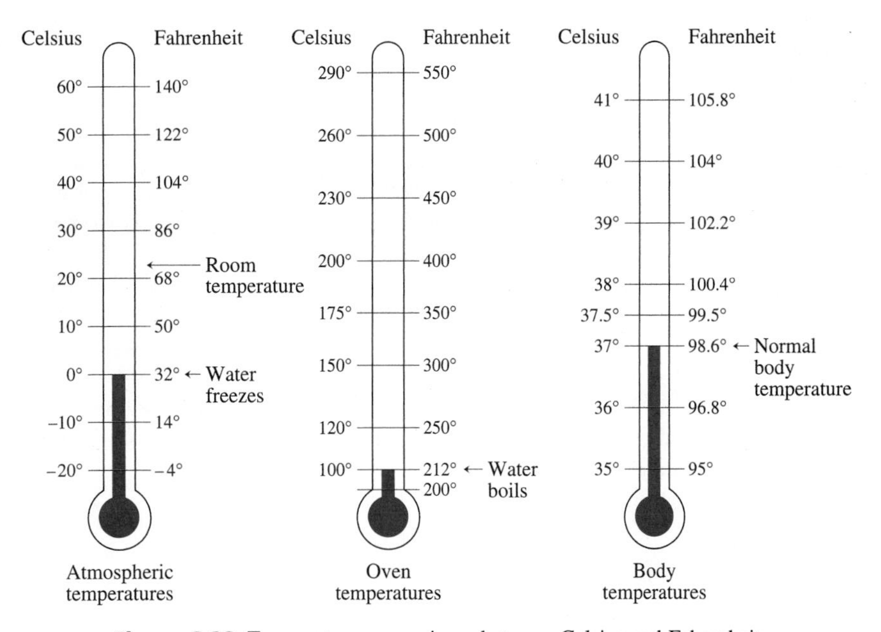

Figure 6.23 Temperature comparisons between Celsius and Fahrenheit

PROBLEM SET 6.5

Essential Ideas

1. **In Your Own Words** Contrast measuring length, area, and volume.
2. **In Your Own Words** State the formulas for the volume of each of the given solids:
   a. right circular cone **b.** sphere
   c. right rectangular prism **d.** pyramid
   e. right circular cylinder
3. **In Your Own Words** What are the freezing point, boiling point, and room temperature in both the metric and U.S. measurement systems?

Level 1

Without measuring, pick the best choice in Problems 4–23 by estimating.

4. A hamburger patty would weigh about
   A. 170 g B. 240 mg C. 2 kg
5. A can of carrots at the grocery store most likely weighs about
   A. 40 kg B. 4 kg C. 0.4 kg
6. A newborn baby would weigh about
   A. 490 mg B. 4 kg C. 140 kg
7. John tells you he weighs 150 kg. If John is an adult, he is
   A. underweight B. about average C. overweight
8. You have invited 15 people for Thanksgiving dinner. You should buy a turkey that weighs about
   A. 795 mg B. 4 kg C. 12 kg
9. Water boils at
   A. 0°C B. 100°C C. 212°C
10. If it is 32°C outside, you would most likely find people
    A. ice skating B. water skiing
11. If the doctor says that your child's temperature is 37°C, you child's temperature is
    A. low B. normal C. high
12. You would most likely broil steaks at
    A. 120°C B. 500°C C. 290°C
13. A kilogram is ___?___ a pound.
    A. more than B. about the same as C. less than
14. The prefix used to mean 1,000 is
    A. centi- B. milli- C. kilo-
15. The prefix used to mean $\frac{1}{1,000}$ is
    A. centi- B. milli- C. kilo-
16. The prefix used to mean $\frac{1}{100}$ is
    A. centi- B. milli- C. kilo-
17. 15 kg is a measure of
    A. length B. capacity C. weight D. temperature
18. 28.5 m is a measure of
    A. length B. capacity C. weight D. temperature
19. 6 L is a measure of
    A. length B. capacity C. weight D. temperature
20. 38°C is a measure of
    A. length B. capacity C. weight D. temperature
21. 7 mL is a measure of
    A. length B. capacity C. weight D. temperature
22. 68 km is a measure of
    A. length B. capacity C. weight D. temperature
23. 14.3 cm is a measure of
    A. length B. capacity C. weight D. temperature

Level 1—Drill

Name the metric unit you would use to measure each of the quantities in Problems 24–35.

24. The distance from New York to Chicago
25. The distance around your waist
26. Your height
27. The height of a building
28. The capacity of a wine bottle
29. The amount of gin in a martini
30. The capacity of a car's gas tank
31. The amount of water in a swimming pool
32. The weight of a pencil

33. The weight of an automobile

34. The outside temperature

35. The temperature needed to bake a cake

Level 2—Drill

Find the volumes of the solids in Problems 36–44 *correct to the nearest cubic unit.*

36.

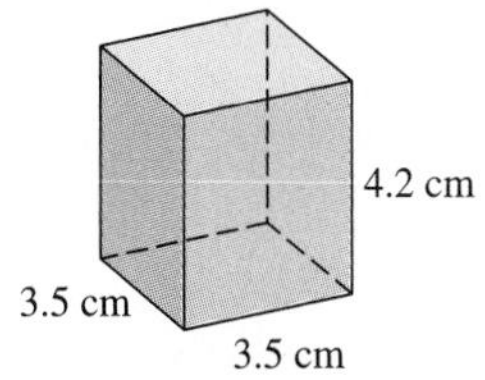

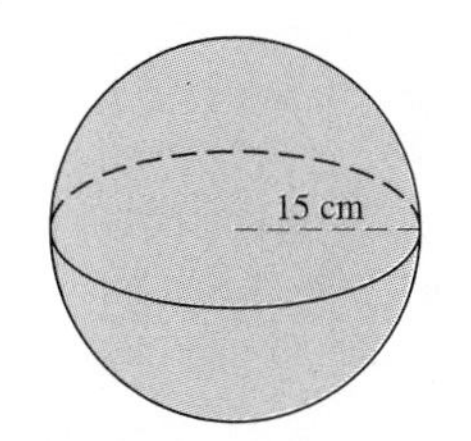

37.

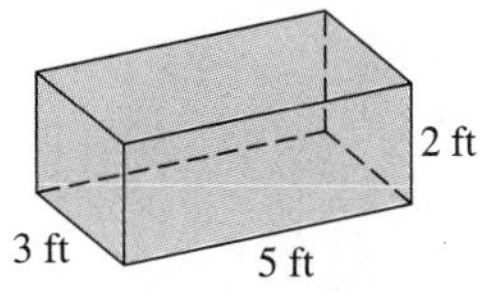

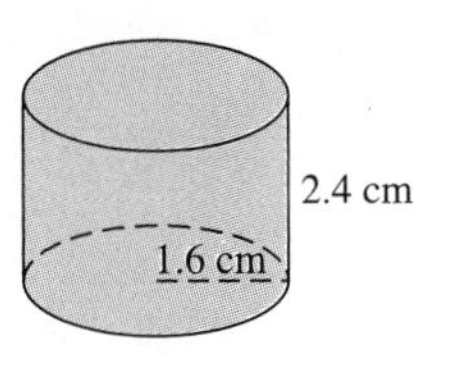

38.

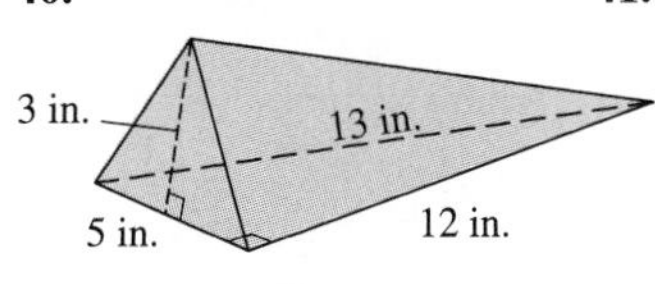

39.

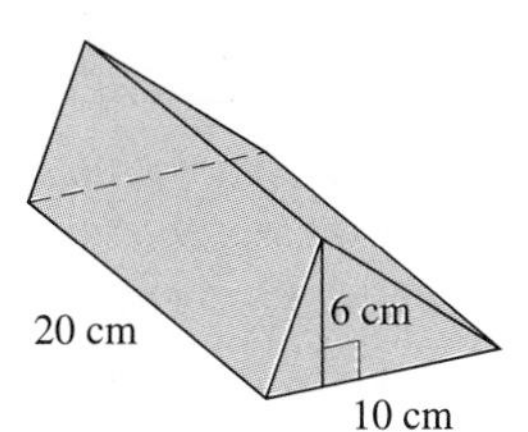

40.

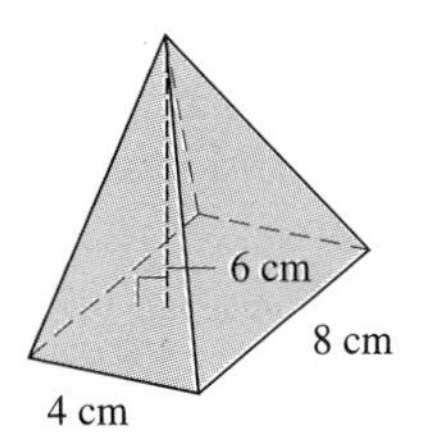

41.

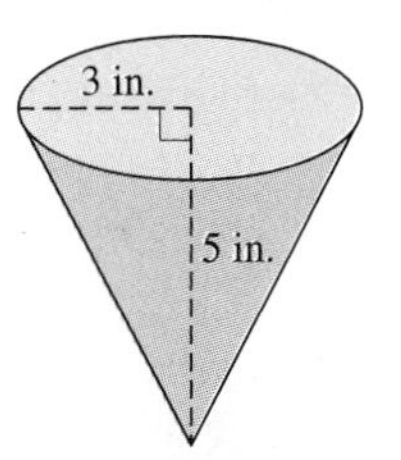

42.

43.

44.

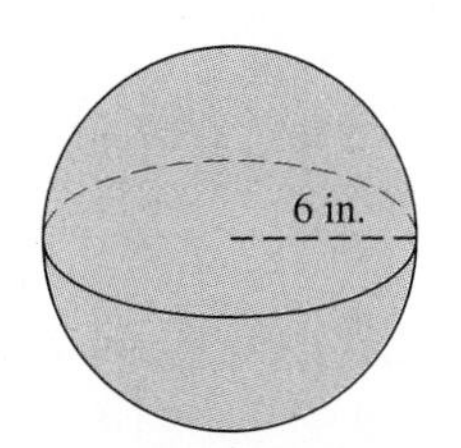

Level 3—Mind Bogglers

45. A polyhedron is a simple closed surface in space whose boundary is composed of polygonal regions (see Figure 6.24). A rather surprising relationship exists among the vertices, edges, and sides of polyhedra. See if you can discover it by looking for patterns in the figures and filling in the blanks.

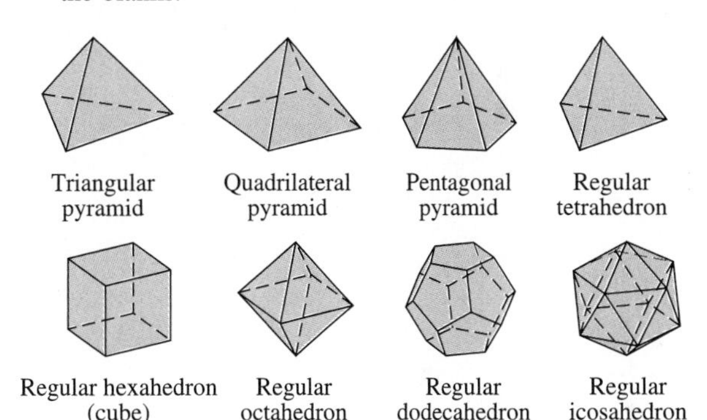

Figure 6.24 Some common polyhedra

FIGURE	*NUMBER OF* *SIDES*	*VERTICES*	*EDGES*
a. Triangular pyramid	4	4	6
b. Quadrilateral pyramid	5	____	8
c. Pentagonal pyramid	____	6	10
d. Regular tetrahedron	4	4	____
e. Cube	____	____	12
f. Regular octahedron	____	6	____
g. Regular dodecahedron	____	____	30
h. Regular icosahedron	____	____	30

*6.6 CONVERTING UNITS

Historical Perspective

Cleveland
94 MILES
151 KILOMETERS

The most difficult problem in changing from the customary (U.S.) system to the metric system in the United States is not mathematical, but psychological. Many people fear that changing to the metric system will require complex multiplying and dividing and the use of confusing decimal points. For example, in a recent popular article, James Collier states:

> *For instance, if someone tells me it's 250 miles up to Lake George, or 400 out to Cleveland, I can pretty well figure out how long it's going to take and plan accordingly. Translating all of this into kilometers is going to be an awful headache. A kilometer is about 0.62 miles, so to convert miles into kilometers you divide by six and multiply by ten, and even that isn't accurate. Who can do that kind of thing when somebody is asking me are we almost there, the dog is beginning to drool and somebody else is telling you you're driving too fast?*
>
> *Of course, that won't matter, because you won't know how fast you're going anyway. I remember once driving in a rented car on a superhighway in France, and every time I looked down at the speedometer we were going 120. That kind of thing can give you creeps. What's it going to be like when your wife keeps shouting, "Slow down, you're going almost 130"?*
>
> *But if you think kilometers will be hard to calculate . . .*

The author of this article has missed the whole point. Why are kilometers hard to calculate? How does he know that it's 400 miles to Cleveland? He knows because the odometer on his car or a road sign told him. Won't it be just as easy to read an odometer calibrated to kilometers or a metric road sign telling him how far it is to Cleveland?

When working with either the metric or the U.S. measurement system, you sometimes must change units of measurement within a particular system.

This section is divided into two parts so that you can work with either the U.S. system or the metric system, or with both if you prefer. The conversion between these systems is optional and is given as an appendix to this section.

METRIC MEASUREMENT CONVERSIONS

One of the advantages (if not the chief advantage) of the metric system is the ease with which you can remember and convert units of measurement. As you've seen, the three basic metric units are related as shown in Figure 6.25.

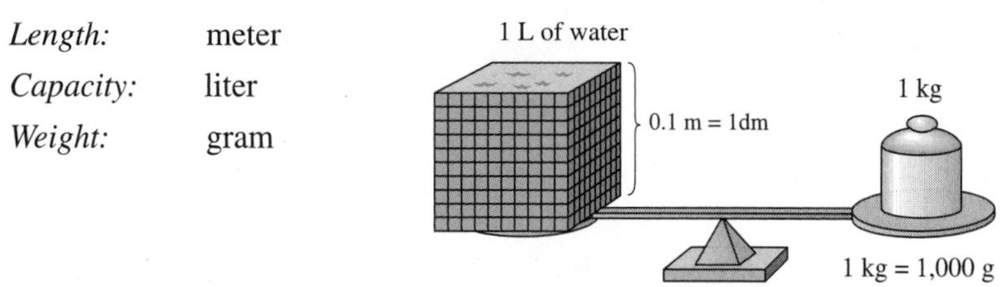

Figure 6.25 Relationship among meter, liter, and gram

*Optional section.

These units are combined with certain prefixes:

milli- means $\frac{1}{1,000}$ *centi-* means $\frac{1}{100}$ *kilo-* means 1,000

Other metric units are used less frequently, but they should be mentioned:

deci- means $\frac{1}{10}$ *deka-* means 10 *hecto-* means 100

A listing of metric units (in order of size) is given in Table 6.3.

Table 6.3 *Metric Measurements*

	Length	Capacity	Weight	Meaning	Memory aid
	kilometer (km)	**kilo**liter (kL)	**kilo**gram (kg)	1,000 units	**K**arl
	hectometer (hm)	**hecto**liter (hL)	**hecto**gram (hg)	100 units	**H**as
	dekameter (dkm)	**deka**liter (dkL)	**deka**gram (dkg)	10 units	**D**eveloped
Basic Unit:	**meter** (m)	**liter** (L)	**gram** (g)	1 unit	**M**y
	decimeter (dm)	**deci**liter (dL)	**deci**gram (dg)	0.1 unit	**D**ecimal
	centimeter (cm)	**centi**liter (cL)	**centi**gram (cg)	0.01 unit	**C**raving for
	millimeter (mm)	**milli**liter (mL)	**milli**gram (mg)	0.001 unit	**M**etrics

To convert from one metric unit to another, you **simply move the decimal point.**

EXAMPLE 1

Converting Metric Capacity

Write 5 L, using all the other prefixes.

Solution **Step 1** Place the decimal point on the given number if it is not shown: 5. L.

Step 2 Set up a chart as shown here and write the decimal point on the line corresponding to the given unit (the line just to the right of liter for this example):

Step 3 Write the number in the chart so that the decimal point is correctly aligned, putting one digit in each column:

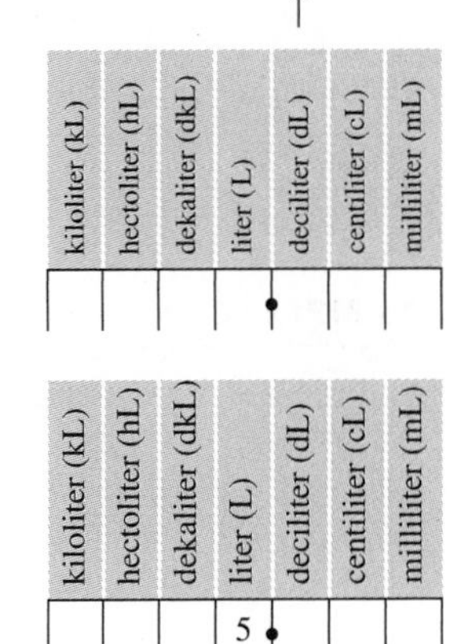

Step 4 Assume that there are zeros in all the other columns. To change to another unit, simply move the decimal point to the column corresponding to the unit to which you are changing.

To change to mL:

kiloliter (kL)	hectoliter (hL)	dekaliter (dkL)	liter (L)	deciliter (dL)	centiliter (cL)	milliliter (mL)
0	0	0	5 •	0	0	0
			Move point to mL column			
0	0	0	5	0	0	0 •

To change to kL:

kiloliter (kL)	hectoliter (hL)	dekaliter (dkL)	liter (L)	deciliter (dL)	centiliter (cL)	milliliter (mL)
0	0	0	5 •	0	0	0
	Move point to kL column					
0 •	0	0	5	0	0	0

These, as well as other possible conversions, are summarized in the following single chart:

kiloliter (kL)	hectoliter (hL)	dekaliter (dkL)	liter (L)	deciliter (dL)	centiliter (cL)	milliliter (mL)	
0	0	0	5 •	0	0	0	5 L This is given
0	0	0	5	0 •	0	0	50 dL
0	0	0	5	0	0 •	0	500 cL
0	0	0	5	0	0	0 •	5,000 mL
0	0	0 •	5	0	0	0	0.5 dkL
0	0 •	0	5	0	0	0	0.05 hL
0 •	0	0	5	0	0	0	0.005 kL

■

In practice, only the necessary zeros are filled in, as shown in Example 2.

EXAMPLE 2

Converting Metric Mass

Write 34.71 kg, using all the other prefixes.

Solution

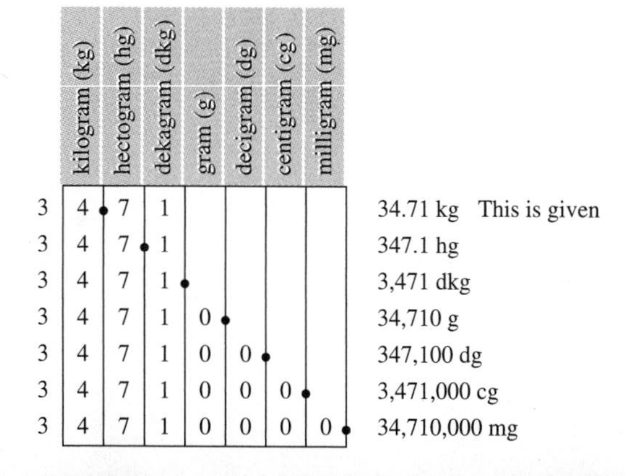

	kilogram (kg)	hectogram (hg)	dekagram (dkg)	gram (g)	decigram (dg)	centigram (cg)	milligram (mg)	
3	4 •	7	1					34.71 kg This is given
3	4	7 •	1					347.1 hg
3	4	7	1 •					3,471 dkg
3	4	7	1	0 •				34,710 g
3	4	7	1	0	0 •			347,100 dg
3	4	7	1	0	0	0 •		3,471,000 cg
3	4	7	1	0	0	0	0 •	34,710,000 mg

To remember this chart, try to memorize a short saying with the same first letters:

Karl	*Kilometer*
Has	*Hectometer*
Developed	*Dekameter*
My	*Meter*
Decimal	*Decimeter*
Craving for	*Centimeter*
Metrics	*Millimeter*

EXAMPLE 3 **Metric Conversions**

Make the indicated conversions.

a. 46.4 cm to m **b.** 6.32 km to m **c.** 503 mL to L
d. 0.031 kL to L **e.** 14 kg to dg

Solution **a.** 46.4 cm = __?__ m

Mentally picture a chart similar to the one shown in Example 1. Then count the number of decimal places involved in this conversion. Remember, "**K**arl **H**as **D**eveloped **M**y **D**ecimal **C**raving for **M**etrics."

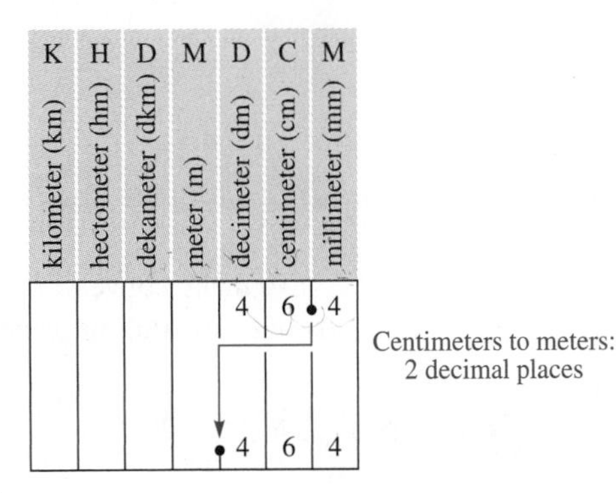

46.4 cm = **0.464 m**

b. 6.32 km = __?__ m

Three decimal places on chart:

K H D M D C M
6 . 3 2

6.32 km = **6,320 m**

c. 503 mL __?__ L

From memory aid change to L:

L
K H D M D C M
5 0 3 .

Three decimal places left (←) on chart: 503 mL = **0.503 L**

d. 0.031 kL = ___?___ L
Three decimal places right (→) on chart: 0.031 kL = **31 L**

e. 14 kg = ___?___ dg
Four decimal places right (→) on chart: 14 kg = **140,000 dg** ■

U.S. MEASUREMENT CONVERSION

Conversion within the U.S. measurement system is more difficult because it is not a decimal system and so involves more than moving the decimal point. First, you must remember the U.S. conversion factors, which are given in the following box.

U.S. Measurement Conversions

LENGTH	*CAPACITY*	*WEIGHT*
12 in. = 1 ft	3 tsp = 1 tbsp	16 oz = 1 lb
3 ft = 1 yd	1 oz = 2 tbsp	2,000 lb = 1 T
1,760 yd = 1 mi	16 tbsp = 1 c	
5,280 ft = 1 mi	16 oz = 1 pt	
	8 oz = 1 c	
	2 c = 1 pt	
	2 pt = 1 qt	
	4 qt = 1 gal	

To make conversions within the U.S. measurement system, you must use the following substitution principle.

Substitution Principle

If two quantities are equal, then one may be substituted (replaced) by an equal quantity without changing the value of the expression.

This means, for example, that

12 in. may be replaced by 1 ft;
1 ft may be replaced by 12 in.;
4 qt may be replaced by 1 gal; and so on.

It also means that if

4 qt = 1 gal then 1 qt = $\frac{1}{4}$ gal   Divide both sides by 4.

EXAMPLE 4

Converting U.S. Length Measurements

Make the following conversions.

a. 8 ft to inches **b.** 8 ft to yards **c.** 1,500 yd to feet
d. 86 in. to feet

Solution

a. $8 \text{ ft} = 8 \times 1 \text{ ft}$

$= 8 \times \mathbf{(12 \text{ in.})}$ Since 1 ft = 12 in., substitute 12 in. for 1 ft.
$= \mathbf{96 \text{ in.}}$

b. $8 \text{ ft} = 8 \times 1 \text{ ft}$

$= 8 \times \mathbf{(\frac{1}{3} \text{ yd})}$ Since 1 ft = $\frac{1}{3}$ yd
$= \frac{8}{3} \text{ yd}$
$= \mathbf{2\frac{2}{3} \text{ yd}}$

c. $1{,}500 \text{ yd} = 1{,}500 \times 1 \text{ yd}$

$= 1{,}500 \times \mathbf{(3 \text{ ft})}$ Since 1 yd = 3 ft
$= \mathbf{4{,}500 \text{ ft}}$

d. $86 \text{ in.} = 86 \times 1 \text{ in.}$

$= 86 \times \mathbf{(\frac{1}{12} \text{ ft})}$ Since 1 in. = $\frac{1}{12}$ ft
$= \frac{86}{12} \text{ ft}$
$= 7\frac{2}{12} \text{ ft}$
$= \mathbf{7\frac{1}{6} \text{ ft}}$

■

EXAMPLE 5

Converting U.S. Capacity Measurements

Make the following conversions.

a. 12 tsp to ounces **b.** 5 c to pints **c.** 2 qt 1 pt to ounces

Solution

a. $12 \text{ tsp} = 12 \times 1 \text{ tsp}$

$= 12 \times \mathbf{(\frac{1}{3} \text{ tbsp})}$ Since 1 tsp = $\frac{1}{3}$ tbsp
$= 12 \times \frac{1}{3} \times 1 \text{ tbsp}$

$= 12 \times \frac{1}{3} \times \mathbf{(\frac{1}{2} \text{ oz})}$ Since 1 tbsp = $\frac{1}{2}$ oz
$= \frac{12}{6} \text{ oz}$
$= \mathbf{2 \text{ oz}}$

b. $5 \text{ c} = 5 \times 1 \text{ c}$

$= 5 \times \mathbf{(\frac{1}{2} \text{ pt})}$ Since 1 c = $\frac{1}{2}$ pt
$= \mathbf{2\frac{1}{2} \text{ pt}}$

c. $2 \text{ qt } 1 \text{ pt} = 2 \times 1 \text{ qt} + 1 \text{ pt}$

$$\begin{aligned} &= 2 \times (\mathbf{2\ pt}) + 1 \text{ pt} && \text{Since 1 qt = 2 pt} \\ &= 4 \text{ pt} + 1 \text{ pt} \\ &= 5 \text{ pt} \\ &= 5 \times 1 \text{ pt} \\ &= 5 \times (\mathbf{16\ oz}) && \text{Since 1 pt = 16 oz} \\ &= \mathbf{80\ oz} \end{aligned}$$

■

EXAMPLE 6 Converting U.S. Weight Measurements

Make the following conversions.

a. 48 oz to pints **b.** 5 lb to ounces **c.** 48 oz to pounds
d. $2\frac{1}{4}$ lb to ounces **e.** 8.5 T to pounds

Solution

a. $48 \text{ oz} = 48 \times (1 \text{ oz})$

$$\begin{aligned} &= 48 \times (\tfrac{1}{16}\ \mathbf{pt}) && \text{Since 1 oz} = \tfrac{1}{16} \text{ pt} \\ &= \tfrac{48}{16} \text{ pt} \\ &= \mathbf{3\ pt} \end{aligned}$$

b. $5 \text{ lb} = 5 \times 1 \text{ lb}$

$$\begin{aligned} &= 5 \times (\mathbf{16\ oz}) && \text{Since 1 lb = 16 oz} \\ &= \mathbf{80\ oz} \end{aligned}$$

c. $48 \text{ oz} = 48 \times 1 \text{ oz}$

$$\begin{aligned} &= 48 \times (\tfrac{1}{16}\ \mathbf{lb}) && \text{Since 1 oz} = \tfrac{1}{16} \text{ lb} \\ &= \tfrac{48}{16} \text{ lb} \\ &= \mathbf{3\ lb} \end{aligned}$$

d. $2\frac{1}{4} \text{ lb} = 2\frac{1}{4} \times 1 \text{ lb}$

$$\begin{aligned} &= 2\tfrac{1}{4} \times (\mathbf{16\ oz}) && \text{Since 1 lb = 16 oz} \\ &= \frac{9}{4} \times \frac{16}{1} \text{ oz} \\ &= \mathbf{36\ oz} \end{aligned}$$

e. $8.5 \text{ T} = 8.5 \times 1 \text{ T}$

$$\begin{aligned} &= 8.5 \times (\mathbf{2{,}000\ lb}) && \text{Since 1 T = 2,000 lb} \\ &= \mathbf{17{,}000\ lb} \end{aligned}$$

PROBLEM SET 6.6

Odd-numbered problems *are U.S. measurement conversions, and* ***even-numbered problems*** *are metric conversions. Use reduced fractions for U.S. conversions, and use decimals for metric conversions.*

LEVEL 1—Drill

In Problems 1–14 make the indicated conversions.

1. a. 1 in. = _____ ft **b.** 1 in. = _____ yd
c. 1 in. = _____ mi

2. a. 1 cm = _____ mm **b.** 1 cm = _____ m
c. 1 cm = _____ km

3. a. 1 ft = _____ in. **b.** 1 ft = _____ yd
c. 1 ft = _____ mi

4. a. 1 mm = _____ cm **b.** 1 mm = _____ m
c. 1 mm = _____ km

5. a. 1 yd = _____ in. **b.** 1 yd = _____ ft
c. 1 yd = _____ mi

6. a. 1 m = _____ mm **b.** 1 m = _____ cm
c. 1 m = _____ km

7. a. 1 tsp = _____ tbsp **b.** 1 tsp = _____ oz
c. 1 tsp = _____ c

8. a. 1 mL = _____ dL **b.** 1 mL = _____ L
c. 1 mL = _____ kL

9. a. 1 oz = _____ c **b.** 1 oz = _____ pt
c. 1 c = _____ pt

10. a. 1 L = _____ mL **b.** 1 L = _____ dL
c. 1 L = _____ kL

11. a. 1 pt = _____ qt **b.** 1 pt = _____ gal
c. 1 qt = _____ gal

12. a. 1 kL = _____ mL **b.** 1 kL = _____ dL
c. 1 kL = _____ L

13. a. 1 oz = _____ lb **b.** 1 oz = _____ T
c. 1 lb = _____ T

14. a. 1 g = _____ mg **b.** 1 g = _____ cg
c. 1 g = _____ kg

LEVEL 2—Drill

In Problems 15–38, write each measurement in terms of all the other units listed in the problem. U.S. units will be changed to other U.S. units, and metric units will be changed to other metric units.

	mile (mi)	yard (yd)	foot (ft)	inch (in.)
15.		9		
17.				6
19.	4			
21.				150

	kilometer (km)	hectometer (hm)	dekameter (dkm)	meter (m)	decimeter (dm)	centimeter (cm)	millimeter (mm)
16.				9 •			
18.						6 •	
20.	4 •						
22.				1	5	0 •	

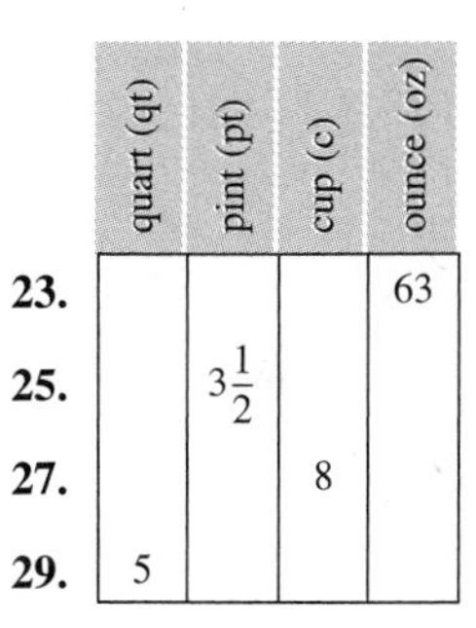

	quart (qt)	pint (pt)	cup (c)	ounce (oz)
23.				63
25.		$3\frac{1}{2}$		
27.			8	
29.	5			

	kiloliter (kL)	hectoliter (hL)	dekaliter (dkL)	liter (L)	deciliter (dL)	centiliter (cL)	milliliter (mL)
24.						6	3 •
26.				3 •	5		
28.			8 •				
30.		3 •	1				

	ton (T)	pound (lb)	ounce (oz)
31.		4	
33.			96
35.		$5\frac{1}{4}$	
37.	$4\frac{1}{2}$		

	kilogram (kg)	hectogram (hg)	dekagram (dkg)	gram (g)	decigram (dg)	centigram (cg)	milligram (mg)
32.	6 •	5					
34.			9	6 •			
36.	5 •	2	5				
38.	4 •	5					

Appendix U.S.–Metric Conversions

This section may be considered as an appendix for reference use. The emphasis in this chapter has been on everyday use of the metric system. However, certain specialized applications require more precise conversions than we've considered. On the other hand, you don't want to become bogged down with arithmetic to the point where you say "nuts to the metric system."

The most difficult obstacle in the change from the U.S. system to the metric system is not mathematical but psychological. If you have understood the chapter to this point, you realize that this should not be the case. In fact, as you saw in the last section, *working within the metric system is much easier than working within the U.S. system.*

With these ideas firmly in mind, and realizing that your everyday work with the metric system is discussed in the other sections of this chapter, we present an appendix of conversion factors between these measurement systems. Many calculators will make these conversions for you; check your owner's manual.

LENGTH CONVERSIONS

U.S. to Metric

When you know	Multiply by	To find
in.	2.54	cm
ft	30.48	cm
ft	0.3048	m
yd	0.9144	m
mi	1.60934	km

Metric to U.S.

When you know	Multiply by	To find
cm	0.39370	in.
m	39.37	in.
m	3.28084	ft
m	1.09361	yd
km	0.62137	mi

CAPACITY CONVERSIONS

U.S. to Metric

When you know	Multiply by	To find
tsp	4.9289	mL
tbsp	14.7868	mL
oz	29.5735	mL
c	236.5882	mL
pt	473.1765	mL
qt	946.353	mL
qt	0.9464	L
gal	3.7854	L

Metric to U.S.

When you know	Multiply by	To find
mL	0.20288	tsp
mL	0.06763	tbsp
mL	0.03381	oz
mL	0.00423	c
mL	0.00211	pt
mL	0.00106	qt
L	1.05672	qt
L	0.26418	gal

WEIGHT CONVERSIONS

U.S. to Metric

When you know	Multiply by	To find
oz	28.3495	g
lb	453.59237	g
lb	0.453592	kg
T	907.18474	kg

Metric to U.S.

When you know	Multiply by	To find
g	0.0352739	oz
g	0.0022046	lb
kg	2.2046226	lb

TEMPERATURE CONVERSIONS

U.S. to Metric

1. Subtract 32 from degrees Fahrenheit.
2. Multiply this result by $\frac{5}{9}$ to get Celsius.

Metric to U.S.

1. Multiply Celsius degrees by $\frac{9}{5}$.
2. Add 32 to this result to get Fahrenheit.

6.7 SUMMARY AND REVIEW

Important Terms

Accuracy [6.1]
Acre [6.3]
Area [6.3]
Capacity [6.4]
Celsius [6.5]
Centi- [6.1, 6.6]
Centimeter [6.1, 6.6]
Circle [6.2, 6.3]
Circular cone [6.5]
Circular cylinder [6.5]
Circumference [6.2]
Cone [6.5]
Cube [6.4]
Cubic centimeter [6.4]
Cubic inch [6.4]
Cup [6.4]
Cylinder [6.5]
Deci- [6.1, 6.6]
Deka- [6.1, 6.6]
Diameter [6.2]
Equilateral triangle [6.2]
Fahrenheit [6.5]
Foot [6.1]
Gallon [6.4]
Gram [6.5, 6.6]
Hecto- [6.1, 6.6]
Inch [6.1]
Kilo- [6.1, 6.6]
Kilogram [6.5]
Kiloliter [6.4]
Kilometer [6.1]
Length [6.1]
Liter [6.4, 6.6]
Mass [6.5]
Measure [6.1]
Meter [6.1, 6.6]
Metric system [6.1]
Mile [6.1]
Milli- [6.1, 6.6]
Milligram [6.5]
Milliliter [6.4]
Ounce [6.4, 6.5]
Parallelepiped [6.4]
Parallelogram [6.3]
Perimeter [6.2]
Pi (π) [6.2]
Pound [6.5]
Precision [6.1]
Prism [6.5]
Pyramid [6.5]
Quart [6.4]
Radius [6.2]
Rectangle [6.2]
Rectangular parallelepiped [6.4]
Right circular cone [6.5]
Right circular cylinder [6.5]
Right rectangular prism [6.5]
Semicircle [6.2]
SI system [6.1]
Sphere [6.5]
Square [6.2]
Square centimeter [6.3]
Square foot [6.3]
Square inch [6.3]
Square unit [6.3]
Temperature [6.5]
Ton [6.5]
Trapezoid [6.3]
United States system [6.1]
Volume [6.4]
Weight [6.5]
Yard [6.1]

Chapter Objectives

The material in this chapter is reviewed in the following list of objectives. A self-test (with answers and suggestions for additional study) is given. This self-test is constructed so that each problem number corresponds to a related objective. This self-test is followed by a practice test with the questions in mixed order.

[6.1]	*Objective 6.1*	Estimate the length of segments without using any measuring device.
[6.1]	*Objective 6.2*	Measure segments with an indicated precision.
[6.1]	*Objective 6.3*	Know your own size in U.S. and metric measurements.
[6.2]	*Objective 6.4*	Find the perimeter of a given figure.
[6.2]	*Objective 6.5*	Find the circumference or distance around a given figure.
[6.3]	*Objective 6.6*	Find the area of a given figure (rectangle, square, parallelogram, triangle, trapezoid, circle, or a combination of these).
[6.4]	*Objective 6.7*	Find the volume of a box.
[6.4]	*Objective 6.8*	Find the capacity of a given container.

[6.5]	*Objective 6.9*	Name a unit in the U.S. and metric measurement systems that you might use to measure everyday objects.
[6.5]	*Objective 6.10*	Find the volume of a solid (right prism, right circular cylinder, pyramid, right circular cone, or sphere).
[6.5]	*Objective 6.11*	State whether a given measurement measures length, capacity, mass, or temperature.
[6.1–6.5]	*Objective 6.12*	Estimate a given quantity in both the U.S. and metric measurement systems.
*[6.6]	*Objective 6.13*	Arrange the metric prefixes from largest to smallest. Know what each prefix means.
*[6.6]	*Objective 6.14*	Make conversions within the metric system.
*[6.6]	*Objective 6.15*	Make conversions within the U.S. system.
[6.2; 6.3; 6.4]	*Objective 6.16*	Solve applied problems involving perimeter, area, and volume.

Self-Test

Each question of this self-test is related to the corresponding objective listed above.

1. Without any measuring device, draw a segment 10 cm long.
2. Measure this segment to the nearest tenth cm. ————————
3. What is the width of your index finger in both the metric and U.S. systems?
4. Find the perimeter in Figure 6.26.

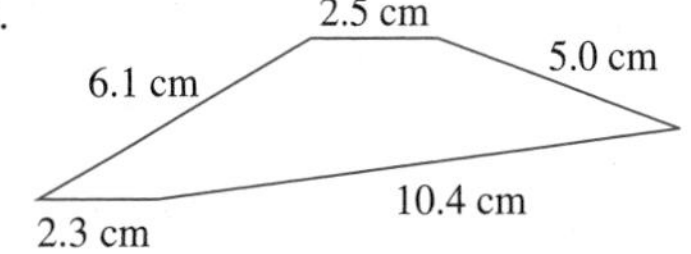

Figure 6.26

5. Find the distance around the region in Figure 6.27 (rounded to the nearest inch).

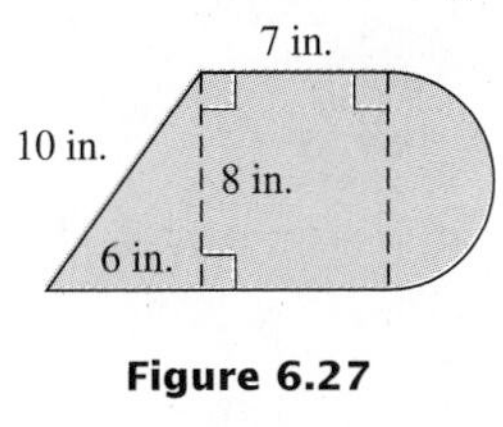

Figure 6.27

6. Find the area in Figure 6.28.

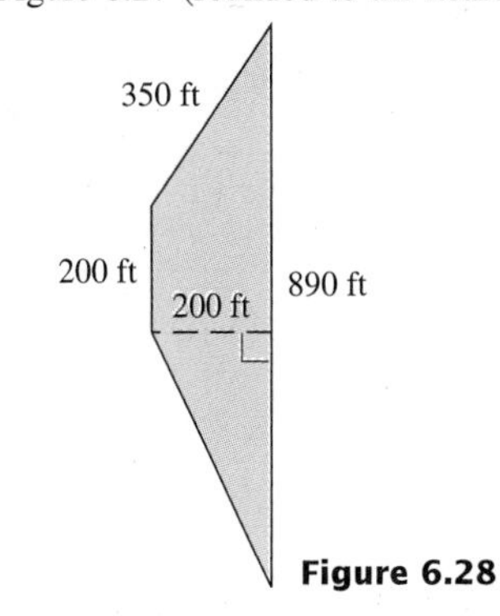

Figure 6.28

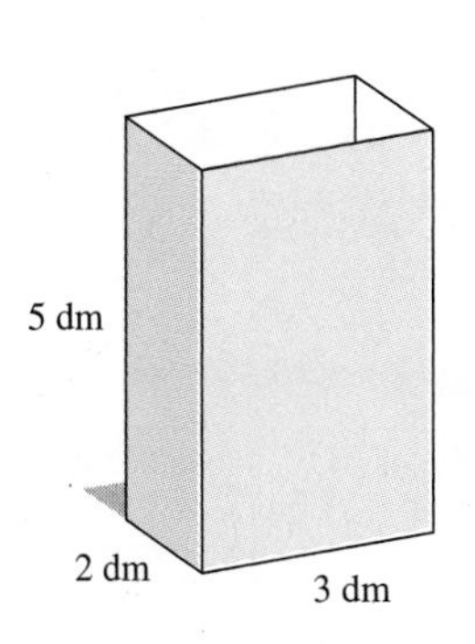

Figure 6.29

7. Find the volume of the box in Figure 6.29.
8. What is the capacity of the box in Figure 6.29?
9. What are the units of measurement in both the U.S. and metric measurement systems that you would use to measure the height of a building?
10. Find the volume (to the nearest tenth of a cubic foot) of a sphere with radius 1 ft.
11. What does the measurement 8.6 mL measure?

12. Estimate the temperature on a hot summer day in both the U.S. and metric measurement systems.

***13.** What does the prefix *milli-* mean?

***14.** How many centimeters are equivalent to 10 km?

***15.** How many tablespoons are equivalent to 3 oz?

16. How many square yards of carpet are necessary to carpet an 11-ft by 16-ft room? (Assume that you cannot purchase part of a square yard.)

STUDY HINTS *Compare your solutions and answers to the self-test. For each problem you missed, work some additional problems in the section listed in the margin. After you have worked these problems, you can test yourself with the practice test.*

Complete Solutions to the Self-Test

Additional Problems

[6.1] *Problems 18–25*

1. ____________________

[6.1] *Problems 26–36*

2. 1.9 cm

[6.1] *Problem 37*

3. Answers vary; $\frac{1}{2}$ in; 1.5 cm

[6.2] *Problems 20–28; 35–37; 47–52*

4. The perimeter is the distance around:

$$2.5 + 5.0 + 10.4 + 2.3 + 6.1 = 26.3$$

The perimeter is 26.3 cm.

[6.2] *Problems 29–34; 38–46*

5. First find the distance around the semicircle:

$$\tfrac{1}{2}\pi d = 4\pi \approx 12.56637061$$

The distance around is

$$10 + 7 + 4\pi + 7 + 6$$

To the nearest inch, the perimeter is 43 in.

[6.3] *Problems 23–52*

6. This figure is a trapezoid, so

$$\begin{aligned} A &= \tfrac{1}{2}h(b + B) \\ &= \tfrac{1}{2}(200)(200 + 890) \\ &= 109{,}000 \end{aligned}$$

The area is 109,000 ft^2. (This is about $2\frac{1}{2}$ acres.)

[6.4] *Problems 11–21*

7.
$$\begin{aligned} V &= \ell wh \\ &= 2(3)(5) \\ &= 30 \end{aligned}$$

The volume is 30 dm^3.

[6.4] *Problems 22–33; 47–56*

8. Since 1 L = 1 dm^3, we see from Problem 7 that the box holds 30 liters.

*Optional section.

[6.5] *Problems 24–35*	**9.** feet and meters
[6.5] *Problems 36–44*	**10.** $V = \frac{4}{3}\pi r^3 = \frac{4}{3}\pi(1)^3 = \frac{4}{3}\pi \approx 4.188790205$ The volume is 4.2 ft^3.
[6.5] *Problems 17–23*	**11.** capacity
[6.1] *Problems 3–15* [6.2] *Problems 4–13* [6.3] *Problems 5–16* [6.4] *Problems 34–43* [6.5] *Problems 4–12*	**12.** Answers vary; 40°C; 100°F
[6.1] *Problems 16–17* [6.4] *Problems 44–46* [6.5] *Problems 13–16* *[6.6] *Table 6.3*	**13.** one-thousandth
*[6.6] *Problems 2–38, even*	**14.** K H D M D C M 10.0 0 0 0 0 0 10 km = 1,000,000 cm
*[6.6] *Problems 1–37, odd*	**15.** 3 oz = 3 × 1 oz = 3 × (2 tbsp) = 6 tbsp
[6.2] *Problems 47–52* [6.3] *Problems 53–79* [6.4] *Problems 57–66*	**16.** $11 \text{ ft} \times 16 \text{ ft} = 3\frac{2}{3} \text{ yd} \times 5\frac{1}{3} \text{ yd} = \frac{11}{3} \times \frac{16}{3} \text{ yd}^2 = \frac{176}{9} \text{ yd}^2 \approx 19.6 \text{ yd}^2$ You must purchase 20 square yards.

Chapter 6 Practice Test

1. Without any measuring device, draw a segment with the indicated length.
   a. 2 in. **b.** 4 in. **c.** 3 cm **d.** 1 cm

2. Measure the segment to the specified precision. ————
   a. the nearest centimeter
   b. the nearest tenth centimeter
   c. the nearest inch
   d. the nearest eighth of an inch

3. Name a unit in the U.S. and metric measurement systems that you might use to measure the named objects.
   a. size of a notebook paper **b.** weight of a dime
   c. capacity of a can of Pepsi **d.** outside temperature

*Optional section.

4. State whether the measurements are measuring length, capacity, mass, or temperature.
a. 4.2 mL **b.** 24 km **c.** 9.3 kg **d.** 23°C

5. Give the requested information about yourself in the U.S. and metric measurement systems.
a. height **b.** weight **c.** normal body temperature

6. **a.** Estimate the distance between San Francisco and New York.
b. Estimate the capacity of a can of 7-Up.
c. Estimate the weight (mass) of a penny.
d. Estimate a comfortable room temperature.

7. Arrange from largest to smallest.
a. mg, cg, dkg **b.** cL, mL, kL **c.** c, tsp, qt **d.** mm, hm, m

8. The following two articles appeared in the *New York Times.*

> Indianapolis, Nov. 20 (AP) Burglars broke into an elementary school here and passed up computer equipment, going instead for 80,000 pennies that pupils had spent months collecting.
> A maintenance worker at Greenbriar Elementary School entered the cafeteria on Friday morning and found that a see-through glass well holding the $800 in pennies, weighing 450 pounds, had been smashed and emptied.

The *New York Times,* November 21, 1994

> Ray Amoroso, collection manager for the Steel Valley Bank in Dillonvale, Ohio, sifted through some of the eight million pennies turned in by a man, now 70, who had collected pennies since he was 5 years old. The pennies, weighing an estimated 48,000 pounds, are worth about $80,000.

The *New York Times,* December 22, 1994

a. What is the weight of a penny, according to the first article?
b. What is the weight of a penny, according to the second article?
c. According to which article does a penny weigh more?
d. Suppose that 330 students in the first article saved for 6 months. If each student contributed equally, how many pennies did each student bring each school day?

9. Find the perimeter and area of each figure.

a.

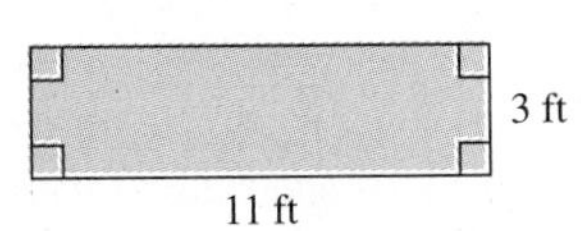

b.

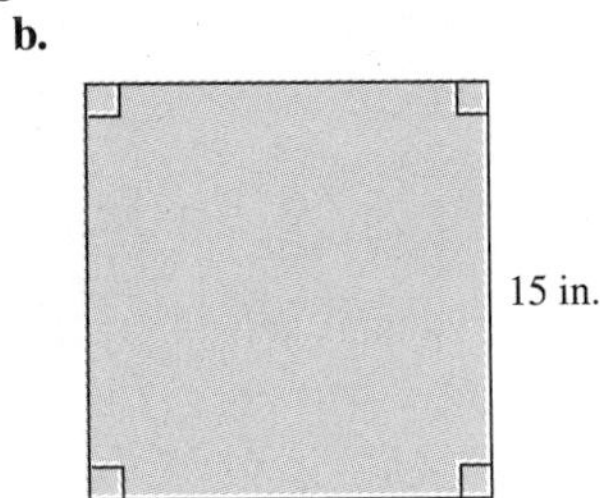

c.

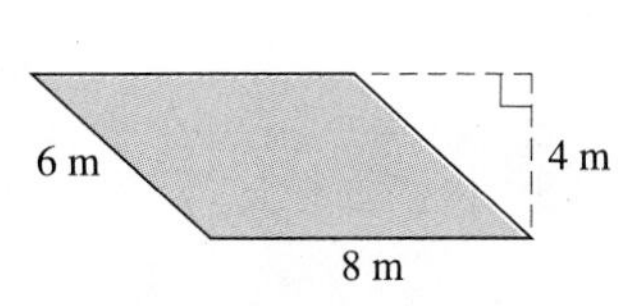

d.

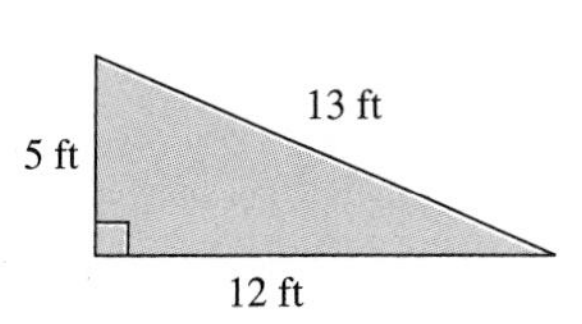

10. Find the area and circumference or the distance around the shaded portions (correct to two decimal places).

a.

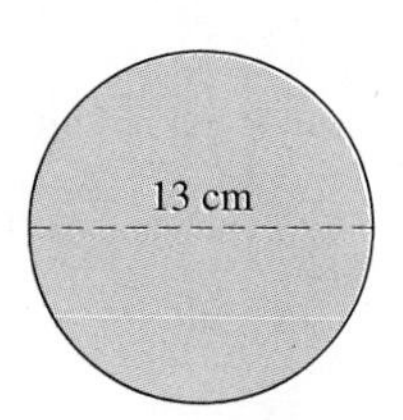

b.

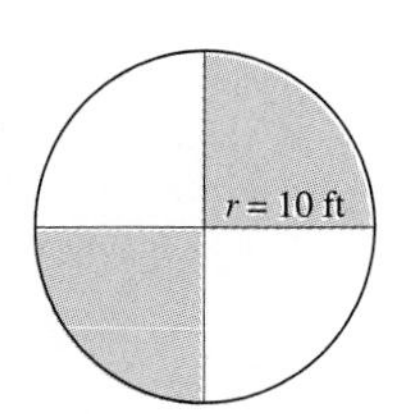

c.

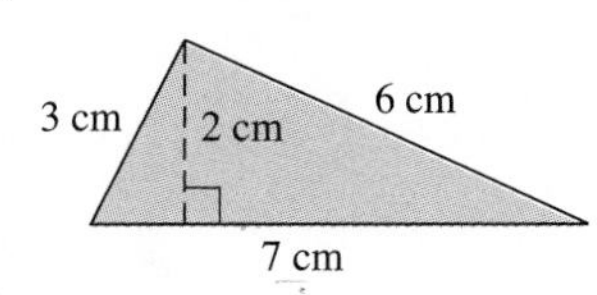

d.

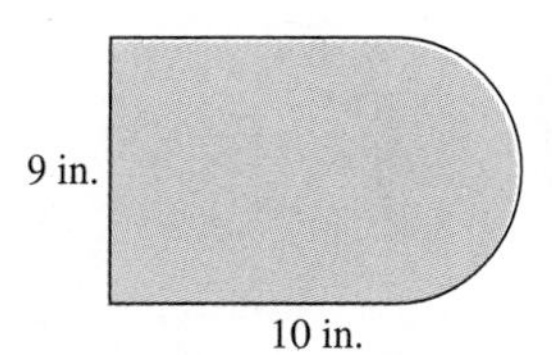

11. Find the volume of each solid as well as its capacity (to the nearest gallon or to the nearest liter).

a.

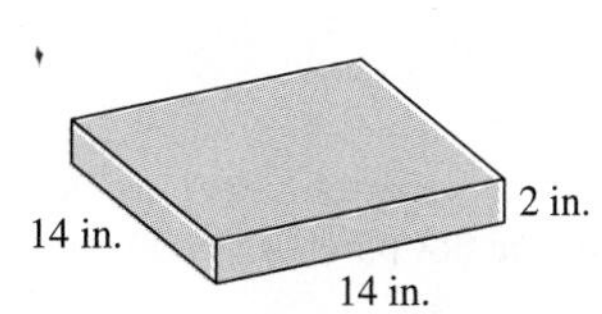

b.

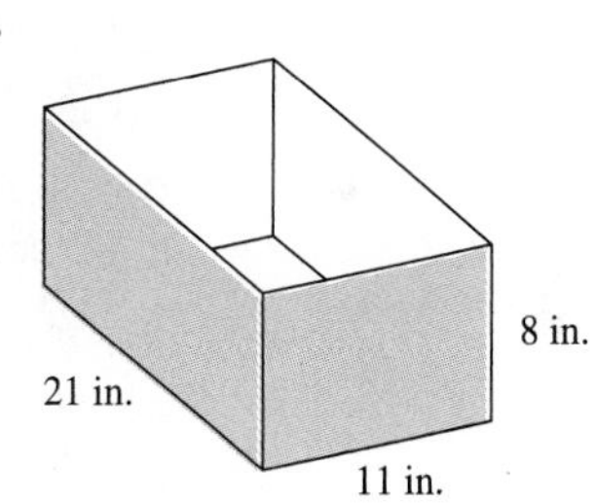

c.

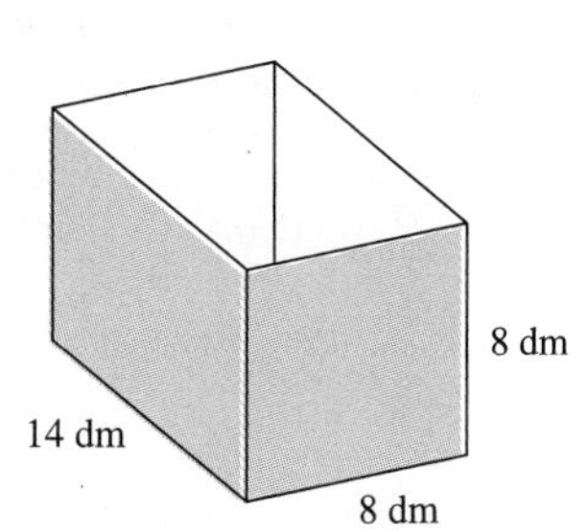

d.

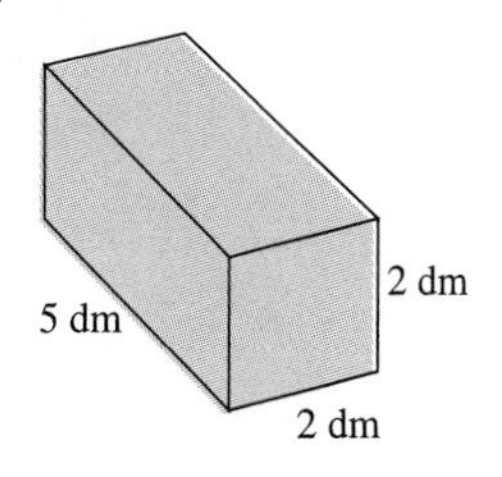

12. Find the exact volume.

a.

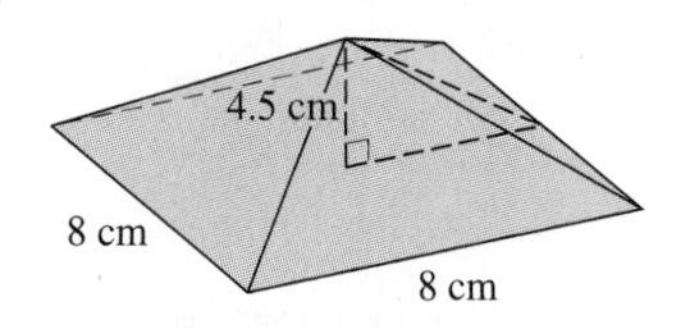

b.

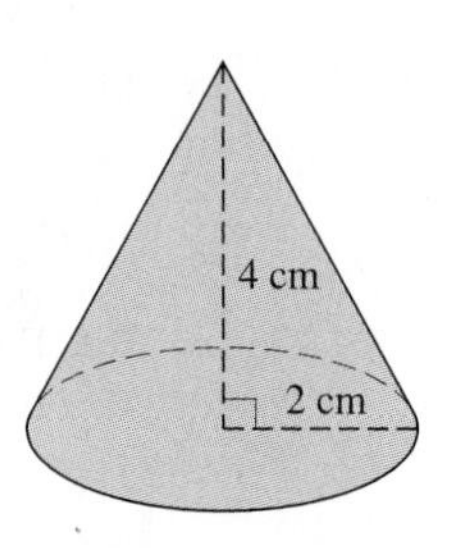

13. Find the volume to the nearest cubic unit.

a.

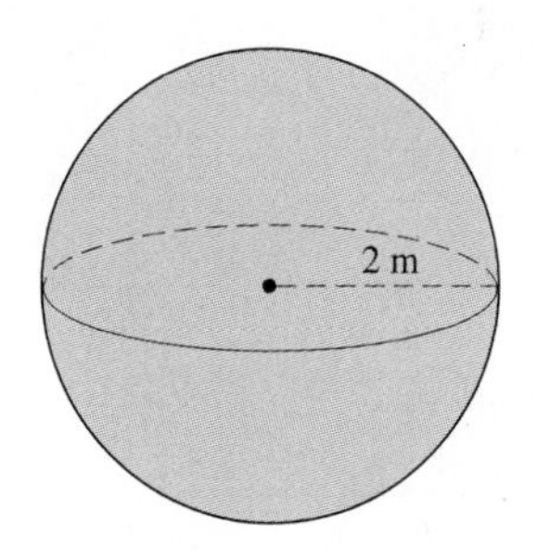

b.

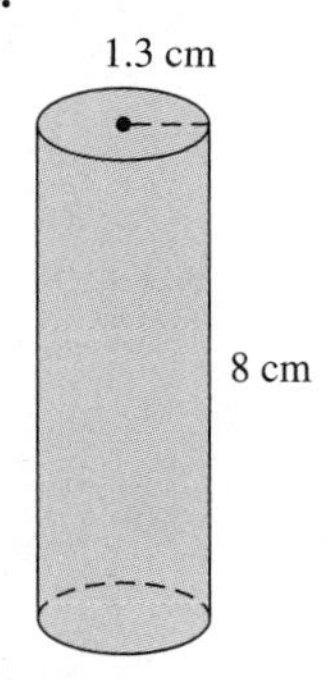

***14.** Fill in the blanks.

a. 4.8 km = _____ m

b. 450 mL = _____ L

c. 480 g = _____ kg

d. 5.2 dm = _____ cm

e. 20 ft = _____ yd

f. 4 in. = _____ yd

g. 2 c = _____ pt

h. $4\frac{1}{2}$ lb = _____ oz

15. How many square yards of carpet are necessary to carpet a 13-ft by 14-ft room? (Assume that you cannot purchase part of a square yard.)

16. If you are charged $100 per square foot for custom stained glass windows, how much would you pay for a window that is 2 ft by 2 ft? How much would you pay for a circular window with a 2-ft diameter?

17. How much water (to the nearest gallon) is necessary to fill to a depth of 2 ft a rectangular bathtub that measures 5 ft by 3 ft?

18. How many cubic yards of concrete are necessary for a rectangular driveway that is 20 ft by 25 ft by 4 in.? Assume that you can order in units of $\frac{1}{2}$ cubic yard.

*Optional section.

19. The Pythagorean theorem (Chapter 2) states that, for a right triangle with sides a and b and hypotenuse c, $a^2 + b^2 = c^2$. In this chapter, we saw that a^2 is the area of a square of length a. We can, therefore, illustrate the Pythagorean theorem for a triangle with sides 3, 4, and 5, as shown in Figure 6.30. Illustrate the Pythagorean theorem for a 5, 12, 13 right triangle.

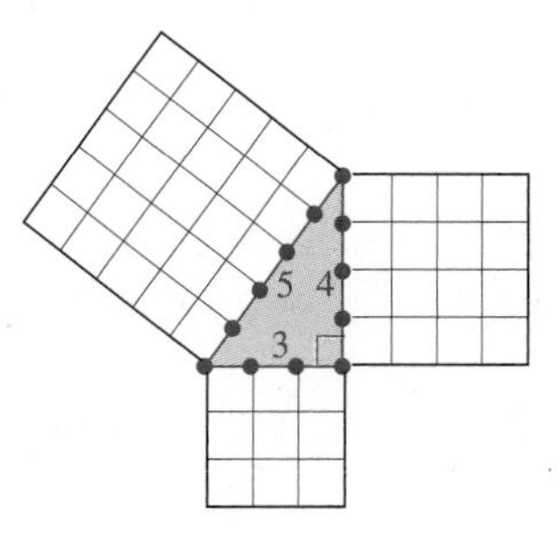

Figure 6.30 Geometric interpretation of the Pythagorean theorem

20. The distributive law (Chapter 1) states that

$$ab + ac = a(b + c)$$

In this chapter, we say ab and ac represent areas of two rectangles, one with sides a and b, and the other with sides a and c. Use the diagram in Figure 6.31 to give a geometric justification of the distributive law.

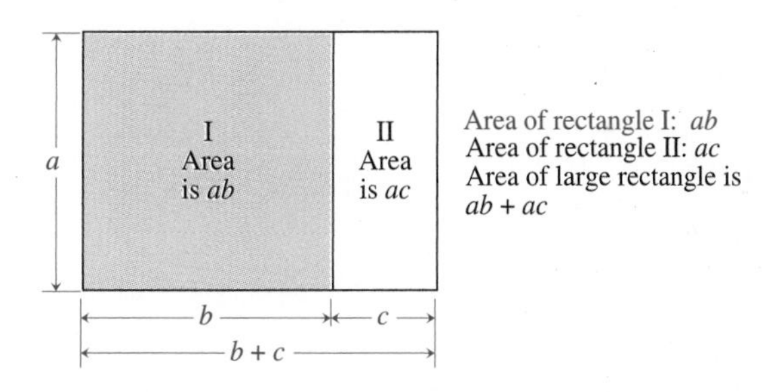

Figure 6.31 Area of large rectangle calculated two ways illustrates the distributive property: $a(b + c) = ab + ac$

PART TWO

APPLICATIONS: THE UTILITY OF MATHEMATICS

"Knowing" mathematics is "doing" mathematics. A person gathers, discovers, or creates knowledge in the course of some activity having a purpose. This active process is different from mastering concepts and procedures. We do not assert that informational knowledge has no value, only that its value lies in the extent to which it is useful in the course of some purposeful activity. It is clear that the fundamental concepts and procedures from some branches of mathematics should be known by all students; established concepts and procedures can be relied on as fixed variables in a setting in which other variables may be unknown. But instruction should persistently emphasize "doing" rather than "knowing that."

Curriculum and Evaluation Standards for School Mathematics, National Council of Teachers of Mathematics, 1989.

How are you doing so far?

Remember the things we talked about in the first section of this book? Keep them in mind as you are confronted with everyday situations that require mathematics. The first part of the book was concerned with problem solving and the mechanics of mathematics. This second part, The Utility of Mathematics, is quite different from the first part. Each section of this part begins with a real-life SITUATION and ends with a question related to that SITUATION. It is my hope that you will be able to put yourself into the setting described, and think about how you could change the facts of the SITUATION to fit yourself.

You will also find something in the problems of this part of the book that you may not be used to seeing in a mathematics book: essay questions. My goal is to get you thinking about how you can *use* mathematics in your own life. It is legitimate to ask, "Why should I be required to take mathematics?"

Mathematics, the science of the ideal, becomes the means of investigating, understanding, and making known the world of the real. The complex is expressed in terms of the simple. From one point of view mathematics may be defined as the science of successive substitutions of simpler concepts for more complex . . . .

Many students have a particular fear and avoidance of applied problems, but I hope you do not turn away from this material without giving it an honest try. I've tried to keep each section as practical as possible without using theory, but many ideas from Part I come into play. However, without being able to *use* the mathematics you have learned, you will be like a cyclist without a cycle or a musician without an instrument. I have made special efforts in this part of the book to allay your fears and give you some insight into how mathematics can be *used* in practical ways.

CHAPTER SEVEN

APPLICATIONS OF PERCENT

7.1 DISCOUNT, SALE PRICE, AND SALES TAX

Situation

Bud came home and told Sue that there was a big home improvement sale at Breuners. Perhaps now was the time they should buy that new living room set. Sue was not sure they had saved enough yet, so she wanted to know what the total cost would be. Bud showed Sue the newspaper ad: "ALL ITEMS REDUCED BY 40% (if you have the appropriate coupon)!!" The cost of the living room set was $2,550 the last time they looked, and Sue had told Bud that they had saved $1,600 for this purchase. "Great news," said Bud, "we can get it!"

In this section, we will see whether Bud and Sue really have enough money for this purchase by looking at discounts, sale prices, and the sales tax that is added after the sale.

DISCOUNT

There are some very common uses of percents for which we can build special models and focus on solving particular types of problems. As consumers, we are all familiar with sales. To encourage us to buy items at one time rather than at another, to introduce us to a particular store, to entice us into a store to buy nonsale items, to sell old or damaged merchandise, or to meet or beat the competition, a retailer will often sell merchandise ON SALE at a *reduced price.* The amount an item is reduced from the regular, or original, price is called the **discount.**

Discount

DISCOUNT = (ORIGINAL PRICE) × (PERCENT MARKDOWN)

The **percent markdown** is written as a decimal to calculate the discount.

EXAMPLE 1

Amount of Discount with a Percent Markdown Given

Find the discount, given the original price and the percent markdown.

a. Hardware, \$175; 20% off **b.** Shoes, \$95; 15% discount

Solution

a. DISCOUNT = (ORIGINAL PRICE) × (PERCENT MARKDOWN)
= 175 × 0.20
= 35

The discount is \$35.

b. DISCOUNT = 95 × 0.15
= 14.25

The discount is \$14.25. ■

Sometimes the markdown is given as a fraction.

EXAMPLE 2

Amount of Discount with a Fractional Markdown Given

Blouses with a regular price of \$59 are on sale marked $\frac{1}{3}$ OFF. What is the discount?

Solution

DISCOUNT $= 59 \times \frac{1}{3} \approx 19.67$ Round money answers to the nearest cent.

The discount is \$19.67. ■

SALE PRICE

As consumers, we are usually concerned with the sale price rather than the amount of discount. The **sale price** can be found by subtracting the discount from the original price. For example, the hardware in part **a** of Example 1 had an original price of \$175 and a discount of \$35, so the sale price is

\$175 − \$35 = \$140

However, there is an easier way to find the sale price, which involves something called the *complement.*

Complement

Two positive numbers less than 1 are called **complements** if their sum is 1.

EXAMPLE 3

Finding Complements

Find the complements.

a. 0.7 **b.** 0.15 **c.** 40% **d.** $\frac{1}{4}$

Solution To find the complement, subtract the given number from 1.

a. $1 - 0.7 = 0.3$; the complement of 0.7 is **0.3.**
b. $1 - 0.15 =$ **0.85**
c. Convert percents to decimals to carry out the arithmetic:

$$1 - 0.4 = 0.6$$

Alternatively, subtract the percent from 100%:

$$100\% - 40\% = 60\%$$

The complement is **60%.**

d. $1 - \frac{1}{4} = \mathbf{\frac{3}{4}}$ ■

The sale price can be found by multiplying the original price by the complement of the markdown.

Sale Price

SALE PRICE = (ORIGINAL PRICE) × (COMPLEMENT OF MARKDOWN)

EXAMPLE 4 **Finding the Sale Price by Using the Complement**

Find the sale price. (See Example 1.)

a. \$175 **b.** \$95

Solution **a.** Complement of a 20% discount is 80%:

SALE PRICE $= 175 \times 0.8 = 140$ The sale price is **\$140.**

b. Complement of 15% discount is 85%:

SALE PRICE $= 95 \times 0.85 = 80.75$ The sale price is **\$80.75.** ■

The same formula for finding the sale price of an item can be used for finding the original price or the percent markdown. Let

$p =$ ORIGINAL PRICE

$s =$ SALE PRICE

$c =$ COMPLEMENT OF THE PERCENT MARKDOWN

Then, the sale price formula is $s = pc$. Also,

$p = \dfrac{s}{c}$ Divide both sides of the formula $s = pc$ by c.

$c = \dfrac{s}{p}$ Divide both sides of the formula $s = pc$ by p.

Furthermore, if we let

$d =$ AMOUNT OF DISCOUNT and $m =$ PERCENT MARKDOWN

we can summarize these formulas.

Discount Formulas

DISCOUNT: $d = pm$

SALE PRICE: $s = p - d$ or $s = pc$

ORIGINAL PRICE: $p = \frac{s}{c}$

COMPLEMENT: $c = \frac{s}{p}$

When solving problems with discounts, markdowns, and sale prices, you can begin with the formula $s = pc$ and solve the equation for the unknown value, or you can pick the appropriate discount formula, as shown in the following examples.

EXAMPLE 5 **Finding the Original Price**

During a 20% OFF sale, an item is marked \$52. What is the regular price?

Solution

$s = pc$ — This is the basic discount formula.

$52 = p(0.80)$ — Given that $s = 52$ and the percent markdown is 20%, the complement of the percent markdown is 80%.

$\frac{52}{0.8} = p$ — Divide both sides by 0.8.

$65 = p$

The regular price is \$65. ■

EXAMPLE 6 **Finding the Percent Markdown**

If the sale price of an item is \$2,288 and the regular price is \$3,520, what is the percent markdown?

Solution

$s = pc$ — This is the basic discount formula.

$2{,}288 = 3{,}520c$ — Given that $s = 2{,}288$ and $p = 3{,}520$

$\frac{2{,}288}{3{,}520} = c$ — Divide both sides by 3,520.

$0.65 = c$

Since this is the complement, the percent markdown is $1 - 0.65 = 0.35$.
The percent markdown is 35%. ■

SALES TAX

Sales tax is levied by almost every state, as shown in Table 7.1.

Table 7.1 *State Sales Tax Rates**

State	Rate	State	Rate	State	Rate	State	Rate	State	Rate
Alabama	4%	Hawaii	4%	Massachusetts	5%	New Mexico	5%	South Dakota	4%
Alaska	0%	Idaho	5%	Michigan	6%	New York	4%	Tennessee	6%
Arizona	5%	Illinois	$6\frac{1}{4}$%	Minnesota	$6\frac{1}{2}$%	North Carolina	6%	Texas	6.25%
Arkansas	4.5%	Indiana	5%	Mississippi	7%	North Dakota	5%	Utah	$4\frac{7}{8}$%
California	$7\frac{1}{4}$%	Iowa	5%	Missouri	5.725%	Ohio	5%	Vermont	5%
Colorado	3%	Kansas	4.9%	Montana	0%	Oklahoma	4.5%	Virginia	4.5%
Connecticut	6%	Kentucky	6%	Nebraska	5%	Oregon	0%	Washington	$6\frac{1}{2}$%
Delaware	0%	Louisiana	4%	Nevada	6.5%	Pennsylvania	6%	West Virginia	6%
Florida	6%	Maine	6%	New Hampshire	0%	Rhode Island	7%	Wisconsin	5%
Georgia	4%	Maryland	5%	New Jersey	6%	South Carolina	5%	Wyoming	4%

*Does not include local sales taxes. *Source: 1996 Information Please Almanac.*

The procedure for finding the amount of sales tax is identical to the procedure for finding the discount.

Sales Tax Formula

SALES TAX = (ORIGINAL PRICE) × (TAX RATE)

EXAMPLE 7 **Finding the Sales Tax**

Find the Pennsylvania sales tax for building materials that cost \$84.65.

Solution

SALES TAX = (ORIGINAL PRICE) × (TAX RATE)
= 84.65(0.06)   Find the sales tax rate in Table 7.1.
= 5.079

The sales tax is \$5.08. ■

To find the total price, add the sales tax to the original price; this is the way it is done on most sales slips you receive at retail stores. For Example 7, this means that the price (including tax) is

$$\$84.65 + \$5.08 = \$89.73$$

However, if you want to find the total price, you can shorten the process by making the following observation:

TOTAL PRICE = (ORIGINAL PRICE) + (SALES TAX)
= (ORIGINAL PRICE) + (ORIGINAL PRICE) × (TAX RATE)
= (ORIGINAL PRICE)(1 + TAX RATE)   Distributive property

Price (Including Tax)

TOTAL AMOUNT = (ORIGINAL PRICE)(1 + TAX RATE)

EXAMPLE 8 **Finding the Price (Including Tax)**

In New Jersey, a truck tire at Sears costs $159.98. What is the total cost of the tire (including tax)?

Solution In Table 7.1, we find that the New Jersey sales tax rate is 6%. Write this as a decimal and add 1: $1 + 0.06 = 1.06$. Multiply this by the original price to find the total amount:

$$\begin{aligned}\text{TOTAL AMOUNT} &= (\text{ORIGINAL PRICE})(1 + \text{TAX RATE})\\ &= 159.98(1.06)\\ &= 169.5788\end{aligned}$$

Round money answers to the nearest cent. **The total amount (including tax) is $169.58.** ■

PROBLEM SET 7.1

ESSENTIAL IDEAS

1. **In Your Own Words** How do you find the discount for an item for sale?

2. **In Your Own Words** What is the complement of a number?

3. **In Your Own Words** How do you find a sale price if the original price and percent discount are known?

4. **In Your Own Words** How do you find a price (including tax) if you know the original price and the tax rate?

LEVEL 1—Drill

Find the discount of the items in Problems 5–14, given the original price and the percent markdown.

5. $450; 5%
6. $65; 20%
7. $55; 25%
8. $250; 35%
9. $25; 15%
10. $14.95; 30%
11. $12.49; 40%
12. $45.50; 50%
13. $9.95; 10%
14. $16.95; 15%

Find the complement of each of the numbers in Problems 15–20.

15. **a.** 0.1 **b.** 0.2 **c.** 0.3 **d.** 0.4 **e.** 0.5

16. **a.** 90% **b.** 80% **c.** 70% **d.** 60% **e.** 50%

17. **a.** $\frac{1}{3}$ **b.** $\frac{2}{3}$ **c.** $\frac{3}{4}$ **d.** $\frac{3}{5}$ **e.** $\frac{1}{6}$

18. **a.** $\frac{1}{5}$ **b.** $\frac{1}{10}$ **c.** $\frac{3}{20}$ **d.** $\frac{4}{15}$ **e.** $\frac{99}{100}$

19. **a.** 0.83 **b.** 17% **c.** $\frac{2}{5}$ **d.** 0.19 **e.** 27%

20. **a.** 16% **b.** $\frac{1}{8}$ **c.** 83% **d.** 39% **e.** 0.42

LEVEL 1

Use estimation to select the best response in Problems 21–26. Do not calculate.

21. The discount for an item marked 20% off with a price of $1,256.95 is about
A. $850 B. $25 C. $250 D. $85

22. An item costs $85.45 and the sales tax rate is 7%. The tax is about
A. $2 B. $4 C. $6 D. $8

23. An item costs $85.45 and the sales tax rate is 7%. The total amount (including tax) is about
A. $87 B. $90 C. $80 D. $104

24. The 5% sales tax on an automobile costing $15,850 is about
A. $800 B. $8,000 C. $80 D. $8

25. In a 15% OFF sale, an item with an original price of $179.95 has a sale price of about
A. $150 B. $300 C. $100 D. $30

26. The sale price of a skirt is $55, and that includes 30% OFF the regular price. The regular price is about
A. $15 B. $40 C. $180 D. $80

WHAT IS WRONG, *if anything, with each of the statements in Problems 27–32? Explain your reasoning.*

27. If an item with an original price of \$50 is on sale for \$40, then the percent markdown is \$10.

28. If an item with an original price of \$50 is on sale for 20% OFF, then the markdown is 20%.

29. In the formula $s = pc$, c represents the percent markdown.

30. To find the percent markdown, use the formula $s = pc$, and solve for c.

31. The complement of 30% is 0.30.

32. To find the price, including tax, add 1 to the tax rate and multiply this answer by the original price.

LEVEL 2—Drill

Find the sales tax for the items in Problems 33–38. Refer to Table 7.1.

33. \$230 golf clubs in Michigan

34. \$95 dress in Louisiana

35. \$85 coat in Washington state

36. \$35 book in Pennsylvania

37. \$18,350 automobile in Wisconsin

38. \$18,350 automobile in Texas

Find the sales tax for the items in Problems 39–44 using the rate in your own state. If your state has no sales tax, use the California rate.

39. Shoes, \$67.99

40. Set of pots and pans, \$109.99

41. Stereo, \$559.95

42. Electronic game, \$349.95

43. Purse, \$46.89

44. Automobile, \$21,955

Find the total amount, including tax, for the items whose price and tax rate are given in Problems 45–50.

45. \$12.50; 6%

46. \$1,925; 5%

47. \$4,312; $3\frac{3}{4}$%

48. \$91.60; 3%

49. \$25.36; $4\frac{1}{2}$%

50. \$365; $4\frac{1}{8}$%

LEVEL 2—Applications

Find the sale price of the items in Problems 51–60, given the original price and the percent markdown. These items were all found in one issue of a local newspaper.

51.

Golf clubs, \$230

52.

Dress, \$85

53.

Coat, \$95

54.

Stereo amplifier, \$805

55.

Jeans, \$45

56.

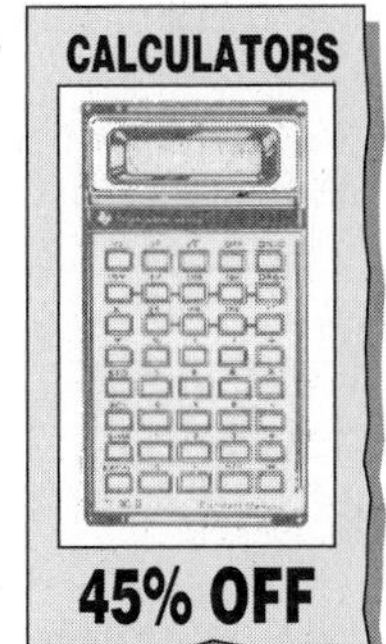

Calculator, \$25

57. \$350 skis; $\frac{1}{2}$ OFF

58. \$875 go-cart; $\frac{1}{2}$ OFF

59. \$32.95 sweater; $\frac{1}{4}$ OFF

60. \$9.50 belt; $\frac{1}{3}$ OFF

61. A dealer selling an automobile for $12,830 offers a $500 rebate. What is the percent markdown (to the nearest tenth of a percent)?

62. If a swimming pool is marked down to $695 from $1,150, what is the percent markdown (to the nearest tenth of a percent)?

63. If an item is marked $420 during a 20% OFF sale, what is the regular price?

64. If an item is marked $165 during a 1/3 OFF sale, what is the regular price?

65. The advertisement shown here claims to offer services at $\frac{1}{2}$ price. However, the sale price is $5.25 and the regular price is $10.00. What is the actual percent markdown (to the nearest tenth of a percent)?

66. What is the percent markdown (to the nearest tenth of a percent) for a cable reconnect as shown in the advertisement?

67. A radio advertisement for a grocery store claimed, "At other stores a bagger bags your groceries, but at *Food For Less* you bag the groceries and save a bundle."

a. Define what you think is meant by "save a bundle".

b. If a bagger earns minimum wage of $6.00/hr, and bags 120 bags in a hour, and you shop and take home 8 bags (the contents of each bag worth $85), how much is your savings in terms of bagger wages?

c. Write your savings from part **b** as a percent of the cost of your groceries.

68. You are employed part-time at a department store that offers an employee discount of 10% that may be applied to purchases. You purchase a 13-in. color television set for $300 that is on sale at 25% OFF. The sales tax is 5.5%. What is the total amount you must pay for this purchase?

69. You are employed at a department store that offers an employee discount of 10% that may be applied to purchases. You purchase some clothes originally marked at $155 that are on sale at 40% OFF. The sales tax is 7%. What is the total amount you must pay for this purchase?

70. Situation Bud came home and told Sue that there was a big home improvement sale at Breuners. Perhaps now was the time they should buy that new living room set. Sue was not sure they had saved enough yet, so she wanted to know what the total cost would be. Bud showed Sue the newspaper ad: "ALL ITEMS REDUCED BY 40% (if you have the appropriate coupon)!!" The cost of the living room set was $2,550 the last time they looked, and Sue had told Bud that they had saved $1,600 for this purchase. "Great news," said Bud, "we can get it!" Is Bud correct?

71. In Your Own Words Write a short paper comparing the SITUATION with your own situation. (That is, list similarities and/or differences.)

LEVEL 3—Team Project

72. Suppose you are given $1,000 to invest. With other members of your team, select an investment strategy, such as savings account, certificate of deposit, stocks, bonds (or even keep the money in a shoe box). Document your investment using your local newspaper. Track your investment for the next 60 days, and then present a report on the results.

7.2 SIMPLE INTEREST

Situation

Jerry has just received an inheritance of $25,000, and he would like to use it to help with his retirement. Since Jerry is 25 years old, he figures that the $25,000 can be invested for 40 years before he will need to use it for retirement. If he would like to have a monthly income of $5,000, how much money will he need to have in his retirement account to provide this income from interest only? He also wants to know what interest rate would be necessary for the $25,000 to grow to this amount.

Both of Jerry's questions are answered by working with the simple interest formula introduced in this section. Jerry will also see how to find the principal, the rate, and the time, given the other components of the interest formula.

AMOUNT OF SIMPLE INTEREST

Certain arithmetic skills enable us to make intelligent decisions about how we spend the money we earn. One of the most fundamental mathematical concepts that consumers, as well as business people, must understand is *interest.* Simply stated, **interest** is money paid for the use of money. We receive interest when we let others use our money (when we deposit money in a savings account, for example), and we pay interest when we use the money of others (for example, when we borrow from a bank).

The amount of the deposit or loan is called the **principal** or **present value,** and the interest is stated as a percent of the principal, called the **interest rate.** The **time** is the length of time for which the money is borrowed or lent. The interest rate is usually an *annual interest rate,* and the time is stated in years unless otherwise given.

Simple Interest Formula

This is a crucial formula to remember.

INTEREST = PRESENT VALUE × RATE × TIME

$I = Prt$

I = AMOUNT OF INTEREST

P = PRESENT VALUE (or PRINCIPAL)

r = ANNUAL INTEREST RATE

t = TIME (in years)

EXAMPLE 1 **Finding the Interest Paid on a Savings Deposit**

Suppose that you decide to save 20¢ per day, but only for a year. At the end of a year, you will have saved $73. If you then put your money into a savings account paying 8.5% interest, how much interest will the bank pay you after one year? After three years?

Solution The present value (P) is 73, the rate (r) is 8.5% = 0.085, and the time (t, in years) is 1. Therefore,

$$\begin{aligned} I &= Prt \\ &= 73(0.085)(1) \\ &= 6.205 \end{aligned}$$

Round money answers to the nearest cent: **After one year, the interest is $6.21.** For three years, $I = 73(0.085)(3) = 18.615$. **After three years, the interest is $18.62.** ■

Sometimes you know the amount of interest but need to find the principal, the rate, or the time. These can be easily found by solving the formula $I = Prt$ for the unknown.

These are all variations of the simple interest formula. Using algebra, you can derive them all by solving for the desired variable.

WHAT YOU WANT	*WHAT YOU KNOW*	*FORMULA*
I, amount of interest	P, r, and t	$I = Prt$
P, present value (principal)	I, r, and t	$P = \dfrac{I}{rt}$
r, rate	I, P, and t	$r = \dfrac{I}{Pt}$
t, time	I, P, and r	$t = \dfrac{I}{Pr}$

EXAMPLE 2 **Using the Simple Interest Formula**

Fill in the blanks.

	INTEREST (I)	*PRINCIPAL* (P)	*RATE* (R)	*TIME* (T)
a.	_____	$1,000	12%	2 yr
b.	$225	_____	9%	1 yr
c.	$112	$800	_____	2 yr
d.	$150	$500	6%	_____

Solution **a.**

$$\begin{aligned} I &= Prt \\ &= 1{,}000(0.12)(2) && \text{Given } P = 1{,}000,\ r = 0.12, \text{ and } t = 2. \\ &= 240 \end{aligned}$$

The amount of interest is $240.

b.

$$\begin{aligned} I &= Prt \\ 225 &= P(0.09)(1) && \text{Given } I = 225,\ r = 0.09, \text{ and } t = 1. \\ \frac{225}{0.09} &= P && \text{Divide both sides by 0.09.} \\ 2{,}500 &= P \end{aligned}$$

The principal is $2,500.

c.
$$I = Prt$$
$$112 = 800(r)(2) \quad \text{Given } I = 112,\ P = 800, \text{ and } t = 2.$$
$$112 = 1{,}600r$$
$$\frac{112}{1{,}600} = r$$
$$0.07 = r$$

The rate is 7%.

d.
$$I = Prt$$
$$150 = 500(0.06)t \quad \text{Given } I = 150,\ P = 500, \text{ and } r = 0.06.$$
$$150 = 30t$$
$$5 = t$$

The time is 5 years. ■

FUTURE VALUE

There is a difference between asking for the amount of interest, as illustrated in Example 2, and asking for the **future value.** The future amount is the amount you will have after the interest is added to the principal, or present value. Let A = FUTURE VALUE.

Once again, all these are algebraic variations of the same formula shown in boldface.

$$\mathbf{A = P + I} \quad \text{or} \quad P = A - I \quad \text{or} \quad I = A - P$$

EXAMPLE 3

Amount of Interest if the Monthly Payments Are Known

Suppose you see a car with a price of $12,436 that is advertised at $290 per month for 5 years. What is the amount of interest paid?

Solution The present value is $12,436. The future value is the total amount of all the payments:

$$\underbrace{\$290}_{\text{Monthly payment}} \times \underbrace{12}_{\text{Number of payments per year}} \times \underbrace{5}_{\text{Number of years}} = \$17{,}400 \quad \text{Use your calculator for this calculation.}$$

Therefore, the amount of interest is

$$I = A - P$$
$$= 17{,}400 - 12{,}436$$
$$= 4{,}964$$

The amount of interest is $4,964. ■

EXAMPLE 4

Problem Solving with the Simple Interest Formula

If a business borrows $18,000 and repays $26,100 in 3 years, what is the simple interest rate?

Solution The present value (P) is $18,000, the future value (A) is $26,100, and the time ($t$) is 3 years. We want to find r, so we use the formula

$I = Prt$ — $P = 18{,}000$; $t = 3$; we can find I by using $I = A - P = 26{,}100 - 18{,}000 = 8{,}100$.

$8{,}100 = 18{,}000(r)(3)$

$8{,}100 = 54{,}000r$ — Use a calculator: [26100] [−] [18000] [=] [÷] [18000] [÷] [3] [=]

$\frac{8{,}100}{54{,}000} = r$

$0.15 = r$

The rate is 15%. ■

EXAMPLE 5 Problem Solving with the Simple Interest Formula

Suppose you wish to save $3,720. If you have $3,000 and invest it at 8% simple interest, how long will it take you to obtain $3,720?

Solution

$I = Prt$ — $A = 3{,}720$; $P = 3{,}000$; we can find I by using $I = A - P = 3{,}720 - 3{,}000 = 720$.

$720 = 3{,}000(0.08)t$

$720 = 240t$

$3 = t$

It takes you 3 years. ■

INTEREST FOR PART OF A YEAR

The numbers in Example 5 were constructed to give a "nice" answer, but the length of time for an investment is not always a whole number of years. There are two ways to convert a number of days to a portion of a year.

Exact interest: 365 days per year

Ordinary interest: 360 days per year

Most applications and businesses use ordinary interest. So in this book, unless it is otherwise stated, assume ordinary interest; that is, use 360 for the number of days in a year.

Ordinary Interest

$$t = \frac{\text{ACTUAL NUMBER OF DAYS}}{360}$$

EXAMPLE 6

Problem Solving with the Simple Interest Formula and Part of a Year

Reconsider Example 5, but this time suppose that you want to save $3,650:

$$I = Prt$$

$$650 = 3{,}000(0.08)t \quad I = 3{,}650 - 3{,}000 = 650;\ P = 3{,}000;\ r = 0.08$$

$$650 = 240t$$

$$\frac{650}{240} = t$$

Calculator display: 2.7803333333

This is 2 years plus some part of a year. To convert the fractional part to days, do not clear your calculator, but subtract 2 (the number of years); then multiply the fractional part by 360:

[650] [÷] [240] [−] [2] [=] [×] [360] *Display:* 255

The time is 2 years 255 days. ■

EXAMPLE 7

Simple Interest Formula for Part of a Year

Suppose that you borrow $1,200 on March 25 at 21% simple interest. How much interest accrues to September 15 (174 days later)? What is the total amount that must be repaid?

Solution We are given $P = 1{,}200$, $r = 0.21$, and $t = \frac{174}{360}$. ← Actual number of days ← Assume ordinary interest.

$$I = Prt$$
$$= 1{,}200(0.21)\left(\tfrac{174}{360}\right)$$
$$= 121.8$$

A calculator is almost essential when working with part of a year:

[1200] [×] [.21] [×] [174] [÷] [360] [=] *Display:* 121.8

The amount of interest is $121.80. To find the amount that must be repaid, find the future value:

$$A = P + I = 1{,}200 + 121.80 = 1{,}321.80$$

The amount that must be repaid is $1,321.80. ■

It is worthwhile to derive a formula for future value because sometimes we will not calculate the interest separately as we did in Example 7.

This is a formula which will be used frequently when working with finances.

FUTURE VALUE = PRESENT VALUE + INTEREST

$$A = P + I$$
$$= P + Prt \quad \text{Substitute } I = Prt.$$
$$= P(1 + rt) \quad \text{Distributive property}$$

You might notice that this is the same procedure we used to find the sales tax and total price in the previous section.

Future Value Formula
(Simple Interest)

$$A = P(1 + rt)$$

EXAMPLE 8 **Finding the Future Value of a Savings Deposit**

If \$10,000 is deposited in an account earning $5\frac{3}{4}\%$ simple interest, what is the future value in 5 years?

Solution We identify $P = 10{,}000$, $r = 0.0575$, and $t = 5$.

$$\begin{aligned} A &= P(1 + rt) \\ &= 10{,}000(1 + 0.0575 \times 5) \\ &= 10{,}000(1 + 0.2875) \\ &= 10{,}000(1.2875) \\ &= 12{,}875 \end{aligned}$$

Don't forget order of operations: multiplication first.

The future value in 5 years is \$12,875. ■

EXAMPLE 9 **Determining the Sum Necessary for Retirement**

Suppose that you have decided that you will need \$4,000 per month on which to live in retirement. If the rate of interest is 8%, how much must you have in the bank when you retire so that you can live on interest only?

Solution We are given $I = 4{,}000$, $r = 0.08$, and $t = \frac{1}{12}$ (one month $= \frac{1}{12}$ year):

$$\begin{aligned} I &= Prt \\ 4{,}000 &= P(0.08)\left(\tfrac{1}{12}\right) && \text{Substitute.} \\ 48{,}000 &= (0.08)P && \text{Multiply both sides by 12.} \\ 600{,}000 &= P && \text{Divide both sides by 0.08.} \end{aligned}$$

You must have \$600,000 on deposit to earn \$4,000 per month at 8%. ■

PROBLEM SET 7.2

ESSENTIAL IDEAS

1. What is interest?
2. What is the simple interest formula? Tell what each variable represents.
3. How do you change a given number of days to years?
4. What is the future value formula for simple interest?

LEVEL 1

Use estimation to select the best response in Problems 5–14. Do not calculate.

5. If you deposit \$100 in a bank account for a year, the amount of interest is likely to be
   A. \$1 B. \$5 C. \$105
   D. impossible to estimate

6. If you deposit \$100 in a bank account for a year, then the future value is likely to be
   A. \$1 B. \$5 C. \$105
   D. impossible to estimate

7. If you purchase a new automobile and finance it for 4 years, the amount of interest you might pay is
   A. \$400 B. \$100 C. \$4,000
   D. impossible to estimate

8. What is a reasonable monthly income when you retire?
   A. \$300 B. \$10,000 C. \$500,000
   D. impossible to estimate

9. To be able to retire and live on the interest only, what is a reasonable amount to have in the bank?
   A. \$300 B. \$10,000 C. \$500,000
   D. impossible to estimate

10. If $I = Prt$ and $P = \$49{,}236.45$, $r = 10.5\%$, and $t = 2$ years, estimate I.
    A. \$10,000 B. \$600 C. \$50,000 D. \$120,000

11. If $I = Prt$ and $I = \$398.90$, $r = 9.85\%$, and $t = 1$ year, estimate P.
    A. \$400 B. \$40 C. \$40,000 D. \$4,000

12. If $t = 3.52895$, then the time is about 3 years and how many days?
    A. 30 B. 300 C. 52 D. 152

13. If a loan is held for 450 days, then t is about
    A. 450 B. 3 C. $1\frac{1}{4}$ D. 5

14. If a loan is held for 180 days, then t is about
    A. 180 B. $\frac{1}{2}$ C. $\frac{1}{4}$ D. 3

LEVEL 1—Drill

Fill in the blanks in Problems 15–44.

	Interest	Principal	Rate	Time	Future Value
15.	____	\$1,000	8%	1 yr	____
16.	____	\$5,000	9%	2 yr	____
17.	____	\$800	12%	3 yr	____
18.	____	\$10,000	15%	6 yr	____
19.	____	\$500	$8\frac{1}{2}\%$	1 yr	____
20.	____	\$300	$12\frac{1}{2}\%$	3 yr	____
21.	____	\$400	$12\frac{1}{2}\%$	4 yr	____
22.	____	\$1,000	$5\frac{3}{4}\%$	5 yr	____
23.	____	\$1,200	$8\frac{1}{4}\%$	10 yr	____
24.	____	\$50,000	9%	40 yr	____
25.	____	\$100,000	10%	40 yr	____
26.	\$360	____	12%	1 yr	____
27.	\$600	\$4,000	____	1 yr	____
28.	\$350	\$2,500	7%	____	____
29.	\$66	____	11%	6 yr	____
30.	\$588	\$700	____	7 yr	____
31.	\$432	\$600	____	8 yr	____
32.	\$180	____	15%	6 yr	____
33.	____	\$3,000	14%	3 yr	____
34.	\$960	\$1,500	16%	____	____
35.	\$432	____	6%	8 yr	____
36.	\$270	\$300	____	9 yr	____
37.	\$624	\$1,200	____	4 yr	____
38.	\$1,800	____	9%	5 yr	____
39.	\$5,850	\$6,500	____	5 yr	____
40.	\$3,240	____	12%	3 yr	____
41.	\$320	\$400	8%	____	____
42.	____	\$512.50	10%	180 days	____
43.	____	\$236.50	8%	255 days	____
44.	____	\$814.90	9%	213 days	____

LEVEL 2

WHAT IS WRONG, *if anything, with each of the statements in Problems 45–50? Explain your reasoning.*

45. Since $I = Prt$, we know that $P = Irt$.

46. To find the rate, use the formula $r = Prt$.

47. If the time is 315 days, then in this book use $t = \frac{315}{365}$.

48. If the time is 5 months, then in this book, use $t = \frac{5}{360}$.

49. If a desk is advertised at a monthly payment of \$45.50 for 3 years, then the future value is \$45.50 $\times$ 3 = \$136.50.

50. The future value formula for simple interest is $A = P(1 + rt)$.

LEVEL 2—Applications

51. If a business borrows \$12,500 and repays \$23,125 in 5 years, what is the simple interest rate?

52. If a business borrows \$18,000 and repays \$26,100 in 4 years, what is the simple interest rate?

53. Sharon wants to save \$3,900. If she has \$3,000 and invests it at 5% simple interest, how long will it take her to obtain \$3,900?

54. Jerry wants to save \$1,000. If he has \$750 and invests it at 12% simple interest, how long (to the nearest day) will it take him to obtain \$1,000?

55. If a friend tells you she earned \$5,075 interest for the year on a 5-year certificate of deposit paying 5% simple interest, what is the amount of the deposit?

56. If Rita receives \$45.33 interest for a deposit earning 3% simple interest for 240 days, what is the amount of her deposit?

57. If John wants to retire with \$10,000 per month,* how much principal is necessary to generate this amount of monthly interest income if the interest rate is 15%?

58. If Melissa wants to retire with \$50,000 per month,* how much principal is necessary to generate this amount of monthly income if the interest rate is 12%?

59. If Jack wants to retire with \$1,000 per month, how much principal is necessary to generate this amount of monthly interest income if the interest rate is 6%?

60. If John from Problem 57 has 30 years before retirement and has \$100,000 to invest today, what simple interest rate does he need between now and then to achieve the amount of principal needed in Problem 57?

61. If Melissa from Problem 58 has 40 years before retirement and has \$100,000 to invest today, what simple interest rate does she need to achieve the amount of principal needed in Problem 58?

62. If Jack from Problem 59 has 5 years before retirement and has \$25,000 to invest today, what simple interest rate does he need to achieve the amount of principal needed in Problem 59?

63. Situation Jerry has just received an inheritance of \$25,000, and he would like to use it to help him with his retirement. Since Jerry is 25 years old, he figures that the \$25,000 can be invested for 40 years before he will need to use it for retirement. If he would like to have a monthly income of \$5,000, how much money will he need to have in his retirement account to provide this income from interest only? Assume that Jerry is able to invest in real estate and can earn a 15% return on his investment.

64. Situation Jerry wants to know what interest rate would be necessary for the \$25,000 to grow to provide an amount so that he can have a monthly income of \$5,000 earned from the interest only. Use the information in Problem 63.

65. In Your On Words Write a short paper comparing the SITUATION with your own situation. (That is, list similarities and/or differences.)

*You might say these are exorbitant monthly incomes, but if you assume 10% average inflation for 40 years, a monthly income of \$220 today will be equivalent to about \$10,000 per month in 40 years. If we assume 2% inflation, that amount is about \$4500. We will discuss inflation in Section 7.5.

7.3 BUYING ON CREDIT

Situation

Karen and Wayne must buy a refrigerator because theirs just broke. Unfortunately, their savings account is depleted, and they will need to borrow money to buy a new one. The bank offers them a personal loan at 21% (APR), and Sears offers them an installment loan at 15% (add-on rate). Karen says that she has heard something about APR rates but doesn't really know what the term means. Wayne says he thinks it has something to do with the prime rate, but he isn't sure.

In this section, Karen and Wayne will learn about installment loans and, in particular, about add-on interest and APR.

CONSUMER LOANS

Two types of consumer credit allow you to make installment purchases. The first, called **closed-end,** is the traditional installment loan. An **installment loan** is an agreement to pay off a loan or a purchase by making equal payments at regular intervals for some specific period of time. In this book, it is assumed that all installment payments are made monthly. The loan is said to be **amortized** if it is completely paid off by these payments; and the payments are called **installments.** If the loan is not amortized, there is a larger final payment, called a **balloon payment.** With an **interest-only loan,** there is a monthly payment equal to the interest, with a final payment equal to the amount received when the loan was obtained.

The second type of consumer credit is called **open-end, revolving credit,** or more commonly, a **credit card** loan. MasterCard, VISA, and Discover cards, as well as those from department stores and oil companies, are examples of open-ended loans. This type of loan allows for purchases or cash advances up to a specified maximum **line of credit** and has a flexible repayment schedule. We will discuss credit cards in Section 7.4.

ADD-ON INTEREST

The most common method for calculating interest on installment loans is by a method known as **add-on interest.** It is nothing more than an application of the simple interest formula. It is called *add-on interest* because the interest is *added* to the amount borrowed so that both interest and the amount borrowed are paid for over the length of the loan. You should be familiar with the following variables:

Spend some time reviewing these variables and the installment loan formulas.

P = AMOUNT TO BE FINANCED (present value)
r = ADD-ON INTEREST RATE
t = TIME (in years) TO REPAY THE LOAN
I = AMOUNT OF INTEREST
A = AMOUNT TO BE REPAID (future value)
m = AMOUNT OF THE MONTHLY PAYMENT
N = NUMBER OF PAYMENTS

Installment Loan Formulas

AMOUNT OF INTEREST: $I = Prt$
AMOUNT TO BE REPAID: $A = P + I$ or $A = P(1 + rt)$
NUMBER OF PAYMENTS: $N = 12t$
AMOUNT OF EACH PAYMENT: $m = \frac{A}{N}$

Notice this agreement about rounding problems dealing with money.

When figuring monthly payments in everyday life, most businesses and banks round up for any fraction of a cent. However, for consistency in this book, we will continue to use the rounding procedures developed in Chapter 1. This means we will round money answers to the nearest cent.

EXAMPLE 1 **Monthly Payments for a Closed-End Loan**

You want to purchase a computer that has a price of \$1,399, and you decide to pay for it with installments over 3 years. The store tells you that the interest rate is 15%. What is the amount of each monthly payment?

Solution You ask the clerk how the interest is calculated, and you are told that the store uses add-on interest. Thus,

$$P = 1{,}399 \quad r = 0.15 \quad t = 3 \quad N = 36$$

Also,

$$I = Prt = 1{,}399(0.15)(3) = 629.55$$
$$A = P + I = 1{,}399 + 629.55 = 2{,}028.55$$

or in one step:

$$A = P(1 + rt) = 1{,}399[1 + (0.15)3] = 2{,}028.55$$

For the monthly payment, divide A by the number of payments:

$$m = \frac{2{,}028.55}{36} \approx 56.35$$

These calculations are best done with a calculator. You can carry out the entire calculation in one step:

[1399] [×] [(] [1] [+] [.15] [×] [3] [)] [÷] [36] [=]   *Display:* 56.3486111

The amount of each monthly payment is $56.35. ■

Among the most common applications of installment loans are the purchase of a car and the purchase of a home. Interest for purchasing a car is add-on interest, but interest for purchasing a home is not. We will, therefore, delay our discussion of home loans until after we have discussed compound interest. The next example shows a calculation for a car loan.

EXAMPLE 2

Problem Solving when Purchasing a Car

Suppose that you have decided to purchase a Honda Civic and want to determine the monthly payment if you pay for the car in 4 years.

Solution Not enough information is given, so you need to ask some questions of the car dealer:

Sticker price of the car (as posed on the window): $7,695.90

Dealer's preparation charges (as posed on the window): $850.00

Tax rate (determined by the state; see Table 7.1): 7%

Add-on interest rate: 8%

You must make an offer. If you are serious about getting the best price, find out the **dealer's cost**—the price the dealer paid for the car you want to buy. In this book, we will tell you the dealer's cost, but in the real world you will need to do some homework to find it (consult a reporting service, an automobile association, a credit union, or the April issue of *Consumer Reports*). Assume that the dealer's cost for this car is $6,925.50. You decide to offer the dealer 5% *over* this cost. We will call this a **5% offer:**

$$\$6{,}925.50(1 + 0.05) = \$7{,}271.775$$

You will notice that we ignored the sticker price and the dealer's preparation charges. Our offer is based only on the *dealer's cost.* Most car dealers will accept an offer that is between 3% and 10% over what they actually paid for the car. For this example, we will assume that the dealer accepted a price of $7,300. We also assume that we have a trade-in with a value of $2,500. Here is a list of calculations shown on the sales contract:

Sales price of Honda:	$7,300.00
Destination charges:	200.00
Subtotal:	7,500.00
Tax (7% rate):	525.00
Less trade-in:	2,500.00
Amount to be financed:	5,525.00

We now calculate several key amounts:

Interest: $I = Prt = 5{,}525(0.08)(4) = 1{,}768$

Amount to be repaid: $A = P + I = 5{,}525 + 1{,}768 = 7{,}293$

Monthly payment: $m = \dfrac{7{,}293}{48} = 151.9375$

The monthly payment for the car is $151.94. ■

ANNUAL PERCENTAGE RATE (APR)

An important aspect of add-on interest is that you are paying a rate that exceeds the quoted add-on interest rate. The reason for this is that you are not keeping the entire amount borrowed for the entire time. For the car payments calculated in Example 2, the principal used was $5,525, but you do not *owe* this entire amount for 4 years. After the first payment, you owe *less* than this amount. In fact, after you make 47 payments, you will owe only $151.94; but the calculation shown in Example 2 assumes that the principal remains constant for 4 years.

To see this a little more clearly, consider a simpler example. Suppose you borrow $2,000 for 2 years with 10% add-on interest. The amount of interest is

$$\$2{,}000 \times 0.10 \times 2 = \$400$$

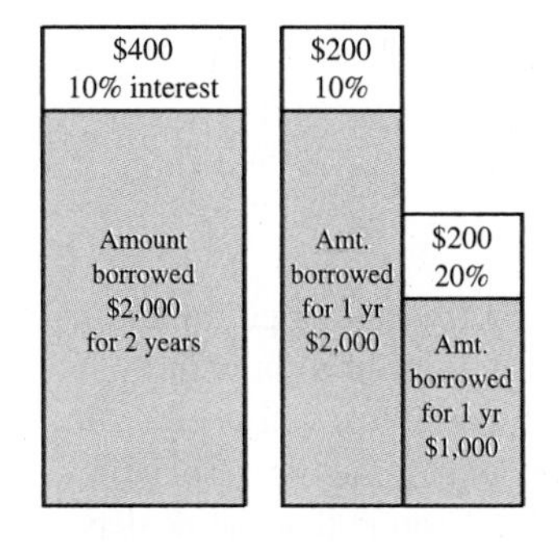

Figure 7.1 Interest on a $2,000 two-year loan

Now, if you pay back $2,000 + $400 at the end of two years, the annual interest rate is 10%. However, if you make a partial payment of $1,200 at the end of the first year and $1,200 at the end of the second year, your total paid back is still the same ($2,400), but you have now paid a higher annual interest rate. Why? Take a look at Figure 7.1. On the left we see that the interest on $2,000 is $400. But if you make a partial payment (figure on the right), we see that $200 for the first year is the correct interest, but the remaining $200 interest piled on the remaining balance of $1,000 is 20% interest (not the stated 10%). Note that since you did not owe $2,000 for 2 years, the interest rate, r, necessary to give $400 interest can be calculated using $I = Prt$:

$$(2{,}000)r(1) + (1{,}000)r(1) = 400$$
$$3{,}000r = 400$$
$$r = \frac{400}{3{,}000} = 0.13333 \text{ or } 13.3\%$$

This number, 13.3%, is called the *annual percentage rate.* This number is too difficult to calculate, as we have just done here, if the number of months is very large. We will, instead, use the formula given in the following box.

APR Formula

The **annual percentage rate,** or **APR,** is the rate paid on a loan when that rate is based on the actual amount owed for the length of time that it is owed. It can be found for an add-on interest rate, r, with N payments by using the formula

$$\text{APR} = \frac{2Nr}{N + 1}$$

Use this formula to compare interest rates.

In 1969 a Truth-in-Lending Act was passed by Congress; it requires all lenders to state the true annual interest rate, which is called the *annual percentage rate (APR)* and is based on the actual amount owed. Regardless of the rate quoted, when you ask a salesperson what the APR is, the law requires that you be told this rate. This regulation enables you to compare interest rates *before* you sign a contract, which must state the APR even if you haven't asked for it.

EXAMPLE 3 **Finding an APR**

In Example 1, we considered the purchase of a computer with a price of \$1,399, paid for in installments over 3 years at an add-on rate of 15%. What is the APR (rounded to the nearest tenth of a percent)?

Solution Knowing the amount of the purchase is not necessary when finding the APR. We only need to know N and r. Since N is the number of payments, we have $N = 12(3) = 36$, and r is given as 0.15:

$$\text{APR} = \frac{2(36)(0.15)}{36 + 1} \approx 0.292 \quad \text{Use a calculator.}$$

The APR is 29.2%. ■

EXAMPLE 4 **Using a Calculator to Find the APR for a Car Purchase**

Consider a Blazer with a price of \$18,436 that is advertised at a monthly payment of \$384.00 for 60 months. What is the APR (to the nearest tenth of a percent)?

Solution We are given

$$P = 18{,}436$$
$$m = 384$$
$$t = 5 \text{ (60 months is 5 years)}$$
$$\text{and } N = 60$$

The APR formula requires that we know the rate r.

Think about the problem before using a calculator. We will use the formula $I = Prt$ to find r by first substituting the values for I, P, and t. We know P and t, but need to calculate I. The future value is the total amount to be repaid ($A = P + I$) and we know $A = 384(60) = 23{,}040$, so

$$\begin{aligned} I &= A - P \\ &= 23{,}040 - 18{,}436 \\ &= 4{,}604 \end{aligned}$$

Now we are ready to use the formula $I = Prt$:

$$\begin{aligned} I &= Prt \\ 4{,}604 &= 18{,}436(r)(5) && \text{Substitute known values.} \\ 920.8 &= 18{,}436r && \text{Divide both sides by 5.} \\ 0.0499457583 &\approx r && \text{Divide both sides by 18,436.} \end{aligned}$$

Finally, for the APR formula,

$$\begin{aligned} \text{APR} &= \frac{2Nr}{N+1} \\ &= \frac{2(60)(0.0499457583)}{61} && \text{Don't round until the last step.} \end{aligned}$$

This is a calculation for our calculator:

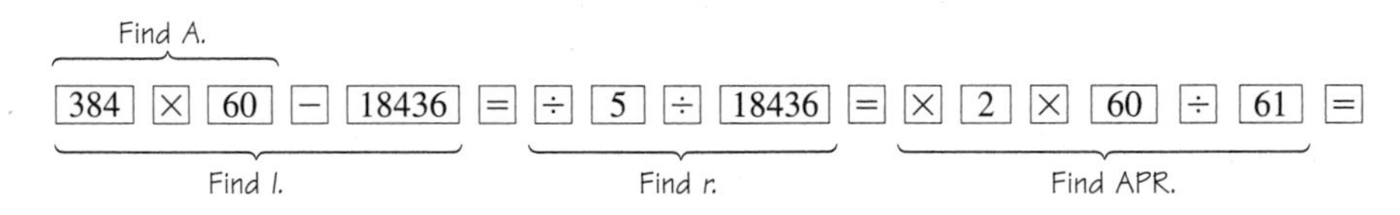

Display: 0982539508

The APR is 9.8%. ■

PROBLEM SET 7.3

Essential Ideas

1. **In Your Own Words** What is add-on interest, and how is it calculated? How do you find the monthly payment for add-on interest?

2. What is the APR formula?

Level 1

Use estimation to select the best response in Problems 3–10. Do not calculate.

3. If you purchase a $2,400 item and pay for it with monthly installments for 2 years, the monthly payment is
   A. $100 per month B. more than $100 per month
   C. less than $100 per month

4. If you purchase a \$595.95 item and pay for it with monthly installments for 1 year, the monthly payment is
A. about \$50 B. more than \$50 C. less than \$50

5. If you purchase an item for \$1,295 at an interest rate of 9.8%, and you finance it for 1 year, then the amount of add-on interest is about
A. \$13.00 B. \$500 C. \$130

6. If you purchase an item for \$1,295 at an interest rate of 9.8%, and you finance it for 4 years, then the amount of add-on interest is about
A. \$13.00 B. \$500 C. \$130

7. If you purchase a new car for \$10,000 and finance it for 4 years, the amount of interest you would expect to pay is about
A. \$4,000 B. \$400 C. \$24,000
D. Can't be estimated

8. A reasonable APR to pay for a 3-year installment loan is
A. 1% B. 12% C. 32%

9. A reasonable APR to pay for a 3-year automobile loan is
A. 6% B. 40% C. \$2,000

10. If you wish to purchase a car with a sticker price of \$10,000, a reasonable offer to make to the dealer is:
A. \$10,000 B. \$9,000 C. \$11,000

WHAT IS WRONG, *if anything, with each of the statements in Problems* 11–16*? Explain your reasoning.*

11. If you purchase a refrigerator for \$895 and make monthly installments for 3 years with an add-on interest rate of 5.8%, then $N = 3$.

12. If you purchase a refrigerator for \$895 and make monthly installments for 3 years with an add-on interest rate of 5.8%, then the monthly payment (m) is found by

$$m = \$895 \div 36 \approx \$24.86$$

13. If you purchase a refrigerator for \$895 and make monthly installments for 3 years with an add-on interest rate of 5.8%, then the amount of interest is found by

$$I = Prt = \$895(5.8)(3) = \$15,573$$

14. If you purchase a sports car for \$36,500 and make monthly payments for 5 years with an add-on interest rate of 9.5%, then $A = \$36,500$.

15. If you purchase a sports car for \$36,500 and make monthly payments for 5 years with an add-on interest rate of 9.5%, then the amount to be repaid is found by

$$A = P(1 + rt) = \$36,500(1 + 0.095 \times 5) = \$53,837.50$$

16. If you purchase a sports car for \$36,500 and make monthly payments for 5 years with an add-on interest rate of 9.5%, then the monthly payment is found by

$$m = \frac{A}{N} = \frac{\$53,837.50}{5} = \$10,767.50$$

LEVEL 2—Applications

Round your answers in Problems 17–22 *to the nearest dollar.*

17. Make a 5% offer on a Honda Civic that has a sticker price of \$16,480 and a dealer cost of \$14,997.

18. Make a 5% offer on a Volkswagen Beetle that has a sticker price of \$16,475 and a dealer cost of \$15,313.

19. Make a 5% offer on a Volvo S90 that has a sticker price of \$35,850 and a dealer cost of \$32,982.

20. Make a 7% offer on a BMW 540i that has a sticker price of \$53,300 and a dealer cost of \$46,904.

21. Make a 6% offer on a Chevrolet Blazer that has a sticker price of \$30,512 and a dealer cost of \$27,460.

22. Make a 5% offer on a Porsche Boxer that has a sticker price of \$41,000 and a dealer cost of \$36,080.

Find the amount of interest and the monthly payment for each of the loans described in Problems 23–32.

23. Purchase a living room set for \$3,600 at 12% add-on interest for 3 years.

24. Purchase a stereo for \$2,500 at 13% add-on interest for 2 years.

25. A \$1,500 loan at 11% add-on rate for 2 years

26. A $2,400 loan at 15% add-on rate for 2 years

27. A $4,500 loan at 18% add-on rate for 5 years

28. A $1,000 loan at 12% add-on rate for 2 years

29. Purchase an oven for $650 at 11% add-on interest for 2 years.

30. Purchase a refrigerator for $2,100 at 14% add-on interest for 3 years.

31. Purchase a car for $5,250 at 9.5% add-on rate for 2 years.

32. Purchase a car for $42,700 at 2.9% add-on rate for 5 years.

Find the APR (rounded to the nearest percent) for each loan listed in Problems 33–42. *These are for the purchases described in Problems* 23–32.

33. Purchase a living room set for $3,600 at 12% add-on interest for 3 years.

34. Purchase a stereo for $2,500 at 13% add-on interest for 2 years.

35. A $1,500 loan at 11% add-on rate for 2 years

36. A $2,400 loan at 15% add-on rate for 2 years

37. A $4,500 loan at 18% add-on rate for 5 years

38. A $1,000 loan at 12% add-on rate for 2 years

39. Purchase an oven for $650 at 11% add-on interest for 2 years.

40. Purchase a refrigerator for $2,100 at 14% add-on interest for 3 years.

41. Purchase a car for $5,250 at 9.5% add-on rate for 2 years.

42. Purchase a car for $42,700 at 2.9% add-on rate for 5 years.

43. A newspaper advertisement offers a $9,000 car for nothing down and 36 easy monthly payments of $317.50. What is the total amount paid for both car and financing?

44. A newspaper advertisement offers a $4,000 used car for nothing down and 36 easy monthly payments of $141.62. What is the total amount paid for both car and financing?

45. A newspaper advertisement offers a $14,350 car for nothing down and 48 easy monthly payments of $488.40. What is the total amount paid for both car and financing?

46. A car dealer will sell you the $16,450 car of your dreams for $3,290 down and payments of $339.97 per month for 48 months. What is the total amount paid for both car and financing?

47. A car dealer will sell you a used car for $6,798 with $798 down and payments of $168.51 per month for 48 months. What is the total amount paid for both car and financing?

In Problems 48–57, *round your answer to the nearest tenth of a percent.*

48. A newspaper advertisement offers a $9,000 car for nothing down and 36 easy monthly payments of $317.50. What is the simple interest rate?

49. A newspaper advertisement offers a $4,000 used car for nothing down and 36 easy monthly payments of $141.62. What is the simple interest rate?

50. A newspaper advertisement offers a $14,350 car for nothing down and 48 easy monthly payments of $488.40. What is the simple interest rate?

51. A car dealer will sell you the $16,450 car of your dreams for $3,290 down and payments of $339.97 per month for 48 months. What is the simple interest rate?

52. A car dealer will sell you a used car for $6,798 with $798 down and payments of $168.51 per month for 48 months. What is the simple interest rate?

53. A newspaper advertisement offers a $9,000 car for nothing down and 36 easy monthly payments of $317.50. What is the APR?

54. A newspaper advertisement offers a $4,000 used car for nothing down and 36 easy monthly payments of $141.62. What is the APR?

55. A newspaper advertisement offers a $14,350 car for nothing down and 48 easy monthly payments of $488.40. What is the APR?

56. A car dealer will sell you the $16,450 car of your dreams for $3,290 down and payments of $339.97 per month for 48 months. What is the APR?

57. A car dealer will sell you a used car for $6,798 with $798 down and payments of $168.51 per month for 48 months. What is the APR?

58. A car dealer carries out the following calculations:

List price	$5,368.00
Options	$1,625.00
Destination charges	$200.00
Subtotal	$7,193.00
Tax	$431.58
Less trade-in	$2,932.00
Amount to be financed	$4,692.58
8% interest for 48 months	$1,501.63
Total	$6,194.21
MONTHLY PAYMENT	$129.05

What is the annual percentage rate?

59. A car dealer carries out the following calculations:

List price	$15,428.00
Options	$3,625.00
Destination charges	$350.00
Subtotal	$19,403.00
Tax	$1,164.18
Less trade-in	$7,950.00
Amount to be financed	$12,617.18
5% interest for 60 months	$3,154.30
Total	$15,771.48
MONTHLY PAYMENT	$262.86

What is the annual percentage rate?

60. A car dealer carries out the following calculations:

List price	$9,450.00
Options	$1,125.00
Destination charges	$300.00
Subtotal	$10,875.00
Tax	$652.50
Less trade-in	$0.00
Amount to be financed	$11,527.50
11% interest for 48 months	$5,072.10
Total	$16,599.60
MONTHLY PAYMENT	$345.83

What is the annual percentage rate?

61. Situation Karen and Wayne must buy a refrigerator because theirs just broke. Unfortunately, their savings account is depleted, and they will need to borrow money to buy a new one. The bank offers them a personal loan at 21% (APR), and Sears offers them an installment loan at 15% (add-on rate). Suppose that the refrigerator at Sears costs $1,598 plus 5% sales tax, and Karen and Wayne plan to pay for the refrigerator for 3 years. Should they finance it with the bank or with Sears?

62. Situation Karen and Wayne must buy a refrigerator because theirs just broke. Unfortunately, their savings account is depleted, and they will need to borrow money to buy a new one. The bank offers them a personal loan at 21% (APR), and Sears offers them an installment loan at 15% (add-on rate). If the refrigerator at Sears costs $1,598 plus 5% sales tax, and Karen and Wayne plan to pay for the refrigerator for 3 years, using the Sears add-on rate, what is the monthly payment?

63. Situation Karen says that she has heard something about APR rates but doesn't really know what the term means. Wayne says he thinks it has something to do with the prime rate, but he isn't sure what. Write a short paper explaining APR to Karen and Wayne.

64. In Your Own Words Write a short paper comparing the SITUATION with your own situation. (That is, list similarities and/or differences.)

LEVEL 3—Team Projects

65. With a team of two or three classmates, investigate the costs of obtaining a $10,000 business loan repaid in a lump sum in 18 months. List specific sources in your area for such a loan, as well as specific options. Include the rate and different methods for calculating the rate. Also include the requirements for the loan (such as a required cosigner or collateral).

Each member of the team should individually obtain information from a different source, and then the team should decide on the best source.

66. With your team, select a particular car and options. Research the cost of the car, as well as sources for purchasing the car. Decide on your best offer and then interview one of the sources to decide if your offer would be acceptable. Prepare a report on your results.

67. What is a bank debit card? Investigate and report on the advantages and disadvantages of using a bank debit card. Some aspects to consider: convenience, privacy, safety, record keeping, acceptance, and liability in case of loss or theft. Individually answer these questions, and then together read, critique, and edit each other's work. Organize and submit a single report.

7.4 CREDIT CARD INTEREST

Situation

Marsha must have surgery, and she does not have the $3,000 cash necessary for the operation. Talking to an administrator at the hospital, she finds that it will accept MasterCard, Visa, and Discover credit cards. All of these credit cards have an APR of 18%, so she figures that it does not matter which card she uses, even though she plans to take a year to pay off the loan.

In this section, Marsha will learn that it does matter which credit card she uses, that not all credit cards are the same—even among credit cards with the same APR.

OPEN-END CREDIT

The most common types of open-end credit used today are through credit cards issued by VISA, MasterCard, Discover, American Express, department stores, and oil companies. Because you don't have to apply for credit each time you want to charge an item, this type of credit is very convenient.

When comparing the interest rates on loans, you should use the APR. In the previous section, we introduced a formula for add-on interest; but for credit cards, the stated interest rate *is* the APR. However, the APR on credit cards is often stated as a daily or a monthly rate. For credit cards, we use a 365-day year rather than a 360-day year.

EXAMPLE 1 **Finding the APR for Credit Cards**

Convert the given credit card rate to APR (rounded to the nearest tenth of a percent).

a. $1\frac{1}{2}\%$ per month **b.** Daily rate of 0.05753%

Solution

a. Since there are 12 months per year, multiply a monthly rate by 12 to get the APR:

$$1\tfrac{1}{2}\% \text{ monthly rate} \times 12 = \mathbf{18\%\ APR}$$

b. Multiply the daily rate by 365 to obtain the APR:

$$0.05753\% \times 365 = 20.99845$$

Rounded to the nearest tenth, this is equivalent to **21.0% APR.** ■

Many credit cards charge an annual fee; some charge $1 every billing period the card is used; others are free. These charges affect the APR differently, depending on how much the credit card is used during the year and on the monthly balance. If you always pay your credit card bill in full as soon as you receive it, the card

Here are some examples of different credit cards:
People's Bank, Bridgeport, CN 13.96% fixed rate; $25 fee (negotiable); average daily balance method.
NationsBank, Dallas, TX 18.15% variable rate; $0 fee; average daily balance method.
Union Bank, San Diego, CA 15.80%; $0 fee; average daily balance method

with no yearly fee would obviously be the best for you. On the other hand, if you use your credit card to stretch out your payments, the APR is more important than the flat fee. For our purposes, we won't use the yearly fee in our calculations of APR on credit cards.

Like annual fees, the interest rates or APRs for credit cards vary greatly. Because VISA and MasterCard are issued by many different banks, the terms even in one locality can vary greatly. Some common examples are shown in the margin.

CALCULATING CREDIT CARD INTEREST

The finance charges can vary greatly even on credit cards that show the *same* APR, depending on the way the interest is calculated. There are three generally accepted methods for calculating these charges: *previous balance, adjusted balance,* and *average daily balance.*

Methods of Calculating Interest on Credit Cards

For credit card interest, use the simple interest formula, $I = Prt$.

Previous balance method: Interest is calculated on the previous month's balance. With this method, P = previous balance, r = annual rate, and $t = \frac{1}{12}$.

Adjusted balance method: Interest is calculated on the previous month's balance *less* credits and payments. With this method, P = adjusted balance, r = annual rate, and $t = \frac{1}{12}$.

Average daily balance method: Add the outstanding balance *each day* in the billing period, and then divide by the number of days in the billing period to find what is called the *average daily balance.* With this method, P = average daily balance, r = annual rate, and t = number of days in the billing period divided by 365.

STOP

If you use credit cards, you should be familiar with these methods for calculating interest. They are all based on the simple interest formula.

In Examples 2–4, we compare the finance charges on a $1,000 credit card purchase, using these three different methods. Assume that a bill for $1,000 is received on April 1 and a payment is made. Then another bill is received on May 1, and this bill shows some finance charges. This finance charge is what we are calculating in Examples 2–4.

EXAMPLE 2

Credit Card Interest by Using the Previous Balance Method

Calculate the interest on a $1,000 credit card bill that shows an 18% APR, using the previous balance method and assuming that $50 is sent and recorded by the due date.

Solution The payment doesn't affect the finance charge for this month unless the bill is paid in full. We have $P = \$1{,}000$, $r = 0.18$, and $t = \frac{1}{12}$. Thus,

$$\begin{aligned} I &= Prt \\ &= \$1{,}000(0.18)(\tfrac{1}{12}) \\ &= \$15 \end{aligned}$$

The interest is $15 for this month. ■

EXAMPLE 3

Credit Card Interest by Using the Adjusted Balance Method

Calculate the interest on a $1,000 credit card bill that shows an 18% APR, using the adjusted balance method and assuming that $50 is sent and recorded by the due date.

Solution First, find the adjusted balance:

Previous balance:	$1,000
Less credits and payments:	50
Adjusted balance:	$950 ← This is P.

We have $P = \$950$, $r = 0.18$, and $t = \frac{1}{12}$. Thus,

$$\begin{aligned} I &= Prt \\ &= \$950(0.18)(\tfrac{1}{12}) \\ &= \$14.25 \end{aligned}$$

The interest is $14.25 for this month. ■

EXAMPLE 4

Interest by Using the Average Daily Balance Method

Calculate the interest on a $1,000 credit card bill that shows an 18% APR, using the average daily balance method. Assume that you sent a payment of $50 on April 1 and that it takes 10 days for this payment to be received and recorded.

Solution For the first 10 days the balance is $1,000, but then the balance drops to $950. Add the balance for *each day:*

10 days @$1,000:	$10,000
20 days @$950:	$19,000
Total:	$29,000

Divide by the number of days in the month (30 in April):

$\$29{,}000 \div 30 = \$966.67$ ← This is the **average daily balance.**

For this problem, we have $P = \$966.67$, $r = 0.18$, and $t = \frac{30}{365}$. Thus,

$$\begin{aligned} I &= Prt \\ &= \$966.67(0.18)(\tfrac{30}{365}) \\ &\approx \$14.30 \end{aligned}$$

You can do this calculation with one calculator session, starting at the top. When using your calculator, think of what you are doing, rather than thinking in terms of individual buttons pressed:

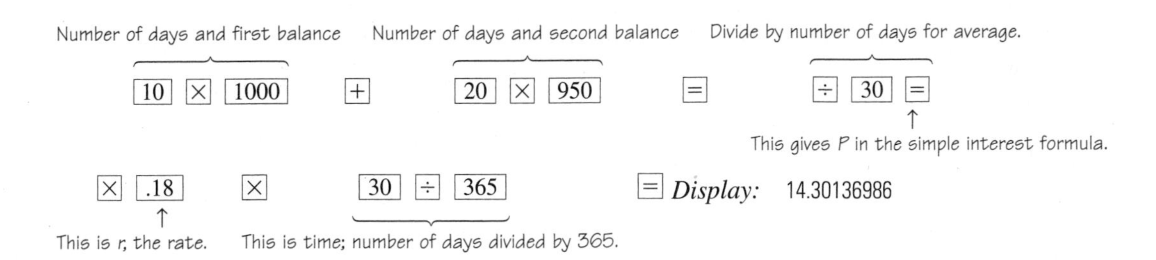

The interest is $14.30 for this month. ■

You can sometimes make good use of credit cards by taking advantage of the period during which no finance charges are levied. Many credit cards charge no interest if you pay in full within a certain period of time (usually 20 or 30 days). This is called the **grace period.** On the other hand, if you borrow cash on your credit card, you should know that many credit cards have an additional charge for cash advances—and these can be as high as 4%. This 4% is *in addition to* the normal finance charges.

Historical Perspective

At the start of the 20th century, a few hotels began to issue credit cards; and as early as 1914, large department stores and gasoline chains were issuing credit cards. In those days, credit cards had three basic functions:

1. They were often prestige items offered only to valued customers.
2. They were more convenient to use than cash.
3. They provided some degree of safety.

During World War II, the use of credit cards virtually ceased, due to government restraint on consumer spending and credit. After wartime restrictions were lifted, however, many plans were reinstated, and railroads and airlines began to issue their own travel cards. In 1949, the Diners Club was established, and it was followed a short time later by American Express and Carte Blanche.

PROBLEM SET 7.4

ESSENTIAL IDEAS

1. What is the formula used to calculate credit card interest?

2. What are the three meanings we give to P when calculating credit card interest?

LEVEL 1

Estimate the answer for Problems 3–12 *by picking the best choice without calculating.*

3. A reasonable APR for a credit card is
   A. 1% B. 30% C. 12%

4. If I do not pay off my credit card each month, the most important cost factor is
A. the annual fee B. the APR

5. If I pay off my credit card balance each month, the most important cost factor is
A. the annual fee B. the APR

6. The method of calculation most advantageous to the consumer is the
A. previous balance method
B. adjusted balance method
C. average daily balance method

7. In an application of the average daily balance method for the month of August, t is
A. $\frac{1}{12}$ B. $\frac{30}{365}$ C. $\frac{31}{365}$

8. When using the average daily balance method for the month of September, t is
A. $\frac{1}{12}$ B. $\frac{30}{365}$ C. $\frac{31}{365}$

9. If your credit card balance is $650 and the interest rate is 12% APR, then the credit card interest charge is
A. $6.50 B. $65 C. $8.25

10. If your credit card balance is $952, you make a $50 payment, the APR is 12%, and the interest is calculated according to the previous balance method, then the finance charge is
A. $9.52 B. $9.02 C. $9.06

11. If your credit card balance is $952, you make a $50 payment, the APR is 12%, and the interest is calculated according to the adjusted balance method, then the finance charge is
A. $9.52 B. $9.02 C. $9.06

12. If your credit card balance is $952, you make a $50 payment, the APR is 12%, and the interest is calculated according to the average daily balance method, then the finance charge is
A. $9.52 B. $9.02 C. $9.06

LEVEL 1—Drill

*Convert each credit card rate in Problems 13–18 to the APR.**

13. Oregon, $1\frac{1}{4}$% per month

14. Arizona, $1\frac{1}{3}$% per month

15. New York, $1\frac{1}{2}$% per month

16. Tennessee, 0.02740% daily rate

17. Ohio, 0.02192% daily rate

18. Nebraska, 0.03014% daily rate

Calculate the monthly finance charge for each credit card transaction in Problems 19–33. Assume that it takes 10 days for a payment to be received and recorded and that the month is 30 days long.

	Balance	*Rate*	*Payment*	*Method*
19.	$300	18%	$50	Previous balance
20.	$300	18%	$50	Adjusted balance
21.	$300	18%	$50	Average daily balance
22.	$300	18%	$250	Previous balance
23.	$300	18%	$250	Adjusted balance
24.	$300	18%	$250	Average daily balance
25.	$500	20%	$50	Previous balance
26.	$500	20%	$50	Adjusted balance
27.	$500	20%	$50	Average daily balance
28.	$3,000	21%	$150	Previous balance
29.	$3,000	21%	$150	Adjusted balance
30.	$3,000	21%	$150	Average daily balance
31.	$3,000	21%	$1,500	Previous balance
32.	$3,000	21%	$1,500	Adjusted balance
33.	$3,000	21%	$1,500	Average daily balance

*These rates were the listed finance charges on purchases of less than $500 on a Citibank VISA statement.

LEVEL 2

WHAT IS WRONG, *if anything, with each of the statements in Problems 34–41? Explain your reasoning.*

34. When calculating credit card interest, use $t = \frac{1}{12}$ because credit cards are billed monthly.

35. From the consumer's point of view, the best method of calculating interest is the average daily balance method.

36. From the bank's point of view, the best method of calculating interest is the average daily balance method.

37. If you owe $1,200 for 7 days in January and $1,000 for 24 days, then the average daily balance is

$$\frac{(\$1{,}200 \times 7) + (\$1{,}000 \times 24)}{30} = \$1{,}080$$

38. If you owe $355 in February 2001 and pay $300 so that your account is credited in 10 days, the average daily balance is

$$\frac{(\$355 \times 10) + (\$55 \times 20)}{30} = \$155$$

39. Suppose that you receive a credit card bill on December 1 for $850. You mail in a payment of $450 the same day, and it is received in 7 days. If the interest is charged as 15% APR and is calculated according to the previous balance method, the amount of interest is

$$\begin{aligned} I &= Prt \\ &= \$450(0.15)(\tfrac{1}{12}) \\ &= \$5.625 \end{aligned}$$

or $5.63.

40. Suppose that you receive a credit card bill on December 1 for $850. You mail in a payment of $450 the same day, and it is received in 7 days. If the interest is charged at 15% APR and is calculated according to the adjusted balance method, the amount of interest is

$$\begin{aligned} I &= Prt \\ &= \$850(0.15)(\tfrac{1}{12}) \\ &= \$10.625 \end{aligned}$$

or $10.63.

41. Suppose that you receive a credit card bill on December 1 for $850. You mail in a payment of $450 the same day, and it is received in 7 days. If the interest is charged at 15% APR and is calculated according to the average daily balance method, the amount of interest is

$$\begin{aligned} I &= Prt \\ &= \$501.6129032(0.15)(\tfrac{1}{12}) \\ &= \$6.27016129 \end{aligned}$$

or $6.27.

Level 2—Applications

42. Most credit cards provide for a minimum finance charge of 50¢ per month. Suppose that you buy a $30 item, make five monthly payments of $5, and then pay the remaining balance. What are the payments and the total interest for this purchase if the interest rate is 12% but you pay the minimum finance charge?

43. The finance charge disclosure on a Sears Revolving Charge Card statement is shown here. Why do you suppose that the limitation on the 50¢ finance charge is from $1.00 to $33.00?

FINANCE CHARGE is based upon account activity during the billing period preceding the current billing period, and is computed upon the "previous balance" ("new balance" outstanding at the end of the preceding billing period) before deducting payments and credits or adding purchases made during the current billing period. The *FINANCE CHARGE* will be the greater of 50¢ (applied to previous balances of $1.00 through $33.00) or an amount determined as follows:

PREVIOUS BALANCE	PERIODIC RATE	*ANNUAL PERCENTAGE RATE*
$1,000 or under	***1.5%*** per month	*18%*
Excess over ***$1,000***	***1.0%*** per month	*12%*

44. Situation Marsha must have surgery, and she does not have the $3,000 cash necessary for the operation. Talking to an administrator at the hospital, she finds that it will accept MasterCard, VISA, and Discover credit cards. All of these credit cards have an APR of 18%, so she figures that it does not matter which card she uses, even though she plans to take a year to pay off the loan. Assume that Marsha makes a payment of $300 and then receives a bill. Show the interest from credit cards of 18% APR according to the previous balance, adjusted balance, and average daily balance methods. Assume that the month has 31 days and that it takes 14 days for Marsha's payment to be mailed and recorded.

45. In Your Own Words Write a short paper comparing the SITUATION with your own situation. (That is, list similarities and/or differences.)

7.5 COMPOUND INTEREST

Situation

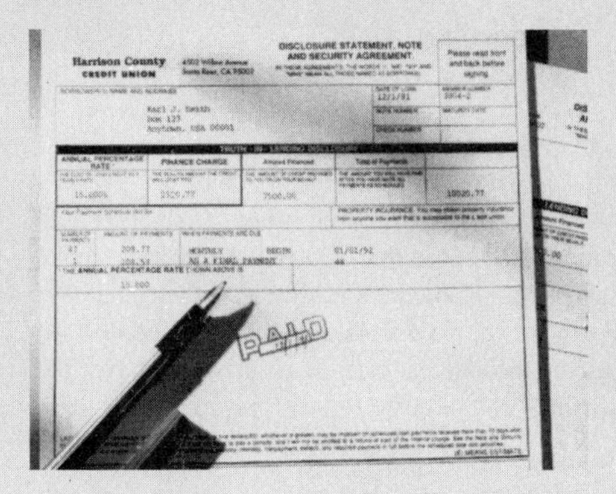

Ron and Lorraine purchased a home some time ago, and they still owe $20,000. Their payments are $195 per month, and they have 10 years left to pay. They want to free themselves of the monthly payments by paying off the loan. They go to the bank and are told that, to pay off the home loan, they must pay $20,000 plus a "prepayment penalty" of 3%. (Older home loans often had prepayment clauses ranging from 1% to 5% of the amount to be paid off. Today, most lenders will waive this penalty, if requested.) This means that, to own their home outright, Ron and Lorraine would have to pay $20,600 ($20,000 + 3% of $20,000). They want to know whether this is a wise financial move.

In this section, Ron and Lorraine will learn about the effects of compound interest, and they will see that with their $20,600 they could not only pay off the home loan but also end up with over $35,000 in cash!

ANNUAL COMPOUNDING

Most banks do not pay interest according to the simple interest formula; instead, after some period of time, they add the interest to the principal and then pay interest on this new, larger amount. When this is done, it is called **compound interest.**

EXAMPLE 1 **Comparing Simple and Compound Interest**

Compare simple and compound interest for a $1,000 deposit at 8% interest for 3 years.

Solution First, calculate the future value of the simple interest:

$$\begin{aligned} A &= P(1 + rt) \\ &= 1{,}000(1 + 0.08 \times 3) \\ &= 1{,}000(1.24) \quad \text{Order of operations; multiplication first} \\ &= 1{,}240 \end{aligned}$$

Using simple interest, the future value in three years is **$1,240.**

Next, assume that the interest is **compounded annually.** This means that the interest is added to the principal after 1 year has passed. This new amount then becomes the principal for the following year. Since the time period for each calculation is 1 year, we let $t = 1$ for each calculation.

First year ($t = 1$):

$$\begin{aligned} A &= P(1 + r) \\ &= 1{,}000(1 + 0.08) \\ &= 1{,}080 \quad \text{This year's balance is next year's principal.} \end{aligned}$$

Second year ($t = 1$): $A = P(1 + r)$ One year's principal is previous year's balance.

$$= 1{,}080(1 + 0.08)$$
$$= 1{,}166.40$$
$$\downarrow$$

Third year ($t = 1$): $A = P(1 + r)$

$$= 1{,}166.40(1 + 0.08)$$
$$= 1{,}259.71$$

Using interest compounded annually, the future value in 3 years is **$1,259.71. The earnings from compounding are $19.71 more than from simple interest.** ■

> **It All Adds UP**
>
> The boy that by addition grows
> And suffers no subtraction
> Who multiplies the thing he knows
> And carries every fraction
> Who well divides the precious time
> The due proportion given
> To sure success aloft will climb
> Interest compound receiving.

The problem with compound interest relates to the difficulty of calculating it. Notice that, to simplify the calculations in Example 1, the variable representing time t was given as 1, and the process was repeated three times. Also notice that, after the future value was found, it was used as the principal in the next step. What if we wanted to compound annually for 20 years instead of for 3 years? Look at Example 1 to discover the following pattern:

Simple interest (20 years) $A = P(1 + rt)$

$$= 1{,}000(1 + 0.08 \times 20)$$
$$= 1{,}000(1 + 1.6)$$
$$= 1{,}000(2.6)$$
$$= 2{,}600$$

Annual compounding (20 years)

$A = P(1 + r)$ First year

$= P(1 + r)(1 + r)$ Second year

$= P(1 + r)^2$ Second year simplified

$= P(1 + r)^2(1 + r)$ Third year

$= P(1 + r)^3$ Third year simplified

$\vdots$

$= P(1 + r)^{20}$ Twentieth year simplified

For a period of 20 years, starting with $1,000 at 8% compounded annually, we have

$$A = 1{,}000(1.08)^{20}$$

You might wish to review exponents and how to use a calculator with exponents, which we presented in Section 1.5.

The difficulty lies in calculating this number. You will need to have a calculator with an exponent key. These are labeled in different ways, depending on the brand. It might be $\boxed{y^x}$ or $\boxed{x^y}$ or $\boxed{\wedge}$. In this book we will show exponents by using $\boxed{\wedge}$, but you should press the appropriate key on your own brand of calculator.

$\boxed{1000}$ $\boxed{\times}$ $\boxed{1.08}$ $\boxed{\wedge}$ $\boxed{20}$ $\boxed{=}$ *Display:* 4660.957144

Round money answers to the nearest cent: **$4,660.96** is the future value of $1,000 compounded annually at 8% for 20 years. This compares with $2,600 from simple interest. The effect of compounding yields $2,060.96 *more* than simple interest.

COMPOUNDING PERIODS

Most banks compound interest more frequently than once a year. For instance, a bank may pay interest as follows:

Semiannually: twice a year or every 180 days
Quarterly: 4 times a year or every 90 days
Monthly: 12 times a year or every 30 days
Daily: 360 times a year

To write a formula for various compounding periods, we must introduce three new variables. First, let

$n =$ NUMBER OF TIMES INTEREST IS CALCULATED EACH YEAR

That is,

Remember these compounding period names.

$n = 1$ for *annual* compounding
$n = 2$ for *semiannual* compounding
$n = 4$ for *quarterly* compounding
$n = 12$ for *monthly* compounding
$n = 360$ for *daily* compounding

Second, let

$N =$ NUMBER OF COMPOUNDING PERIODS

That is,

$$N = nt$$

Third, let

$i =$ RATE PER PERIOD

That is,

$$i = \frac{r}{n}$$

We can now summarize the variables we use for interest.

Variables Used with Interest Formulas

It is tempting to skip past this box, but you must take some time to study what these variables represent.

$A =$ FUTURE VALUE	This is the principal plus interest.
$P =$ PRESENT VALUE	This is the same as the principal.
$r =$ INTEREST RATE	This is the *annual* interest rate.
$t =$ TIME	This is the time *in years*.
$n =$ NUMBER OF COMPOUNDING PERIODS EACH YEAR	
$N = nt$	This is the number of periods.
$i = \frac{r}{n}$	This is the rate per period.

EXAMPLE 2 **Identifying the Variables Used with Interest Formulas**

Fill in the blanks, given an annual rate of 12% and a time period of 3 years for the following compounding periods.

Solution **a.** annual **b.** semiannual **c.** quarterly **d.** monthly **e.** daily

	n	r	t	$N = nt$	$i = \frac{r}{n}$
Compounding Period		*Yearly Rate*	*Time*	*Number of Periods*	*Rate per Period*
a. Annual	**1**	**12%**	**3 yr**	**3**	**12%**
b. Semiannual	**2**	**12%**	**3 yr**	**6**	**6%**
c. Quarterly	**4**	**12%**	**3 yr**	**12**	**3%**
d. Monthly	**12**	**12%**	**3 yr**	**36**	**1%**
e. Daily	**360**	**12%**	**3 yr**	**1,080**	**0.03%**

To find the number of periods, use $N = nt$; for part **e**: $N = 360(3) = 1{,}080$

To find the rate per period, use $i = \frac{r}{n}$; for part **e**; $i = \frac{0.12}{360} = 0.000333\ldots$. As a percent, $i \approx 0.03\%$. ■

We are now ready to state the future value formula for compound interest, which is sometimes called the **compound interest formula.**

Future Value (Compound Interest)

$$A = P(1 + i)^N$$

where A is future value, P is present value, $i = r/n$, and $N = nt$

Whether you use the formula (calculator required) or the table, you will follow the same procedure.

Procedure for Using the Future Value Formula

Identify the present value, P.
Identify the rate, r.
Identify the time, t.
Identify the number of times interest is compounded each year, n.
Calculate $N = nt$.
Calculate $i = \frac{r}{n}$.
Use the future value formula using the calculated values of i and N.

EXAMPLE 3 **Finding Future Value**

Find the future value of $1,000 invested for 10 years at 8% interest compounded

a. annually **b.** semiannually **c.** quarterly

Solution Identify the variables:

$P = 1{,}000$

$r = 0.08$

$t = 10$

a. $n = 1$ Calculate $N = nt = 1(10) = 10$ and $i = \dfrac{r}{n} = \dfrac{0.08}{1} = 0.08$

We use the future value formula, $A = P(1 + i)^N$ and a calculator:

[1000] [×] [(] [1] [+] [.08] [)] [∧] [10] [=] *Display:* 2158.924997

The future value is $2,158.92.

b. $n = 2$ Calculate $N = nt = 2(10) = 20$ and $i = \dfrac{r}{n} = \dfrac{0.08}{2} = 0.04$

[1000] [×] [(] [1] [+] [.04] [)] [∧] [20] [=] *Display:* 2191.123143

The future value is $2,191.12.

c. $n = 4$ Calculate $N = nt = 4(10) = 40$ and $i = \dfrac{r}{n} = \dfrac{0.08}{4} = 0.02$

[1000] [×] [(] [1] [+] [.02] [)] [∧] [40] [=] *Display:* 2208.039664

The future value is $2,208.04. ■

EXAMPLE 4 **Finding Future Value Using a Calculator**

Find the future value of $815 invested for 10 years and 6 months at 8.25% interest compounded

a. quarterly **b.** daily

Solution $P = \$815$; $r = 0.0825$; $t = 10.5$

a. $n = 4$ Calculate:

$$N = nt = 4(10.5) = 42 \qquad i = \frac{r}{n} = \frac{0.0825}{4}$$

$$A = P(1 + i)^N = 815\left(1 + \frac{0.0825}{4}\right)^{42}$$

You will need the parentheses keys on your calculator:

[815] [(] [1] [+] [.0825] [÷] [4] [)] [∧] [42] [=]

Display: 1921.047468

The future value is $1,921.05.

b. $n = 360$; $N = nt = 360(10.5) = 3{,}780$; $i = \dfrac{0.0825}{360}$

$$A = P(1 + i)^N = 815\left(1 + \frac{0.0825}{360}\right)^{3{,}780}$$

[815] [(] [1] [+] [.0825] [÷] [360] [)] [∧] [3780] [=]

Display: 1937.858685

The future value is $1,937.86. ■

EXAMPLE 5

Problem Solving with Compound Interest

Consider the situation of Ron and Lorraine described at the beginning of this section. They wish to know whether it is a wise financial move to use $20,600 to pay off their home mortgage.

Solution The first question is: How much principal do Ron and Lorraine need to generate $195 per month in income (to make their mortgage payments) if they are able to invest it at 15% interest? "Wait! Where can you find 15% interest? My bank pays only 5% interest!" There are investments you could make (not a deposit into a savings account) that could yield a 15% return on your money. Problem solving often requires that you make certain assumptions to have sufficient information to answer a question that you might have. This first question is an application of simple interest, because the interest is withdrawn each month and is not left to accumulate:

I is the amount withdrawn each month.
↓

$$I = Prt \quad \leftarrow t = \tfrac{1}{12} \text{ because it is monthly income.}$$

$$195 = P(0.15)(\tfrac{1}{12})$$

$$2{,}340 = 0.15P \quad \text{Multiply both sides by 12.}$$

$$15{,}600 = P \quad \text{Divide both sides by 0.15.}$$

This means that a deposit of $15,600 will be sufficient to pay off their home loan, with the added advantage that, when their home is eventually paid for, they will still have the $15,600!

Now, since they have $20,600 to invest, they can also allow the remaining $5,000 to grow at 15% interest. This is an example of compound interest because the interest is not withdrawn, but instead accumulates. Since there are 10 years left on the home loan, we have $P = 5{,}000$; $r = 0.15$, $t = 10$, and $n = 1$ (assume annual compounding). Calculate:

$$N = nt = 10 \qquad i = \frac{r}{n} = \frac{0.15}{1} = 0.15$$

Thus,

$$A = 5{,}000(1 + 0.15)^{10} \approx 20{,}227.79$$

If you ever said "When will I ever use this?" here is a real-life example that could save you a great deal of money.

This means that Ron and Lorraine have two options:

1. Use their $20,600 to pay off the home loan. They will have the home paid for and will not need to make any further payments.

2. Dispose of their \$20,600 as follows:

   Deposit \$15,600 into an account to make their house payments. They will not need to make any further payments. At the end of 10 years they will still have the \$15,600.

   Deposit the excess \$5,000 into an account and let it grow for 10 years. The accumulated value will be \$20,227.79.

   Under Option 2 they will have \$35,827.79 in 10 years in addition to all the benefits gained under Option 1. ■

INFLATION

Any discussion of compound interest is incomplete without a discussion of inflation. When there is an increase in the amount of money in circulation, there is a fall in its value which, in turn, causes a rise in prices. This is called **inflation.** The same procedure we used to calculate compound interest can be used to calculate the effects of inflation. The government releases reports of monthly and annual inflation rates. In 1981, the inflation rate was nearly 9%, but in 1999 it was less than 2%. Keep in mind that inflation rates can vary tremendously and that the best we can do in this section is to assume different constant inflation rates. For our purposes in this book, we will assume $n = 1$ (annual compounding) when working inflation problems.

EXAMPLE 6 **Effect of Inflation on a Salary**

Richard is earning \$30,000 per year and would like to know what salary he could expect in 20 years if inflation continues at an average of 9%.

Solution $P = 30{,}000$, $r = 0.09$, and $t = 20$. In this book, assume $n = 1$ for inflation problems. This means that $N = 20$ and $i = 0.09$. Find

$$\begin{aligned} A &= P(1 + i)^N \\ &= 30{,}000(1 + 0.09)^{20} \\ &\approx 168{,}132.32 \quad \text{Use a calculator.} \end{aligned}$$

The answer means that, if inflation continues at a constant 9% rate, an annual salary of **\$168,000** will have about the same purchasing power in 20 years as a salary of \$30,000 today. ■

PRESENT VALUE

Sometimes we know the future value of an investment and wish to know its present value. Such a problem is called a **present value** problem. The formula follows directly from the future value formula (by division).

Present Value Formula

This is the same as the future value formula, algebraically solved for P.

$$P = \frac{A}{(1+i)^N}; \text{ on a calculator this is } A \div (1+i)^N$$

where P is present value, A is future value, $i = r/n$, and $N = nt$.

EXAMPLE 7

Finding a Present Value

Suppose that you want to take a trip to Tahiti in 5 years and you decide that you will need $5,000. To have that much money set aside in 5 years, how much money should you deposit into a bank account paying 6% compounded quarterly?

Solution In this problem, P is unknown and A is given: $A = 5{,}000$. Also, $r = 0.06$; $t = 5$; and $n = 4$. Calculate:

$$N = nt = 4(5) = 20 \qquad i = \frac{r}{n} = \frac{0.06}{4}$$

We use the formula $P = \dfrac{A}{(1+i)^N}$.

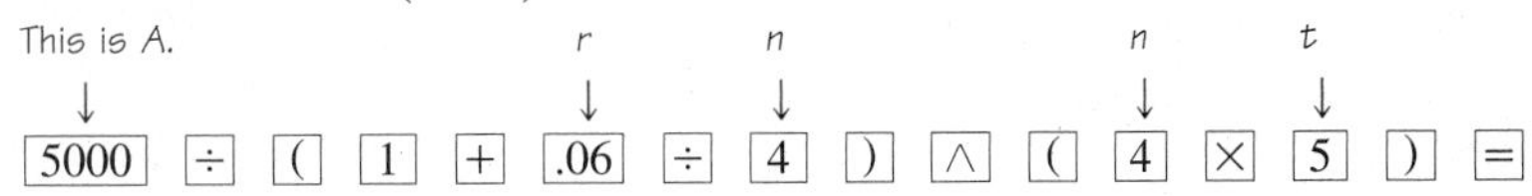

Display: 3712.352091

You should deposit $3,712.35. ■

EXAMPLE 8

Using Present Value with a Retirement Decision

An insurance agent wishes to sell you a policy that will pay you $100,000 in 30 years. What is the value of this policy in today's dollars, if we assume a 9% inflation rate?

Solution This is a present value problem for which $A = 100{,}000$, $r = 0.09$, $n = 1$, and $t = 30$. We calculate

$$N = nt = 1(30) = 30 \quad i = \frac{r}{n} = \frac{0.09}{1} = 0.09$$

To find the present value, calculate

$$P = \frac{100{,}000}{(1+0.09)^{30}} \approx \mathbf{\$7{,}537.11}$$

This means that the agent is offering you an amount comparable to $7,537.11 in terms of today's dollar. ■

PROBLEMS SET 7.5

ESSENTIAL IDEAS

1. What is n for the following compounding periods?
   a. quarterly **b.** semiannually
   c. monthly **d.** daily

2. Tell what each of the following variables represents when used in connection with interest formulas.
   a. A **b.** P
   c. r **d.** t
   e. n

3. **a.** How do you calculate N?
   b. How do you calculate i?

4. What is the future value formula for compound interest?

5. What is the present value formula for compound interest?

6. What formula do you use to calculate inflation?

LEVEL 1

WHAT IS WRONG, *if anything, with each of the statements in Problems 7–12? Explain your reasoning.*

7. If an investment is compounded monthly, then $N = 12$.

8. If an investment is compounded daily, then $n = 365$.

9. If the rate of an investment is 12% compounded monthly, then $r = 1\%$.

10. If the rate of an investment is 6% compounded monthly, then $i = 0.05$.

11. To find the amount of interest I for compound interest, first find A and then subtract P.

12. If \$500 is deposited into an account for 4 years paying 4.5% compounded daily, then $P = 500$, $N = 4$, and $i = 0.045$, so

$$A = P(1 + i)^N = 500(1 + 0.045)^4$$

Using a calculator, we find $A \approx \$596.26$.

LEVEL 1—Drill

In Problems 13–18, compare the future amounts (A) you would have if the money were invested at simple interest and if it were invested with annual compounding.

13. \$1,000 at 8% for 5 years

14. \$5,000 at 10% for 3 years

15. \$2,000 at 12% for 3 years

16. \$2,000 at 12% for 5 years

17. \$5,000 at 12% for 20 years

18. \$1,000 at 14% for 30 years

Fill in the blanks in Problems 19–34.

Compounding Period	P *Present Value*	r *Annual Rate*	t *Time*	n	i *Period Rate*	N *Number of Periods*	A *Future Value*
19. Annual	\$1,000	9%	5 yr	____	____	____	____
20. Semiannual	\$1,000	9%	5 yr	____	____	____	____
21. Annual	\$500	8%	3 yr	____	____	____	____
22. Semiannual	\$500	8%	3 yr	____	____	____	____
23. Quarterly	\$500	8%	3 yr	____	____	____	____

Compounding Period	*P* Present Value	*r* Annual Rate	*t* Time	*n*	*i* Period Rate	*N* Number of Periods	*A* Future Value
24. Semiannual	\$3,000	18%	3 yr	____	____	____	______
25. Quarterly	\$5,000	18%	10 yr	____	____	____	______
26. Quarterly	\$624	16%	5 yr	____	____	____	______
27. Quarterly	\$5,000	20%	10 yr	____	____	____	______
28. Monthly	\$350	12%	5 yr	____	____	____	______
29. Monthly	\$4,000	24%	5 yr	____	____	____	______
30. Quarterly	\$800	12%	90 days	____	____	____	______
31. Quarterly	\$900	12%	180 days	____	____	____	______
32. Quarterly	\$1,900	12%	270 days	____	____	____	______
33. Quarterly	\$1,250	16%	450 days	____	____	____	______
34. Quarterly	\$1,000	12%	900 days	____	____	____	______

Level 2—Drill

In Problems 35–48, find the future value, using the future value formula and a calculator.

35. \$35 at 17.65% compounded annually for 20 years

36. \$155 at 21.25% compounded annually for 25 years

37. \$835 at 3.5% compounded semiannually for 6 years

38. \$9,450 at 7.5% compounded semiannually for 10 years

39. \$575 at 5.5% compounded quarterly for 5 years

40. \$3,450 at 4.3% compounded quarterly for 8 years

41. \$9,730.50 at 7.6% compounded monthly for 7 years

42. \$3,560 at 9.2% compounded monthly for 10 years

43. \$45.67 at 3.5% compounded daily for 3 years

44. \$34,500 at 6.9% compounded daily for 2 years

45. \$89,500 at 6.2% compounded monthly for 30 years

46. \$119,400 at 7.5% compounded monthly for 30 years

47. \$225,500 at 8.65% compounded daily for 30 years

48. \$355,000 at 9.5% compounded daily for 30 years

Level 2—Applications

49. Find the cost of each item in 5 years, assuming an inflation rate of 9%.

a. cup of coffee, \$0.75 **b.** Sunday paper, \$1.25

c. Big Mac, \$1.95 **d.** gallon of gas, \$1.35

50. Find the cost of each item in 10 years, assuming an inflation rate of 5%.

a. movie admission, \$5.00 **b.** CD, \$14.95

c. textbook, \$40.00 **d.** electricity bill, \$65

51. Find the cost of each item in 10 years, assuming an inflation rate of 12%.

a. phone bill, \$45 **b.** pair of shoes, \$65

c. new suit, \$370 **d.** monthly rent, \$600

52. Find the cost of each item in 20 years, assuming an inflation rate of 6%.

a. TV set, $600 **b.** small car, $8,000

c. car, $18,000 **d.** tuition, $16,000

53. How much would you have in 5 years if you purchased a $1,000 5-year savings certificate that paid 4% compounded quarterly?

54. What is the future value after 15 years if you deposited $1,000 for your child's education and the interest was guaranteed at 16% compounded quarterly?

55. Find the cost of a home in 30 years, assuming an annual inflation rate of 10%, if the present value of the house is $125,000.

56. Find the cost of the monthly rent for a two-bedroom apartment in 30 years, assuming an annual inflation rate of 10%, if the current rent is $650.

57. Suppose that an insurance agent offers you a policy that will provide you with a yearly income of $50,000 in 30 years. What is the comparative annual salary today, assuming an inflation rate of 6%?

58. Suppose that an insurance agent offers you a policy that will provide you with a yearly income of $50,000 in 30 years. What is the comparable annual salary today, assuming an inflation rate of 10%?

59. Suppose an insurance agent offers you a policy that will provide you with a yearly income of $50,000 in 30 years. What is the comparable annual salary today, assuming an inflation rate of 4%?

60. In Your Own Words Conduct a survey of banks, savings and loan companies, and credit unions in your area. Prepare a report on the different types of savings accounts available and the interest rates they pay. Include methods of payment as well as interest rates.

61. Do you expect to live long enough to be a millionaire? Suppose that your annual salary today is $19,000. If inflation continues at 12%, how long will it be before $19,000 increases to an annual salary of a million dollars?

62. In Your Own Words Consult an almanac or some government source, and then write a report on the current inflation rate. Project some of these results to the year of your own expected retirement.

63. Situation Ron and Lorraine purchased a home some time ago, and they still owe $20,000. Their payments are $195 per month and they have 10 years left to pay. They want to free themselves of the monthly payments by paying off the loan. They go to the bank and are told that, to pay off the home loan, they must pay $20,000 plus a "prepayment penalty" of 3%. (Older home loans often had prepayment clauses ranging from 1% to 5% of the amount to be paid off. Today, most lenders will waive this penalty, if requested.) This means that, to own their home outright, Ron and Lorraine would have to pay $20,600 ($20,000 + 3% of $20,000). They want to know whether this is a wise financial move. In Example 5, we assumed a rather high 15% investment rate. Repeat the analysis using an 8% interest rate.

64. Situation Rework Problem 63, assuming an 11% interest rate.

65. In Your Own Words Write a short paper comparing the SITUATION with your own situation. (That is, list similarities and/or differences.)

7.6 BUYING A HOME

Situation

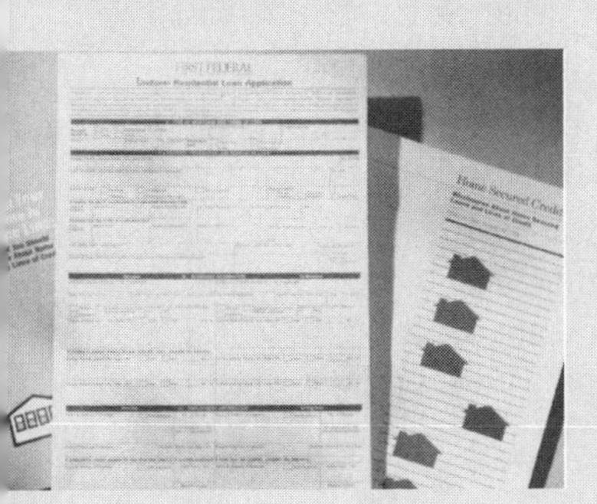

Dorothy and Wes have found the "home of their dreams," and it costs only $240,000. They have never purchased a home before—or anything else involving so much money—and they are very excited. There seems to be so much to do, and so many new words are being used by the realtor and the banker that they are also confused. If their combined salaries are $39,500 and if the current interest rate is 7.5% APR, how much of a down payment will they need to be able to afford their "dream home," and what is the monthly payment?

In this section, Dorothy and Wes will be introduced to the terminology of buying a house, as well as the steps in negotiating a sales contract and obtaining a mortgage.

HOME LOANS

To purchase a home, you will probably need a lender to agree to provide the money you need. You, in turn, promise to repay the money based on terms set forth in an agreement, or loan contract, called a **mortgage.** As the borrower, you pledge your home as security. It remains pledged until the loan is paid off. If you fail to meet the terms of the contract, the lender has the right to **foreclose,** which means that the lender may take possession of the property.

I. TYPE OF MORTGAGE AND TERMS OF LOAN
Mortgage Applied for: V.A. / Conventional / Other: / FHA / FmHA
Agency Case Number
Lender Case Number
Amount $
Interest Rate %
No. of Months
Amortization Type: Fixed Rate / Other (explain): / GPM / ARM (type):
II. PROPERTY INFORMATION AND PURPOSE OF LOAN
Subject Property Address (street, city, state, ZIP)
No. of Uni
Legal Description of Subject Property (attach description if necessary)
Year Built
Purpose of Loan: Purchase / Construction / Other (explain): / Refinance / Construction-Permanent
Property will be: Primary Residence / Secondary Residence / Investment
Complete this line if construction or construction-permanent loan.
Year Lot Acquired | Original Cost $ | Amount Existing Liens $ | (a) Present Value of Lot $ | (b) Cost of Improvements $ | Total (a + b) $
Complete this line if this is a refinance loan.
Year Acquired | Original Cost $ | Amount Existing Liens $ | Purpose of Refinance | Describe Improvements: made / to be mad | Cost: $
Title will be held in what Name(s)
Manner in which Title will be held
Estate will be held in: Fee Simple / Leasehold (show expiration date)
Source of Down Payment, Settlement Charges and/or Subordinate Financing (explain)

There are three types of mortgage loans: (1) conventional loans made between you and a private lender; (2) VA loans made to eligible veterans (these are guaranteed by the Veterans Administration, so they cost less than the other types of loans); and (3) FHA loans made by private lenders and insured by the Federal Housing Administration. Regardless of the type of loan you obtain, you will pay certain lender costs. By *lender costs,* we mean all the charges required by the lender: closing costs plus interest.

When you shop around for a loan, certain rates will be quoted:

1. **Interest rate.** This is the annual interest rate for the loan; it fluctuates on a daily basis. The APR, as stated on the loan agreement, is generally just a little higher than the quoted interest rate. This is because the quoted interest rate is usually based on ordinary interest (360-day year), whereas the APR is based on exact interest (365-day year).
2. **Origination fee.** This is a one-time charge to cover the lender's administrative costs in processing the loan. It may be a flat \$100 to \$300 fee, or it may be expressed as a percentage of the loan.
3. **Points.** This refers to discount points, a one-time charge used to adjust the yield on the loan to what the market conditions demand. It offsets constraints placed on the yield by state and federal regulations. Each point is equal to 1% of the amount of the loan.

EXAMPLE 1 **Finding the Charges for a Loan**

If you are obtaining a \$120,000 loan and the bank charges 8.5% plus \$250 and $2\frac{1}{2}$ points, what are the one-time charges?

Solution The APR probably affects the cost of a loan the most of the three quoted rates, but it does not affect the one-time charges. The \$250 origination fee is added to the fee called *points.* Since 1 point = 1%, we have

$$\text{POINTS} = \$120{,}000 \times 2.5\% = \$120{,}000 \times 0.025 = \mathbf{\$3{,}000}$$
$$\text{FEE} = \mathbf{\$250}$$
$$\text{TOTAL ONE-TIME CHARGES} = \$3{,}000 + \$250 = \mathbf{\$3{,}250}$$

■

Suppose that Lender A quotes 8.5% + 2.5 points + \$250, and Lender B quotes 8.8% + 1 point + \$100. Which lender should you choose? It is desirable to incorporate these three charges—interest rate, origination fee, and points—into one formula so that you can decide which lending institution is offering you the best terms on your loan. This could save you thousands of dollars over the life of your loan. The **comparison rate** formula shown in the following box can be used to calculate the combined effects of these fees. Even though it is not perfectly accurate, it is usually close enough to permit meaningful comparison among lenders.*

Comparison Rate

This is not the actual rate charged, but a rate you can use to compare rates offered by different lenders.

$$\text{COMPARISON RATE} = \text{APR} + 0.125\left(\text{POINTS} + \frac{\text{ORIGINATION FEE}}{\text{AMOUNT OF LOAN}}\right)$$

*The 0.125 is a weighting factor. Since the points and origination fee are one-time factors, they should not be weighted as heavily as the interest rate. Also since the decimal 0.125 is equivalent to $\frac{1}{8}$, we see that the one-time fees are weighted to be one-eighth as important as the interest rate.

EXAMPLE 2 **Finding the Best Loan**

Suppose that you wish to obtain a home loan for \$120,000, and you obtain the following quotations from lenders:

Lender A: 8.5% + 2.5 points + \$250

Lender B: 8.8% + 1 point + \$100

Which lender is making you the better offer?

Solution Calculate the comparison rate for both lenders (a calculator is helpful for this calculation).

Lender A:

$$\text{COMPARISON RATE} = \text{APR} + 0.125\left(\text{POINTS} + \frac{\text{FEE}}{\text{AMT OF LOAN}}\right)$$

$$= \underbrace{0.085}_{\text{APR}} + 0.125\left(\underset{\uparrow}{0.025} + \frac{250}{120{,}000}\right)$$

Write points as a decimal.

$$\approx 0.0884 \quad \text{or} \quad 8.84\%$$

Use a calculator. Display: 0.0883854167

Lender B:

$$\text{COMPARISON RATE} = \text{APR} + 0.125\left(\text{POINTS} + \frac{\text{FEE}}{\text{AMT OF LOAN}}\right)$$

$$= 0.088 + 0.125\left(0.01 + \frac{100}{120{,}000}\right)$$

$$\approx 0.0894 \quad \text{or} \quad 8.94\%$$

Display: 0.0893541667

Lender A is making the better offer, because its comparison rate of 8.84% is less than the comparison rate of Lender B (8.94%). ■

MONTHLY PAYMENTS

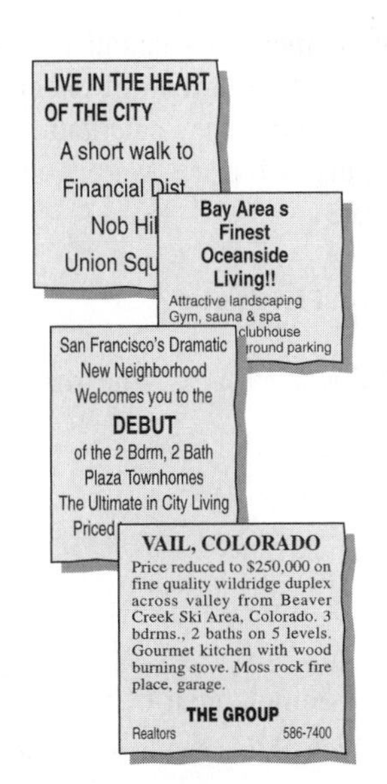

Now imagine that you have selected a lender; next you need to calculate the monthly payments. Several factors influence the amount of the **monthly payment:**

1. **Length of the loan.** Most lenders assume a 30-year period. The shorter the period, the greater the monthly payment; however, the greater the monthly payment, the less the finance charges.
2. Amount of the **down payment.** The greater the down payment, the less the amount to finance; and the less the amount to finance, the smaller the monthly payment.
3. The **APR.** The smaller the interest rate (APR), the smaller the finance charges and consequently the smaller the monthly payment.

The effect of some of these factors is shown in Table 7.2 (page 342). Some financial counselors suggest making as large a down payment as you can afford, whereas others suggest making as small a down payment as is allowed. You will have to as-

Table 7.2 *Effect of Down Payment on the Cost of a $120,000 Home with Interest at 8%*

Down payment	Percent	Monthly payment			Total payments (to the nearest dollar)		
		20 Years	25 Years	30 Years	20 Years	25 Years	30 Years
$ 0	0%	$1,003.73	$926.18	$880.52	$240,895	$277,854	$316,987
$ 6,000	5%	$ 953.54	$879.87	$836.49	$228,850	$263,961	$301,136
$12,000	10%	$ 903.36	$833.56	$792.47	$216,806	$250,068	$285,289
$24,000	20%	$ 802.98	$740.94	$704.41	$192,715	$222,282	$253,588
$30,000	25%	$ 752.80	$694.63	$660.39	$180,672	$208,389	$237,740
$36,000	30%	$ 702.61	$648.33	$616.36	$168,626	$194,499	$221,890

sess many factors, such as your tax bracket and your investment potential, to determine how large a down payment you should make.

EXAMPLE 3 **Comparing the Effect of a Down Payment**

If you are obtaining a $120,000 30-year 8% loan, what would you save in monthly payment and in total payments by increasing your down payment from $12,000 to $24,000?

Solution From Table 7.2, we see that the monthly payments for a 30-year 8% loan with a 10% ($12,000) down payment are $792.47, and with a 20% ($24,000) down payment the monthly payments are $704.41. This is a *monthly* savings of **$88.06.** The total savings are also found by subtracting the amounts shown in Table 7.2:

$$\$285{,}289 - \$253{,}588 = \mathbf{\$31{,}701}$$

■

The amount of the down payment and the monthly payments are calculated in Table 7.2. However, it is unlikely that you will buy a home with a loan of exactly $120,000 and an interest rate of exactly 8%. Let's now consider these calculations.

Since the down payment is simply a percent of the purchase price, we encountered the necessary procedures in the first part of this text.

EXAMPLE 4 **Finding the Down Payment and Amount to Be Financed**

The home you select costs $145,500 and you pay 20% down. What is your down payment? How much will be financed?

Solution

$$\text{DOWN PAYMENT} = \$145{,}500 \times 0.20 = \mathbf{\$29{,}100}$$

$$\begin{aligned}\text{AMOUNT TO BE FINANCED} &= \text{TOTAL AMOUNT} - \text{DOWN PAYMENT}\\ &= \$145{,}500 - \$29{,}100\\ &= \$116{,}400\end{aligned}$$

You can also find the amount to be financed by using the complement of the down payment. The complement of a 20% down payment is 0.80.

$$\begin{aligned}\text{AMOUNT TO BE FINANCED} &= \text{TOTAL AMOUNT} \times \text{COMPLEMENT}\\ &= \$145{,}500(0.80)\\ &= \mathbf{\$116{,}400}\end{aligned}$$

■

We will use Table II in Appendix D (at the back of the book) to calculate the amount of the monthly payment for a loan. You will need to know the amount to be financed, the interest rate, and the length of time the loan is to be financed.

Monthly Payment

$$\text{MONTHLY PAYMENT} = \frac{\text{AMOUNT OF LOAN}}{1{,}000} \times \text{TABLE II ENTRY}$$

EXAMPLE 5 **Finding the Monthly Payment for a Loan**

What is the monthly payment for a home loan of $54,800 if the interest rate is $7\frac{1}{2}\%$, financed for 30 years?

Solution Notice that Table II is expressed in thousands of dollars. This means that you need to divide the amount to be financed by 1,000 (you can do this mentally). For this example, we obtain 54.8. Next, find the entry in Table II for 7.5% for 30 years: 6.99. This represents the cost to repay $1,000. Since there are 54.8 thousands, we multiply

$$6.99 \times 54.8 = 383.052$$

The monthly payments for this loan are $383.05. ■

MAXIMUM HOUSE PAYMENT

For many people, buying a home is the single most important financial decision of a lifetime. There are three steps in buying and financing a home. The first step consists of finding a home you would like to buy and then reaching an agreement with the seller on the price and the terms. For this step, you will need to negotiate a *purchase-and-sale agreement* or *sales contract.* The second step involves finding a lender to finance the purchase. You will need to understand interest as you shop around to obtain the best terms. Finally, the third step involves paying certain *closing costs* in a process called *settlement* or **closing,** where the deal is finalized. The term **closing costs** refers to money exchanged at the settlement, above and beyond the down payment on the property. These costs may include an attorney's fee, the lender's administration fee, taxes, and points.

The first step in buying a home consists of finding a house you can afford and then coming to an agreement with the seller. A real estate agent can help you find the type of home you want for the money you can afford to pay. A great deal depends on the amount of down payment you can make. It is nice to look at something like Table 7.2, but in the real world the house you can afford and the down payment you can make are determined by your income. A useful rule of thumb in determining the monthly payment you can afford is given here:

1. Subtract any monthly bills (not paid off in the next 6 months) from your gross monthly income.
2. Multiply by 36%.

Your house payment should not exceed this amount. Another way of determining whether you can afford a house is to multiply your annual salary by 4; the purchase price should not exceed this amount. Today, the 36% factor is more common, and it is the one we will use in this book.

EXAMPLE 6 **Finding the Maximum House Payment**

Suppose that your monthly income is \$3,500 and your current monthly payments on bills total \$290. What is the maximum amount you should plan to spend for house payments?

Solution

Step 1: \$3,500 − \$290 = \$3,210

Step 2: \$3,210 × 0.36 = \$1,155.60

The maximum house payment should be \$1,155.60 per month. ■

In everyday life, the maximum loan we can obtain is important in deciding what price home to seek.

You now know how much you can afford (maximum house payment), but how does this relate to the amount of the loan? Finding the answer involves using Table II "backward."

Maximum Amount of a Loan

$$\text{MAXIMUM LOAN} = \frac{\text{MONTHLY PAYMENT YOU CAN AFFORD}}{\text{TABLE II ENTRY}} \times 1{,}000$$

EXAMPLE 7 **Finding the Maximum Loan**

Suppose that the maximum house payment you can afford is \$1,155.60 per month. What is the maximum loan you should seek if you will finance your purchase with an 8.5% 30-year loan?

Solution We find the entry in Table II corresponding to 8.5% over 30 years: 7.69. Then

$$\text{MAXIMUM LOAN} = \frac{1{,}155.60}{7.69} \times 1{,}000 \approx 150{,}273.08 \qquad \textit{Display:} \ 150273.0819$$

The maximum loan is \$150,273.08. ■

We conclude this section with two applied examples about buying a house.

EXAMPLE 8 **Financial Considerations in Buying a House**

Suppose that you want to purchase a home for \$110,000 with a 30-year mortgage at 8% interest. Suppose that you can put 20% down.

a. What is the amount of the down payment?
b. What is the amount to be financed?
c. What are the monthly payments?
d. What is the total amount of interest paid on this loan?
e. What is the necessary monthly income to be able to afford this loan?

Solution

a. DOWN PAYMENT = \$110,000 × 0.20 = **\$22,000**
b. AMOUNT TO BE FINANCED = \$110,000 × 0.80 = **\$88,000**
c. Use Table II; the entry for 8% on a 30-year loan is 7.34. Thus,

MONTHLY PAYMENTS = 88 × \$7.34 = **\$645.92**

d. The total interest paid can be found by subtracting the amount financed from the total paid:

$$\begin{aligned} \text{TOTAL PAID} &= (\text{NUMBER OF PAYMENTS})(\text{AMOUNT OF EACH PAYMENT}) \\ &= 360 \times \$645.92 \\ &= \$232{,}531.20 \end{aligned}$$

$$\begin{aligned} I &= A - P \\ &= \$232{,}531.20 - \$88{,}000 \\ &= \mathbf{\$144{,}531.20} \end{aligned}$$

Don't forget to subtract the amount of the loan and NOT the price of the house.

e. Remember, the amount of the house payment should be no more than 36% of the gross monthly income less monthly bills. Consequently, you must answer the question:

36% of WHAT NUMBER is \$645.92?

$$\begin{aligned} \frac{36}{100} &= \frac{645.92}{W} \\ 36W &= 64{,}592 \\ W &\approx 1{,}794.22 \end{aligned}$$

This means that the monthly salary remaining after monthly bills would need to be about **\$1,794.22.** This is an annual salary (assuming no other monthly bills) of about \$21,500. ■

EXAMPLE 9 **Finding the Necessary Down Payment for a Home Loan**

Suppose that your gross monthly salary is \$2,500 and your spouse's gross salary is \$2,000 per month. Your monthly bills are \$800. The home you wish to purchase costs \$178,000, and the loan is a 10% 30-year loan. How much down payment (rounded to the nearest hundred dollars) is necessary for you to be able to afford this home?

Solution First, determine the monthly payment you can afford:

$$\$2,500 + \$2,000 - \$800 = \$3,700$$

You can afford 36% of this for the house payment:

$$0.36(\$3,700) = \$1,332 \quad \text{This is what you can afford.}$$

Next, determine the Table II entry for 10% over 30 years: 8.78. Then

$$\text{MAXIMUM AMOUNT OF LOAN} = \frac{1,332}{8.78} \times 1,000$$
$$\approx 151,708.43$$

Since the house costs \$178,000, it would be necessary to make a down payment of \$178,000 − \$151,708 = \$26,292. **The necessary down payment is \$26,300.** (This is about a 15% down payment.) ■

PROBLEM SET 7.6

Essential Ideas

1. What factors are important when looking for a home loan?

2. What are points on a home loan?

3. What factors influence the amount of monthly payment for a home loan?

4. What is the rule of thumb for determining the monthly payment you can afford?

5. **In Your Own Words** Why would you use a formula for comparing interest rates on home loans?

6. **In Your Own Words** How do you find the monthly payment for a home loan?

7. **In Your Own Words** How do you find the maximum amount of home loan?

Level 1

Use estimation to select the best response in Problems 8–13. Do not calculate.

8. A \$98,000 30-year loan at 10% + 0.5 point + \$150 would have a fee for points of about
   A. \$10,000 B. \$500 C. \$650 D. \$850

9. A \$98,000 30-year loan at 10% + 0.5 point + \$150 would have total fees of about
   A. \$10,000 B. \$500 C. \$650 D. \$850

10. A \$98,000 30-year loan at 10% + 0.5 point + \$150 would have a down payment of about
    A. \$10,000 B. \$500 C. \$650 D. \$850

11. A \$98,000 30-year loan at 10% + 0.5 point + \$150 would have a monthly payment of about
    A. \$10,000 B. \$500 C. \$650 D. \$850

12. A \$98,000 30-year loan at 10% + 0.5 point + \$150 would have a comparison rate of
    A. 10% B. more than 10% C. less than 10%

13. A \$98,000 30-year loan at 10% + 0.5 point + \$150 would require what monthly income for the buyer?
    A. \$8,000 B. \$28,600 C. \$2,500 D. \$850

Level 1—Drill

Determine the maximum monthly payment for a house (to the nearest dollar), given the information in Problems 14–19.

14. Gross monthly income of \$985; current monthly payments of \$147

15. Gross monthly income of \$1,240; current monthly payments of \$215

16. Gross monthly income of \$1,480; current monthly payments of \$520

17. Gross monthly income of \$2,300; current monthly payments of \$350

18. Gross monthly income of \$2,800; current monthly payments of \$540

19. Gross monthly income of \$3,600; current monthly payments of \$370

Determine the down payment for each home described in Problems 20–25.

20. \$48,500; 5% down

21. \$69,900; 20% down

22. \$53,200; 10% down

23. \$64,350; 10% down

24. \$85,000; 20% down

25. \$112,000; 30% down

What are the bank charges for the points indicated in Problems 26–31?

26. \$48,500; 2 points

27. \$69,900; 3 points

28. \$53,200; 3 points

29. \$64,350; 5 points

30. \$85,000; $\frac{1}{2}$ point

31. \$112,000; $\frac{3}{4}$ point

For Problems 32–37, *determine the comparable interest rate (to two decimal places) for a* \$50,000 *loan when the quoted information is given.*

32. 11.5% + 2 pts + \$450

33. 7.25% + 3 pts + \$250

34. 8.7% + $\frac{1}{2}$ pt + \$250

35. 6.5% + $\frac{3}{4}$ pt + \$350

36. 9.3% + $\frac{3}{4}$ pt + \$150

37. 10.8% + $\frac{1}{2}$ pt + \$200

Use Table II to estimate the monthly payment for each loan described in Problems 38–51.

38. \$48,500; 30 years; 6%

39. \$48,500; 20 years; 8%

40. \$48,500; 15 years; 14%

41. \$48,500; 25 years; 14%

42. \$48,500; 30 years; 14%

43. \$85,000; 20 years; 10%

44. \$85,000; 25 years; 10%

45. \$85,000; 30 years; 10%

46. \$69,900; 15 years; $6\frac{1}{2}$%

47. \$69,900; 20 years; $6\frac{1}{2}$%

48. \$69,900; 30 years; $6\frac{1}{2}$%

49. \$112,000; 15 years; 7%

50. \$112,000; 25 years; 7%

51. \$112,000; 30 years; 7%

Level 2

WHAT IS WRONG, *if anything, with each of the statements in Problems* 52–57? *Explain your reasoning.*

52. For a \$120,000 loan with interest rate 8% and 30% down, the amount that can be saved monthly by changing the term from 30 years to 25 years is \$31.97, because the monthly payments change from \$648.33 to \$616.36. You can see this by looking at Table 7.2.

53. For a \$120,000 loan with interest rate 8% and 30% down, the total amount that can be saved by changing the term from 30 years to 25 years is a total of \$53,264, because the total amount of payments changes from \$221,890 to \$168,626. You can see this by looking at Table 7.2.

54. If you obtain a quotation for a home costing \$435,000, obtain a 20-year loan at 6% + 1.5 points + \$200, and make a 20% down payment, you can find the comparison rate as follows:

$$\text{COMPARISON RATE} = 0.06 + \left(0.015 + \frac{200}{348{,}000}\right)$$
$$= 0.0755747126$$

The comparison rate is 7.56%.

55. If you obtain a quotation for a home costing \$315,000 and obtain a 30-year loan at 5.5% + $\frac{3}{4}$ point + \$100, and make a 20% down payment, you can find the comparison rate as follows:

$$\text{COMPARISON RATE} = 5.5 + 0.125\left(0.75 + \frac{100}{252{,}000}\right)$$
$$\approx 5.59\%$$

56. If you obtain a quotation for a home costing \$435,000 and obtain a 20-year loan at 6% + 1.5 points + \$200, and make a 20% down payment, you can find the monthly payment as follows: \$435,000 × 0.8 = \$348,000; the Table II entry for a 6%, 20-year loan is 6.00, so the monthly payment is \$6.00 × 348 = \$2,088.

57. If you obtain a quotation for a home costing \$315,000, obtain a 30-year loan at 6.5% + $\frac{3}{4}$ point + \$100, and make a 20% down payment, you can find the total amount of interest paid on this loan as follows: \$315,000 × 0.8 = \$252,000; the Table II entry for a 6.5%, 30-year loan is 6.32, so the monthly payment is \$6.32 × 252 = \$1,592.64. Then,

$$I = A - P = \$1{,}592.64(360) - \$315{,}000 = \$258{,}350.40.$$

Level 2—Applications

In Problems 58–63, find the amount of the down payment (rounded to the nearest hundred dollars) necessary for the buyer to afford the monthly payments for the described home.

58. Monthly salary of \$2,900, with monthly payments of \$400; \$125,000 home with a 30-year 10% loan

59. Monthly salary of \$1,900, with monthly payments of \$250; \$89,000 home with a 30-year 9% loan

60. Monthly salary of \$4,900, with monthly payments of \$700; \$195,000 home with a 30-year 11% loan

61. Monthly salary of \$4,900, with monthly payments of \$700; \$195,000 home with a 30-year 9% loan

62. Monthly salary of \$3,500, with monthly payments of \$500; \$148,000 home with a 30-year 12% loan

63. Monthly salary of \$3,500, with monthly payments of \$500; \$225,000 home with a 30-year 10% loan

64. Suppose that you want to purchase a home for \$75,000, with a 30-year mortgage at 7% interest. Suppose that you can put 20% down.
 a. What is the amount of the down payment?
 b. What is the amount to be financed?
 c. What are the monthly payments?
 d. What is the total amount of interest paid on the 30-year loan?
 e. What is the necessary monthly income for you to be able to afford this home?

65. Suppose that you want to purchase a home for \$100,000, with a 20-year mortgage at 7.5% interest. Suppose that you can put 20% down.
 a. What is the amount of the down payment?
 b. What is the amount to be financed?

 c. What are the monthly payments?
 d. What is the total amount of interest paid on the 20-year loan?
 e. What is the necessary monthly income for you to be able to afford this home?

66. Suppose that you want to purchase a home for \$150,000, with a 30-year mortgage at 6% interest. Suppose that you can put 20% down.
 a. What is the amount of the down payment?
 b. What is the amount to be financed?
 c. What are the monthly payments?
 d. What is the total amount of interest paid on the 30-year loan?
 e. What is the necessary monthly income for you to be able to afford this home?

67. Suppose that you want to purchase a home for \$225,000, with a 20-year mortgage at 11% interest. Suppose that you can put 20% down.
 a. What is the amount of the down payment?
 b. What is the amount to be financed?
 c. What are the monthly payments?
 d. What is the total amount of interest paid on the 20-year loan?
 e. What is the necessary monthly income for you to be able to afford this home?

68. In this book we use 36% as a factor of net income in calculating the amount of the maximum monthly payment for a home loan. This is not the only criterion. A booklet from Fannie Mae called "A Guide to Home Ownership" computes the maximum monthly payment as follows:

Method 1 Calculate 33% of the gross monthly income.

Method 2 Calculate 38% of the gross monthly income. Subtract the monthly debt payments.

Use the lesser of (1) or (2) to compute the maximum monthly house payment. Use this criterion to assess the information given in Example 6.

69. Situation Dorothy and Wes have found the "home of their dreams," and it costs only $240,000. They have never purchased a home before—or anything else involving so much money—and they are very excited. There seems to be so much to do, and so many new words are being used by the realtor and the banker that they are also confused. If their combined salaries are $39,500 and if the current interest rate is 7.5% APR, how much of a down payment will they need to be able to afford their "dream home," and what is the monthly payment if the loan is financed for 30 years?

LEVEL 3—Team Project

70. Along with other members of your team, look in a local newspaper and select a home to purchase. Write a report on necessary income, costs, and monthly payments for various options when buying this home. You might consider some of the questions described in the SITUATION for this section.

7.7 SUMMARY AND REVIEW

Important Terms

Add-on interest [7.3]
Adjusted balance method [7.4]
Amortized loan [7.3]
Annual compounding [7.5]
Annual percentage rate [7.3]
APR [7.3; 7.6]
Average daily balance method [7.4]
Balloon payment [7.3]
Closed-end loan [7.3]
Closing [7.6]
Closing costs [7.6]
Comparison rate for home loans [7.6]
Complement [7.1]
Compound interest [7.5]
Compound interest formula [7.5]
Credit card [7.3; 7.4]
Daily compounding [7.5]
Dealer's cost [7.3]
Discount [7.1]
Down payment [7.6]
Exact interest [7.2]
Five-percent offer [7.3]
Foreclose [7.6]
Future value [7.2]
Grace period [7.4]
Inflation [7.5]
Installment loan [7.3]
Installments [7.3]
Interest [7.2]
Interest-only loan [7.3]
Interest rate [7.2; 7.6]
Length of loan [7.6]
Line of credit [7.3]
Monthly compounding [7.5]
Monthly payment [7.6]
Mortgage [7.6]
Open-end loan [7.3]
Ordinary interest [7.2]
Origination fee [7.6]
Percent markdown [7.1]
Points [7.6]
Present value [7.2; 7.5]
Previous balance method [7.4]
Principal [7.2]
Quarterly compounding [7.5]
Revolving credit [7.3]
Sale price [7.1]
Sales tax [7.1]
Semiannual compounding [7.5]
Sticker price [7.3]
Time (for a loan) [7.2]

Chapter Objectives

The material in this chapter is reviewed in the following list of objectives. A self-test (with answers and suggestions for additional study) is given. The self-test is constructed so that each problem number corresponds to a related objective. This self-test is followed by a practice test with the questions in mixed order.

[7.1]	*Objective 7.1*	Find the discount, given the original price and the percent markdown.
[7.1]	*Objective 7.2*	Find the complement of a given number.
[7.1]	*Objective 7.3*	Find the sales tax or price including tax.
[7.1]	*Objective 7.4*	Find the sale price, original price, or markdown, given two of these three values.
[7.2]	*Objective 7.5*	Find the future value, using simple interest.
[7.2]	*Objective 7.6*	Use the simple interest formula to find principal, rate, or time.
[7.3]	*Objective 7.7*	Make an offer for an automobile.
[7.3]	*Objective 7.8*	Find the amount of interest and the monthly payment for an installment loan using add-on interest.
[7.3]	*Objective 7.9*	Explain what APR means, and find it for a given loan.
[7.3]	*Objective 7.10*	Find the simple interest rate and APR, given the price and information about the installments.
[7.4]	*Objective 7.11*	Convert credit card rates to APRs.
[7.4]	*Objective 7.12*	Calculate the monthly finance charge for credit card transactions, using the previous balance, adjusted balance, and average daily balance methods.
[7.5]	*Objective 7.13*	Find the future value, using compound interest.
[7.5]	*Objective 7.14*	Find the cost of an item, taking into account the effect of inflation.
[7.5]	*Objective 7.15*	Find the present value, using compound interest.
[7.6]	*Objective 7.16*	Determine the maximum monthly payment for a house.
[7.6]	*Objective 7.17*	Determine the down payment and charges for points for a home loan.
[7.6]	*Objective 7.18*	Determine the comparable interest rate (to two decimal places) for a given bank loan.
[7.6]	*Objective 7.19*	Estimate the monthly payment for a home loan.
[7.6]	*Objective 7.20*	Find the amount of down payment necessary to afford a given home.
[7.1–7.6]	*Objective 7.21*	Estimate answers to financial problems.
[7.1–7.6]	*Objective 7.22*	Work applied interest problems.

Self-Test

Each question of this self-test is related to the corresponding objective listed above.

1. If an item is priced at $286 and the discount is 35%, what is the amount of the discount?

2. Find the complement of each given number.
a. 0.73 **b.** 38% **c.** $\frac{1}{8}$

3. What is the price, including tax, for a set of four tires that cost $89.95 each, if the sales tax is $5\frac{1}{2}$%?

4. a. If a billiard table you have been wanting costs $3,500 but now is on sale for 20% OFF, what is the sale price for the billiard table?

b. If the sale price of an item is $384 and the sale is 20% OFF the regular price, what is the regular price?

c. The sale price of a golf cart is $1,365, and the original price is $1,950. What is the percent markdown?

5. What is the future value for a deposit of $35,500 at 8% simple interest for 6 years?

6. Fill in the blanks.

Interest	*Principal*	*Rate*	*Time*
a. $560	_____	8%	2 years
b. $819	$1,950	_____	6 years
c. $510	$ 850	15%	_____

7. Suppose that the car you wish to purchase has a sticker price of $22,730 with a dealer cost of $18,579. Make a 5% offer for this car (rounded to the nearest hundred dollars).

8. Suppose that the amount to be financed for a car purchase is $13,500 at an add-on interest rate of 2.9% for 2 years. What are the monthly installment and the amount of interest that you will pay?

9. Find the APR for the loan described in Problem 8.

10. If a car with a cash price of $11,450 is offered for nothing down with 48 monthly payments of $353.04, what is the APR?

11. If a credit card has a daily rate (365-day year) of 0.0547945%, what is the APR (rounded to the nearest tenth of a percent)?

12. From the consumer's point of view, which method of calculating interest on a credit card is most advantageous? Illustrate the three types of calculating interest for a purchase of $525 with 9% APR for a 31-day month in which it takes 7 days for your $100 payment to be received and recorded.

13. What is the future value of a deposit of $855 at 9% for 3 years compounded semiannually?

14. If you purchase a home today for $185,000, what would you expect it to be worth in 30 years if you assume an inflation rate of 8%?

15. Suppose that you want to have $1,000,000 in 50 years. To achieve this goal, how much do you need to deposit today if you can earn 9% interest compounded monthly?

16. What is the maximum monthly payment (to the nearest dollar) you can afford for a house if your gross monthly income is $3,500 and you have current monthly payments of $1,400?

17. What is a 30% down payment for a house costing $220,000, and what amount must be paid for $1\frac{1}{2}$ points?

18. What is the comparison rate (rounded to the nearest tenth of a percent) for a home loan of 11% with fees of 3 points and $350? Assume that the loan value is $154,000.

19. What is the monthly payment for a home loan of $154,000 if the rate is 11% and the time is 20 years?

20. If you have a gross monthly income of $3,500 (with no outstanding bills) and the home you wish to purchase costs $154,000, what is the necessary down payment (rounded to the nearest hundred dollars) if the loan is for 20 years at 11%?

21. Estimate the minimum amount of money necessary in a retirement account at the time of retirement. (Pick the best response.)
A. $100,000 B. $1,000,000 C. $100,000,000

22. Suppose that you expect to receive a $100,000 inheritance when you reach 21 in 3 years and 4 months. What is the present value of your inheritance if the current interest rate is 6.4% compounded monthly?

STUDY HINTS *Compare your solutions and answers to the self-test. For each problem you missed, work some additional problems in the section listed in the margin. After you have worked these problems, you can test yourself with the practice test.*

Complete Solutions to the Self-Test

Additional Problems

[7.1] *Problems 5–14*

1. DISCOUNT = ORIGINAL PRICE × PERCENT MARKDOWN
= 286 × 0.35
= 100.1
The discount is $100.10.

[7.1] *Problems 15–20*

2. a. $1 - 0.73 = 0.27$ **b.** $100\% - 38\% = 62\%$ **c.** $1 - \frac{1}{8} = \frac{7}{8}$

[7.1] *Problems 33–50*

3. TOTAL PRICE = ORIGINAL PRICE × (TAX RATE + 1)
= 4(89.95)(1 + 0.055)
= 379.589
The total price of the tires is $379.59.

[7.1] *Problems 51–64*

4. a. SALE PRICE = ORIGINAL PRICE × COMPLEMENT
= 3,500 × 0.80
= 2,800
The sale price is $2,800.

b. ORIGINAL PRICE = SALE PRICE ÷ COMPLEMENT
= 384 ÷ 0.80
= 480
The original price is $480.

c. COMPLEMENT = SALE PRICE ÷ ORIGINAL PRICE
= 1,365 ÷ 1,950
= 0.7
The percent markdown is 30%.

[7.2] *Problems 15–44*

5. $A = P(1 + rt) = 35{,}500(1 + 0.08 \times 6) = 52{,}540$
The future value is $52,540.

[7.2] *Problems 45–50*

6. a.
$I = Prt$ — Simple interest formula
$560 = P(0.08)(2)$ — Substitute values.
$280 = 0.08P$ — Divide both sides by 2.
$3{,}500 = P$ — Divide both sides by 0.08.
The principal is $3,500.

b. $I = Prt$ — Simple interest formula
$819 = 1{,}950(r)(6)$ — Substitute values.
$136.5 = 1{,}950r$ — Divide both sides by 6.
$0.07 = r$ — Divide both sides by 1,950.
The rate is 7%.

c. $I = Prt$ — Simple interest formula
$510 = 850(0.15)t$ — Substitute values.
$510 = 127.5t$ — Multiply.
$4 = t$ — Divide both sides by 127.5.
The time is 4 years.

[7.3] *Problems 17–22*

7. $18{,}579(1 + 0.05) = \$19{,}507.95$
You should offer $19,500 for the car.

[7.3] *Problems 23–32*

8. $I = Prt = 13{,}500(0.029)(2) = 783$
$A = P + I = 13{,}500 + 783 = 14{,}283$
Monthly payment is $\$14{,}283 \div 24 = \$595.125$
The total interest is $783, and the monthly payment is $595.13.

[7.3] *Problems 2; 33–42*

9. $\text{APR} = \dfrac{2Nr}{N+1} = \dfrac{2(24)(0.029)}{25} = 0.05568$
The APR is 5.568%.

[7.3] *Problems 48–60*

10. $A = 48(353.04) = 16{,}945.92$
$I = A - P = 16{,}945.92 - 11{,}450.00 = 5{,}495.92$
Also,

$I = Prt$
$5{,}495.92 = 11{,}450(r)(4)$ — 48 months is 4 years, so $t = 4$.
$1{,}373.98 = 11{,}450r$ — Divide both sides by 4.
$0.12 \approx r$ — Divide both sides by 11,450.

Finally,

$$\text{APR} = \frac{2Nr}{N+1} = \frac{2(48)r}{49} \approx 0.235$$ *Display:* 0.2350986187

The APR is about 23.5%.

[7.4] *Problems 13–18*

11. $\text{APR} = 0.0547945\% \times 365 = 19.999993\%$
The APR is about 20.0%.

[7.4] *Problems 19–33*

12. The adjusted balance method is most advantageous to the consumer.

PREVIOUS BALANCE METHOD:
$I = Prt = 525(0.09)(\frac{1}{12}) = 3.9375$
The finance charge is $3.94.

ADJUSTED BALANCE METHOD:
$I = Prt = (525 - 100)(0.09)(\frac{1}{12}) = 3.1875$
The finance charge is $3.19.

AVERAGE DAILY BALANCE METHOD:

Average balance: $(525 \times 7 + 425 \times 24) \div 31 \approx 447.58$

$I = Prt = P(0.09)(\frac{31}{365}) \approx 3.42$

The finance charge is \$3.42. *Display:* 3.421232877

[7.5] *Problems 13–48; 53–54*

13. By calculator: $A = P(1 + i)^N = 855\left(1 + \frac{0.09}{2}\right)^6 \approx 1{,}113.432407$

The future value is \$1,113.43.

[7.5] *Problems 49–52; 55*

14. For inflation, use future value where $n = 1$; for this problem, we have $P = 185{,}000$, $t = 30$, and $r = 0.08$.
By calculator, $185{,}000(1.08)^{30} \approx 1{,}861{,}591.52$.
The future value is \$1,861,591.52.

[7.5] *Problems 56–58*

15. $A = \$1{,}000{,}000$; $t = 50$; $r = 0.09$; $n = 12$; $N = nt = 50(12) = 600$; $i = r/n = 0.09/12$.
Thus,

$$P = A \div (1 + i)^N$$
$$= 1{,}000{,}000 \div \left(1 + \frac{0.09}{12}\right)^{600}$$
$$\approx 11{,}297.10$$

Display: 11297.10362

Deposit \$11,297.10 to have a million dollars in 50 years.

[7.6] *Problems 14–19*

16. $0.36(3{,}500 - 1{,}400) = 756$
The maximum monthly payment is \$756.

[7.6] *Problems 20–31*

17. DOWN PAYMENT: $0.30(\$220{,}000) = \$66{,}000$
POINTS: $0.015(154{,}000) = 2{,}310$
The fee for points is \$2,310.

[7.6] *Problems 32–37*

18. COMPARISON RATE $= 0.11 + 0.125\left(0.03 + \frac{350}{154{,}000}\right)$
Display: 0.1140340909
The comparison rate is 11.4%.

[7.6] *Problems 38–51*

19. MONTHLY PAYMENT $= \frac{154{,}000}{1{,}000}$ (TABLE II ENTRY)
$= 154(10.32)$
$= 1{,}589.28$

The monthly payment is \$1,589.28.

[7.6] *Problems 58–63*

20. AMOUNT YOU CAN AFFORD $= 0.36(3{,}500) = 1{,}260$
Table II entry is 10.32. Thus,

MAXIMUM LOAN $= \frac{1{,}260}{10.32} \times 1{,}000 \approx 122{,}093.02$

The necessary down payment is $154{,}000 - 122{,}093 = 31{,}907$.
The down payment should be \$31,900.

[7.1] *Problems 21–26*
[7.2] *Problems 5–14*
[7.3] *Problems 3–10*
[7.4] *Problems 3–12*
[7.6] *Problems 8–13*

21. Estimate 10% annual interest:
A. $\$100{,}000(0.1) = \$10{,}000$; seems too little
B. $\$1{,}000{,}000(0.1) = \$100{,}000$; seems too high, but remember that you need to take into account inflation between now and when you retire.
Estimate that B is the best choice.

[7.1] *Problems 51–70*
[7.2] *Problems 51–64*
[7.3] *Problems 17–63*
[7.4] *Problems 42–44*
[7.5] *Problems 49–64*
[7.6] *Problems 58–69*

22. $A = \$100{,}000$; $r = 0.064$; $t = 3\frac{4}{12}$; $n = 12$. Calculate:

$N = nt = 12\left(\frac{40}{12}\right) = 40 \qquad i = \frac{r}{n} = \frac{0.064}{12}$

$$P = A \div (1 + i)^N$$
$$= 100{,}000 \div \left(1 + \frac{0.064}{12}\right)^{40}$$
$$\approx 80{,}834.49$$

The present value is \$80,834.49. *Display:* 80834.48968

Chapter 7 Practice Test

1. **a.** Find a 5% discount on an item costing \$135.
b. Find a 30% discount on an item costing \$30.
c. Find a 20% discount on an item selling for \$89.95.
d. Find a 45% discount on an item selling for \$99.

2. Find the complement.
a. 0.18 **b.** $\frac{1}{7}$ **c.** 13% **d.** $33\frac{1}{3}\%$

3. Convert the credit card rates to APR rates (rounded to the nearest percent).
a. monthly rate of $1\frac{1}{2}\%$ **b.** monthly rate of $1\frac{1}{4}\%$
c. daily rate of 0.05753% **d.** daily rate of 0.0246575%

4. Assume that the month is 30 days long and that it takes 8 days to receive a payment. Suppose that a \$1,200 purchase is made and the current billing shows this charge. What is the finance charge in the next billing period if the credit card uses 20% APR?
a. \$100 payment; previous balance method
b. \$100 payment; adjusted balance method
c. \$100 payment; average daily balance method
d. \$1,100 payment; average daily balance method

5. Assume that \$2,500 is invested at 18% for five years. Find the future value.
a. simple interest **b.** compounded annually
c. compounded quarterly **d.** compounded monthly
e. compounded semiannually

6. How much must be deposited today to have \$100,000 in 30 years if it is invested at 12%, compounded daily?

7. Suppose that your present salary is \$22,000 per year. Predict what your salary will be (to the nearest thousand dollars) when you retire in 40 years, if you assume an annual rate of inflation as specified.
a. 8% **b.** 2% **c.** 12% **d.** 7%

8. Fill in the blanks.

	Interest	*Principal*	*Rate*	*Time*
a.	____	\$4,500	19%	3 years
b.	\$1,650	____	15%	2 years
c.	\$6,210	\$4,500	____	6 years
d.	\$456	\$600	19%	____

9. Determine the comparable interest rate (to two decimal places) for the described bank loans.
a. \$42,200 30-year loan: 10.5% + 3 points + \$500
b. \$82,000 20-year loan: 7% + 1 point + \$200
c. \$105,000 25-year loan: 13.2% + 4 points + \$100
d. \$160,000 30-year loan: 8.4% + $\frac{3}{4}$ point + \$150

10. What is the monthly payment for the described loans?
a. \$45,800 30-year loan: 10.5% + 3 points + \$500
b. \$45,800 20-year loan: 10.5% + 3 points + \$500
c. \$245,000 25-year loan: 7% + 1 point + \$200
d. \$245,000 30-year loan: 7% + 1 point + \$200

11. If a coat you have been wanting normally costs \$254, but the store now has a special 20% OFF sale, what is the sale price for this coat?

12. If a table has a regular price of \$120 but is marked $\frac{1}{3}$ OFF, what is the sale price?

13. If a dress is marked down from \$145 to \$87, what is the percent markdown?

14. If an item is marked $\frac{1}{4}$ OFF and is on sale for \$187.50, what is its regular price?

15. What is the tax on a suit that sells for \$250 if the tax rate is 6.5%?

16. If the price for a car is \$10,950 and the tax rate is 4.5%, what is the total amount, including tax?

17. What is the future value of \$12,000 at 17% simple interest for 149 days?

18. If a Ford Taurus LX has a sticker price of \$18,355 and a dealer cost of \$16,152.40, make a 7% offer.

19. Find the amount of interest, the monthly payment, and the APR (rounded to the nearest tenth of a percent) for a bedroom set costing \$2,350 with 18% add-on interest for 3 years.

20. Suppose a washer and dryer are advertised at a cash price of \$895.95 or \$39 per month for 36 months. What is the APR for the financed price (rounded to the nearest tenth of a percent)?

21. If a car with a cash price of \$15,450 is offered for nothing down with 60 monthly payments of \$360.50, what is the APR (rounded to the nearest tenth of a percent)?

22. What is the present value of an investment worth \$5,000 in 3 years with an 8% rate compounded quarterly?

23. Determine the maximum affordable monthly payment for a house (to the nearest dollar) for someone with a gross monthly income of \$4,580 and current monthly payments of \$1,200.

24. What is a 10% down payment for a condominium that costs \$82,000?

25. What are the fees for a \$205,000 house purchased with a loan with terms of 30% down, 6% + $\frac{1}{2}$ point + \$250?

26. If you decide to buy your newborn daughter a \$1,000 long-term bond that pays 18% compounded quarterly, how much will your child have on her 20th birthday?

27. An insurance agent offers you a policy that will provide you with $20,000 in 30 years. What is the present value of this policy if you assume an inflation rate of 7%?

28. Suppose that you purchase a $1,000 3-year certificate paying 4.58% simple interest. How much will you have when the certificate matures?

29. To make a $45 per month payment, you decide to deposit a lump sum into an account paying 12% interest. How much must you deposit to make the monthly payments from the interest on your deposit?

30. If you have a gross monthly income of $2,850 with no monthly bills and the home you wish to purchase costs $175,000, what is the necessary down payment (rounded to the nearest hundred dollars) for the indicated loans?
a. 25 years at 8%
b. 25 years at 7%
c. 30 years at 7%

CHAPTER EIGHT

SETS AND LOGIC

8.1 INTRODUCTION TO SETS

Situation

"What is there in a vacuum to make one afraid?" said the flea. "There is nothing in it," I said, "and that is what makes one afraid to contemplate it. A person can't think of a place with nothing at all in it without going nutty, and if he tries to think that nothing is something after all, he gets nuttier." "But that is only part of it," retorted the flea. "What about the infinity of the universe?" "Infinity!" said I. "Infinity is an expression of your imagination." Is the infinite plausible? Can you contemplate the vastness of infinity?*

This section lays the groundwork for some fundamental concepts in mathematics. Sets are a unifying idea and will lead us to a means of discussing not only the meaning of something with nothing at all in it, but also the meaning of something that is infinite.

DENOTING SETS

A fundamental concept in mathematics—and in life, for that matter—is the sorting of objects into similar groupings. Every language has an abundance of words that mean "a collection" or "a grouping." For example, we speak of a *herd* of cattle, a *flock* of birds, a *school* of fish, a track *team*, a stamp *collection,* and a *set* of dishes. All these grouping words serve the same purpose, and in mathematics we use the word *set* to refer to any collection of objects. Since the beginning of this book, we have assigned problem *sets*; and in Section 1.2, we defined the *sets* of natural numbers, counting numbers, and whole numbers.

In mathematics we do not define the word **set.** It is what, in mathematics, is called an **undefined term.** It is impossible to define all terms, because every defi-

*The first part is quoted from Don Marquis's *Archy and Mehitabel* (The Merry Flea). The second part is presented with apologies to Mr. Marquis.

nition requires other terms, so some *undefined terms* are necessary to get us started. To illustrate this idea, let's try to define the word *set* by using dictionary definitions:

"*Set*: a *collection* of objects." What is a collection?

"*Collection*: an *accumulation*." What is an accumulation?

"*Accumulation*: a *collection,* a *pile,* or a *heap.*" We see that the word *collection* gives us a **circular definition.** What is a *pile*?

"*Pile*: a *heap.*" What is a heap?

"*Heap*: a *pile.*"

Do you see that a dictionary leads us in circles? In mathematics, we do not allow circular definitions, and this forces us to accept some words without definition. The term *set* is undefined. Remember, the fact that we do not define set does not prevent us from having an intuitive grasp of how to use the word.

Sets are usually specified in one of two ways. The first is by *description,* and the other is by the *roster* method. In the **description method,** we specify the set by describing it in such a way that we know exactly which elements belong to it. An example is the set of 50 states in the United States of America. We say that this set is **well defined,** since there is no doubt that the state of California belongs to it and that the state of Germany does not; neither does the state of confusion. Lack of confusion, in fact, is necessary in using sets. The distinctive property that determines the inclusion or exclusion of a particular element is called the *defining property* of the set.

Consider the example of *the set of good students in this class.* This set is not well defined, since it is a matter of opinion whether a student is a "good" student. If we agree, however, on the meaning of the words *good students,* then the set is said to be *well defined.* A better (and more precise) formulation is usually required—for example, *the set of all students in this class who received a C or better on the first examination.* This is well defined, since it can be clearly determined exactly which students received a C or better on the first test.

In the **roster method,** the set is defined by listing the members. The objects in a set are called **members** or **elements** of the set and are said to **belong to** or **be contained** in the set. For example, instead of defining a set as *the set of all students in this class who received a C or better on the first examination,* we might simply define the set by listing its members: {Howie, Mary, Larry}.

Sets are usually denoted by capital letters, and the notation used for sets is braces. Thus, the expression

$$A = \{4, 5, 6\}$$

means that A is the name for the set whose members are the numbers 4, 5, and 6.

Sometimes we use braces with a defining property, as in the following examples:

{states in the United States of America}

{all students in this class who received an A on the first test}

EXAMPLE 1 **Changing a Description Definition to a Roster Definition**

Specify the given sets by roster. If the set is not well defined, say so.

a. {counting numbers between 10 and 20}
b. {distinct letters in the word *happy*}
c. {counting numbers greater than 1,000}
d. {U.S. presidents arranged in chronological order}
e. {good U.S. presidents}

Solution

a. {11, 12, 13, 14, 15, 16, 17, 18, 19}
Notice that *between* does not include the first and last numbers.
b. {*h, a, p, y*}
c. {1001, 1002, 1003, . . .}
Notice that it is sometimes impossible or impractical to write *all* the elements of a particular set using the roster method. We use *three dots* to indicate that some elements have been omitted. You must be careful, however, to list enough elements so that someone looking at the set can see the intended pattern.
d. {Washington, Adams, Jefferson, . . . , Reagan, Bush, Clinton}
e. Not well defined ■

EXAMPLE 2 **Changing a Roster Definition to a Description Definition**

Specify the given sets by description.

a. $\{1, 2, 3, 4, 5, \ldots\}$
b. $\{0, 1, 2, 3, 4, 5, \ldots\}$
c. $\{\ldots, -3, -2, -1, 0, 1, 2, 3, \ldots\}$
d. $\{12, 14, 16, \ldots, 98\}$
e. $\{4, 44, 444, 4444, \ldots\}$
f. $\{m, a, t, h, e, i, c, s\}$

Solution Answers may vary.

a. Counting (or natural) numbers
b. Whole numbers
c. Integers
d. {Even numbers between 10 and 100}
e. {Counting numbers whose digits consist of fours only}
f. {Distinct letters in the word *mathematics*} ■

If we try to list the set of rational numbers by roster, we will find that this is a difficult task (see Problem 50 in Problem Set 8.1). A new notation called **set-builder notation** was invented to allow us to combine both the roster and the description methods. Consider:

The set of all x

$\{x \mid x \text{ is an even counting number}\}$

such that

We now use this notation for the set of rational numbers:

$\{\frac{a}{b} \mid a \text{ is an integer and } b \text{ is a nonzero integer}\}$

Read this as: "The set of all $\frac{a}{b}$ such that a is an integer and b is a nonzero integer."

We can further shorten this notation. In Chapter 2, we gave names to the common sets of numbers:

$\mathbb{N} = \{1, 2, 3, 4, \ldots\}$	**Set of natural, or counting, numbers**
$\mathbb{W} = \{0, 1, 2, 3, 4 \ldots\}$	**Set of whole numbers**
$\mathbb{I} = \{\ldots, -2, -1, 0, 1, 2, \ldots\}$	**Set of integers**
$\mathbb{Q} = \{\frac{a}{b} \mid a \in \mathbb{I}, b \in \mathbb{I}, b \neq 0\}$	**Set of rational numbers**

Notice that we used a new symbol in the set-builder notation for the set $\mathbb{Q}$. If S is a set, we write $a \in S$ if a is a member of the set S, and we write $b \notin S$ if b is not a member of the set S. Thus, "$a \in \mathbb{I}$" means that the variable a is an integer, and the statement "$b \in \mathbb{I}, b \neq 0$" means that the variable b is a nonzero integer.

EXAMPLE 3

Using Set Inclusion Notation

Let C = cities in California; a = city of Anaheim, b = city of Berlin. Use set membership notation to describe relations among a, b, and C.

Solution $\boldsymbol{a \in C; b \notin C}$ ■

EQUAL AND EQUIVALENT SETS

We say that two sets are **equal** if they contain exactly the same elements. Thus, if $E = \{2, 4, 6, 8, \ldots\}$,

$$\{x \mid x \text{ is an even counting number}\} = \{x \mid x \in E\}$$

The order in which you represent elements in a set has no effect on set membership. Thus,

$$\{1, 2, 3\} = \{3, 1, 2\} = \{2, 1, 3\} = \cdots$$

Also, if an element appears in a set more than once, it is not generally listed more than a single time. For example,

$$\{1, 2, 3, 3\} = \{1, 2, 3\}$$

Another possible relationship between sets is that of *equivalence.* Two sets are **equivalent** if they have the same *number* of elements. Don't confuse this concept with equality. Equivalent sets do not need to be equal sets, but equal sets are always equivalent.

EXAMPLE 4 **Recognizing Equivalent Sets**

Which of the following sets are equivalent? Are any equal?
{○, △, □}, {5, 8, 11}, {1, ⌜, ⊓}, {•, ⊙, ★}, {1, 2, 3}

Solution **All of the given sets are equivalent.** Notice that no two of them are equal, but all share the property of "threeness." ■

The number of elements in a set is often called its **cardinality.** The cardinality of the sets in Example 4 is 3; that is, the common property of the sets is the **cardinal number** of the set. Equivalent sets with four elements each have in common the property of "fourness," and thus we would say that their cardinality is 4.

EXAMPLE 5 **Finding the Cardinality of a Given Set**

Find the cardinality of

a. $R = \{5, \triangle, \Upsilon, \pi\}$ **b.** $S = \{\ \}$
c. $T = \{\text{states of the United States}\}$

Solution
a. The cardinality of R is **4.**
b. The cardinality of S (the empty set) is **0.**
c. The cardinality of T is **50.** ■

Certain sets such as $\mathbb{N}$, $\mathbb{I}$, $\mathbb{W}$, or $A = \{1000, 2000, 3000, \ldots\}$ have a common property. We call these *infinite sets.* If the cardinality of a set is 0 or a counting number, we say it is **finite.** Otherwise, we say it is **infinite.** We can also say that a set is finite if it has a cardinality less than some counting number, even though we may not know its precise cardinality. For example, we can safely assert that the set of students attending the University of Hawaii is finite even though we may not know its cardinality, because the cardinality is certainly less than a million.

You use the ideas of cardinality and of equivalent sets every time you count something, even though you don't use the term *cardinality.* For example, a set of sheep is counted by finding a set of counting numbers that has the same cardinality as the set of sheep. These equivalent sets are found by using an idea called *one-to-one correspondence.* If the set of sheep is placed into a one-to-one correspondence with a set of pebbles, the set of pebbles can then be used to represent the set of sheep.

One-to-One Correspondence

Two sets A and B are said to be in a **one-to-one correspondence** if we can find a pairing so that

1. Each element of A is paired with precisely one element of B;

and

2. Each element of B is paired with precisely one element of A.

EXAMPLE 6 **Showing That Sets Are Equivalent**

Show which of the following sets are equivalent.

$M = \{3\}, \quad N = \{\text{three}\}, \quad P = \{\triangle, \square, \circ, \triangledown\}, \quad Q = \{t, h, r, e\}$

Solution Notice that

$$\begin{array}{ccc} M = \{3\} & & P = \{\triangle, \square, \circ, \triangledown\}, \\ \updownarrow & & \quad\ \ \updownarrow\ \ \updownarrow\ \ \updownarrow\ \ \updownarrow \\ N = \{\text{three}\} & & Q = \{\,t,\ \ h,\ \ r,\ \ e\} \end{array}$$

so we say "M is equivalent to N" and we write $\boldsymbol{M \leftrightarrow N}$. Note that $\boldsymbol{P \leftrightarrow Q}$. We also see that $M = N$. ■

Since infinity is not a number, we cannot correctly say that the cardinality of the counting numbers is infinity. In the late 18th century, Georg Cantor assigned a cardinal number $\aleph_0$ (pronounced "aleph-null") to the set of counting numbers. That is, $\aleph_0$ is the cardinality of the set of counting numbers

$\mathbb{N} = \{1, 2, 3, 4, \ldots\}$

The set

$E = \{2, 4, 6, 8, \ldots\}$

also has cardinality $\aleph_0$, since it can be put into a one-to-one correspondence with set $\mathbb{N}$:

$$\begin{array}{l} \mathbb{N} = \{1,\ 2,\ 3,\ 4,\ \ldots,\ n,\ \ldots\} \\ \quad\ \ \ \updownarrow\ \ \updownarrow\ \ \updownarrow\ \ \updownarrow \qquad\ \ \updownarrow \\ E = \{2,\ 4,\ 6,\ 8,\ \ldots,\ 2n,\ \ldots\} \end{array}$$

Historical Perspective

Georg Cantor (1845–1918)

The concept of sets and what we call *set theory* is attributed to Georg Cantor, a great German mathematician. Cantor's ideas were not adopted as fundamental underlying concepts in mathematics until the 20th century, but today the idea of a set is used to unify and explain nearly every other concept in mathematics. Cantor believed that "The essence of mathematics lies in its freedom," but in the final analysis he was not able to withstand the rejection of his ideas. He underwent a series of mental breakdowns and died in a mental hospital in 1918.

UNIVERSAL AND EMPTY SETS

We conclude this section by considering two special sets. The first is the set that contains every element under consideration, and the second is the set that contains no elements. A **universal set,** denoted by U, contains all the elements under consideration in a given discussion; and the **empty set** contains no elements, and thus has cardinality 0. The empty set is denoted by { } or ∅. Do not confuse the notations ∅, 0, and {∅}. The symbol ∅ denotes a *set* with no elements; the symbol 0 denotes a *number*; and the symbol {∅} is a set with one element (namely, the set containing ∅).

For example, if $U = \{1, 2, 3, 4, 5, 6, 7, 8, 9\}$, then all sets we would be considering would have elements only among the elements of U. No set could contain the number 10, since 10 is not in that agreed-upon universe.

For every problem, a universal set must be specified or implied, and it must remain fixed for that problem. However, when a new problem is begun, a new universal set can be specified.

Notice that we defined *a* universal set and *the* empty set; that is, a universal set may vary from problem to problem, but there is only one empty set. After all, it doesn't matter whether the empty set contains no numbers or no people—it is still empty. The following are examples of descriptions of the empty set:

{living saber-toothed tigers}   {counting numbers less than 1}

PROBLEM SET 8.1

Essential Ideas

1. **In Your Own Words** Why do you think mathematics accepts the word *set* as an undefined term?
2. **In Your Own Words** Distinguish between equal and equivalent sets.
3. **In Your Own Words** What is a universal set?
4. **In Your Own Words** What is the empty set?
5. **In Your Own Words** Give three examples of the empty set.
6. **In Your Own Words** Give three examples of an infinite set.

Level 1

WHAT IS WRONG, *if anything, with each of the statements in Problems 7–12? Explain your reasoning.*

7. A set is defined to be a collection of objects.
8. The empty set does not have any elements, so $\varnothing = 0$.
9. The cardinality of {0} is zero.
10. $\{0\} = \varnothing$
11. The number of grains of sand on all the beaches on earth can't be counted, so the set of grains of sand on beaches is infinite.
12. The set of counting numbers has a cardinality that is infinite.

Level 1—Drill

Tell whether each set in Problems 13–20 is well defined. If it is not well defined, change it so that it is well defined.

13. The set of students attending the University of California yes
14. {Grains of sand on earth}
15. The set of counting numbers between 3 and 4 yes
16. The set of happy people in your country
17. The set of people with pointed ears no

18. {Counting numbers less than 0}

19. {Good bets on the next race at Hialeah}

20. {Years that will be bumper years for growing corn in Iowa}

Specify the sets in Problems 21–28 by roster.

21. {Distinct letters in the word *mathematics*}

22. {Current U.S. president}

23. {Odd counting numbers less than 11}

24. {Positive multiples of 3}

25. {Distinct letters in the word *pipe*}

26. {Counting numbers greater than 150}

27. {Even counting numbers between 5 and 15}

28. {Counting numbers containing only 1s}

Specify the sets in Problems 29–34 by description

29. {1, 2, 3, 4, 5, 6, 7, 8, 9}

30. {1, 11, 121, 1331, 14641, . . .}

31. {10, 20, 30, . . ., 100} **32.** {50, 500, 5000, . . .}

33. {101, 103, 105, . . . 169} **34.** {m, i, s, p}

Write out in words the description of the sets given in Problems 35–40, and then list each set in roster form.

35. $\{x \mid x$ is an odd counting number$\}$

36. $\{x \mid x$ is a natural number between 1 and 10$\}$

37. $\{x \mid x \in \mathbb{N}, x \neq 8\}$

38. $\{x \mid x \in \mathbb{W}, x \leq 8\}$

39. $\{x \mid x \in \mathbb{W}, x < 8\}$

40. $\{x \mid x \in \mathbb{W}, x \notin E\}$ where $E = \{2, 4, 6, \ldots\}$

Level 2

41. In Your Own Words Give an example of a set
a. with cardinality 2.
b. with cardinality 10.

42. In Your Own Words Give an example of a set
a. with cardinality 0.
b. whose cardinality is larger than a million.

43. a. Which of the following sets can be placed into a one-to-one correspondence?

A = {distinct letters in the word *pipe*}
$B = \{4\}$
$C = \{p, i, e\}$
$D = \{2 + 1\}$
E = {three}
$F = \{3\}$

b. Which of the given sets are equal?

44. a. Which of the following sets can be placed into a one-to-one correspondence?

$A = \{16\}$
$B = \{10 + 6\}$
$C = \{10, 6\}$
$D = \{2^5\}$
$E = \{2, 5\}$

b. Which of the given sets are equal?

45. Show that there is more than one one-to-one correspondence between the sets {m, a, t} and {1, 2, 3}.

46. Show that the following sets have the same cardinality, by placing the elements into a one-to-one correspondence.

{1, 2, 3, 4, . . . , 999, 1000}
{7964, 7965, 7966, 7967, . . . , 8962, 8963}

47. Show that the following sets do not have the same cardinality.

{1, 2, 3, 4, . . . , 586, 587}
{550, 551, 552, 553, . . . , 902, 903}

48. Do the following sets have the same cardinality?

{48, 49, 50, . . . , 783, 784}
{485, 487, 489, . . . , 2053, 2055}

Level 2—Applications

49. Classify the following sets as finite or infinite.
  a. The set of stars in the Milky Way
  b. The set of counting numbers greater than one million
  c. The set of drops of water in all the oceans on earth
  d. The set of grains of sand on all the beaches on earth

50. List the set of rational numbers between 0 and 1 by roster. List the elements in order, and do not list elements that have been previously listed.

51. **Situation** "What is there in a vacuum to make one afraid?" said the flea. "There is nothing in it," I said, "and that is what makes one afraid to contemplate it. A person can't think of a place with nothing at all in it without going nutty, and if he tries to think that nothing is something after all, he gets nuttier." "But that is only part of it," retorted the flea. "What about the infinity of the universe?" "Infinity!" said I. "Infinity is an expression of your imagination." Is the infinite plausible? Can you contemplate the vastness of infinity?

52. **In Your Own Words** Write a short paper comparing the SITUATION with your own situation. (That is, can you think of *any* relationship between the idea of infinity and something that relates to your life?)

8.2 SUBSETS

Situation

"Niels, where are you going?" asked Sammy. "I'm going to the barber for a shave," answered Niels. "You know, Niels, Darryl, the barber, does a very good business. He shaves those men and only those men who do not shave themselves." "Come on, Sammy, that's impossible! Who do you think shaves him?"

The study of sets and relationships among sets can lead us to a consideration of questions such as this. Consider the following *Barber's Rule: Darryl shaves those men and only those men who do not shave themselves.*

If he shaves himself, according to the Barber's Rule, he does not shave himself. On the other hand, if he does not shave himself, then, according to the Barber's Rule, he shaves himself!

SUBSETS

To deal effectively with sets, you must understand certain relationships among sets. A set A is a **subset** of a set B, denoted by $A \subseteq B$, if every element of A is an element of B. Consider the following sets:

$$U = \{1, 2, 3, 4, 5, 6, 7, 8, 9\}$$
$$A = \{2, 4, 6, 8\} \quad B = \{1, 3, 5, 7\} \quad C = \{5, 7\}$$

Now, A, B, and C are subsets of the universal set U (*all* sets we consider are subsets of the universe, by definition of a universal set). We also note that $C \subseteq B$ since

every element of C is also an element of B. However, C is not a subset of A, written $C \not\subseteq A$. In this case, we merely note that $5 \in C$ and $5 \notin A$.

Do not confuse the notions of "element" and "subset." In the present example, 5 is an *element* of C, since it is listed in C. By the same token, $\{5\}$ is a *subset* of C, but it is not an element, since we do not find $\{5\}$ contained in C; if we did, C might look like this: $\{5, \{5\}, 7\}$.

Consider all possible subsets of $C = \{5, 7\}$:

$\{5\}$, $\{7\}$ are obvious subsets.

$\{5, 7\}$ is a also a subset, since both 5 and 7 are elements of C.

$\{\quad\}$ is also a subset of C. It is a subset because all of its elements belong to C. Stated a different way, if it were not a subset of C, we would have to be able to find an element of $\{\quad\}$ that is not in C. Since we cannot find such an element, we say that the empty set is a subset of C (and in fact, the *empty set is a subset of every set*).

We see that there are four subsets of C, even though C has only two elements. We must therefore be careful to distinguish between a subset and an element. Remember, 5 and $\{5\}$ mean different things.

The subsets of C can be classified into two categories: *proper* and *improper*. Since every set is a subset of itself, we immediately know one subset for *any* given set: the set itself. A *proper subset* is a subset that is not equal to the original set; that is, A is a **proper subset** of a set B, written $A \subset B$, if A is a subset of B and $A \neq B$. An **improper subset** of a set A is the set A.

We see there are three proper subsets of $C = \{5, 7\}$: $\varnothing$, $\{5\}$, and $\{7\}$. There is one improper subset of C: $\{5, 7\}$.

EXAMPLE 1

Classifying Proper and Improper Subsets

Find the proper and improper subsets of $A = \{2, 4, 6, 8\}$. What is the cardinality of A?

Solution

The cardinality of A is 4 (because there are 4 elements in A). There is one improper subset: $\{2, 4, 6, 8\}$. The proper subsets are:

$\{\quad\}$,
$\{2\}, \{4\}, \{6\}, \{8\}$,
$\{2, 4\}, \{2, 6\}, \{2, 8\}, \{4, 6\}$ $\{4, 8\}, \{6, 8\}$,
$\{2, 4, 6\}, \{2, 4, 8\}, \{2, 6, 8\}, \{4, 6, 8\}$ ■

VENN DIAGRAMS

A useful way to depict relationships among sets is to let the universal set be represented by a rectangle, with the proper sets in the universe represented by circular or oval regions, as shown in Figure 8.1 (page 368). These figures are called **Venn diagrams,** after John Venn (1834–1923).

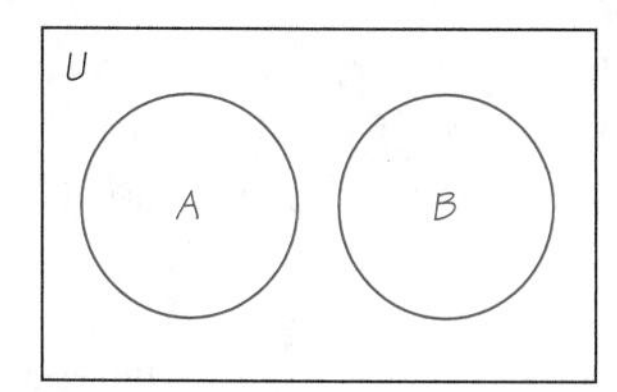

a. Disjoint sets

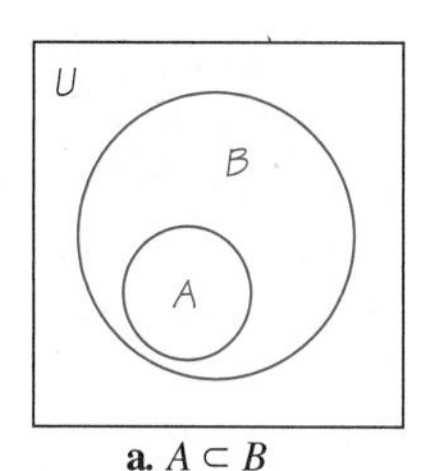

a. $A \subset B$

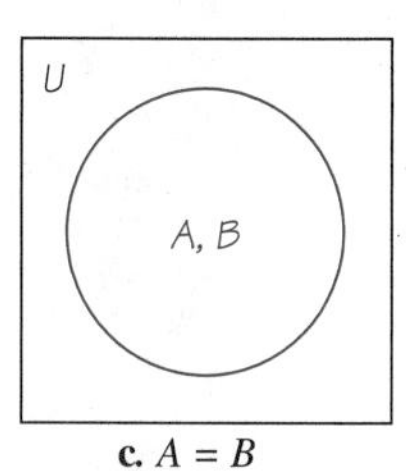

b. $B \subset A$

c. $A = B$

Figure 8.1 Venn diagrams for subset

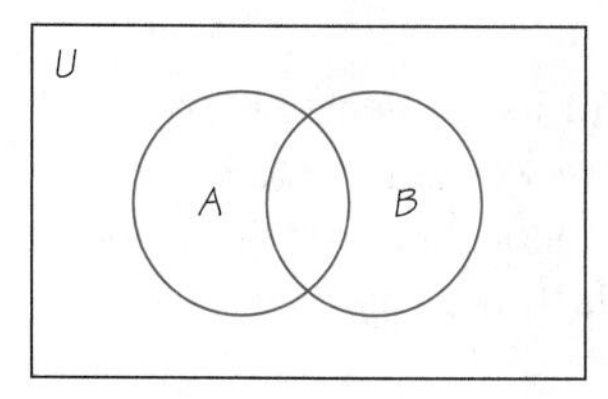

b. Overlapping sets

Figure 8.2 Set relationships

We can also illustrate other relationships between two sets: A and B may have no elements in common, in which case they are **disjoint** (as depicted by Figure 8.2a), or they may be **overlapping sets** that have some elements in common (as depicted by Figure 8.2b).

Sometimes we are given two sets X and Y, and we know nothing about how they are related. In this situation, we draw a general figure, such as the accompanying one. In it, there are four regions, labeled I, II, III, and IV.

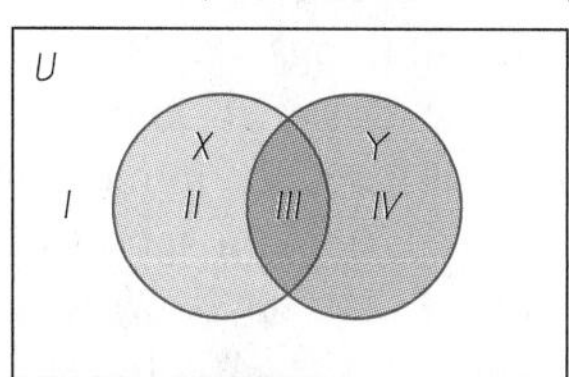

If $X \subseteq Y$, then region II is empty.
If $Y \subseteq X$, then region IV is empty.
If $X = Y$, then regions II and IV are empty.
If X and Y are disjoint, then region III is empty.

PROBLEM SET 8.2

Essential Ideas

1. **In Your Own Words** Explain the difference between the terms *element of a set* and *subset of a set.* Give examples.

2. **In Your Own Words** Explain the difference between the terms *subset* and *proper subset.*

3. **In Your Own Words** Explain the difference between *universal set* and *empty set.*

4. **In Your Own Words** What is a Venn diagram?

Level 1

WHAT IS WRONG, *if anything, with each of the statements in Problems 5–6? Explain your reasoning.*

5. A proper subset of the set of even numbers is $\{2, 4, 6, 8, 10, 12, \ldots\}$.

6. The list of all subsets of $\{1, 2, 3\}$ is $\{1\}$, $\{2\}$, $\{3\}$, $\{1, 2\}$, $\{2, 3\}$, and $\{1, 3\}$.

Level 1—Drill

List all subsets of each set given in Problems 7–12.

7. $\{m, y\}$
8. $\{4, 5\}$
9. $\{y, o, u\}$
10. $\{3, 6, 9\}$
11. $\{m, a, t, h\}$
12. $\{1, 2, 3, 4\}$

Level 2

13. a. List all possible subsets of the set $A = \varnothing$.
 b. Repeat part **a** for $B = \{1\}$.
 c. Repeat part **a** for $C = \{1, 2\}$.
 d. Repeat part **a** for $D = \{1, 2, 3\}$.

e. Repeat part **a** for $E = \{1, 2, 3, 4\}$.

f. Look for a pattern. Can you guess how many subsets the set $F = \{1, 2, 3, 4, 5\}$ has?

14. Using Problem 13, make a conjecture about the number of subsets that can be formed from a set consisting of n elements.

Level 2—Applications

15. Situation "Niels, where are you going?" asked Sammy. "I'm going to the barber for a shave," answered Niels. "You know, Niels, Darryl, the barber, does a very good business. He shaves those men and only those men who do not shave themselves." "Come on, Sammy, that's impossible! Who do you think shaves him?" Write a paper on who you think shaves the barber.

16. In Your Own Words Write a short paper comparing the SITUATION with your own situation. (That is, can you think of *any* relationships between sets, subsets, and operations among sets that relate to something in your own situation?)

Level 3—Team Project

17. With your team, draw Venn diagrams showing all possible regions for the following number of intersecting sets:

a. one set **b.** two sets **c.** three sets
d. four sets **e.** five sets

8.3 OPERATIONS WITH SETS

Situation

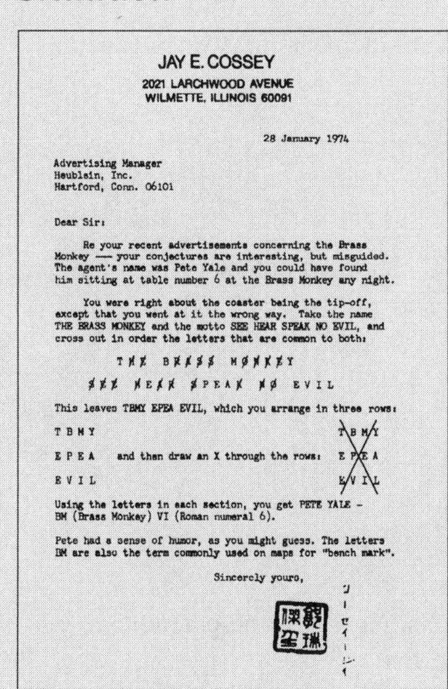

JAY E. COSSEY
2021 LARCHWOOD AVENUE
WILMETTE, ILLINOIS 60091

28 January 1974

Advertising Manager
Heublein, Inc.
Hartford, Conn. 06101

Dear Sir:

Re your recent advertisements concerning the Brass Monkey — your conjectures are interesting, but misguided. The agent's name was Pete Yale and you could have found him sitting at table number 6 at the Brass Monkey any night.

You were right about the coaster being the tip-off, except that you went at it the wrong way. Take the name THE BRASS MONKEY and the motto SEE HEAR SPEAK NO EVIL, and cross out in order the letters that are common to both:

T H E B R A S S M O N K E Y

S E E H E A R S P E A K N O E V I L

This leaves TBMY EPEA EVIL, which you arrange in three rows:

T B M Y

E P E A and then draw an X through the rows: T B M Y / E P E A / E V I L

E V I L

Using the letters in each section, you get PETE YALE - BM (Brass Monkey) VI (Roman numeral 6).

Pete had a sense of humor, as you might guess. The letters BM are also the term commonly used on maps for "bench mark".

Sincerely yours,

"Do you know anything about a Brass Monkey, Jay?" asked Tom. "Do I ever!" exclaimed Jay. "Let me show you a letter I wrote to Heublein over two decades ago. They had this advertising campaign, which was a puzzle. It received a lot of response. Anyway, I solved the puzzle by finding the intersection of two sets. . . ."

In this section, Tom will learn not only about a Brass Monkey, but also about three operations on sets, namely, union, intersection, and complementation.

There are three common operations with sets: union, intersection, and complementation.

UNION

Union is an operation for sets A and B in which a set is formed that consists of all the elements that are in A or B or both. The symbol for the operation of union is $\cup$, and we write $A \cup B$.

Union

The **union** of sets A and B, denoted by $A \cup B$, is the set consisting of all elements of A or B or both.

EXAMPLE 1 **Finding the Union of Sets**

Let

$$U = \{1, 2, 3, 4, 5, 6, 7, 8, 9\}$$
$$A = \{2, 4, 6, 8\} \quad B = \{1, 3, 5, 7\} \quad C = \{5, 7\}$$

Find the following unions of sets.

a. $A \cup C$ **b.** $B \cup C$ **c.** $A \cup B$ **d.** $(A \cup B) \cup \{9\}$

Solution **a.** The union of A and C is the set consisting of all elements in A or in C or in both:

$$A \cup C = \mathbf{\{2, 4, 5, 6, 7, 8\}}$$

b. $B \cup C = \mathbf{\{1, 3, 5, 7\}}$

Notice that, even though the elements 5 and 7 appear in both sets, they are listed only once; that is, the sets $\{1, 3, 5, 7\}$ and $\{1, 3, 5, 5, 7, 7\}$ are equal (exactly the same). Sometimes we write

$$B \cup C = B$$

c. $A \cup B = \mathbf{\{1, 2, 3, 4, 5, 6, 7, 8\}}$

d.
$$\begin{aligned}(A \cup B) \cup \{9\} &= \{1, 2, 3, 4, 5, 6, 7, 8\} \cup \{9\} \\ &= \mathbf{\{1, 2, 3, 4, 5, 6, 7, 8, 9\}} \\ &= U\end{aligned}$$

Here we are considering the union of three sets. However, the parentheses indicate the operation that should be performed first. ■

We can use Venn diagrams to illustrate union. In Figure 8.3, we first shade A (horizontal lines) and then shade B (vertical lines). *The union is all parts that have been shaded* (either horizontal or vertical) *at least once*; this region is shown as a color screen.

Figure 8.3 Venn diagram for union

Intersection

A second operation is called intersection.

Intersection

The **intersection** of sets A and B, denoted by $A \cap B$, is the set consisting of all elements common to both A and B.

EXAMPLE 2 **Finding the Intersection of Sets**

Let $U = \{a, b, c, d, e\}$, and let

$$A = \{a, c, e\} \quad B = \{c, d, e\} \quad C = \{a\} \quad D = \{e\}$$

Find the following intersections of sets.

a. $A \cap B$ **b.** $A \cap C$ **c.** $B \cap C$ **d.** $(A \cap B) \cap D$

Solution

a. The intersection of A and B is the set consisting of elements in both A and B: $\mathbf{A \cap B = \{c, e\}}$

b. $\mathbf{A \cap C = C}$ $A \cap C = \{a\}$, and $\{a\} = C$.

c. $\mathbf{B \cap C = \varnothing}$ B and C have no elements in common (they are disjoint).

d. $(A \cap B) \cap D = \{c, e\} \cap \{e\}$ Parentheses first

$= \{e\}$ Intersection

$= \mathbf{D}$ ■

The intersection of sets can also be easily shown using a Venn diagram, as in Figure 8.4. The intersection is all parts shaded twice (both horizontal and vertical), as shown with a color screen.

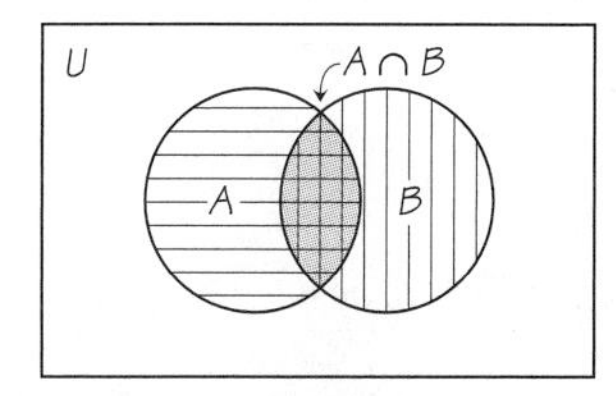

Figure 8.4 Venn diagram for intersection

Complementation

CAUTION

Complementation is an operation on a set that must be performed in reference to a universal set.

Complement

The **complement** of a set A, denoted by $\overline{A}$, is the set of all elements in U that are not in the set A.

EXAMPLE 3 **Finding Complements**

Let U = {people in California}, and let
A = {people who are over 30};
B = {people who are 30 or under};
C = {people who own a car}
Find the following complements of sets.

a. $\overline{C}$ **b.** $\overline{A}$ **c.** $\overline{B}$ **d.** $\overline{U}$ **e.** $\overline{\varnothing}$

Solution

a. $\overline{C}$ = **{Californians who do not own a car}**
b. $\overline{A}$ = $\overline{\{\text{people who are over 30}\}}$ = {people who are 30 or under} = ***B***
c. $\overline{B}$ = ***A***
d. $\overline{U} = \varnothing$
e. $\overline{\varnothing}$ = ***U*** ■

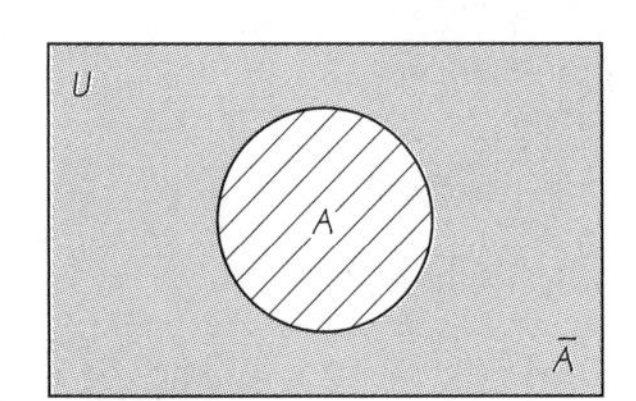

Figure 8.5 Venn diagram for complementation

Complementation, too, can be shown using a Venn diagram. The shaded part (color screen) of Figure 8.5 shows the complement of A. In a Venn diagram, *the complement is everything in U that is not in the set under consideration* (in this case, everything not in A).

PROBLEM SET 8.3

ESSENTIAL IDEAS

1. **In Your Own Words** Distinguish between the union and the intersection of two sets.

2. **In Your Own Words** Define *union* and illustrate with examples.

3. **In Your Own Words** Define *intersection* and illustrate with examples.

4. **In Your Own Words** Define *complement* and illustrate with examples.

LEVEL 1

WHAT IS WRONG, *if anything, with each of the statements in Problems 5–8? Explain your reasoning. Let* $U = \mathbb{N}$, $A = \{1, 2, 3, 4, 5\}$, $B = \{4, 5, 6, 7, 8, 9\}$, C = *{odd numbers}, and* E = *{even numbers}.*

5. $A \cup B = \{1, 2, 3, 4, 5\} \cup \{4, 5, 6, 7, 8, 9\}$
$= \{4, 5\}$

6. $A \cap B = \{1, 2, 3, 4, 5\} \cap \{4, 5, 6, 7, 8, 9\}$
$= \{1, 2, 3, 4, 5, 4, 5, 6, 7, 8, 9\}$

7. $\overline{A} = \overline{\{1, 2, 3, 4, 5\}} = \{6, 7, 8, 9\}$

8. $\overline{C} = E$

Perform the given set operations in Problems 9–18.

Let $U = \{1, 2, 3, 4, 5, 6, 7, 8, 9, 10\}$.

9. $\{2, 6, 8\} \cup \{6, 8, 10\}$

10. $\{2, 6, 8\} \cap \{6, 8, 10\}$

11. $\{1, 2, 3, 4, 5\} \cap \{3, 4, 5, 6, 7\}$

12. $\{1, 2, 3, 4, 5\} \cup \{3, 4, 5, 6, 7\}$

13. $\{2, 5, 8\} \cup \{3, 6, 9\}$

14. $\{2, 5, 8\} \cap \{3, 6, 9\}$

15. $\overline{\{2, 8, 9\}}$

16. $\overline{\{1, 2, 5, 7, 9\}}$

17. $\overline{\{8\}}$

18. $\overline{\{6, 7, 8, 9, 10\}}$

Let $U = \{1, 2, 3, 4, 5, 6, 7\}$*;* $A = \{1, 2, 3, 4\}$*;*
$B = \{1, 2, 5, 6\}$*; and* $C = \{3, 5, 7\}$.
List all members of each of the sets in Problems 19–28.

19. $A \cup B$

20. $A \cup C$

21. $B \cup C$

22. $A \cap B$

23. $A \cap C$

24. $B \cap C$

25. $\overline{A}$

26. $\overline{B}$

27. $\overline{C}$

28. $\overline{\varnothing}$

Draw Venn diagrams for each of the relationships in Problems 29–34.

29. $X \cup Y$

30. $X \cup Z$

31. $X \cap Y$

32. $X \cap Z$

33. $\overline{X}$

34. $\overline{Y}$

35. If $A \subseteq B$, describe:

a. $A \cup B$ **b.** $A \cap B$ **c.** $A \cup B$ **d.** $A \cap \overline{B}$

36. Situation "Do you know anything about a Brass Monkey, Jay?" asked Tom. "Do I ever!" exclaimed Jay. "Let me show you a letter I wrote to Heublein over two decades ago. They had this advertising campaign, which was a puzzle. It received a lot of response. Anyway, I solved the puzzle by finding the intersection of two sets:

{T, H, E, B, R, A, S, S, M, O, N, K, E, Y}

{S, E, E, H, E, A, R, S, P, E, A, K, N, O, E, V, I, L}"

a. Write these sets properly, without repeated letters.

b. Find the intersection of these sets.

8.4 VENN DIAGRAMS

Situation

James knows that half the students from his school are accepted at the public university nearby. Also, half are accepted at the local private college. James thinks that this adds up to 100%, so he will surely be accepted at one or the other institution. Explain why James may be wrong. If possible, use a diagram in your explanation.

A summary of the results of this examination question was printed in 1989 in *A Question of Thinking: A First Look at Students' Performance on Open-Ended Questions in Mathematics.* The findings are shown in the clipping on the left.

Of the approximately 500 papers scored, only 20 percent of the responses showed that students were able to explain, with or without the help of diagrams, that an overlap can exist between the groups admitted to the college and the university. Approximately 25 percent of the students in the sample digressed from the point of the problem. Another 40 percent were unable to interpret the situation sensibly and gave inappropriate answers.

In the previous section, we defined union, intersection, and complementation of sets. However, the real payoff for studying these relationships comes when dealing with combined operations or with several sets at the same time.

EXAMPLE 1 **Combined Operations with Sets**

Draw a Venn diagram for each of the following.

a. $\overline{A \cup B}$ **b.** $\overline{A} \cup \overline{B}$

Verify your diagrams for the following particular example.

Let $U = \{1, 2, 3, 4, 5, 6, 7, 8, 9, 10\}$ and let $A = \{1, 2\}$, $B = \{4, 5, 6, 7, 8\}$.

Solution These are combined operations; in part **a,** first find the union, and then find the complement; and in part **b,** first find the complements and then find the union.

a. For the Venn diagram, first draw the union:

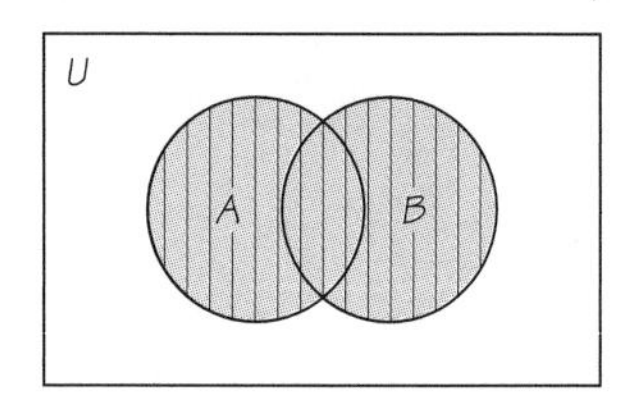

Then find the complement:

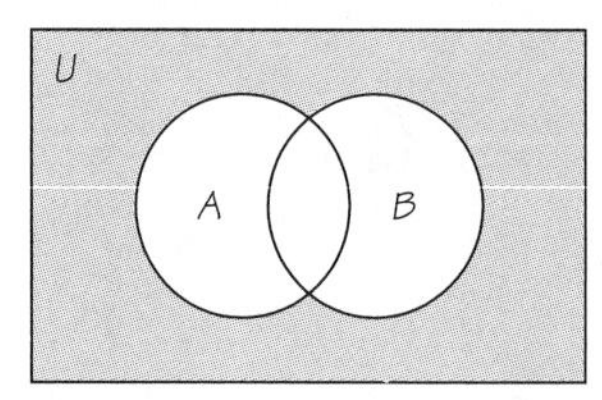

These steps are generally combined in one diagram:

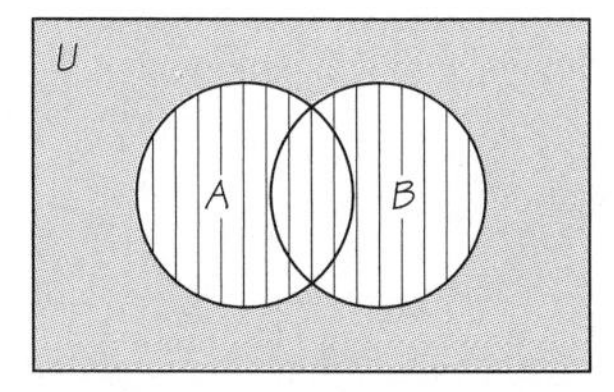

The answer is shown as the shaded portion; the lines show the intermediate steps. In your own work, you will generally find it easier to show the *final answer* (the part shaded here) in a second color (pen or highlighter).

For the particular example at hand, we find

$$\begin{aligned}\overline{A \cup B} &= \overline{\{1, 2\} \cup \{4, 5, 6, 7, 8\}}\\ &= \overline{\{1, 2, 4, 5, 6, 7, 8\}}\\ &= \{3, 9, 10\}\end{aligned}$$

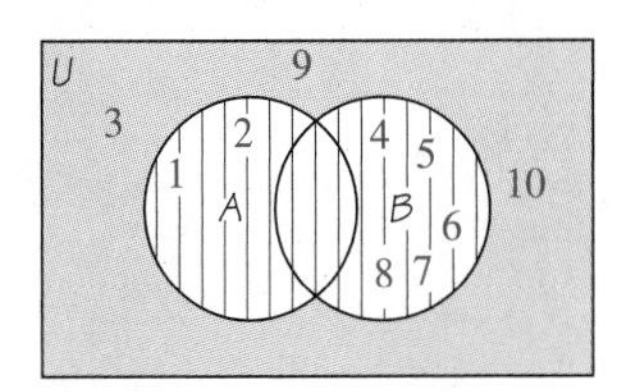

b. First find $\overline{A}$ and $\overline{B}$:

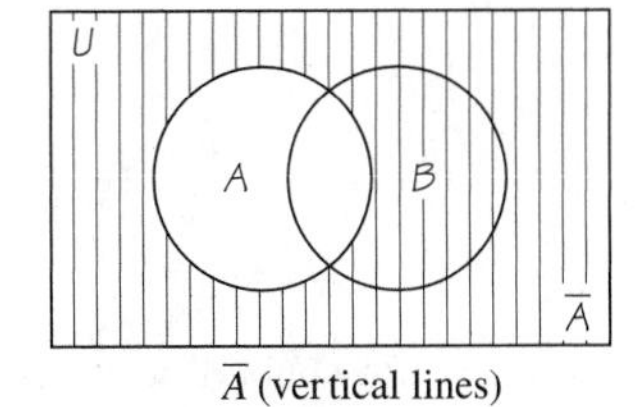

$\overline{A}$ (vertical lines)

$\overline{A}$ with $\overline{B}$ (horizontal lines)

For $\overline{A} \cup \overline{B}$, the union is all parts that have horizontal or vertical lines (or both).

For the particular example at hand, we find

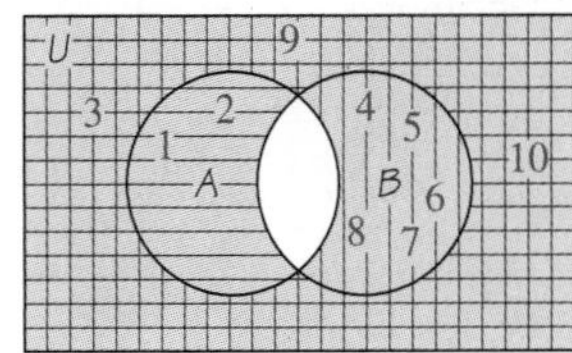

$$\begin{aligned}\overline{A} \cup \overline{B} &= \overline{\{1, 2\}} \cup \overline{\{4, 5, 6, 7, 8\}}\\ &= \{3, 4, 5, 6, 7, 8, 9, 10\} \cup \{1, 2, 3, 9, 10\}\\ &= \{1, 2, 3, 4, 5, 6, 7, 8, 9, 10\}\end{aligned}$$

Notice from the Venn diagram that the shaded portion is not the entire universe. *For this example, $\overline{A} \cup \overline{B} = U$, but this is not true in general, since (as the Venn diagram shows) the entire region is not shaded.* ■

Venn diagrams can be useful in making general statements about sets. (See part **b** of Example 1, for example.) We also see for the Venn diagrams in Example 1 that $\overline{A \cup B} \neq \overline{A} \cup \overline{B}$. If they were equal, the final *shaded* portions of the Venn diagrams would be the same.

EXAMPLE 2 Using Venn Diagrams to Prove a Statement about Sets

Prove that $\overline{A \cup B} = \overline{A} \cap \overline{B}$.

Solution The procedure is to draw a Venn diagram for the left side and a separate diagram for the right side, and then to look to see whether they are the same.

Venn diagrams: Left side of equal sign | Right side of equal sign

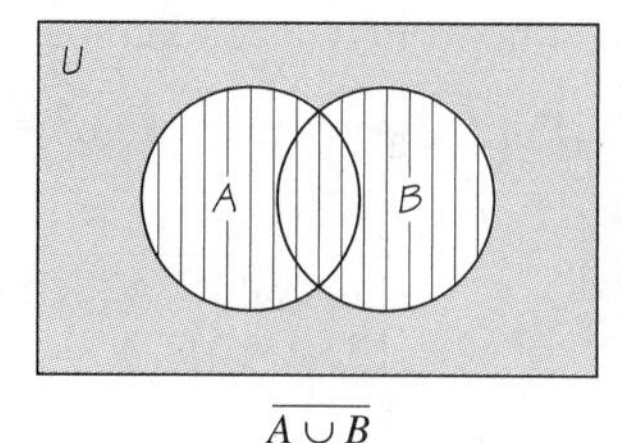

$\overline{A \cup B}$

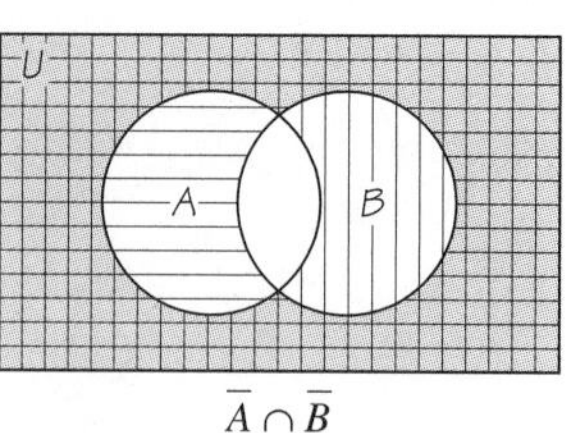

$\overline{A} \cap \overline{B}$

Detail: See Example 1.

$\overline{A}$ (vertical lines)
$\overline{B}$ (horizontal lines)
$\overline{A} \cap \overline{B}$ is the intersection of the horizontal and vertical lines; it is shown as the shaded color screen.

Compare the shaded portions of the two Venn diagrams. They are the same, so we have proved that $\overline{A \cup B} = \overline{A} \cap \overline{B}$. ■

The result proved in Example 2 is one part of a result known as **De Morgan's law.** (In Problem 41 of Problem Set 8.4, you are asked to prove the second part.)

De Morgan's Laws

$$\overline{X \cup Y} = \overline{X} \cap \overline{Y} \quad \text{and} \quad \overline{X \cap Y} = \overline{X} \cup \overline{Y}$$

The *order* of operations for sets is from left to right, unless there are parentheses. Operations within parentheses are performed first, as shown in Example 3.

EXAMPLE 3

Combined Operations with Venn Diagrams

Illustrate $(A \cup C) \cap \overline{C}$ with a Venn diagram.

Solution

First, draw $A \cup C$ (vertical lines). Next, draw $\overline{C}$ (horizontal lines). The result is the intersection of the vertical and horizontal parts and is the portion shaded in color.

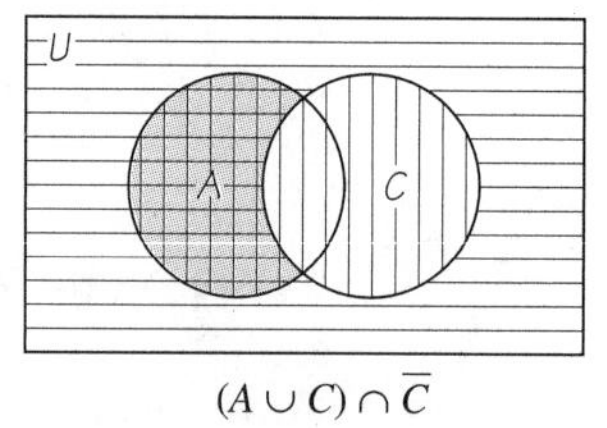

$(A \cup C) \cap \overline{C}$

VENN DIAGRAMS WITH THREE SETS

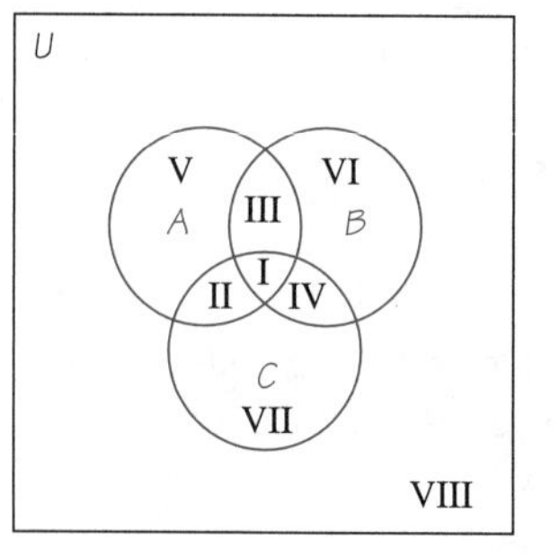

Figure 8.6 Three general sets

Sometimes you will be asked to consider the relationships among three sets. The general Venn diagram is shown in Figure 8.6. Notice that the three sets divide the universe into eight regions. Can you number each part?

EXAMPLE 4

Venn Diagrams for Three Sets

Find the results of the following operations among three sets:

a. $A \cup B$ **b.** $A \cap C$ **c.** $B \cap C$ **d.** $\overline{A}$ **e.** $\overline{A \cup B}$ **f.** $(A \cap B) \cap C$

Solution

a.

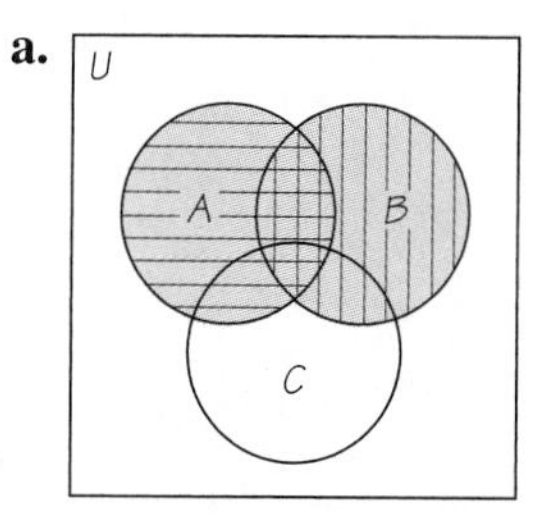

b.

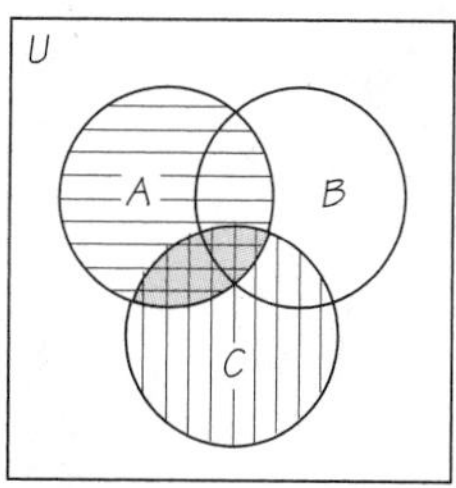

c.

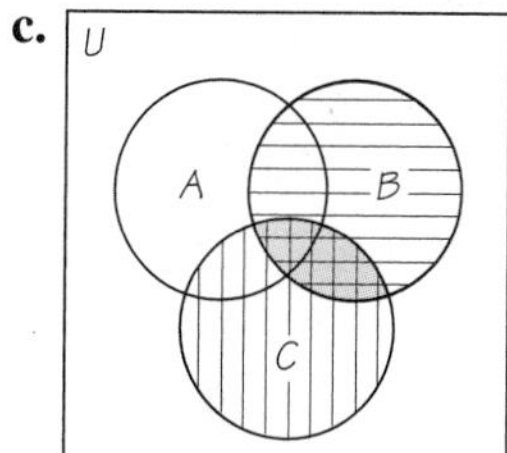

d.

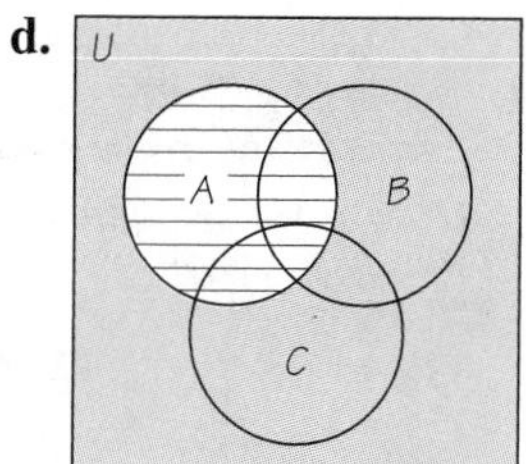

e.

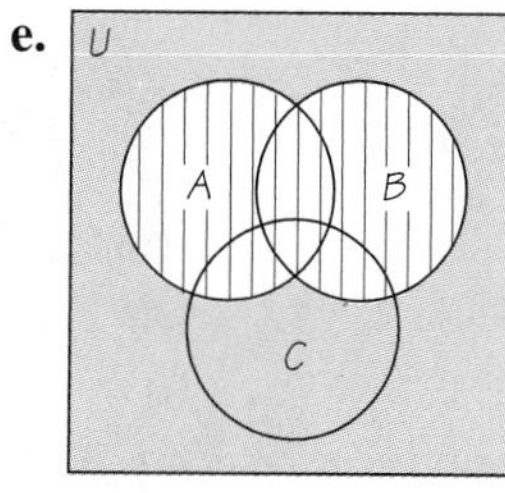

f.

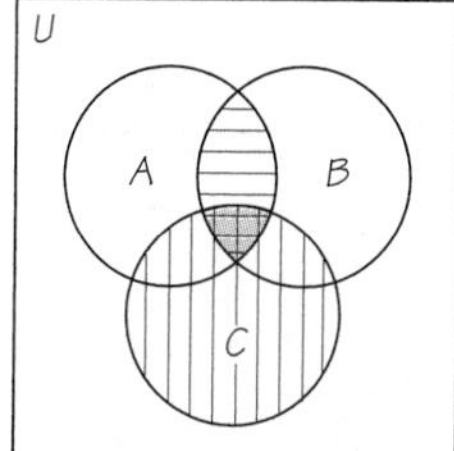

■

EXAMPLE 5

Venn Diagrams with Three Sets and Parentheses

Perform the following combinations of operations among three sets.

a. $A \cup (B \cap C)$ **b.** $\overline{A \cup B} \cap C$

Solution

a. Do parentheses first (vertical lines); then shade A (horizontal lines).

The union is all parts that show either vertical or horizontal lines (or both); the answer is the part that is shaded.

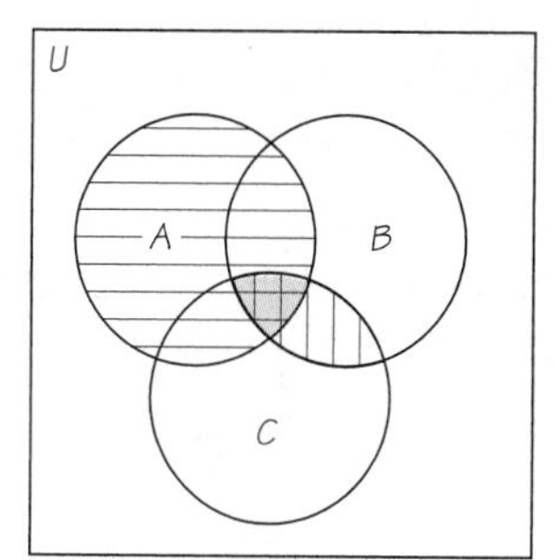

b. Overbars "act like" parentheses, so do $\overline{A \cup B}$ first (vertical); then shade C (horizontal).

The intersection is all parts that show *both* vertical and horizontal lines; the answer is the part that is shaded.

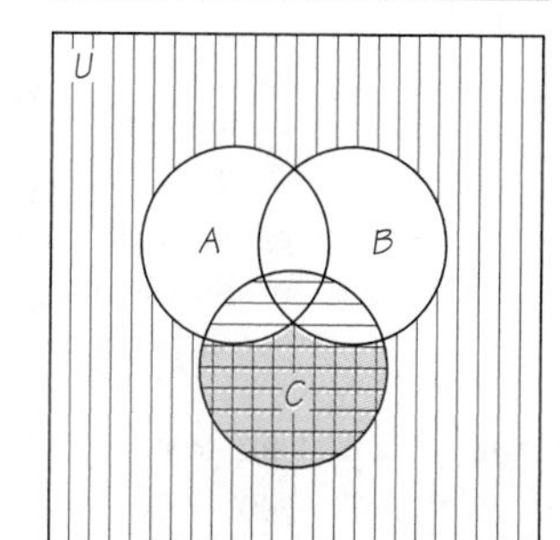

■

PROBLEM SET 8.4

ESSENTIAL IDEAS

1. Draw a Venn diagram for $A \cup B$.

2. Draw a Venn diagram for $A \cap B$.

3. Draw a Venn diagram for $\overline{A}$.

4. Write in symbols: the complement of the union of sets A and B. Draw a Venn diagram.

5. Write in symbols: the union of the complements of sets A and B. Draw a Venn diagram.

LEVEL 1

WHAT IS WRONG, *if anything, with each of the statements in Problems 6–10? Explain your reasoning.*

6. For $\overline{A \cup B}$, first find the complement, then find the union.

7. $\overline{A \cup B} = \overline{A} \cup \overline{B}$

8. $\overline{A \cap B} = \overline{A} \cap \overline{B}$

9. To draw the Venn diagram for $(A \cap B) \cup \overline{C}$, we would carry out (in order) the following steps:
   (1) Draw a rectangle (for the universe) containing three overlapping circles. Label the circles A, B, and C.
   (2) Shade in all parts that are in both A and B.
   (3) Shade in all parts that are not in C.
   (4) Use a highlighter to mark all parts that have been shaded in both parts (2) and (3).

10. To draw the Venn diagram for $\overline{A \cup B} \cap C$, we would carry out (in order) the following steps:
   (1) Draw a rectangle (for the universe) containing three overlapping circles. Label the circles A, B, and C.
   (2) Shade in all parts that are not in either A or B.
   (3) Shade in C.
   (4) Use a highlighter to mark all parts that have been shaded in both parts (2) and (3).

LEVEL 1—Drill

Draw a Venn diagram to illustrate each relationship given in Problems 11–25.

11. $A \cup \overline{B}$
12. $\overline{A} \cup B$
13. $\overline{A} \cap C$
14. $\overline{B \cap C}$
15. $\overline{A \cap C}$
16. $\overline{A \cup B}$
17. $\overline{B \cup C}$
18. $A \cap (B \cup C)$
19. $A \cup (B \cup C)$
20. $\overline{(A \cup B) \cup C}$
21. $(A \cap B) \cap (A \cap C)$
22. $(A \cap B) \cup (A \cap C)$
23. $\overline{(A \cap B) \cup C}$

24. $A \cap \overline{B \cup C}$

25. $\overline{A \cup B} \cup C$

Level 2—Drill

Let $U = \{1, 2, 3, 4, 5, 6, 7, 8, 9, 10\}$, *and let*
$A = \{2, 4, 6, 8\}$, $B = \{5, 9\}$, $C = \{2, 5, 8, 9, 10\}$.
List all the members of each set in Problems 26–35.

26. $(A \cup B) \cup C$

27. $(A \cap B) \cap C$

28. $A \cup (B \cap C)$

29. $A \cap (\overline{B} \cup C)$

30. $\overline{A} \cap (B \cup C)$

31. $A \cap (B \cap \overline{C})$

32. $\overline{A} \cup (B \cap C)$

33. $A \cap \overline{B \cup C}$

34. $\overline{A \cap (B \cup C)}$

35. $\overline{A \cup (B \cap C)}$

In Problems 36–41, *use Venn diagrams to prove or disprove each expression. For each of these problems, draw one Venn diagram for the left side of the equation and another for the right. If the final shaded portions are the same, then you have proved the result. If the final shaded portions are not identical, then you have disproved the result.*

36. $\overline{A \cup B} = \overline{A} \cup \overline{B}$

37. $\overline{A} \cap \overline{B} = \overline{A \cup B}$

38. $(A \cup B) \cup C = A \cup (B \cup C)$

39. $A \cup (B \cup C) = (A \cup B) \cup (A \cup C)$

40. $A \cap (B \cup C) = (A \cap B) \cap (A \cap C)$

41. Prove De Morgan's law: $\overline{X \cap Y} = \overline{X} \cup \overline{Y}$

Level 2—Applications

42. Draw a Venn diagram showing the relationship among cats, dogs, and animals.

43. Draw a Venn diagram showing the relationship among trucks, buses, and cars.

44. Draw a Venn diagram showing people who are over 30, people who are 30 or under, and people who drive a car.

45. Draw a Venn diagram showing people in a classroom wearing some black, people wearing some blue, and people wearing some brown.

46. Draw a Venn diagram showing that all Chevrolets are automobiles.

47. Draw a Venn diagram showing birds, bees, and living creatures.

48. **Situation** James knows that half the students from his school are accepted at the public university nearby. Also, half are accepted at the local private college. James thinks that this adds up to 100%, so he will surely be accepted at one or the other institution. Explain why James may be wrong. If possible, use a diagram in your explanation.

8.5 SURVEY PROBLEMS USING SETS

Situation

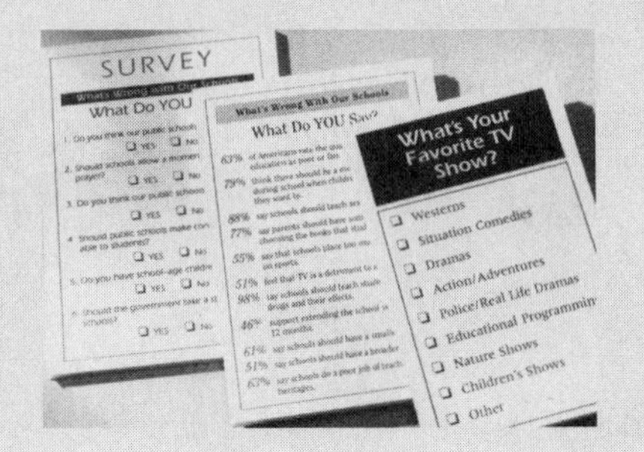

"Why are you so gloomy, Matt?" asked Dave. "Oh, I didn't get the job. The personnel manager sent me out to poll 100 people about their favorite types of TV shows. I didn't have time to complete the survey, so I made up some responses. I don't know how she found out, but she didn't give me the job. I guess I'll need to learn more about surveys before I try that again."

In this section, Matt will learn how to use sets to reach conclusions about surveys.

Venn diagrams can sometimes be used to solve survey problems. Suppose that a survey indicates that 45 students are taking mathematics and that 41 are taking English. How many students are taking math or English? At first, it might seem that all we need to do is add 41 and 45, but such is not the case, as you can see by looking at Figure 8.7.

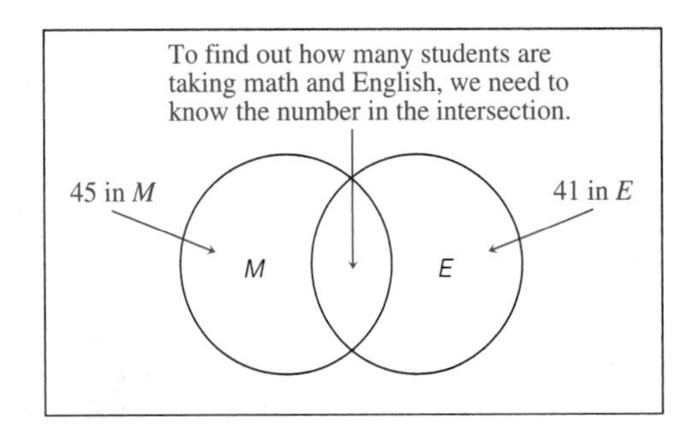

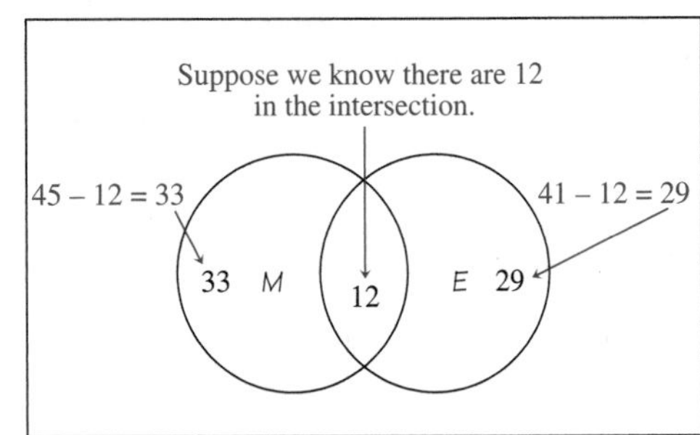

Figure 8.7 Using a Venn diagram to find the numbers in regions

We see that there are $33 + 29 + 12 = 74$ students enrolled in mathematics or in English or in both. (See Figure 8.7.) We see that the proper procedure is to fill in the number in the intersection and then to find the numbers in the other regions by subtraction.

For three sets, the situation is a little more involved, as illustrated by Example 1. The overall procedure is to fill in the number in the innermost set first and work your way out through the rest of the Venn diagram by using subtraction.

EXAMPLE 11 **Survey Problem**

A survey of 100 randomly selected students gave the following information:

45 students are taking mathematics.
41 students are taking English.
40 students are taking history.
15 students are taking math and English.
18 students are taking math and history.
17 students are taking English and history.
7 students are taking all three.

a. How many are taking only mathematics?
b. How many are taking only English?
c. How many are taking only history?
d. How many are not taking any of these courses?
e. How many are not taking mathematics?

Solution We draw a Venn diagram and fill in the various regions.

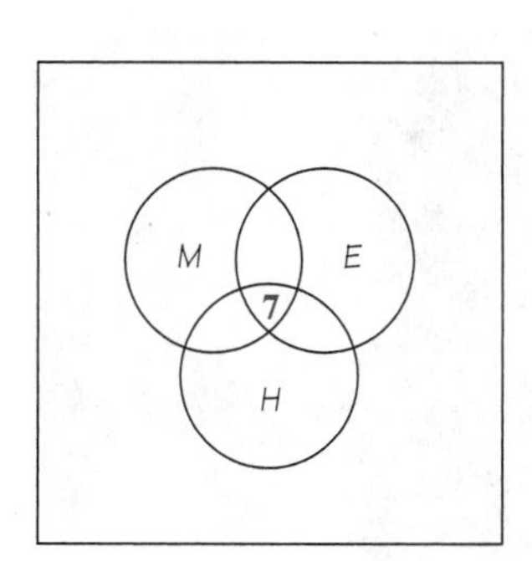

Let $M = \{\text{students taking mathematics}\}$
$E = \{\text{students taking English}\}$
$H = \{\text{students taking history}\}$

Start by filling in the innermost section; there are 7 in the region $M \cap E \cap H$. Notice that we have filled in a 7 in this region.

Now fill in the portions of intersection of each pair of two sets, using subtraction:

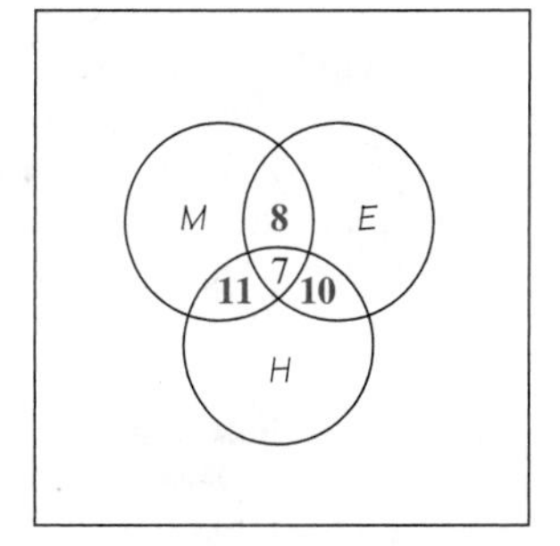

$E \cap H$ is given as 17, but 7 have previously been accounted for, so only the *remaining* 10 are added to the Venn diagram. (See the figure at the right: $17 - 7 = 10$.)

$M \cap H$ is given as 18; so fill in an additional 11 students ($18 - 7 = 11$).

$M \cap E$ is given as 15; so fill in 8 students ($15 - 7 = 8$).

The next step is to fill in the regions of the Venn diagram representing single sets:

H is given as 40, but 28 ($7 + 10 + 11$) have previously been accounted for, so only an *additional* 12 members are needed.

E is given as 41; but here we subtract 25 ($7 + 10 + 8$) and fill in 16.

M is given as 45; but in this case we subtract 26 ($7 + 11 + 8$), leaving us with 19 new students.

Finally, add all the numbers in the diagram and confirm that 83 persons have been accounted for: $19 + 8 + 16 + 11 + 7 + 10 + 12 = 83$. Since 100 students

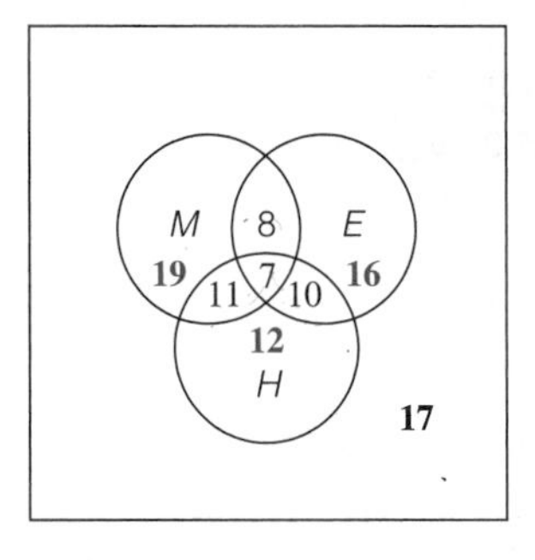

Figure 8.8 Survey problem

were surveyed, we see that $100 - 83 = 17$ are not taking any of the three courses. We fill in 17 as the number outside the three interlocking circles.

We now have all the information we need to answer the questions simply by looking at Figure 8.8.

a. **19** (the number in M only, and not in the other sets)
b. **16** (the number in E only, and not in the other sets)
c. **12** (the number in H only, and not in the other sets)
d. **17** (the number inside the universe but outside all the other sets)
e. **55** ($100 - 45 = 55$; you do not need the Venn diagram for this answer.) ■

PROBLEM SET 8.5

Essential Ideas

1. How many regions are there with two overlapping sets?

2. How many regions are there with three overlapping sets?

Level 1

3. **WHAT IS WRONG,** *if anything, with the following statement? Explain your reasoning.* If there are 145 elements in set A and 100 elements in set B, then there are $145 + 100 = 245$ elements in $A \cup B$.

Level 1—Applications

4. Santa Rosa Junior College enrolled 29,000 students in the fall of 1999. It was reported that of that number, 58% were female and 42% were male. In addition, 62% were over the age of 25. How many students were there in each category? Draw a Venn diagram showing these relationships.

5. In a sample of defective tires, 72 have defects in materials, 89 have defects in workmanship, and 17 have defects of both types. How many tires are in the sample?

6. Montgomery College has a 50-piece band and a 36-piece orchestra. If 14 people are members of both the band and the orchestra, can the band and orchestra travel in two 40-passenger buses?

Level 2—Applications

7. In 1995 the United States population was approximately 263 million. It was reported that of that number, 72% have two white parents, 11.5% have two black parents, and 9% have two Hispanic parents. If $\frac{1}{2}$% have one black and one white parent, 2% have one black and one Hispanic parent, and 1% have one white and one Hispanic parent, how many people are there in each category? Draw a Venn diagram.

8. Listed here are five female and five male Wimbledon tennis champions, along with their country of citizenship and handedness.

Female	*Male*
Steffi Graf, Germany, right	Michael Stich, Germany, right
Martina Navratilova, U.S., left	Stefan Edberg, Sweden, right
Chris Evert Lloyd, U.S., right	Boris Becker, Germany, right
Evonne Goolagong, Australia, right	Pat Cash, Australia, right
Virginia Wade, Britain, right	John McEnroe, U.S., left

Using Figure 8.9, indicate in which region each of these individuals would be placed.

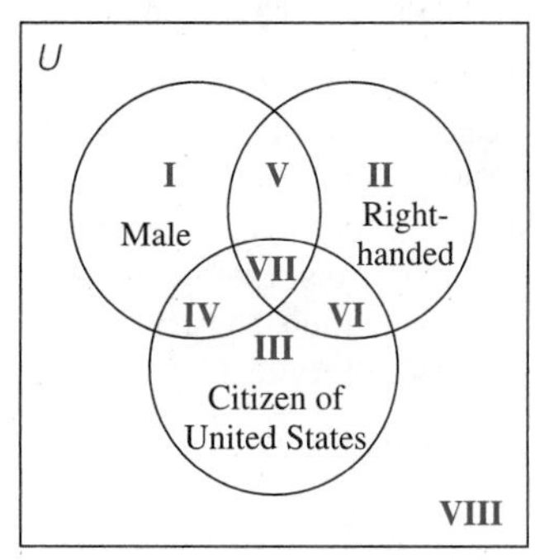

Figure 8.9 Problem 8

9. From a survey of 100 college students, a marketing research company found that 75 students owned stereos, 45 owned cars, and 35 owned both cars and stereos.

a. How many students owned either a car or a stereo (but not both)?

b. How many students do not own either a car or a stereo?

10. In a survey of a TriDelt chapter with 50 members, 18 were taking mathematics and 35 were taking English; 6 were taking both. How many were not taking either of these subjects?

11. The senior class at Rancho Cotati High School included 25 football players and 16 basketball players. If 7 persons played both sports, how many different people played in these sports?

12. The fire department wants to send booklets on fire hazards to all teachers and homeowners in town. How many booklets does it need, using these statistics?

50,000 homeowners
4,000 teachers
3,000 teachers who own their own homes

13. A survey of 100 women finds that 40 jog, 25 swim, 16 cycle, 15 swim and jog, 10 swim and cycle, 8 jog and cycle, and 3 jog, swim, and cycle. Let $J = \{\text{people who jog}\}$, $S = \{\text{people who swim}\}$, and $C = \{\text{people who cycle}\}$. Use a Venn diagram to show how many are in each of the eight possible categories.

14. In a recent survey of 100 women, the following information was gathered:

59 use shampoo A.
51 use shampoo B.
35 use shampoo C.
24 use shampoos A and B.
19 use shampoos A and C.
13 use shampoos B and C.
11 use all three.

Let $A = \{\text{women who use shampoo A}\}$; $B = \{\text{women who use shampoo B}\}$; and $C = \{\text{women who use shampoo C}\}$. Use a Venn diagram to show how many women are in each of the eight possible categories.

15. A poll was taken of 100 students at a commuter campus to find out how they got to campus. The results were:

42 drove alone.
28 rode in a carpool.
31 rode public transportation.
9 used both carpools and public transportation.
10 used both a carpool and sometimes their own cars.
6 used buses as well as their own cars.
4 used all three methods.

How many used none of these means of transportation?

16. In an interview of 50 students,

12 liked Proposition 8 and Proposition 13.
18 liked Proposition 8 but not Proposition 2.
4 liked Proposition 8, Proposition 13, and Proposition 2.
25 liked Proposition 8.
15 liked Proposition 13.
10 liked Proposition 2 but not Proposition 8 or Proposition 13.
1 liked Proposition 13 and Proposition 2 but not Proposition 8.

a. Of those surveyed, how many did not like any of the three propositions?
b. How many liked Proposition 8 and Proposition 2?
c. Show the completed Venn diagram.

17. Human blood is typed Rh^+ (positive blood) or Rh^- (negative blood). This Rh factor is called an *antigen.* There are two other antigens known as A and B types. Blood lacking both A and B types is called type O. Sketch a Venn diagram showing the three types of antigens A, B, and Rh, and label each of the eight regions. For example, O^+ is inside the Rh set (anything in Rh is positive), but outside both A and B. On the other hand, O^- is that region outside all three circles.

18. Each of the circles in Figure 8.10 is identified by a letter, each having a number value from 1 to 9. Where the circles overlap, the number is the sum of the values of the letters in the overlapping circles. What is the number value for each letter?*

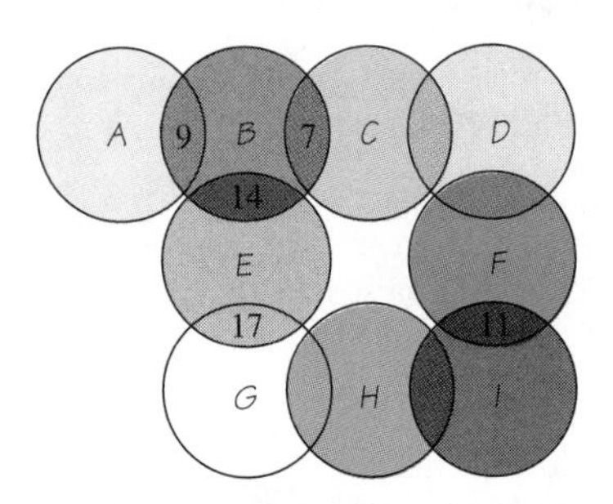

Figure 8.10 Circle intersection puzzle

19. On the *NBC Nightly News* on Thursday, May 25, 1995, Tom Brokaw read a brief report on computer use in the United States. The story compared computer users by ethnic background, and Brokaw reported that 14% of blacks and 13% of Hispanics use computers, but 27% of whites use computers. Brokaw then commented that computer use by whites was equal to that of blacks and Hispanics combined.

a. Draw a Venn diagram showing the white, black, Hispanic, Asian, and native American populations of the United States, along with U.S. computer users.

b. Use the Venn diagram from part **a** to show that percentages cannot be added as was done by Brokaw.

20. Situation "Why are you so gloomy, Matt?" asked Dave. "Oh, I didn't get the job. The personnel manager sent me out to poll 100 people about their favorite types of TV shows. I didn't have time to complete the survey, so I made up some responses. I don't know how she found out, but she didn't give me the job. I guess I'll need to learn more about surveys before I try that again."

Matt's data follow:

59 preferred comedies.
38 preferred variety shows.
42 preferred serious drama.
18 preferred comedies and variety programs.
12 preferred variety and serious drama.
16 preferred comedies and serious drama.
7 preferred all types.
2 didn't like any TV shows.

If you were the personnel manager, would you hire Matt on the basis of this survey?

21. In Your Own Words Write a short paper comparing the SITUATION with your own situation. (That is, can you imagine *any* situation in which you would need to use the information of this section?)

*From "Perception Puzzles," by Jean Moyer, *Sky,* January 1995, p. 120.

8.6 INDUCTIVE AND DEDUCTIVE REASONING

Situation

"Sarah, do you remember the old 'Ode to Billy Joe'? I just saw the movie on my VCR and it was great. I finally figured out what Billy Joe threw off the Tallahatchee Bridge," said Bob. "Oh, yeah? I'll bet you didn't!" answered Sarah. "Anyway," continued Sarah, "the movie is not faithful to the song. From the song you cannot figure out for sure what was thrown from the bridge." Is it possible to use deductive reasoning to answer this question?

In this section, Sarah and Bob will find out what is meant by deductive reasoning and how to use Venn diagrams to reach valid conclusions.

In Example 2 of Section 8.4 we used Venn diagrams to prove De Morgan's laws. There is a close association between sets and **logic.** In this section, we will look at two types of logical reasoning and then apply Venn diagrams to reach certain conclusions.

INDUCTIVE REASONING

One type of logical reasoning, called *inductive reasoning,* reaches conclusions by observing patterns and then predicting other results based on those observations. It is the type of reasoning used in much of scientific investigation, and it involves reasoning from particular facts or individual cases to a general *conjecture.* A **conjecture** is a guess or generalization predicted from incomplete or uncertain evidence.

Inductive Reasoning

Inductive reasoning is reasoning from the particular to the general.

The more individual occurrences you observe, the better able you are to make a correct generalization. The cartoon at the top of page 386 shows an incorrect generalization based on only two questions.

Born Loser reprinted by permission of Newspaper Enterprise Association, Inc.

The child in the cartoon reached the conclusion that the pattern was a numerical sequence of garbage trucks. You must keep in mind that an inductive conclusion is always tentative and may have to be revised in light of new evidence. For example, consider this pattern:

$1 \times 9 = 9$
$2 \times 9 = 18$
$3 \times 9 = 27$
$4 \times 9 = 36$
$5 \times 9 = 45$
$6 \times 9 = 54$
$7 \times 9 = 63$
$8 \times 9 = 72$
$9 \times 9 = 81$
$10 \times 9 = 90$

Conjecture: The sum of the digits of the answer of any product involving 9 is always 9.
Evidence: The first ten examples shown here substantiate this speculation.

The conjecture based on this pattern is suspect because it is based on so few cases. In fact, it is shattered by the very next case, where we get $11 \times 9 = 99$. An example that contradicts a conjecture is called a **counterexample.**

EXAMPLE 1

Finding Counterexamples

Tell whether you believe the given conjecture is true or false. If you believe it to be false, see if you can find a counterexample.

a. The sum of the digits in any product of 3 and another number is always divisible by 3.
b. All primes are odd.
c. All numbers between 1 and 2 are rational.

Solution

a. Look for a pattern:

$$\left.\begin{array}{l} 1 \times 3 = 3 \\ 2 \times 3 = 6 \\ 3 \times 3 = 9 \\ 4 \times 3 = 12 \\ 5 \times 3 = 15 \\ 6 \times 3 = 18 \\ 7 \times 3 = 21 \\ 8 \times 3 = 24 \\ 9 \times 3 = 27 \\ 10 \times 3 = 30 \end{array}\right\}$$

The sum of the digits in each of these answers is divisible by 3.

Try other examples: $8{,}340 \times 3 = 25{,}020$

Here $2 + 5 + 0 + 2 + 0 = 9$, which is divisible by 3. **It seems to be a valid conjecture.**

b. Try some examples:

3, 5, 7, 11, 13, 17, 19, 23, . . .

are all odd. **However, 2 is a prime and it is even, so it serves as a counterexample for the conjecture.**

c. Some examples: $1\frac{1}{2}$, 1.66666, and $1.\overline{6}$ are all between 1 and 2, and they are all rational. Examples, however, do not prove the conjecture; they simply support the conjecture. We can find a counterexample: **1.6066066606666066666606 . . . is not rational.** ■

Sometimes you must formulate a conjecture based on inductive reasoning.

EXAMPLE 2 Formulating a Conjecture

What is the 100th consecutive positive odd number? Use this information to find the sum of the first 100 consecutive positive odd numbers.

Solution Positive odd numbers are 1, 3, 5, ... What is the 100th consecutive odd number?

1 is the 1st odd number;

3 is the 2nd odd number;

5 is the 3rd odd number;

7 is the 4th odd number;

9 is the 5th odd number;

↑

This seems to be one less than twice the term number.

$2(\mathbf{6}) - 1 = 11$ is the 6th odd number;

$2(\mathbf{7}) - 1 = 13$ is the 7th odd number;

⋮

$2(\mathbf{100}) - 1 = 199$ is the **100**th odd number.

We now look for a pattern to find $1 + 3 + 5 + \cdots + 199$; certainly it is too large an equation to do by brute force (even with a calculator).

$$
\begin{aligned}
1 &= 1 && \text{One term} \\
1 + 3 &= 4 && \text{Two terms} \\
1 + 3 + 5 &= 9 && \text{Three terms} \\
1 + 3 + 5 + 7 &= 16 && \text{Four terms} \\
1 + 3 + 5 + 7 + 9 &= 25 && \text{Five terms} \\
&\vdots
\end{aligned}
$$

It appears that the sum of 2 terms is $4 = 2^2$; of 3 terms is $9 = 3^2$; of 4 terms is $16 = 4^2$; and of 5 terms is $25 = 5^2$. Thus, the sum of the first 100 consecutive odd numbers seems to be 100^2.

Conjecture: $\mathbf{1 + 3 + 5 + \cdots + 199 = 100^2 = 10{,}000}$ ■

Historical Perspective

Aristotle (384–322 B.C.)

Logic began to flourish during the classical Greek Period. Aristotle was the first person to study the subject systematically, and he and many other Greeks searched for universal truths that were irrefutable. The logic of this period, referred to as Aristotelian logic, is still used today and is based on the syllogism, which is discussed in this section.

The second great period for logic came with the use of symbols to simplify complicated arguments; it began with the great German mathematician Gottfried Leibniz (1646–1716). The world took little notice of Leibniz's logic until George Boole (1815–1864) formalized logic and led it into a third and most important period, characterized by the use of symbolic logic. Symbolic logic is a course that is required in many degree programs today.

DEDUCTIVE REASONING

Although mathematicians often proceed by inductive reasoning to formulate new ideas, they are not content to stop at the "probable" stage. Often they formulate their predictions into conjectures and then try to prove these *deductively. Deductive reasoning* is a formal structure based on a set of *unproved* statements and a set of *undefined* terms, as well as being a logical process of deriving new results. The unproved statements are called **premises** or **axioms,** and the proved new results are called **theorems.** For example, consider the following argument:

1. If you read the *Times,* then you are well informed.
2. You read the *Times.*
3. Therefore, you are well informed.

If you accept statements 1 and 2 as true, then you *must* accept statement 3 as true. Statements 1 and 2 are called the *premises* of the argument, and statement 3 is called the **conclusion.** Such reasoning is called *deductive reasoning*; and if the conclusion follows from the premises, the reasoning is said to be **valid.**

Deductive Reasoning

Deductive reasoning consists of reaching a conclusion by using a formal structure based on a set of *undefined terms* and on a set of accepted unproved *axioms* or *premises.* The conclusions are said to be **proved** and are called **theorems.**

Logic accepts no conclusions except those that are inescapable. This is possible because of the strict way in which concepts are defined. Difficulty in simplifying arguments may arise because of their length, the vagueness of the words used, the literary style, or the possible emotional impact of the words used. Consider the following two arguments:

1. If George Washington was assassinated, then he is dead.
   Therefore, if he is dead, he was assassinated.
2. If you use heroin, then you first used marijuana.
   Therefore, if you use marijuana, then you will use heroin.

Logically, these two arguments are exactly the same, and both are **invalid** forms of reasoning. Nearly everyone would agree that the first is invalid, but many people see the second as valid because of the emotional appeal of the words used.

To avoid these difficulties, we look at the *form* of the arguments and not at the independent truth or falsity of the statements. One type of logic problem is called a **syllogism.** A syllogism has three parts: two *premises,* or hypotheses, and a *conclusion.* The premises give us information from which we form a conclusion. With the syllogism, we are interested in knowing whether the conclusion *necessarily follows* from the premises. If it does, it is called a *valid syllogism;* if not, it is called *invalid.* Consider the following examples:

VALID FORMS OF REASONING		*INVALID FORMS OF REASONING*
All Chevrolets are automobiles.	**Premise**	Some people are nice.
All automobiles have four wheels.	**Premise**	Some people are broke.
All Chevrolets have four wheels.	**Conclusion**	There are some nice broke people.
All teachers are crazy.	**Premise**	All dodos are extinct.
Karl Smith is a teacher.	**Premise**	No dinosaurs are dodos.
Therefore, Karl Smith is crazy.	**Conclusion**	Therefore, all dinosaurs are extinct.

To analyze such arguments, we need to have a systematic method of approach. We will use Venn diagrams.* For two sets p and q, we make the interpretations shown in Figure 8.11.

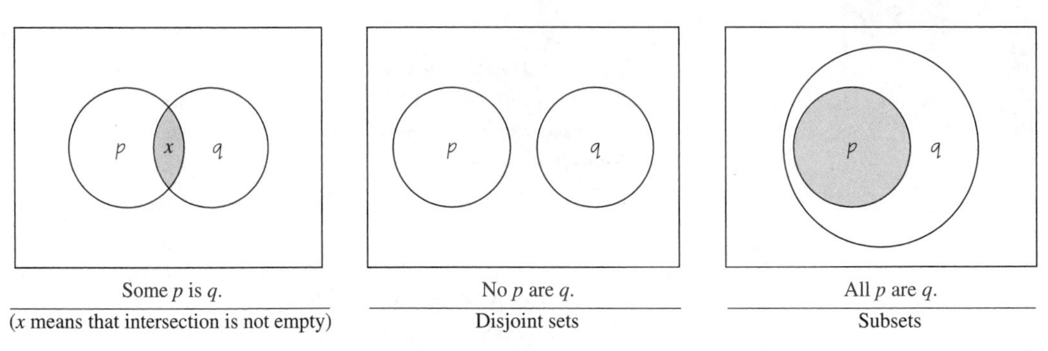

Figure 8.11 Venn diagrams for syllogisms

EXAMPLE 3

Using Venn Diagrams in a Syllogism

Test the validity of the following arguments.

a. All dictionaries are books.
This is a dictionary.
Therefore, this is a book.

b. If you like potato chips, then you will like Krinkles.
You do not like potato chips.
Therefore, you do not like Krinkles.

Solution

a. Begin with a Venn diagram showing the first premise:
All dictionaries are books.

Let p: dictionaries
q: books

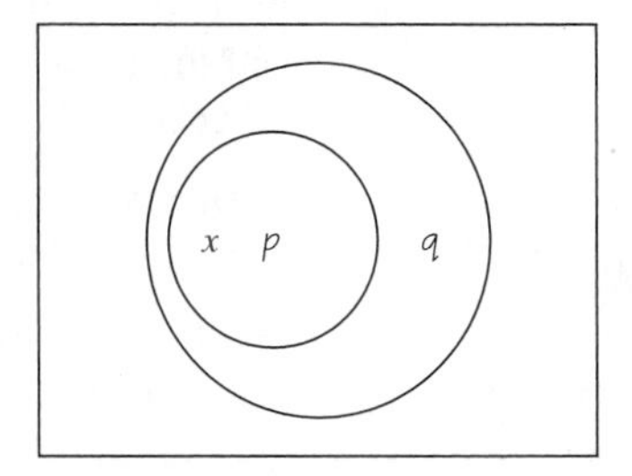

For the second premise, we place x (this object) inside the set of dictionaries (set p). The conclusion, "This object is a book," cannot be avoided (since x *must* be in q) so it is **valid.**

*In logic, Venn diagrams are often called *Euler circles,* after the famous mathematician Leonhard Euler, who used circles and ovals to analyze this type of argument. However, Venn used these circles in a more general way.

b. Again, begin with a Venn diagram:
If you like potato chips, then you will like Krinkles.

The first premise is the same as:

All people who like potato chips like Krinkles.

Let p: people who like potato chips
q: people who like Krinkles

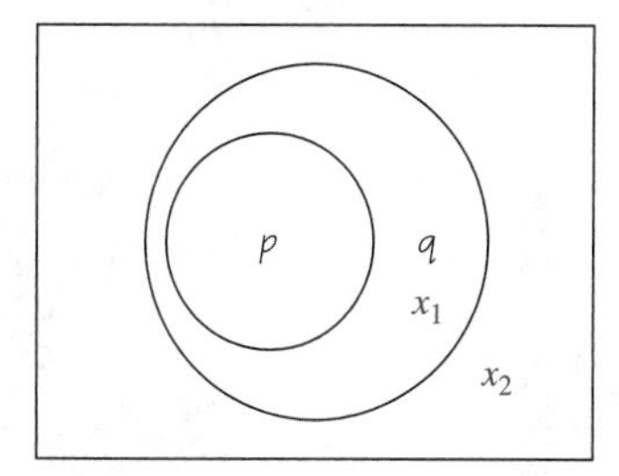

For the second premise, you will place the x (you) outside the circle labeled p. Notice that you are not forced to place x into a single region; it could be placed in either of two places—those labeled x_1 and x_2. Since the stated conclusion is not forced, the argument is **not valid.** ■

Many syllogisms require more than two sets. The next example requires that you label all eight parts of the Venn diagram.

EXAMPLE 4

Naming the Parts of a Venn Diagram with Three Sets

If the regions of a Venn diagram are labeled as shown in Figure 8.12, describe each numbered region in words.

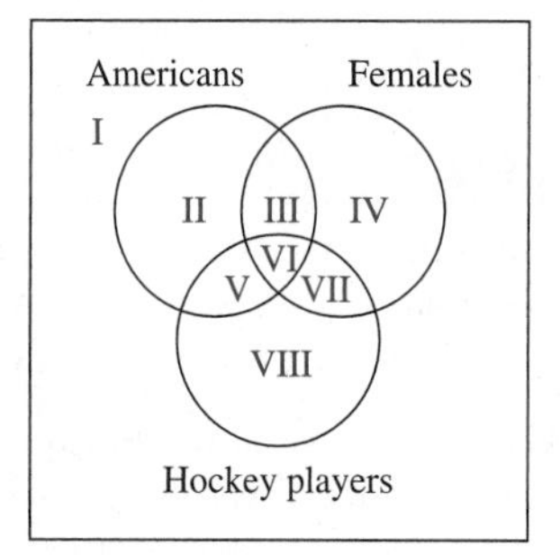

Figure 8.12 Sets of Americans, females, and hockey players

Solution We note (from common usage) that a nonfemale is a male.

Region I:	Non-American males who are not hockey players
Region II:	Male Americans who are not hockey players
Region III:	American females who are not hockey players
Region IV:	Non-American females who are not hockey players
Region V:	American male hockey players
Region VI:	American female hockey players
Region VII:	Non-American female hockey players
Region VIII:	Non-American male hockey players

■

EXAMPLE 5 **Using Venn Diagrams to Test the Validity of an Argument**

Test the validity of the following argument.

Premises: 1. All dictionaries are useful books.
2. All useful books are valuable.

Conclusion: All dictionaries are valuable.

Solution Let U be the set of all books.

d: set of dictionaries
u: set of useful books
v: set of valuable books

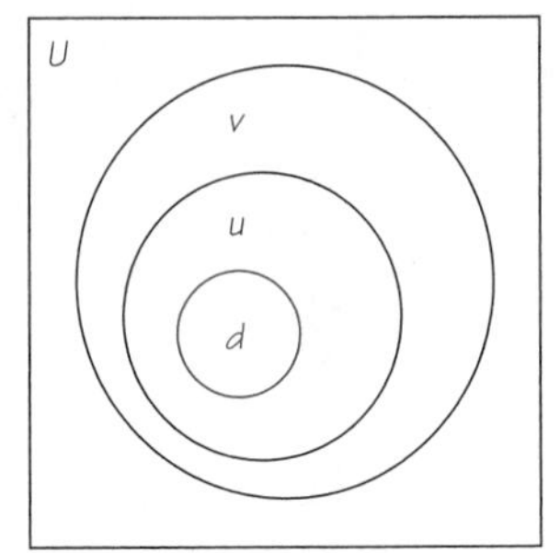

We draw a Venn diagram. The first premise identifies d as a subset of u. The second premise identifies u as a subset of v.

We now check to see whether we can avoid the conclusion; in this case, we cannot. The set d *must* be inside the set v, so the conclusion is **valid.** ■

PROBLEM SET 8.6

Essential Ideas

1. **In Your Own Words** Explain inductive reasoning. Give an original example of an occasion when you have used inductive reasoning or heard it being used.

2. **In Your Own Words** In your own words, explain deductive reasoning. Give an original example of an occasion when you have used deductive reasoning or heard it being used.

Level 1

3. **In Your Own Words** Does the following *B.C.* cartoon illustrate inductive or deductive reasoning? Explain your answer.

B.C. By permission of Johnny Hart and Creators Syndicate, Inc.

4. **In Your Own Words** Does the news story illustrate inductive or deductive reasoning? Explain your answer.

> The young man in charge of the checkroom in a fancy restaurant was known for his memory. He never used the usual markers to identify the hats and coats. One day a guest decided to put him to a test, and said to the clerk, "How do you know this is my hat?" "I don't," was the response. "But why did you return it to me?" asked the guest. "Because," said the clerk, "it's the one you gave to me."

5. **In Your Own Words** What is a syllogism?

6. Consider the successive products of 8 ($1 \times 8, 2 \times 8, 3 \times 8, 4 \times 8, \ldots$). If the product has more than 1 digit, add the digits successively to obtain a single digit. What is the pattern that you observe?

7. Consider the successive products of 3 ($1 \times 3,\ 2 \times 3,\ 3 \times 3,\ 4 \times 3, \ldots$). If the product has more than 1 digit, add the digits successively to obtain a single digit. What pattern do you observe?

8. What is the sum of the first 25 consecutive odd numbers?

9. What is the sum of the first 1,000 consecutive odd numbers?

10. What is the 100th consecutive even number?

11. Give the 100th number in the pattern: 1, 4, 9, 16, 25, . . .

12. Compute the following.
    a. 11^2 **b.** 111^2 **c.** $1{,}111^2$
    d. Describe a pattern from parts **a–c.**

13. Compute the following.
    a. $9 \times 1 - 1$ **b.** $9 \times 21 - 1$
    c. $9 \times 321 - 1$
    d. Describe a pattern from parts **a–c.**

14. If the regions of a Venn diagram are labeled as shown, describe each region in words.

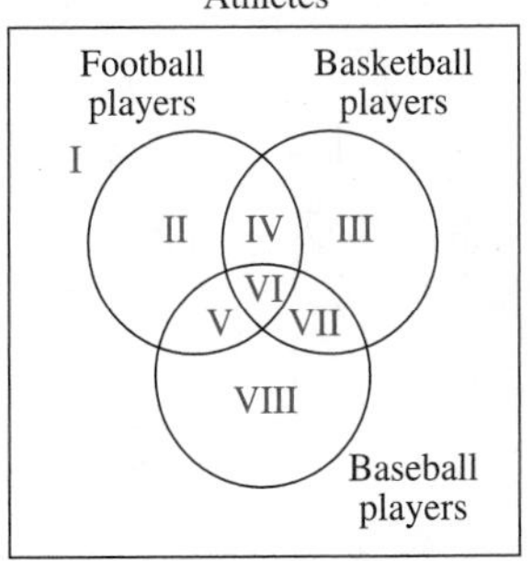

15. If the regions of a Venn diagram are labeled as shown, describe each region in words.

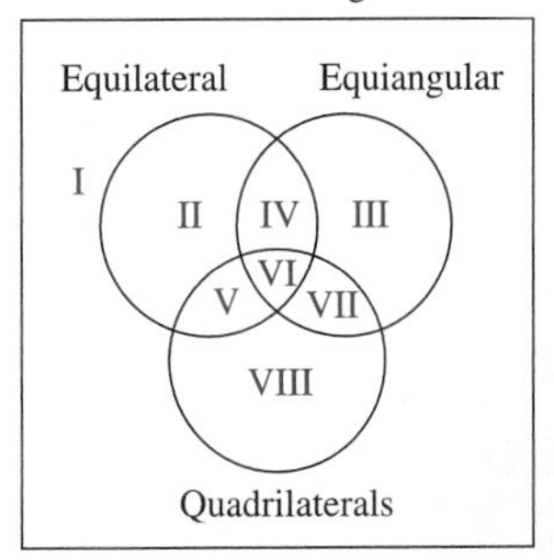

16. If the regions of a Venn diagram are labeled as shown, describe each region in words.

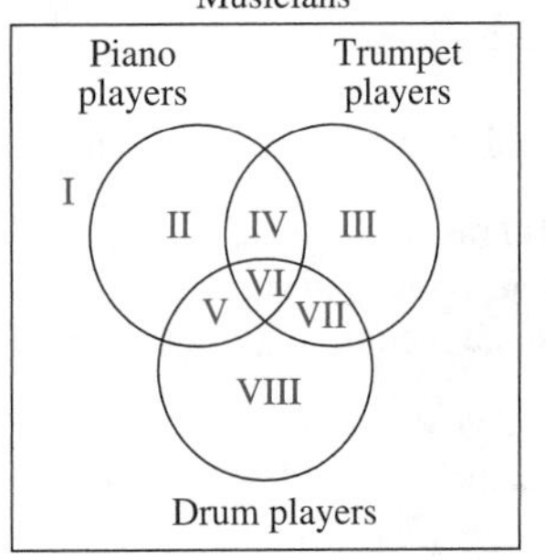

Tell whether you believe that the conjecture in Problems 17–26 *is true or false. If you believe it to be false, see if you can find a counterexample.*

17. The sum of two even numbers is even.

18. The sum of two odd numbers is odd.

19. The sum of two primes is prime.

20. The product of two even numbers is even.

21. The product of two odd numbers is odd.

22. The product of two primes is prime.

23. The area of a rectangle (number of square units) is larger than its perimeter (number of units).

24. The basic metric units for length, capacity, and weight are larger than their U.S. counterparts.

25. All of the subsets of $S = \{a, b, c, d, e\}$ have cardinality less than the cardinality of S.

26. Every equilateral triangle has one angle of 60°.

Level 2

Use Venn diagrams to check the validity of the arguments in Problems 27–37.

27. All mathematicians are eccentrics.
All eccentrics are rich.
Therefore, all mathematicians are rich.

28. Some beautiful women are blond.
Blonds have more fun.
Therefore, some beautiful women have more fun.

29. All bachelors are handsome.
Some bachelors do not drink lemonade.
Therefore, some handsome men do not drink lemonade.

30. No students are enthusiastic.
You are enthusiastic.
Therefore, you are not a student.

31. Some women are tall.
No men are strong.
Therefore, some tall people are not strong.

32. No politicians are honest.
Some dishonest people are found out.
Therefore, some politicians are found out.

33. All candy is fattening.
All candy is delicious.
Therefore, all fattening food is delicious.

34. No professors are ignorant.
All ignorant people are vain.
Therefore, no professors are vain.

35. No monkeys are soldiers.
All monkeys are mischievous.
Therefore, some mischievous creatures are not soldiers.

36. All lions are fierce.
Some lions do not drink coffee.
Therefore, some creatures that drink coffee are not fierce.

37. All red hair is pretty.
No pretty things are valuable.
Therefore, no red hair is valuable.

Level 2—Applications

Problems 38–41 *refer to the lyrics of "By the Time I Get to Phoenix." Tell whether each answer you give is arrived at inductively or deductively.*

By the Time I Get to Phoenix

By the time I get to Phoenix she'll be risin'.
She'll find the note I left hangin' on her door.
She'll laugh when she reads the part that says I'm leavin',
'Cause I've left that girl so many times before.

By the time I make Albuquerque she'll be workin'.
She'll probably stop at lunch and give me a call.
But she'll just hear that phone keep on ringin'
Off the wall, that's all.

By the time I make Oklahoma she'll be sleepin'.
She'll turn softly and call my name out low.
And she'll cry just to think I'd really leave her,
'tho' time and time I've tried to tell her so,
She just didn't know
I would really go.

38. In what basic direction (north, south, east, or west) is the person traveling?

39. What method of transportation or travel is the person using?

40. What is the probable starting point of this journey?

41. List five facts you know about each person involved.

Problems 42–45 *refer to the lyrics of "Ode to Billy Joe." Tell whether each answer you give is arrived at inductively or deductively.*

Ode to Billy Joe

It was the third of June, another sleepy, dusty, delta day.
I was choppin' cotton and my brother was balin' hay.
And at dinnertime we stopped and walked back to the house to eat,
And Mama hollered at the back door, "Y'all remember to wipe your feet."
Then she said, "I got some news this mornin' from Choctaw Ridge,
Today Billy Joe McAllister jumped off the Tallahatchee Bridge."

Papa said to Mama, as he passed around the black-eyed peas,
"Well, Billy Joe never had a lick o' sense, pass the biscuits please,
There's five more acres in the lower forty I've got to plow,"
And Mama said it was a shame about Billy Joe anyhow.
Seems like nothin' ever comes to no good up on Choctaw Ridge,
And now Billy Joe McAllister's jumped off the Tallahatchee Bridge.

Brother said he recollected when he and Tom and Billy Joe,
Put a frog down my back at the Carroll County picture show,
And wasn't I talkin' to him after church last Sunday night,
I'll have another piece of apple pie, you know, it don't seem right,
I saw him at the sawmill yesterday on Choctaw Ridge,
And now you tell me Billy Joe's jumped off the Tallahatchee Bridge."

Mama said to me, "Child, what's happened to your appetite?
I been cookin' all mornin' and you haven't touched a single bite,
That nice young preacher Brother Taylor dropped by today,
Said he'd be pleased to have dinner on Sunday, Oh, by the way,
He said he saw a girl that looked a lot like you up on Choctaw Ridge
And she an' Billy Joe was throwin' somethin' off the Tallahatchee Bridge."

A year has come and gone since we heard the news 'bout Billy Joe,
Brother married Becky Thompson, they bought a store in Tupelo,
There was a virus goin' round, Papa caught it and he died last spring.
And now Mama doesn't seem to want to do much of anything.
And me I spend a lot of time pickin' flowers up on Choctaw Ridge,
And drop them into the muddy water off the Tallahatchee Bridge.

Music and lyrics by Bobbie Gentry. © 1967

42. How many people are involved in this story? List them by name and/or description.

43. Who "saw him at the sawmill yesterday"?

44. In which state is the Tallahatchee Bridge located?

45. On what day or days of the week could the death not have taken place? On what day of the week was the death most probable?

Problems 46–51 *are included to develop critical thinking and are not directly related to the material in the text. Some will seem very easy, whereas others will seem very difficult. You are not expected to be able to answer them all, but you are expected to try.*

46. If you had only one match and entered a room in which there was a kerosene lamp, an oil burner, and a wood-burning stove, which would you light first?

47. Do they have a 4th of July in England?

48. Some months have 30 days and some have 31 days. How many have 28 days?

49. A man built a house that has four sides and it is rectangular. Each side has a southern exposure. A big bear came wandering by. What color is the bear?

50. If a doctor gave you three pills and told you to take one every half-hour, how long would they last you?

51. A woman gives a beggar 50¢. The woman is the beggar's sister, but the beggar is not the woman's brother. How come?

52. Situation "Sarah, do you remember the old song 'Ode to Billy Joe'? I just saw the movie on my VCR and it was great. I finally figured out what Billy Joe threw off the Tallahatchee Bridge," said Bob. "Oh, yeah? I'll bet you didn't!" answered Sarah. "Anyway," continued Sarah, "the movie is not faithful to the song. From the song you cannot figure out for sure what was thrown from the bridge." Is it possible to use deductive reasoning to answer this question?

53. In Your Own Words Write a short paper comparing the SITUATION with your own situation. (That is, find some examples from your life in which you have used inductive or deductive reasoning. Give examples, sources, and classifications as inductive or deductive.)

8.7 SUMMARY AND REVIEW

Important Terms

Axiom [8.6]
Belong to a set [8.1]
Cardinal number [8.1]
Cardinality [8.1]
Circular definition [8.1]
Complement [8.3]
Conclusion [8.6]
Conjecture [8.6]
Contained in a set [8.1]
Counterexample [8.6]
Deductive reasoning [8.6]
De Morgan's laws [8.4]
Description method [8.1]
Disjoint sets [8.2]
Element [8.1]
Empty set [8.1]
Equal sets [8.1]
Equivalent sets [8.1]
Finite set [8.1]
Improper subset [8.2]
Inductive reasoning [8.6]
Infinite set [8.1]
Intersection [8.3]
Logic [8.6]
Member [8.1]
One-to-one correspondence [8.1]
Overlapping sets [8.2]
Premise [8.6]
Proper subset [8.2]
Proved result [8.6]
Roster method [8.1]
Set [8.1]
Set-builder notation [8.1]
Subset [8.2]
Syllogism [8.6]
Theorem [8.6]
Undefined terms [8.1]
Union [8.3]
Universal set [8.1]
Valid [8.6]
Venn diagram [8.2; 8.4]
Well-defined set [8.1]

Chapter Objectives

The material in this chapter is reviewed in the following list of objectives. A self-test (with answers and suggestions for additional study) is given. The self-test is constructed so that each problem number corresponds to a related objective. This self-test is followed by a practice test with the questions in mixed order.

[8.1] *Objective 8.1* Decide whether a given set is well defined.

[8.1] *Objective 8.2* Specify a set in roster form, given a description or set-builder notation.

[8.1] *Objective 8.3* Specify a set given by roster in description form.

[8.1] *Objective 8.4* Decide whether two sets can be placed into a one-to-one correspondence; if so, show at least one such correspondence.

[8.1] *Objective 8.5* Classify a given set as finite or infinite.

[8.2] *Objective 8.6* List all the subsets of a given set.

[8.3] *Objective 8.7* Find the union of given sets.

[8.3] *Objective 8.8* Find the intersection of given sets.

[8.3] *Objective 8.9* Find the complement of a given set.

[8.3] *Objective 8.10* Draw Venn diagrams for union, intersection, and complement.

[8.4] *Objective 8.11* Draw a Venn diagram for combined operations with sets.

[8.4] *Objective 8.12* Prove or disprove set statements by using Venn diagrams.

[8.4] *Objective 8.13* Find members of a set given by combined operations among sets.

[8.4] *Objective 8.14* Draw a Venn diagram to represent given sets.

[8.5] *Objective 8.15*	Answer questions about a survey by using a Venn diagram.
[8.6] *Objective 8.16*	Distinguish between inductive and deductive reasoning.
[8.6] *Objective 8.17*	Use inductive reasoning to find a pattern, or find a counterexample if the statement is not true.
[8.6] *Objective 8.18*	Use Venn diagrams to check the validity of a given argument.

Self-Test

Each question of this self-test is related to the corresponding objective listed above.

1. Is the following set well defined?

The set of persons in this classroom who are under 20 years of age

2. Write the following set in roster form:

$\{x \mid 2 < x \leq 5; x \in \mathbb{N}\}$

3. Write the following set in description form:

$\{101, 103, 105, \ldots, 197, 199\}$

4. Can the following sets be placed in a one-to-one correspondence?

$\mathbb{N} = \{1, 2, 3, 4, \ldots\}$ and $E = \{2, 4, 6, 8, \ldots\}$

If so, show one such correspondence; if not, give a counterexample.

5. Is the number of reruns of *I Love Lucy* finite or infinite? Explain your answer.

6. List all subsets of the set $\{0, 1\}$.

7. Find $\{1, 2, 3, 4, \ldots\} \cup \{2, 4, 6, 8, \ldots\}$.

8. Let $A = \{\text{multiples of } 2\}$ and $B = \{\text{multiples of } 3\}$. Find $A \cap B$.

9. If the universe is $\mathbb{W}$ (whole numbers), find $\overline{\mathbb{N}}$ where $\mathbb{N}$ is the set of natural numbers.

10. Draw a Venn diagram for: **a.** $A \cap C$ **b.** $A \cup B$

11. Draw a Venn diagram for $\overline{A \cup \overline{B}}$.

12. Prove or disprove that $\overline{A \cup B} = \overline{A} \cup \overline{B}$.

13. Let $U = \{m, a, t, h, i, s, f, u, n\}$, $A = \{m, a\}$, $B = \{t, h, i, s\}$, $C = \{i, s\}$, and $D = \{n, u, t, s\}$; find $\overline{A \cup B} \cup (D \cap C)$.

14. Draw a Venn diagram showing people who read *MAD* magazine or *Rolling Stone.* Also, suppose 19 people like to read *MAD* and 25 read *Rolling Stone,* but 8 like to read both. How many like to read either one or the other or both?

15. A survey of 70 college students showed the following data: 42 had a car; 50 had a TV; 30 had a bicycle; 17 had a car and a bicycle; 35 had a car and a TV; 25 had a TV and a bicycle; and 15 had all three.

16. Your brother watches *Saturday Night Live* every Saturday evening, and tonight is Saturday evening, so you conclude that he must be watching *Saturday Night Live.* What type of reasoning are you using?

17. What is the sum of the first 99 counting numbers?

18. Use a Venn diagram to check the validity of the following argument:

All squares are rectangles.

All rectangles are polygons.

Therefore, all squares are polygons.

STUDY HINTS *Compare your solutions and answers to the self-test. For each problem you missed, work some additional problems in the section listed in the margin. After you have worked these problems, you can test yourself with the practice test.*

Complete Solutions to the Self-Test

Additional Problems

Additional Problems	Solution
[8.1] *Problems 13–20*	**1.** Yes, it is well defined because it is clear what it means to be under 20 years of age.
[8.1] *Problems 21–28; 35–40*	**2.** {3, 4, 5}
[8.1] *Problems 29–34*	**3.** Answers vary; the set of odd counting numbers between 100 and 200.
[8.1] *Problems 43–48*	**4.** Yes; answers vary.

$$\begin{array}{l} \mathbb{N} = \{1, 2, 3, 4, \ldots\} \\ \quad\ \ \updownarrow\ \updownarrow\ \updownarrow\ \updownarrow \\ E = \{2, 4, 6, 8, \ldots\} \end{array}$$

Additional Problems	Solution
[8.1] *Problem 49*	**5.** Finite; answers vary.
[8.2] *Problems 7–14*	**6.** $\varnothing$, {0}, {1}, {0, 1}
[8.3] *Problems 9; 12–13; 19–21*	**7.** $\mathbb{N} \cup E = \{1, 2, 3, 4, \ldots\} = \mathbb{N}$, where E is the set of even numbers.
[8.3] *Problems 10–11; 14; 22–24*	**8.** $A \cap B$ = {numbers that are multiples of both 2 and 3} = {multiples of 6}
[8.3] *Problems 15–18; 25–28*	**9.** {0}
[8.3] *Problems 29–34*	**10. a.**

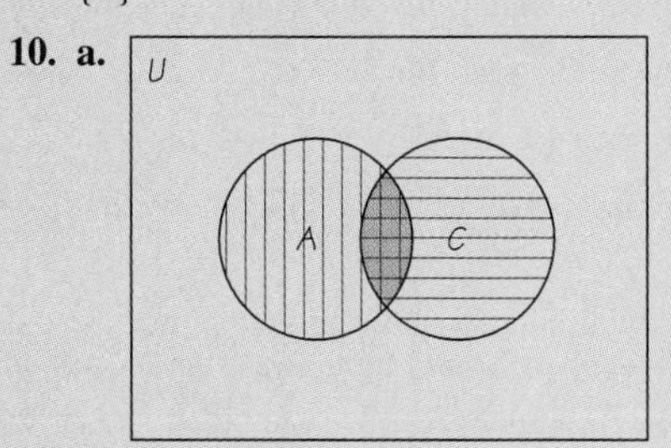

b.

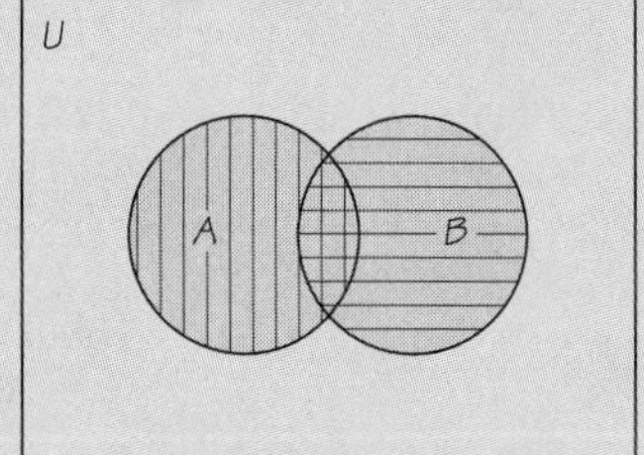

[8.4] *Problems 11–25*

11.

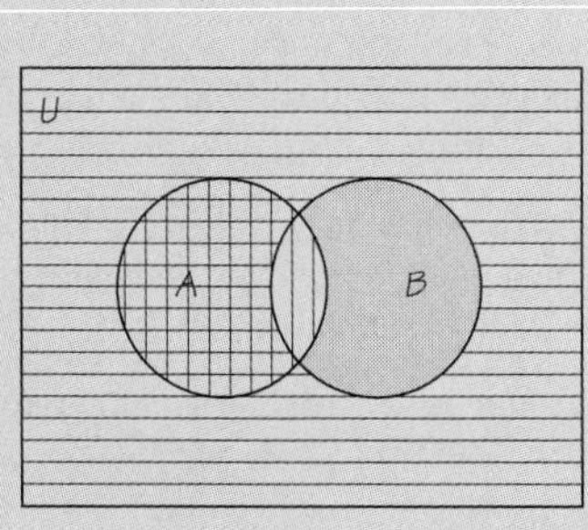

[8.4] *Problems 36–41*

12. Not valid

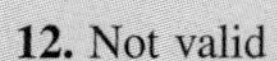

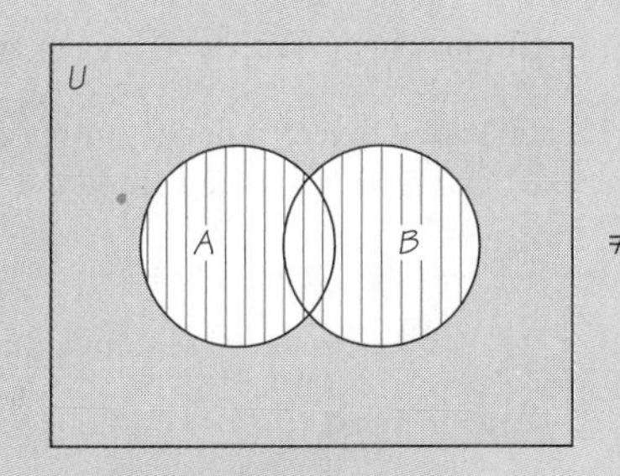

$\neq$

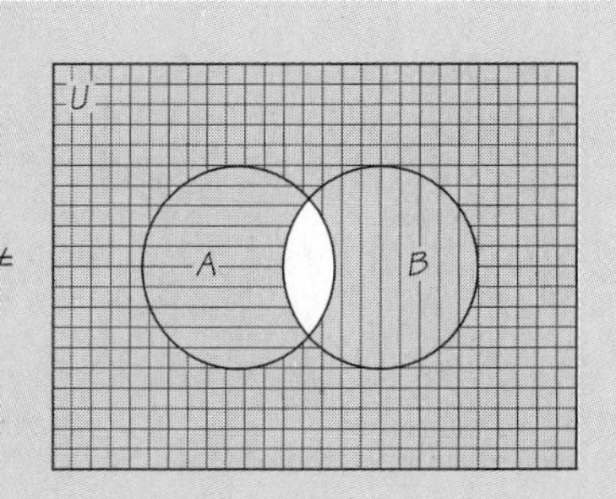

[8.4] *Problems 26–35*

13. $\overline{A \cup B} \cup (D \cap C) = \overline{\{m, a, t, h, i, s\}} \cup \{s\}$
$= \{f, u, n\} \cup \{s\}$
$= \{f, u, n, s\}$

[8.4] *Problems 42–48*

14.

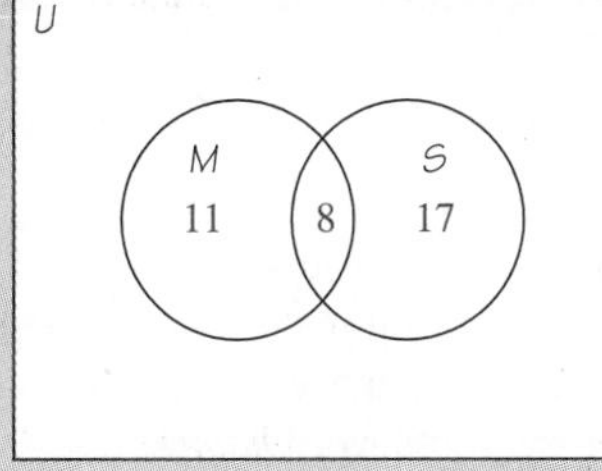

$19 + 25 - 8 = 36$

[8.5] *Problems 4–20*

15. Draw a Venn diagram; begin with the innermost part first:
15 have all three.
Next, use the information for two overlapping sets to fill in 2, 20, and 10.
Fill in the regions in each circle not yet completed to fill in 5, 5, and 3.
Finally, total all the numbers to see that 10 of the students have none of these items.

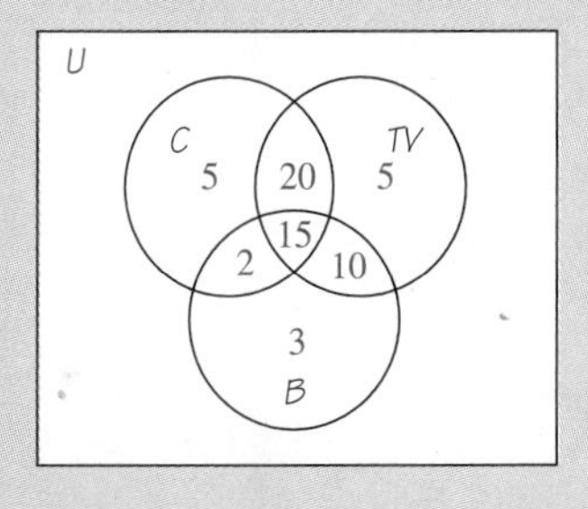

[8.6] *Problems 3–4; 38–45*

16. It is inductive reasoning because it is based on past observations.

[8.6] *Problems 6–13; 17–26*

17. Look for a pattern:
$$1 + 2 + 3 + \cdots + 49 + 50 + 51 + \cdots + 97 + 98 + 99$$
Pair up first and last on list $(1 + 99 = 100)$
next pair $(2 + 98 = 100)$
next pair $(3 + 97 = 100)$
$\vdots$
There are 49 pairs with middle number of 50 not paired. Thus,

$$49 \times 100 + 50 = 4{,}950$$

[8.6] *Problems 14–16; 27–37*

18.

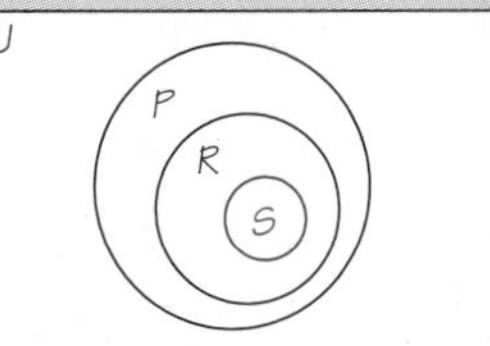

It is valid.

Chapter 8 Practice Test

1. **In Your Own Words** Compare and contrast inductive and deductive reasoning.

2. **In Your Own Words** If you make a conjecture based on a series of observations, what type of reasoning are you using?

3. True or false? (If false, give a counterexample.)

   There are no irrational numbers between 0 *and* 1.

4. True or false? (If false, give a counterexample.)

   $x^2 > x$ for all x

5. Are the given sets well defined?
   a. The set of good baseball players
   b. {baseball players earning more than \$1 million per year}
   c. $\{x \mid x < 0, x \in \mathbb{N}\}$
   d. $\{x \mid x$ is a happy number$\}$

6. Write the given sets by roster.
   a. {even counting numbers less than 10}
   b. $\{x \mid x$ is a positive multiple of 3 that is less than 30$\}$
   c. $\{x \mid x$ is a counting number between 453 and 490$\}$
   d. {distinct letters in the word *happy*}

7. Write the given sets by description.
   a. {10, 12, 14, 16, 18}
   b. {2001, 2003, 2005, . . . , 2999}
   c. {1, 2, 3, 4, . . .}
   d. {1, 4, 9, 16, 25, 36, . . . , 100, 121, 144}

8. Do the given pairs of sets have the same cardinality? If so, show a possible one-to-one correspondence.
   a. $A = \{3, 5, 7, 9\}$; $B = \{a, b, c, d\}$
   b. $C = \{1, 2, 3, \ldots, 588\}$; $D = \{185, 187, 189, \ldots, 1361\}$
   c. $E = \{1, 2, 3, \ldots, 40\}$; $F = \{988, 978, , 508\}$
   d. $G = \{1, 2, 3, \ldots\}$; $H = \{2, 4, 6, \ldots\}$

9. Tell whether the given sets are finite or infinite.
   a. The set of counting numbers greater than 5 million
   b. The set of counting numbers less than 5 million
   c. The set of cells in your body at a particular time
   d. The set of all possible books that have been written or could ever be written.

10. List the subsets of the given sets.
    a. $\{H, I\}$
    b. $\{m, a, d\}$
    c. {4, 5, 6, 7}
    d. {0}

For Problems 11–13, *let* $U = \mathbb{N}$ *(counting numbers); and let*

$A = \{1, 2, 3, 4, 5, 6, 7, 8, 9, 10\}$
$B = \{x \mid x \text{ is a multiple of 3 that is less than } 31\}$
$E = \{x \mid x \text{ is an even counting number}\}$

11. Find the union as requested.
a. $\{1, 3, 5, 7, 9\} \cup \{2, 4, 5, 9, 10\}$
b. $\{10, 11, \ldots, 24, 25\} \cup B$
c, $A \cup B$
d. $\varnothing \cup U$

12. Find the intersection as requested.
a. $\{1, 3, 5, 7, 9\} \cap \{2, 4, 5, 9, 10\}$
b. $\{10, 11, \ldots, 24, 25\} \cap B$
c. $A \cap B$
d. $\varnothing \cap U$

13. Find the complements.
a. $\overline{\{1, 3, 5, 7, 9\}}$
b. $\overline{E}$
c. $\overline{A}$
d. $\overline{\varnothing}$

14. Draw Venn diagrams for the given sets.
a. $\overline{A \cap B}$ **b.** $A \cup B$ **c.** $\overline{A}$ **d.** $\overline{B}$

15. Draw Venn diagrams for the given sets.
a. $A \cup \overline{B}$ **b.** $\overline{A} \cup \overline{B}$ **c.** $(A \cup B) \cap \overline{C}$ **d.** $\overline{(A \cup B)} \cup C$

16. Prove or disprove.
a. $(A \cup B) \cup C = A \cup (B \cup C)$
b. $A \cap (B \cup C) = (A \cap B) \cup C$
c. $(A \cup B) \cap C = (A \cap C) \cup (B \cap C)$

17. Prove one of De Morgan's laws.

18. Use the results of the following survey. The survey included 500 motorists.

140 had received a ticket for speeding.
120 had received a ticket for failure to yield the right-of-way.
98 had received a ticket for failure to stop at a stop sign.
41 had received tickets for speeding and failure to yield.
35 had received tickets for speeding and failure to stop at a stop sign.
47 had received tickets for failure to yield and failure to stop at a stop sign.
24 had received tickets for all three violations.

a. How many motorists did not receive any tickets?
b. How many motorists received only one ticket?
c. How many motorists received exactly two tickets?
d. How many motorists received only a ticket for speeding?

19. Does the following story illustrate inductive or deductive reasoning?

Q: What has 18 legs and catches flies?
A: I don't know, what?
Q: A baseball team. What has 36 legs and catches flies?
A: I don't know that either.
Q: Two baseball teams. If the United States has 100 senators and 50 states, what does . . . ?
A: I know this one!
Q: Good. What does each state have?
A: Three baseball teams!

20. Consider the successive products of 6. If the product has more than one digit, add the digits successively to obtain a single digit. What is the pattern you observe?

21. What is the 100th number in the pattern 2, 5, 10, 17, 26, . . . ?

Use a Venn diagram to check the validity of each of the arguments given in Problems 22–25.

22. All artists are creative.
Some musicians are artists.
Therefore, some musicians are creative.

23. No apples are bananas.
All apples are fruit.
Therefore, no bananas are fruit.

24. All birds have wings.
All flies have wings.
Therefore, some flies are birds.

25. All rectangles are polygons.
All squares are rectangles.
Therefore, all squares are polygons.

CHAPTER NINE

PROBABILITY

9.1 INTRODUCTION TO PROBABILITY

Situation

We see examples of probability every day. Weather forecasts, stock market analyses, contests, children's games, television game shows, and gambling all involve ideas of probability. Probability is the mathematics of uncertainty. "Wait a minute," interrupts Sammy. "I thought that mathematics was absolute and that there was no uncertainty about it!" Well, Sammy, suppose that you are playing a game of Monopoly. Do you know what you'll land on in your next turn? "No, but neither do you!" That's right, but mathematics can tell us what property you are most likely to land on. There is nothing uncertain about probability; rather, probability is a way of describing uncertainty. For example, what is the probability of tossing a coin and obtaining heads? "That's easy; it's one-half," answers Sammy. Right, but why do you say it is one-half?

In this section, we'll investigate the definition of probability and some of the procedures for dealing with probability.

Historical Perspective

Irving Hertzel, a professor from Iowa State University, used a computer to determine the most probable squares in the game of Monopoly™: 1. Illinois Avenue; 2. Go; 3. B & O Railroad; 4. Free Parking.

TERMINOLOGY

An **experiment** is an observation of any physical occurrence. The **sample space** of an experiment is the set of all its possible outcomes. An **event** is a subset of the sample space. If an event is the empty set, it is called the **impossible event;** and if it has only one element, it is called a **simple event.**

EXAMPLE 1 **Listing a Sample Space**

a. What is the sample space for the experiment of tossing a coin and then rolling a die?

b. List the following events for the sample space in part **a:**

$E = \{\text{rolling an even number on the die}\}$

$H = \{\text{tossing a head}\}$

$X = \{\text{rolling a six and tossing a tail}\}$

Which of these (if any) are simple events?

Solution A coin is considered **fair** if the outcomes of head and tail are **equally likely.** We also note that a fair die is one for which the outcomes from rolling it are *equally likely.* A die for which one outcome is more likely than the others is called a **loaded die.** In this book, we will assume fair dice and fair coins unless otherwise noted.

Figure 9.1 Sample space for tossing a coin and rolling a die

a. You can visualize the sample space as shown in Figure 9.1. If the sample space is very large, it is sometimes worthwhile to make a direct listing, as shown in Figure 9.1. For this reason, it is sometimes helpful to build sample spaces by using what is called a **tree diagram:**

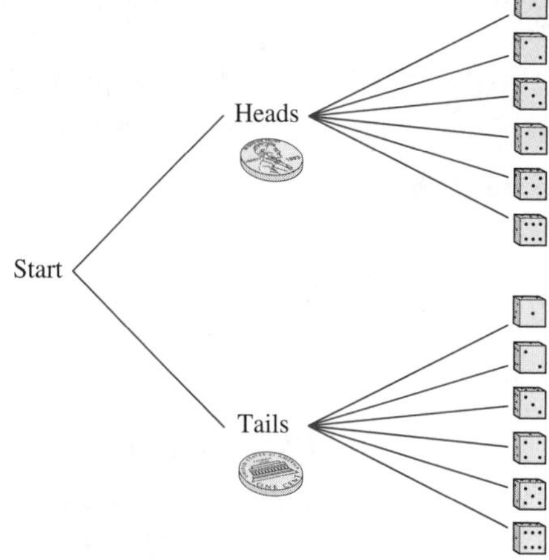
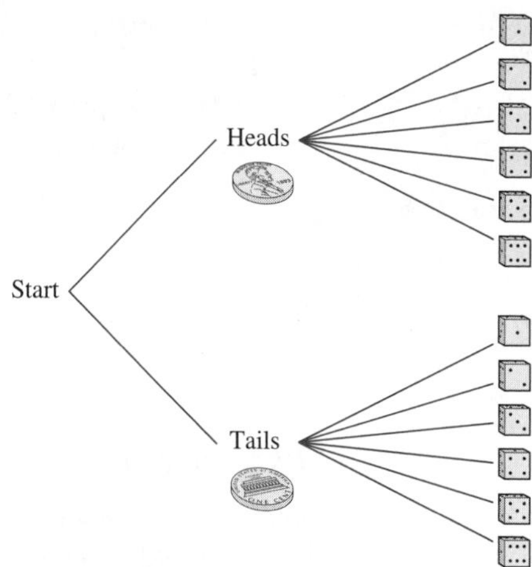

We see that S (the sample space) is

$S =$ **{H1, H2, H3, H4, H5, H6, T1, T2, T3, T4, T5, T6}**

b. An event must be a *subset* of the sample space. Notice that E, H, and X are all subsets of S. Therefore,

$E =$ **{H2, H4, H6, T2, T4, T6}**

$H =$ **{H1, H2, H3, H4, H5, H6}**

$X =$ **{T6}**

***X* is a simple event,** because it has only one element. ■

Two events E and F are said to be **mutually exclusive** if $E \cap F = \varnothing$.

EXAMPLE 2 **Mutually Exclusive Events**

Suppose that you perform an experiment of rolling a die. Find the sample space, and then let $E = \{1, 3, 5\}$, $F = \{2, 4, 6\}$, $G = \{1, 3, 6\}$, and $H = \{2, 4\}$. Which of these are mutually exclusive?

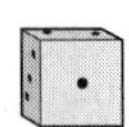
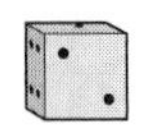

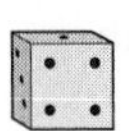

Figure 9.2 Sample space for a single die

Solution The **sample space for a single die,** as shown in Figure 9.2, is

$$S = \{1, 2, 3, 4, 5, 6\}$$ We look only at the number of spots on the top.

Evidently,

***E* and *F* are mutually exclusive,** since $E \cap F = \varnothing$.

***G* and *H* are mutually exclusive,** since $G \cap H = \varnothing$.

***E* and *H* are mutually exclusive,** since $E \cap H = \varnothing$.

But

***F* and *H* are *not* mutually exclusive** (elements 2 and 4 in common).

***E* and *G* are *not* mutually exclusive** (elements 1 and 3 in common).

***F* and *G* are *not* mutually exclusive** (element 6 in common). ■

PROBABILITY

If the sample space can be divided into *mutually exclusive* and *equally likely* outcomes, we can define the probability of an event. Let's consider the experiment of tossing a single coin. A suitable sample space is

$$S = \{\text{heads, tails}\}$$

Suppose we wish to consider the event of obtaining heads; we'll call this event A. Then,

$$A = \{\text{heads}\}$$

and this is a simple event.

We wish to define the probability of event A, which we denote by $P(A)$. Notice that the outcomes in the sample space are mutually exclusive; that is, if one occurs, the other cannot occur. If we flip a coin, there are two possible outcomes, and *one and only one* outcome can occur on a toss. If each outcome in the sample space is equally likely, we define the probability of A as

$$P(A) = \frac{\text{NUMBER OF SUCCESSFUL RESULTS}}{\text{NUMBER OF POSSIBLE RESULTS}}$$

A "successful" result is a result that corresponds to the event whose probability we are seeking—in this case, {heads}. Since we can obtain a head (success) in only one way, and the total number of possible outcomes is two, the probability of heads is given by this definition as

$$P(\text{heads}) = P(A) = \tfrac{1}{2}$$

This must correspond to the empirical results you would obtain if you repeated the experiment a large number of times. In Problem 58 of Problem Set 9.1, you are asked to repeat this experiment 100 times.

Definition of Probability

This is the foundational definition for this chapter.

If an experiment can occur in any of n mutually exclusive and equally likely ways, and if s of these ways are considered favorable, then the **probability** of an event E, denoted by $P(E)$, is

$$P(E) = \frac{s}{n} = \frac{\text{NUMBER OF OUTCOMES FAVORABLE TO } E}{\text{NUMBER OF ALL POSSIBLE OUTCOMES}}$$

EXAMPLE 3 **Using the Definition of Probability**

Use the definition of probability to find first the probability of white and second the probability of black, using the spinner shown and assuming that the arrow will never lie on a border line.

Solution Looking at the spinner, we note that it is divided into three areas of the same size. We assume that the spinner is equally likely to land in any of these three areas.

$P(\text{white}) = \dfrac{\mathbf{2}}{\mathbf{3}}$ ← Two sections are white. ← Three sections altogether

$P(\text{black}) = \dfrac{\mathbf{1}}{\mathbf{3}}$ ← One section is black. ← Three sections altogether

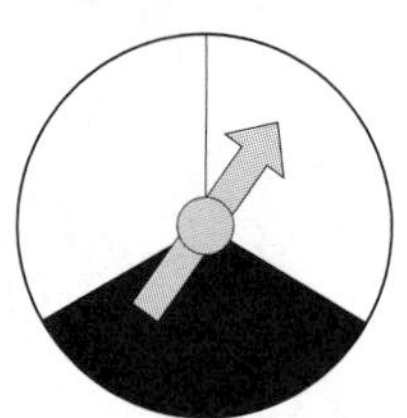

■

EXAMPLE 4 **Using the Definition of Probability**

Consider a jar that contains marbles as shown. Suppose that each marble has an equal chance of being picked from the jar. Find:

a. $P(\text{black})$ **b.** $P(\text{blue})$ **c.** $P(\text{white})$

Solution

a. $P(\text{black}) = \dfrac{4}{12}$ ← 4 black marbles in jar ← 12 marbles in jar

$= \dfrac{\mathbf{1}}{\mathbf{3}}$ Reduce fractions.

b. $P(\text{blue}) = \dfrac{\mathbf{7}}{\mathbf{12}}$

c. $P(\text{white}) = \dfrac{\mathbf{1}}{\mathbf{12}}$

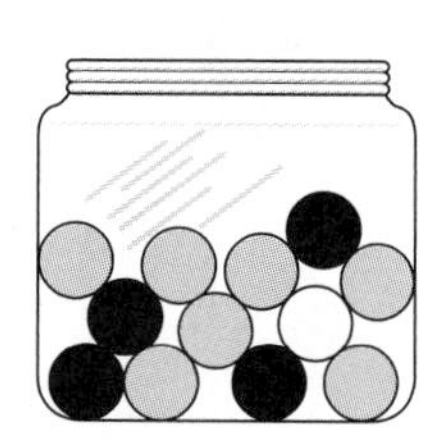

■

Reduced fractions are used to state probabilities when the fractions are fairly simple. If, however, the fractions are not simple, and you have a calculator, it is acceptable to state the probabilities as decimals, as shown in Example 5.

EXAMPLE 5

Using the Definition of Probabilities with Decimals

Suppose that, in a certain study, 46 out of 155 people showed a certain kind of behavior. Assign a probability to this behavior.

Solution $P(\text{behavior shown}) = \dfrac{46}{155} \approx \mathbf{0.3}$ *Calculator display:* .2967741935 ■

EXAMPLE 6

Using Probabilities in Playing a Game

Consider two spinners as shown. You and an opponent are to spin a spinner simultaneously, and the one with the higher number wins. Which spinner should you choose, and why?

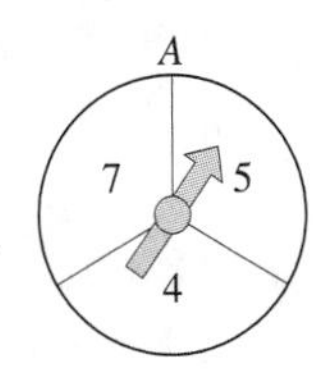

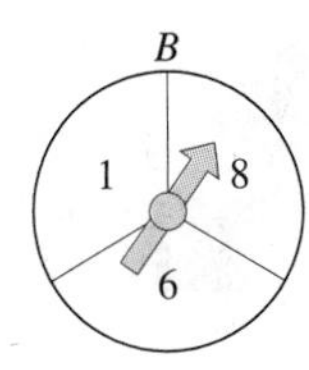

Solution We begin by listing the sample space:

A \ B	1	6	8
4	[(4, 1)]	(4, 6)	(4, 8)
5	[(5, 1)]	(5, 6)	(5, 8)
7	[(7, 1)]	[(7, 6)]	(7, 8)

The times that A wins are enclosed in boxes.

$$P(A \text{ wins}) = \frac{4}{9}; \quad P(B \text{ wins}) = \frac{5}{9}$$

We would choose spinner B because it has a greater probability of winning. ■

EXAMPLE 7

Probabilities with a Deck of Cards

Suppose that a single card is selected from an ordinary deck of 52 cards. Find:

a. $P(\text{ace})$
b. $P(\text{heart})$
c. $P(\text{face card})$

Solution The **sample space for a deck of cards** is shown in Figure 9.3.

Spades (black cards)

Hearts (red cards)

Clubs (black cards)

Diamonds (red cards)

Figure 9.3 Sample space for a deck of cards

a. An ace is a card with one spot. $P(\text{ace}) = \dfrac{4}{52} = \dfrac{\mathbf{1}}{\mathbf{13}}$

b. $P(\text{heart}) = \dfrac{13}{52} = \dfrac{\mathbf{1}}{\mathbf{4}}$

c. $P(\text{face card}) = \dfrac{12}{52}$ ← A face card is a jack, queen, or king. ← Number of cards in the sample space

$= \dfrac{\mathbf{3}}{\mathbf{13}}$ ∎

You will need to be familiar with a deck of cards for many examples in this chapter.

EXAMPLE 8 **Probabilities with a Pair of Dice**

Suppose you are just beginning a game of Monopoly.™ You roll a pair of dice. What is the probability that you land on a railroad on the first roll of the dice?

Solution You need to know the locations of the railroads on a Monopoly board (see SITUATION at the opening of this section). There are four railroads altogether, and these are positioned so that only one can be reached on one roll of a pair of dice. The required number to roll is a 5. We begin by listing the sample space.

You might try {2, 3, 4, 5, 6, 7, 8, 9, 10, 11, 12}, but these possible outcomes are not equally likely, which you can see by considering a tree diagram. The roll of the first die has 6 possibilities, and then *each* of these in turn can combine with any of 6 possibilities for a total of 36 possibilities. The **sample space for a pair of dice** is summarized in Figure 9.4.

We will refer to this sample space for many of the examples that follow.

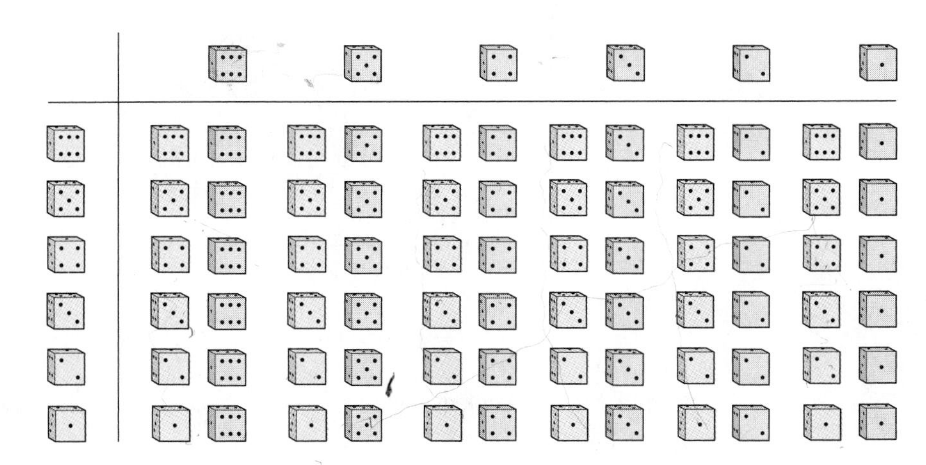

Figure 9.4 Sample space for a pair of dice

Thus, $n = 36$ for the definition of probability. We need to look at Figure 9.4 to see how many possibilities there are for obtaining a 5. We find (1, 4), (2, 3), (3, 2), and (4, 1), so $s = 4$. Then

$$P(\text{five}) = \frac{4}{36} \begin{array}{l} \leftarrow \text{4 ways to obtain a 5} \\ \leftarrow \text{36 ways to roll a pair of dice} \end{array}$$

$$= \frac{1}{9}$$

■

You should use Figure 9.4 when working probability problems that deal with rolling a pair of dice. You will be asked to perform an experiment (Problem 59 in Problem Set 9.1) in which you roll a pair of dice 100 times and then compare the results you obtain with the probabilities you calculate by using Figure 9.4.

PROBABILITIES OF UNIONS AND INTERSECTIONS

The word *or* is translated as ∪ (union), and the word *and* is translated as ∩ (intersection). We will find the probabilities of combinations of events involving the words *or* and *and* by finding unions and intersections of events.

EXAMPLE 9 **Finding Probabilities of Unions and Intersections**

Suppose that a single card is selected from an ordinary deck of cards.

a. What is the probability that it is a two or a king?
b. What is the probability that it is a two and a heart?
c. What is the probability that it is a two or a heart?
d. What is the probability that it is a two and a king?

Solution

a. $P(\text{two or a king}) = P(\text{two} \cup \text{king})$
Look at Figure 9.3.

two = {two of hearts, two of spades, two of diamonds, two of clubs}

king = {king of hearts, king of spades, king of diamonds, king of clubs}

two $\cup$ king = {two of hearts, two of spades, two of diamonds, two of clubs, king of hearts, king of spades, king of diamonds, king of clubs}

There are 8 possibilities for success. It is usually not necessary to list all of these to know that there are 8 possibilities—simply look at Figure 9.3.

$$P(\text{two} \cup \text{king}) = \frac{8}{52} = \mathbf{\frac{2}{13}}$$

b. Look at Figure 9.3.

two = {**two of hearts,** two of spades, two of diamonds, two of clubs}

heart = {ace of hearts, **two of hearts,** three of hearts, . . . , king of hearts}

two $\cap$ heart = {**two of hearts**}

There is one element in common (as shown in boldface), so

$$P(\text{two} \cap \text{heart}) = \mathbf{\frac{1}{52}}$$

c. This is very similar to parts **a** and **b,** but there is one important difference. Look at the sample space and notice that although there are 4 twos and 13 hearts, the total number of successes is *not* 4 + 13 = 17, *but rather* 16.

two = {**two of hearts,** two of spades, two of diamonds, two of clubs}

heart = {ace of hearts, **two of hearts,** three of hearts, . . . , king of hearts}

two $\cup$ heart = {**two of hearts,** two of spades, two of diamonds, two of clubs, ace of hearts, three of hearts, four of hearts, five of hearts, six of hearts, seven of hearts, eight of hearts, nine of hearts, ten of hearts, jack of hearts, queen of hearts, king of hearts}

It is not necessary to list these possibilities. The purpose of doing so in this case was to reinforce the fact that there are *actually* 16 (not 17) possibilities. The reason for this is there is one common element (as found in part **b**):

$$P(\text{two} \cup \text{heart}) = \frac{16}{52} = \mathbf{\frac{4}{13}}$$

d. two $\cap$ king = $\varnothing$, so there are no elements in the intersection.

$$P(\text{two} \cap \text{king}) = \frac{0}{52} = \mathbf{0}$$

■

The probability of the empty set is 0, which means that the event cannot occur. In Problem Set 9.1, you will be asked to show that the probability of an event that *must* occur is 1. These are the two extremes. All other probabilities fall somewhere in between. The closer a probability is to 1, the more likely the event is to occur; the closer a probability is to 0, the less likely the event is to occur.

Now let's summarize the procedure for finding the probability of an event E when all simple events in the sample space are equally likely:

1. Describe and identify the sample space, S. The number of elements in S is n.
2. Count the number of occurrences that interest us; call this the number of successes and denote it by s.
3. Compute the probability of the event using the formula

$$P(E) = \frac{s}{n}$$

The procedure just outlined will work only when the simple events in S are equally likely. If the equally likely model does not apply to your experiments, you need a more complicated model, or else you must proceed experimentally. This model will, however, be sufficient for the problems you will find in this book.

PROBLEM SET 9.1

Essential Ideas

1. **In Your Own Words** What is a sample space?

2. **In Your Own Words** What is an event?

3. **In Your Own Words** Describe the concept of probability.

4. **In Your Own Words** What is the formula for finding probability? What are the necessary conditions for using this formula?

5. **In Your Own Words** How is the probability of a union translated?

6. **In Your Own Words** How is the probability of an intersection translated?

Level 1

WHAT IS WRONG, *if anything, with each of the statements in Problems 7–14? Explain your reasoning.*

7. The sample space for rolling a single die is $S = \{1, 2, 3, 4, 5, 6\}$, so $P(\text{two}) = \frac{1}{6}$.

8. The sample space for rolling a pair of dice is $S = \{2, 3, 4, 5, 6, 7, 8, 9, 10, 11, 12\}$, so $P(\text{two}) = \frac{1}{11}$.

9. The sample space for drawing a card from a deck of cards has 52 elements, so

$$P(\text{two}) = \frac{1}{52}$$

10. The definition of the probability of an event E is

$$P(E) = \frac{\text{NUMBER OF OUTCOMES FAVORABLE TO } E}{\text{NUMBER OF ALL POSSIBLE OUTCOMES}}$$

11. $P(A \text{ or } B)$ means $P(A \cap B)$.

12. $P(A \text{ and } B)$ means $P(A \cup B)$.

13. $P(A \cup B) = P(A) + P(B)$

14. $P(A \cap B) = P(A) \cdot P(B)$

Level 1—Drill

For the spinners in Problems 15–18, assume that the pointer can never lie on a border line.

15. a. $P(A)$
b. $P(B)$
c. $P(C)$

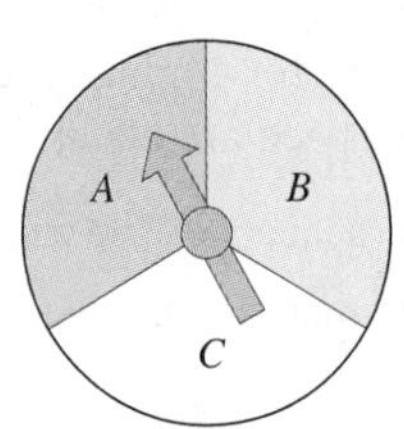

16. a. $P(D)$
b. $P(E)$
c. $P(F)$

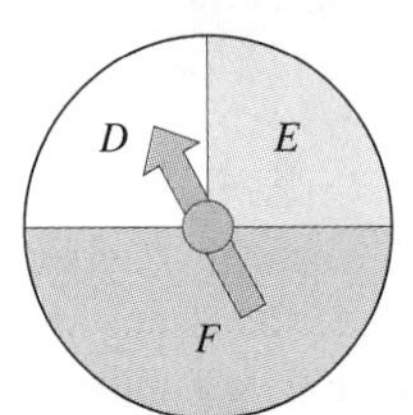

17. a. $P(G)$
b. $P(H)$
c. $P(I)$
d. $P(J)$

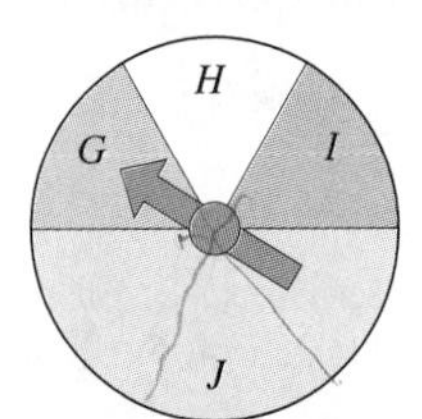

18. a. $P(\text{white})$
b. $P(\text{black})$
c. $P(\text{gray})$

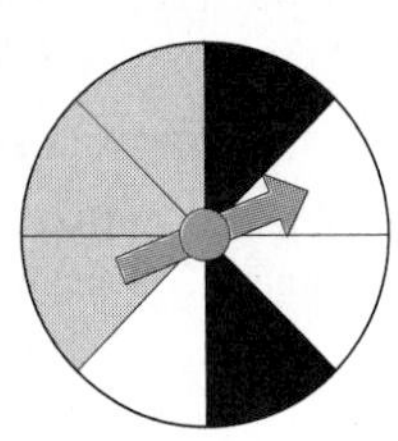

Level 1—Applications

Give the probabilities in Problems 19–22 in decimal form (correct to two decimal places). A calculator may be helpful with these problems.

19. Last year, 1,485 calculators were returned to the manufacturer. If 85,000 were produced, assign a number to specify the probability that a particular calculator would be returned.

20. Last semester, a certain professor gave 13 As out of 285 grades. If one of the professor's students were selected randomly, what is the probability of the student receiving an A?

21. Last year in Ferndale, CA, it rained on 75 days. What is the probability of rain on a day selected at random?

22. The campus vets club is having a raffle and is selling 1,500 tickets. If the people on your floor of the dorm bought 285 of those tickets, what is the probability that someone on your floor will hold the winning ticket?

Consider the jar containing marbles shown here. Suppose each marble has an equal chance of being picked from the jar. Find the probabilities in Problems 23–25.

23. $P(\text{white})$

24. $P(\text{black})$

25. $P(\text{blue})$

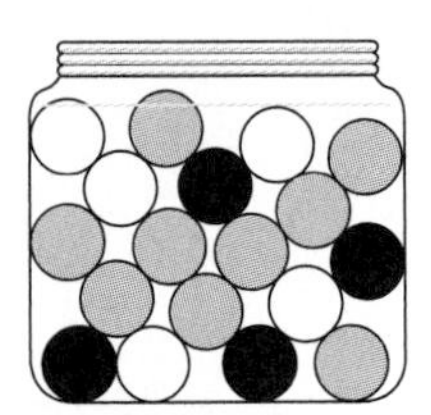

Level 2—Applications

Poker is a common game in which players are dealt five cards from a deck of cards. It can be shown that there are 2,598,960 different possible poker hands. The winning hands (from highest to lowest) are shown in Table 9.1. Find the requested probabilities in Problems 26–34. Use a calculator, and show your answers to whatever accuracy possible on your calculator.

26. $P(\text{royal flush})$
27. $P(\text{straight flush})$
28. $P(\text{four of a kind})$
29. $P(\text{full house})$
30. $P(\text{flush})$
31. $P(\text{straight})$
32. $P(\text{three of a kind})$
33. $P(\text{two pair})$
34. $P(\text{one pair})$

A single card is selected from an ordinary deck of cards. The sample space is shown in Figure 9.3. Find the probabilities in Problems 35–44.

35. $P(\text{five of hearts})$
36. $P(\text{five})$
37. $P(\text{heart})$
38. $P(\text{jack})$
39. $P(\text{diamond})$
40. $P(\text{even number})$
41. $P(\text{five and a jack})$
42. $P(\text{five or a jack})$
43. $P(\text{heart and a jack})$
44. $P(\text{heart or a jack})$

Suppose that you toss a coin and roll a die in Problems 45–48. The sample space is shown in Figure 9.1.

Table 9.1
Poker Hands

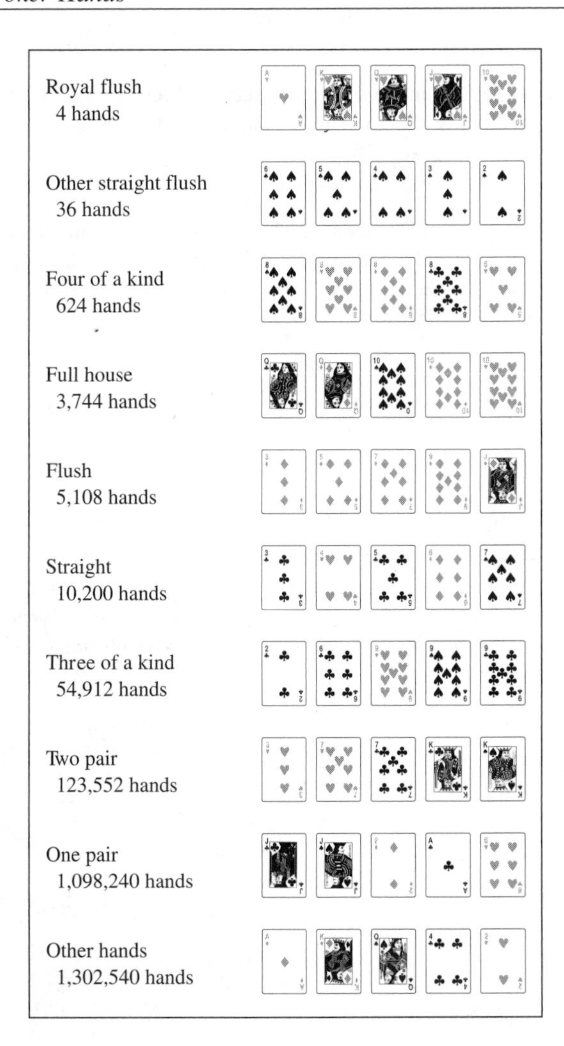

45. What is the probability of obtaining:
  a. Tails *and* a five?
  b. Tails *or* a five?
  c. Heads *and* a two?

46. What is the probability of obtaining:
  a. Tails?
  b. One, two, three, *or* four?
  c. Heads *or* a two?

47. What is the probability of obtaining:
  a. Heads *and* an odd number?
  b. Heads *or* an odd number?

48. What is the probability of obtaining:
  a. Heads *and* a five? **b.** Heads *or* a five?

Use the sample space shown in Figure 9.4 *to find the probabilities in Problems* 49–57 *for the experiment of rolling a pair of dice.*

49. *P*(five) **50.** *P*(six) **51.** *P*(seven)

52. *P*(eight) **53.** *P*(nine) **54.** *P*(two)

55. *P*(four *or* five) **56.** *P*(even number)

57. *P*(eight *or* ten)

Perform the experiments in Problems 58–61, *tally your results, and calculate the probabilities (to the nearest hundredth).*

58. Toss a coin 100 times. Make sure that, each time the coin is flipped, it rotates several times in the air and lands on a table or on the floor. Keep a record of the results of this experiment. Based on your experiment, what is *P*(heads)?

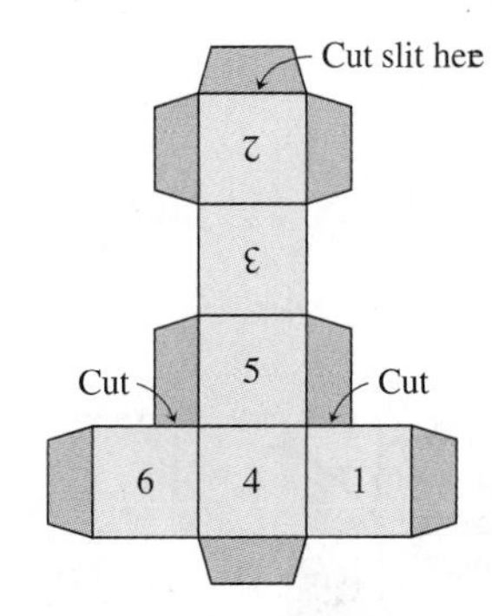

If you do not have any dice, you can use this pattern to construct your own.

59. Roll a pair of dice 100 times. Keep a record of the results. Based on your experiment, find:
  a. *P*(two) **b.** *P*(three)
  c. *P*(four) **d.** *P*(five)
  e. *P*(six) **f.** *P*(seven)
  g. *P*(eight) **h.** *P*(nine)
  i. *P*(ten) **j.** *P*(eleven)
  k. *P*(twelve)

60. Flip three coins simultaneously 100 times, and note the results. The possible outcomes are:
  a. three heads **b.** two heads and one tail
  c. two tails and one head **d.** three tails
  Based on your experiment, find the probabilities of each of these events. Do these appear to be equally likely?

61. Simultaneously toss a coin and roll a die 100 times, and note the results. The possible outcomes are H1, H2, H3, H4, H5, H6, T1, T2, T3, T4, T5, and T6. Do these appear to be equally likely outcomes?

62. Dice is a popular game in gambling casinos. Two dice are tossed, and various amounts are paid according to the outcome. If a seven or eleven occurs on the first roll, the player wins. What is the probability of winning on the first roll?

63. In dice, the player loses if the outcome of the first roll is a two, three, or twelve. What is the probability of losing on the first roll?

64. In dice, a pair of ones is called *snake eyes.* What is the probability of losing a dice game by rolling snake eyes?

B.C. By permission of Johnny Hart and Creators Syndicate, Inc.

65. Consider a die with only four sides, marked 1, 2, 3, and 4. Write out a sample space similar to the one in Figure 9.4 for rolling a pair of these dice.

66. Using the sample space you found in Problem 65, find the probability that the sum of the dice is the given number. Assume equally likely outcomes.
a. P(two) **b.** P(three) **c.** P(four)

67. Using the sample space you found in Problem 65, find the probability that the sum of the dice is the given number. Assume equally likely outcomes.
a. P(five) **b.** P(six) **c.** P(seven)

68. The game of Dungeons and Dragons uses nonstandard dice. Consider a die with eight sides marked 1, 2, 3, 4, 5, 6, 7, and 8. Write a sample space similar to the one in Figure 9.4 for rolling a pair of these dice.

Suppose you and an opponent each pick one of the spinners shown here. A "win" means spinning a higher number. Construct a sample space to answer each question, and tell which of the two spinners given in Problems 69–74 you would choose in each case.

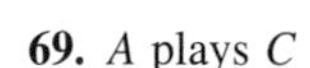

69. *A* plays *C*

70. *A* plays *B*

71. *B* plays *C*

72. *D* plays *E*

73. *E* plays *F*

74. *D* plays *F*

A
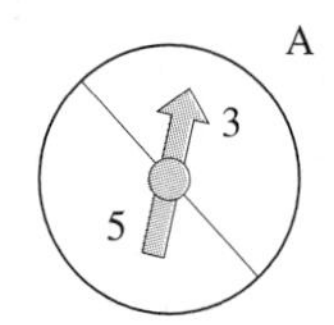

B
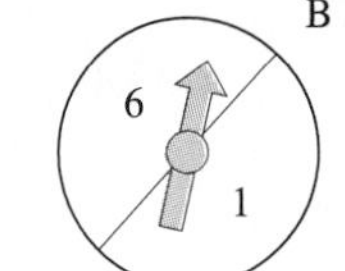

C
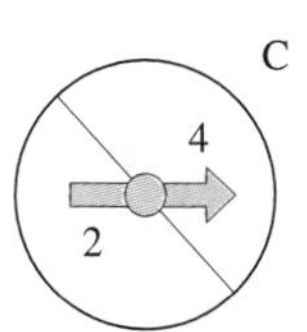

D
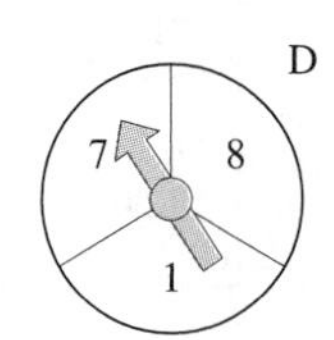

E
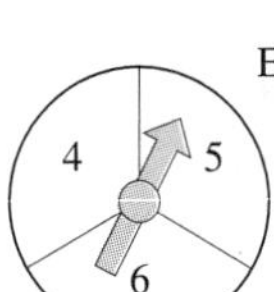

F

2 3 9

75. The game of WIN Construct a set of nonstandard dice as shown in Figure 9.5.

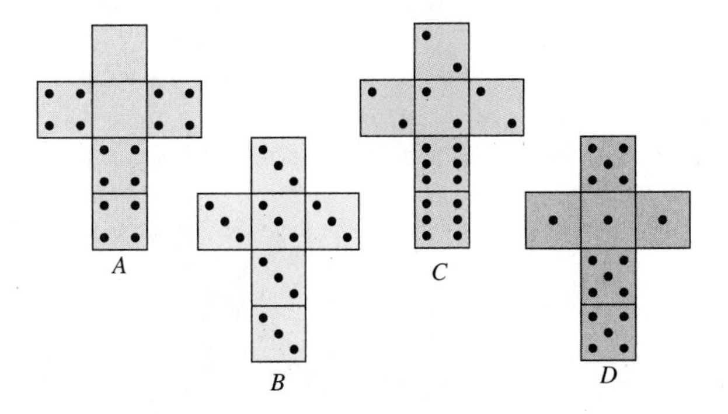

Figure 9.5 Faces on the dice for a game of Win

Suppose that your opponent picks die A and that you pick die B. Then we can enumerate the sample space as shown here.

A: / **B:**	**0**	**0**	**4**	**4**	**4**	**4**
3	(3, 0)	(3, 0)	(3, 4)	(3, 4)	(3, 4)	(3, 4)
3	(3, 0)	(3, 0)	(3, 4)	(3, 4)	(3, 4)	(3, 4)
3	(3, 0)	(3, 0)	(3, 4)	(3, 4)	(3, 4)	(3, 4)
3	(3, 0)	(3, 0)	(3, 4)	(3, 4)	(3, 4)	(3, 4)
3	(3, 0)	(3, 0)	(3, 4)	(3, 4)	(3, 4)	(3, 4)
3	(3, 0)	(3, 0)	(3, 4)	(3, 4)	(3, 4)	(3, 4)
	B wins		A wins			

We see that the probability of A winning is $\frac{24}{36}$, or $\frac{2}{3}$. If you were to play the game of WIN, would you choose your die first or second?

76. Show that the probability of some event E falls between 0 and 1. Show that $P(E) = 0$ if the event cannot occur and $P(E) = 1$ if the event must occur.

77. Situation We see examples of probability every day. Weather forecasts, stock market analyses, contests, children's games, television game shows, and gambling all involve ideas of probability. Probability is the mathematics of uncertainty. "Wait a minute," interrupts Sammy. "I thought that mathematics was absolute and that there was no uncertainty about it!" Well, Sammy, suppose you are playing a game of Monopoly™. Do you know what you'll land on in your next turn? "No, but neither do you!" Suppose you are just beginning a Monopoly game. What space are you most likely to land on with your first roll of the dice?

78. Situation Probability is a way of describing uncertainty. For example, what is the probability of tossing a coin and obtaining heads? "That's easy, it's one-half," answers Sammy. Right, but *why* do you say it is one-half? What should Sammy say?

79. In Your Own Words Write a short paper comparing the SITUATION with your own situation. (That is, list similarities and/or differences.)

LEVEL 3—Team Project

80. We see examples of probability every day. Weather forecasts, stock market analyses, contests, children's games, television game shows, and gambling all involve ideas of probability. With your team members, search current newspapers and magazines to provide examples of probability. After looking for at least a week, collate your material to present a portfolio of the examples you have found.

9.2 PROBABILITY MODELS

Situation

DEAR ABBY: My husband and I just had our eighth child. Another girl, and I am really one disappointed woman. I suppose I should thank God she was healthy, but, Abby, this one was supposed to have been a boy. Even the doctor told me the law of averages was in our favor 100 to one. What is the probability of our next child being a boy?

"We need to talk, Ben," said Melissa, "I know we have eight children, but I want a boy!" "I know, dear, so do I, but I'm not sure we can afford another child. The doctor told us that the law of averages for having a boy is in our favor if we try again, but look what happened last time." Melissa quickly cut in, "My dad is a math teacher, and he said that the chances of having nine girls in a row is 1 out of 512." A worried look swept over Ben's face as he said, "Right, and the doctor said 1 out of 100 last time, too. I'll tell you what, write to Ann Landers or Dear Abby and see what advice the experts can give, then . . . "

COMPLEMENTARY PROBABILITIES

In Section 9.1, we looked at the probability of an event E. Now we wish to expand our discussion. Let

s = NUMBER OF GOOD OUTCOMES (successes)

f = NUMBER OF BAD OUTCOMES (failures)

n = TOTAL NUMBER OF POSSIBLE OUTCOMES $(s + f = n)$

Then the probability that event E occurs is

$$P(E) = \frac{s}{n}$$

The probability that event E does not occur is

$$P(\overline{E}) = \frac{f}{n}$$

An important property of probability is found by adding these probabilities:

This says that some event either happens or does not.

$$P(E) + P(\overline{E}) = \frac{s}{n} + \frac{f}{n} = \frac{s+f}{n} = \frac{n}{n} = 1$$

Probabilities whose sum is 1 are called **complementary probabilities,** and the following box shows what is known as the **property of complements.**

Property of Complements

$$P(E) = 1 - P(\overline{E}) \quad \text{or} \quad P(\overline{E}) = 1 - P(E)$$

EXAMPLE 1

Using the Property of Complements

Use Table 9.1 (on page 413) to find the probability of not obtaining one pair with a poker hand.

Solution From Table 9.1, we see that $P(\text{pair}) = \dfrac{1{,}098{,}240}{2{,}598{,}960} \approx 0.42$, so that

$$P(\text{no pair}) = P(\overline{\text{pair}}) = 1 - P(\text{pair}) \approx 1 - 0.42 = \mathbf{0.58}$$ ■

EXAMPLE 2

Finding a Probability

What is the probability of obtaining at least one head when a coin is flipped three times?

Solution Let F = {receive at least one head when a coin is flipped three times}

Method I Work directly; use a tree diagram to find the possibilities.

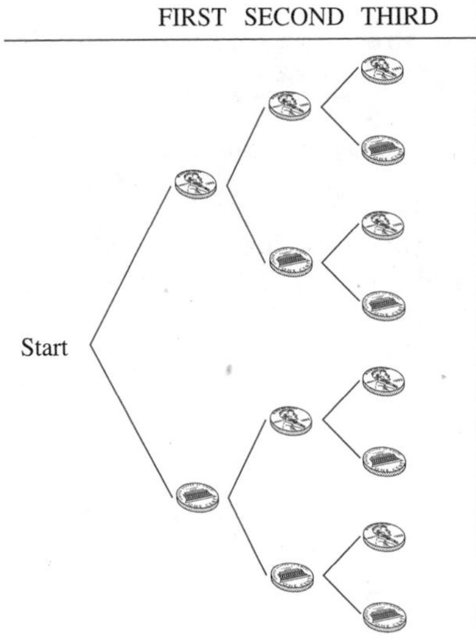

FIRST	SECOND	THIRD	SUCCESS
H	H	H	yes
H	H	T	yes
H	T	H	yes
H	T	T	yes
T	H	H	yes
T	H	T	yes
T	T	H	yes
T	T	T	no

$$P(F) = \tfrac{7}{8}$$

Method II For one coin there are 2 outcomes (head and tail); for two coins there are 4 outcomes (HH, HT, TH, TT); and for three coins there are 8 outcomes. We answer the question by finding the complement; $\overline{F}$ is the event of receiving no heads (that is, of obtaining all tails). *Without* drawing the tree diagram, we note that there is only one way of obtaining all tails (TTT). Thus

$$P(F) = 1 - P(\overline{F}) = 1 - \tfrac{1}{8} = \tfrac{7}{8}$$ ■

FUNDAMENTAL COUNTING PRINCIPLE

Read this counting principle several times.

For one coin there are 2 possible outcomes and for 2 coins there are 4 possible outcomes. In Example 2, we considered flipping a coin three times, and by drawing a tree diagram we found 8 possibilities. These are applications of a counting technique called the **fundamental counting principle,** which can be understood by looking at tree diagrams.

Fundamental Counting Principle

If task A can be performed in m ways, and if, after task A is performed, a second task B can be performed in n ways, then task A followed by task B can be performed in $m \cdot n$ ways.

EXAMPLE 3 **Verifying the Fundamental Counting Principle with a Tree Diagram**

What is the probability that, in a family with two children, the children are of opposite sexes?

Solution The fundamental counting principle tells us that, since there are 2 ways of having a child (B or G), for 2 children there are

$$2 \cdot 2 = 4 \text{ ways}$$

We verify this by looking a tree diagram:

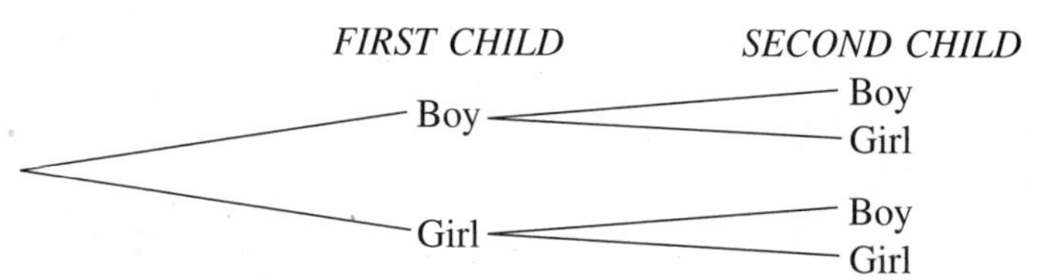

There are 4 equally likely outcomes: BB, BG, GB, and GG. Thus, the probability of having a boy and a girl in a family of two children is

$$\frac{\text{NUMBER OF SUCCESSFUL OUTCOMES}}{\text{TOTAL NUMBER OF POSSIBLE OUTCOMES}} = \frac{2}{4} = \frac{1}{2}$$ ■

EXAMPLE 4

Probability Using the Fundamental Counting Principle

What is a family's probability of having two boys and two girls, if it has four children?

Solution The fundamental counting principle tells us the number of possibilities is

$$2 \times 2 \times 2 \times 2 = 16$$

Thus, $n = 16$. To find s we list the event (by using a tree diagram for four children) that brings success:

$$\{BBGG, BGBG, BGGB, GBGB, GBBG, GGBB\}$$

Since there are 6 elements in this set, we see that $s = 6$. Thus, the desired probability is

$$\frac{6}{16} = \mathbf{\frac{3}{8}}$$

■

EXAMPLE 5

License Plate Problem

What is the probability (rounded to the nearest percent) of getting a license plate that has a repeated letter or digit if you live in a state where the scheme is three letters followed by three numerals?

Solution A license plate bears three letters followed by three digits. We begin by using the fundamental counting principle to find n (the total number of possibilities):

Letters in the alphabet ↓ Number of digits ↓

$$\underbrace{26 \times 26 \times 26}_{\text{Three letters}} \times \underbrace{10 \times 10 \times 10}_{\text{Three digits}} = 17{,}576{,}000$$

To count the number of successes, you must understand the problem. To confirm this, which of the following plates would be considered a success?

ABC123	Failure
ABC122	Success; repeated digit
MMA456	Success; repeated letter
TTT111	Success; repeated letter and repeated numeral
XYZ 890	Failure

Success is one repetition or two repetitions or three repetitions . . . . Let R be the event that a repetition is received. It is difficult to count all possible repetitions, but we can use the fundamental counting principle to count the number of license plates that *do not* have a repetition:

Letters in the alphabet ↓ 26 × 25 ↑ Letters left after the first one (no repetitions) × 24 × 10 × 9 × 8 = 11,232,000 (↓ Number of digits; ↑ Digits left)

$$26 \times 25 \times 24 \times 10 \times 9 \times 8 = 11{,}232{,}000$$

$$P(R) = 1 - P(\overline{R}) = 1 - \frac{11{,}232{,}000}{17{,}576{,}000} \approx 0.36 \qquad \textit{Display:} \quad .3609467456$$

This means that **about 36% of all license plates in the state have a repeated letter or digit.** ■

PROBLEM SET 9.2

ESSENTIAL IDEAS

1. **In Your Own Words** Explain what we mean by "complementary probabilities."
2. **In Your Own Words** What is the fundamental counting principle?

LEVEL 1

Use estimation to select the best response in Problems 3– 8. *Do not calculate.*

3. According to the National Safety Council, which of these three means of travel is the safest?
   A. car (1.12 deaths per 100 billion passenger miles)
   B. plane (0.04 death per 100 billion passenger miles)
   C. train (0.06 death per 100 billion passenger miles)
4. Which of the following is most probable?
   A. winning the grand prize in a state lottery (1 chance in 5,000,000)
   B. being struck by lightning (1 chance in 600,000)
   C. appearing on the *Tonight Show* (1 chance in 490,000)
5. Which of the following is more probable?
   A. obtaining at least 2 heads in 3 flips of a coin
   B. obtaining at least 2 heads in 4 flips of a coin
6. Which of the following is more probable?
   A. obtaining a six 2 times in 3 rolls of a die
   B. obtaining a six at least 2 times in 3 rolls of a die
7. Which of the following is more probable?
   A. guessing all the correct answers on a 20-question true–false examination
   B. obtaining all heads in 20 tosses of a coin
8. Which of the following is more probable?
   A. Guessing all the correct answers on a 10-question 5-part multiple-choice test
   B. Your living room is filled with white ping pong balls. There is also one red ping pong ball in the room. You reach in and select a ping pong ball at random and select the red ping pong ball.

WHAT IS WRONG, *if anything, with each of the statements in Problems* 9–13*? Explain your reasoning.*

9. If $P(E) = \frac{3}{7}$, then $P(\overline{E}) = \frac{3}{5}$.
10. $P(E)$ and $P(\overline{E})$ are reciprocals.
11. The probability of obtaining a head in one toss of a coin is $\frac{1}{2}$, and the probability of obtaining two heads in two tosses of a coin is $\frac{1}{4}$. It then follows that the probability of obtaining five heads in five tosses is $\frac{1}{32}$. Suppose I flip a coin four times and obtain four heads. Then it is better to bet on tails for the fifth flip because the probability of five heads in a row is very small.
12. This example is from the book *How to Take a Chance* by Darrell Huff. "On August 18, 1913 at a casino in Monte Carlo, black came up 26 times in a row on a roulette wheel. If you had bet $1 on black and continued to let your bet ride for the entire run of blacks, you would have won $67,108,863.00." If you were to bet on the next spin of the wheel, you should bet on red because the chances of black on the 27th spin are very, very, small.
13. Draw five cards from a deck of cards. The probability of drawing a heart on the fifth card is 13/48.

Level 1—Drill

Find the requested probabilities in Problems 14–19.

14. $P(\overline{A})$ if $P(A) = 0.6$

15. $P(\overline{B})$ if $P(B) = \frac{4}{5}$

16. $P(C)$ if $P(\overline{C}) = \frac{9}{13}$

17. $P(D)$ if $P(\overline{D}) = 0.005$

18. Obtaining at least one head in 4 flips of a coin

19. Obtaining a sum of at least four when rolling a pair of dice

List the outcomes considered a failure, as well as the probability of failure, for the experiments named in Problems 20–25.

	EXPERIMENT	*SUCCESS*	*P(success)*
20.	Tossing a coin	head	$\frac{1}{2}$
21.	Rolling a die	four or six	$\frac{1}{3}$
22.	Guessing an answer on a 5-choice multiple-choice test	correct guess	$\frac{1}{5}$
23.	Card game	drawing a heart	0.18
24.	Baseball game	White Sox win	0.57
25.	Football game	your school wins	0.83

26. Three fair coins are tossed. What is the probability that at least one is a head?

27. A card is selected from an ordinary deck. What is the probability that it is not a face card?

28. Choose a natural number between 1 and 100, inclusive. What is the probability that the number chosen is not a multiple of 5?

Level 2—Applications

29. Experiment. Consider the following birth dates of some famous mathematicians:

Abel	August 5, 1802
Cardano	September 21, 1576 (died; birth date unknown)
Descartes	March 31, 1596
Euler	April 15, 1707
Fermat	August 20, 1601 (baptized; birth date unknown)
Galois	October 25, 1811
Gauss	April 30, 1777
Newton	December 25, 1642
Pascal	June 19, 1623
Riemann	September 17, 1826

Add to this list the birth dates of the members of your class. *But before you compile this list, guess the probability that at least two people in this group will have exactly the same birthday* (not counting the year). Be sure to make your guess *before* finding out the birth dates of your classmates. The answer, of course, depends on the number of people on the list. Ten mathematicians are listed and you may have 20 people in your class, giving 30 names on the list.

30. Birthday problem. This problem is related to Problem 29. Suppose that you select 23 people out of a crowd. The probability that two or more of them will have the same birthday is greater than 50%! This seemingly paradoxical situation will fool most people. Figure 9.6 is a chart showing these probabilities.

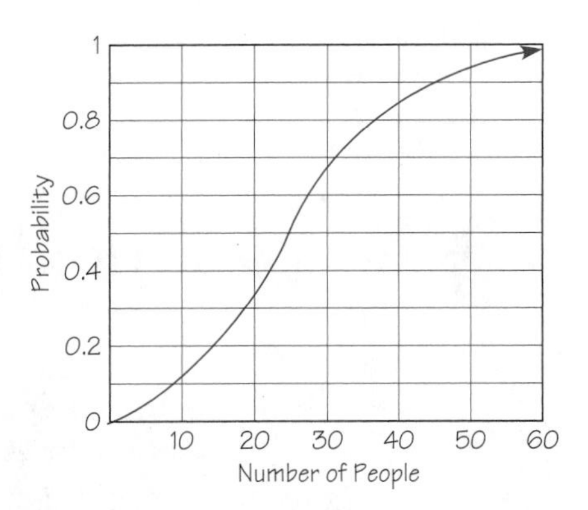

Figure 9.6 Probabilities for the birthday problem

a. Pick 23 names at random from a biographical dictionary or a *Who's Who,* and verify some of the probabilities of the table. The further you go past 23, the greater the probability of finding two people with the same birthday.

b. The graph (Figure 9.6) is approaching a probability of 1 as the number of people increases. How many people are necessary for the probability actually to reach 1?

31. One roulette system is to bet $1 on black. If black comes up on the first spin of the wheel, you win $1 and the game is over. If black does not come up, double your bet ($2). If you win, the game is over and your net winnings for two spins is still $1 (show this). If black does not come up, double your bet again ($4). If you win, the game is over and your net winnings for three spins is still $1 (show this). Continue this doubling procedure until you eventually win; in every case your net winnings amount to $1 (show this). What is the fallacy with this betting system?

32. **Situation** "We need to talk, Ben," said Melissa, "I know we have eight children, but I want a boy!" "I know, dear, so do I, but I'm not sure we can afford another child. The doctor told us that the law of averages for having a boy is in our favor if we try again, but look what happened last time." Melissa quickly cut in, "My dad is a math teacher, and he said that the chances of having nine girls in a row is 1 out of 512." A worried look swept over Ben's face as he said, "Right, and the doctor said 1 out of 100 last time, too. I'll tell you what, write to Ann Landers or Dear Abby and see what advice the experts can given, then . . . . " What is the probability that the next child for Melissa and Ben is a boy?

33. **In Your Own Words** Write a short paper comparing the SITUATION with your own situation. (That is, list similarities and/or differences.) Answer the question asked in the situation: "Can probability help us decide on the gender of an unborn child or children?"

9.3 ODDS AND CONDITIONAL PROBABILITY

Situation

"Last week I won a free Big Mac at McDonald's. I sure was lucky!" exclaimed Charlie. "Do you go there often?" asked Pat. "Only twenty or thirty times a month. And the odds of winning a Big Mac were only 20 to 1." "Don't you mean 1 to 20?" queried Pat. "Don't confuse me with details. I don't even understand odds at the racetrack, and I go all the time. Why do I need to know anything about odds anyway?"

In this section, Charlie will learn about odds, the calculation of odds, and their relationship to probability. He will also learn about two models to help in calculating probabilities: one to use in finding the probability that something will not occur, and the other to use for calculating the probability when the sample space has been altered because of some additional information.

ODDS

Related to probability is the notion of odds. Instead of forming ratios

$$P(E) = \frac{s}{n} \quad \text{and} \quad P(\overline{E}) = \frac{f}{n}$$

we form the following ratios:

Odds in favor of an event E: $\frac{s}{f}$ (ratio of successes to failures);

Odds against an event E: $\frac{f}{s}$ (ratio of failures to successes);

where

$s =$ NUMBER OF SUCCESSES

$f =$ NUMBER OF FAILURES

$n =$ NUMBER OF POSSIBILITIES

EXAMPLE 1 **Finding the Odds of an Event**

If a jar has 2 quarters, 200 dimes, and 800 pennies, what are the odds against picking a quarter if a coin is chosen at random?

Solution We are interested in picking a quarter, so let $s = 2$; a failure is not obtaining a quarter, so $f = 1{,}000$. (Find f by adding 200 and 800.)

Odds in favor of obtaining a quarter: $\frac{s}{f} = \frac{2}{1{,}000} = \frac{1}{500}$

Odds against obtaining a quarter: $\frac{f}{s} = \frac{1{,}000}{2} = \frac{500}{1}$

Do not write $\frac{500}{1}$ as 500.

The odds against choosing a quarter at random are 500 to 1. ■

EXAMPLE 2 **Visualizing Odds**

The odds against winning a lottery are 50,000,000 to 1. Make up an example to help visualize these odds.

Solution Imagine one red ping pong ball and 50,000,000 white ping pong balls. Winning a lottery is equivalent to reaching into a container containing these 50,000,001 ping pong balls and obtaining the red one. To visualize this, consider the size container you would need. Assume that a ping pong ball takes up a volume of 1 in.^{3}. A home of 1,200 ft^2 with an 8-ft ceiling has a volume of

$$\begin{aligned} 1{,}200 \text{ ft}^2 \times 8 \text{ ft} &= 9{,}600 \text{ ft}^3 \\ &= 9{,}600 \times (12 \text{ in.} \times 12 \text{ in.} \times 12 \text{ in.}) = 16{,}588{,}800 \text{ in.}^3 \end{aligned}$$

This means that we would need to *fill* about 3 homes with ping pong balls and then choose one ball out of *one* of the houses. If the red one is chosen, we win the lottery! ■

Sometimes you know the probability and want to find the odds, or you may know the odds and want to find the probability. These relationships are easy if you remember:

$$s + f = n$$

Finding the Odds Given the Probability

Pay attention here. It is easy to confuse odds and probability.

Suppose that you *know* $P(E)$. Then

Odds in favor of an event E: $\dfrac{P(E)}{P(\overline{E})}$

Odds against an event E: $\dfrac{P(\overline{E})}{P(E)}$

You can show that these formulas are true:

$$\frac{P(E)}{P(\overline{E})} = \frac{\frac{s}{n}}{\frac{f}{n}} = \frac{s}{n} \cdot \frac{n}{f} = \frac{s}{f} = \text{odds in favor}$$

The verification of the second formula is left for the problem set (Problem 30).

Finding the Probability Given Odds in Favor or Odds Against an Event

Suppose that you *know* the odds in favor of an event E:

s to f

or the odds against an event E:

f to s

Then,

$$P(E) = \frac{s}{s+f} \quad \text{and} \quad P(\overline{E}) = \frac{f}{s+f}$$

EXAMPLE 3 **Finding the Odds when the Probability Is Known**

If the probability of an event is 0.45, what are the odds in favor of the event?

Solution $P(E) = 0.45 = \frac{45}{100} = \frac{9}{20}$ This is given.

$P(\overline{E}) = 1 - \frac{9}{20} = \frac{11}{20}$

Then the odds in favor of E are $\frac{P(E)}{P(\overline{E})} = \frac{\frac{9}{20}}{\frac{11}{20}} = \frac{9}{20} \cdot \frac{20}{11} = \frac{9}{11}$

The odds in favor are 9 to 11. ■

EXAMPLE 4 **Finding the Probability when the Odds Against Are Known**

If the odds against you are 20 to 1, what is the probability of the event (rounded to three decimal places)?

Solution Odds against are f to s, so $f = 20$ and $s = 1$; thus,

$$P(E) = \frac{1}{20 + 1} = \frac{1}{21} \approx \mathbf{0.048}$$ ■

EXAMPLE 5 **Finding the Probability when the Odds in Favor Are Known**

If the odds in favor of some event are 2 to 5, what is the probability of the event (rounded to three decimal places)?

Solution Odds in favor are s to f, so $s = 2$ and $f = 5$; thus,

$$P(E) = \frac{2}{2 + 5} = \frac{2}{7} \approx \mathbf{0.286}$$ ■

CONDITIONAL PROBABILITY

Frequently, we wish to compute the probability of an event but we have additional information that will alter the sample space. For example, suppose that a family has two children. What is the probability that the family has two boys?

$$P(2 \text{ boys}) = \frac{1}{4}$$ Sample space: BB, BG, GB, GG; 1 success out of 4 possibilities

Now, let's complicate the problem a little. Suppose that we know that the older child is a boy. We have altered the sample space as follows:

Original sample space: BB, BG, GB, GG; but we need to cross out the last two possibilities because we *know* that the older child is a boy.

Success ↓ These are crossed out.

Altered sample space: BB, BG, ~~GB~~, ~~GG~~, therefore,

Altered sample space has two elements.

$$P(\text{2 boys given the older is a boy}) = \frac{1}{2}$$

This is a problem involving a **conditional probability**—namely, *the probability of an event* **given** *that another event F has occurred.* We denote this by

$P(E \mid F)$ Read this as "probability of E given F."

EXAMPLE 6

Finding a Conditional Probability

Suppose that you toss two coins (or a single coin twice). What is the probability that two heads are obtained if you know that at least one head is obtained?

Solution Consider an altered sample space: HH, HT, TH, ~~TT~~. The probability is $\frac{1}{3}$. ■

EXAMPLE 7

Finding Conditional Probabilities

Suppose that you draw two cards from a deck of cards. Find the following probabilities (correct to three decimal places). Let H = {the second card drawn is a heart}.

a. $P(H)$
b. $P(H \mid \text{a heart is drawn on the first draw})$
c. $P(H \mid \text{a heart is not drawn on the first draw})$

Solution

a. $P(H) = \frac{13}{52} = \frac{\mathbf{1}}{\mathbf{4}} = \mathbf{0.250}$

Remember, it does not matter what happened on the first draw because we do not know what happened on that draw. The second card "does not remember" what is drawn on the first draw.

b. This time we know what happened on the first draw, so $n = 51$ and $s = 12$ (because a heart was drawn on the first draw):

$$P(H \mid \text{a heart is drawn on the first draw}) = \frac{\mathbf{12}}{\mathbf{51}} \approx \mathbf{0.235}$$

c. We still have $n = 51$, but this time $s = 13$:

$$P(H \mid \text{a heart is not drawn on the first draw}) = \frac{\mathbf{13}}{\mathbf{51}} \approx \mathbf{0.255}$$ ■

Historical Perspective

The engraving depicts gambling in 18th-century France. The mathematical theory of probability arose in France in the 17th century when a gambler, Chevalier de Méré, became interested in adjusting the stakes so that he could win more often than he lost. In 1654, he wrote to Blaise Pascal, who in turn sent his questions to Pierre de Fermat. Together they developed the first theory of probability.

Engraving by Darcis: *Le Trente-et-un*

PROBLEM SET 9.3

ESSENTIAL IDEAS

1. **In Your Own Words** Explain what we mean by "conditional probability."

2. **In Your Own Words** Contrast probability and odds.

LEVEL 1

WHAT IS WRONG, *if anything, with each of the statements in Problems 3–5? Explain your reasoning.*

3. If $P(E) = \frac{2}{3}$, then the odds in favor of E are 2 to 3.

4. If the odds in favor of an event are 3 to 5, then the odds against the event are 2 to 5.

5. The odds in favor of an event and the odds against an event are complements.

LEVEL 1—Drill

6. What are the odds in favor of drawing an ace from an ordinary deck of cards?

7. What are the odds in favor of drawing a heart from an ordinary deck of cards?

8. What are a family's odds against having four boys, if it has four children?

9. Suppose that the odds that a man will be bald by the time he is 60 are 9 to 1. State this as a probability.

10. Suppose the odds are 33 to 1 that someone will lie to you at least once in the next seven days. State this as a probability.

11. Racetracks quote the approximate odds for each race on a large display board called a *tote board.* Here's what it might say for a particular race:

HORSE NUMBER	*ODDS*
1	2 to 1
2	15 to 1
3	3 to 2
4	7 to 5
5	1 to 1

What would be the probability of winning for each of these horses? [*Note:* The odds stated are for the horse's *losing.* Thus, $P(\text{horse 1 losing}) = \frac{2}{2+1} = \frac{2}{3}$, so $P(\text{horse 1 winning}) = 1 - \frac{2}{3} = \frac{1}{3}$.]

Level 2—Drill

12. Suppose that a family wants to have four children.
- **a.** What is the sample space?
- **b.** What is the probability of 4 girls? 4 boys?
- **c.** What is the probability of 1 girl and 3 boys? 1 boy and 3 girls?
- **d.** What is the probability of 2 boys and 2 girls?
- **e.** What is the sum of your answers in parts **b–d?**

13. What is a family's probability of having exactly two boys, given that at least one of their three children is a boy?

14. What is the probability of obtaining exactly three heads in four flips of a coin, given that at least two are heads?

A single card is drawn from a standard deck of cards. In Problems 15–20, find the probabilities if the given information is known about the chosen card. A face card is a jack, queen, or king.

15. $P(\text{face card} \mid \text{jack})$ **16.** $P(\text{jack} \mid \text{face card})$

17. $P(\text{heart} \mid \text{not a spade})$ **18.** $P(\text{two} \mid \text{not a face card})$

19. $P(\text{black} \mid \text{jack})$ **20.** $P(\text{jack} \mid \text{black})$

Two cards are drawn from a standard deck of cards, and one of the two cards is noted. Find the probabilities of the second card, given the information about the noted card provided in Problems 21–26.

21. $P(\text{ace} \mid \text{two})$ **22.** $P(\text{king} \mid \text{king})$

23. $P(\text{heart} \mid \text{heart})$ **24.** $P(\text{heart} \mid \text{spade})$

25. $P(\text{black} \mid \text{red})$ **26.** $P(\text{black} \mid \text{black})$

For Problems 27–30, see Figure 9.4 on page 409.

27. On a single roll of a pair of dice, what is the probability that the sum is seven, given that at least one die came up two?

28. On a single roll of a pair of dice, suppose that we are given that at least one die came up two. What is the probability that the sum is
a. Three? **b.** Four? **c.** Two?

29. What are the odds in favor of rolling a seven or eleven on a single roll of a pair of dice?

30. Show that the odds against an event E can be found by computing $\dfrac{P(\overline{E})}{P(E)}$.

Level 2—Applications

31. Situation "Last week I won a free Big Mac at McDonald's. I sure was lucky!" exclaimed Charlie. "Do you go there often?" asked Pat. "Only twenty or thirty times a month. And the odds of winning a Big Mac were only 20 to 1." "Don't you mean 1 to 20?" queried Pat. "Don't confuse me with details. I don't even understand odds at the racetrack, and I go all the time." Is Charlie or Pat correct about the odds?

32. In Your Own Words Write a short paper comparing the SITUATION with your own situation. (That is, list similarities and/or differences.) Answer the question asked in the situation: "Why do I need to know anything about odds anyway?"

9.4 MATHEMATICAL EXPECTATION

Situation

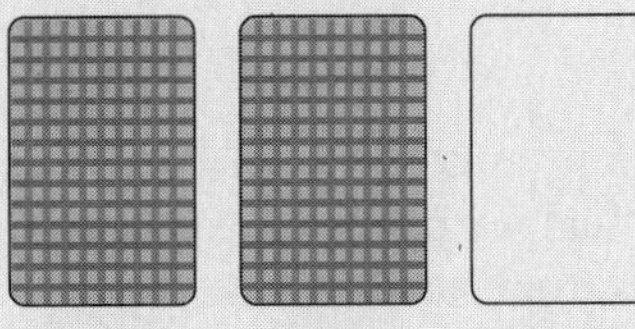

Black on both sides | Black on one side, white on the other | White on both sides

Suppose that your friend George shows you three cards. One card is white on both sides, one is black on both sides, and the last one is black on one side and white on the other. He mixes the cards and tells you to select one at random and place it on the table. Suppose that the upper side turns out to be black. It is not the white–white card; it must be either the black–black or the black–white card. "Thus," says George, "I'll bet you $1 that the other side is black." Would you play? Perhaps you hesitate. Now George says he feels generous. You need to pay him only 75¢ if you lose, and he will still pay you $1 if he loses. Would you play now?

In this section, you'll learn how to analyze a variety of gambling situations. Whether you enjoy gambling and games of chance or are opposed to them and would never play a gambling game, you should find some valuable information in this section. Gambling situations range from dice, cards, and slot machines, to buying insurance and selling a home. By analyzing these games, you can show that without proper analysis a person could be destined for financial ruin, given enough time and limited resources. You can also find situations that should not be considered gambling—situations in which you can't lose.

EXPECTATION

Smiles toothpaste is giving away \$10,000. All you must do to have a chance to win is send a postcard with your name on it (the fine print says you do not need to buy a tube of toothpaste). Is it worthwhile to enter?

Suppose the contest receives 1 million postcards (a conservative estimate). We wish to compute the **expected value** (or your **expectation**) of entering this contest. The expected value of this contest is obtained by multiplying the amount to win by the probability of winning:

$$
\begin{aligned}
E &= (\text{AMOUNT TO WIN}) \times (\text{PROBABILITY OF WINNING}) \\
&= \$10{,}000 \times \frac{1}{1{,}000{,}000} \\
&= \$0.01
\end{aligned}
$$

What does this expected value mean? It means if you were to play this "game" a large number of times, you would expect your *average winnings per game* to be \$0.01. A game is said to be **fair** if the expected value equals the cost of playing the game. If the expected value is positive, then the game is in your favor; if the expected value is negative, then the game is not in your favor. Is this game fair? If the toothpaste company charges you 1¢ to play the game, then it is fair. But how much does the postcard cost? We see that this is not a fair game.

EXAMPLE 1 **Finding the Expected Value**

Suppose that you draw a card from a deck of cards and are paid \$10 if the card is an ace. What is the expected value?

Solution $E = \$10 \times \frac{4}{52} \approx \mathbf{\$0.77}$ Since expected value involves a dollar amount, round your answer to the nearest cent. ■

Sometimes there is more than one possible payoff, and we define the expected value (or expectation) as the sum of the expected values from each separate payoff.

EXAMPLE 2 **Expected Value for a Contest**

A recent contest offered one grand prize worth \$10,000, two second prizes worth \$5,000 each, and ten third prizes worth \$1,000 each. What is the expected value if you assume that there are 1 million entries and that the winners' names are replaced after being drawn?

Solution $P(\text{1st prize}) = \frac{1}{1{,}000{,}000}$; $P(\text{2nd prize}) = \frac{2}{1{,}000{,}000}$; $P(\text{3rd prize}) = \frac{10}{1{,}000{,}000}$

$$E = \overbrace{\$10{,}000}^{\text{Amount of 1st prize}} \times \underbrace{\frac{1}{1{,}000{,}000}}_{P(\text{1st prize})} + \overbrace{\$5{,}000 \times \frac{2}{1{,}000{,}000}}^{\text{2nd prize}} + \underbrace{\$1{,}000 \times \frac{10}{1{,}000{,}000}}_{\text{3rd prize}}$$

$$= \$0.01 + \$0.01 + \$0.01$$

$$= \mathbf{\$0.03}$$ ■

EXAMPLE 3 **Using Expectation to Make a Decision**

You are offered two games:

Game A: Two dice are rolled. You will be paid \$3.60 if you roll two ones, and you will not receive anything for any other outcome.

Game B: Two dice are rolled. You will be paid \$36.00 if you roll any pair, but you must pay \$3.60 for any other outcome.

Which game should you play?

Solution You might say, "I'll play the first game because, if I play that game, I cannot lose anything." This strategy involves *minimizing your losses.* On the other hand, you can use a strategy that *maximizes your winnings.* In this book, we will base our decisions on maximizing the winnings—that is, we wish to select the game that provides the larger expectation.

Game A: $E = \$3.60 \times \frac{1}{36} = \$0.10$

Game B: When calculating the expected value with a charge (a loss), write that charge as a negative number (a negative payoff is a loss).

$$\begin{aligned} E &= \$36.00 \times \frac{6}{36} + (-\$3.60) \times \frac{30}{36} \\ &= \$6.00 + (-\$3.00) \\ &= \$3.00 \end{aligned}$$

This means that, if you were to play each game 100 times, you would expect your winnings for Game A to be about 100($0.10) or $10 and those from playing Game B to be about 100($3.00) or $300. **You should choose to play Game B.** ■

Now we give a formal definition of expectation.

Mathematical Expectation

If an event E has several possible outcomes with probabilities $p_1, p_2, p_3, \ldots,$ and if for each of these outcomes the amount that can be won is $a_1, a_2, a_3, \ldots,$ respectively, then the **mathematical expectation** (or expected value) of E is

$$E = a_1p_1 + a_2p_2 + a_3p_3 + \cdots$$

Use this definition to help you decide whether to place a bet, play a game, or enter a business venture.

EXAMPLE 4 **Mathematical Expectation of a Contest**

A contest offered the prizes shown below. What is the expected value for this contest?

Solution We note the following values:

$a_1 = \$15{,}000;\ p_1 = 0.000008$

$a_2 = \$1{,}000;\ p_2 = 0.000016$

$a_3 = \$625;\ p_3 = 0.000016$

$a_4 = \$525;\ p_4 = 0.000016$

$a_5 = \$390;\ p_5 = 0.000032$

$a_6 = \$250;\ p_6 = 0.000032$

$ *WIN • WIN • WIN* $

PRIZE	VALUE	*PROBABILITY*
Grand Prize Trip	$15,000	0.000008
Samsonite Luggage	$1,000	0.000016
Magic Chef Range	$625.00	0.000016
Murray Bicycle	$525.00	0.000016
Lawn Boy Mower	$390.00	0.000032
Weber Kettle	$250.00	0.000032

$$\begin{aligned} E = \$15{,}000(0.000008) &+ \$1{,}000(0.000016) + \$625(0.000016) \\ &+ \$525(0.000016) + \$390(0.000032) + \$250(0.000032) \\ \approx \mathbf{\$0.17} \end{aligned}$$

■

Life insurance is also a form of gambling. When you purchase a policy, you are betting that you will die during the term of the policy, and the company is betting that you will live. The probability that you will die in any particular year of your life is an empirical probability and is listed in what is called a **mortality table,** such as Table III in Appendix D at the back of this book.

EXAMPLE 5 Expected Value of Buying an Insurance Policy

Suppose that you turned 21 years old today, and you wish to take out a $100,000 insurance policy for 1 year. How much should you be willing to pay for this policy?

Solution To answer this question, we need to know the probability that a 21-year-old person will die during his or her next year. By consulting Table III, we can see that, out of 96,478 persons of age 21, we could expect 177 to die within a year. Thus,

$$\begin{aligned} E &= (\text{AMOUNT TO WIN}) \times (\text{PROBABILITY OF WINNING}) \\ &= \$100{,}000 \times \frac{177}{96{,}478} \\ &\approx \mathbf{\$183.46} \end{aligned}$$

This means that you should be willing to pay $183.46. However, if the company charged $183.46 for such a policy, their gain (in the long run) would be zero. The actual premium, then, should be fixed at $183.46 plus a set amount for administrative costs and profit. ■

EXPECTATION WITH A COST OF PLAYING

Many games charge you a fee to play. If you must pay to play, this cost of playing should be taken into consideration when you calculate the expected value. Remember, if the expected value is 0, it is a fair game; if the expected value is positive, you should play; but if it is negative, you should not.

EXAMPLE 6 Expected Value with a Cost of Playing

Consider a game consisting of drawing a card from a deck of cards. If it is a face card, you win $20. Should you play the game if it costs $5 to play?

Subtract the cost of play from your winnings.

Solution

$$E = (\$20 - \$5)\underbrace{\left(\frac{12}{52}\right)}_{\text{Probability of winning}} + (-\$5)\underbrace{\left(1 - \frac{12}{52}\right)}_{\text{Probability of losing}}$$

$$\begin{aligned} &= \$15\left(\tfrac{3}{13}\right) - (5)\left(\tfrac{10}{13}\right) \\ &= \frac{45 - 50}{13} \end{aligned}$$

$$= \frac{-5}{13}$$
$$\approx -\$0.38$$

You should not play this game, because it has a negative expectation. ■

You must understand the nature of the game to know whether to subtract the cost of playing, as we did in Example 6. If you surrender your money to play, then it must be subtracted, but if you "leave it on the table," then you do not subtract it. Consider the following example of a U.S. roulette game. In this game, your bet is placed on the table but is not collected until after the play of the game and it is determined that you lost.

EXAMPLE 7 **Expected Value when Playing Roulette**

What is the expectation for playing roulette if you bet $1 on number 5?

Solution This is classified as problem solving because we need to know more information. A U.S. roulette wheel has 38 numbered slots (1–36, 0, and 00), as shown in Figure 9.7. Some of the more common bets and payoffs are shown. If the payoff is listed as 6 to 1, you would receive $6 for each $1 bet. In addition, you would keep the $1 you originally wagered. One play consists of having the dealer spin the wheel and a little ball in opposite directions. As the ball slows to a stop, it lands in one of the 38 numbered slots, which are colored black, red, or green. A single-number bet has a payoff of 35 to 1. The $1 you bet is collected only if you lose.

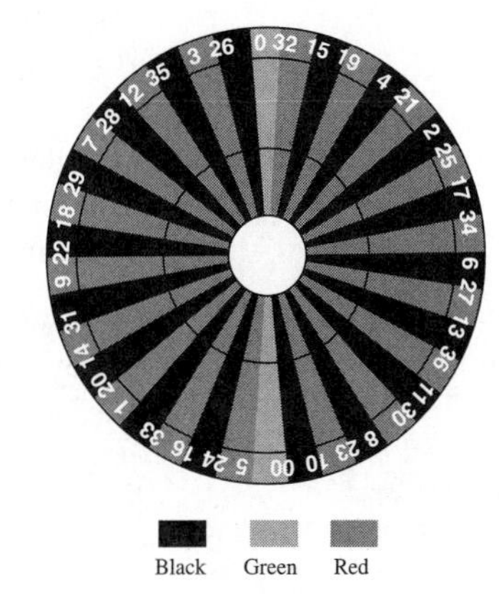

Here is how bets are placed on the roulette table

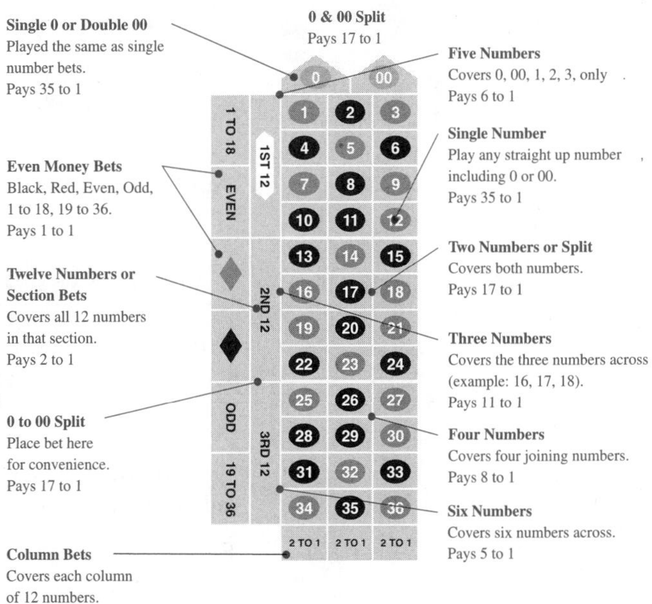

Figure 9.7 U.S. roulette wheel and board

Now, you can calculate the expected value:

$$E = 35(\tfrac{1}{38}) + (-1)(\tfrac{37}{38})$$
$$\approx -\$0.05$$

The expected loss is about 5¢ per play. ■

EXAMPLE 8

Expected Value for Selling a House

Walt, who is a realtor, knows that if he takes a listing to sell a house, it will cost him $1,000. However, if he sells the house, he will receive 6% of the selling price. If another realtor sells the house, Walt will receive 3% of the selling price. If the house remains unsold after 3 months, he will lose the listing and receive nothing. Suppose that the probabilities for selling a particular $200,000 house are as follows: the probability that Walt will sell the house is 0.4; the probability that another agent will sell the house is 0.2; and the probability that the house will remain unsold is 0.4. What is Walt's expectation if he takes this listing?

Solution First, we must decide whether this is an "entrance fee" problem. Is Walt required to pay the $1,000 before the "game" of selling the house is played? The answer is yes, so the $1,000 must be subtracted from the payoffs. Now let's calculate those payoffs:

$$6\% \text{ of } \$200{,}000 = 0.06(\$200{,}000) = \$12{,}000$$
$$3\% \text{ of } \$200{,}000 = 0.03(\$200{,}000) = \$6{,}000$$

$$E = \underbrace{(\$12{,}000 - \$1{,}000)(0.4)}_{\text{Walt sells the house.}} + \underbrace{(\$6{,}000 - \$1{,}000)(0.2)}_{\text{Another agent sells.}} + \underbrace{(-\$1{,}000)(0.4)}_{\text{House doesn't sell.}}$$
$$= \$11{,}000(0.4) + \$5{,}000(0.2) - \$1{,}000(0.4)$$
$$= \$4{,}400 + \$1{,}000 - \$400$$
$$= \$5{,}000$$

Walt's expectation is $5,000. ■

EXAMPLE 9

Using the Expected Value in Making a Decision

Remember the situation from the beginning of this section? Your friend George shows you three cards. One is white on both sides, one is black on both sides, and the last is black on one side and white on the other. He mixes the cards and lets you select one at random and place it on the table. George then calls out the color on the underside. If he selects the right color, you lose $1; if he does not select the correct color on the underside, you win $1. Should you play?

Solution First, decide whether there is an "entry fee to play." Since you and George are each putting up $1, there is no entry fee. However, suppose that you are not sure you want to play, so George explains:

Suppose you select a card, and we see that it's black on top. We know it's not the white–white card, so it must be either the black–black card or the black–white card. This gives you a 50–50 chance of winning, so it's a fair game. Come on, let's play!

Before you agree to play, consider the sample space. Let's start by distinguishing between the front (side 1) and the back (side 2) of each card. The sample space of *equally likely* events is as follows:

RESULT	*CARD 1*		*CARD 2*		*CARD 3*	
Side showing (side 1)	B_1	B_2	B	W	W_1	W_2
Side not showing (side 2)	B_2	B_1	W	B	W_2	W_1

Let's also assume that, after we select the card, we see that a black side is face up. (If it is a white side that we see, we can repeat the same argument, with colors reversed.) We also see that

$$P(\text{black is face down}) = \frac{3}{6} = \frac{1}{2}$$ (This is George's **incorrect** argument.)

But we have additional information. We wish to compute the *conditional probability* of black on the underside, given that a black card is face up. Alter the sample space to take into account the additional information:

RESULT	*CARD 1*		*CARD 2*		*CARD 3*	
Side showing (side 1)	B_1	B_2	B	~~W~~	~~W_1~~	~~W_2~~
Side not showing (side 2)	B_2	B_1	W	~~B~~	~~W_2~~	~~W_1~~
				↑	↑	↑

Cross these out (black on top).

$$P(\text{black is face down} \mid \text{black on top}) = \frac{2}{3}$$

This means that, if George picks the color on the bottom to match the color on top, he will have a probability of winning of 2/3, so the probability that we will win is 1/3. We now calculate the expectation:

$$E = (\$1)\left(\frac{1}{3}\right) + (-\$1)\left(\frac{2}{3}\right) \approx -\$0.33$$

Since the expectation is negative, **you should not play.** ■

PROBLEM SET 9.4

Essential Ideas

1. What is the formula for mathematical expectation?

2. **In Your Own Words** What is mathematical expectation, and what can it be used for?

Level 1

Use estimation to select the best response in Problems 3–8. Do not calculate.

3. The expectation from playing a game in which you win \$950 by correctly calling heads or tails when you flip a coin is about
   A. \$500 B. \$50 C. \$950

4. The expectation from playing a game in which you win \$950 by correctly calling heads or tails on each of five flips of a coin is about
   A. \$500 B. \$50 C. \$950

5. If the expected value of playing a \$1 game of blackjack is \$0.04, then after playing the game 100 times you should have netted about
   A. \$104 B. −\$4 C. \$4

6. If your expected value when playing a \$1 game of roulette is −\$0.05, then after playing the game 100 times you should have netted about
   A. −\$105 B. −\$5 C. \$5

7. The probability of correctly guessing a telephone number is about
   A. 1 out of 100
   B. 1 out of 1,000
   C. 1 out of 10,000,000

8. Winning the grand prize in a state lottery is about as probable as
   A. having a car accident
   B. having an item fall out of the sky into your yard
   C. being a contestant on *Jeopardy*

9. A box contains one each of \$1, \$5, \$10, \$20, and \$100 bills. You reach in and withdraw one bill. What is the expected value?

10. A box contains one each of \$1, \$5, \$10, \$20, and \$100 bills. It costs \$20 to reach in and withdraw one bill. What is the expected value?

11. Suppose that you roll two dice. You will be paid \$5 if you roll a double. You will not receive anything for any other outcome. How much should you be willing to pay for the privilege of rolling the dice?

12. A magazine subscription service is having a contest in which the prize is \$80,000. If the company receives 1 million entries, what is the expectation of the contest?

13. Suppose that you have 5 quarters, 5 dimes, 10 nickels, and 5 pennies in your pocket. You reach in and choose a coin at random. What is the expectation? What is the most likely to be picked?

14. A game involves tossing two coins and receiving 50¢ if they are both heads. What is a fair price to pay for the privilege of playing?

15. Krinkles potato chips is having a "Lucky Seven Sweepstakes." The one grand prize is \$70,000; 7 second prizes each pay \$7,000; 77 third prizes each pay \$700; and 777 fourth prizes each pay \$70. What is the expectation of this contest (rounded to the nearest cent), if there are 10 million entries?

16. A punch-out card contains 100 spaces. One space pays \$100, five spaces pay \$10, and the others pay nothing. How much should you pay to punch out one space?

WHAT IS WRONG, *if anything, with each of the statements in Problems 17–22? Explain your reasoning.*

17. In roulette, if you bet on black, the probability of winning is $\frac{1}{2}$ because there are the same numbers of black and red spots.

18. An expected value of \$5 means that you should expect to win \$5 each time you play the game.

19. If the expected value of a game is positive, then it is a game you should play.

20. If you were asked to choose between a sure \$10,000, or an 80% chance of winning \$15,000 and a 20% chance of winning nothing, which would you take?

    Game A: $E = \$10{,}000(1) = \$10{,}000$ This is a sure thing.

    Game B: $E = \$15{,}000(0.8) + \$0(0.2) = \$12{,}000$

The better choice, according to expected value, is to take the 80% chance of winning $15,000.

21. Suppose that you buy a lottery ticket for $1. The payoff is $50,000, with a probability of winning of 1/1,000,000. Therefore, the expected value is

$$E = \$50{,}000\left(\frac{1}{1{,}000{,}000}\right) = \$0.05$$

22. Suppose you pay $1. The payoff is $50 with odds against winning 99 to 1. Then the expected value is

$$E = (\$50 - \$1)\left(\frac{1}{99}\right) + (-\$1)\left(\frac{98}{99}\right) = -\$0.49$$

Level 2—Applications

Use the mortality table (Table III *in Appendix D) to answer the questions in Problems* 23–28.

23. What is the expected value of a 1-year, $10,000 policy issued at age 10?

24. What is the expected value of a 1-year, $10,000 policy issued at age 38?

25. What is the expected value of a 1-year, $10,000 policy issued at age 65?

26. What is the expected value of a 1-year, $20,000 policy issued at age 18?

27. What is the expected value of a 1-year, $25,000 policy issued at age 19?

28. What is the expected value of a 1-year, $50,000 policy issued at age 23?

29. In old gangster movies on TV, you often hear of "numbers runners" or the "numbers racket." This numbers game, which is still played today, involves betting $1 on the last three digits of the number of stocks sold on a particular day in the future as reported in *The Wall Street Journal.* If the payoff is $500, what is the expectation for this numbers game?

30. A realtor who takes the listing on a house to be sold knows that she will spend $800 trying to sell the house. If she sells it herself, she will earn 6% of the selling price. If another realtor sells a house from her list, the first realtor will earn only 3% of the price. If the house remains unsold after 6 months, she will lose the listing. Suppose that probabilities are as follows:

EVENT	*PROBABILITY*
Sell by herself	0.50
Sell by another realtor	0.30
Not sell in 6 months	0.20

What is the expected profit from listing a $185,000 house?

31. An oil-drilling company knows that it costs $25,000 to sink a test well. If oil is hit, the income for the drilling company will be $425,000. If only natural gas is hit, the income will be $125,000. If nothing is hit, there will be no income. If the probability of hitting oil is $\frac{1}{40}$ and if the probability of hitting gas is $\frac{1}{20}$, what is the expectation for the drilling company? Should the company sink the test well?

32. In Problem 31, suppose that the income for hitting oil is changed to $825,000 and the income for gas to $225,000. Now what is the expectation for the drilling company? Should the company sink the test well?

33. Suppose that you roll one die. You are paid $5 if you roll a one, and you pay $1 otherwise. What is the expectation?

34. A game involves drawing a single card from an ordinary deck. If an ace is drawn, you receive 50¢; if a heart is drawn, you receive 25¢; if the queen of spades is drawn, you receive $1. If the cost of playing is 10¢, should you play?

35. Consider the following game in which a player rolls a single die. If a prime (2, 3, or 5) is rolled, the player wins $2. If a square (1 or 4) is rolled, the player wins $1. However, if the player rolls a perfect number (6), it costs the player $11. Is this a good deal for the player or not?

Consider the spinners in Problems 36–39. Determine which represent fair games. Assume that the cost to spin the wheel once is \$5.00 *and that you will receive the amount shown on the spinner after it stops.*

36.

\$8.00
\$2.00

37.

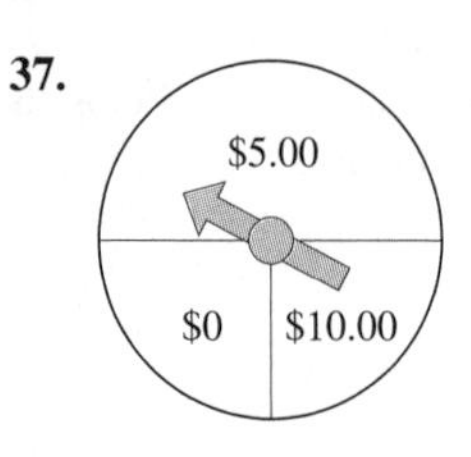

38.

\$2.00
\$8.00
\$5.00

39.

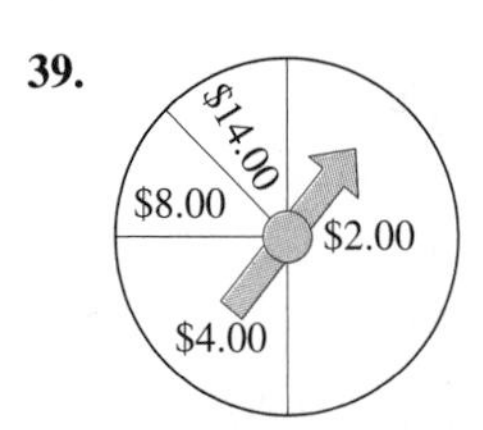

40. Assume that a dart is randomly thrown at the following dart board and that it strikes the board every time. The payoffs are listed on the board. How much should you be willing to pay for the opportunity to play this game?

\$1.00 | \$6.00 | \$8.00 | \$4.00 | \$10.00

41. A company held a contest, and the following information was included in the fine print:

\$ WIN\$ WIN\$

PRIZE	NUMBER OF PRIZES	PROBABILITY OF WINNING INDICATED PRIZE
\$10,000	13	0.000005
\$1,000	52	0.00002
\$100	520	0.0002
\$10	28,900	0.010886
TOTAL	**29,485**	**0.011111**

Read this information carefully, and calculate the expectation (to the nearest cent) for this contest.

42. A company held a bingo contest for which the following chances of winning were given:

PLAY ONE CARD AND WIN
YOUR CHANCES OF WINNING ARE AT LEAST:

	1 TIME	*7 TIMES*	*13 TIMES*
\$25 prize	1 in 21,252	1 in 3,036	1 in 1,630
\$3 prize	1 in 2,125	1 in 304	1 in 163
\$1 prize	1 in 886	1 in 127	1 in 68
Any prize	1 in 609	1 in 87	1 in 47

What is the expectation (to the nearest cent) from playing one card 13 times?

43. Calculate the expectation (to the nearest cent) for the *Reader's Digest* sweepstakes described below. Assume there are 197,000,000 entries.

Official Disclosure of Dates
To be eligible to win the *FIVE MILLION DOLLAR* Grand Prize, you must return the attached Sweepstakes Entry Document by August 19, 1991. Failure to respond by that date will result in the forfeiture of Grand Prize eligibility. To be eligible to win any of 58,567 other prizes (but not the Grand Prize) return your Sweepstakes Entry Document by March 2, 1992.

Grand Prize Distribution Information
If you are chosen Grand Prize winner, you will receive *FIVE MILLION DOLLARS* in your choice of payment options: Either 30 equal yearly payments of \$167,000.00 each *OR* 360 equal monthly payments of \$14,000.00 each. You must specify your choice of payment option now by detaching the appropriate card at left and affixing it to the box provided on your Sweepstakes Entry Document.

OFFICIAL PRIZE LIST

1 First Prize \$100,000.00
2 Second Prizes \$50,000.00
3 Third Prizes \$20,000.00
4 Fourth Prizes \$5,000.00
10 Fifth Prizes \$500.00
400 Sixth Prizes \$100.00
58,147 Seventh Prizes.....Winner's choice of a "Special Edition" Men's or Women's wristwatch, at an \$89.00 approximate retail value.

Sponsor D. P. Burr
SWEEPSTAKES DIRECTOR

\$10,500,000.00
SWEEPSTAKES ENTRY DOCUMENT

In Problems 44–53, *what is the expectation for the* \$1 *bets on a U.S. roulette wheel? See Figure* 9.7 *on page* 433.

44. Black

45. Odd

46. Single-number bet

47. Double-number bet

48. Three-number bet

49. Four-number bet

50. Five-number bet

51. Six-number bet

52. Twelve-number bet

53. Column bet

54. St. Petersburg paradox. Suppose that you toss a coin and will win \$1 if it comes up heads. If it comes up tails, you toss again. This time you will receive \$2 if it comes up heads. If it comes up tails, toss again. This time you will receive \$4 if it is heads and nothing if it comes up tails. What is the mathematical expectation for this game?

55. St. Petersburg paradox. Suppose that you toss a coin and will win \$1 if it comes up heads. If it comes up tails, you toss again. This time you will receive \$2 if it comes up heads. If it comes up tails, toss again. This time you will receive \$4 if it is heads. Continue in this fashion for a total of 10 flips of the coin, after which you receive nothing if it comes up tails. What is the mathematical expectation for this game?

56. St. Petersburg paradox. Suppose that you toss a coin and will win \$1 if it comes up heads. If it comes up tails, you toss again. This time you will receive \$2 if it comes up heads. If it comes up tails, toss again. This time you will receive \$4 if it is heads. Continue in this fashion for a total of 1,000 flips of the coin after which you receive nothing if it comes up tails. What is the mathematical expectation for this game?

57. St. Petersburg paradox. Suppose that you toss a coin and will win \$1 if it comes up heads. If it comes up tails, you toss again. This time you will receive \$2 if it comes up heads. If it comes up tails, toss again. This time you will receive \$4 if it is heads. You continue in this fashion until you finally toss a head. Would you pay \$100 for the privilege of playing this game? What is the mathematical expectation for this game?

58. Suppose that you are in class and your instructor makes you the following legitimate offer. Take out a piece of paper and without communicating with your classmates, write one of the following messages under your name:

☐ I share the wealth and will receive \$1,000 if *everyone* in the class checks this box.

☐ I will not share the wealth and want a certain \$100.

If *everyone* checks the first box, then all will receive \$1,000. If at least *one* person checks the second box, then only those who check the second box will receive \$100. Which box would you check, and why?

59. Repeat Problem 58 except change the stakes to \$110 and \$100, respectively.

60. Repeat Problem 58 except change the stakes to \$10,000 and \$10, respectively.

61. Repeat Problem 58 except change the stakes to be an A grade in the class if *everyone* checks the first box, but if anyone checks the second box, the following will occur: Those who check it will have a 50% chance of an A and a 50% chance of a F, but those who do not check the second box will be given an F in this course.

62. Situation Suppose that your friend George shows you three cards. One card is white on both sides, one is black on both sides, and the last is black on one side and white on the other. He mixes the cards and tells you to select one at random and place it on the table. Suppose the upper side turns out to be black. It is not the white–white card; it must be either the black–black or the black–white card. "Thus," says George, "I'll bet you \$1 that the other side is black." Would you play? Perhaps you hesitate. Now George says he feels generous. You need to pay him only 75¢ if you lose, and he will still pay you \$1 if he loses. Would you play now?

63. In Your Own Words Write a short paper comparing the SITUATION with your own situation. (That is, where could you use the ideas of expected value in your own life?)

9.5 SUMMARY AND REVIEW

Important Terms

Cards (sample space) [9.1]
Complementary probabilities [9.2]
Conditional probability [9.3]
Dice (sample space) [9.1]
Equally likely outcomes [9.1]
Event [9.1]
Expectation [9.4]
Expected value [9.4]
Experiment [9.1]
Fair coin [9.1]
Fair game [9.4]
Fundamental counting principle [9.2]
Impossible event [9.1]
Loaded die [9.1]
Mathematical expectation [9.4]
Mortality table [9.4]
Mutually exclusive [9.1]
Odds against [9.3]
Odds in favor [9.3]
Probability [9.1]
Property of complements [9.2]
Sample space [9.1]
Simple event [9.1]
Tree diagram [9.1]

Chapter Objectives

The material of this chapter is reviewed in the following list of objectives. A self-test (with answers and suggestions for additional study) is given at the end of this problem set.

[9.1] *Objective 9.1* Understand and apply the definition of probability.

[9.1] *Objective 9.2* Find probabilities by looking at the sample space. In particular, be able to use Figure 9.2 (single die), Figure 9.3 (deck of cards), and Figure 9.4 (pair of dice).

[9.1] *Objective 9.3* Find probabilities involving the union and intersection of events by looking at the sample space.

[9.2] *Objective 9.4* Work probability problems involving complementary events.

[9.3] *Objective 9.5* Work problems involving odds.

[9.3] *Objective 9.6* Find conditional probabilities.

[9.4] *Objective 9.7* Find the expected value of a given event.

[9.2, 9.4] *Objective 9.8* Estimate probabilities, odds, and expectations.

Self-Test

Each question of this self-test is related to the corresponding objective listed above.

1. If a sample from an assembly line reveals four defective items out of 1,000 sample items, what is the probability that any one item is defective?

2. **a.** What is the probability of rolling a single die and obtaining a prime?
b. A pair of dice is rolled. What is the probability that the resulting sum is eight?
c. A card is selected from an ordinary deck of cards. What is the probability that it is a jack or better? (A jack or better is a jack, queen, king, or ace.)

3. Roll a pair of dice. Find the following.
a. P(5 on one of the dice)
b. P(5 on one die or 4 on the other)
c. P(5 on one die and 4 on the other)

4. If the probability of dropping an egg is 0.01, what is the probability of not dropping the egg?

5. a. If $P(E) = 0.9$, what are the odds in favor of E?
 b. If the odds against you are 1,000 to 1, what is the probability of the event?

6. A single card is drawn from a standard deck of cards. What is the probability that it is an ace, if you know that one ace has already been removed from the deck?

7. a. A game consists of rolling a die and receiving \$12 if a one is rolled and nothing otherwise. What is the mathematical expectation?
 b. What is the expected value for a \$100,000 life insurance policy issued at age 24?

8. Which one of the following events has odds against of about a million to one?
 A. Selecting a particular Social Security number by picking the digits at random
 B. Selecting a particular phone number at random if the first digit is known
 C. Winning first prize in a state lottery

STUDY HINTS *Compare your solutions and answers to the self-test. For each problem you missed, work some additional problems in the section listed in the margin. After you have worked these problems, you can test yourself with the Practice Test.*

Complete Solutions to the Self-Test

Additional Problems

[9.1] *Problems 3–4; 26–34; 58–61; 76*

1. $P(\text{defective}) = \dfrac{4}{1{,}000} = 0.004$

[9.1] *Problems 15–25; 35–40; 49–54; 62–75*

2. a. The possible primes are 2, 3, and 5, so $P(\text{prime}) = \dfrac{3}{6} = \dfrac{1}{2}$.
 b. Look at Figure 9.4. $P(\text{eight}) = \dfrac{5}{36}$
 c. $P(\text{jack or better}) = \dfrac{16}{52} = \dfrac{4}{13}$

[9.1] *Problems 41–48; 55–57*

3. Look at Figure 9.4.
 a. $P(5 \text{ on one of the dice}) = \dfrac{11}{36}$
 b. $P(5 \text{ on one die or 4 on the other}) = \dfrac{20}{36} = \dfrac{5}{9}$
 c. $P(5 \text{ on one die and 4 on the other}) = \dfrac{2}{36} = \dfrac{1}{18}$

[9.2] *Problems 1; 14–30*

4. $P(E) = 0.01$; $P(\overline{E}) = 1 - P(E) = 1 - 0.01 = 0.99$

[9.3] *Problems 6–11; 29–30*

5. a. $P(E) = \dfrac{9}{10}$; $P(\overline{E}) = 1 - \dfrac{9}{10} = \dfrac{1}{10}$

$$\text{Odds in favor} = \frac{P(E)}{P(\overline{E})} = \frac{\frac{9}{10}}{\frac{1}{10}} = \frac{9}{10} \cdot \frac{10}{1} = \frac{9}{1}\text{; 9 to 1}$$

b. Let E be the event. Given $f = 1{,}000$ and $s = 1$,

$$P(E) = \frac{s}{s+f} = \frac{1}{1{,}000+1} = \frac{1}{1{,}001}$$

[9.3] *Problems 12–28*

6. $P(\text{ace}) = \frac{3}{51} = \frac{1}{17}$

[9.4] *Problems 9–16; 23–61*

7. a. $E = P(\text{one}) \cdot \$12 = \frac{1}{6}(12) = 2$

The expected value is $2.

b. $E = \$100{,}000\left(\frac{183}{95{,}940}\right)$ From Table III in Appendix D

$\approx \$190.74$

The expected value is $190.74.

[9.2] *Problems 3–8*
[9.4] *Problems 3–18*

8. A typical Social Security number is 555–55–5555, so by the fundamental counting principle, the odds are about 1 in 10^9.

A typical phone number is 555–1234; if the first digit is known, the odds are about 1 in 10^6.

A state lottery almost always has more than a million entries; the best choice is B.

Chapter 9 Practice Test

1. **In Your Own Words** Define probability; explain the words "equally likely and mutually exclusive."

2. If a spinner has 18 red, 18 black, and 2 green compartments, what is the probability of landing on a green in one spin of the wheel?

3. If a jar contains 25 blue marbles, 55 cat-eye marbles, and 20 clear marbles, what is the probability of obtaining a clear marble when one marble is chosen from the jar? Assume that all marbles have an equal chance of being drawn.

4. A die is rolled. What is the probability that it comes up:
   a. a three? **b.** a four? **c.** a three and a four?
   d. even? **e.** odd? **f.** an even or an odd?

5. A pair of dice is tossed. What is the probability that the sum is:
   a. six? **b.** seven?

6. A card is selected from an ordinary deck of cards. What is the probability that it comes up
   a. a diamond? **b.** a two? **c.** a diamond and a two?

7. A card is selected from an ordinary deck of cards. What is the probability that it comes up
   a. a king? **b.** a heart? **c.** a king or a heart?

8. Consider a pair of dice in which each die has three green faces, two red faces, and one yellow face. Roll the pair of dice. Write out the sample space and find the probability of obtaining two faces of the same color with one roll of the pair of dice.

9. Using the sample space from Problem 8, find the following.
a. P(both green) **b.** P(both red) **c.** P(both yellow)

10. If $P(A) = \frac{5}{11}$, what is $P(\overline{A})$?

11. If the probability of drawing a defective item from an assembly line is 0.09, what is the probability of not drawing a defective item?

12. What is the probability of having a phone number with at least one repeated digit? (Assume the phone numbers are assigned randomly and also that the first digit cannot be a 0 or a 1.)

13. A coin is tossed four times. What is the probability of obtaining at least one tail?

14. What are the odds in favor of drawing a heart from an ordinary deck of cards?

15. If the odds in favor of a particular event are 20 to 1, what is the probability of that event?

16. According to the Internal Revenue Service, the odds against having a corporate tax return being audited are 15 to 1. What is the probability that a corporate tax return will be audited?

17. If Ferdinand the Frog has twice as good a chance of winning the jumping contest as either one of the other two champion frogs, what is the probability of Ferdinand winning (stated as a percent)? What are the odds in favor of Ferdinand winning?

18. A jar contains three orange balls and two purple balls. Two balls are chosen at random. What is the probability of obtaining two orange balls if we know that the first ball drawn was orange and if we draw the second ball without replacing the first ball?

19. Repeat Problem 18, except replace the first ball before drawing the second ball.

20. What is the probability that a family of four children will contain two boys and two girls, given that the first child was a boy?

21. A single card is drawn from a standard deck of cards. What is the probability that it is a king, if you know that it is a face card?

22. A game consists of cutting a standard deck of cards. You win $2 if a face card turns up, but you lose $1 otherwise. Should you play?

23. What is the expected value of a $50,000 life insurance policy issued at age 25?

24. A lottery offers a prize of a color TV set (value $500), and 800 tickets are sold.
What is the expectation if you buy three tickets?
If the tickets cost $1 each, should you buy them?
How much should you be willing to pay for the three tickets?

25. Explain mathematical expectation using your own words. Tell why it is a useful concept, and give some examples of when you might use mathematical expectation.

CHAPTER TEN

STATISTICS

10.1 FREQUENCY DISTRIBUTIONS AND GRAPHS

Situation

"Kids Prefer Del Monte Fruit Cups." "Nine out of ten dentists recommend Trident for their patients who chew gum." "Crest has been shown to be . . . ." "You can clearly see that Bufferin is the most effective . . . ." "Penzoil is better suited . . . ." "Sylvania was preferred by . . . ." "How can anyone analyze the claims of the commercials we see and hear on a daily basis?" asked Betty. "I even subscribe to Consumer Reports, *but so many of the claims seem to be unreasonable. I don't like to buy items by trial and error, and I really don't believe all the claims in advertisements."*

The first step in dealing with and understanding statistics is to think critically and understand some of the ways in which data can be organized. In this section, Betty will learn how to make a frequency distribution and then how to organize the data into a bar graph or a line graph.

Statistics is used in a variety of ways. Think about how you have seen statistics used.

FREQUENCY DISTRIBUTIONS

Undoubtedly, you have some idea about what is meant by the term *statistics*. For example, we hear about:

1. Statistics on population growth
2. Information about the depletion of the rain forests or the ozone layer
3. The latest statistics on the cost of living
4. The Gallup Poll's use of statistics to predict election outcomes
5. The Nielsen ratings, which indicate that one show has 30% more viewers than another
6. Baseball or sports statistics

We could go on and on, but you can see from these examples that there are two main uses for the word **statistics**. First, we use the term to mean a mass of data, in-

Table 10.1
***Frequency Distribution** for 50 rolls of a pair of dice*

Outcome	Tally	Frequency
2	|	1
3	|||	3
4	𝍸	5
5	||	2
6	𝍸 ||||	9
7	𝍸 |||	8
8	𝍸 |	6
9	𝍸 |	6
10	||||	4
11	𝍸	5
12	|	1

cluding charts and tables. This is the everyday, nontechnical use of the word. Second, the word refers to a methodology for collecting, analyzing, and interpreting data. In this chapter we'll examine some of these statistical methods.

Computers, spreadsheets, and simulation programs have done a lot to help us deal with hundreds or thousands of pieces of information at the same time. We can deal with large batches of data by organizing them into groups, or **classes.** The difference between the lower limit of one class and the lower limit of the next class is called the **interval** of the class. After determining the number of values within a class, termed the **frequency,** you can use this information to summarize the data. The end result of this classification and tabulation is called a **frequency distribution.** For example, suppose that you roll a pair of dice 50 times and obtain these outcomes:

3, 2, 6, 5, 3, 8, 8, 7, 10, 9, 7, 5, 12, 9, 6, 11, 8,
11, 11, 8, 7, 7, 7, 10, 11, 6, 4, 8, 8, 7, 6, 4, 10, 7,
9, 7, 9, 6, 6, 9, 4, 4, 6, 3, 4, 10, 6, 9, 6, 11

We can organize these data in a convenient way by using a frequency distribution, as shown in Table 10.1.

EXAMPLE 1 Making a Frequency Distribution

Make a frequency distribution for the information in Table 10.2.

Table 10.2 *States Sales Tax Rates**

Alabama	4%	Louisiana	4%	Ohio	5%
Alaska	0%	Maine	6%	Oklahoma	4.5%
Arizona	5%	Maryland	5%	Oregon	0%
Arkansas	4.5%	Massachusetts	5%	Pennsylvania	6%
California	$7\frac{1}{4}\%$	Michigan	6%	Rhode Island	7%
Colorado	3%	Minnesota	$6\frac{1}{2}\%$	South Carolina	5%
Connecticut	6%	Mississippi	7%	South Dakota	4%
Delaware	0%	Missouri	5.725%	Tennessee	6%
Florida	6%	Montana	0%	Texas	6.25%
Georgia	4%	Nebraska	5%	Utah	$4\frac{7}{8}\%$
Hawaii	4%	Nevada	6.5%	Vermont	5%
Idaho	5%	New Hampshire	0%	Virginia	4.5%
Illinois	$6\frac{1}{4}\%$	New Jersey	6%	Washington	$6\frac{1}{2}\%$
Indiana	5%	New Mexico	5%	West Virginia	6%
Iowa	5%	New York	4%	Wisconsin	5%
Kansas	4.9%	North Carolina	6%	Wyoming	4%
Kentucky	6%	North Dakota	5%		

*Does not include local sales taxes.

Solution

Sales tax	Tally	Frequency
0%	𝍸	5
3%	𝍷	1
4%	𝍸 𝍷𝍷	7
$4\frac{1}{2}\%$	𝍷𝍷𝍷	3
$4\frac{7}{8}\%$	𝍷	1
4.9%	𝍷	1
5%	𝍸 𝍸 𝍷𝍷𝍷	13
5.725%	𝍷	1
6%	𝍸 𝍸	10
$6\frac{1}{4}\%$	𝍷𝍷	2
$6\frac{1}{2}\%$	𝍷𝍷𝍷	3
7%	𝍷𝍷	2
$7\frac{1}{4}\%$	𝍷	1

Make three columns. First, list the sales tax categories; next, tally; and finally, count the tallies to determine the frequency of each (see margin). To account for irregularities (such as $4\frac{7}{8}\%$), the categories are often grouped. For this example, the data are divided into 8 categories (or groups) in the **grouped frequency distribution,** as shown here. The number of categories is usually arbitrary and chosen for convenience.

Sales tax	Tally	Frequency
0–1%	𝍸	5
1^{+}–2%		0
2^{+}–3%	𝍷	1
3^{+}–4%	𝍸 𝍷𝍷	7
4^{+}–5%	𝍸 𝍸 𝍸 𝍷𝍷𝍷	18
5^{+}–6%	𝍸 𝍸 𝍷	11
6^{+}–7%	𝍸 𝍷𝍷	7
Over 7%	𝍷	1

■

BAR GRAPHS

Frequency distribution for 50 rolls of a pair of dice

Outcome	Tally	Frequency
2	𝍷	1
3	𝍷𝍷𝍷	3
4	𝍸	5
5	𝍷𝍷	2
6	𝍸 𝍷𝍷𝍷𝍷	9
7	𝍸 𝍷𝍷𝍷	8
8	𝍸 𝍷	6
9	𝍸 𝍷	6
10	𝍷𝍷𝍷𝍷	4
11	𝍸	5
12	𝍷	1

A **bar graph** compares several related pieces of data using horizontal or vertical bars of uniform width. There must be some type of scale or measurement on both the horizontal and vertical axes. An example of a bar graph is shown in Figure 10.1, which shows the data from Table 10.1.

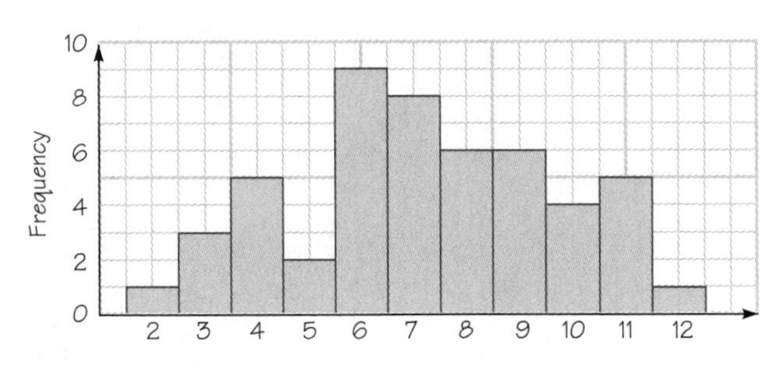

Figure 10.1 Outcomes of experiment of rolling a pair of dice

EXAMPLE 2

Constructing a Bar Graph

Construct a bar graph for the data given in Example 1. Use the grouped categories.

Solution To construct a bar graph, draw and label the horizontal and vertical axes, as shown in Figure 10.2a. It is helpful (although not necessary) to use graph paper. Next, draw marks indicating the frequencies as shown in Figure 10.2b. Finally, complete the bars and shade them as shown in Figure 10.2c.

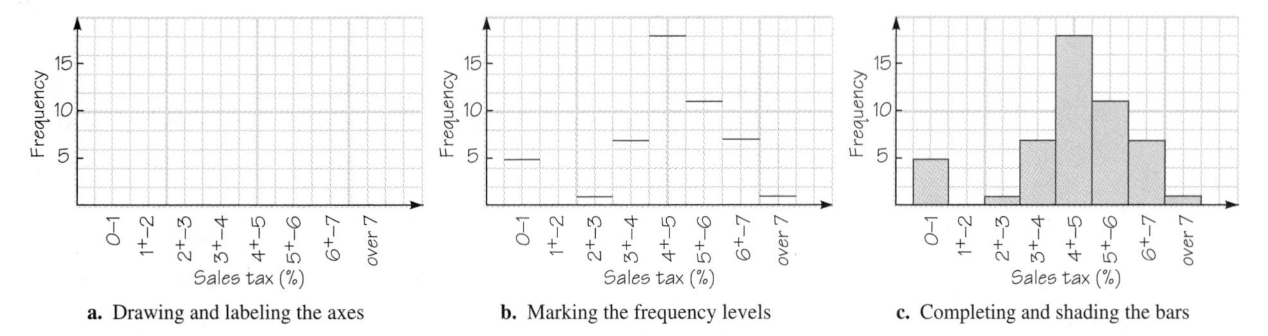

a. Drawing and labeling the axes **b.** Marking the frequency levels **c.** Completing and shading the bars

Figure 10.2 Bar graph for sales tax data ■

If a bar graph represents a frequency distribution (as in Example 1), it is called a **histogram.** A histogram consists of a series of bars drawn with the same width on a horizontal axis, with each bar drawn in proportion to the frequency of values that occur.

You will frequently need to look at and interpret bar graphs in which bars of different lengths are used for comparison.

EXAMPLE 3

Reading a Bar Graph

Refer to Figure 10.3 to answer the following questions.

a. Who is the favorite in the election poll?
b. What percent of the voters favored Becker in July 2000?
c. In which month was the undecided vote the greatest?

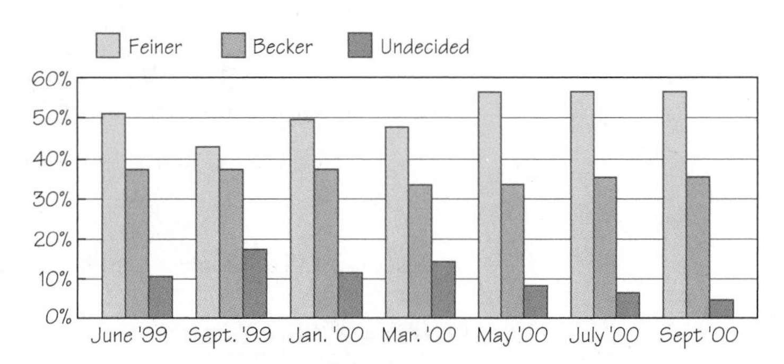

Figure 10.3 Example of a bar graph from an election

Solution

a. Feiner was the favorite, since a larger percentage of voters preferred her throughout the polling period.
b. The bar for Becker for July 2000 looks like it comes midway between the lines marking 30% and 40%, so we estimate 35%.
c. The undecided vote is shown by the darkest bars, and the darkest bar is the tallest for the month of September 1999. ■

LINE GRAPHS

A graph that uses a broken line to illustrate how one quantity changes with respect to another is called a **line graph.** A line graph is one of the most widely used kinds of graph.

EXAMPLE 4

Drawing a Line Graph

Draw a line graph for the data given in Example 1. Use the previously grouped categories.

Solution The line graph uses points instead of bars to designate the locations of the frequencies. These points are then connected by line segments, as shown in Figure 10.4. To plot the points, use the frequency distribution (from Example 1) to find the category (0–1%, for example); then plot a point showing the frequency (5, in this example). This step is shown in Figure 10.4a. The last step is to connect the dots with line segments as shown in part b.

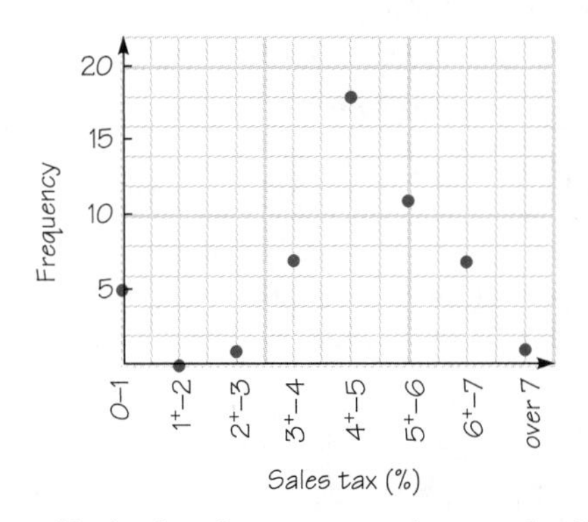

a. Plotting the points to represent frequency levels

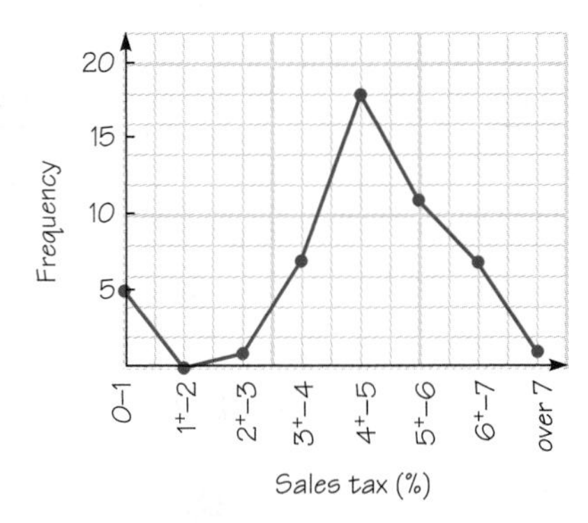

b. Connecting the dots with line segments

Figure 10.4 Constructing a line graph for sales tax data ■

Just as with bar graphs, you must be able to read and interpret line graphs.

EXAMPLE 5

Reading a Line Graph

The monthly rainfalls for Honolulu and New Orleans are shown in Figure 10.5.

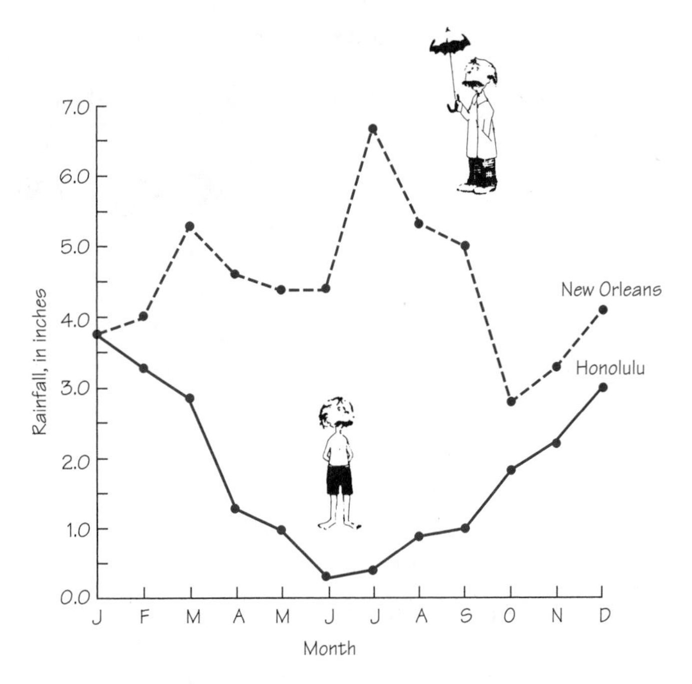

Figure 10.5 Monthly normal rainfall for two U.S. cities

a. During which month does it rain the most in New Orleans? In Honolulu?
b. During which month does it rain the least in New Orleans? In Honolulu?

Solution **a.** It rains the most in New Orleans (dashed line graph) in July, and in Honolulu (solid line graph) in January.
b. The least rain falls in New Orleans in October, and in Honolulu in June. ■

CIRCLE GRAPHS

Another type of commonly used graph is the **circle graph,** also known as a **pie chart.** This graph is particularly useful in illustrating how a whole quantity is divided into parts—for example, income or expenses in a budget.

To create a circle graph, first write the number in each category as a percent of the total. Then convert this percent to an angle in a circle. Remember that a circle is divided into 360°, so we multiply the percent by 360 to find the number of degrees for each category. You can use a protractor to construct a circle graph, as shown in Example 6.

EXAMPLE 6 **Constructing a Circle Graph**

The 1996 expenses for Karlin Enterprises are shown in Figure 10.6.

KE KARLIN ENTERPRISES — EXPENSE REPORT FY 1999

Salaries	$ 72,000
Rents, taxes, insurance	$ 24,000
Utilities	$ 6,000
Advertising	$ 12,000
Shrinkage	$ 1,200
Materials and supplies	$ 1,200
Depreciation	$ 3,600
TOTAL	**$120,000**

Figure 10.6 Expenses for Karlin Enterprises

Construct a circle graph showing the expenses for Karlin Enterprises.

Solution The first step in constructing a circle graph is to write the ratio of each entry to the total of the entries, as a percent. This is done by finding the total ($120,000) and then dividing each entry by that total.

$$\text{Salaries: } \frac{72{,}000}{120{,}000} = 60\% \qquad \text{Rents: } \frac{24{,}000}{120{,}000} = 20\%$$

$$\text{Utilities: } \frac{6{,}000}{120{,}000} = 5\% \qquad \text{Advertising: } \frac{12{,}000}{120{,}000} = 10\%$$

$$\text{Shrinkage: } \frac{1{,}200}{120{,}000} = 1\% \qquad \text{Materials/supplies: } \frac{1{,}200}{120{,}000} = 1\%$$

$$\text{Depreciation: } \frac{3{,}600}{120{,}000} = 3\%$$

A circle has 360°, so the next step is to multiply each percent by 360°:

Salaries: $360° \times 0.60 = 216°$ Rents: $360° \times 0.20 = 72°$
Utilities: $360° \times 0.05 = 18°$ Advertising: $360° \times 0.10 = 36°$
Shrinkage: $360° \times 0.01 = 3.6°$ Depreciation: $360° \times 0.03 = 10.8°$
Materials/supplies: $360° \times 0.01 = 3.6°$

Finally, use a protractor to construct the circle graph, as shown in Figure 10.7.

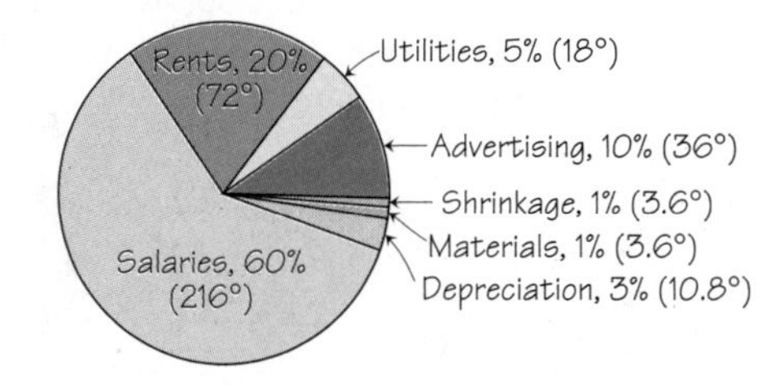

Figure 10.7 Circle graph showing the expenses for Karlin Enterprises ■

PICTOGRAPHS

A **pictograph** is a representation of data that uses pictures to show quantity. Consider the raw data shown in Table 10.3, and the bar graph of those data shown in Figure 10.8.

Table 10.3 *Marital Status of Persons Age 65 and Older**

	Married	Widowed	Divorced	Never Married
Women	5.5	7.2	0.5	0.8
Men	7.5	1.4	0.5	0.6

*Figures are in millions, rounded to the nearest 100,000.

Figure 10.8 Bar graph showing the marital status of persons age 65 and older

A pictograph uses a picture to illustrate data; it is normally used only in popular publications, rather than for scientific applications. For these data, suppose that we draw pictures of a woman and a man so that each picture represents 1 million persons, as shown in Figure 10.9.

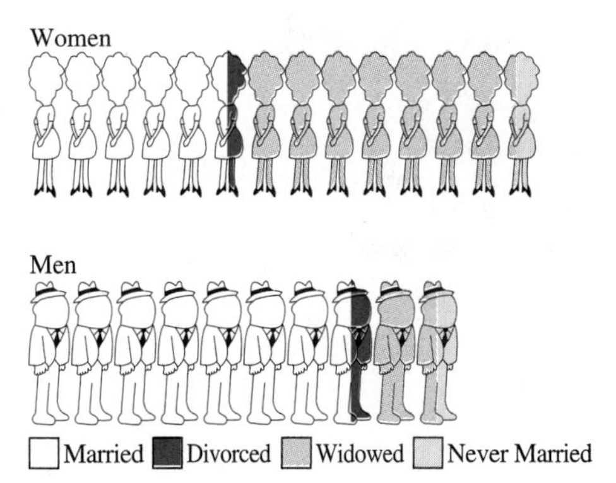

Figure 10.9 Pictograph showing the marital status of persons age 65 and older

MISUSES OF GRAPHS

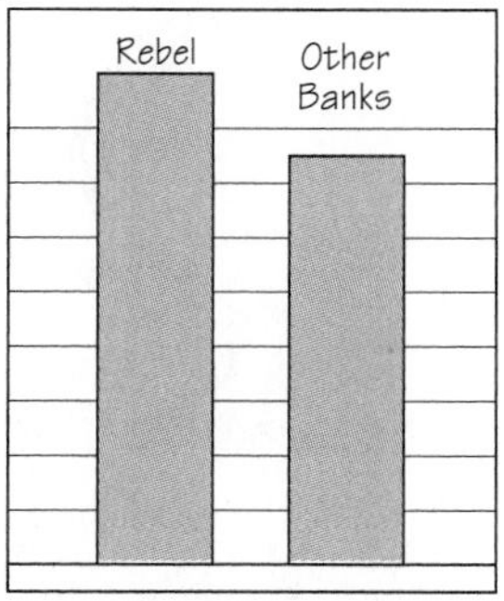

The scales on the axes of bar or line graphs are frequently chosen so as to exaggerate or diminish the real difference. Even worse, graphs are often presented with no scale whatsoever. For instance, Figure 10.10 shows a graphical "comparison" between rates paid by banks clipped from a newspaper advertisement; only the name of the bank has been changed.

Figure 10.10 Misuse of a bar graph

The most misused type of graph is the pictograph. Consider the data from Table 10.3. Such data can be used to determine the height of a three-dimensional object, as in Figure 10.11a. When an object (such as a person) is viewed as three-dimensional, differences seem much larger than they actually are. Look at part **b** of Figure 10.11 and notice that, as the height and width are doubled, the volume is actually increased eightfold.

a. Pictograph showing the number of widowed persons of age 65 and older

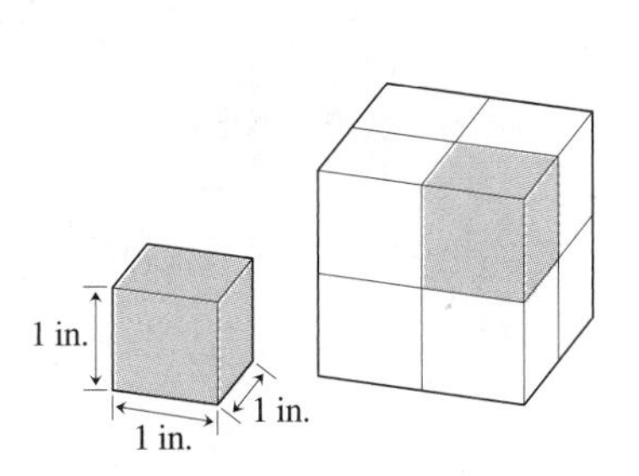

b. Change in volume as a result of changes in length and width

Figure 10.11 Examples of misuses in pictographs

PROBLEM SET 10.1

Essential Ideas

1. What is a histogram?

2. What is a line graph?

3. What is a circle graph?

4. What is a pictograph?

Level 1—Drill

Use the following information in Problems 5–7.

The heights of 30 students are as follows (rounded to the nearest inch):

66, 68, 65, 70, 67, 67, 68, 64, 64, 66, 64, 70, 72, 71, 69,
64, 63, 70, 71, 63, 68, 67, 67, 65, 69, 65, 67, 66, 69, 69

5. Prepare a frequency distribution.

6. Draw a histogram.

7. Draw a line graph.

Use the following information in Problems 8–10.

The wages of employees of a small accounting firm are as follows:

$20,000, $25,000, $25,000, $25,000, $30,000, $30,000, $35,000, $50,000, $60,000, $18,000, $18,000, $16,000, $14,000

8. Prepare a frequency distribution.

9. Draw a histogram.

10. Rearrange the salaries into the following groups and draw a bar graph.

$0–$20,000
$20,001–$30,000
$30,001–$40,000
$40,001–$50,000
Over $50,000

Use the following information in Problems 11–13.

The waiting times, in days, for a marriage license in the 50 *states are:*

Alabama, 0; Alaska, 3; Arizona, 0; Arkansas, 0; California, 0; Colorado, 0; Connecticut, 4; Delaware, 0; Florida, 3; Georgia, 3; Hawaii, 0; Idaho, 0; Illinois, 0; Indiana, 3; Iowa, 3; Kansas, 3; Kentucky, 0; Louisiana, 3; Maine, 3; Maryland, 2; Massachusetts, 3; Michigan, 3; Minnesota, 5; Mississippi, 3; Missouri, 0; Montana, 0; Nebraska, 0; Nevada, 0; New Hampshire, 3; New Jersey, 3; New Mexico, 0; New York, 0; North Carolina, 0; North Dakota, 0; Ohio, 5; Oklahoma, 0; Oregon, 3; Pennsylvania, 3; Rhode Island, 0; South Carolina, 1; South Dakota, 0; Tennessee, 3; Texas, 0; Utah, 0; Vermont, 3; Virginia, 0; Washington, 3; West Virginia, 3; Wisconsin, 5; Wyoming, 0.

11. Prepare a frequency distribution.

12. Draw a histogram.

13. Draw a line graph.

Use the following information in Problems 14–15.

The purchasing power of the dollar (1967 = $1) *is (to the nearest cent):*

1964, $1.07; 1966, $1.03; 1968, $0.96; 1970, $0.86; 1972, $0.80; 1974, $0.68; 1976, $0.59; 1978, $0.49; 1980, $0.38; 1982, $0.33; 1984, $0.31; 1986, $0.30; 1988, $0.29; 1990, $0.28; 1992, $0.27; 1994, $0.27; 1996, $0.26; 1998, $0.26

14. Draw a line graph.

15. Draw a bar graph.

16. Figure 10.12 shows a line graph.

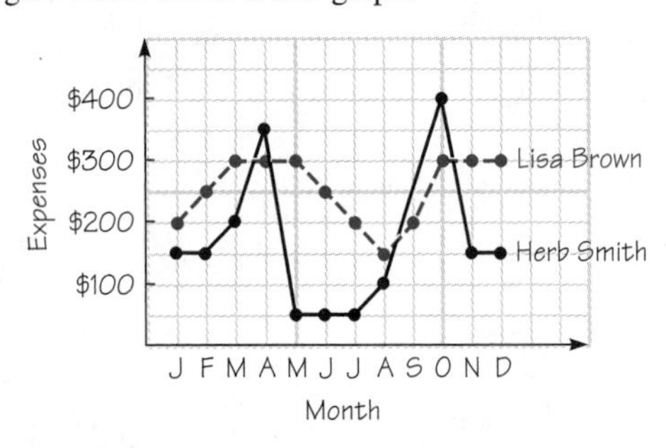

Figure 10.12 Expenses for two salespeople of the Leadwell Pencil Company

a. During which month did Herb incur the most expenses?
b. During which month did Lisa incur the least expenses?

17. Use the bar graph in Figure 10.13.

a. How many candidates were running for office?
b. Who was the favorite in the election poll?
c. What percent of the voters favored Thompson in July 2000?
d. What percent of the voters favored Morton in September 1999?

"If the election for U.S. Senate were being held today, for whom would you vote?"

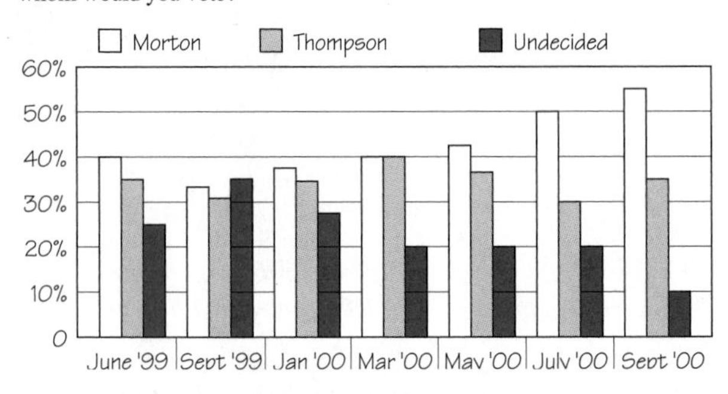

Figure 10.13 Political bar graph from the 2000 election

e. In which month was the undecided vote the greatest?
f. In which month was Morton's lead the greatest?

18. Use the bar graph in Figure 10.14.

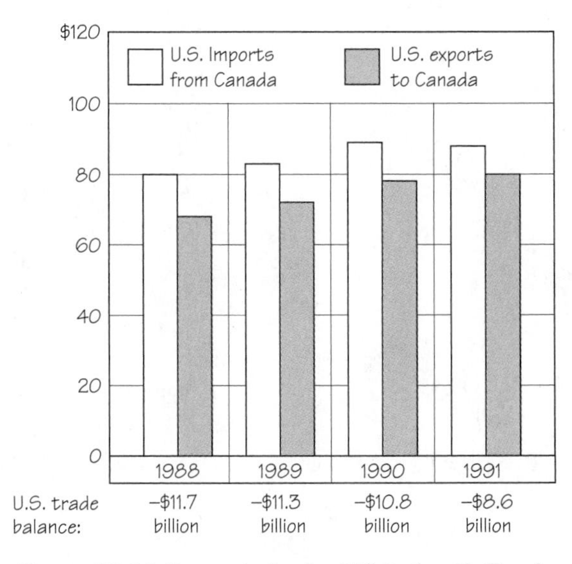

Figure 10.14 Bar graph showing U.S. trade with Canada

a. What is being illustrated in this graph?
b. Did the United States have more imports or exports for the years 1988–1991?
c. What was the approximate balance of trade with Canada in 1991?
d. What was the approximate dollar amount of United States imports from Canada in 1990?

Level 2—Applications

19. Use the bar graph in Figure 10.15.

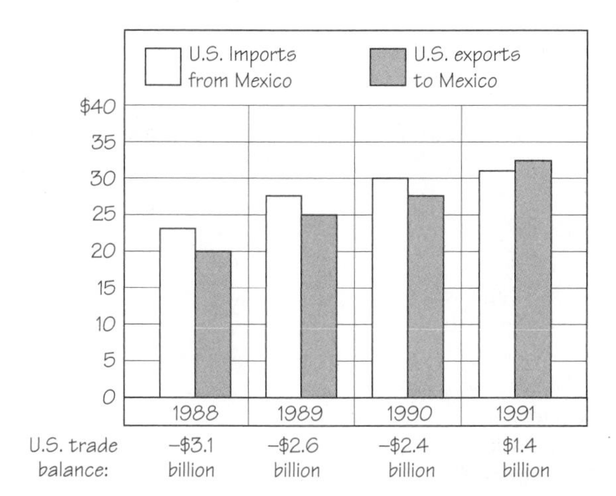

Figure 10.15 Bar graph showing U.S. trade with Mexico

a. What is being illustrated in the graph?

b. Did the United States have more imports or exports for the years 1988–1991?

c. What was the approximate balance of trade with Mexico in 1991?

d. What was the approximate dollar amount of United States exports to Mexico in 1989?

When a person in California renews the registration for an automobile, the bar graph shown in Figure 10.16 *is included with the bill. Use this bar graph to answer the questions in Problems* 20–25. *The following statement is included with the graph:*

> There is no safe way to drive after drinking. These charts show that a few drinks can make you an unsafe driver. They show that drinking affects your BLOOD ALCOHOL CONCENTRATION (BAC). The BAC zones for various numbers of drinks and time periods are printed in white, gray, and black. HOW TO USE THESE CHARTS: First, find the chart that includes your weight. For example, if you weigh 160 lbs., use the "150 to 169" chart. Then look under "Total Drinks" at the "2" on this "150 to 169" chart. Now look below the "2" drinks, in the row for 1 hour. You'll see your BAC is in the gray shaded zone. This means that if you drive after 2 drinks in 1 hour, you could be arrested. In the gray zone, your chances of having an accident are 5 times higher than if you had no drinks. But if you had 4 drinks in 1 hour, your BAC would be in the black shaded area . . . and your chances of having an accident 25 times higher.

BAC Zones: 90 to 109 lbs. | 110 to 129 lbs. | 130 to 149 lbs. | 150 to 169 lbs.

TIME FROM 1st DRINK — Total Drinks 1 2 3 4 5 6 7 8 (for each chart)

1 hr
2 hrs
3 hrs
4 hrs

BAC Zones: 170 to 189 lbs. | 190 to 209 lbs. | 210 to 229 lbs. | 230 lbs. & Up

TIME FROM 1st DRINK — Total Drinks 1 2 3 4 5 6 7 8 (for each chart)

1 hr
2 hrs
3 hrs
4 hrs

SHADINGS IN THE CHARTS ABOVE MEAN:

(.01% - .04%) Seldom illegal
(.05% - .09%) May be illegal
(.05% - .09%) Illegal if under 18 yrs. old
(.10% Up) Definitely illegal

Prepared by the Department of Motor Vehicles in cooperation with the California Highway Patrol.

Figure 10.16 Blood Alcohol Concentration (BAC) charts

20. Suppose that you weigh 115 pounds and that you have two drinks in two hours. If you then drive, how much more likely are you to have an accident than if you had refrained from drinking?

21. Suppose that you weigh 115 pounds and that you have four drinks in three hours. If you then drive, how much more likely are you to have an accident than if you had refrained from drinking?

22. Suppose that you weigh 195 pounds and have two drinks in two hours. According to Figure 10.16, are you seldom illegal, maybe illegal, or definitely illegal?

23. Suppose that you weigh 195 pounds and that you have four drinks in three hours. According to Figure 10.16, are you seldom illegal, maybe illegal, or definitely illegal?

24. If you weigh 135 pounds, how many drinks in three hours would you need to be definitely illegal?

25. If you weigh 185 pounds and drink a six-pack of beer during a 3-hour baseball game, can you legally drive home?

Use the following information in Problems 26–29. The amount of electricity used in a typical all-electric home is shown in the circle graph in Figure 10.17. If, in a certain month, a home used 1,100 kwh (kilowatt-hours), find the amounts of electricity used from the graph.

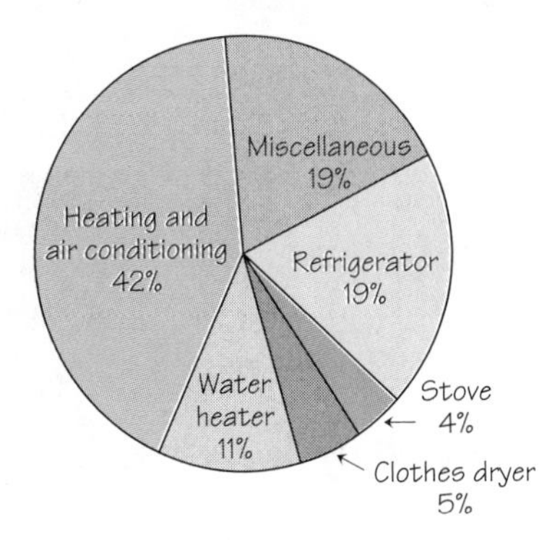

Figure 10.17 Home electricity usage

26. The amount of electricity used by the water heater

27. The amount of electricity used by the stove

28. The amount of electricity used by the refrigerator

29. The amount of electricity used by the clothes dryer

Use the following information in Problems 30–32.
The gross profits for January at Tower Center for the five departments were:

Appliance Department, $20,000;
Automotive Department, $10,000;
Clothing Department, $55,000;
Grocery Department, $260,000;
Nursery Department, $15,000.

30. Change each of the profit amounts to percents (to the nearest tenth of a percent).

31. Change each of the percents in Problem 30 to degrees.

32. Draw a circle graph of these data.

In Problems 33–36, draw a circle graph to represent the given information.

33. The breakdown for the cost of serving a meal in a restaurant is:

12.5% for food,
22.0% for rent,
10.0% for marketing and administrative costs,
15.0% for dishes and equipment,
30.0% for labor,
10.5% for profit.

34. The amount of electricity used in a typical all-electric home includes:

42% for heating and air conditioning,
11% for water heater,
19% for refrigerator,
4% for stove,
5% for clothes dryer,
19% miscellaneous.

35. According to the Specialty Coffee Association, the cost of a $1.75 latte includes:

$0.26 for the coffee beans and milk,
$0.20 for dishes and equipment,
$0.44 for rent,
$0.18 for marketing and administrative costs,
$0.58 for labor,
$0.09 profit.

36. The breakdown for the electricity consumption of a home using 8,000 kwh of electricity is:

3,840 kwh for heating,
960 kwh for hot water,
800 kwh for refrigerator,
560 kwh for freezer,
560 kwh for lighting,
1,280 kwh for other uses.

37. The cost to mail a 1-oz, first-class letter in the United States during this century is given in the following list:

1900–1918	3 cents
1919–1931	2 cents
1932–1957	3 cents
1958–1962	4 cents
1963–1967	5 cents
1968–1970	6 cents
1971–1973	8 cents
1974	10 cents
1975–1977	13 cents
1978–1980	15 cents
1981–1984	20 cents
1985–1987	22 cents
1988–1991	25 cents
1992–1994	29 cents
1995–1998	32 cents
1999	33 cents

Summarize this information, using a line graph.

38. The time required for three leading pain relievers to reach your bloodstream is as follows: Brand A, 480 seconds; Brand B, 500 seconds; Brand C, 490 seconds.

a. Draw a bar graph using the scale shown in Figure 10.18a.

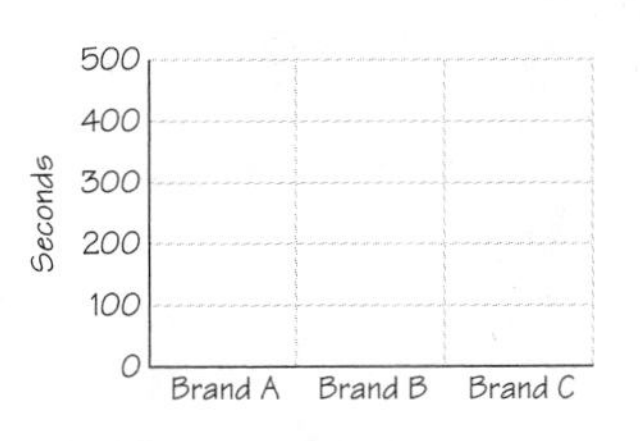

a. Scale from 0 to 500 seconds

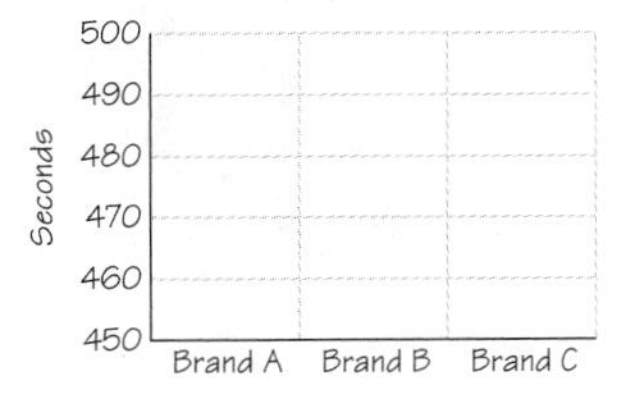

b. Scale from 450 to 500 seconds

Figure 10.18 Time for pain reliever to reach your bloodstream

b. Draw a bar graph using the scale shown in Figure 10.18b.

c. If you were an advertiser working on a promotion campaign for Brand A, which graph from Figure 10.18 would seem to give your product a more distinct advantage?

WHAT IS WRONG, *if anything, with each of the statements in Problems* 39–43*? Explain your reasoning.*

39. Consider the graph shown in Figure 10.19. Clearly, Brand B is better.

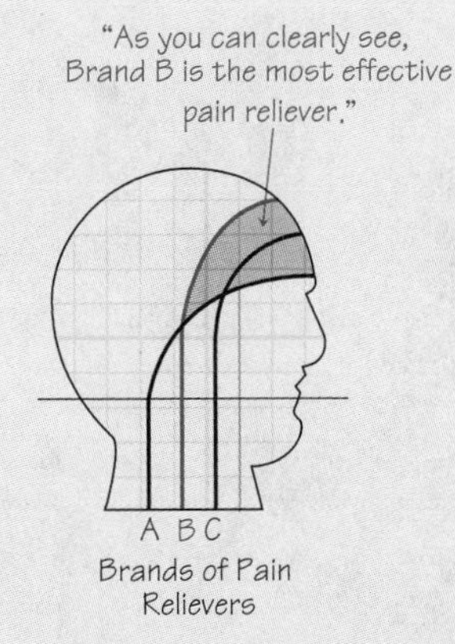

Figure 10.19 From an advertisement for pain reliever

40. Consider the graph shown in Figure 10.20 on page 458. The potential commercial forest growth, as compared with the current commercial forest growth, is almost double (44.9 to 74.2).

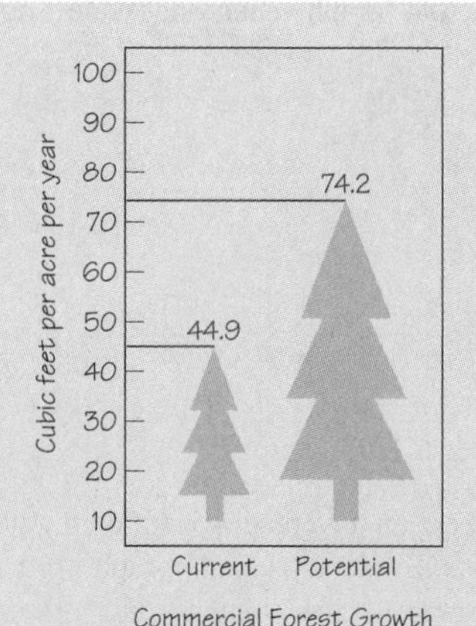

Figure 10.20 From the U.S. Forest Service

41. In 1998 it was reported on the evening news that the number of deaths since the lifting of the 55 mph speed limit was up by 350, representing a 22% increase. From this study, the conclusion reached was that there was a connection between the rise in the death rate and the raising of the speed limit.

42. Consider the graph shown in Figure 10.21.

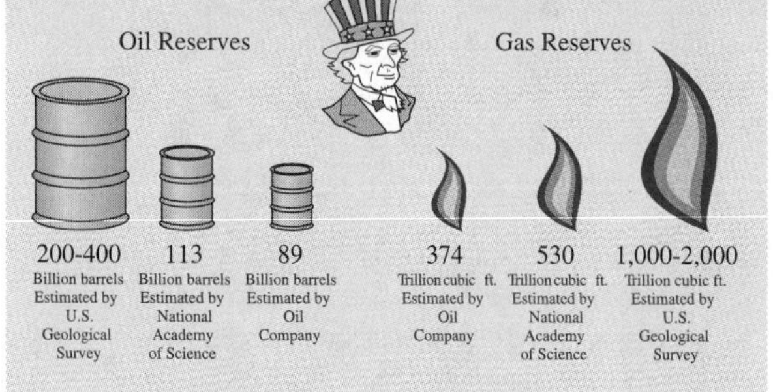

Figure 10.21 Pictograph showing oil and gas reserves

43. Consider the Saab advertisement.

There are two cars built in Sweden. Before you buy theirs, drive ours.

When people who know cars think about Swedish cars, they think of them as being strong and durable. And conquering some of the toughest driving conditions in the world.

But, unfortunately, when most people think about buying a Swedish car, the one they think about usually isn't ours. (Even though ours doesn't cost any more.)

Ours is the SAAB 99E. It's strong and durable. But it's a lot different from their car.

Our car has Front-Wheel Drive for better traction, stability and handling.

It has a 1.85 liter, fuel-injected, 4 cylinder, overhead cam engine as standard in every car. 4-speed transmission is standard too. Or you can get a 3-speed automatic (optional).

Our car has four-wheel disc brakes and dual-diagonal braking system so you can stop straight and fast every time.

It has a wide stance. (About 55 inches.) So it rides and handles like a sports car.

Outside, our car is smaller than a lot of "small" cars. 172" overall length, 57" overall width.

Inside, our car has bucket seats up front and a full five feet across in the back so you can easily accommodate five adults.

It has more headroom than a Rolls Royce and more room from the brake pedal to the back seat than a Mercedes 280. And it has factory air conditioning as an option.

There are a lot of other things that make our car different from their car. Like roll cage construction and a special "hot seat" for cold winter days.

So before you buy their car, stop by your nearest SAAB dealer and drive our car. The SAAB 99E. We think you'll buy it instead of theirs.

SAAB 99E

44. Situation "Nine out of ten dentists recommend Trident for their patients who chew gum." "Crest has been shown to be . . . ." "You can clearly see that Bufferin is the most effective . . . ." "Penzoil is better suited . . . ." "Sylvania was preferred by . . . ." "How can anyone analyze the claims of the commercials we see and hear on a daily basis?" asked Betty. "I even subscribe to *Consumer Reports,* but so many of the claims seem to be unreasonable. I don't like to buy items by trial and error, and I really don't believe all the claims in advertisements." Collect examples of good statistical graphs and examples of misleading graphs. Use some of the leading newspapers and national magazines.

45. In Your Own Words Read *How to Lie with Statistics* by Darrell Huff (New York: Norton, 1954). Then write a short paper relating its material to your own SITUATION.

10.2 MEASURES OF CENTRAL TENDENCY

Situation

"My dad is better than your dad! Na, na, na-na na!"

Do you suppose that Violet's dad bowled better on Monday nights (185 avg.) than on Thursday nights (170 avg.)? Don't be too hasty to say yes.

In Section 10.1, we organized data into a frequency distribution and then discussed their presentation in graphical form. However, some properties of data can help us interpret masses of information. We will use the situation to introduce the notion of *average.*

Let's begin by looking at some possible bowling scores:

	MONDAY NIGHT	*THURSDAY NIGHT*
Game 1	175	180
Game 2	150	130
Game 3	160	161
Game 4	180	185
Game 5	160	163
Game 6	183	185
Game 7	287	186
Totals	1,295	1,190

To find the averages used by Violet in the *Peanuts* cartoon, we divide these totals by the number of games:

MONDAY NIGHT $\frac{1,295}{7} = 185$ *THURSDAY NIGHT* $\frac{1,190}{7} = 170$

If we consider the averages, Violet's dad did better on Mondays; but if we consider the games separately, we see that Violet's dad typically did better on Thursdays (five out of seven games). Would any other properties of the bowling scores tell us this fact?

The average is the most commonly used statistical measure.

The average used by Violet is only one kind of statistical measure that can be used. It is the measure that most of us think of when we hear someone use the word *average.* It is called the *mean.* Other statistical measures, called **averages** or **measures of central tendency,** are defined in the following box.

Measures of Central Tendency: Mean, Median, Mode

The *mean* is the most sensitive average. It reflects the entire distribution and is the most common average.
The *median* gives the middle value. It is useful when there are a few extraordinary values to distort the mean.
The *mode* is the average that measures "popularity." It is possible to have no mode or more than one mode.

1. **Mean.** The number found by adding the data and then dividing by the number of values in the data set
2. **Median.** The middle number when the numbers in the data set are arranged in order of size. If there are two middle numbers (in the case of an even number of values in the data set), the median is the mean of these two middle numbers.
3. **Mode.** The value that occurs most frequently. If no number occurs more than once, there is no mode. It is possible to have more than one mode.

Consider these other measures of central tendency for Violet's dad's bowling scores.

Median Rearrange the data from smallest to largest when finding the median:

MONDAY NIGHT		*THURSDAY NIGHT*
150		130
160		161
160		163
175	← Middle number →	**180**
180	is the median.	185
183		185
287		186

Be sure you can distinguish among mean, median, and mode.

Mode Look for the number that occurs most frequently:

MONDAY NIGHT		*THURSDAY NIGHT*	
150		130	
160	Most frequent is the mode.	161	
160		163	
175		180	
180		**185**	Most frequent is the mode.
183		**185**	
287		186	

If we compare the three measures of central tendency for the bowling scores, we find the following:

	MONDAY NIGHT	*THURSDAY NIGHT*
Mean	185	170
Median	175	180
Mode	160	185

We are no longer convinced that Violet's dad did better on Monday nights than on Thursday nights.

EXAMPLE 1 **Finding Measures of Central Tendency**

Find the mean, median, and mode for the following sets of numbers.

a. 3, 5, 5, 8, 9 **b.** 4, 10, 9, 8, 9, 4, 5 **c.** 6, 5, 4, 7, 1, 9

Solution

a. *Mean:* $\frac{\text{SUM OF TERMS}}{\text{NUMBER OF TERMS}} = \frac{3+5+5+8+9}{5} = \frac{30}{5} = \mathbf{6}$

Median: Arrange in order: 3, 5, **5,** 8, 9. The middle term is the median: **5.**

Mode: The most frequently occurring term is the mode: **5.**

b. *Mean:* $\frac{4+10+9+8+9+4+5}{7} = \frac{49}{7} = \mathbf{7}$

Median: 4, 4, 5, **8,** 9, 9, 10; the median is **8.**

Mode: The data set has two modes: **4** and **9.** If a data set has two modes, we say it is **bimodal.**

c. *Mean:* $\frac{6+5+4+7+1+9}{6} = \frac{32}{6} \approx \mathbf{5.33}$

Median: 1, 4, $\underbrace{5, 6}_{\frac{5+6}{2}=\frac{11}{2}}$, 7, 9; the median is $\frac{11}{2} = \mathbf{5.5}$.

Mode: There is **no mode** because no term appears more than once. ■

A rather nice physical model illustrates the idea of the mean. Consider a seesaw that consists of a plank and a movable support (called a *fulcrum*). We assume that the plank has no weight and is marked off into units as shown in Figure 10.22.

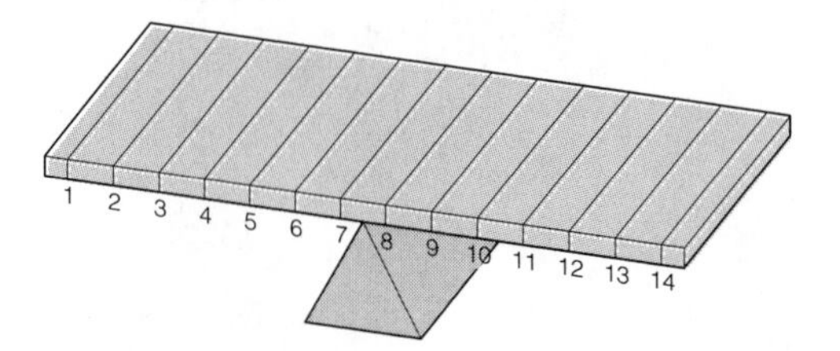

Figure 10.22 Fulcrum and plank model for mean

Now let's place some 1-lb weights in the positions of the numbers in some given distribution. The balance point for the plank is the mean. For example, consider the data from part **a** of Example 1: 3, 5, 5, 8, 9. If these weights are placed on the plank, the balance point is 6, as shown in Figure 10.23.

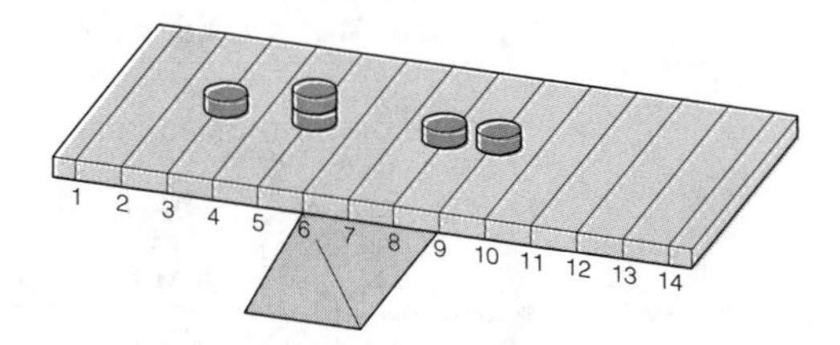

Figure 10.23 The balance point for the data set 3, 5, 5, 8, 9 is 6.

EXAMPLE 2

Measures of Central Tendency when Data Are Presented with a Frequency Distribution

Consider the number of days one must wait for a marriage license in the various states in the United States.

DAYS' WAIT	*FREQUENCY*
0	25
1	1
2	1
3	19
4	1
5	3
Total	50

What are the mean, the median, and the mode for these data?

Solution *Mean:* To find the mean, we could, of course, add all 50 individual numbers, but instead, notice that

0 occurs 25 times, so write	0×25
1 occurs 1 time, so write	1×1
2 occurs 1 time, so write	2×1
3 occurs 19 times, so write	3×19
4 occurs 1 time, so write	4×1
5 occurs 3 times, so write	5×3

Thus, the mean is

$$\frac{0 \times 25 + 1 \times 1 + 2 \times 1 + 3 \times 19 + 4 \times 1 + 5 \times 3}{50} = \frac{79}{50} = \mathbf{1.58}$$

Median: Since the median is the middle number, and since there are 50 values, the median is the mean of the 25th and 26th numbers (when they are arranged in order).

$$\left.\begin{array}{l}\text{25th term is } 0\\ \text{26th term is } 1\end{array}\right\} \frac{0+1}{2} = \frac{\mathbf{1}}{\mathbf{2}}$$

Mode: The mode is the value that occurs most frequently, which is **0.** ■

PROBLEM SET 10.2

Essential Ideas

1. **In Your Own Words** What is the mean?
2. **In Your Own Words** What is the median?
3. **In Your Own Words** What is the mode?
4. **In Your Own Words** Compare and contrast measures of central tendency.

Level 1

WHAT IS WRONG, *if anything, with each of the statements in Problems 5–10? Explain your reasoning.*

5. The average of the numbers 11, 50, 50, and 9 is found as follows:

$$\frac{11 + 50 + 50 + 9}{4} = \frac{120}{4} = 30$$

6. In 1989, Andy Van Slyke's batting average was better than Dave Justice's; and in 1990 Andy once again beat Dave. Therefore, it follows that Andy's combined 1989–1990 batting average is better than Dave's.

	Andy			Dave		
	Hits	AB	Avg	Hits	AB	Avg
1989	113	476	0.236	12	51	0.235
1990	140	493	0.284	124	439	0.282

7. For the data 6, 7, 10, 10, and 15, the median is

$$\frac{6 + 7 + 10 + 10 + 15}{5} = \frac{48}{5}$$

8. For the data 6, 7, 10, and 15, the median is $\frac{7 + 10}{2} = \frac{17}{2}$.

9. For the data 6, 7, 10, 10, and 15,

the mean is $\frac{6 + 7 + 10 + 10 + 15}{5} = 9.6$

the median is $\frac{7 + 10}{2} = \frac{17}{2}$

the mode is 10

10. For the data 6, 7, 10, 10, 15, and 15, the mode is 10.

Level 1—Drill

In Problems 11–22, find the three measures of central tendency (the mean, median, and mode).

11. 1, 2, 3, 4, 5

12. 17, 18, 19, 20, 21

13. 103, 104, 105, 106, 107

14. 765, 766, 767, 768, 769

15. 4, 7, 10, 7, 5, 2, 7

16. 15, 13, 10, 7, 6, 9, 10

17. 3, 5, 8, 13, 21

18. 1, 4, 9, 16, 25

19. 79, 90, 95, 95, 96

20. 70, 81, 95, 79, 85

21. 1, 2, 3, 3, 3, 4, 5

22. 0, 1, 1, 2, 3, 4, 16, 21

23. By looking at Problems 11–14, discover a pattern to find the mean of the numbers 613, 614, 615, 616, 617.

Level 2—Applications

24. **In Your Own Words** A professor gives five exams. Two students' scores have the same mean, although one student did better on all the tests except one. Give an example of such scores.

25. Find the mean, the median, and the mode of the following salaries of employees of the Moe D. Lawn Landscaping Company:

SALARY	*FREQUENCY*
$25,000	4
28,000	3
30,000	2
45,000	1

26. G. Thumb, the leading salesperson for the Moe D. Lawn Landscaping Company, turned in the following summary of sales for the week of October 23–28:

DATE	*NUMBER OF CLIENTS CONTACTED BY G. THUMB*
Oct. 23	12
Oct. 24	9
Oct. 25	10
Oct. 26	16
Oct. 27	10
Oct. 28	21

Find the mean, the median, and the mode of the number of clients.

27. Find the mean, the median, and the mode of the following scores:

TEST SCORE	*FREQUENCY*
90	1
80	3
70	10
60	5
50	2

28. A class obtained the following scores on a test:

SCORE	*FREQUENCY*
90	1
80	6
70	10
60	4
50	3
40	1

Find the mean, the median, and the mode for the class.

29. A class obtained the following test scores:

SCORE	*FREQUENCY*
90	2
80	4
70	9
60	5
50	3
40	1
30	2
0	4

Find the mean, the median, and the mode for the class.

30. The following salaries for the executives of a certain company are known:

POSITION	*SALARY*
President	$170,000
1st VP	140,000
2nd VP	120,000
Supervising manager	54,000
Accounting manager	40,000
Personnel manager	40,000
Department manager	30,000
Department manager	30,000

Find the mean, the median, and the mode. Which measure seems to best describe the average executive salary for the company?

31. The 1998 prices, per ton, of wine grapes for different kinds of wine are given in the following table.

VARIETY	*PRICE*
Cabernet Sauvignon	$1,650
Merlot	$1,800
Zinfandel	$1,600
Pinot Noir	$1,500
Chardonnay	$1,600
Sauvignon Blanc	$1,100

Find the mean, the median, and the mode. Which measure seems to best describe the average price?

32. The 1999 monthly cost for health care for an employee and two dependents is given in the following table.

PROVIDER	*COST*
Maxicare	$415.24
Cigna	$424.77
Health Net	$427.48
Pacific Care	$428.05
Health Plan of the Redwoods	$431.52
Kaiser	$433.50
Aetna U.S. Healthcare	$436.11
Blue Shield HMO	$442.28
Omni Healthcare	$457.86
Lifeguard	$457.94

Find the mean, the median, and the mode. Which measure seems to best describe the average cost?

33. The California counties with population more than one million are given in the following table.

COUNTY	*1997 POPULATION*
Alameda	1,371,067
Los Angeles	9,145,219
Orange	2,674,091
Riverside	1,447,791
Sacramento	1,125,976
San Bernardino	1,615,817
San Diego	2,722,650
Santa Clara	1,609,037

Find the mean, the median, and the mode. Which measure seems to best describe the average population?

34. Roll a single die until all six numbers occur at least once. Repeat the experiment 20 times. Find the mean, the median, and the mode of the number of tosses.

35. Roll a pair of dice until all 11 numbers occur at least once. Repeat the experiment 20 times. Find the mean, the median, and the mode of the number of tosses.

LEVEL 3—Team Project

36. Your team has been hired to conduct a survey, and prepare a report of your findings. Begin by determining a suitable multiple-choice question, for example, "Do you watch any soap opera, and if so, which one is your favorite?" Identify five to ten possible responses for your question, including "None of the above."

Determine your team's survey methods; for example, specify the number of responses and where you will obtain your data. Discuss how location and time of day can affect the results. Conduct the survey, tabulate the results, and construct bar, line, and circle graphs to display the results.

Your report should include your methods, results, and conclusions, as well as tabulated responses and summary graphs.

10.3 MEASURES OF POSITION

Situation

"I can't make up my mind which candidate to hire," sighed Jane. "Both Gustavo and Shannon have stunning recommendations, which are equal in all respects." "How will you decide—flip a coin?" asks Hannah. "No, since they both came from equally prestigious schools, I think I will use their standing in their graduating classes," responded Jane.

Which candidate should Jane hire?

Transcript Records

Be hereby notified that

Gustavo G. Rivas

graduated 10th in his class of 50 graduates.

It is also hereby confirmed that

Shannon J. Sovndal

graduated 25th in his class of 625 graduates.

These notices are confirmed without prejudice and acknowledge the good work of both candidates.

In the last section, we considered measures of central tendency, and we noted that the median divides the data into two equal parts; half the values are above the median and half are below the median. Sometimes we use benchmark positions, sometimes called **measures of position,** that divide the data into more than two parts. **Quartiles,** denoted by Q_1 (first quartile), Q_2 (second quartile), and Q_3 (third quartile), divide the data into four equal parts.

EXAMPLE 1 Finding Quartiles

On a recent examination the following scores were obtained:

98, 95, 92, 90, 88, 83, 82, 80, 77, 77, 74, 69, 65, 60, 59, 37

Divide these data into quartiles.

Solution The median is the measure that determines the middle of a given data set. This data set has 16 scores (an even number), so we must find the mean of the 8th and 9th scores:

$$\frac{80 + 77}{2} = 78.5$$

The median is 78.5.

The first quartile is the value that has 25% of the scores below it, and since there are 16 scores, 25% (one-fourth) of 16 is 4. The value of the first quartile is the fifth score from the bottom (four scores below it), namely, 69.

The third quartile is the value that has 75% of the scores below it (or 25% above), so that it is the fifth score from the top, namely, 88.

We see that **$Q_1 = 69$, $Q_2 = 78.5$, and $Q_3 = 88$ are the quartiles.** These scores divide the data into four parts, by position. ■

It is often desirable to divide the data into more than four parts. For example, if John tells you he received a test score of 83, this does not really tell you how well he did on the exam. If John says his score is one of the scores listed in Example 1, then we have a bit more information, and we can say he scored in the third quartile. Recall, $Q_1 = 69$, $Q_2 = 78.5$, and $Q_3 = 88$, so these numbers divide the data into four parts:

1st quartile scores:	0–69
2nd quartile scores:	70–78
3rd quartile scores:	79–88
4th quartile scores:	89–100

We could also say that John's score is the sixth highest score in the class, but unless we know the number of scores, we do not have a meaningful measure of position. In other words, to know position we would need to know that John scored sixth out of 16 taking the test.

Deciles are nine values that divide the data into ten equal parts, and **percentiles** are 99 values that divide the data into 100 equal parts. For example, when you take the Scholastic Assessment Test (SAT), your score is recorded as a percentile score. If you scored in the 92nd percentile, it means that you scored better than approximately 92% of those who took the test.

EXAMPLE 2 Finding Percentile Rank

Lisa says her score was 23, and that score was the 85th score from the top in a class of 240 scores. Calculate Lisa's percentile rank.

Solution In Lisa's class, there are $240 - 85 = 155$ scores ranked below her. Thus,

$$\frac{155}{240} \approx 0.65$$

This means that approximately 65% of Lisa's classmates were ranked lower than she, so we would say **Lisa is in the 65th percentile.** ■

EXAMPLE 3 Finding the Class Rank

Lee has received a percentile rank of 85% in a class of 50 students. What is Lee's rank in the class?

Solution Lee's percentile rank means that 85% of the students have scores less than his, so

$$0.85 \text{ of } 50 = 42.5$$

This means that 42 students scored lower than Lee, so Lee's rank in the class is

$$50 - 42 = 8$$

or **eighth in a class of 50.** ■

EXAMPLE 4 Finding Decile Rank

The 1999 monthly cost for health care for an employee and two dependents is given in the following table.

PROVIDER	*COST*
Maxicare	$415.24
Cigna	$424.77
Health Net	$427.48
Pacific Care	$428.05
Health Plan of the Redwoods	$431.52
Kaiser	$433.50
Aetna U.S. Healthcare	$436.11
Blue Shield HMO	$442.28
Omni Healthcare	$457.86
Lifeguard	$457.94

What is the decile rank for Kaiser?

Solution Since the data are arranged from lowest to highest, we see that there are five below Kaiser, and there are ten providers. **Kaiser is the 5th decile.** You say, "How convenient it was that there were 10 health care providers." Suppose that Omni and Lifeguard are eliminated from the list. Now, in which percentile is Kaiser? There are still five ranked below Kaiser out of a total of eight providers, so

$$\frac{5}{8} = 0.625 \quad \text{or} \quad \mathbf{62.5\%}$$

We would say that Kaiser is in the 62nd percentile or in the 6th decile. ■

Historical Perspective

Florence Nightingale (1820–1910)

Even though the origins of statistics are ancient, the development of modern statistical techniques began in the 19th century. Adolph Quetelet (1796–1874) was the first person to apply statistical methods to accumulated data. He correctly predicted (to his own surprise) the crime and mortality rates from year to year. Florence Nightingale was an early proponent of the use of statistics in her work. Since the advent of the computer, statistical methods have come within the reach of almost everyone. Today, statistical methods strongly influence agriculture, biology, business, chemistry, economics, education, electronics, medicine, physics, psychology, and sociology. If you decide to enter one of these disciplines for a career, you will no doubt be required to take at least one statistics course.

PROBLEM SET 10.3

Essential Ideas

1. What is a quartile?

2. What is a decile?

3. What is a percentile?

Level 1—Drill

Find the percentile rank in Problems 4–9.

4. 10th in a class of 50
5. 12th in a class of 200
6. 14th out of 75
7. 58th out of 450
8. 83rd out of 690
9. 106th out of 17,500

Find the class ranking in Problems 10–15.

10. 90th percentile out of 20

11. 80th percentile out of 30

12. 55th percentile out of 140

13. 99th percentile out of 5,000

14. 70th percentile out of 35

15. 73rd percentile out of 38

Level 2—Drill

16. California ranks 39th (out of 50) in graduation rate for public high schools. What is California's quartile rank?

17. New Jersey ranks 8th (out of 50) in graduation rate for public high schools. What is New Jersey's decile rank?

18. Florida ranks 48th (out of 50) in graduation rate for public high schools. What is Florida's percentile rank?

19. Columbia University ranks as the 13th most expensive out of 50 listed universities. What is Columbia's quartile rank?

20. Given a class of examination scores:

$$\{93, 86, 83, 75, 71, 67, 65, 63, 60, 53\}$$

   a. What is the percentile rank of a score of 93?
   b. What is the median?
   c. What score is the first quartile?
   d. What score is the fourth decile?
   e. What score is at the 70th percentile?

21. Given a class of examination scores:

$$\{93, 86, 83, 75, 71, 67, 65, 63, 60, 53, 45, 38\}$$

   a. What is the percentile rank of a score of 93?
   b. What is the median?
   c. What score is the first quartile?
   d. What score is the fourth decile?
   e. What score is at the 70th percentile?

Level 2—Applications

22. If there are 36 students in a class, and Amy's score is at the third quartile, how many students scored lower than Amy?

23. Jerry is ranked at the first quartile of his graduating class of 1400 students. What is his rank in the class?

24. James is ranked at the median of his math class of 45 students. What is his rank in the class?

25. James is ranked 80th in his class of 450, and John, who is in the same class, has a percentile rank of 80. Who has a higher standing in the class?

26. In 1997, in terms of population, Contra Costa County ranked 9th out of 58 counties. What was its percentile rank?

27. In 1997, in terms of population, Modoc County ranked 55th out of 58 counties. What was its decile rank?

28. The top 18 American Kennel Club 1996 Registrations are given:

BREED	*NUMBER*	*BREED*	*NUMBER*
Labrador Retriever	149,505	Pomeranian	40,216
Rottweiler	89,867	Shih Tzu	38,055
German Shepherd	79,076	Chihuahua	36,562
Golden Retriever	68,993	Boxer	36,398
Beagle	56,803	Shetland Sheepdog	33,577
Poodle	56,803	Dalmation	32,972
Dachshund	48,426	Miniature Schnauzer	31,834
Cocker Spaniel	45,305	Doberman Pinscher	20,355
Yorkshire Terrier	40,316	Pug	18,398

Use these data to answer the following questions.

   a. What is the rank of the Poodle? the Dachshund?
   b. What is the percentile rank of the Golden Retriever?
   c. What is the decile rank of the Chihuahua?
   d. What is the quartile rank of the Cocker Spaniel?

29. In 1996, the number of cars imported into the United States was as follows:

COUNTRY	*NUMBER*	*COUNTRY*	*NUMBER*
Canada	1,589,980	South Korea	140,572
Japan	1,012,785	Sweden	86,593
Mexico	550,620	United Kingdom	43,890
Germany	234,381	Italy	1,125

Use these data to answer the following questions.
a. What is the rank of Mexico?
b. What is the percentile rank of United Kingdom?
c. What is the decile rank of Germany?
d. What is the quartile rank of Sweden?

30. Situation "I can't make up my mind which candidate to hire," sighed Jane. "Both Gustavo and Shannon have stunning recommendations, which are equal in all respects." "How will you decide—flip a coin?" asks Hannah. "No, since they both came from equally prestigious schools, I think I will use their standing in their graduating classes," responded Jane. Which candidate should Jane hire? Assume that Gustavo G. Rivas graduated 10th in his class of 50 graduates, and Shannon J. Sovndal graduated 25th in his class of 625 graduates.

31. In Your Own Words Prepare a report or exhibit showing how statistics are used in baseball.

32. In Your Own Words Prepare a report or exhibit showing how statistics are used in educational testing.

33. In Your Own Words Prepare a report or exhibit showing how statistics are used in psychology.

LEVEL 3—Team Projects

34. With your team, prepare a report or exhibit showing how statistics are used in business. Use a daily report of transactions on the New York Stock Exchange. What inferences can you make from the information reported?

35. With your team, investigate the work of Adolph Quetelet, François Galton, Karl Pearson, R. A. Fisher, and Florence Nightingale. Prepare a report or an exhibit of their work in statistics.

10.4 MEASURES OF DISPERSION

Situation

"We need privacy and a consistent wind," said Wilbur. "Did you write to the Weather Bureau to find a suitable location?" "Well," replied Orville, "I received this list of possible locations and Kitty Hawk, North Carolina, looks like just what we want. Look at this . . . ." However, Orville and Wilbur spent many days waiting in frustration after they arrived in Kitty Hawk, because the winds weren't suitable. The Weather Bureau's information gave the averages, but the Wright brothers didn't realize that an acceptable average can be produced by unacceptable extremes.

In this section, we'll discuss not only averages, but also measures of dispersion that would have helped Wilbur and Orville find a more suitable location for their experiments.

We have considered measures of central tendency and measures of position, but these measures do not give the entire story. For example, consider these sets of data:

Set A: $\{8, 9, 9, 9, 10\}$ *Mean:* $\dfrac{8+9+9+9+10}{5} = 9$

Median: 9

Mode: 9

Set B: $\{2, 9, 9, 12, 13\}$ *Mean:* $\dfrac{2+9+9+12+13}{5} = 9$

Median: 9

Mode: 9

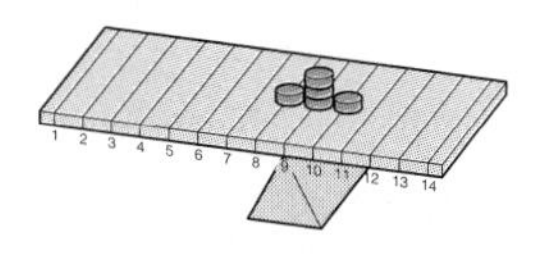

a. $A = \{8, 9, 9, 9, 10\}$

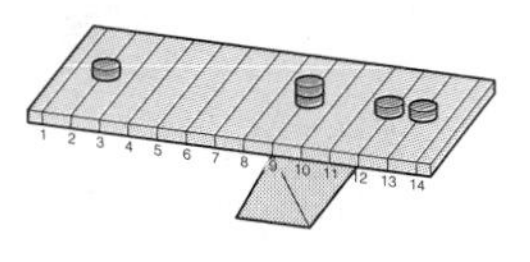

b. $B = \{2, 9, 9, 12, 13\}$

Figure 10.24 Visualization of dispersion of sets of data

Notice that, for sets A and B, the measures of central tendency do not distinguish the data. However, if you look at the data placed on planks, as shown in Figure 10.24, you will see that the data in Set B are relatively widely dispersed along the plank, whereas the data in Set A are clumped around the mean. In statistics, a **population** is the total set of items defined by some characteristic. Thus, we might refer to sets A and B as *populations.*

We'll consider three **measures of dispersion:** the *range,* the *standard deviation,* and the *variance.*

Range

The **range** of a set of data is the difference between the largest and the smallest numbers in the set.

EXAMPLE 1 Finding the Range

Find the ranges for the following populations:

a. Set $A = \{8, 9, 9, 9, 10\}$ **b.** Set $B = \{2, 9, 9, 12, 13\}$

Solution Notice from Figure 10.24 that the means for these sets of data are the same. The range is found by computing the difference between the largest and smallest values in the set.

a. $10 - 8 = \mathbf{2}$ **b.** $13 - 2 = \mathbf{11}$ ■

The range is used, along with quartiles, to construct a statistical tool called a *box plot.* For a given set of data, a **box plot** consists of a rectangular box positioned above a numerical scale, drawn from Q_1 (the first quartile) to Q_3 (the third quartile). The median (Q_2, or second quartile) is shown as a dashed line, and a segment is extended to the left to show the distance to the minimum value; another segment is extended to the right for the maximum value.

EXAMPLE 2 **Drawing a Box Plot**

The 1999 monthly cost for health care for an employee and two dependents is given in the following table.

PROVIDER	*COST*
Maxicare	$415.24
Cigna	$424.77
Health Net	$427.48
Pacific Care	$428.05
Health Plan of the Redwoods	$431.52
Kaiser	$433.50
Aetna U.S. Healthcare	$436.11
Blue Shield HMO	$442.28
Omni Healthcare	$457.86
Lifeguard	$457.94

Draw a box plot for the monthly costs of health care.

Solution We find the quartiles by first finding Q_2, the median. Since there are 10 entries, the median is the mean of the two middle items:

$$Q_2 = \frac{431.52 + 433.50}{2} = 432.51$$

Next, we find Q_1 (first quartile), which is the median of all items below Q_2:

$$Q_1 = 427.48$$

Finally, find Q_3 (third quartile), which is the median of all items above Q_2:

$$Q_3 = 442.28$$

The minimum cost is $415.24 and the maximum cost is $457.94, so we have a box plot, as shown in Figure 10.25.

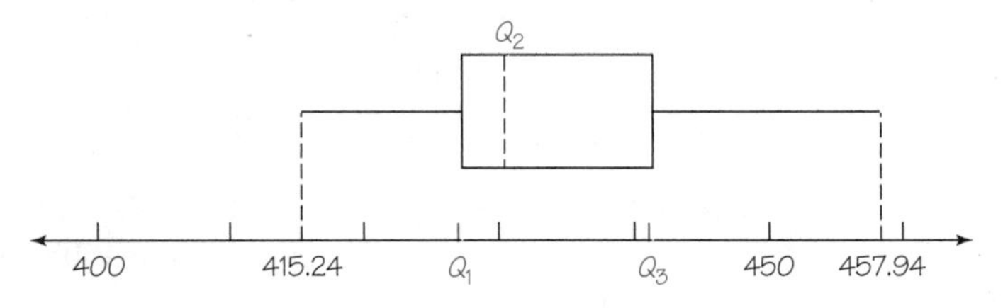

Figure 10.25 Box plot ■

Sometimes a box plot is called a *box-and-whisker plot.* Its usefulness should be clear when you look at Figure 10.25. It shows:

1. the median (labeled Q_2, a measure of central tendency);
2. the location of the middle half of the data (represented by the extent of the box; the endpoints are Q_1 and Q_3);

3. the range (a measure of dispersion, the ends of the whiskers sticking out of the box);
4. the skewness (the nonsymmetry of both the box and the whiskers).

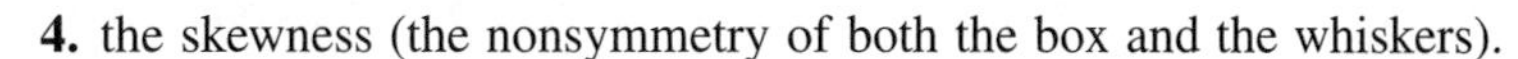

Notice that the range is determined by only the largest and the smallest numbers in the set; it does not give us any information about the other numbers. It thus seems reasonable to invent another measure of dispersion that takes into account all the numbers in the data. We will consider the *deviations from the mean* and then the *standard deviation.* We proceed here as we did with the range, but instead of finding the difference between the largest and smallest values, we find the difference between *each number* and the mean.

Again, let's use the data sets we worked with in Example 1.

Set $A = \{8, 9, 9, 9, 10\}$	Set $B = \{2, 9, 9, 12, 13\}$
Mean is 9.	Mean is 9.

Find the deviation of each term from the mean by subtracting the mean from each member in the set:

$8 - 9 = -1$	$2 - 9 = -7$
$9 - 9 = 0$	$9 - 9 = 0$
$9 - 9 = 0$	$9 - 9 = 0$
$9 - 9 = 0$	$12 - 9 = 3$
$10 - 9 = 1$	$13 - 9 = 4$
↑ mean	↑ mean

If we total these deviations (to obtain a measure of the total deviation), in each case we obtain 0. The reason is that the positive and negative differences "cancel each other out." Thus, to find a measure of total dispersion of a population, we calculate the mean of the *square of each of these deviations:*

Set $A = \{8, 9, 9, 9, 10\}$	Set $B = \{2, 9, 9, 12, 13\}$
$(8 - 9)^2 = (-1)^2 = 1$	$(2 - 9)^2 = (-7)^2 = 49$
$(9 - 9)^2 = 0^2 = 0$	$(9 - 9)^2 = 0^2 = 0$
$(9 - 9)^2 = 0^2 = 0$	$(9 - 9)^2 = 0^2 = 0$
$(9 - 9)^2 = 0^2 = 0$	$(12 - 9)^2 = (3)^2 = 9$
$(10 - 9)^2 = (1)^2 = 1$	$(13 - 9)^2 = (4)^2 = 16$

$$\text{Mean: } \frac{1 + 0 + 0 + 0 + 1}{5} = \frac{2}{5} = 0.4$$

$$\text{Mean: } \frac{49 + 0 + 0 + 9 + 16}{5} = \frac{74}{5} = 14.8$$

We sometimes must refer to this mean of the squares of the deviations, so we give it a name: the *variance.* Still, we need to continue to develop a true picture of the dispersion. Since we squared each difference (to eliminate the effect of positive and negative differences), it seems reasonable that we should find the square root of the variance as a more meaningful measure of dispersion. This number, called the *standard deviation,* is denoted by the lowercase Greek letter sigma (σ). You will need a calculator to find square roots:

Set A: $\sigma = \sqrt{0.4}$
≈ 0.632 *Display:* .632455532

Set B: $\sigma = \sqrt{14.8}$
≈ 3.85 *Display:* 3.847076812

We summarize these steps in the following box.

Standard Deviation

You might want to spend some time reading the steps in this box.

The **standard deviation** of a population, denoted by σ, is the square root of the variance. To find it, carry out these steps:

1. Determine the mean of the set of numbers.
2. Subtract the mean from each number in the set.
3. Square each of these differences.
4. Find the sum of the squares of these differences.
5. Divide this sum by the number of pieces of data.

   The result is the ***variance*** *of a population:*
   the mean of the squares of the deviations from the mean.

6. Take the square root of the variance.

EXAMPLE 3 **Finding the Standard Deviation of a Population**

Suppose that Missy received the following test scores in a math class: 92, 85, 65, 89, 96, and 71. What is the standard deviation for her test scores?

Solution First, we calculate the mean:

Step 1: $\dfrac{92 + 85 + 65 + 89 + 96 + 71}{6} = 83$ This is the mean.

We summarize steps 2 through 4 by using a table format:

SCORE	*(DEVIATION FROM THE MEAN)*^{2}
92	$(92 - 83)^2 = 9^2 = 81$
85	$(85 - 83)^2 = 2^2 = 4$
65	$(65 - 83)^2 = (-18)^2 = 324$
89	$(89 - 83)^2 = 6^2 = 36$
96	$(96 - 83)^2 = 13^2 = 169$
71	$(71 - 83)^2 = (-12)^2 = 144$

Mean of squares of deviations:

$$\frac{81 + 4 + 324 + 36 + 169 + 144}{6} = \frac{758}{6} = 126\tfrac{1}{3}$$

We note that this number, $126\frac{1}{3}$, is called the variance. If you do not have access to a calculator, you can use the variance as a measure of dispersion. However, we assume that you have a calculator and can find the standard deviation:

$$\sigma = \sqrt{\frac{758}{6}} \approx \mathbf{11.24} \qquad \textit{Display:} \text{ 11.2398102}$$ ■

How can we use the standard deviation? We will begin the discussion in Section 10.5 with this question. For now, however, we will give one example. Suppose that Missy obtained 65 on an examination for which the mean was 50 and the standard deviation was 15, whereas in another class Shannon scored 76 on an examination for which the mean was 80 and the standard deviation was 3. Did Shannon or Missy do better in her respective class? We see that Missy scored one standard deviation above the mean ($50 + 15 = 65$), whereas Shannon scored two standard deviations *below* the mean ($80 - 2 \times 3 = 76$); so Missy did better compared to her classmates than did Shannon.

PROBLEM SET 10.4

Essential Ideas

1. What is the range?

2. What is the variance?

3. What is the standard deviation?

4. In Your Own Words Compare and contrast measures of dispersion.

Level 1

WHAT IS WRONG, *if anything, with each of the statements in Problems 5–6? Explain your reasoning.*

5. The standard deviation is always smaller than the variance, because the standard deviation is the square root of the variance.

6. For the data 70, 80, and 90, the standard deviation is found as follows:

The mean is $\frac{70 + 80 + 90}{3} = \frac{240}{3} = 80.$

$(70 - 80)^2 = (-10)^2 = 100$

$(80 - 80)^2 = 0^2 = 0$

$(90 - 80)^2 = 10^2 = 100$

The mean of the deviations is $\frac{100 + 0 + 100}{3} = \frac{200}{3} = 66\frac{2}{3}.$

The standard deviation is $\sqrt{66\frac{2}{3}} \approx 8.16.$

Level 1—Drill

In Problems 7–18, find the range and the standard deviation (correct to two decimal places). If you do not have a calculator, find the range and the variance.

7. 1, 2, 3, 4, 5

8. 17, 18, 19, 20, 21

9. 103, 104, 105, 106, 107

10. 765, 766, 767, 768, 769

11. 4, 7, 10, 7, 5, 2, 7

12. 15, 13, 10, 7, 6, 9, 10

13. 3, 5, 8, 13, 21

14. 1, 4, 9, 16, 25

15. 79, 90, 95, 95, 96

16. 70, 81, 95, 79, 85

17. 1, 2, 3, 3, 3, 4, 5

18. 0, 1, 1, 2, 3, 4, 16, 21

19. By looking at Problems 7–10, and discovering a pattern, find the mean and the variance of the numbers 217,849, 217,850, 217,851, 217,852, and 217,853.

20. Suppose that a variance is zero. What can you say about the data?

Level 2—Applications

21. A class obtained the following test scores:

SCORE	*FREQUENCY*
90	2
80	4
70	9
60	5
50	3
40	1
30	2
0	4

Draw a box-and-whisker plot for these data.

22. The salaries of employees of the Moe D. Lawn Landscaping Company are given:

SALARY	*FREQUENCY*
$25,000	4
28,000	3
30,000	2
45,000	1

Draw a box-and-whisker plot for these data.

23. Find the standard deviation of the salaries of employees in Problem 22.

24. G. Thumb, the leading salesperson for the Moe D. Lawn Landscaping Company, turned in the following summary of sales for the week of October 23–28:

DATE	*NUMBER OF CLIENTS CONTACTED BY G. THUMB*
Oct. 23	12
Oct. 24	9
Oct. 25	10
Oct. 26	16
Oct. 27	10
Oct. 28	21

Find the standard deviation rounded to the nearest unit.

25. Find the standard deviation for the following scores (rounded to the nearest unit):

TEST SCORE	*FREQUENCY*
90	1
80	3
70	10
60	5
50	2

26. A class obtained the following scores on a test:

SCORE	*FREQUENCY*
90	1
80	6
70	10
60	4
50	3
40	1

Find the standard deviation rounded to the nearest unit.

27. A class obtained the following test scores:

SCORE	*FREQUENCY*
90	2
80	4
70	9
60	5
50	3
40	1
30	2
0	4

Find the standard deviation rounded to the nearest unit.

28. In Your Own Words A professor gives six exams. Two students' scores have the same mean, although one student's scores have a small standard deviation and the other student's scores have a large standard deviation. Give an example of such scores.

29. The number of miles driven on each of five tires was 17,000, 19,000, 19,000, 20,000, and 21,000. Find the mean, the range, and the standard deviation (rounded to the nearest unit) for these mileages.

30. Roll a single die until all six numbers occur at least once. Repeat the experiment 20 times. Find the mean, the median, the mode, and the range of the number of tosses.

31. Roll a pair of dice until all 11 numbers occur at least once. Repeat the experiment 20 times. Find the mean, the median, the mode, and the range of the number of tosses.

32. If you roll a pair of dice 36 times, the expected number of times for rolling each of the numbers is as given in the accompanying table. A graph of these data is shown in Figure 10.26. Find the mean, the variance, and the standard deviation for this model.

OUTCOME	*EXPECTED FREQUENCY*
2	1
3	2
4	3
5	4
6	5
7	6
8	5
9	4
10	3
11	2
12	1

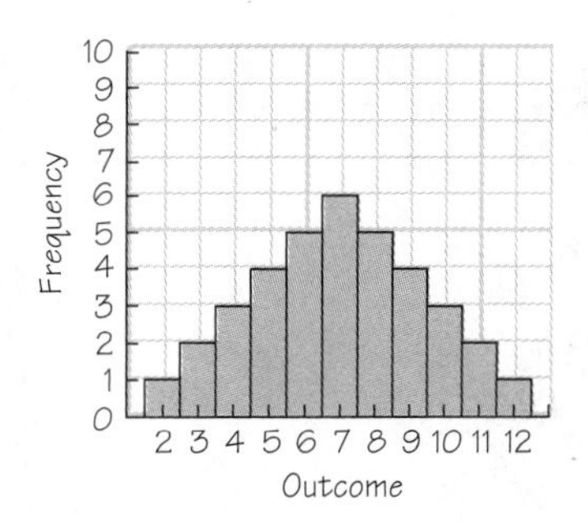

Figure 10.26 Distribution for rolling a pair of dice

33. Continuation of Problem 32. Roll a pair of dice 36 times. Construct a table and a graph similar to the ones shown for Problem 32. Find the mean, the variance, and the standard deviation for your experiment.

34. Continuation of Problems 32 and 33. Compare the results of Problems 32 and 33. If this is a class problem, you might wish to pool the results for the entire class for Problem 33 before making the comparison.

35. Situation "We need privacy and a consistent wind," said Wilbur. "Did you write to the Weather Bureau to find a suitable location?" "Well," replied Orville, "I received this list of possible locations and Kitty Hawk, North Carolina, looks like just what we want. Look at this . . . ." However, Orville and Wilbur spent many days waiting in frustration after they arrived in Kitty Hawk, because the winds weren't suitable. The Weather Bureau's information gave the averages, but the Wright brothers didn't realize that an acceptable average can be produced by unacceptable extremes. Write a paper explaining how it is possible to have an acceptable average produced by unacceptable extremes.

Wright Brothers' first flight at Kitty Hawk, North Carolina

10.5 THE NORMAL CURVE

Situation

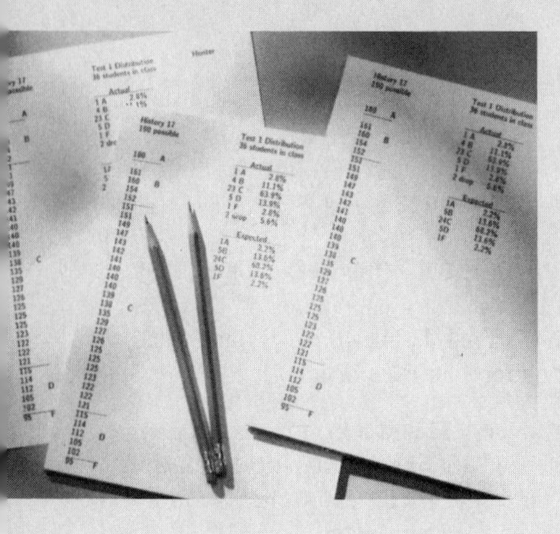

"I signed up for Hunter in History 17," said Ben. "Why did you sign up for her?" asked Ted. "Don't you know she has given only three A's in the last 14 years?!" "That's just a rumor. Hunter grades on a curve," Ben replied. "Don't give me that 'on a curve' stuff," continued Ted, "I'll bet you don't even know what that means. Anyway, my sister-in-law had her last year and said it was so bad that . . . ."

In this section, Ted and Ben will learn what it means to grade on a curve. Many everyday examples can be represented by bell-shaped, or normal, curves. Ted and Ben will see how to make certain predictions based on this normal distribution.

The cartoon in the margin suggests that most people do not like to think of themselves or their children as having "normal intelligence." But what do we mean by *normal* or *normal intelligence?*

Suppose we survey the results of 20 children's scores on an IQ test. The scores (rounded to the nearest 5 points) are 115, 90, 100, 95, 105, 95, 105, 105, 95, 125, 120, 110, 100, 100, 90, 110, 100, 115, 105, and 80. A frequency graph of these data is shown in Figure 10.27a. If we consider 10,000 scores instead of only 20, we might obtain the frequency distribution shown in Figure 10.27b.

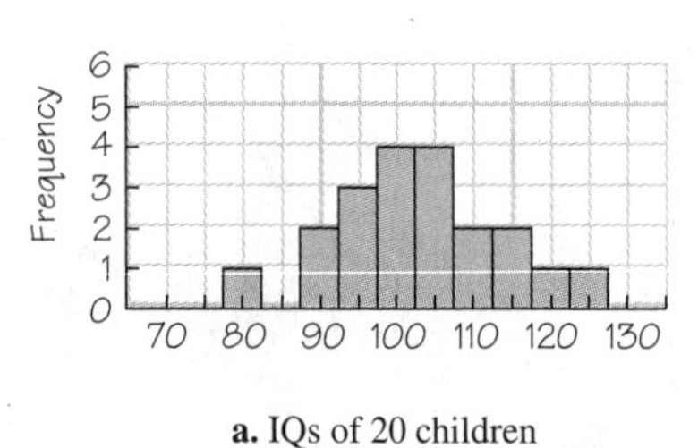

a. IQs of 20 children

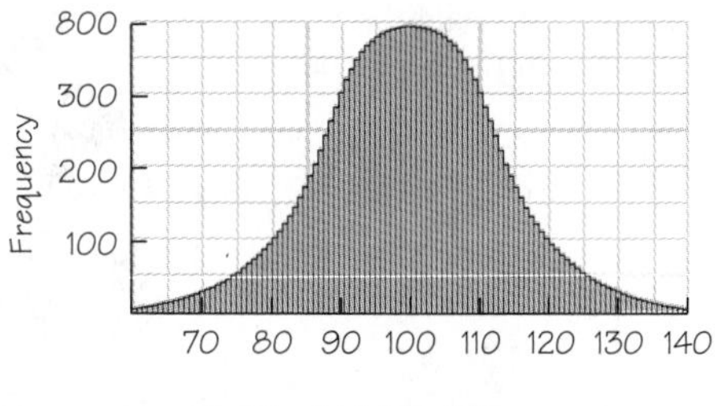

b. IQs of 10,000 children

Figure 10.27 Frequency distributions for IQ scores

CAUTION

This curve is important in many different applications and when observing many natural phenomena.

The data illustrated in Figure 10.27 approximate a commonly used curve called a *normal frequency curve,* or simply a **normal curve.** (See Figure 10.28.)

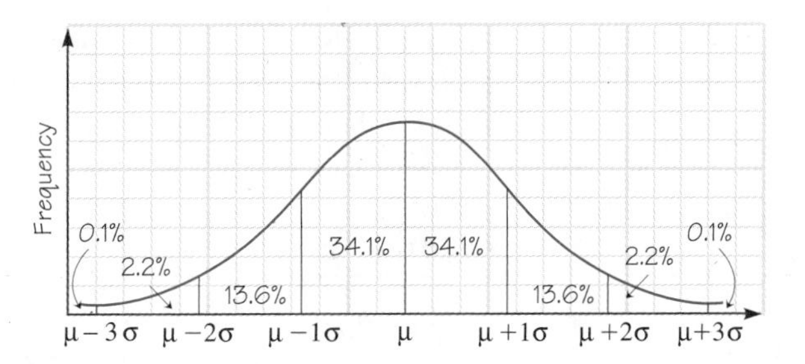

Figure 10.28 A normal curve

If we obtain the frequency distribution of a large number of measurements (as with IQ), the corresponding graph tends to look normal, or **bell-shaped.** The normal curve has some interesting properties. In it, the mean, the median, and the mode all have the same value, and all occur exactly at the center of the distribution; we denote this value by the Greek letter mu (μ). The standard deviation for this distribution is σ (sigma). Roughly 68% of all values lie within the region from one standard deviation below to one standard deviation above the mean. About 96% lie within two standard deviations on either side of the mean, and virtually all (99.8%) values lie within three standard deviations on either side. These percentages are the same regardless of the particular mean or standard deviation.

EXAMPLE 1 Using a Normal Curve to Predict Occurrence of IQ Scores

Predict the distribution of IQ scores of 1,000 people if we assume that IQ scores are normally distributed, and if we know that the mean is 100 and the standard deviation is 15.

Solution First, find the breaking points around the mean: $\mu = 100$ and $\sigma = 15$.

$$\mu + \sigma = 100 + 15 = 115 \qquad \mu - \sigma = 100 - 15 = 85$$
$$\mu + 2\sigma = 100 + 2 \cdot 15 = 130 \qquad \mu - 2\sigma = 100 - 2 \cdot 15 = 70$$
$$\mu + 3\sigma = 100 + 3 \cdot 15 = 145 \qquad \mu - 3\sigma = 100 - 3 \cdot 15 = 55$$

We use Figure 10.28 to find that 34.1% of the scores will be between 100 and 115 (i.e., between μ and $\mu + \sigma$):

$$0.341 \times 1{,}000 = 341$$

About 13.6% will be between 115 and 130 (i.e., between $\mu + \sigma$ and $\mu + 2\sigma$):

$$0.136 \times 1{,}000 = 136$$

About 2.2% will be between 130 and 145 (i.e., between $\mu + 2\sigma$ and $\mu + 3\sigma$):

$$0.022 \times 1{,}000 = 22$$

About 0.1% will be above 145 (more than $\mu + 3\sigma$):

$$0.001 \times 1{,}000 = 1$$

SCORES	*%*	*EXPECTED NUMBER*
Below 55	0.1	1
55–69	2.2	22
70–84	13.6	136
85–99	34.1	341
100–114	34.1	341
115–129	13.6	136
130–144	2.2	22
Above 144	0.1	1
Totals	100.0	1,000

The intervals for standard deviations below the mean repeat these expected numbers, since the normal curve is symmetric to the left and to the right of the mean. The distribution is shown in the margin. ■

EXAMPLE 2 Grading on a Curve

Suppose that an instructor "grades on a curve." Show the grading distribution on an examination of 45 students, if the scores are normally distributed with a mean of 73 and a standard deviation of 9.

Solution Grading on a curve means determining students' grades according to the percentages shown in Figure 10.29.

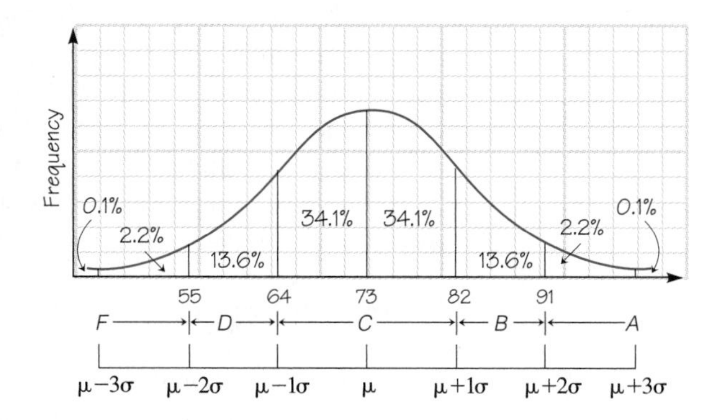

Figure 10.29 Grade distribution for a class "graded on a curve"

We calculate these numbers as shown:

Calculation	*Scores*	*Grade*	*Calculation*	*Number*
Two or more standard deviations above the mean	91–100	A	$0.023 \times 45 = 1.035$	1
$\mu + 2\sigma = 73 + 2 \cdot 9 = 91$	82–90	B	$0.136 \times 45 = 6.12$	6
$\mu + 1\sigma = 73 + 1 \cdot 9 = 82$				
Mean: $\mu = 73$	64–81	C	$0.682 \times 45 = 30.69$	31
$\mu - 1\sigma = 73 - 1 \cdot 9 = 64$				
$\mu - 2\sigma = 73 - 2 \cdot 9 = 55$	55–63	D	same as grade B	6
Two or more standard deviations below the mean	0–54	F	same as grade A	1

Grading on a curve means that the person with the top score in the class receives an A; the next 6 ranked persons (from the top) receive B grades; the bottom score in the class receives an F; the next 6 ranked persons (from the bottom) receive D grades; and finally, the remaining 31 persons receive C grades. Notice the majority of the class (34.1% + 34.1% = 68.2%) will receive an "average" C grade. ■

EXAMPLE 3 **Problem Solving with a Normal Distribution**

The Eureka Light Bulb Company tested a new line of light bulbs and found their lifetimes to be normally distributed, with a mean life of 98 hours and a standard deviation of 13 hours.

a. What percentage of bulbs will last less than 72 hours?

b. What is the probability that a bulb selected at random will last longer than 111 hours?

Solution Draw a normal curve with mean 98 and standard deviation 13, as shown in Figure 10.30.

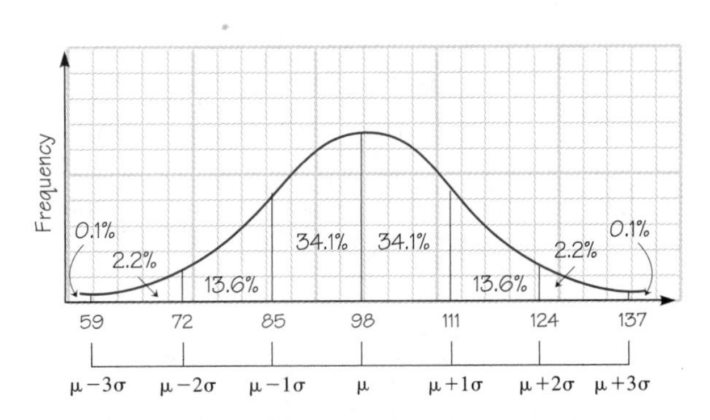

Figure 10.30 Light bulb life is normally distributed.

a. About **2.3%** (2.2% + 0.1% = 2.3%) will last less than 72 hours.

b. We know that about 15.9% (13.6% + 2.2% + 0.1% = 15.9%) of the bulbs will last longer than 111 hours, so

$$P(\text{bulb life} > 111 \text{ hours}) \approx \mathbf{0.159}$$

■

PROBLEM SET 10.5

Essential Ideas

1. Draw a normal curve, label the mean, and standard deviations on each side of the mean. Fill in the frequency percents for each standard deviation.

Level 1—Applications

In Problems 2–6, suppose that people's heights (in centimeters) are normally distributed, with a mean of 170 and a standard deviation of 5. We find the heights of 50 people.

2. How many would you expect to be between 165 and 175 cm tall?
3. How many would you expect to be taller than 160 cm?
4. How many would you expect to be taller than 175 cm?
5. What is the probability that a person selected at random is taller than 165 cm?
6. What is the variance for this experiment?

In Problems 7–11, suppose that, for a certain exam, a teacher grades on a curve. It is known that the grades follow a normal distribution with a mean of 50 and a standard deviation of 5. There are 45 students in the class.

7. How many students should receive a C?
8. How many students should receive an A?
9. What score would be necessary to obtain an A?
10. If an exam paper is selected at random, what is the probability that it will be a failing paper?
11. What is the variance for this exam?

Level 2—Applications

12. The diameter of an electric cable is normally distributed, with a mean of 0.9 inch and a standard deviation of 0.01 inch. What is the probability that the diameter will exceed 0.91 inch?

13. Suppose that the breaking strength of a rope (in pounds) is normally distributed, with a mean of 100 pounds and a standard deviation of 16. What is the probability that a certain rope will break when subjected to a force of 132 pounds or less?
14. Suppose that the annual rainfall in Ferndale, California, is known to be normally distributed, with a mean of 35.5 inches and a standard deviation of 2.5 inches. About 2.3% of the time, the annual rainfall will exceed how many inches?
15. In Problem 14, what is the probability that the rainfall in a given year will exceed 30.5 inches in Ferndale?
16. The diameter of a pipe is normally distributed, with a mean of 0.4 inch and a variance of 0.0004. What is the probability that the diameter will exceed 0.44 inch?
17. The breaking strength (in pounds) of a certain new synthetic is normally distributed, with a mean of 165 and a variance of 9. The material is considered defective if the breaking strength is less than 159 pounds. What is the probability that a single, randomly selected piece of material will be defective?
18. Suppose the neck size of men is normally distributed, with a mean of 15.5 inches and a standard deviation of 0.5 inch. A shirt manufacturer is going to introduce a new line of shirts. How many of each of the following sizes should be included in a batch of 1,000 shirts?

    a. 14 **b.** 14.5 **c.** 15 **d.** 15.5
    e. 16 **f.** 16.5 **g.** 17 **h.** 17.5

19. A package of Toys Galore Cereal is marked "Net Wt. 12 oz." The actual weight is normally distributed, with a mean of 12 oz and a variance of 0.04.
  a. What percent of the packages will weigh less than 12 oz?
  b. What weight will be exceeded by 2.3% of the packages?

20. Instant Dinner comes in packages with weights that are normally distributed, with a standard deviation of 0.3 oz. If 2.3% of the dinners weigh more than 13.5 oz, what is the mean weight?

21. In Your Own Words Select something that you think might be normally distributed (for example, the ring size of students at your college). Next, select 100 people and make the appropriate measurements (in this example, ring size). Calculate the mean and standard deviation. Illustrate your findings using a bar graph. Do your data appear to be normally distributed?

22. Situation "I signed up for Hunter in History 17," said Ben. "Why did you sign up for her?" asked Ted. "Don't you know she has given only three A's in the last fourteen years?!" "That's just a rumor. Hunter grades on a curve," Ben replied. "Don't give me that 'on a curve' stuff," continued Ted, "I'll bet you don't even know what that means. Anyway, my sister-in-law had her last year and said it was so bad that . . . ." Assume that Hunter does, indeed, grade on a curve. Show the grading distribution if 200 students take an examination, with a mean of 73 and a standard deviation of 9. What are the grades Hunter would give?

23. In Your Own Words Write a short paper comparing the SITUATION with your own situation. (That is, list similarities and/or differences.)

10.6 SAMPLING

Situation

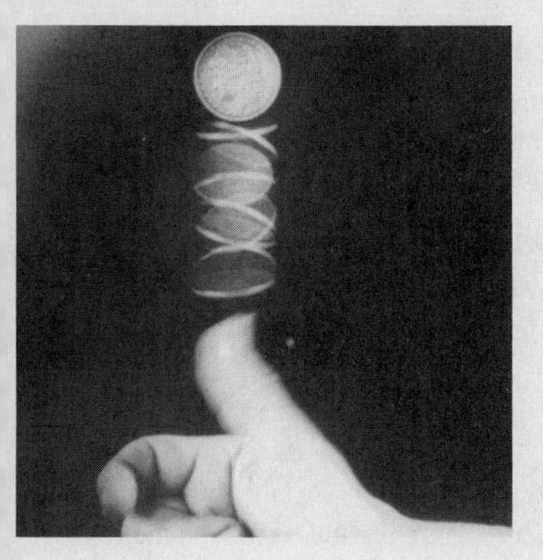

Suppose that John hands you a coin to flip and wants to bet on the outcome. Now, John has tried this sort of thing before, and you suspect that the coin is "rigged." You decide to test this hypothesis by taking a sample. You flip the coin twice, and it is heads both times. You say, "Aha, I knew it was rigged!" John replies, "Don't be silly. Any coin can come up heads twice in a row."

In this section, you'll see how to test a hypothesis, and you'll take a brief look at sampling, polls, and statistical predictions.

The Bettmann Archive

The previous sections of this chapter dealt with the accumulation of data, measures of central tendency, and dispersion. However, a more important part of statistics is its ability to help us make predictions about a population based on a sample from that population. A **sample** is a smaller group of items chosen to represent a larger group, or population. Thus, a sample is a proper subset of the population. The sample is analyzed; then based on this analysis, some conclusion about the entire population is made. Sampling necessarily involves some error, because the sample and the population are not identical.

The inference drawn from a poll can, of course, be wrong, so statistics is also concerned with estimating the error involved in predictions based on samples. In 1936, the *Literary Digest* predicted that Alfred Landon would defeat Franklin D. Roosevelt—who was subsequently reelected President by a landslide. (The magazine ceased publication the following year.) In 1948, the *Chicago Daily Tribune* drew an incorrect conclusion from its polls and declared in a headline that Thomas Dewey had just been elected president over Harry S. Truman. And in 1976, the *Milwaukee Sentinel* printed the erroneous headline shown here about the Wisconsin Democratic primary, again based on the result of its polls.

In an attempt to minimize error in their predictions, statisticians follow very careful procedures:

Step 1 Propose some hypothesis about a population.
Step 2 Gather a sample from the population.
Step 3 Analyze the data.
Step 4 Accept or reject the hypothesis.

Suppose that you want to decide whether a certain coin is a "fair" coin. You decide to test the hypothesis, "This is a fair coin," by flipping the coin 100 times. This provides a *sample.* Suppose the result is

Heads: 55 Tails: 45

Do you accept or reject the hypothesis that "This coin is fair"? The expected number of heads is 50, but certainly a fair coin might well produce the results obtained.

As you can readily see, two types of error are possible:

These types of error are essential for the ideas of sampling.

Type I: Rejection of the hypothesis when it is true.
Type II: Acceptance of the hypothesis when it is false.

How can we minimize the possibility of making either error? Let's carry this example further, and repeat the experiment of flipping the coin 100 times:

Trial number:	1	2	3	4	5	6	. . .
Number of heads:	55	52	54	57	59	55	. . .

If the coin is fair and we repeat the experiment a large number of times, the distribution of the number of heads should be normal, with a mean of 50 and a standard deviation of 5, as shown in Figure 10.31. We assume this information from experimentation with *known* fair coins. The question is whether to accept the *unknown* test coin as fair.

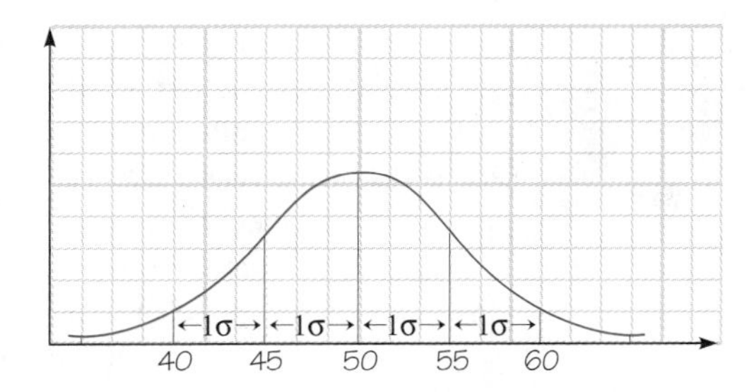

Figure 10.31 Given normal distribution for the number of heads upon flipping a fair coin 100 times

Suppose that you are willing to accept the coin as fair only if the number of heads falls between 45 and 55 (that is, within one standard deviation on either side of the expected number). If you adopt this standard, you know you will be correct 68% of the time, if the coin is fair. How do you know this? Look at Figure 10.31, and note that 34.1% of the results will be within $+1\sigma$ and 34.1% will be within -1σ; the total is 34.1% + 34.1% = 68.2%.

But a friend says, "Yes, you will be correct 68% of the time, but you will also be rejecting a lot of fair coins!" You respond, "But suppose that a coin really is a bad coin (it really favors heads), with a mean number of heads of 60 and a standard deviation of 5. If I adopted the same standard ($\pm 1\sigma$), I'd be accepting all the coins in the shaded region of Figure 10.32."

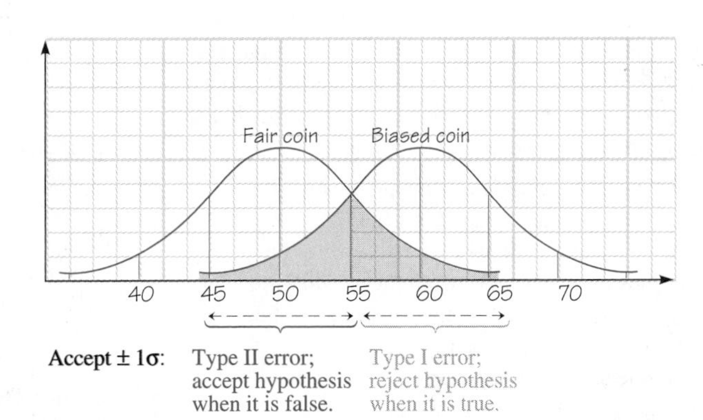

Figure 10.32 Comparison of Type I and Type II errors

As you can see, a decrease in the probability of a Type I error increases the probability of a Type II error, and vice versa. Deciding which type of error to minimize depends on the stakes involved and on some statistical calculations that go beyond the scope of this course.

Poll-taker to boss: "Our latest opinion poll showed that 90% of the people aren't interested in the opinions of others."

Consider a company that produces two types of valves. The first type is used in jet aircraft, and the failure of this valve might cause many deaths. A sample of the valves is taken and tested, and the company must accept or reject the entire shipment on the basis of these test results. Under these circumstances, the company would rather reject many good valves than accept a bad one. On the other hand, the second valve is used in toy airplanes; the failure of this valve would merely cause the crash of the model. In this case, the company wouldn't want to reject too many good valves, so they would minimize the probability of rejecting the good valves.

Many times we read of a poll in the newspaper or hear of it on the evening news, and a percent is given. For example, "The candidate was favored by 48% of those sampled." What is not often stated (or is stated only in small print) is that there is a margin of error and a confidence level. For example, "The margin of error is 4 percent at a confidence level of 95%." This means that we are 95% confident that the actual percentage of people who favor the candidate is between 44% and 52%.

PROBLEM SET 10.6

LEVEL 2—Applications

1. **In Your Own Words** You are interested in knowing the number and ages of children (0–18 years) in a part (or all) of your community. You will need to sample 50 families, finding the number of children in each family and the age of each child. It is important that you select the 50 families at random. How to do this is a subject of a course in statistics. For this problem, however, follow these steps:

   Step 1 Determine the geographic boundaries of the area with which you are concerned.

   Step 2 Consider various methods for selecting the families at random. For example, could you:
   (i) select the first 50 homes at which someone is at home when you call?
   (ii) select 50 numbers from the phone book?
   Using (i) or (ii) could result in a biased sample. Can you guess why this might be true? In a statistics course, you might explore other ways of selecting the homes. For this problem, use one of these methods.

   Step 3 Consider different ways of asking the question. Can the way the family is approached affect the response?

   Step 4 Gather your data.

   Step 5 Organize your data. Construct a frequency distribution for the number of children per family. Also construct a frequency distribution for the ages of the children, with integral values from 0 to 18.

   Step 6 Find out the number of families who actually live in the area you've selected. If you can't do this, assume that the area has 1,000 families.

   a. What is the average number of children per family?
   b. What percent of the children are in the first grade (age 6)?
   c. If all the children age 12–15 are in junior high, how many are in junior high for the geographic area you are considering?
   d. See if you can actually find out the answers to parts **b** and **c,** and compare these answers with your projections.
   e. What other inferences can you make from your data?

2. **In Your Own Words** Five identical containers (shoe boxes, paper cups, etc.) must be prepared for this problem, with contents as follows: There are five boxes containing red and white items (such as marbles, poker chips, or colored slips of paper).

BOX	*CONTENTS*
#1	15 red and 15 white
#2	30 red and 0 white
#3	25 red and 5 white
#4	20 red and 10 white
#5	10 red and 20 white

Select one of the boxes at random so that you don't know its contents.

Step 1 Shake the box.
Step 2 Select one marker, note the result, and return it to the box.
Step 3 Repeat the first two steps 20 times with the same box.

a. What do you think is inside the box you have sampled?
b. Could you have guessed the contents by repeating the experiment five times? Ten times? Do you think you should have more than 20 observations per experiment? Discuss.

3. **Situation** Suppose that John hands you a coin to flip and wants to bet on the outcome. Now, John has tried this sort of thing before, and you suspect that the coin is "rigged." You decide to test this hypothesis by taking a sample. You flip the coin twice, and it is heads both times. You say, "Aha, I knew it was rigged!" John replies, "Don't be silly. Any coin can come up heads twice in a row." The following scheme was devised by mathematician John von Neumann to allow fair results even if the coin is somewhat biased. The coin is flipped twice. If it comes up heads both times or tails both times, it is flipped twice again. If it comes up heads–tails, this will decide the outcome in favor of the first party; and if it comes up tails–heads, this will decide the outcome in favor of the second party. Show that this will result in a fair toss even if the coins are biased.

4. **In Your Own Words** Write a short paper comparing the SITUATION with your own situation. (That is, list similarities and/or differences.)

5. **In Your Own Words** Write a short book report on *Innumeracy* by John Allen Paulos (New York: Hill & Wang, 1988). In particular, comment on the contents of Chapter 5, "Statistics, Trade-Offs, and Society."

10.7 SUMMARY AND REVIEW

Important Terms

Average [10.2]
Bar graph [10.1]
Bell-shaped curve [10.5]
Bimodal [10.2]
Box plot [10.4]
Circle graph [10.1]
Classes [10.1]
Decile [10.3]
Frequency [10.1]
Frequency distribution [10.1]
Grouped frequency distribution [10.1]
Histogram [10.1]
Interval [10.1]
Line graph [10.1]
Mean [10.2]
Measure of central tendency [10.2]
Measure of dispersion [10.4]
Measure of position [10.3]
Median [10.2]
Mode [10.2]
Normal curve [10.5]
Percentile [10.3]
Pictograph [10.1]
Pie chart [10.1]
Population [10.4]
Quartile [10.3]
Range [10.4]
Sample [10.6]
Standard deviation [10.4]
Statistics [10.1]
Type I error [10.6]
Type II error [10.6]
Variance [10.4]

Chapter Objectives

The material in this chapter is reviewed in the following list of objectives. A self-test (with answers and suggestions for additional study) is given. This self-test is constructed so that each problem number corresponds to a related objective. This self-test is followed by a practice test with the questions in mixed order.

[10.1] *Objective 10.1*	Make a frequency distribution and draw a bar or line graph to represent data.
[10.1] *Objective 10.2*	Draw a circle graph.
[10.1] *Objective 10.3*	Be able to read and interpret graphs.
[10.1] *Objective 10.4*	Be able to recognize misuses of statistics.
[10.2] *Objective 10.5*	Find the mean, the median, and the mode.
[10.3] *Objective 10.6*	Find the percentile, decile, and quartile rank, given a data position.
[10.3] *Objective 10.7*	Find a class ranking, given the percentile and number of data points.
[10.4] *Objective 10.8*	Draw a box-and-whisker plot.
[10.4] *Objective 10.9*	Find the range, the variance, and the standard deviation for a set of data.
[10.5] *Objective 10.10*	Understand and use a normal distribution.
[10.6] *Objective 10.11*	Understand the basic ideas of sampling.

Self-Test

Each question of this self-test is related to the corresponding objective listed above.

1. a. Make a frequency table for the following results of tossing a coin:

HTTTT HHTHH TTHHT THTHT HHTTH THHTH HHTHT TTTTT

b. Draw a bar graph for the number of heads and tails given in part **a.**

c. Three coins are tossed onto a table, and the following frequencies are noted:

Number of heads	3	2	1	0
Frequency	18	56	59	17

Draw a line graph to represent these data.

2. Suppose that a student's college expenses are as follows:

Registration fee	$1,800
Education fee	$1,000
Books and supplies	$ 600
Room and board	$6,400
Miscellaneous	$ 200

Make a circle graph showing these data.

3. Consider the graph shown in Figure 10.33, which shows the salary and sex data from a survey of United States company vice-presidents in 1994.

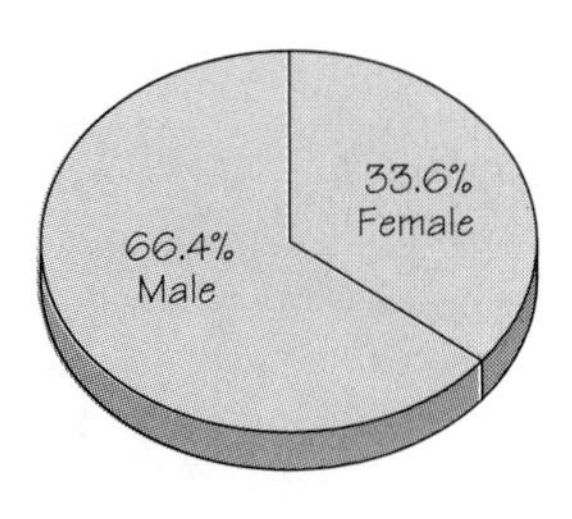

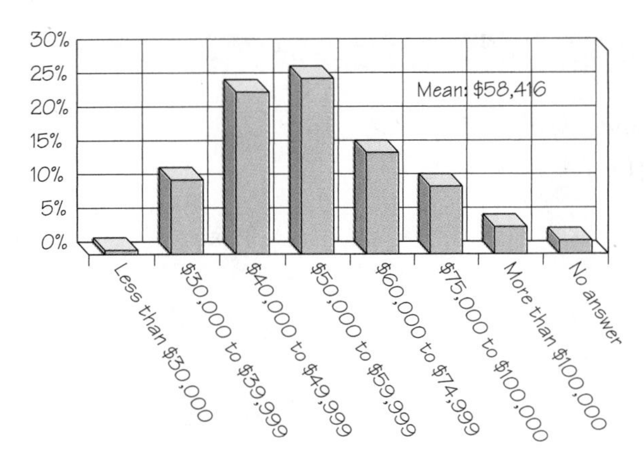

Figure 10.33 Salary survey

a. Which salary had the greatest percent of respondents? What percent of respondents received salaries in that class?

b. What percent of respondents are female? (Choose the best response.)

A. $\frac{1}{3}$ B. $\frac{2}{3}$ C. $\frac{4}{5}$ D. $\frac{1}{2}$ E. Information not available

c. If there were 2,000 respondents to this survey, what would be the approximate number of males?

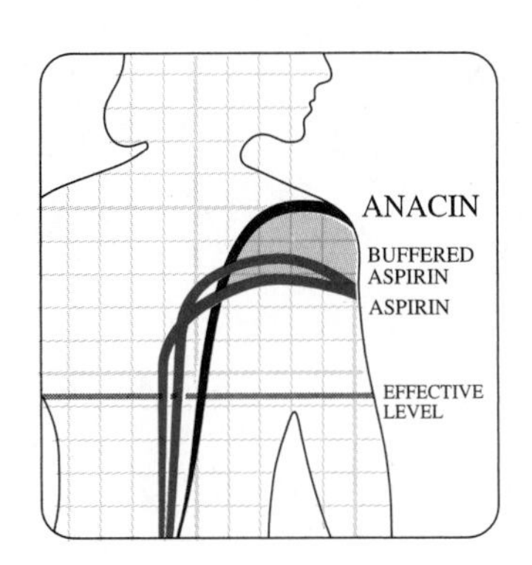

Figure 10.34 Anacin advertisement

4. Consider the graph of an Anacin advertisement, as shown in Figure 10.34. Does it tell you anything at all about the effectiveness of the three pain relievers? Explain your answer.

In Problems 5–8, use the following test scores.

96, 92, 92, 89, 88, 88, 87, 87, 87, 87, 80, 79, 79, 78, 76,
76, 76, 74, 73, 72, 72, 71, 71, 70, 70, 66, 66, 60, 53, 20

5. a. Find the mean (rounded to the nearest unit).

b. Find the median.

c. Find the mode.

d. Which measure of central tendency is most appropriate for these test scores?

6. If Frank's score is 87, what is his percentile rank?

7. If Jane's score is in the 87th percentile, what is her class rank?

8. Draw a box-and-whisker plot for these test scores (ranked highest to lowest).

9. a. Find the variance for the scores 90, 85, 70, 70, and 65.

b. Find the standard deviation (rounded to the nearest tenth) for the scores in part **a.**

10. If grades are assigned to a class of 30 students and the instructor "grades on a curve," how many students each will receive A's, B's, C's, D's, and F's?

11. Explain how you would obtain a sample of people in your community to ask how they feel about a recent crime.

STUDY HINTS *Compare your solutions and answers to the self-test. For each problem you missed, work some additional problems in the section listed in the margin. After you have worked these problems, you can test yourself with the practice test.*

Complete Solutions to the Self-Test

Additional Problems

[10.1] *Problems 5–15; 37*

1. a. Heads: 𝍸 𝍸 𝍸 ||| (18); Tails 𝍸 𝍸 𝍸 𝍸 || (22)

b.

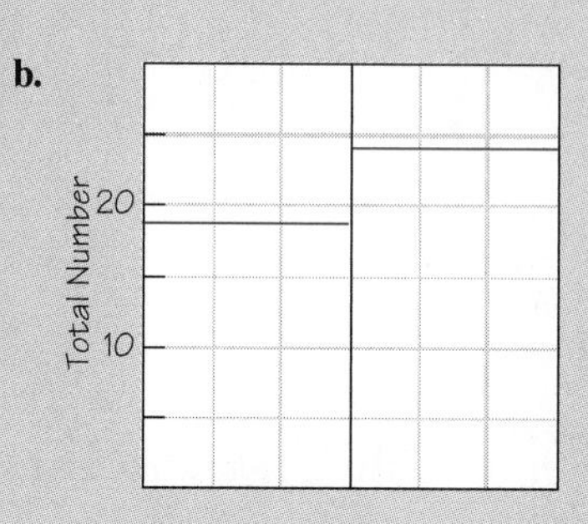

c.

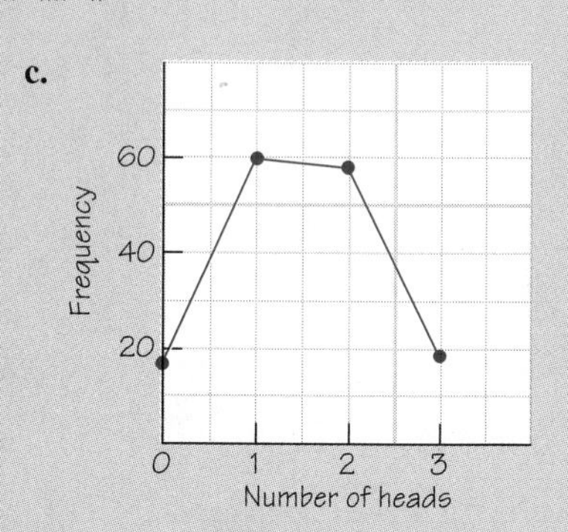

[10.1] *Problems 30–36*

2. The sum of the expenses is $10,000; so the percents and degrees are as follows:

Registration: 0.18(360°) = 64.8°

Education: 0.10(360°) = 36°

Books: 0.06(360°) = 21.6°

Room: 0.64(360°) = 230.4°

Misc.: 0.02(360°) = 7.2°

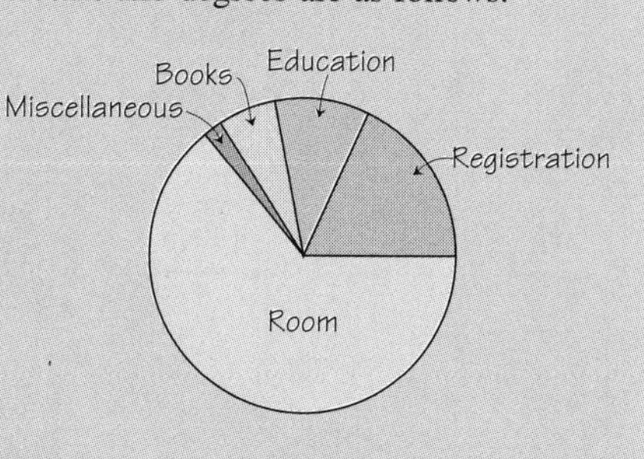

[10.1] *Problems 16–29*

3. a. $50,000 to $59,999; 27% **b.** A **c.** 0.664(2,000) = 1,328

[10.1] *Problems 38–43*

4. No, it does not. No scale is shown.

[10.2] *Problems 11–35*

5. a. Add the numbers and divide by 30: $2,275 \div 30 = 75.8\overline{3}$. The mean is 76.
b. The median is 76.
c. The mode is 87.
d. The mean is the most appropriate measure.

[10.3] *Problems 4–9; 16–30*

6. A score of 87 is higher than 20 of the 30 scores, so 87 is in the 67th percentile.

[10.3] *Problems 10–15*

7. $30 \times 0.87 = 26.1$; the 87th percentile is the score 89, which means that Jane's rank is 4th.

[10.4] *Problems 21–22*

8. Since there are 30 scores, the median (Q_2) is the mean of the 15th and 16th scores; this is 76. The 7th score (Q_1) is 87, and the 7th score from the bottom (Q_3) is 70. The range is 96 − 20 = 76.

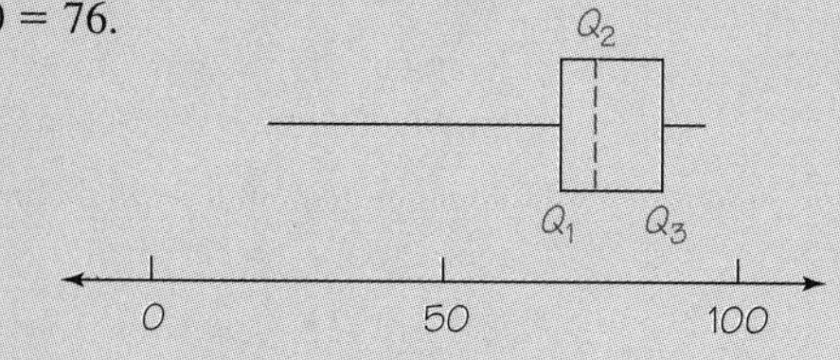

[10.4] *Problems 7–20; 23–34*

9. a. The variance is found as follows:

$$\text{MEAN} = \frac{90 + 85 + 70 + 70 + 65}{5} = 76$$

Data	90	85	70	70	65
Deviation from the mean	−14	−9	6	6	11
Square of deviation	196	81	36	36	121

$$\text{Mean of the squares: } \frac{196 + 81 + 36 + 36 + 121}{5} = 94$$

b. The variance is 94 and the standard deviation is $\sqrt{94} \approx 9.7$.

[10.5] *Problems 2–22*

10. 2.3% should be A's: 0.023(30) = 0.69; 1 A

13.6% should be B's: 0.136(30) = 4.08; 4 B's

68.2% should be C's: 0.682(30) = 20.46; 20 C's

Same number of D's as B's: 4 D's

Same number of F's as A's: 1 F

[10.6] *Problems 1–5*

11. Answers vary.

Chapter 10 Practice Test

1. Make a frequency table for the following data:

7, 8, 8, 7, 8, 9, 6, 9, 8, 6, 8, 9, 10, 7, 7

2. Make a frequency table identifying the political party of the U.S. Speaker of the House for the 103 Congresses. The parties are designated by Democrat (D), Republican (R), Federalist (F), Dem.–Rep. (DR), Whig (W), and American (A):†

F, F, F, F, F, DR, F, DR, DR, DR, DR, DR, DR, DR, DR, DR, DR, DR, DR, DR, DR, DR, D, D, D, D, W, D, D, D, W, D, D, W, D, D, D, A, D, W, R, R, R, R, R, R, R, R, D, D, D, D, D, D, R, D, D, D, R, D, D, R, R, R, R, R, R, R, R, D, D, D, D, R, R, R, R, R, R, D, D, D, D, D, D, D, D, D, D, R, D, D, R, D, D, D, D, D, D, D, D, D, D, D, D, D, D, D, D, D, D, D

3. Make a bar graph for the data in Problem 2.

4. Make a line graph for the data in Problem 2.

†There are more than 103 entries because some Congresses had more than one speaker.

YEAR	EXPENDITURES (billions)
1989	$ 957
1990	$1,150
1991	$1,162
1992	$1,265

5. Consider the data listed in the margin. The table shows the U.S. government's expenditures for social welfare.
a. Draw a bar graph to represent these data.
b. Suppose that you had the following viewpoint: *Social welfare expenditures rose only \$0.3 trillion from 1989 to 1992.* Draw a graph that shows very little increase in the expenditures.
c. Suppose you had the following viewpoint: *Social welfare expenditures rose \$308,000,000,000 from 1989 to 1992.* Draw a graph that shows a tremendous increase in the expenditures.
d. In view of parts **a–c,** discuss the possibilities of using statistics to mislead or support different views.

6. Find the mean, the median, and the mode for the following data: 23, 24, 25, 26, 27.

SIZE	NO. OF CASES
6 oz	5
10 oz	10
12 oz	35
16 oz	50

7. A small grocery store stocked several sizes of Copycat Cola last year. The sales figures are shown in the accompanying table. Find the mean, the median, and the mode for size.

8. A student's scores in a certain math class are 72, 73, 74, 85, and 91. Find the mean, the median, and the mode.

9. The earnings of the part-time employees of a small real estate company for the past year were (in thousands of dollars): 12, 22, 18, 15, 9, 11, 18, 17, and 10. Find the mean, the median, and the mode.

10. In Your Own Words Discuss the differences and similarities among the mean, the median, and the mode. Amplify with your own examples.

11. The manager of the store described in Problem 7 decides to cut back on variety and stock only one size of cola. Which of the measures of central tendency will be most useful in making this decision?

12. Which measure of central tendency would be most appropriate for the situation in Problem 8?

13. Which measure of central tendency is most representative of the employees' earnings as described in Problem 9?

Find the range and the standard deviation (to two decimal places) for the data from the indicated problem. Find the variance if you do not have a calculator.

14. Problem 6

15. Problem 7

16. Problem 8

17. Problem 9

18. The heights of 10,000 women students at Eastern College are known to be normally distributed, with a mean of 67 inches and a standard deviation of 3 inches. How many women would you expect to be between 64 and 70 inches tall?

19. How many women in Problem 18 would you expect to be shorter than 5 ft 1 in.?

20. Determine the upper and lower scores for the middle 68% of scores on a test on which the scores are normally distributed with mean 73 and standard deviation 8.

21. **In Your Own Words** Two instructors gave the same Math 1A test in their classes. Both classes had the same mean score, but one class had a standard deviation twice as large as that of the other class. Which class do you think would be easier to teach, and why?

22. **In Your Own Words** What do we mean by Type I error?

23. **In Your Own Words** What do we mean by Type II error?

24. **In Your Own Words** What is the difference between a sample and a population?

25. **In Your Own Words** How would you obtain a sample of one-syllable words used in this text? How would you apply those results to the whole book?

CHAPTER ELEVEN

GRAPHS

11.1 ORDERED PAIRS AND THE CARTESIAN COORDINATE SYSTEM

Situation

Jack had not been to San Francisco before, and he was looking forward to seeing Fisherman's Wharf, North Beach, Nob Hill, Chinatown, and the Civic Center. However, the first thing he wanted to do was look up his old friend, Charlie, who lives on Main Street in San Francisco. He bought a map and looked up Charlie's street in the index. It was listed as (7, B). What does (7, B) mean?

In mathematics, such a representation is called an *ordered pair* and has many useful applications. The first one we'll consider in this section is called a *Cartesian coordinate system* and sets the stage for the remainder of this chapter.

ORDERED PAIRS

Have you ever drawn a picture by connecting the dots or found a city or street on a map? If you have, then you've used the ideas we'll be discussing in this section.

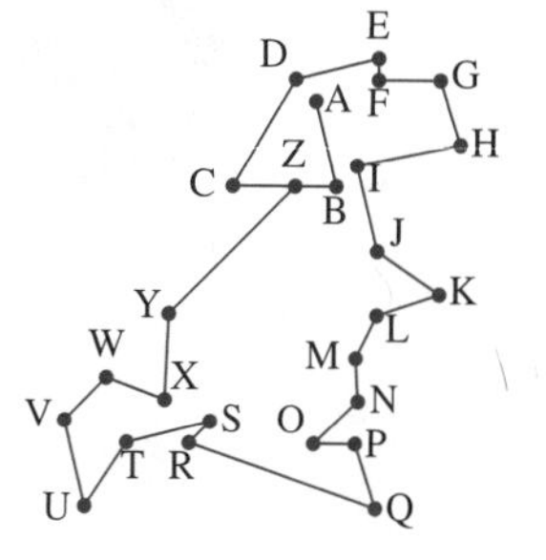

Can you connect the dots?

Suppose we wish to find Fisherman's Wharf in San Francisco. We look in the index for the map in Figure 11.1 and find Fisherman's Wharf listed as (6, D).

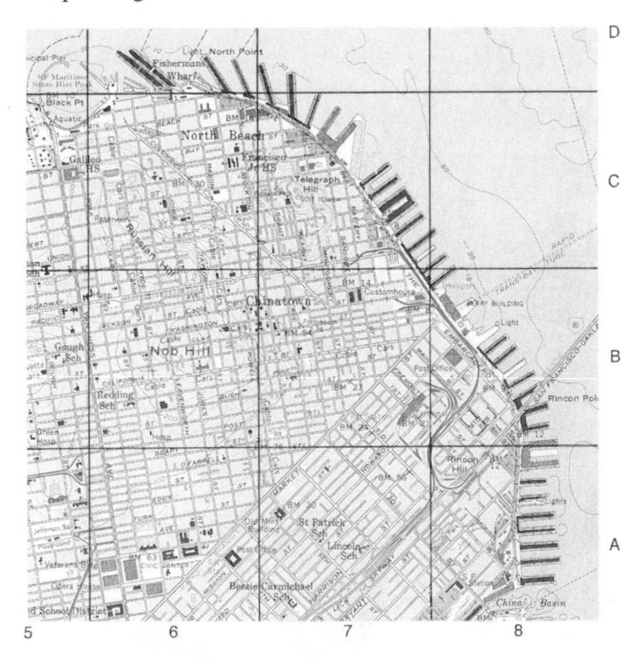

Figure 11.1 Map of San Francisco: Can you locate section (6, D)?

Even with the index listing of, say, Main Street at (7, B), the task of finding the street is not an easy one. How could we improve the map? Let's create a smaller grid, as shown in Figure 11.2.

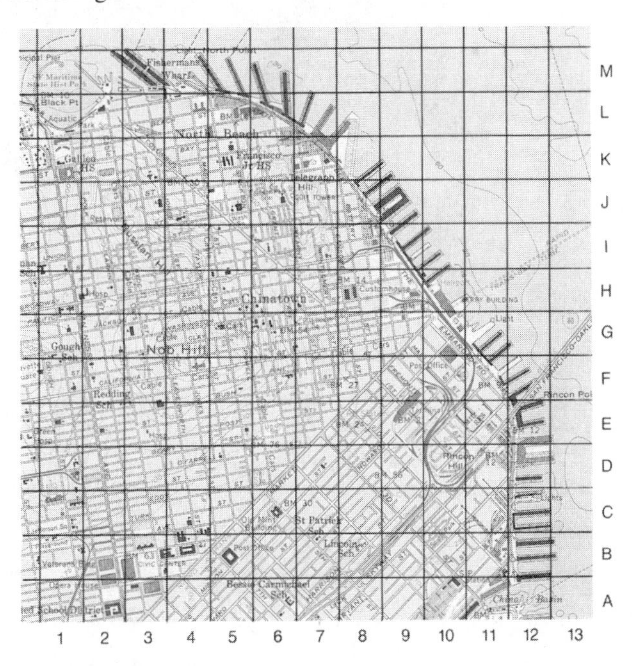

Figure 11.2 Map of San Francisco with a refined grid

first component
↓
(x, y) is an ordered pair.
↑
second component

Now Main Street is located at (9, G) and is easier to find because of the smaller grid. However, a smaller grid means that a lot of letters are needed for the vertical scale, and we might need more letters than there are in the alphabet. So let's use a notation that will allow us to represent points on the map with pairs of numbers. A pair of numbers written as (2, 3) is called an **ordered pair** to remind you that the order in which the numbers 2 and 3 are listed is important. That is, (2, 3) specifies a different location than does (3, 2). For the ordered pair (2, 3), 2 is the **first component** and 3 is the **second component.**

For our map, suppose that we relabel the vertical scale with numbers and change both scales so that we label the *lines instead of the spaces,* as shown in Figure 11.3. (By the way, most technical maps number lines instead of spaces.) Now we can fix the location of any street on the map quite precisely. Notice that, if we use an ordered pair of numbers (instead of a number and a letter), it is important to know which component of the ordered pair represents the horizontal distance and which component represents the vertical distance. If we use Figure 11.3, we see that the coordinates of Main Street are about (9.5, 6.3); we can even say that Main Street

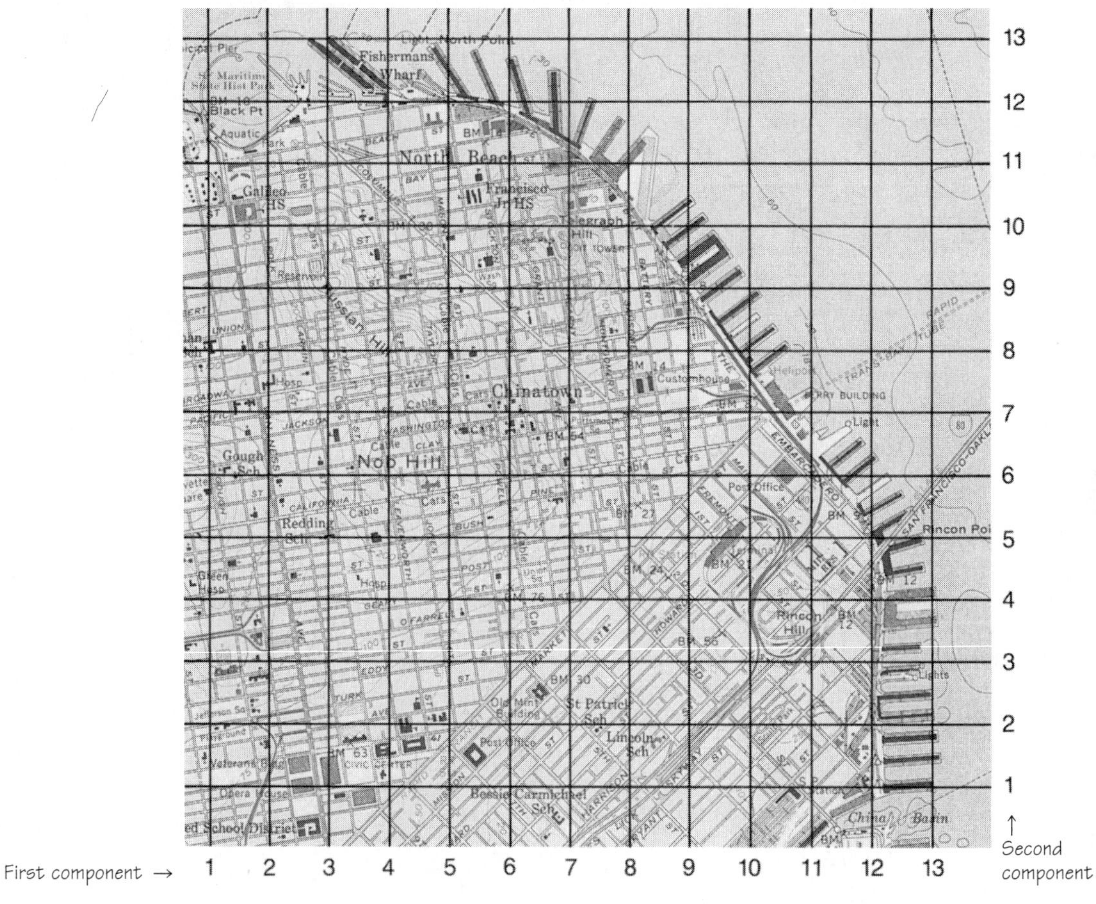

Figure 11.3 Map of San Francisco with a coordinate grid

runs from about (9.5, 6.3) to (11.2, 4.5). Notice that, by using ordered pairs and numbering the lines instead of the spaces, we have refined our grid. We've refined it even more with decimal components. When using pairs of numbers instead of numbers and letters, remember that the first component is on the horizontal axis and the second component is on the vertical axis.

EXAMPLE 1 **Using Ordered Pairs on a Map**

Find the major landmarks with the following coordinates in Figure 11.3.

a. (3.5, 1.5) **b.** (3.5, 12.5) **c.** (5, 11) **d.** (4, 6) **e.** (6, 7.3)

Solution **a. Civic Center** **b. Fisherman's Wharf** **c. North Beach**
d. Nob Hill **e. Chinatown** ■

There are many ways to use ordered pairs to find particular locations. For example, a teacher may make a seating chart like the one shown in Figure 11.4.

5	Otis Moorehouse	Susan Reiland		Jeff Clark	Shannon Sovndal
4	Sharon Boschen	Linda Smith	Bob Anderson	Josephine Lee	Milt Hoehn
3	Jeff Atz	Laurie Pederson	Tim Selbo	Todd Humann	Brian Claasen
2	Terry Shell	Jane Morton	Clint Stevenson	Steve Switzer	
1	Rosamond Foley	Beverly Schaap	Niels Pedersen	Melissa Becker	Eva Mikalson
	1	2	3	4	5

Front of class

Figure 11.4 Classroom seating chart

In the grade book, the teacher records

Anderson (3, 4) *Remember, first component is horizontal direction*

Atz (1, 3) *Second component is vertical direction*

Anderson's seat is in column 3, row 4. Can you think of some other ways in which ordered pairs could be used to locate a position?

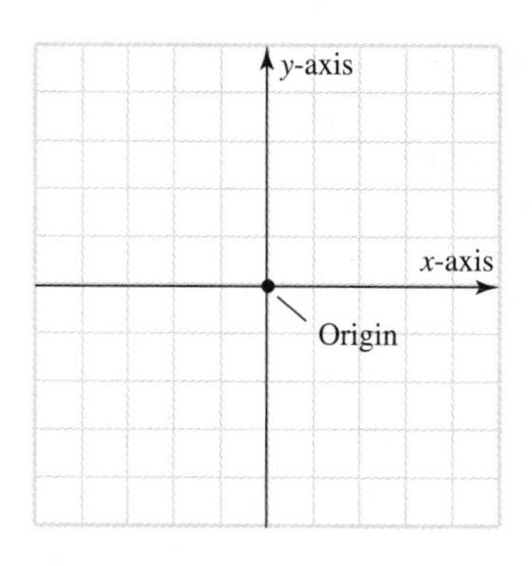

Figure 11.5 Cartesian coordinate system

CARTESIAN COORDINATE SYSTEM

The idea of using an ordered pair to locate a certain position requires particular terminology. **Axes** are two perpendicular real number lines, such as those shown in Figure 11.5. The point of intersection of the axes is called the **origin.** In Chapter 2, we associated direction to the right or up with positive numbers. The upward and rightward arrows in Figure 11.5 are pointing in the positive direction. These perpendicular lines are usually drawn so that one is horizontal and the other is vertical. The horizontal axis is called the ***x*-axis,** and the vertical axis is called the ***y*-axis.**

Notice that the axes of a Cartesian coordinate system divide the plane into four parts. These parts are called **quadrants** and are labeled as shown in Figure 11.6.

We can now label points in the plane by using ordered pairs. The first component of the pair gives the horizontal distance, and the second component gives the vertical distance, as shown in Figure 11.7. If x and y are components of a point, representation of the point (x, y) is called the **rectangular** or **Cartesian coordinates** of the point. To **plot** (or **graph**) a point means to show the coordinates of the ordered pair by drawing a dot at the specified location.

y

QUADRANT II x is negative y is positive	QUADRANT I x is positive y is positive
QUADRANT III x is negative y is negative	QUADRANT IV x is positive y is negative

x

Figure 11.6 Quadrants

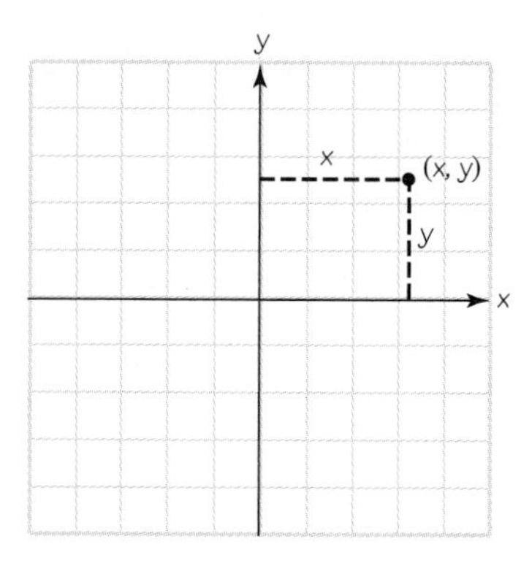

Figure 11.7 (x, y) are coordinates of a point.

EXAMPLE 2

Plotting Points

Plot (graph) the given points.

a. $(5, 2)$ **b.** $(3, 5)$ **c.** $(-2, 1)$ **d.** $(0, 5)$ **e.** $(-6, -4)$
f. $(3, -2)$ **g.** $(\frac{1}{2}, 0)$ **h.** $(0, 0)$

Solution

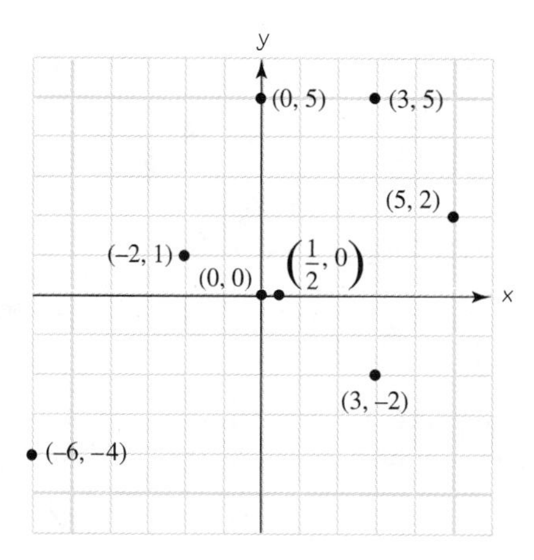

■

PROBLEM SET 11.1

Level 1—Drill

Find the landmarks given by the coordinates in Problems 1–10. Use the map in Figure 11.3.

1. (7, 10)
2. (1, 2)
3. (11, 6)
4. (12, 0.5)
5. (11, 7.3)
6. (1.5, 11.3)
7. (6.5, 2.5)
8. (6.5, 4.5)
9. (6.9, 9.8)
10. (5.5, 1.5)

Using the seating chart in Figure 11.4, name the occupant of the seat given by the coordinates in Problems 11–20.

11. (5, 1)
12. (4, 1)
13. (1, 2)
14. (4, 2)
15. (2, 3)
16. (3, 1)
17. (1, 5)
18. (4, 5)
19. (2, 1)
20. (2, 4)

Plot the points given in Problems 21–25. Use the indicated scale.

21. Scale: 1 square on your paper = 1 unit
    a. (1, 2)
    b. (3, −3)
    c. (−4, 3)
    d. (−2, −3)
    e. (0, 4)

22. Scale: 1 square on your paper = 1 unit
    a. (−1, 4)
    b. (6, 2)
    c. (1, −5)
    d. (3, 0)
    e. (−1, −2)

23. Scale: 1 square on your paper = 10 units
    a. (10, 25)
    b. (−5, 15)
    c. (0, 0)
    d. (−50, −35)
    e. (−30, −40)

24. Scale: 1 square on your paper = 50 units
    a. (100, −225)
    b. (−50, −75)
    c. (0, 175)
    d. (−200, 125)
    e. (50, 300)

25. Scale, x-axis: 1 square on your paper = 1 unit
    Scale, y-axis: 1 square on your paper = 10 units
    a. (4, 40)
    b. (−3, 0)
    c. (2, −40)
    d. (−5, −10)
    e. (0, 0)

Level 2—Drill

26. Plot the following coordinates on graph paper, and connect each point with the preceding one: (−2, 2), (−2, 9), (−7, 2), (−2, 2). Start again: (−6, −6), (−6, −8), (6, −8), (10, −8), (10, −6), (6, −6), (6, 0), (10, 0), (10, 2), (6, 2), (6, 16), (−2, 14), (−11, 2), (−9, −1), (−2, −1), (−2, −6), (−6, −6).

27. Plot the following coordinates on graph paper, and connect each point with the preceding one: (9, 0), (7, −1), (6, −2), (7, −5), (5, −8), (−8, −10), (−2, −5), (0, −2), (−2, −1), (−6, −5), (−5, −3), (−6, −2), (−5, −1), (−6, 0), (−5, 1), (−6, 2), (−5, 3), (−6, 4), (−5, 5), (−6, 7), (−2, 3), (0, 4), (−1, 10), (−6, 15), (0, 14.5), (3, 14), (6, 5), (4, 3), (5, 2), (7, 2), (9, 0). Finally, plot a point at (7, 1).

28. Suppose that we place coordinate axes on the connect-the-dots figure shown here. Let A be (2, 9) and Z be (1, 5). Write directions similar to those of Problems 26 and 27 about how to sketch this figure.

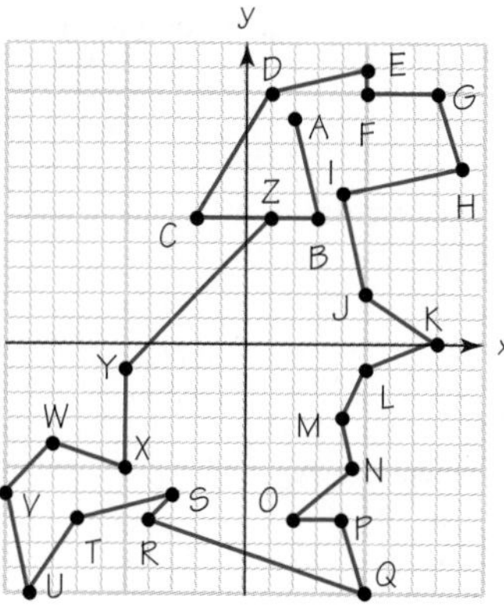

29. Connect the points (0, 2), (1, 5), (2, 8), (−1, −1), (−2, −4). What do you observe?

30. **a.** How many points lie on a line?
b. How many points do you need to plot in order to determine a line?

31. Plot five points, each of which has 2 as the first component. Connect the plotted points.

32. Plot five points, each of which has 3 as the second component. Connect the plotted points.

33. Plot five points in each of which the second component is twice the first component. Connect the plotted points.

34. Plot five points in each of which the second component is 2 less than the first component. Connect the plotted points.

Level 2 — Applications

35. **In Your Own Words** Draw a picture similar to the ones in Problems 26 and 27, and then describe the picture using ordered pairs.

36. **Situation** Jack had not been to San Francisco before and he was looking forward to seeing Fisherman's Wharf, North Beach, Nob Hill, Chinatown, and the Civic Center. However, the first thing he wanted to do was look up his old friend, Charlie, who lives on Main Street in San Francisco. He bought a map and looked up Charlie's street in the index. It was listed as (7, B). What does (7, B) mean?

37. **In Your Own Words** Write a short paper comparing the SITUATION with your own situation. (That is, list similarities and/or differences.)

11.2 FUNCTIONS

Situation

"Look out! That old abandoned well is very dangerous, Huck. I think that it should be capped so that nobody could fall in and kill his self." "Ah, shucks, Tom, let's climb down and see how deep it is. I'll bet it is a mile down to the bottom." If we assume that it is not possible for Tom or Huck to climb down into the well, how can they determine the depth of the well?

Tom and Huck will see in this section that the distance an object falls depends on the length of time it falls. When one number is related to another, it is often beneficial to be able to state that relationship. In this section, Tom and Huck will see that we call this relationship a *function*, if there are no "fickle pickers."

The idea of looking at two sets of variables at the same time was introduced in the previous section. Sets of ordered pairs provide a very compact and useful way to represent relationships between various sets of numbers. To consider this idea, let's look at the following cartoon.

B.C. By permission of Johnny Hart and Creators Syndicate, Inc.

The distance an object will fall depends on (among other things) the length of time it falls. If we let the variable d be the distance the object has fallen and the variable t be the time it has fallen (in seconds), and if we disregard air resistance, the formula is

$$d = 16t^2$$

Therefore, in the *B.C.* cartoon, if the well is 16 seconds deep (and we neglect the time it takes for the sound to come back up), we know that the depth of the well (in feet) is

$$\begin{aligned} d &= 16(16)^2 \\ &= 16(256) \\ &= 4{,}096 \end{aligned}$$

The formula $d = 16t^2$ gives rise to a set of data:

Time (in seconds)	0	1	2	3	4 . . . 15	16
Distance (in ft)	0	16	64	144	256 . . . 3,600	4,096

first component (values for t)
↓
(x, y)
↑
second component (values for d)

For every nonnegative value of t, there is a corresponding value of d. We can represent the data in the table as a set of ordered pairs in which the first component represents a value for t and the second component represents a corresponding value for d. For this example, we have (0, 0), (1, 16), (2, 64), (3, 144), (4, 256), . . . , (15, 3600), (16, 4096).

Whenever we have a situation comparable to the one illustrated by this example—namely, whenever the first component of an ordered pair is associated with exactly one second component—we call the set of ordered pairs a *function*.

Function

This is one of the unifying ideas in all of mathematics.

A **function** is a set of ordered pairs in which the first component is associated with exactly one second component.

Not all sets of ordered pairs are functions, as we can see from the following examples.

EXAMPLE 1 **Distinguishing Functions from Nonfunctions**

Which of the following sets of ordered pairs are functions?

a. $\{(0, 0), (1, 2), (2, 4), (3, 9), (4, 16)\}$
b. $\{(0, 0), (1, 1), (1, -1), (4, 2), (3, -2)\}$
c. $\{(1, 3), (2, 3), (3, 3), (4, 3)\}$
d. $\{(3, 1), (3, 2), (3, 3), (3, 4)\}$

Solution **a.** We see that

$$\begin{aligned} 0 &\rightarrow 0 \\ 1 &\rightarrow 2 \\ 2 &\rightarrow 4 \\ 3 &\rightarrow 9 \\ 4 &\rightarrow 16 \end{aligned}$$

Sometimes it is helpful to think of the first component as the "picker" and the second component as the "pickee." Given an ordered pair (x, y), we find that each replacement for x "picks" a partner, or a second value. We can symbolize this by $x \rightarrow y$.

Since each first component is associated with exactly one second component, the set is **a function.**

b. For this set,

$0 \rightarrow 0$

$1 \rightarrow 1$

$1 \rightarrow -1$

$4 \rightarrow 2$

$3 \rightarrow -2$

Since 1 picks up two values as a partner or second component, we call the number 1 a "fickle picker." But **a function is a set of ordered pairs for which there are no fickle pickers.**

Since the first component can be associated with more than one second component, the set is **not a function.**

c. For this set,

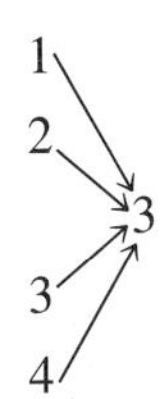

There are no fickle pickers, so it is a function.

This is an example of **a function.**

d. Finally,

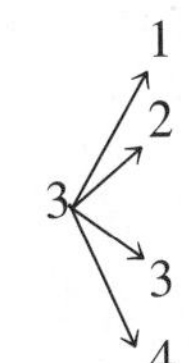

The number 3 is a fickle picker, so this set is not a function.

Since the first component is associated with several second components, the set is **not a function.** ■

Another way to consider functions is with the idea of a function machine. Think of a function machine as shown in Figure 11.8.

Think of this machine as having an input where items are entered and an output where results are obtained, much like a vending machine. If the number 2 is dropped into the input, a function machine will output a single value. If the name of the function machine is f, then the output value is called "f of 2" and is written as $f(2)$. This is called **functional notation.**

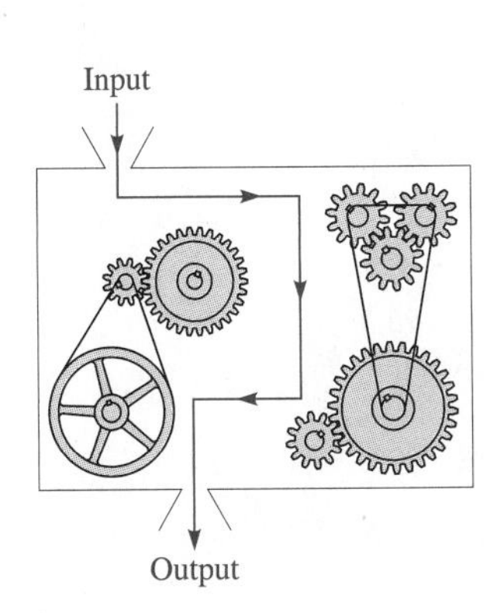

Figure 11.8 Function machine

EXAMPLE 2

Finding Output Values for a Function

If you input each of the given values into a function machine named g, illustrated at the top of page 504, what is the resulting output value?

a. 4 **b.** -3 **c.** π **d.** x

Solution **a.** Input is 4; output is $g(4)$, pronounced "gee of four." To calculate $g(4)$, first add 2, then multiply by 5:

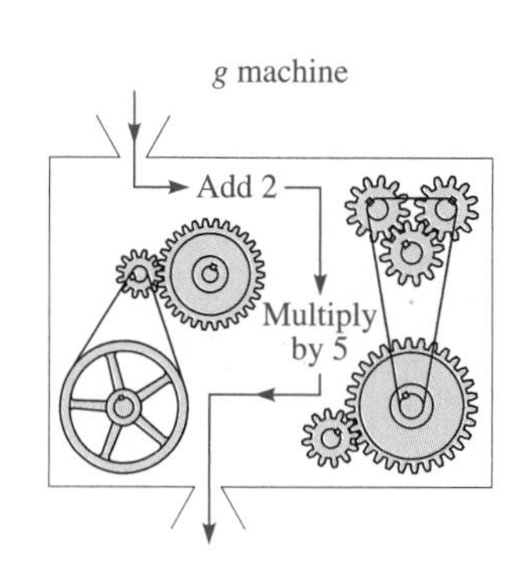

input value ↓ output value

$$(4 + 2) \times 5 = 6 \times 5 = 30$$

We write $\mathbf{g(4) = 30}$.

b. Input is -3; output is $g(-3)$, pronounced "gee of negative three."

$$g(-3) = (-3 + 2) \times 5 = -1 \times 5 = \mathbf{-5}$$

c. Input is π; output is $g(\pi)$, pronounced "gee of pi."

$$g(\pi) = (\pi + 2) \times 5 = \mathbf{5(\pi + 2)}$$

d. Input is x; output is $g(x)$, pronounced "gee of ex."

$$g(x) = (x + 2) \times 5 = \mathbf{5(x + 2)}$$

■

EXAMPLE 3 **Using Functional Notation**

If a function machine f squares the input value, we write $f(x) = x^2$, where x represents the input value. We usually define functions by simply saying "Let $f(x) = x^2$." Identify the value of f for the given value.

a. $f(2)$ **b.** $f(8)$ **c.** $f(-3)$ **d.** $f(t)$

Solution **a.** $f(2) = 2^2 = \mathbf{4}$ **b.** $f(8) = 8^2 = \mathbf{64}$
c. $f(-3) = (-3)^2 = \mathbf{9}$ **d.** $f(t) = \mathbf{t^2}$ ■

PROBLEM SET 11.2

ESSENTIAL IDEAS

1. What is a function?
2. What does it mean to evaluate a function, and what is the notation for this process?

LEVEL 1

WHAT IS WRONG, *if anything, with each of the statements in Problems 3–8? Explain your reasoning.*

3. The set {(1, 1), (2, 1), (3, 1)} is not a function because 1, 2, and 3 all are associated with the same number, namely, 1.
4. If $f(5) = 6$, then the input value is 6.
5. If $f(5) = 6$, then the output value is $f(5)$.
6. If $f(x) = x + 10$, then $f(10) = 10$.
7. If $g(x) = 10x$, then $g(10) = 10$.
8. If $h(x) = 2x$, then $h(\pi) = 6.28$.

LEVEL 1—Drill

Which of the sets in Problems 9–20 are functions?

9. {(1, 4), (2, 5), (4, 7), (9, 12)}
10. {(4, 1), (5, 2), (7, 4), (12, 9)}

11. $\{(1, 1), (2, 1), (3, 4), (4, 4), (5, 9), (6, 9)\}$

12. $\{(1, 1), (1, 2), (4, 3), (4, 4), (9, 5), (9, 6)\}$

13. $\{(4, 3), (17, 29), (18, 52), (4, 19)\}$

14. $\{(13, 4), (29, 4), (5, 4), (9, 4)\}$

15. $\{(19, 4), (52, 18), (29, 17), (3, 4)\}$

16. $\{(4, 9), (4, 4), (4, 29), (4, 19)\}$

17. $\{(5, 0)\}$

18. $\{(0, 0)\}$

19. $\{1, 2, 3, 4, 5\}$

20. $\{69, 82, 44, 37\}$

LEVEL 2—Drill

Tell what the output value is for each of the function machines in Problems 21–26 *for* **(a)** 4, **(b)** 6, **(c)** -8, **(d)** $\frac{1}{2}$, **(e)** x.

21.

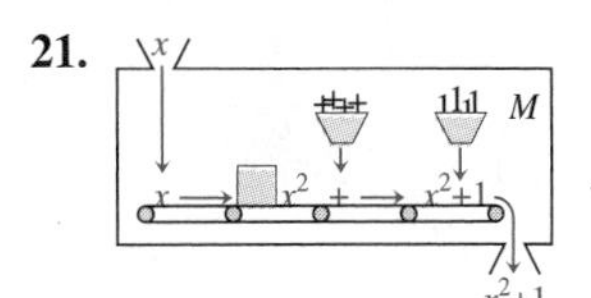

22.

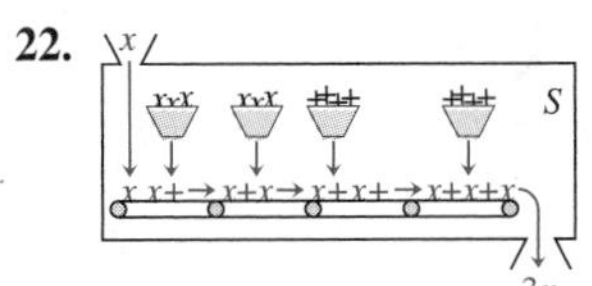

23.

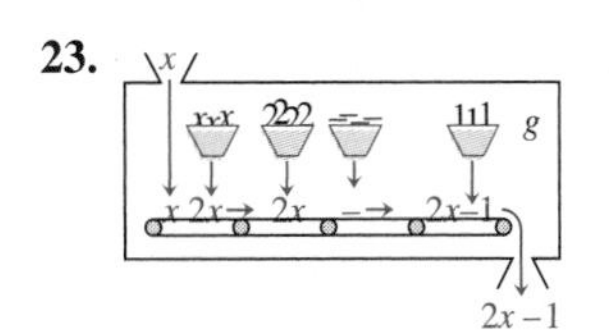

24.

25.

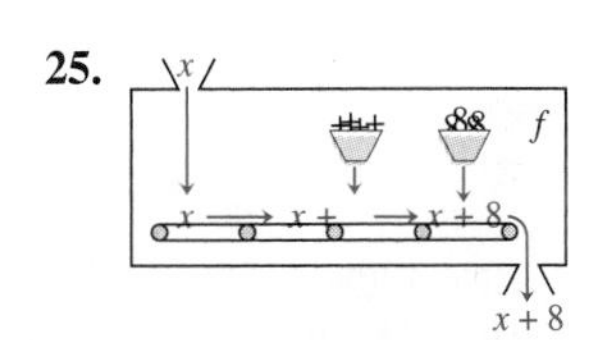

26.

27. If $f(x) = x - 7$, find
a. $f(15)$ **b.** $f(-9)$ **c.** $f(p)$

28. If $g(x) = 2x$, find
a. $g(100)$ **b.** $g(-25)$ **c.** $g(m)$

29. If $h(x) = 3x - 1$, find
a. $h(0)$ **b.** $h(-10)$ **c.** $h(a)$

30. If $f(x) = x^2 + 1$, find
a. $f(-3)$ **b.** $f(\frac{1}{2})$ **c.** $f(b)$

31. If $g(x) = \frac{x}{2}$, find
a. $g(10)$ **b.** $g(-4)$ **c.** $g(3)$

32. If $h(x) = 0.6x$, find
a. $h(4.1)$ **b.** $h(2.3)$ **c.** $h(\frac{5}{2})$

LEVEL 2—Applications

33. The velocity v (in feet per second) of the rock dropped into the well in the *B.C.* cartoon at the beginning of this section is also related to time t (in seconds) by the formula

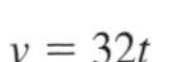

$$v = 32t$$

Complete the table showing the time and velocity of the rock.

Time (in sec)	0	1	2	3	4 . . .	8 . . .	16
Velocity (in ft per sec)	0	32	**a.**	**b.**	**c.**	**d.**	**e.**

34. Using the table of values in Problem 33, find the velocity of the rock at the instant it was released, after 8 seconds, and when it hit the bottom of the well. Write your answers in the form (t, v).

35. An independent distributor bought a new vending machine for \$2,000. It had a probable scrap value of \$100 at the end of its expected 10-year life. The value V at the end of n years is given by

$$V = 2{,}000 - 190n$$

Complete the table showing the year and the value of the machine.

Year	0	1	3	5	7	9	10
Value	2,000	1,810	**a.**	**b.**	**c.**	**d.**	**e.**

36. Using the table of values in Problem 35, find the value of the machine when it is purchased, when it is 5 years old, and when it is scrapped. Write your answers in the form (n, V).

37. Let A be the set of the following cities:

$A = \{$Arm, MI; Bone, ID; Cheek, TX; Doublehead, AL; Elbow Lake, MN$\}$

Suppose that these cities are connected by direct service as shown by a, b, c, d, and e in Figure 11.9. We define a set of ordered pairs (x, y) if and only if x and y are connected by direct service. Find the elements of this set of ordered pairs. Is this set a function?

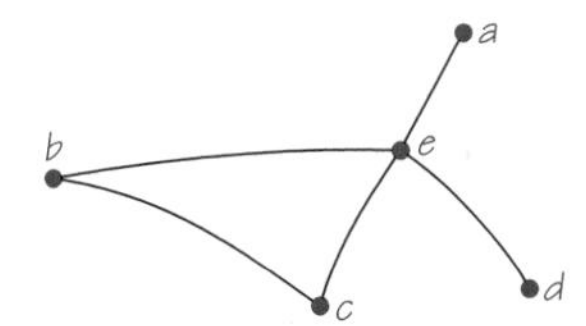

Figure 11.9 City network

38. Suppose that we modify the network of Problem 37 by making the lines one-way, as shown by the arrows in Figure 11.10. We define a set of all pairs (x, y) for which there is one-way service from x to y. Find the elements of this set. Is this set a function?

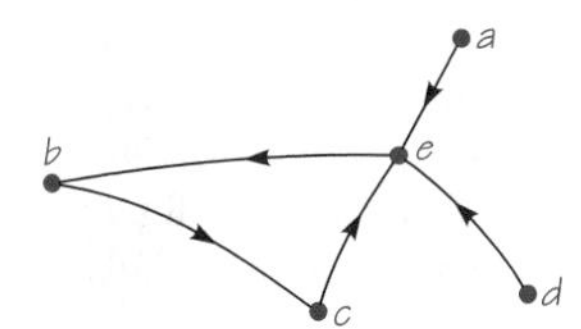

Figure 11.10 Directed network

39. **Situation** "Look out! That old abandoned well is very dangerous, Huck. I think that it should be capped so that nobody could fall in and kill his self." "Ah, shucks, Tom, let's climb down and see how deep it is. I'll bet it is a mile down to the bottom." If we assume that it is not possible for Tom or Huck to climb down into the well, how can they find out the depth of the well? Assume that the well is a mile deep. How long before the rock hits the bottom?

40. **In Your Own Words** Write a short paper comparing the SITUATION with your own situation. (That is, list similarities and/or differences.)

11.3 LINES

Situation

René Descartes (1596–1650)

"René, would you please get out of bed! This is the third time I've called you to come down for breakfast." "But, Mama, I don't feel well today. Must I get up?" said René. "If I can lie here just a bit longer, Mama, I can finish my meditation. I was watching this fly on my ceiling and I have this great idea about representing its position in the room. Can't you let me stay just a few minutes more? Please!"

In this section, we will learn how to represent a point on a flat surface, such as a piece of paper. In more advanced work, the representation of a point in a room (space) is discussed. René Descartes, the person after whom we name the coordinate system, was a person of frail health. He had a lifelong habit of lying in bed until late in the morning or even the early afternoon. It is said that these hours in bed were probably his most productive. During one of these periods, it is presumed, Descartes made the discovery of the coordinate system.

SOLVING EQUATIONS WITH TWO VARIABLES

Let's consider an equation with two variables, say, x and y. If there are two values x and y that make an equation true, then we say that the ordered pair (x, y) **satisfies** the equation and that it is a **solution** of the equation.

EXAMPLE 1

Finding a Solution for a Given Equation

Tell whether each point satisfies the equation $3x - 2y = 7$.

a. $(1, -2)$ **b.** $(-2, 1)$ **c.** $(-1, -5)$

Solution

You should substitute each pair of values of x and y into the given equation to see whether the resulting equation is true or false. If it is true, the ordered pair is a solution; if it is false, the ordered pair is not a solution. If it is a solution, we say that it satisfies the equation.

a. $(1, -2)$ means $x = \mathbf{1}$ and $y = \mathbf{-2}$:

$$\begin{aligned} 3x - 2y &= 3(\mathbf{1}) - 2(\mathbf{-2}) \\ &= 3 + 4 \\ &= 7 \end{aligned}$$

Since $7 = 7$, we see that $(1, -2)$ is **a solution.**

b. $(-2, 1)$ means $x = \mathbf{-2}$ and $y = \mathbf{1}$ (compare with part **a** and notice that the order is important in determining which variable takes which value).

$$\begin{aligned} 3x - 2y &= 3(\mathbf{-2}) - 2(\mathbf{1}) \\ &= -6 - 2 \\ &= -8 \end{aligned}$$

Since $-8 \neq 7$, we see that $(-2, 1)$ is **not a solution.**

c. $(-1, -5)$:

$$\begin{aligned} 3x - 2y &= 3(\mathbf{-1}) - 2(\mathbf{-5}) \\ &= -3 + 10 \\ &= 7 \end{aligned}$$

The ordered pair $(-1, -5)$ is **a solution.** ■

GRAPHING A LINE

The process of graphing a line requires that you find ordered pairs that make an equation true. To do this, *you*, the student, must choose convenient values for x and then solve the resulting equation to find a corresponding value for y.

EXAMPLE 2

Finding Ordered Pairs that Satisfy an Equation

Find three ordered pairs that satisfy the equation $y = -2x + 3$.

Solution

You choose any x value—say, $\mathbf{x = 1}$. Substitute this value into the given equation to find a corresponding value of y:

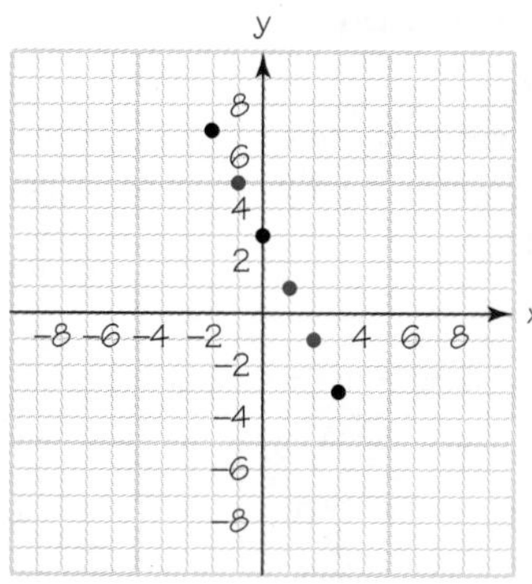
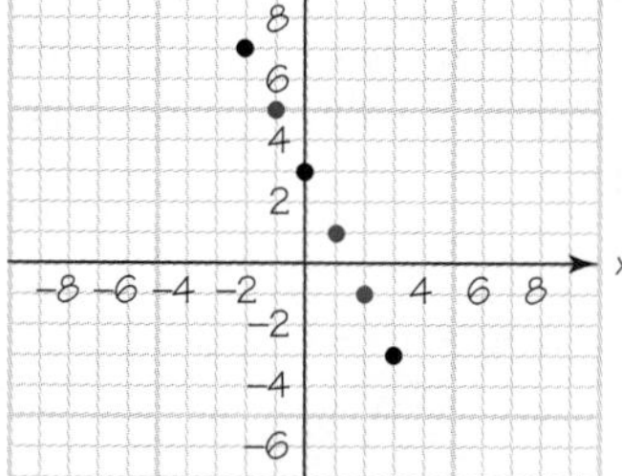

Figure 11.11 Points that satisfy $y = -2x + 3$

$$
\begin{aligned}
y &= -2x + 3 && \text{Given equation} \\
&= -2(\mathbf{1}) + 3 && \text{Substitute chosen value.} \\
&= -2 + 3 \\
&= 1
\end{aligned}
$$

You choose this value.
↓
The first ordered pair is **(1, 1).**
↑
You find this value by substitution into the equation.

Choose a second value—say, $\boldsymbol{x = 2}$. Then

$$
\begin{aligned}
y &= -2x + 3 && \text{Start with given equation.} \\
&= -2(\mathbf{2}) + 3 && \text{Substitute.} \\
&= -1 && \text{Simplify.}
\end{aligned}
$$

The second ordered pair is **(2, −1).**

Choose a third value—say, $\boldsymbol{x = -1}$. Then

$$
\begin{aligned}
y &= -2x + 3 \\
&= -2(\mathbf{-1}) + 3 \\
&= 5
\end{aligned}
$$

The third ordered pair is **(−1, 5).** ■

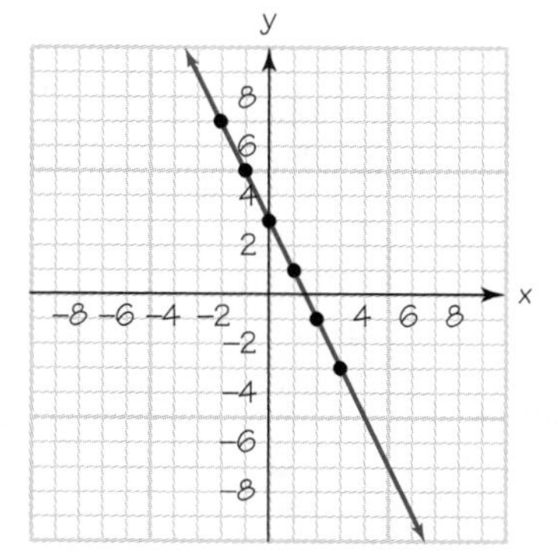

Figure 11.12 Representation of all points satisfying $y = -2x + 3$

Have we found *all* the ordered pairs that satisfy the equation $y = -2x + 3$? Can you find others? Suppose we plot the ordered pairs in Example 2, along with three additional points, as shown in Figure 11.11. Do you notice anything about the arrangement of these points in the plane? Suppose that we draw a line passing through the points, as shown in Figure 11.12. This line represents the set of *all* ordered pairs that satisfy the equation.

If we carry out the process of finding three ordered pairs that satisfy an equation with two first-degree variables, and then we draw a line through those points to find the representation of all ordered pairs satisfying the equation, we say that we are *graphing the line,* and the final set of points we have drawn represents the **graph** of the line.

Procedure for Graphing a Line

To graph a **line,** find two ordered pairs that lie on the line. Two points determine a line. Find a third point as a check. Draw the line (using a straightedge) passing through these three points.

If the three points don't lie on a straight line, then you have made an error.

EXAMPLE 3

Graphing a Line

Graph $y = 2x + 2$.

Solution

It is generally easier to pick x and find y.

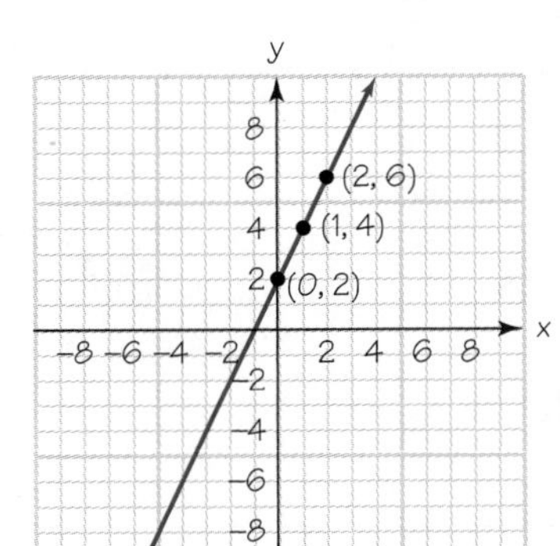

If $x = \mathbf{0}$: $y = 2x + 2$ *Given equation*

$= 2(\mathbf{0}) + 2$ *Substitute.*

$= 2$ *Simplify.*   Plot the point (0, 2).

If $x = \mathbf{1}$: $y = 2x + 2$

$= 2(\mathbf{1}) + 2$

$= 4$ Plot (1, 4).

If $x = \mathbf{2}$: $y = 2x + 2$

$= 2(\mathbf{2}) + 2$

$= 6$ Plot (2, 6).

Draw the line passing through the three plotted points. ■

EXAMPLE 4

Graphing a Line Given in Standard Form

Graph the line $x + 2y = 6$.

Solution

It is generally easier if you solve for y before you substitute in values for x.

$$x + 2y = 6$$

$$2y = -x + 6 \quad \text{Subtract } x \text{ from both sides.}$$

$$y = -\tfrac{1}{2}x + 3 \quad \text{Divide both sides (all terms) by 2.}$$

If $x = \mathbf{0}$: $y = -\frac{1}{2}x + 3$

$= -\frac{1}{2}(\mathbf{0}) + 3$

$= 3$ Plot (0, 3).

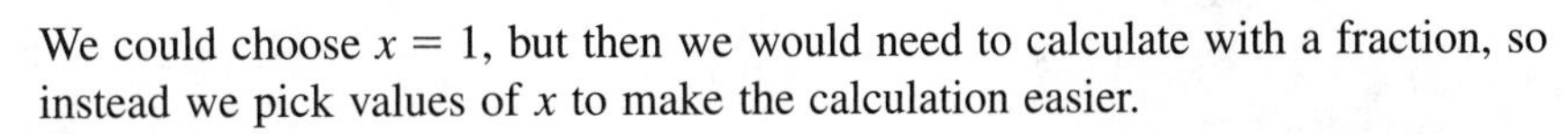

We could choose $x = 1$, but then we would need to calculate with a fraction, so instead we pick values of x to make the calculation easier.

If $x = \mathbf{2}$: $y = -\frac{1}{2}x + 3$

$= -\frac{1}{2}(\mathbf{2}) + 3$

$= -1 + 3$

$= 2$ Plot (2, 2).

If $x = \mathbf{-2}$: $y = -\frac{1}{2}x + 3$

$= -\frac{1}{2}(\mathbf{-2}) + 3$

$= 1 + 3$

$= 4$ Plot (−2, 4).

Figure 11.13 Graph of $x + 2y = 6$

Draw the line connecting these points, as shown in Figure 11.13. ■

EXAMPLE 5 **Graphing a Horizontal Line**

Graph $y = 4$.

Solution Since the only requirement is that y (the second component) equal 4, we see that there is no restriction on the choice for x. Thus, (0, 4), (1, 4), and (−2, 4) all satisfy the equation $y = 4$. If you plot and connect these points, you will see that the line formed is a **horizontal line.**

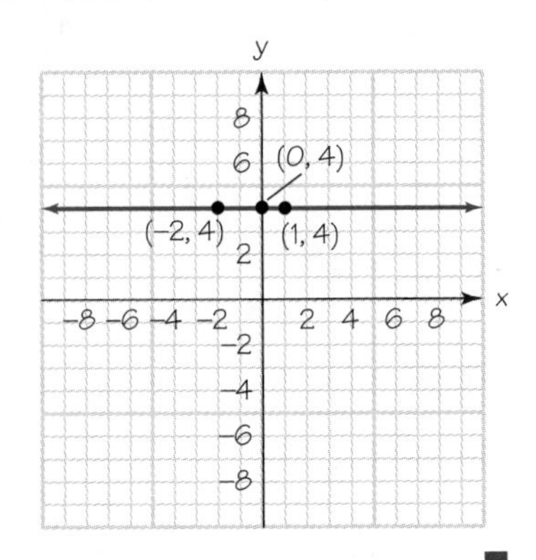

■

EXAMPLE 6 **Graphing a Vertical Line**

Graph $x = 3$.

Solution Since the only requirement is that x (the first component) equal 3, we see that there is no restriction on the choice for y. Thus, (3, 2), (3, −1), and (3, 0) all satisfy the condition that $x = 3$. If you plot and connect these points, you will see that the line formed is a **vertical line.**

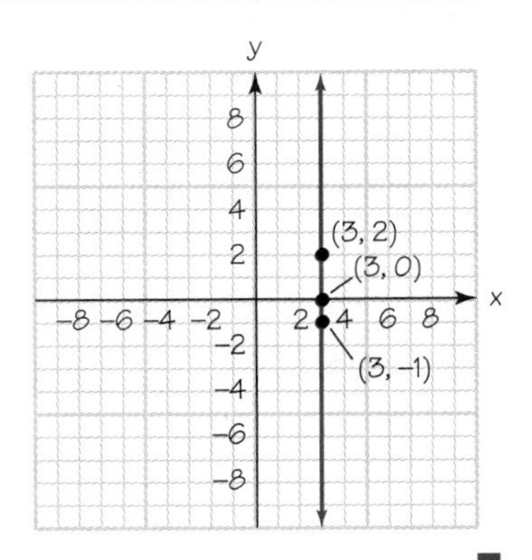

■

PROBLEM SET 11.3

ESSENTIAL IDEAS

1. **In Your Own Words** What does it mean to be a solution to an equation with two variables?

2. **In Your Own Words** Describe a process for graphing a line.

LEVEL 1—Drill

Find three ordered pairs that satisfy each equation in Problems 3–14.

3. $y = x + 5$ **4.** $y = 2x - 1$ **5.** $y = 2x + 5$

6. $y = x - 4$ **7.** $y = x - 1$ **8.** $y = x + 1$

9. $y = -2x + 1$ **10.** $y = -3x + 1$ **11.** $y = 2x + 1$

12. $y - 3x = 1$ **13.** $3x + 4y = 8$ **14.** $x + 2y = 4$

LEVEL 2

WHAT IS WRONG, *if anything, with each of the statements in Problems 15–20? Explain your reasoning.*

15. The point (5, 4) satisfies the equation $3x - 5y = -13$ because

$$\begin{aligned} 3x - 5y &= 3(4) - 5(5) \\ &= 12 - 25 \\ &= -13 \end{aligned}$$

16. To graph $2x - 3y = 12$, we first find three ordered pairs that satisfy the equation:

$$2x - 3y = 12$$
$$-3y = -2x + 12$$
$$y = \frac{2}{3}x + 4$$

If $x = 0$, then $y = 4$; if $x = 3$, then $y = 6$; and if $x = -3$, then $y = 2$.

17. The graph of $y = 5$ does not exist because there is no x-value in the equation.

18. The graph of $x = 15$ passes through the points (15, 0), (15, 15), and (15, −3.5).

19. The graph of $x = 5$ is a horizontal line.

20. The graph of $y = 15$ is a vertical line.

Level 2—Drill

Graph the lines in Problems 21–32. These are the same equations as those in Problems 3–14.

21. $y = x + 5$ **22.** $y = 2x - 1$ **23.** $y = 2x + 5$

24. $y = x - 4$ **25.** $y = x - 1$ **26.** $y = x + 1$

27. $y = -2x + 1$ **28.** $y = -3x + 1$ **29.** $y = 2x + 1$

30. $y - 3x = 1$ **31.** $3x + 4y = 8$ **32.** $x + 2y = 4$

Graph the lines in Problems 33–53.

33. $y = x + 3$ **34.** $y = 1 - x$ **35.** $y = 4 - x$

36. $y = 3x - 2$ **37.** $y = 5x - 2$ **38.** $y = 2x + 3$

39. $x + y = 0$ **40.** $x - y = 0$ **41.** $2x + y = 3$

42. $3x + y = 10$ **43.** $2x + 4y = 16$ **44.** $x + 3y = 6$

45. $x = 5$ **46.** $x = -4$ **47.** $x = -1$

48. $y = -2$ **49.** $y = 6$ **50.** $y = 4$

51. $x + y + 100 = 0$ **52.** $5x - 3y = 27$ **53.** $y = 50x$

Level 2—Applications

54. Situation "René, would you please get out of bed! This is the third time I've called you to come down for breakfast." "But, Mama, I don't feel well today. Must I get up?" said René. "If I can lie here just a bit longer, Mama, I can finish my meditation. I was watching this fly on my ceiling, and I have this great idea about representing its position in the room. Can't you let me stay just a few minutes more? Please!" Use patterns or do some outside research to explain a three-dimensional coordinate system for describing the fly's position in René's room.

11.4 SYSTEMS AND INEQUALITIES

Situation

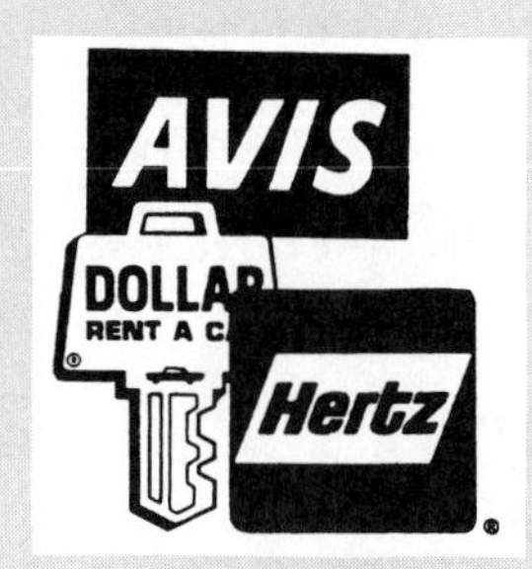

Linda is on a business trip and needs to rent a car for the day. The car rental agency has the following options on the car she wants to rent: Option A: $40 plus 50¢ per mile; Option B: Flat $60 with unlimited mileage. Which car should she rent?

In this section, Linda will learn how to use lines to solve systems of equations to help her with problems like this one.

SOLVING SYSTEMS OF EQUATIONS BY GRAPHING

Many situations involve two variables or unknowns that are related in some specific fashion. The situation relates the cost c of a car rental to the number of miles driven, in the following way:

Option A COST = BASIC CHARGE + MILEAGE CHARGE

50¢ per mile ↓

COST = 40 + 0.5(NUMBER OF MILES)

Let COST = c NUMBER OF MILES = m

$$c = 40 + 0.5m$$

Option B COST = FLAT FEE

$$c = 60$$

Suppose that we represent these relationships in a graph, as described by a Cartesian coordinate system. We will find ordered pairs (m, c) that make these equations true.

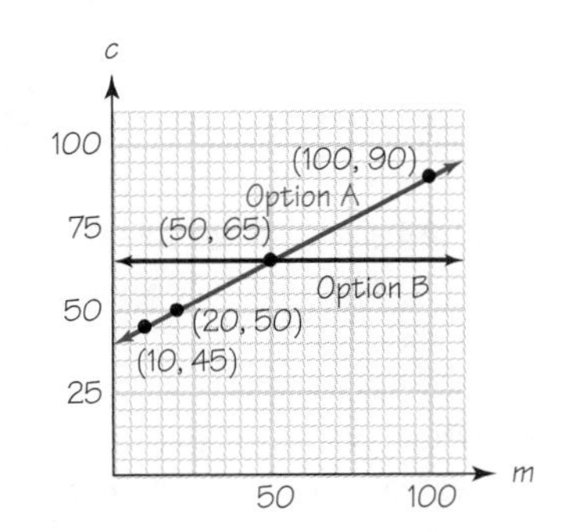

Figure 11.14 Car rental problem: Graphs of $c = 40 + 0.5m$ and $c = 60$

Option A $c = 40 + 0.5m$

Let $m = \mathbf{20}$: $c = 40 + 0.5(\mathbf{20}) = 40 + 10 = 50$

This means that, if Linda drives 20 miles, the cost of the rental is \$50; plot (20, 50).

Let $m = \mathbf{50}$: $c = 40 + 0.5(\mathbf{50}) = 40 + 25 = 65$ Plot (50, 65).

Let $m = \mathbf{100}$: $c = 40 + 0.5(\mathbf{100}) = 40 + 50 = 90$ Plot (100, 90).

From the graph, we can see that these points all lie on the same line. Draw a line through these points, as shown in Figure 11.14.

Option B $c = 60$

The second component of the ordered pair (m, c) is always 60 regardless of the number of miles, m. This is shown as a black line in Figure 11.14.

EXAMPLE 1

Problem Solving Using Lines

As an extension of the situation, Linda is offered two options when renting a car:*

*Problem 55 asks you to answer the question in the SITUATION using a flat fee of \$50 and a variable fee of \$30 plus 25¢/mi.

Option A \$40 plus 50¢ per mile; equation is $c = 40 + 0.5m$ for a cost of c dollars and m miles.

Option B Flat \$60 with unlimited miles; equation is $c = 60$.

Estimate the mileage for which both rates are the same.

Solution The solution is the point of intersection of the graphs shown in Figure 11.14. We estimate the coordinates to be (40, 60), as shown in Figure 11.15.

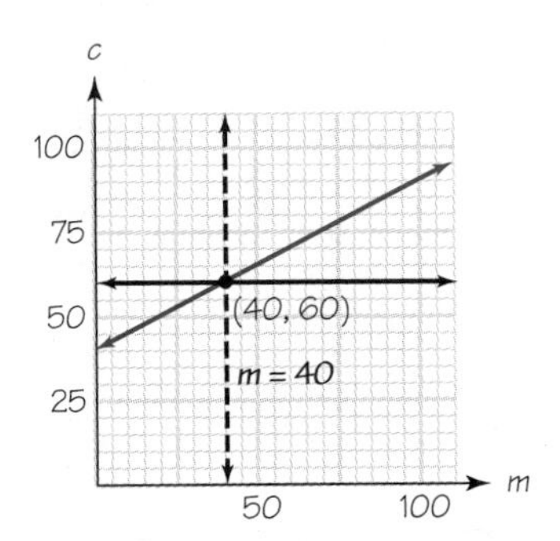

Figure 11.15 Comparing car rental rates

This point of intersection means that, **if Linda drives 40 miles, the rates are the same.** It also means that, if Linda expects to drive more than 40 miles, she should take the fixed rate. ■

In Example 1, we looked at the intersection point of the graphs of two lines. When two or more equations are considered together, we call them a **system of equations.** The intersection point on the graph is called the **simultaneous solution** of a system of equations.

EXAMPLE 2

Finding the Simultaneous Solution of a System of Equations

Solve the following system of equations by graphing:

$$\begin{cases} x + 2y = 5 \\ 3x - y = 8 \end{cases}$$

The brace is used to signify that we want to find the simultaneous solution of the system of equations. The method of graphing involves looking for the point of intersection for the two lines.

Solution Graph both lines on the same coordinate axes.

Line $x + 2y = 5$ | Line $3x - y = 8$

x	y
1	2
3	1
5	0

x	y
0	−8
1	−5
2	−2

Remember, to find these points, you choose the x value and then calculate the corresponding y value.

Plot these points and draw each line passing through them, as shown in Figure 11.16.

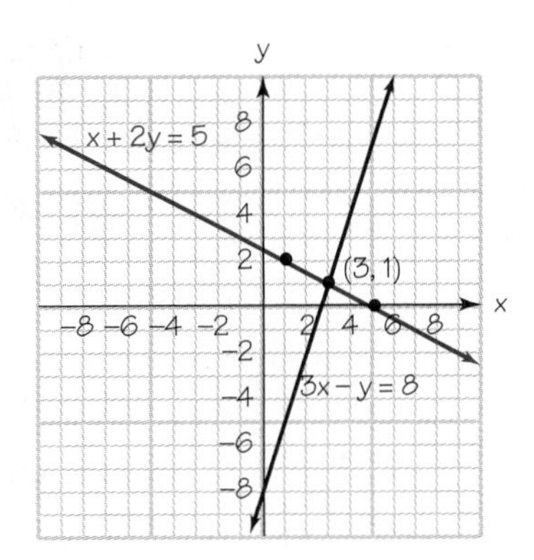

Figure 11.16 Graphs of $x + 2y = 5$ and $3x - y = 8$

The solution to the system of equations is the point of intersection; it looks like the point **(3, 1).** We can check this point to see whether it satisfies *both* equations:

$$\begin{aligned} x + 2y &= 5 \\ (\mathbf{3}) + 2(\mathbf{1}) &= 5 \\ 5 &= 5 \text{ is true} \end{aligned} \qquad \begin{aligned} 3x - y &= 8 \\ 3(\mathbf{3}) - (\mathbf{1}) &= 8 \\ 8 &= 8 \text{ is true} \end{aligned}$$

The point (3, 1) satisfies both of the given equations, so we say that (3, 1) checks. ■

An interesting application of graphing systems of equations has to do with **supply and demand.** If supply greatly exceeds demand, money will be lost because of unsold items. On the other hand, if demand greatly exceeds supply, money will be lost because of insufficient inventory. The most desirable situation is when supply and demand are equal. If we assume that supply and demand are linear, then the solution of the system is called the **equilibrium point** and represents the point at which supply and demand are equal. In more advanced courses, it is shown that the price that maximizes the profit occurs at this equilibrium point.

EXAMPLE 3 **Finding an Equilibrium Point**

Suppose that you have a small product that is marketable to the students on your campus. A little research shows that only 200 people would buy it at \$10, but 2,000 would buy it at \$1. This information represents the *demand.* You also find that a local shop can make the product during slack time and can supply 100 items at a price that allows you to sell the product for \$1 each. To supply more, the shop must use overtime. If the shop supplies 1,500 items, you would have to charge \$6. This information represents the *supply.* What price should you charge for this item to maximize your profit?

Solution First, find the demand points. Let p = PRICE and n = NUMBER OF ITEMS. Because the demand n is determined by the price p, let the price be the independent variable; that is, let the ordered pair be (p, n). From the demand, we have

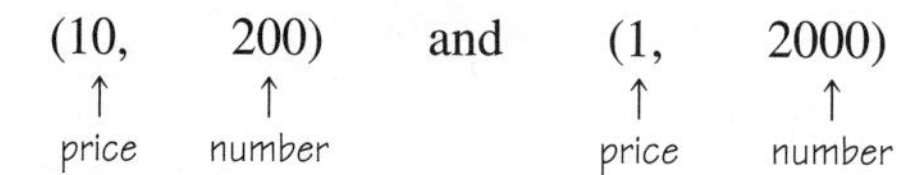

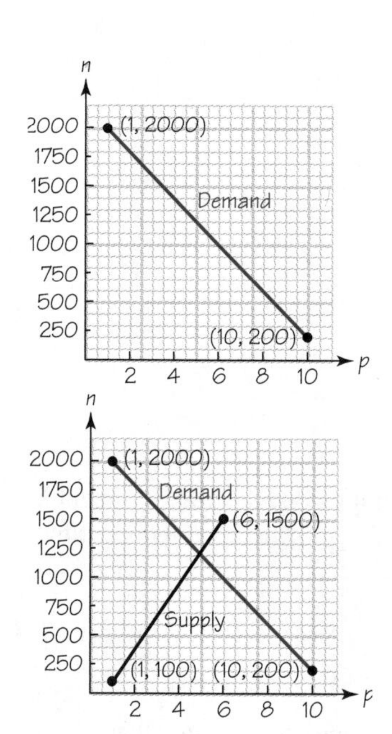

We call these points the *demand points* because they determine the demand line. Draw a line through these points, and label the line "Demand."

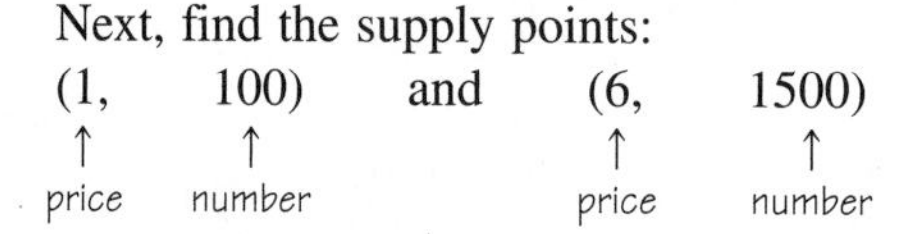

Draw a line passing through these points, and label the line "Supply."

The profit is maximized at the equilibrium point, which is the intersection of the supply and demand lines, as shown in Figure 11.17.

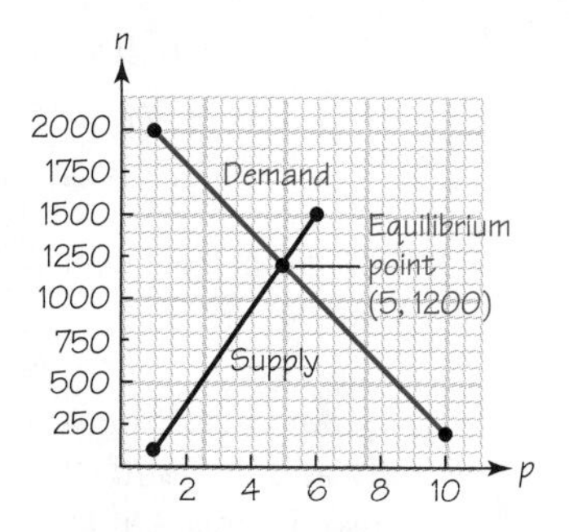

Figure 11.17 Equilibrium point

The price charged should be \$5. At this price, you should expect to sell 1,200 items. ■

GRAPHING LINEAR INEQUALITIES*

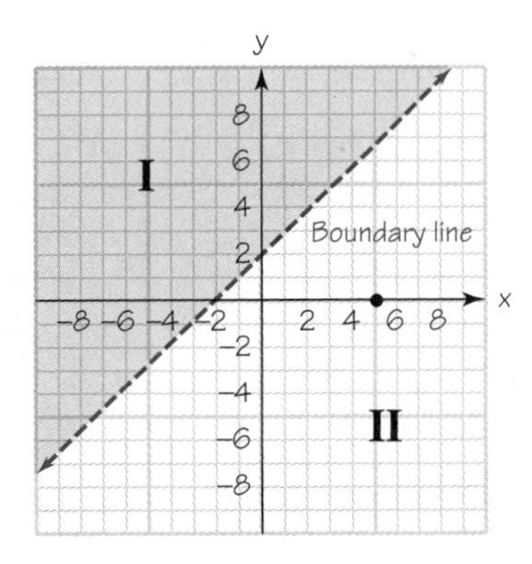

Figure 11.18 Half-planes

A second application of graphing lines involves extending the concept to graphing linear inequalities. We begin by noting that every line divides a plane into three parts, as shown in Figure 11.18. Two parts are labeled I and II; these are called **half-planes.** The third part is called the **boundary** and is the line separating the half-planes. The solution of a first-degree inequality in two unknowns is the set of all ordered pairs that satisfy the given inequality. This solution set is a half-plane. The following table offers some examples of first-degree inequalities with two unknowns, along with associated terminology.

EXAMPLE	*INEQUALITY SYMBOL*	*BOUNDARY INCLUDED*	*TERM*
$3x - y > 5$	$>$	no	**open half-plane**
$3x - y < 5$	$<$	no	open half-plane
$3x - y \geq 5$	$\geq$	yes	**closed half-plane**
$3x - y \leq 5$	$\leq$	yes	closed half-plane

We can now summarize the procedure for graphing a first-degree inequality in two unknowns.

Procedure for Graphing a Linear Inequality

Step 1 Graph the boundary.

Replace the inequality symbol with an equality symbol and draw the resulting line. This is the boundary line.

Use a solid line when the boundary is included ($\leq$ or $\geq$).

Use a dashed line when the boundary is not included ($<$ or $>$).

Step 2 Test a point.

Choose any point in the plane that is not on the boundary line; the point (0, 0) is usually the simplest choice.

If this **test point** makes the *inequality* true, shade in that half-plane for the solution.†

If the test point makes the *inequality* false, shade in the other half-plane for the solution.

This process sounds complicated, but if you know how to draw lines from equations, you will not find it difficult.

*This topic requires Section 3.6.

†A highlighter pen does a nice job of shading your work.

EXAMPLE 4 **Graphing a Linear Inequality**

Graph $3x - y \geq 5$.

Solution Note that the inequality symbol is $\geq$, so the boundary is included.

Step 1 Graph the boundary; draw the (solid) line corresponding to

$3x - y = 5$ Replace the inequality symbol with an equality symbol.

Let $x = \mathbf{0}$; then $3(\mathbf{0}) - y = 5$ or $y = -5$. Plot $(0, -5)$.
Let $x = \mathbf{1}$; then $3(\mathbf{1}) - y = 5$ or $y = -2$. Plot $(1, -2)$.
Let $x = \mathbf{2}$; then $3(\mathbf{2}) - y = 5$ or $y = 1$. Plot $(2, 1)$.

The boundary line is shown in Figure 11.19.

Step 2 Choose a test point; we choose (0, 0).

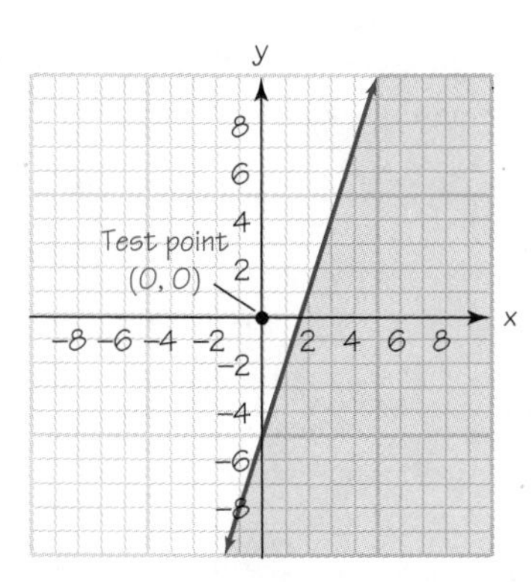

Figure 11.19 Graph of $3x - y \geq 5$

Plot **(0, 0)** in Figure 11.19 and note that it lies in one of the half-planes determined by the boundary line. We now check this test point with the given *inequality.*

$$3x - y \geq 5$$
$$3(\mathbf{0}) - (\mathbf{0}) \geq 5 \quad \text{You can usually test this in your head.}$$
$$0 \geq 5 \quad \text{This is false.}$$

Therefore, shade the half-plane that does *not* contain (0, 0), as shown in Figure 11.19. ■

EXAMPLE 5 **Graphing a Linear Inequality that Passes Through (0, 0)**

Graph $y < x$.

Solution Note that the inequality is $<$, so the boundary line is not included.

Step 1 Draw the (dashed) boundary line, $y = x$, as shown in Figure 11.20.

Step 2 Choose a test point. We can't pick (0, 0), because (0, 0) is on the boundary line. Since we must choose some point not on the boundary, we choose (1, 0):

$$y < x$$
$$0 < 1 \quad \text{This is true.}$$

Therefore, shade the half-plane that contains the test point, as shown in Figure 11.20. ■

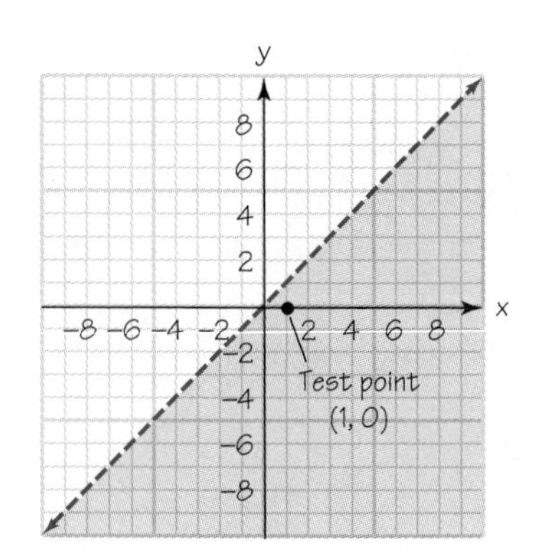

Figure 11.20 Graph of $y < x$

PROBLEM SET 11.4

Level 1

WHAT IS WRONG, *if anything, with each of the statements in Problems 1–12? Explain your reasoning.*

1. (3, 2) is a solution to the system $\begin{cases} 2x - y = 1 \\ x + 2y = 8 \end{cases}$ because $2(\mathbf{2}) - \mathbf{3} = 1$ is true and $\mathbf{2} + 2(\mathbf{3}) = 8$ is true.

2. (3, 2) is a solution to the system $\begin{cases} x - y = 1 \\ 5 - y = x \end{cases}$ because $\mathbf{3} - \mathbf{2} = 1$ is true and $5 - \mathbf{2} = \mathbf{3}$ is true.

3. Nothing seems to work in the system $\begin{cases} x = 1 - y \\ y = 3 - x \end{cases}$ so the solution is (0, 0).

4. The linear inequality $2x + 5y < 2$ does not have a boundary line, because the inequality is $<$.

5. A good test point for the linear inequality $y \geq x$ is (0, 0).

6. To graph the linear inequality $x - y \leq 2$, we write

$$x - y \leq 2$$
$$-y \leq -x + 2$$
$$y \geq x - 2$$

The half-plane to be shaded is the upper half-plane because the inequality is $\geq$.

7. The test point (0, 0) satisfies the inequality $2y - 3x < 2$.

8. The test point (0, 0) satisfies the inequality $3x - 2y \geq -1$.

9. The test point (0, 0) satisfies the inequality $3x > 2y$.

10. The test point $(-2, 4)$ satisfies the inequality $y > 2x - 1$.

11. The test point $(-2, 4)$ satisfies the inequality $5x + 2y \leq 9$.

12. The test point $(-2, 4)$ satisfies the inequality $4x < 3y$.

Solve the systems in Problems 13–27 graphically.

13. $\begin{cases} x + y = 5 \\ x - y = 1 \end{cases}$

14. $\begin{cases} 2x + y = 5 \\ x - y = 4 \end{cases}$

15. $\begin{cases} 2x + y = 1 \\ x - y = 5 \end{cases}$

16. $\begin{cases} 2x - y = -2 \\ 3x + y = 7 \end{cases}$

17. $\begin{cases} 2x - y = 6 \\ 4x + y = 6 \end{cases}$

18. $\begin{cases} y = 4x + 5 \\ y = x + 2 \end{cases}$

19. $\begin{cases} y = 2x + 1 \\ y = 3x + 3 \end{cases}$

20. $\begin{cases} y = x + 1 \\ y = -2x + 4 \end{cases}$

21. $\begin{cases} 4x + y = -3 \\ 2x + y = 1 \end{cases}$

22. $\begin{cases} 3x + 2y = 4 \\ x + y = 1 \end{cases}$

23. $\begin{cases} x + 2y = 2 \\ 3x + y = -4 \end{cases}$

24. $\begin{cases} 4x + y = -3 \\ 3x + 2y = 4 \end{cases}$

25. $\begin{cases} 2x - 3y = 6 \\ x + 2y = 10 \end{cases}$

26. $\begin{cases} x - 2y = -2 \\ 3x - y = 4 \end{cases}$

27. $\begin{cases} x + y = 1 \\ 3x + 2y = -4 \end{cases}$

Graph the first-degree inequalities in two unknowns in Problems 28–42.

28. $y \leq 2x + 1$

29. $y \geq 5x - 3$

30. $y > 5x - 3$

31. $y \geq -2x + 3$

32. $y > 3x - 3$

33. $y < -2x + 5$

34. $3x \leq 2y$

35. $2x < 3y$

36. $x \geq y$

37. $y > x$

38. $y \geq 0$

39. $x \leq 0$

40. $x - 3y \geq 9$

41. $6x - 2y < 1$

42. $3x + 2y > 1$

Level 2—Applications

43. Suppose that a car rental agency gave you the following choices:

Option A $30 per day plus 40¢ per mile

Option B Flat $50 per day

Write equations for options A and B.

44. Graph the equations you found in Problem 43.

45. Use the graph in Problem 44 to estimate the mileage for which both rates are the same.

46. You can rent a paint sprayer for $4 per hour or for $24 per day. Write equations for these options, assuming that you need the sprayer for one day or less.

47. Graph the equations you found in Problem 46.

48. Use the graph in Problem 47 to estimate the number of hours for which both rates are the same.

Find the price that maximizes the profit in Problems 49–52.

49. The demand for a product varies from 200 at a price of $6 each to 800 at a price of $2 each. Also, 800 could be supplied at a price of $7 each, whereas only 200 could be supplied for $1 each.

50. The demand for a product varies from 1,000 at a price of $9 each to 7,000 at a price of $3 each. Also, 6,000 could be supplied at a price of $10 each, whereas only 2,000 could be supplied for $2 each.

51. The demand for a commodity ranges from 5,000 at a one-dollar price, down to 1,000 at a seven-dollar price. It is possible to obtain 2,000 at a two-dollar price but 5,000 can be had at a price of eight dollars each.

52. The demand for a commodity is twenty at $2,000 each, but only five at $8,000 each. It is possible to obtain nine at $2,000 each, but eleven can be available at $10,000 each.

53. By test-marketing its calculators at UCLA, California Instruments, a manufacturer of calculators, finds that it can sell 180 calculators when they are priced at $10, but only 20 calculators when they are priced at $40. On the other hand, it finds that 20 calculators can be supplied at $10 each. If they were supplied at $40 each, overtime shifts could be used to raise the supply to 180 calculators. What is the optimum price for the calculators?

54. A manufacturer of class rings test-marketed a new item at the University of Texas. It found that it could sell 900 items when they are priced at $1, but only 300 items if the price is raised to $7. On the other hand, it found that 600 items can be supplied at $1 each. If they were supplied at $9 each, overtime shifts could be used to raise the supply to 1,000 items. What is the optimum price for the items?

55. Situation Linda is on a business trip and must rent a car for the day. The car rental agency has the following options on the car she wants to rent: Option A: $30 plus 25¢ per mile (instead of $40 and 50¢/mi); Option B: Flat $50 (instead of $60) with unlimited mileage. Which car should she rent?

56. In Your Own Words Write a short paper comparing the SITUATION with your own situation. (That is, list similarities and/or differences.)

11.5 GRAPHING CURVES

Situation

"Well, Billy, can I make it?" asked Evel. "If you can accelerate to the proper speed, and if the wind is not blowing too much, I think you can," answered Billy. "It will be one huge money-maker, but I want some assurance that it can be done!" retorted Evel.

About 20 years ago, a daredevil named Evel Knievel attempted a skycycle ride across the Snake River. In this section, Evel will learn how to plot points to draw curves that are not lines. He will learn about a curve called a *parabola* that traces the path of a projectile fired upward. He will also find that the growth of money (which he, no doubt, earned from this stunt) is modeled by a curve called an *exponential curve.*

GRAPHING CURVES BY PLOTTING POINTS

When cannons were introduced in the 13th century, their primary use was to demoralize the enemy. It was much later that they were used for strategic purposes. In fact, cannons existed nearly three centuries before enough was known about the behavior of projectiles to use them with any accuracy. The cannonball does not travel in a straight line, it was discovered, because of an unseen force that today we know as *gravity.* Consider Figure 11.21. It is a scale drawing (graph) of the path of a cannonball fired in a particular way.

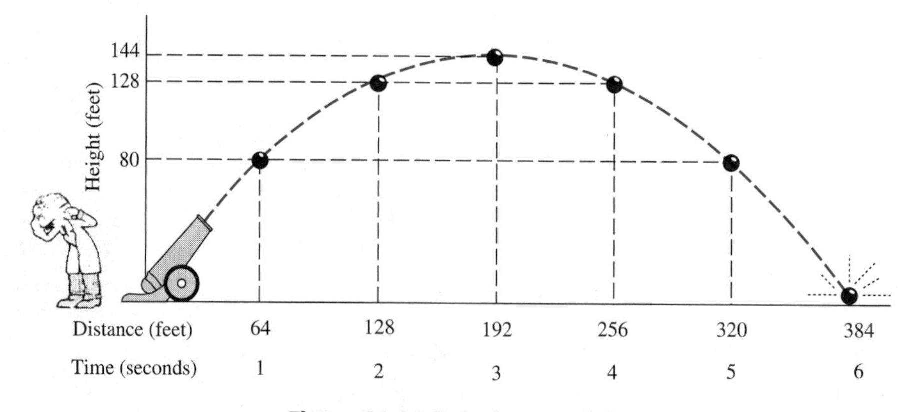

Figure 11.21 Path of a cannonball

The path described by a projectile is called a **parabola.** Any projectile—a ball, an arrow, a bullet, a rock from a slingshot, even water from the nozzle of a hose or sprinkler—will travel a parabolic path. Note that this parabolic curve has a maximum height and is symmetric about a vertical line through that height. In other words, the ascent and descent paths are symmetric.

EXAMPLE 1

Graphing a Nonlinear Curve

Graph $y = x^2$.

Solution We will choose x values and find corresponding y values.

Let $x = \mathbf{0}$:

$y = \mathbf{0}^2 = 0$; plot (0, 0).

Let $x = \mathbf{1}$:

$y = \mathbf{1}^2 = 1$; plot (1, 1).

Let $x = \mathbf{-1}$:

$y = (\mathbf{-1})^2 = 1$; plot $(-1, 1)$.

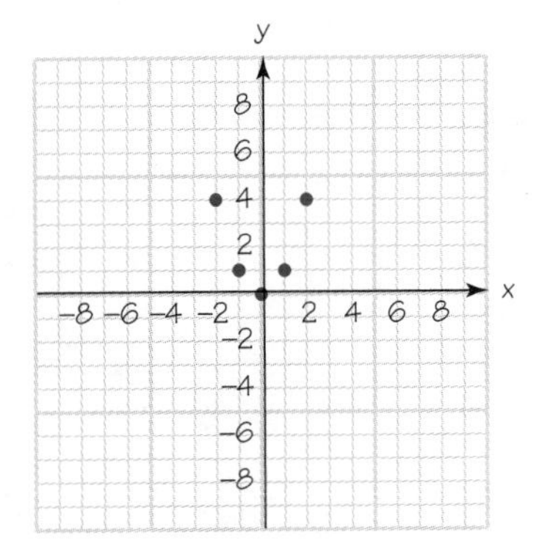

Figure 11.22 Points satisfying the equation $y = x^2$ are not linear

Notice in Figure 11.22 that these points do not fall in a straight line. If we find two more points, we can see the shape of the graph.

Let $x = \mathbf{2}$:

$y = \mathbf{2}^2 = 4$; plot (2, 4).

Let $x = \mathbf{-2}$:

$y = (\mathbf{-2})^2 = 4$; plot $(-2, 4)$.

Connect the points to form a smooth curve, as shown in Figure 11.23.

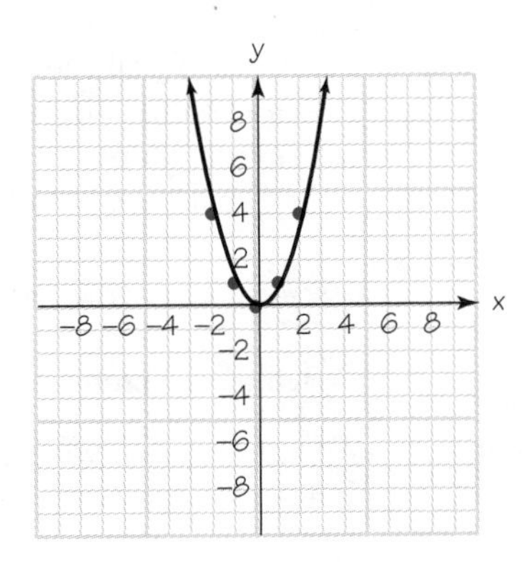

Figure 11.23 Graph of $y = x^2$ ■

The curve shown in Figure 11.23 is a parabola that is said to *open upward.* The lowest point, (0, 0) in Example 1, is called the **vertex.** The following example is a parabola that *opens downward.*

EXAMPLE 2

Graphing a Parabola

Sketch $y = -\frac{1}{2}x^2$.

Solution

Let $x = \mathbf{0}$: $y = -\frac{1}{2}(\mathbf{0})^2 = 0$; plot $(0, 0)$.

Let $x = \mathbf{1}$: $y = -\frac{1}{2}(\mathbf{1})^2 = -\frac{1}{2}$; plot $(1, -\frac{1}{2})$.

Let $x = \mathbf{-1}$: $y = -\frac{1}{2}(\mathbf{-1})^2 = -\frac{1}{2}$; plot $(-1, -\frac{1}{2})$.

Let $x = \mathbf{2}$: $y = -\frac{1}{2}(\mathbf{2})^2 = -2$; plot $(2, -2)$.

Let $x = \mathbf{-2}$: $y = -\frac{1}{2}(\mathbf{-2})^2 = -2$; plot $(-2, -2)$.

Let $x = \mathbf{4}$: $y = -\frac{1}{2}(\mathbf{4})^2 = -8$; plot $(4, -8)$.

Connect these points to form a smooth curve, as shown in Figure 11.24.

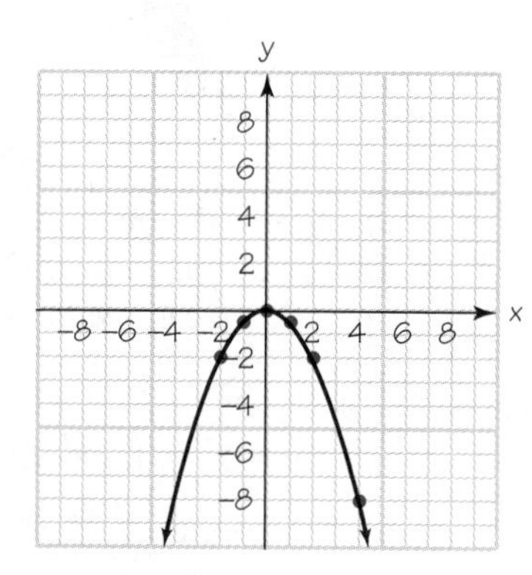

Figure 11.24 Graph of $y = -\frac{1}{2}x^2$ ■

You can sketch many different curves by plotting points. The procedure is to decide whether you should pick x values and find the corresponding y values, or pick y values and find the corresponding x values. Find enough ordered pairs so that you can connect those points with a smooth graph. Many graphs in mathematics are not smooth, but we will not consider those in this course.

EXAMPLE 3

Curve Sketching by Choosing y Values

Sketch the graph of the curve $x = y^2 - 6y + 4$.

Solution

We use the same procedure to find points on the graph, except that in this equation we see that it will be easier to choose y values and find corresponding x values.

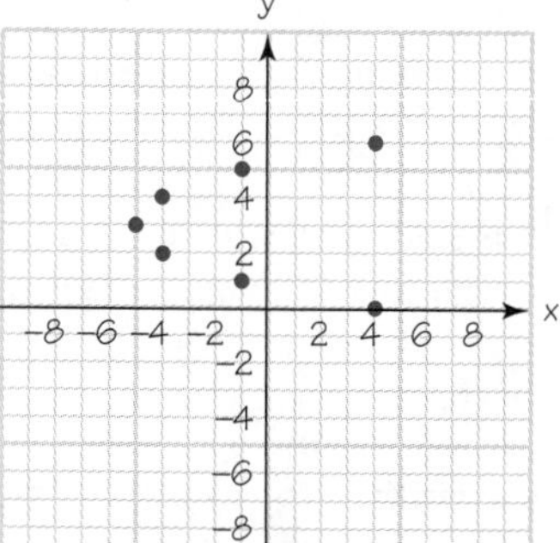

Let $y = \mathbf{0}$: $x = \mathbf{0}^2 - 6 \cdot \mathbf{0} + 4 = 4$; plot $(4, 0)$.
Let $y = \mathbf{1}$: $x = \mathbf{1}^2 - 6 \cdot \mathbf{1} + 4 = 1 - 6 + 4 = -1$; plot $(-1, 1)$.
Let $y = \mathbf{2}$: $x = \mathbf{2}^2 - 6 \cdot \mathbf{2} + 4 = 4 - 12 + 4 = -4$; plot $(-4, 2)$.
Let $y = \mathbf{3}$: $x = \mathbf{3}^2 - 6 \cdot \mathbf{3} + 4 = 9 - 18 + 4 = -5$; plot $(-5, 3)$.
Let $y = \mathbf{4}$: $x = \mathbf{4}^2 - 6 \cdot \mathbf{4} + 4 = 16 - 24 + 4 = -4$; plot $(-4, 4)$.
Let $y = \mathbf{5}$: $x = \mathbf{5}^2 - 6 \cdot \mathbf{5} + 4 = 25 - 30 + 4 = -1$; plot $(-1, 5)$.
Let $y = \mathbf{6}$: $x = \mathbf{6}^2 - 6 \cdot \mathbf{6} + 4 = 36 - 36 + 4 = 4$; plot $(4, 6)$.

How many points should we find? For lines, we had to find two points, with a third point as a check. For curves that are not lines, we do not have a particular number of points to check. We must plot as many points as are necessary to enable us to draw a smooth curve. We connect the points we have plotted to draw the curve shown in Figure 11.25.

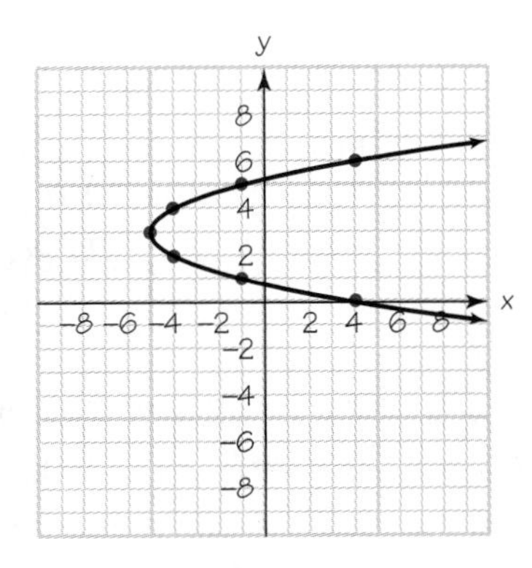

Figure 11.25 Graph of $x = y^2 - 6y + 4$ ■

EXPONENTIAL CURVES

An **exponential equation** is one in which a variable appears as an exponent. Consider the equation $y = 2^x$. This equation represents a doubling process. The graph of such an equation is called an **exponential curve** and is graphed by plotting points.

EXAMPLE 4 **Graphing an Exponential Curve by Plotting Points**

Sketch the graph of $y = 2^x$ for nonnegative values of x.

Solution Choose x values and find corresponding y values. These values form ordered pairs (x, y). Plot enough ordered pairs so that you can see the general shape of the curve, and then connect the points with a smooth curve.

Let $x = \mathbf{0}$: $y = 2^0 = 1$; plot (0, 1).

Let $x = \mathbf{1}$: $y = 2^1 = 2$; plot (1, 2).

Let $x = \mathbf{2}$: $y = 2^2 = 4$; plot (2, 4).

Let $x = \mathbf{3}$: $y = 2^3 = 8$; plot (3, 8).

Let $x = \mathbf{4}$: $y = 2^4 = 16$; plot (4, 16).

Connect the points with a smooth curve, as shown in Figure 11.26.

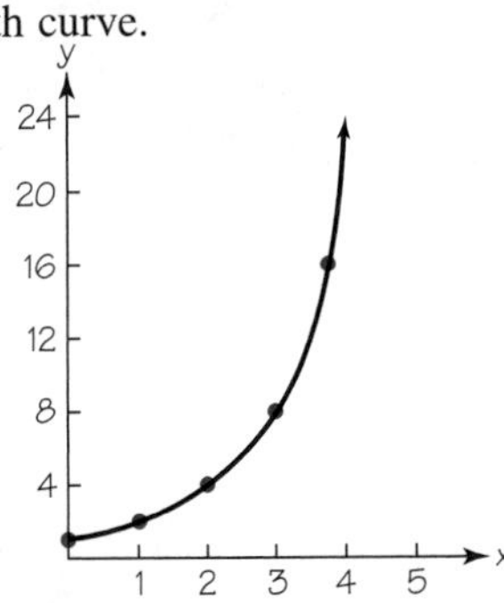

Figure 11.26 Graph of $y = 2^x$ ■

Population growth is described by an exponential equation. The population P at some future date can be predicted if you know the present population, P_0, and the growth rate, r. In more advanced courses, it is shown that the predicted population in t time periods is approximated by the equation $P = P_0(2.72)^{rt}$.

EXAMPLE 5

Problem Solving with an Exponential Equation

Tony calls his local Chamber of Commerce and finds that the growth rate of his town is now 5%. Also, according to the 1990 census, the population is 2,500. Find the growth equation, and draw a graph showing the population between the years 1990 and 2010.

Solution $P_0 = 2{,}500$ and $r = 5\% = 0.05$ Remember, to change a percent to a decimal, move the decimal point two places to the left.

The equation of the graph is $\mathbf{P = 2{,}500(2.72)^{0.05t}}$.

If $t = \mathbf{0}$: $P = 2{,}500(2.72)^{\mathbf{0}} = 2{,}500$ Plot the point (0, 2500). This is the 1990 population; 1990 is called the base year. It is also called the "present time" (even though it is not now 1990). Thus, if $t = 5$, then the population corresponds to the year 1995. If $t = 10$, the population is for the year 2000.

If $t = \mathbf{10}$: $P = 2{,}500(2.72)^{0.05(\mathbf{10})} \approx 4{,}123$

Press: [2500] [×] [2.72] [^] [(] [.05] [×] [10] [)] [=]
Display: 4123.105626

This means that the predicted population in the year 2000 is 4,123. Plot the point (10, 4123).

If $t = \mathbf{20}$: $P = 2{,}500(2.72)^{0.05(\mathbf{20})} = 6800$

Press: [2500] [×] [2.72] [^] [(] [.05] [×] [20] [)] [=]
Display: 6800 Plot the point (20, 6800).

We have plotted these points in Figure 11.27. ■

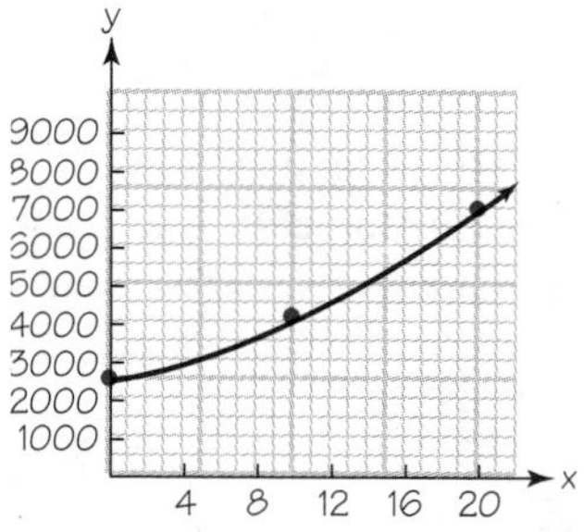

Figure 11.27 Graph showing population P from 1990 to 2010 (base year, $t = 0$, is 1990)

One of the first applications in Part II and one of the most important concepts for intelligent functioning in today's world is *interest.* We can now return to this idea and look at graphs showing the growth of money as the result of interest accumulation. Simple interest is represented by a line, whereas compound interest requires an exponential graph. This means that, for short periods of time, there is very little difference between simple and compound interest, but as time increases, the differences become significant.

EXAMPLE 6 **Graphs Containing Simple and Compound Interest**

Suppose that you deposit \$100 at 10% interest. Graph the total amounts you will have if you invest your money at simple interest and if you invest your money at interest compounded annually. Here are the appropriate formulas:

Simple interest: $A = P(1 + rt)$

Compound interest: $A = P(1 + r)^t$

Solution For this example, $P = 100$ and $r = 0.10$; the variables are t and A. We begin by finding some ordered pairs. A calculator is necessary for calculating compound interest.

YEAR	*SIMPLE INTEREST*	*COMPOUND INTEREST*
$t = \mathbf{0}$	$A = 100(1 + 0.1 \cdot \mathbf{0}) = 100$ Point (0, 100)	$A = 100(1 + 0.1)^{\mathbf{0}} = 100$ Point (0, 100)
$t = \mathbf{1}$	$A = 100(1 + 0.1 \cdot \mathbf{1}) = 110$ Point (1, 110)	$A = 100(1 + 0.1)^{\mathbf{1}} = 110$ Point (1, 110)
$t = \mathbf{10}$	$A = 100(1 + 0.1 \cdot \mathbf{10}) = 200$ Point (10, 200)	$A = 100(1 + 0.1)^{\mathbf{10}} \approx 259$ Point (10, 259)
$t = \mathbf{20}$	$A = 100(1 + 0.1 \cdot \mathbf{20}) = 300$ Point (20, 300)	$A = 100(1 + 0.1)^{\mathbf{20}} \approx 673$ Point (20, 673)

The graphs through these points are shown in Figure 11.28.

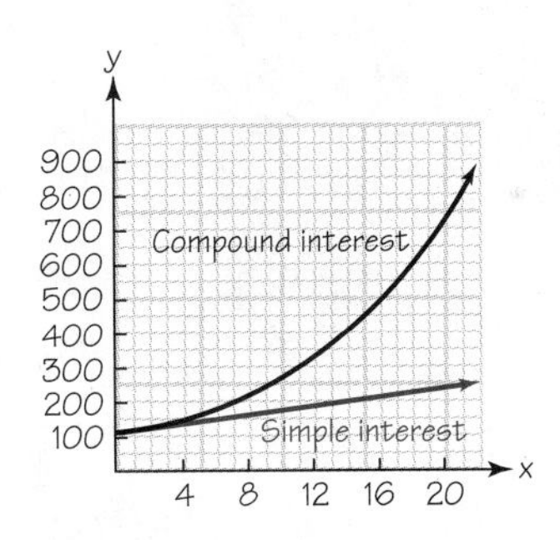

Figure 11.28 Comparison of simple interest and compound interest

■

PROBLEM SET 11.5

ESSENTIAL IDEAS

1. What is a parabola?

2. What is an exponential graph?

LEVEL 1

WHAT IS WRONG, *if anything, with each of the statements in Problems 3–5? Explain your reasoning.*

3. In graphing $y = 2x^2$, we need to plot three points:
If $x = \mathbf{0}$, then $y = 2(\mathbf{0})^2 = 0$; plot $(0, 0)$.
If $x = \mathbf{1}$, then $y = 2(\mathbf{1})^2 = 2$; plot $(1, 2)$.
If $x = \mathbf{-2}$, then $y = 2(\mathbf{-2})^2 = 8$; plot $(-2, 8)$.

Complete the graph by drawing the line passing through these three points.

4. In graphing $x = y^2$, it is easier to pick y values and to find the corresponding x values.

If $y = \mathbf{0}$, then $x = \mathbf{0}^2 = 0$; plot $(0, 0)$.
If $y = \mathbf{1}$, then $x = \mathbf{1}^2 = 1$; plot $(1, 1)$.
If $y = \mathbf{-1}$, then $x = (\mathbf{-1})^2 = 1$; plot $(1, -1)$.
If $y = \mathbf{2}$, then $x = \mathbf{2}^2 = 4$; plot $(4, 2)$.
If $y = \mathbf{3}$, then $x = \mathbf{3}^2 = 9$; plot $(9, 3)$.

Complete the graph by drawing the curve passing through these points. This curve is a parabola.

5. In graphing $y = x^3$, plot points to draw the graph shown in Figure 11.29.

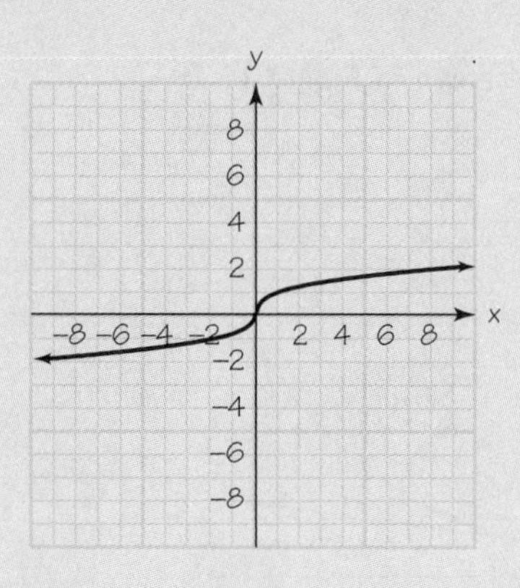

Figure 11.29 Graph of $y = x^3$

Level 2—Drill

Sketch the graph of each equation in Problems 6–32.

6. $y = 3x^2$ **7.** $y = 2x^2$ **8.** $y = 10x^2$

9. $y = -x^2$ **10.** $y = -2x^2$ **11.** $y = -3x^2$

12. $y = -5x^2$ **13.** $y = 5x^2$ **14.** $y = \frac{1}{2}x^2$

15. $x = 2y^2$ **16.** $x = -3y^2$ **17.** $x = \frac{1}{2}y^2$

18. $y = 2x^3$ **19.** $y = -x^3$ **20.** $x = -y^3$

21. $y = \frac{1}{3}x^2$ **22.** $y = -\frac{2}{3}x^2$ **23.** $y = \frac{1}{10}x^2$

24. $y = -\frac{1}{10}x^2$

25. $y = \frac{1}{1,000}x^2$

26. $y = x^2 - 4$

27. $y = x^2 + 4$

28. $y = 9 - x^2$

29. $y = -3x^2 + 4$

30. $y = 2x^2 - 3$

31. $y = x^2 - 2x + 1$

32. $y = x^2 + 4x + 4$

Sketch the graphs of the equations in Problems 33–44 *for non-negative values of x.*

33. $y = 3^x$

34. $y = 4^x$

35. $y = 5^x$

36. $y = -2^x$

37. $y = -6^x$

38. $y = -7^x$

39. $y = 10^x$

40. $y = 100 - 2^x$

41. $y = (\frac{1}{2})^x$

42. $y = (\frac{1}{3})^x$

43. $y = (\frac{1}{10})^x$

44. $y = -(\frac{1}{10})^x$

Draw the graphs in Problems 45–50.

45. $y = -2x^2 + 4x - 2$

46. $y = \frac{1}{4}x^2 - \frac{1}{2}x + \frac{1}{4}$

47. $y = \frac{1}{2}x^2 + x + \frac{1}{2}$

48. $y = x^2 - 2x + 3$

49. $x = 3y^2 + 12y + 14$

50. $x = 4y - y^2 - 4$

51. Change the growth rate in Example 5 to 6%, and graph the population curve.

52. Change the growth rate in Example 5 to 3.5%, and graph the population curve.

53. Change the growth rate in Example 5 to 1.5%, and graph the population curve.

54. Rework Example 6 for a 12% interest rate.

55. Rework Example 6 for an 8% interest rate.

56. Rework Example 6 for an 18% interest rate.

Level 2 — Applications

57. **a.** Graph $h = -\frac{144}{9}(t - 3)^2 + 144$.

b. Does the parabola in part **a** open upward or downward?

c. If we relate the graph in part **a** to the cannonball in Figure 11.21, can t be negative?

d. Draw the graph of the parabola in part **a** for $0 \leq t \leq 6$.

58. **a.** Graph $d = 16t^2$.
b. Does the parabola in part **a** open upward or downward?
c. If we relate the graph in part **a** to the falling object in the *B.C.* cartoon in Section 11.2, can t be negative?
d. Draw the graph of the parabola in part **a** for $0 \leq t \leq 16$.

59. In archaeology, carbon-14 dating is a standard method of determining the age of certain artifacts. Decay rate can be measured by half-life, which is the time required for one-half of a substance to decompose. Carbon-14 has a half-life of approximately 5,600 years. If 100 grams of carbon were present originally, then the amount A of carbon present today is given by the formula

$$A = 100(\tfrac{1}{2})^{t/5,600}$$

where t is the time since the artifact was alive until today. Graph this relationship by letting $t = 5{,}600$, $t = 11{,}200$, . . . , and adjust the scale accordingly.

60. **In Your Own Words** Predict the population of your city or state for the year 2000.

61. **Situation** "Well, Billy, can I make it?" asked Evel. "If you can accelerate to the proper speed, and if the wind is not blowing too much, I think you can," answered Billy. "It will be one huge moneymaker, but I want some assurance that it can be done!" retorted Evel. About 20 years ago, a daredevil named Evel Knievel attempted a skycycle ride across the Snake River (see Figure 11.30).

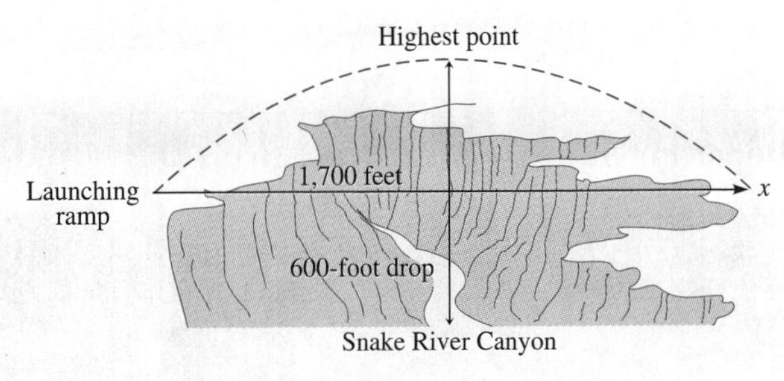

Figure 11.30 Snake River Canyon

Suppose that the path of Evel Knievel's skycycle is

$$y = -0.0005x^2 + 2.39x$$

assuming that the ramp is at the origin and x is the horizontal distance traveled. Graph this relationship. Using your graph, answer Evel's question: Will he make it? Assume that the actual distance the skycycle must travel is 4,700 ft.

62. **In Your Own Words** Write a short paper comparing the SITUATION withyour own situation. (That is, we don't expect that you are planning a skycycle ride; rather, how could you use a graph to assist you in dealing with something that might happen in your life?)

11.6 SUMMARY AND REVIEW

Important Terms

Axes [11.1]
Boundary [11.4]
Cartesian coordinates [11.1]
Closed half-plane [11.4]
Demand [11.4]
Equilibrium point [11.4]
Exponential curve [11.5]
Exponential equation [11.5]
First component [11.1]
Function [11.2]
Functional notation [11.2]
Graph [11.1, 11.3]
Half-plane [11.4]
Horizontal line [11.3]
Line [11.3]
Open half-plane [11.4]
Ordered pair [11.1]
Origin [11.1]
Parabola [11.5]
Plot a point [11.1]
Quadrants [11.1]
Rectangular coordinates [11.1]
Satisfy [11.3]
Second component [11.1]
Simultaneous solution [11.4]
Solution [11.3]
Supply [11.4]
System of equations [11.4]
Test point [11.4]
Vertex [11.5]
Vertical line [11.3]
x-axis [11.1]
y-axis [11.1]

Chapter Objectives

The material in this chapter is reviewed in the following list of objectives. A self-test (with answers and suggestions for additional study) is given. The self-test is constructed so that each problem number corresponds to a related objective. This self-test is followed by a practice test with the questions in mixed order.

[11.1] *Objective 11.1* Plot points on a coordinate system.
[11.2] *Objective 11.2* Decide whether a given set is a function.
[11.2] *Objective 11.3* Find the values of a function.
[11.3] *Objective 11.4* Graph a first-degree equation in two unknowns.
[11.4] *Objective 11.5* Solve a system of equations graphically.
[11.4] *Objective 11.6* Graph a first-degree inequality in two unknowns.
[11.5] *Objective 11.7* Sketch the graph of an equation with two variables that is not linear.
[11.5] *Objective 11.8* Sketch the graph of an exponential equation.
[11.1–11.5] *Objective 11.9* Solve applied problems.

Self-Test

Each question of this self-test is related to the corresponding objective listed above.

1. Plot the points (25, 100), (−50, 50), and (−125, −75).

2. Is {(4, 3), (5, −2), (6, 3)} a function? Why or why not?

3. If $f(x) = 10 - 3x$, find
a. $f(0)$ **b.** $f(10)$ **c.** $f(-4)$

4. Graph $5x - y = 15$.

5. Find the simultaneous solution for the system $\begin{cases} 2x - y = -1 \\ x - 3y = 7 \end{cases}$

6. Graph $x < 3y$.

7. Graph $y = 1 - x^2$.

8. Graph $y = -2^x$ for nonnegative values of x.

9. Graph the earnings for the next 50 years of $50 invested at a simple interest rate of 10%. Use the formula $A = P(1 + rt)$.

STUDY HINTS *Compare your solutions and answers to the self-test. For each problem you missed, work some additional problems in the section listed in the margin. After you have worked these problems, you can test yourself with the practice test.*

Complete Solutions to the Self-Test

Additional Problems

[11.1] *Problems 1–36*

1.

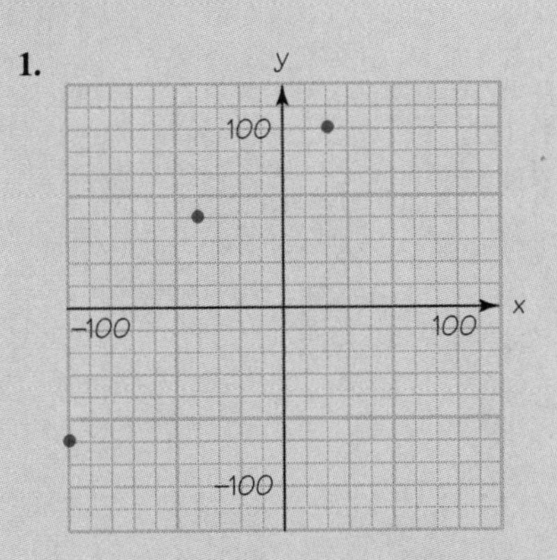

[11.2] *Problems 9–20*

2. Yes, each x value is associated with exactly one y value.

[11.2] *Problems 21–39*

3. a. $f(0) = 10 - 3(0) = 10$
b. $f(10) = 10 - 3(10) = 10 - 30 = -20$
c. $f(-4) = 10 - 3(-4) = 10 + 12 = 22$

[11.3] *Problems 3–14; 21–53*

4.

x	y
0	−15
1	−10
3	0

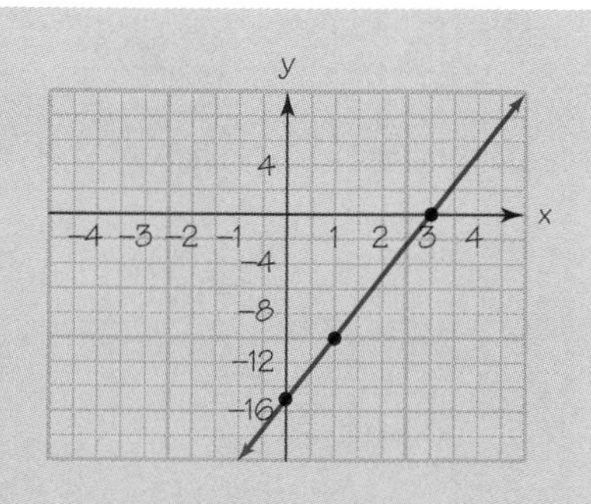

[11.4] *Problems 13–27*

5. *FIRST LINE:*

x	y
0	1
1	3
−1	−1

SECOND LINE:

x	y
−2	−3
7	0
4	−1

The point of intersection is (−2, −3).

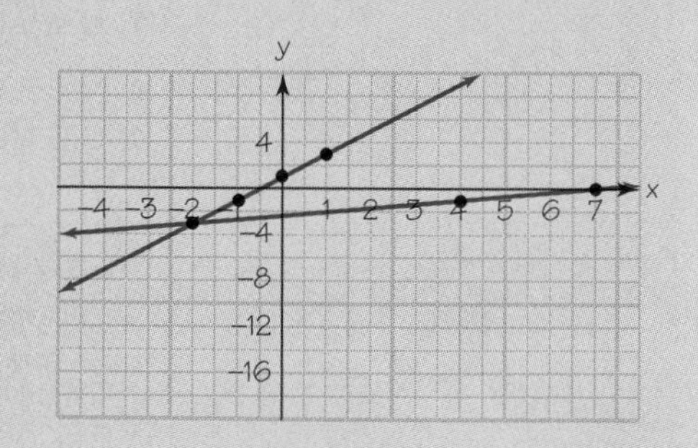

[11.4] *Problems 28–42*

6.

x	y
0	0
3	1
6	2

Test point: (0, 3)

$x < 3y$

$0 < 3(3)$ is true

Shade the half-plane on the same side as the test point.

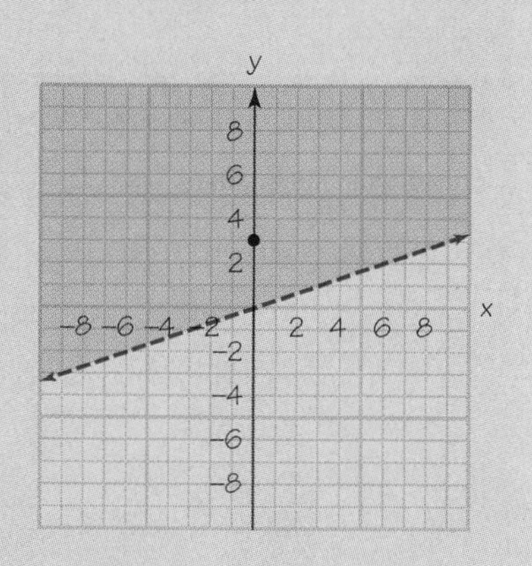

[11.5] *Problems 6–56*

7.

x	y
0	1
1	0
−1	0
2	−3
3	−8

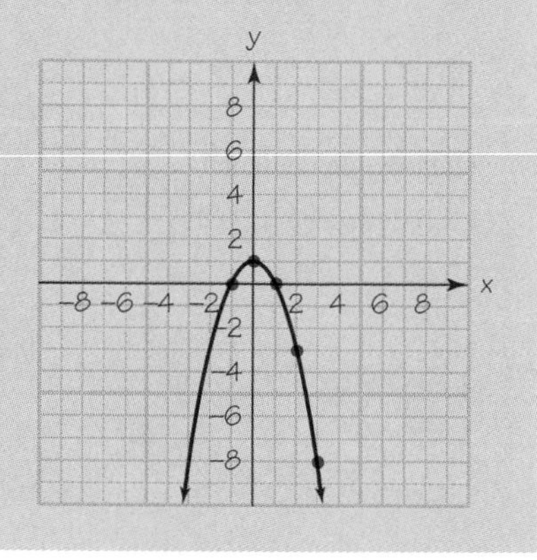

[11.5] *Problems 33–44; 51–56*

8. 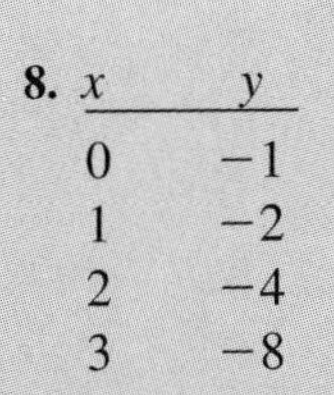

x	y
0	−1
1	−2
2	−4
3	−8

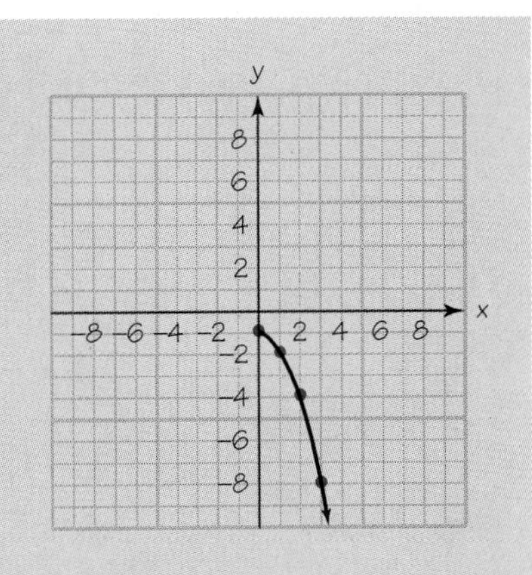

[11.1] *Problems 35–36*
[11.2] *Problems 33–39*
[11.3] *Problem 54*
[11.4] *Problems 43–55*
[11.5] *Problems 57–61*

9. $A = P(1 + rt) = 50(1 + 0.1t)$

t	A
0	50
10	100
30	200
50	300

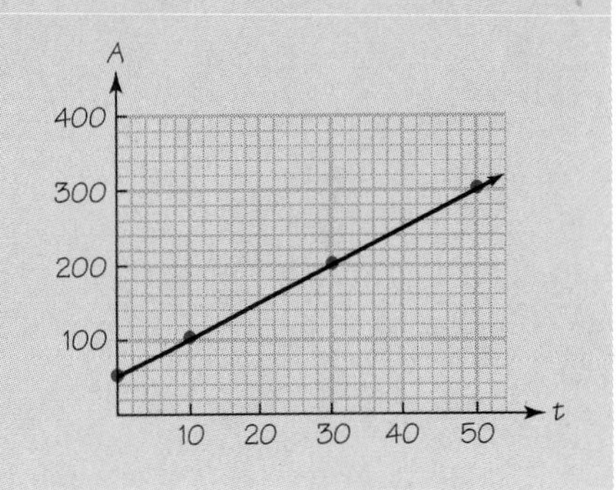

Chapter 11 Practice Test

1. Plot five points so that the first and second components are the same. Draw the line passing through the plotted points.

2. Plot five points so that the first component is 5. Draw the line passing through the plotted points.

3. Decide whether each set is a function.
a. $\{(4, 19), (19, 4), (18, 19)\}$
b. $\{(12, 6), (12, 8), (12, 10)\}$
c. $\{(1, 2), (2, 4), (3, 9), (4, 16), (5, 25)\}$
d. $\{(1, 5, 9, 10\}$

4. Find the value of each function.
a. $f(x) = 3x + 2$; find $f(6)$.
b. $g(x) = x^2 - 3$; find $g(0)$.
c. $F(x) = 5x + 25$; find $F(10)$.
d. $m(x) = 5$; find $m(10)$.

Graph the equations in Problems 5–12.

5. $y = 3x - 2$

6. $3x + y = 0$

7. $x + 2y + 10 = 0$

8. $y = 25$

9. $y = 4x^2$

10. $y = 5 - x^2$

11. $3y^2 + x = 0$

12. $y = x^2 + 6x + 11$

Graph the equations in Problems 13–16 *for nonnegative values of x.*

13. $y = -3^x$

14. $y = (\frac{1}{4})^x$

15. $y = 2^x - 10$

16. $y = 10 - 2^x$

Graph the inequalities in Problems 17–20.

17. $y > 2x - 1$

18. $y > 3x + 2$

19. $y \geq 2x - 3$

20. $y \leq 3x + 4$

Solve the systems of equations in Problems 21–24.

21. $\begin{cases} x + y = 5 \\ 2x - y = 1 \end{cases}$

22. $\begin{cases} 2x + y = 5 \\ x - y = 4 \end{cases}$

23. $\begin{cases} 2x + y = 1 \\ x - y = 5 \end{cases}$

24. $\begin{cases} x + y = 1 \\ 2x - y = 5 \end{cases}$

25. If a cannonball is fired upward with an initial velocity of 128 feet per second, its height can be calculated according to the formula

$$y = 128t - 16t^2$$

where t is the length of time (in seconds) after the cannonball is fired. Sketch this equation by letting $t = 0, 1, 2, \ldots, 7, 8$. Connect these points with part of a parabola.

26. Steve can rent a rototiller for $8 per day or $40 per week. He wants to rent the tiller for one week or less.

a. Write the equation for the cost if he rents it by the day.

b. Write the equation for the cost if he rents it by the week.

c. Graph the equations for parts **a** and **b** on the same axes.

d. At what number of days are both rates the same?

27. **a.** Graph the earnings for the next 40 years of $100 invested at a simple interest rate of 12%. Use the formula

$$A = P(1 + rt)$$

b. Graph the earnings for the next 40 years of $100 invested at 12% interest compounded annually. Use the formula

$$A = P(1 + r)^t$$

28. Draw a population curve for a city whose growth rate is 1.3% and whose present population is 53,000. The appropriate equation is $P = P_0(2.72)^{rt}$. Let $t = 0$, 10, and 20 to help you find points for graphing this curve.

APPENDIX A

HINDU–ARABIC NUMERATION SYSTEM

Historical Perspective

From *Margarita Philophica Nova*, 1512, Courtesy of the Museum of the History of Science, University of Oxford. This picture depicts a contest between one person computing with a form of abacus and another with Hindu–Arabic numerals.

The numeration system in common use today (the one we have been calling the decimal system) has ten symbols—namely, 0, 1, 2, 3, 4, 5, 6, 7, 8, and 9. The selection of ten digits was no doubt a result of our having ten fingers (digits).

The symbols originated in India about 300 B.C. However, because the early specimens do not contain a zero or use a positional system, this numeration system offered no advantage over other systems then in use in India.

The date of the invention of the zero symbol is not known. The symbol did not originate in India but probably came from the late Babylonian period via the Greek world.

By the year 750 A.D., the zero symbol and the idea of a positional system had been brought to Baghdad and translated into Arabic. We are not certain how these numerals were introduced into Europe, but it is likely that they came via Spain in the 8th century. Gerbert, who later became Pope Sylvester II, studied in Spain and was the first European scholar known to have taught these numerals. Because of the origins, these numerals are called the **Hindu–Arabic numerals.** Since ten basic symbols are used, the Hindu–Arabic numeration system is also called the *decimal numeration system,* from the Latin word *decem,* meaning "ten."

Although we now know that the decimal system is very efficient, its introduction met with considerable controversy. Two opposing factions, the "algorists" and the "abacists," arose. Those favoring the Hindu–Arabic system were called algorists, because the symbols were introduced in Europe in a book called (in Latin) *Liber Algorismi de Numero Indorum,* by the Arab mathematician al-Khowârizmî. The word *algorismi* is the origin of our word *algorithm.* The abacists favored the status quo—using Roman numerals and doing arithmetic on the abacus. The battle between the abacists and the algorists lasted for 400 years. The Roman Catholic Church exerted great influence in commerce, science, and theology. The church criticized those using the "heathen" Hindu–Arabic numerals and consequently kept the world using Roman numerals until 1500. Roman numerals were easy to write and learn, and addition and subtraction with them were easier than with the "new" Hindu–Arabic numerals. It seems incredible that our decimal system has been in general use only since about 1500.

Let's examine the Hindu–Arabic or decimal numeration system:

1. It uses ten symbols, called digits.
2. Large numbers are expressed in terms of powers of 10.
3. It is positional.

Consider how we count objects:

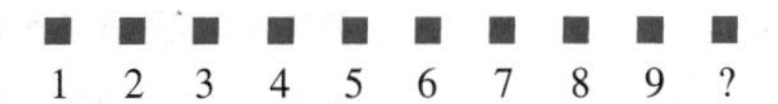

At this point we could invent another symbol as the Egyptians did (you might suggest 0, but remember that 0 represents no objects), or we could reuse the digit symbols by repeating them or by altering their positions. We agree to use 10 to mean 1 group of

The symbol 0 was invented as a placeholder to show that the 1 here is in a different position from the 1 representing ■. We continue to count:

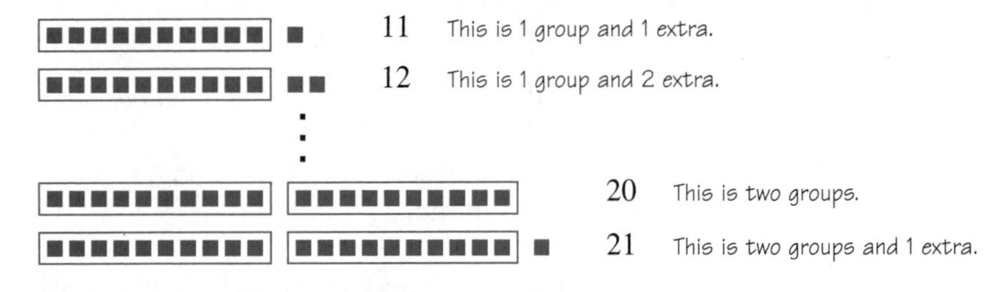

We continue in the same fashion until we have 9 groups and 9 extra. What's next? It is 10 groups or a group of groups:

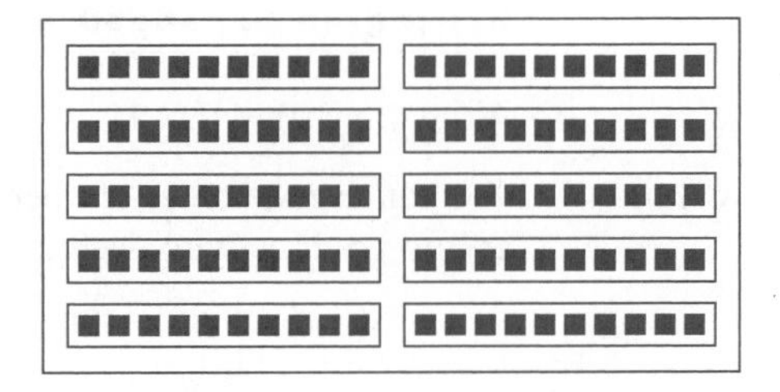

We call this group of groups a $10 \cdot 10$ or 10^2 or a **hundred.** We again use position and repeat the symbol 1 with still different meaning: 100.

EXAMPLE 1 **Meaning of a Number Given in Decimal Form**

What does 134 mean?

Solution 134 means that we have 1 group of 100, 3 groups of 10, and 4 extra. In symbols,

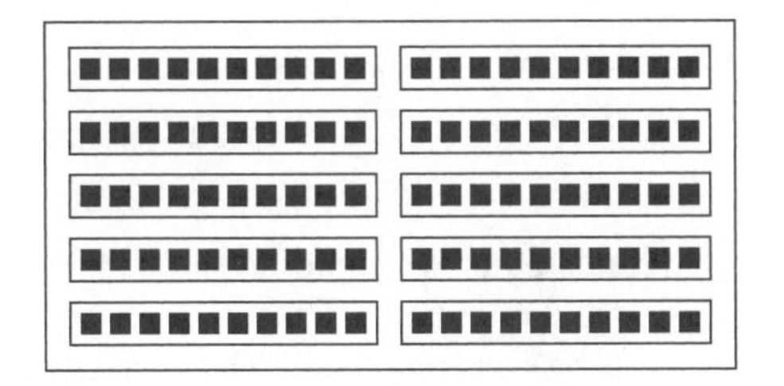

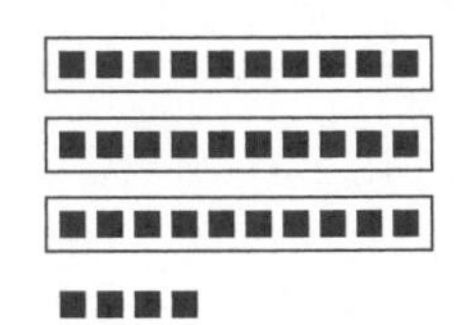

We denote this more simply by writing:

These represent the number in each group.

$$(\mathbf{1} \times \underbrace{10^2}) + (\mathbf{3} \times \underbrace{10}) + 4$$

These are the names of the groups.

This leads us to the meaning *one hundred, three tens, four ones.* ■

The representation, or the meaning, of the number 134 in Example 1 is called **expanded notation.**

EXAMPLE 2 **Writing a Decimal Numeral in Expanded Notation**

Write 52,613 in expanded form.

Solution

$$\begin{aligned} 52{,}613 &= 50{,}000 + 2{,}000 + 600 + 10 + 3 \\ &= 5 \times 10^4 + 2 \times 10^3 + 6 \times 10^2 + 1 \times 10 + 3 \end{aligned}$$

■

EXAMPLE 3 **Writing an Expanded Numeral in Decimal Notation**

Write $4 \times 10^8 + 9 \times 10^7 + 6 \times 10^4 + 3 \times 10 + 7$ in decimal form.

Solution You can use the order of operations and multiply out the digits, but you should be able to go directly to decimal form if you remember what place value means:

49 00600 37 = **490,060,037**

Notice that there were no powers of 10^6, 10^5, 10^3, or 10^2. ■

A period, called a **decimal point** in the decimal system, is used to separate the fractional parts from the whole parts. The positions to the right of the decimal point are fractions:

$$\frac{1}{10} = 10^{-1}, \quad \frac{1}{100} = 10^{-2}, \quad \frac{1}{1{,}000} = 10^{-3}$$

To complete the pattern, we also sometimes write $10 = 10^1$ and $1 = 10^0$.

EXAMPLE 4 **Writing a Decimal Numeral with Fractional Parts in Expanded Form**

Write 479.352 using expanded notation.

Solution

$$\begin{aligned} 479.352 &= 400 + 70 + 9 + 0.3 + 0.05 + 0.002 \\ &= 400 + 70 + 9 + \frac{3}{10} + \frac{5}{100} + \frac{2}{1{,}000} \\ &= \mathbf{4 \times 10^2 + 7 \times 10^1 + 9 \times 10^0 + 3 \times 10^{-1} + 5 \times 10^{-2} + 2 \times 10^{-3}} \end{aligned}$$ ■

PROBLEM SET A

Essential Ideas

1. **In Your Own Words** What is expanded notation?

2. **In Your Own Words** What is a group of groups in:
   a. base 10 **b.** base 8

3. **In Your Own Words** What do we mean by a decimal numeration system?

Level 1—Drill

Explain each of the concepts or procedures in Problems 4–6.

4. Illustrate the meaning of 123.

5. Illustrate the meaning of 145.

6. Illustrate the meaning of 1,234 by showing the appropriate groupings.

Give the meaning of the numeral 5 in each of the numbers in Problems 7–10.

7. 805 **8.** 508 **9.** 0.00567 **10.** 58,000,000

Write the numbers in Problems 11–24 in decimal notation.

11. a. 10^5 **b.** 10^3 **12. a.** 10^6 **b.** 10^4

13. b. 10^{-4} **b.** 10^{-3} **14. a.** 10^{-2} **b.** 10^{-6}

15. a. 5×10^3 **b.** 5×10^2 **16. a.** 8×10^{-4} **b.** 7×10^{-3}

17. a. 6×10^{-2} **b.** 9×10^{-5} **18. a.** 5×10^{-6} **b.** 2×10^{-9}

19. a. $1 \times 10^4 + 0 \times 10^3 + 2 \times 10^2 + 3 \times 10^1 + 4 \times 10^0$

b. $6 \times 10^1 + 5 \times 10^0 + 0 \times 10^{-1} + 8 \times 10^{-2} + 9 \times 10^{-3}$

20. **a.** $5 \times 10^5 + 2 \times 10^4 + 1 \times 10^3 + 6 \times 10^2 + 5 \times 10^1 + 8 \times 10^0$

b. $6 \times 10^7 + 4 \times 10^3 + 1 \times 10^0$

21. **a.** $7 \times 10^6 + 3 \times 10^{-2}$ **b.** $6 \times 10^9 + 2 \times 10^{-3}$

22. $5 \times 10^5 + 4 \times 10^2 + 5 \times 10^1 + 7 \times 10^0 + 3 \times 10^{-1} + 4 \times 10^{-2}$

23. $3 \times 10^3 + 2 \times 10^1 + 8 \times 10^0 + 5 \times 10^{-1} + 4 \times 10^{-2} + 2 \times 10^{-4}$

24. $2 \times 10^4 + 6 \times 10^2 + 4 \times 10^{-1} + 7 \times 10^{-3} + 6 \times 10^{-4} + 9 \times 10^{-5}$

Write each of the numbers in Problems 25–32 in expanded notation.

25. **a.** 741 **b.** 728,407

26. **a.** 0.096421 **b.** 27.572

27. **a.** 47.00215 **b.** 521

28. **a.** 6,245 **b.** 2,305,681

29. **a.** 428.31 **b.** 5,245.5

30. **a.** 0.00000527 **b.** 100,000.001

31. **a.** 893.0001 **b.** 8.00005

32. **a.** 678,000.01 **b.** 57,285.9361

APPENDIX B

DIFFERENT NUMERATION SYSTEMS

Historical Perspective

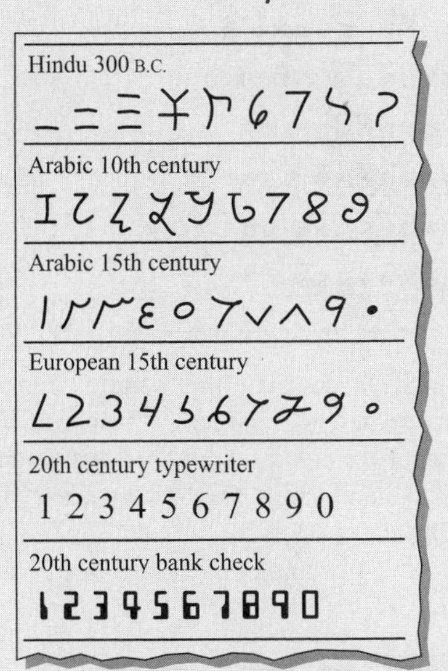

The symbols that we use in the Hindu–Arabic numeration system have changed considerably over the centuries. We have recently seen the digits in bar codes, which can easily be read by computer.

You might ask, "When would we ever use a base other than base 10?" We can cite one obvious example. When photographs are sent back from space, they are sent using a binary numeration system. The images are not usually recorded on photographic film, but instead the image is broken up into tiny dots, called *pixels*. For example, a photograph might be divided into 1,000 pixels horizontally and 500 vertically. Each pixel is then assigned a number representing its brightness: 0 for pure white and 63 for pure black. These numbers are sent back as six-digit binary numbers 000000 to 111111. A computer translates these digits into a photograph. The photo shown here is the scarred face of Triton, Neptune's largest moon. This image was sent by Voyager 2 on August 29, 1989.

In the previous appendix, we discussed the Hindu–Arabic numeration system and grouping by tens. However, we could group by twos, fives, twelves, or any other counting number. In this appendix, we summarize numeration systems with bases other than ten. This not only will help you understand our own numeration system, but will give you insight into the numeration systems used with computers, namely, base 2 (**binary**), base 8 (**octal**), and base 16 (**hexadecimal**).

NUMBER OF SYMBOLS

The number of symbols used in a particular base depends on the method of grouping for that base. For example, in base ten the grouping is by tens, and in base five the grouping is by fives. Suppose we wish to count ■■■■■■■■■■■ in various bases. Let's look for patterns in Table B.1. Note the use of the subscript following the numeral to keep track of the base in which we are working.

Table B.1 *Grouping in Various Bases*

Base	Symbols	Method of grouping	Notation
two	0, 1		1011_{two}
three	0, 1, 2		102_{three}
four	0, 1, 2, 3		23_{four}
five	0, 1, 2, 3, 4		21_{five}
six	0, 1, 2, 3, 4, 5		15_{six}
seven	0, 1, 2, 3, 4, 5, 6		14_{seven}
eight	0, 1, 2, 3, 4, 5, 6, 7		13_{eight}
nine	0, 1, 2, 3, 4, 5, 6, 7, 8		12_{nine}
ten	0, 1, 2, 3, 4, 5, 6, 7, 8, 9		11_{ten}

Do you see any patterns? Suppose we wish to continue this pattern. Can we group by elevens or twelves? We can, provided new symbols are "invented." For base eleven (or higher bases), we use the symbol T to represent ■■■■■■■■■■. For base twelve (or higher bases), we use E to stand for ■■■■■■■■■■■. For bases larger than twelve, other symbols can be invented.

For example, $2T_{twelve}$ means that there are two groupings of twelve and T (ten) extra:

We continue with the pattern from Table B.1 by continuing beyond base ten in Table B.2.

Table B.2 *Grouping in Various Bases*

Base	Symbols	Method of grouping	Notation
eleven	0, 1, 2, . . . , 8, 9, T		10_{eleven}
twelve	0, 1, 2, . . . , 8, 9, T, E		E_{twelve}
thirteen	0, 1, 2, . . . , 8, 9, T, E, U		$E_{thirteen}$
fourteen	0, 1, 2, . . . , 9, T, E, U, V		$E_{fourteen}$

Do you see more patterns? Can you determine the number of symbols in each base system?

CHANGE FROM SOME BASE TO BASE TEN

To change from base b to base ten, we write the numerals in expanded notation. The resulting number is in base ten.

EXAMPLE 1 **Changing into Base Ten**

Change each number to base ten.

a. 1011.01_{two} **b.** 1011.01_{four} **c.** 1011.01_{five}

Solution

a. $1011.01_{two} = 1 \times 2^3 + 0 \times 2^2 + 1 \times 2^1 + 1 \times 2^0 + 0 \times 2^{-1} + 1 \times 2^{-2}$
$= 8 + 0 + 2 + 1 + 0 + 0.25 = \mathbf{11.25}$

b. $1011.01_{four} = 1 \times 4^3 + 0 \times 4^2 + 1 \times 4^1 + 1 \times 4^0 + 0 \times 4^{-1} + 1 \times 4^{-2}$
$= 64 + 0 + 4 + 1 + 0 + 0.0625 = \mathbf{69.0625}$

c. $1011.01_{five} = 1 \times 5^3 + 0 \times 5^2 + 1 \times 5^1 + 1 \times 5^0 + 0 \times 5^{-1} + 1 \times 5^{-2}$
$= 125 + 0 + 5 + 1 + 0 + 0.04 = \mathbf{131.04}$ ■

CHANGE FROM BASE TEN TO SOME BASE

To see how to change from base ten to any other valid base, let's again look for a pattern:

To change from base ten to base two, group by twos.

To change from base ten to base three, group by threes.

To change from base ten to base four, group by fours.

To change from base ten to base five, group by fives.

⋮

The groupings from this pattern are summarized in Table B.3.

Table B.3 *Place-value chart*

Base	Place value					
2	$2^5 = 32$	$2^4 = 16$	$2^3 = 8$	$2^2 = 4$	$2^1 = 2$	$2^0 = 1$
3	$3^5 = 243$	$3^4 = 81$	$3^3 = 27$	$3^2 = 9$	$3^1 = 3$	$3^0 = 1$
4	$4^5 = 1{,}024$	$4^4 = 256$	$4^3 = 64$	$4^2 = 16$	$4^1 = 4$	$4^0 = 1$
5	$5^5 = 3{,}125$	$5^4 = 625$	$5^3 = 125$	$5^2 = 25$	$5^1 = 5$	$5^0 = 1$
7	$7^5 = 16{,}807$	$7^4 = 2{,}401$	$7^3 = 343$	$7^2 = 49$	$7^1 = 7$	$7^0 = 1$
8	$8^5 = 32{,}768$	$8^4 = 4{,}096$	$8^3 = 512$	$8^2 = 64$	$8^1 = 8$	$8^0 = 1$
10	$10^5 = 100{,}000$	$10^4 = 10{,}000$	$10^3 = 1{,}000$	$10^2 = 100$	$10^1 = 10$	$10^0 = 1$
12	$12^5 = 248{,}832$	$12^4 = 20{,}736$	$12^3 = 1{,}728$	$12^2 = 144$	$12^1 = 12$	$12^0 = 1$

The next example shows how we can interpret this grouping process in terms of a simple division.

EXAMPLE 2

Changing to Base Two

Convert 42 to base two.

Solution Using Table B.3, we see that the largest power of two smaller than 42 is 2^5, so we begin with $2^5 = 32$:

$$42 = 1 \times 2^5 + 10$$
$$10 = 0 \times 2^4 + 10$$
$$10 = 1 \times 2^3 + 2$$
$$2 = 0 \times 2^2 + 2$$
$$2 = 1 \times 2^1 + 0$$
$$0 = 0 \times 2^0$$

We could now write out 42 in expanded notation:

$$42 = 1 \times 2^5 + 0 \times 2^4 + 1 \times 2^3 + 0 \times 2^2 + 1 \times 2^1 + 0 \times 2^0 = 101010_{two}$$

Instead of carrying out the steps by using Table B.3, we will begin with 42 and carry out repeated division, saving each remainder as we go. We are changing to base 2, so we do repeated division by 2:

$$2\overline{)42}\ \text{with quotient } \mathbf{21} \quad \text{r. } 0 \leftarrow \textit{Save remainder.}$$

Next, we need to divide 21 by 2, but instead of rewriting our work we work our way up:

$$\begin{array}{rl} \mathbf{10} & \text{r.}1 \leftarrow \text{Save all remainders.} \\ \mathbf{2}\overline{)21} & \text{r.}0 \leftarrow \text{Save remainder.} \\ 2\overline{)42} & \end{array}$$

Continue by doing repeated division.

Stop when you get a zero here.
↓

$$\begin{array}{rl} 0 & \text{r. } 1 \\ 2\overline{)1} & \text{r. } 0 \\ 2\overline{)2} & \text{r. } 1 \\ 2\overline{)5} & \text{r. } 0 \\ 2\overline{)10} & \text{r. } 1 \\ 2\overline{)21} & \text{r. } 0 \\ 2\overline{)42} & \end{array}$$

Answer is found by reading down. ↓

Thus, $\mathbf{42 = 101010_{two}}$.

You can check by using expanded notation:

$$101010_{two} = 1 \times 2^5 + 1 \times 2^3 + 1 \times 2 = 32 + 8 + 2 = 42$$

■

EXAMPLE 3

Changing from Base 10 to Other Bases

Write 42 in

a. base three **b.** base four

Solution **a.** Begin with $3^3 = 27$ (from Table B.3):

$$\begin{aligned} 42 &= 1 \times 3^3 + 15 \\ 15 &= 1 \times 3^2 + 6 \\ 6 &= 2 \times 3^1 + 0 \\ 0 &= 0 \times 3^0 \end{aligned} \quad \text{or} \quad \begin{array}{rl} 0 & \text{r.}1 \\ 3\overline{)1} & \text{r.}1 \\ 3\overline{)4} & \text{r.}2 \\ 3\overline{)14} & \text{r.}0 \\ 3\overline{)42} & \end{array}$$

Thus, **$42 = 1120_{three}$**.

b. Begin with $4^2 = 16$ (from Table B.3):

$$\begin{aligned} 42 &= 2 \times 4^2 + 10 \\ 10 &= 2 \times 4^1 + 2 \\ 2 &= 2 \times 4^0 \end{aligned} \quad \text{or} \quad \begin{array}{rl} 0 & \text{r. }2 \\ 4\overline{)2} & \text{r. }2 \\ 4\overline{)10} & \text{r. }2 \\ 4\overline{)42} & \end{array}$$

Thus, **$42 = 222_{four}$**. ■

EXAMPLE 4

Applied Number Base Problem

Suppose you need to purchase 1,000 name tags and can buy them by the gross (144), the dozen (12), or individually. The name tags cost \$0.50 each, \$4.80 per dozen, and \$50.40 per gross. How should you order to minimize the cost?

Solution If you purchase 1,000 tags individually, the cost is $\$0.50 \times 1{,}000 = \$500$. This is not the least cost possible, because of the bulk discounts. We must find the maximum number of gross, the number of dozens, and then purchase the remainder individually. We will proceed by repeated division by 12, which we recognize as equivalent to changing the number to base twelve. Change 1,000 to base 12:

$$\begin{array}{rl} 0 & \text{r. }6 \\ 12\overline{)\quad 6} & \text{r. }11\text{, or } E \text{ in base twelve} \\ 12\overline{)\quad 83} & \text{r. }4 \\ 12\overline{)1{,}000} & \end{array}$$

Thus, $1{,}000 = 6E4_{twelve}$ so **you must purchase 6 gross, 11 dozen, and 4 individual tags.** Let's check: The cost is $6 \times \$50.40 + 11 \times \$4.80 + 4 \times \$0.50 = \$357.20$. As you can see, this is considerably less expensive than purchasing the individual name tags. ■

PROBLEM SET B

Essential Ideas

1. **In Your Own Words** Explain the process of changing from base eight to base ten.
2. **In Your Own Words** Explain the process of changing from base sixteen to base ten.
3. **In Your Own Words** Explain the process of changing from base ten to base eight.
4. **In Your Own Words** Explain the process of changing from base ten to base sixteen.

Level 1—Drill

5. Count the number of people in the indicated base.

a. base ten **b.** base five **c.** base thirteen
d. base eight **e.** base two

In Problems 6–10, write the numbers in expanded notation.

6. **a.** 643_{eight} **b.** 5387.9_{twelve}
7. **a.** 110111.1001_{two} **b.** 5411.1023_{six}
8. **a.** 64200051_{eight} **b.** 1021.221_{three}
9. **a.** 323000.2_{four} **b.** 234000_{five}
10. **a.** 3.40231_{five} **b.** 2033.1_{four}

Change the numbers in Problems 11–17 to base ten.

11. **a.** 527_{eight} **b.** 527_{twelve}
12. **a.** $25TE_{twelve}$ **b.** 1101.11_{two}
13. **a.** 431_{five} **b.** 65_{eight}
14. **a.** 1011.101_{two} **b.** 11101000110_{two}
15. **a.** 573_{twelve} **b.** 4312_{eight}
16. **a.** 2110_{three} **b.** 4312_{five}
17. **a.** 537.1_{eight} **b.** 3721_{eight}
18. Change 628 to base four.
19. Change 724 to base five.
20. Change 427 to base twelve.
21. Change 256 to base two.
22. Change 615 to base eight.
23. Change 412 to base five.
24. Change 615 to base two.
25. Change 5,133 to base twelve.
26. Change 795 to base seven.
27. Change 512 to base two.
28. Change 4,731 to base twelve.
29. Change 52 to base three.
30. Change 76 to base four.
31. Change 602 to base eight.

Level 2—Applications

Use number bases to answer the questions given in Problems 32–44.

32. Change 52 days to weeks and days.
33. Change 158 hours to days and hours.
34. Change 55 inches to feet and inches.
35. Change 39 ounces to pounds and ounces.

36. Change 500 to gross, dozens, and units.

37. Change \$4.59 to quarters, nickels, and pennies.

38. Using only quarters, nickels, and pennies, what is the minimum number of coins needed to make \$0.84?

39. Suppose you have two quarters, four nickels, and two pennies. Use base five to write a numeral to indicate your financial status.

40. A bookstore ordered 9 gross, 5 dozen, and 4 pencils. Write this number in base twelve and in base ten.

41. Change \$8.34 to the smallest number of coins consisting of quarters, nickels, and pennies.

42. Change 44 days to weeks and days.

43. Change 54 months to years and months.

44. Change 29 hours to days and hours.

45. In Your Own Words The *duodecimal numeration system* refers to the base twelve system, which uses the symbols 0, 1, 2, 3, 4, 5, 6, 7, 8, 9, *T*, *E*. Historically, a numeration system based on 12 is not new. There were 12 tribes in Israel and 12 Apostles of Christ. In Babylon, 12 was used as a base for the numeration system before it was replaced by 60. In the 18th and 19th centuries, Charles XII of Sweden and Georg Buffon (1707–1788) advocated the adoption of the base twelve system. Even today there is a Duodecimal Society of America that advocates the adoption of this system. According to the Society's literature, no one "who thought long enough—three to 17 minutes—to grasp the central idea of the duodecimal system ever failed to concede its superiority." Study the duodecimal system from 3 to 17 minutes, and comment on whether you agree with the Society's statement.

46. In Your Own Words Discuss the binary (base two) numeration system.

APPENDIX C

GLOSSARY

Absolute value [2.2] The absolute value of a number is the distance of that number from the origin. Symbolically,

$$|n| = \begin{cases} n & \text{if } n \geq 0 \\ -n & \text{if } n < 0 \end{cases}$$

Accuracy [6.1] One speaks of an *accurate statement* in the sense that it is true and correct, or of an accurate computation in the sense that it contains no numerical error. Accurate to a certain decimal place means that all digits preceding and including the given one are correct.

Acre [6.3] A unit commonly used in the United States system for measuring land. It contains 43,560 ft^2.

Acute angle [5.2] An angle whose measure is smaller than a right angle

Addition of integers [2.2] If the integers to be added have the same sign, the answer will also have that same sign and will have a magnitude equal to the sum of the absolute values of the given integers. If the integers to be added have opposite signs, the answer will have the sign of the integer with the larger absolute value, and will have a magnitude equal to the difference of the absolute values. Finally, if one or both of the given integers is 0, use the property $n + 0 = n$ for any integer n.

Addition law of exponents [3.1] To multiply two numbers with like bases, add the exponents. That is, $b^m \cdot b^n = b^{m+n}$.

Addition property of equations [3.4] The solution of an equation is unchanged by adding the same number to both sides of the equation.

Addition property of inequality [3.7] The solution of an inequality is unchanged if you add the same number to both sides of the inequality.

Add-on interest [7.3] It is a method of calculating interest and installments on a loan. The amount of interest is calculated according to the formula $I = Prt$ and is then added to the amount of the loan. This sum divided by the number of payments is the amount of monthly payment.

Adjacent side [5.4] In a right triangle, an acute angle is made up of two sides, one called the *hypotenuse* (the side opposite the right angle) and the other side called the adjacent side.

Adjusted balance method [7.4] A method of calculating credit card interest using the formula $I = Prt$ in which P is the balance owed after the current payment is subtracted

Algebra [2.1] A generalization of arithmetic that uses letters, or variables, to denote numbers

Algebraic expression Any meaningful combination of numbers, variables, and signs of operation

Amortization The process of paying off a debt by systematically making partial payments until the debt (principal) and interest are repaid

Amortization schedule A table showing the schedule of payments of a loan detailing the amount of each payment that goes to repay the principal and how much goes to pay interest

Amortized loan [7.3] A loan that is fully paid off with the last periodic payment

Angle [5.2] Two rays or segments with a common endpoint

Annual compounding [7.5] In the compound interest formula, it is when $n = 1$.

Annual percentage rate [7.3] The percentage rate charged on a loan based on the actual amount owed and the actual time it is owed. The approximation formula for annual percentage rate (APR) is

$$\text{APR} = \frac{2Nr}{N+1}$$

APR [7.3; 7.6] Abbreviation for annual percentage rate. See *Compound interest.*

Area [6.3] A number describing the two-dimensional content of a set. Specifically, it is the number of square units enclosed in a plane figure.
Area formulas Square, s^2; rectangle, ℓw; circle, πr^2; trapezoid, $\frac{1}{2}h(b_1 + b_2)$
Associative property A property of grouping that applies to addition and multiplication, but not to subtraction or division: If a, b, and c are real numbers, then

$$(a + b) + c = a + (b + c) \text{ and } (ab)c = a(bc)$$

Average [2.5; 10.2] A single number that is used to typify or represent a set of numbers. In this book, it refers to the *mean, median,* or *mode.*
Average daily balance method [7.4] A method of calculating credit card interest using the formula $I = Prt$ in which P is the average daily balance owed for a current month, and t is the number of days in the month divided by 365
Axes [11.1] The intersecting lines of a Cartesian coordinate system. The horizontal axis is called the x-axis, and the vertical axis is called the y-axis. The axes divide the plane into four parts called *quadrants.*
Axiom [5.1; 8.6] A statement that is accepted without proof

Balloon payment [7.3] A single larger payment made at the end of the time period of an installment loan that is not amortized
Bar graph [10.1] See *Graph.*
Base [4.4] The whole quantity in a percent problem
Base of an exponential [1.5] In $y = b^x$, the *base* is b $(b \neq 1)$.
Bell-shaped curve [10.5] See *Normal curve.*
Belong to a set [8.1] To be an element of a set
Bimodal [10.2] A data set that contains two modes
Binomial [3.1] A polynomial with exactly two terms
Bisect To divide into two equal or congruent parts
Bond An interest-bearing certificate issued by a government or business, promising to pay the holder a specified amount (usually $1,000) on a certain date
Boundary [11.4] See *Half-plane.*
Box plot [10.4] A rectangular box positioned above a numerical scale, drawn from Q_1 (the first quartile) to Q_3 (the third quartile). The median (Q_2) is shown as a dashed line and a segment is extended to show both the maximum and minimum values.
Braces See *Grouping symbols.*
Brackets See *Grouping symbols.*

Canceling [1.6] The process of reducing a fraction by dividing the same number into both the numerator and the denominator
Capacity [6.4] A measurement for the amount of liquid a container holds
Cardinal number [8.1] A number that designates the manyness of a set; the number of units, but not the order in which they are arranged
Cardinality [8.1] The number of elements in a set
Cartesian coordinate system [11.1] Two intersecting lines, called *axes*, used to locate points in a plane. Ordered pairs used to locate points in this coordinate system are called *Cartesian coordinates.* If the intersecting lines are perpendicular, the system is called a *rectangular coordinate system.*
Cartesian plane See *Cartesian coordinate system.*
Celsius [6.5] A metric measurement for temperature for which the freezing point of water is 0° and the boiling point of water is 100°
Center See *Circle.*
Centi- [6.1; 6.6] A prefix that means 1/100
Centimeter [6.1; 6.6] One hundredth of a meter
Circle [6.2; 6.3] The set of points in a plane that are a given distance from a given point. The given point is called the *center,* and the given distance is called the *radius.* The diameter is twice the radius. The *unit circle* is the circle with center at (0, 0) and $r = 1$.
Circle graph [10.1] See *Graph.*
Circular cone [6.5] A cone with a circular base
Circular cylinder [6.5] A cylinder with a circular base
Circular definition [8.1] A definition that relies on the use of the word being defined, or other words that rely on the word being defined
Circumference [6.2] The distance around a circle
Classes [10.1] One of the groupings when organizing data. The difference between the lower limit of one class and the lower limit of the next class is called the *interval* of the class. The number of values within a class is called the *frequency.*
Closed-end loan [7.3] An installment loan

Closed half-plane [11.4] See *Half-plane.*

Closing [7.6] The process of settlement on a real estate loan

Closing costs [7.6] Costs paid at the closing of a real estate loan

Coefficient [3.1] Any factor of a term is said to be the coefficient of the remaining factors. Generally, the word *coefficient* is taken to be the numerical coefficient of the variable factors.

Common denominator [1.7] For two or more fractions, a common multiple of the denominators

Common factor A factor that two or more terms of a polynomial have in common

Common fraction [1.3; 1.6] Fractions written in the form of one integer divided by a whole number are common fractions. For example, 1/10 is common fraction representation and 0.1 is the decimal representation of the same number.

Commutative property A property of order that applies to addition and multiplication, but not to subtraction and division. If a and b are real numbers, then

$$a + b = b + a \text{ and } ab = ba$$

Comparison property [3.7] For any two numbers x and y, exactly one of the following is true: 1. $x = y$; x is equal to y (the same as) 2. $x > y$; x is greater than y (bigger than) 3. $x < y$; x is less than y (smaller than). This is sometimes known as the *trichotomy* property.

Comparison rate for home loans [7.6] A formula for comparing terms of a home loan. The formula is

$$\text{APR} + 0.125\left(\text{POINTS} + \frac{\text{ORIGINATION FEE}}{\text{AMOUNT OF LOAN}}\right)$$

Compass [5.1] An instrument for describing circles or for measuring distances between two points

Complement [7.1; 8.3] (1) Two numbers less than 1 are called complements if their sum is 1. (2) The complement of a set is everything not in the set relative to a given universe.

Complementary probabilities [9.2]

$$P(E) = 1 - P(\overline{E})$$

Complex decimal [1.6] A form that mixes decimal and fractional form, such as $0.12\frac{1}{2}$

Complex fraction A rational expression a/b where a or b (or both) have fractional form

Components See *Ordered pair.*

Composite [1.5] A number that has two or more prime factors

Compound interest [7.5] A method of calculating interest by adding the interest to the principal at the end of the compounding period so that this sum is used in the interest calculation for the next period

Compound interest formula [7.5]

$A = P(1 + i)^N$, where A = future value;

P = present value (or principal);

r = annual interest rate (APR);

t = number of years;

n = number of times compounded per year;

$i = \dfrac{r}{n}$; and $N = nt$.

Compounded See *Interest.*

Compounded annually [7.5] $n = 1$ in the compound interest formula

Conclusion [8.6] The statement that follows (or is to be proved to follow) as a consequence of the hypothesis of the theorem

Conditional equation [3.4] See *Equation.*

Conditional inequality See *Inequality.*

Conditional probability [9.3] A probability that is found on the condition that a certain event has occurred. The notation $P(E \mid F)$ is the probability of event E *on the condition* that event F has occurred.

Cone [6.5] A solid bounded by a region for its base in a plane and a surface formed by the line segments that join points on the base to a fixed point (called the *vertex*) not in the plane of the base.

Congruent [5.1] Of the same size and shape; if one is placed on top of the other, the two figures will coincide exactly in all their parts.

Congruent triangles [5.3] Two triangles with the same size and shape

Conjecture [8.6] A guess or prediction based on incomplete or uncertain evidence

Consecutive numbers Counting numbers that differ by 1

Consistent system If a system of equations has at least one solution, it is consistent; otherwise it is said to be *inconsistent.*

Constant Symbol with exactly one possible value

Construction [5.1] The process of drawing a figure that will satisfy certain given conditions

Contained in a set [8.1] An element is contained in a set if it is a member of the set.

Contradiction An equation for which the solution set is empty

Coordinate A numerical description for a point. Also see *Ordered pair.*

Coordinate plane See *Cartesian coordinate system.*

Corresponding angles [5.4] Angles in different triangles that are similarly related to the rest of the triangle

Corresponding parts [5.3] Points, angles, lines, etc., in different figures, similarly related to the rest of the figures

Corresponding sides [5.4] Sides of different triangles that are similarly related to the rest of the triangle

Cosine [5.4] In a right triangle ABC with right angle C, $\cos A =$ (adjacent side of A)/hypotenuse.

Counterclockwise In the direction of rotation opposite to that in which the hands move around the dial of a clock

Counterexample [8.6] An example used to disprove a proposition

Counting numbers [1.2] See *Natural numbers.*

Credit card [7.3; 7.4] A card signifying that the person or business issued the card has been approved open-ended credit. It can be used at certain restaurants, airlines, and stores accepting that card.

Cube [1.5; 6.4] (1) A solid with six equal square sides. (2) In an expression such as x^3, which is pronounced "x cubed," it means xxx.

Cube root See *Root of a number.*

Cubic centimeter [6.4] A cube with all sides equal to 1 cm

Cubic inch [6.4] A cube with all sides equal to 1 in.

Cup [6.4] A standard unit of capacity equal to 8 oz; abbreviated as c

Cursor A mark indicating the location on a computer screen or a calculator display

Cylinder [6.5] A closed surface consisting of two simple closed curves in parallel planes and a *lateral surface* that is the union of all line segments joining corresponding points on the given closed curves.

Daily compounding [7.5] In the compound interest formula, it is when $n = 365$ (exact interest) or when $n = 360$ (ordinary interest). In this book, use ordinary interest unless otherwise indicated.

Dealer's cost [7.3] The actual amount that a dealer pays for the goods sold

Decagon [5.2] A polygon having ten sides

Deci- [6.1; 6.6] A prefix that means 1/10

Deciles [10.3] Nine values that divide the data into ten parts

Decimal [1.3] Any number written in decimal notation. Sometimes called a Hindu–Arabic numeral.

Decimal fraction [1.3] A number in decimal notation that has fractional parts, such as 23.25. If a common fraction p/q is written as a decimal fraction, the result will be either a *terminating decimal* as with

$\frac{1}{4} = 0.25$ or a *repeating decimal* as with
$\frac{2}{3} = 0.6666. . . .$

Deductive reasoning [8.6] A formal structure based on a set of axioms and a set of undefined terms. New terms are defined in terms of the given undefined terms and new statements, or *theorems*, are derived from the axioms by proof.

Degree [3.1; 5.2] (1) The degree of a term in one variable is the exponent of the variable, or it is the sum of the exponents of the variables if there are more than one. The degree of a polynomial is the degree of its highest-degree term. (2) A unit of measurement of an angle that is equal to 1/360 of a revolution

Deka- [6.1; 6.6] A prefix that means 10

Deleted point A single point that is excluded from the domain

Demand [11.4] The number of items that can be sold at a given price

De Morgan's laws [8.4] For sets X and Y,

$$\overline{X \cup Y} = \overline{X} \cap \overline{Y} \text{ and } \overline{X \cap Y} = \overline{X} \cup \overline{Y}$$

Denominator [1.3] See *Rational number.*

Dependent system If *every* ordered pair satisfying one equation in a system of equations also satisfies every other equation of the given system, then we describe the system as dependent.

Dependent variable The variable associated with the second component of an ordered pair

Derived equation See *Equation.*

Description method [8.1] A method of defining a set by describing the set (as opposed to listing its elements)

Diameter [6.2] See *Circle.*

Difference [1.2; 2.1] The result of a subtraction

Dimension A configuration having length only is said to be of one dimension; area and not volume, two dimensions; volume, three dimensions.

Discount [7.1] A reduction from a usual or list price

Disjoint sets [8.2] Sets that have no elements in common

Distributive law (for multiplication over addition) [1.2; 3.2] If a, b, and c are real numbers, then $a(b + c) = ab + ac$ and $(a + b)c = ac + bc$ for the basic operations. That is, the number outside the parentheses indicating a sum or difference is distributed to each of the numbers inside the parentheses.

Dividend [2.1] The number or quantity to be divided. In a/b, the dividend is a.

Division $a/b = x$ is $a \div b = x$ and means $a = bx$.

Division by zero [1.3; 2.5] For $a \div b$ we insist $b \neq 0$ because if $b = 0$, then $bx = 0$, regardless of the value of x, and therefore could not equal a nonzero number a. On the other hand, if $a = 0$, then $0/0 = 1$ checks from the definition, and so also does $0/0 = 2$, which means that $1 = 2$, another contradiction. Thus, division by 0 is not ever possible.

Division of integers [2.5] The quotient of two integers is the quotient of the absolute values, and is positive if the given integers have the same sign, and negative if the given integers have opposite signs. Furthermore, division by zero is not possible and division into 0 gives the answer 0.

Division property of equations [3.4] The solution of an equation is unchanged by dividing both sides of the equation by the same nonzero number.

Division property of inequality See *Multiplication property of inequality.*

Dodecagon [5.2] A polygon with 12 sides

Divisor [2.1] The quantity by which the dividend is to be divided. In a/b, b is the divisor.

Domain [2.1] The *domain* of a variable is the set of replacements for the variable. The *domain* of a graph of an equation with two variables x and y is the set of permissible real-number replacements for x.

Double negative $-(-a) = a$

Down payment [7.6] An amount paid at the time a product is financed. The purchase price minus the down payment is equal to the amount financed.

Element [8.1] One of the individual objects that belong to a set

Elementary operations [1.2] Refers to the operations of addition, subtraction, multiplication, and division

Empty set [8.1] See *Set.*

Equal [2.1; 2.2] Two numbers are equal if they represent the same quantity or are identical. In mathematics, this refers to a relationship that satisfies the axioms of equality.

Equal angles [5.2] Two angles that have the same measure

Equal sets [8.1] Sets that contain the same elements

Equality, axioms of For $a, b, c \in \mathbb{R}$,

- **reflexive:** $a = a$
- **substitution:** If $a = b$, then a may be replaced throughout by b (or b by a) in any statement without changing the truth or falsity of the statement.
- **symmetric:** If $a = b$, then $b = a$.
- **transitive:** If $a = b$ and $b = c$, then $a = c$.

Equally likely outcomes [9.1] Outcomes whose probabilities of occurring are the same

Equation [3.4] A statement of equality. If always true, an equation is called an *identity;* if always false, it is called a *contradiction.* If it is sometimes true and sometimes false, it is called a *conditional equation.* Values that make an equation true are said to *satisfy* the equation and are called *solutions* or *roots* of the equation. Equations with the same solutions are called *equivalent equations.*

Equation of a graph Every point on the graph has coordinates that satisfy the equation, and every ordered pair that satisfies the equation has coordinates that lie on the graph.

Equilateral triangle [6.2] A triangle whose three sides all have the same length

Equilibrium point [11.4] A point for which the supply and demand are equal

Equivalent equations See *Equation.*

Equivalent sets [8.1] Sets that have the same cardinality

Equivalent systems Systems that have the same solution set

Estimation [1.2] An approximation (usually mental) of size or value used to form an opinion

Euclidean geometry [5.1] The study of geometry based on the assumptions of Euclid. These basic assumptions are called Euclid's postulates.

Euclid's postulates [5.1] The five postulates are: 1. A straight line can be drawn from any point to any other point. 2. A straight line extends infinitely far in either direction. 3. A circle can be described with any point as center and with a radius equal to any finite straight line drawn from the center. 4. All right angles are equal to each other. 5. Given a straight line and any point not on this line, there is one and only one line through that point that is parallel to the given line.

Evaluate [2.1] To *evaluate* an expression means to replace the variables by given numerical values and then simplify the resulting numerical expression. To *evaluate* a trigonometric ratio means to find its approximate numerical value.

Event [9.1] A subset of a sample space.

Exact interest [7.2] The calculation of interest assuming that there are 365 days in a year

Expectation [9.4] See *Mathematical expectation.*

Expected value [9.4] See *Mathematical expectation.*

Experiment [9.1] An observation of any physical occurrence

Exponent [1.5] Where b is any nonzero real number and n is any natural number, exponent is defined as follows:

$$b^n = \underbrace{b \cdot b \cdot \cdots \cdot b}_{n \text{ factors}} \qquad b^0 = 1 \qquad b^{-n} = \frac{1}{b^n}$$

b is called the *base*, n is called the *exponent*, and b^n is called a *power*.

Exponential curve [11.5] The graph of an exponential equation. It indicates an increasingly steep rise, and passes through the point (0, 1).

Exponential equation [11.5] An equation of the form $y = b^x$ where b is positive and not equal to 1

Exponential notation [1.5] A notation involving exponents

Expression Numbers, variables, and operations involving numbers and variables

Extended order of operations [1.7] 1. First, perform any operations enclosed in parentheses. 2. Next, perform any operations that involve raising to a power. 3. Perform multiplication and division, reading from left to right. 4. Do addition and subtraction, reading from left to right.

Exterior angle [5.3] An exterior angle of a triangle is the angle on the other side of an extension of one side of the triangle.

Extraneous root A number obtained in the process of solving an equation that is not a root of the equation to be solved

Extremes [4.1] See *Proportion.*

Factor [1.5; 2.1] Each of the numbers multiplied to form a product is called a factor of the product.

Factoring [1.5] The process of determining the factors of a product

Factorization [1.5] The result of factoring a number or an expression

Fahrenheit [6.5] A unit of measurement in the United States system for measuring temperature where the freezing point of water is 32° and the boiling point of water is 212°.

Fair coin [9.1] A coin for which heads and tails are equally likely

Fair game [9.4] A game for which the mathematical expectation is zero

Finite set [8.1] See *Set.*

First component [11.1] See *Ordered pair.*

Five-percent offer [7.3] An offer made that is 105% of the price paid by the dealer. That is, it is an offer that is 5% over the cost.

Foot [6.1] A unit of linear measure in the United States system that is equal to 12 inches

Foreclose [7.6] If the scheduled payments are not made, the lender takes the right to redeem the mortgage and keeps the collateral property.

Formula [3.6] A general answer, rule, or principle stated in mathematical notation

Fraction [1.3] See *Rational number.*

Frequency [10.1] See *Classes.*

Frequency distribution [10.1] For a collection of data, the tabulation of the number of elements in each class

Function [11.2] A rule that assigns to each element in the domain a single (unique) element

Functional notation [11.2] The representation of a function f using the notation $f(x)$ to denote the output value for f when x is the input value

Fundamental counting principle [9.2] If one task can be performed in m ways and a second task can be performed in n ways, then the number of ways that the tasks can be performed one after the other is mn.

Fundamental property of equations If P and Q are algebraic expressions, and k is a real number, then each of the following is equivalent to $P = Q$:

Addition	$P + k = Q + k$
Subtraction	$P - k = Q - k$
Nonzero multiplication	$kP = kQ,\ k \neq 0$
Nonzero division	$\frac{P}{k} = \frac{Q}{k},\ k \neq 0$

Fundamental property of fractions [1.6] If both the numerator and denominator are multiplied or divided by the same nonzero number, the resulting fraction will be the same.

Fundamental property of inequalities If P and Q are algebraic expressions, and k is a real number, then each of the following is equivalent to $P < Q$:

Addition	$P + k < Q + k$
Subtraction	$P - k < Q - k$
Positive multiplication	$kP < kQ,\ k > 0$
Positive division	$\frac{P}{k} < \frac{Q}{k},\ k > 0$
Negative multiplication	$kP > kQ,\ k < 0$
Negative division	$\frac{P}{k} > \frac{Q}{k},\ k < 0$

This property also applies for $\leq$, $>$, and $\geq$.

Fundamental property of rational expressions

$$\frac{PK}{QK} = \frac{P}{Q}, \quad Q, K \neq 0$$

Future value [7.2] See *Compound interest.*

Future value formula [7.2, 7.5] For simple interest: $A = P(1 + rt)$; for compound interest: $A = P(1 + i)^N$

Gallon [6.4] A measure of capacity in the United States system that is equal to 4 quarts or 231 cubic inches

Googol [1.5] The number with 1 followed by 100 zeros—that is, 10,000,000,000,000,000,000,000,000,000,000,-000,000,000,000,000,000,000,000,000,000,000,000,000,-000,000,000,000,000,000,000,000,000,000,000,000,000

Grace period [7.4] A period of time between when an item is purchased and when it is paid for during which no interest is charged

Gram [6.5, 6.6] A unit of weight in the metric system. It is equal to the weight of one cubic centimeter of water at 4°C.

Graph [10.1; 11.1; 11.3] (1) In statistics, it is a drawing that shows the relation between certain sets of numbers. Common forms are *bar graphs, line graphs, pictographs,* and *pie charts (circle graphs).* (2) A drawing that shows the relation between certain sets of numbers. It may be one-dimensional ($\mathbb{R}$), two-dimensional ($\mathbb{R}^2$), or three-dimensional ($\mathbb{R}^3$).

Graph of an equation See *Equation of a graph.*

Greater than [2.2] If a lies to the right of b on a number line, then a is greater than b, $a > b$. Formally, $a > b$ if and only if $a - b$ is positive.

Greater than or equal to Written $a \geq b$, means $a > b$ or $a = b$

Grouped frequency distribution [10.1] If the data are grouped before they are tallied, then the resulting distribution is called a *grouped frequency distribution.*

Grouping symbols Parentheses ( ), brackets [ ], and braces { } indicate the order of operations and are also sometimes used to indicate multiplication, as in (2)(3) = 6. Also called *symbols of inclusion.*

Growth formula $A = A_0e^{rt}$ or $A = A_0(2.72)^{rt}$

Half-plane [11.4] The part of a plane that lies on one side of a line in the plane. It is a *closed* half-plane if the line is included. It is an *open* half-plane if the line is not included. The line is the *boundary* of the half-plane in either case.

Hecto- [6.1; 6.6] A prefix meaning 100

Heptagon [5.2] A polygon having seven sides

Hexagon [5.2] A polygon having six sides

Histogram [10.1] A bar graph that represents a frequency diagram

Horizontal line [11.3] A line with zero slope; that is, a line that is level, usually drawn so that it is parallel to the top edge of your paper. It is a line parallel to the horizon. Its equation has the form y = constant.

Hypotenuse [2.6; 5.4] The longest side in a right triangle

Hypothesis An assumed proposition used as a premise in proving something else

Implication A statement that follows from other statements. It is also a proposition formed from two given propositions by connecting them with an "if . . . , then . . . " form. It is symbolized by $p \rightarrow q$.

Impossible event [9.1] An event for which the probability is zero—that is, an event that cannot happen

Improper fraction [1.3] A fraction for which the numerator is greater than the denominator

Improper subset [8.2] See *Subset.*

Inch [6.1] A linear measurement in the United States system equal in length to the following segment:

———————

Inconsistent system See *Consistent system.*

Independent variable The variable associated with the first component of an ordered pair

Inductive reasoning [8.6] A type of reasoning accomplished by first observing patterns and then predicting answers for more complicated similar problems

Inequality A statement of order. If always true, an inequality is called an *absolute inequality;* if always false, an inequality is called a *contradiction.* If sometimes true and sometimes false, it is called a *conditional inequality.* Values that make the statement true are said to *satisfy* the inequality. A *string of inequalities* may be used to show the order of three or more quantities.

Inequality symbols [3.7] The symbols $>$, $\geq$, $<$, and $\leq$

Infinite set [8.1] See *Set.*

Inflation [7.5] An increase in the amount of money in circulation, resulting in a fall in its value and a rise in prices. In this book, we assume annual compounding with the future value formula; that is, use $A = P(1 + r)^n$, where r is the projected annual inflation rate, n is the number of years, and P is the present value.

Installment loan [7.3] A financial problem in which an item is paid for over a period of time. It is calculated using add-on or compound interest.

Installments [7.3] Part of a debt paid at regular intervals over a period of time

Integers [2.2] $\mathbb{Z} = \{\ldots, -3, -2, -1, 0, 1, 2, 3, \ldots\}$, composed of the natural numbers, their opposites, and 0

Intercepts The point or points where a line or a curve crosses a coordinate axis. The x-intercepts are sometimes called the *zeros* of the equation.

Interest [7.2] An amount of money paid for the use of another's money. See *Compound interest.*

Interest-only loan [7.3] A loan in which periodic payments are for interest only so that the principal amount of the loan remains the same

Interest rate [7.2; 7.6] The percentage rate paid on financial problems. In this book it is denoted by r and is assumed to be an annual rate unless otherwise stated.

Intersection [8.3] The *intersection* of sets A and B, denoted by $A \cap B$, is the set consisting of elements in *both* A and B.

Interval [10.1] See *Classes.*

Invert [1.6] In relation to the fraction a/b, it means to interchange the numerator and the denominator to obtain the fraction b/a.

Irrational number [2.6] A number that can be expressed as a nonrepeating, nonterminating decimal; the set of irrational numbers is denoted by $\mathbb{Q}'$.

Juxtaposition [1.2] When two variables, a number and a variable, or a symbol and a parenthesis are written next to each other with no operation symbol, as is xy, $2x$, or $3(x + y)$. Juxtaposition is used to indicate multiplication.

Kilo- [6.1, 6.6] A prefix that means 1,000

Kilogram [6.5] 1,000 grams

Kiloliter [6.4] 1,000 liters

Kilometer [6.1] 1,000 meters

LCD [1.7] An abbreviation for least common denominator

Least common denominator (LCD) The smallest number that is exactly divisible by each of the given numbers

Length [6.1] A measurement of an object from end to end

Less than [2.2] If a is to the left of b on a number line, then a is less than b, $a < b$. Formally, $a < b$ if and only if $b > a$.

Less than or equal to Written $a \leq b$, means $a < b$ or $a = b$

Like terms [3.2] Terms that differ only in their numerical coefficients. Also called *similar terms.*

Line [5.1; 11.3] In mathematics, it is an undefined term. It is a curve that is straight, so it is sometimes referred to as a *straight line.* It extends in both directions, and is considered one-dimensional, so it has no thickness.

Line graph [10.1] See *Graph.*

Line of credit [7.3] The maximum amount of credit to be extended to a borrower. That is, it is a promise by a lender to extend credit up to some predetermined amount.

Line of symmetry [5.1] A line with the property that for a given curve, any point P on the curve has a corresponding point Q (called the reflection point of P) so that the perpendicular bisector of $\overline{PQ}$ is on the line of symmetry.

Line segment [5.1] A part of a line between two points on the line

Linear equation A first-degree equation with one or two variables. For example, $x + 5 = 0$ is a linear equation, as is $x + y + 5 = 0$. An equation is linear in a certain variable if it is first degree in that variable. For example, $x + y^2 = 0$ is linear in x, but not y.

Linear function A function whose equation can be written in the form $f(x) = mx + b$.

Linear inequality A first-degree inequality with one or two variables

Linear polynomial A first-degree polynomial

Linear system A system of equations, each of which is first degree

Liter [6.4; 6.6] The basic unit of capacity in the metric system. It is the capacity of 1 cubic decimeter.

Literal equation [3.6] An equation with more than one variable

Loaded die [9.1] A die in which the faces do not have an equal chance of occurring

Logic [8.6] The science of correct reasoning

Logical conclusion The statement that follows logically as a consequence of the hypotheses of a theorem

Lowest common denominator [1.7] For two or more fractions, the smallest common multiple of the denominators

Mass [6.5] In this course, it is the amount of matter an object comprises. Formally, it is a measure of the tendency of a body to oppose changes in its velocity.

Mathematical expectation [9.4] A calculation defined as the product of an amount to be won and the probability that it is won. If there is more than one amount to be won, it is the sum of the expectations of all the prizes. It is also called the *expected value.*

Mathematical modeling An iterative procedure that makes assumptions about real-world problems to formulate the problem in mathematical terms. After the mathematical problem is solved, it is tested for accuracy in the real world and revised for the next step in the iterative process.

Mathematical system A set with at least one defined operation and some developed properties

Maximum loan [7.6] In this book, it refers to the maximum amount of loan that can be obtained for a home with a given amount of income and a given amount of debt

Mean [2.5; 10.2] The number found by adding the data and dividing by the number of values in the data set

Means [4.1] See *Proportion.*

Measure [6.1] Comparison to some unit recognized as standard

Measure of central tendency [10.2] Refers to the averages of mean, median, and mode

Measure of dispersion [10.4] Refers to the measures of range, standard deviation, and variance

Measure of position [10.3] A benchmark position to describe data sets, such as quartiles, deciles, or percentiles

Median [10.2] The middle number when the numbers in the data are arranged in order of size. If there are two middle numbers (in the case of an even number of data values), the median is the mean of these two middle numbers.

Members [8.1] See *Set.*

Meter [6.1; 6.6] The basic unit for measuring length in the metric system

Metric system [6.1] A decimal system of weights and measures in which the gram, the meter, and the liter are the basic units of mass, length, and capacity, respectively. One gram is the mass of one cm^3 of water and one liter is the same as 1,000 cm^3. In this book, the metric system refers to the SI metric system as revised in 1960.

Mile [6.1] A unit of linear measurement in the United States system that is equal to 5,280 ft

Milli- [6.1; 6.6] A prefix that means 1/1,000

Milligram [6.5] 1/1,000 of a gram

Milliliter [6.4] 1/1,000 of a liter

Minus [2.1; 2.3] Refers to the operation of subtraction. The symbol "$-$" means minus only when it appears between two numbers, between two variables, or between numbers and variables.

Mixed number [1.3] A number that has both a counting number part and a proper fraction part; for example, $3\frac{1}{2}$

Mode [10.2] The value in a data set that occurs most frequently. If no number occurs more than once, there is no mode. It is possible to have more than one mode.

Monomial [3.1] A polynomial with one and only one term

Monthly compounding [7.5] In the compound interest formula, it is when $n = 12$.

Monthly payment [7.6] In an installment loan, it is a periodic payment that is made once every month.

Mortality table [9.4] A table showing the probability of a person living or dying during a particular year of his or her life

Mortgage [7.6] An agreement, or loan contract, in which a borrower pledges a home or other real estate as security.

Multiplication of integers [2.4] If the integers to be multiplied both have the same sign, the result is positive and the magnitude of the answer is the product of the absolute values of the integers. If the integers to be multiplied have opposite signs, the product is negative and has magnitude equal to the product of the absolute values of the given integers. Finally, if one or both of the given integers is 0, the product is 0.

Multiplication principle See *Fundamental counting principle.*

Multiplication property of equations [3.4] Both sides of an equation may be multiplied or divided by any nonzero number to obtain an equivalent equation.

Multiplication property of inequality [3.7] Both sides of an inequality may be multiplied or divided by a positive number, and the order of the inequality will remain unchanged. The order is reversed if both sides are multiplied or divided by a negative number.

Multiplicative identity The number 1, with the property that $1 \cdot a = a$ for any real number a

Multiplicative inverse See *Reciprocal.*

Multiplicative law of inequality If $a < b$, then $ac < bc$ if $c > 0$, and $ac > bc$ if $c < 0$. This also applies to $\leq$, $>$, and $\geq$.

Mutually exclusive [9.1] Events are *mutually exclusive* if their intersection is empty.

Natural numbers [1.2] $\mathbb{N} = \{1, 2, 3, 4, 5, \ldots\}$, the positive integers, also called the *counting numbers*

Negative number [2.2] A number less than zero

Negative sign [2.2; 2.3] The symbol "$-$" when used in front of a number, as in -5. Do not confuse with the same symbol used for subtraction, as in $8 - 5$, or with the symbol for opposite.

***n*-gon** [5.2] A polygon with n sides

Nonagon [5.2] A polygon with 9 sides

Normal curve [10.5] A graphical representation of a normal distribution. Its high point occurs at the mean, and it is symmetric with respect to this mean. On each side of the mean, it has an area that includes 34.1% of the population within one standard deviation, 13.6% from one to two standard deviations, and about 2.3% of the population more than two standard deviations from the mean.

Null set See *Set.*

Number line [2.2] A line used to display a set of numbers graphically (the axis for a one-dimensional graph)

Numerator [1.3] See *Rational number.*

Numerical coefficient [3.1] See *Coefficient.*

Obtuse angle [5.2] An angle that is greater than a right angle and smaller than a straight angle

Octagon [5.2] A polygon with eight sides

Odds [9.2] If $s + f = n$, where s is considered favorable to an event E and n is the total number of possibilities, then the *odds in favor of E* are s/f and the *odds against E* are f/s.

One-dimensional coordinate system A real number line

One-to-one correspondence [8.1] Between two sets A and B, this means each element of A can be matched with exactly one element of B and also each element of B can be matched with exactly one element of A.

Open-end loan [7.3] A preapproved line of credit that the borrower can access as long as timely payments are made and the credit line is not exceeded. It is usually known as a credit card loan.

Open half-plane [11.4] See *Half-plane.*

Opposite [2.2; 2.3; 2.4; 3.4] Opposites x and $-x$ are the same distance from 0 on the number line but in opposite directions; $-x$ is also called the *additive inverse* of x. Do not confuse the symbol "$-$" meaning opposite with the same symbol as used to mean subtraction or negative.

Opposite side [5.4] In a right triangle, an acute angle is made up of two sides. The opposite side of the angle refers to the third side that is not used to make up the sides of the angle.

Order of an inequality Refers to a $>$, $\geq$, $<$, or $\leq$ relationship

Order of operations [1.2; 1.7] If no grouping symbols are used in a numerical expression, first perform all multiplication and division from left to right, and then perform all addition and subtraction from left to right.

Order symbols [2.2] Refers to $>$, $\geq$, $<$, $\leq$ in an inequality

Ordered pair [11.1] A pair of numbers, written (x, y), in which the order of naming is important. The numbers x and y are sometimes called the *first* and *second components* of the pair and are called the *coordinates* of the point designated by (x, y).

Ordinary interest [7.2] The calculation of interest assuming a year has 360 days. In this book, we assume ordinary interest unless otherwise stated.

Origin [11.1] The point designating 0 on a number line. In two dimensions, the point of intersection of the coordinate axes; the coordinates are (0, 0).

Origination fee [7.6] A fee paid to obtain a real estate loan

Ounce [6.4, 6.5] (1) A unit of capacity in the United States system that is equal to 1/128 of a gallon (2) A unit of mass in the United States system that is equal to 1/16 of a pound

Overlapping sets [8.2] Sets whose intersection is not empty

Parabola [11.5] A set of points in the plane equidistant from a given point (called the *focus*) and a given line (called the *directrix*). It is the path of a projectile. The *axis of symmetry* is the axis of the parabola. The point where the axis cuts the parabola is the *vertex.*

Parallel lines [5.1] Two nonintersecting straight lines in the same plane

Parallelepiped [6.4] A polyhedron, all of whose faces are parallelograms

Parallelogram [6.3] A quadrilateral with its opposite sides parallel

Parentheses See *Grouping symbols.*

Pentagon [5.2] A polygon with five sides

Percent [4.3] The ratio of a given number to 100; hundredths; denoted by %; that is, 5% means 5/100.

Percent markdown [7.1] The percent of an original price used to find the amount of discount

Percent problem [4.4] The percent problem is one that can be restated as "*A* is *P*% of *W*," and is formulated as a proportion

$$\frac{P}{100} = \frac{A}{W}$$

Percentage [4.4] The total amount in a percentage problem

Percentiles [10.3] Ninety-nine values that divide the data into one hundred parts

Perfect square [2.6] Since $1^2 = 1$, $2^2 = 4$, $3^2 = 9$, . . . , the perfect squares are 1, 4, 9, 16, 25, 36, 49, ...

Perimeter [6.2] The distance around a polygon

Perpendicular Two lines or line segments are perpendicular if their intersection forms right angles.

Pi (π) [6.2] A number that is defined as the ratio of the circumference to the diameter of a circle. It cannot be represented exactly as a decimal, but it is a number between 3.1415 and 3.1416.

Pictograph [10.1] See *Graph.*

Pie chart [10.1] See *Graph.*

Place-value names [1.3] From large to small, the place value names are: trillions, hundred billions, ten billions, billions, hundred millions, ten millions, millions, hundred thousands, ten thousands, thousands, hundreds, tens, units, tenths, hundredths, thousandths, ten-thousandths, hundred-thousandths, and millionths.

Plane [5.1] In mathematics, it is an undefined term. It is flat and level and extends infinitely in horizontal and vertical directions. It is considered two-dimensional.

Plot a point [11.5] To mark the position of a point

Point [1.3; 5.1; 7.6] (1) In the decimal representation of a number, it is a mark that divides the whole number part of a number from its fractional part. (2) In geometry, it is an undefined word that signifies a position, but that has no dimension or size. (3) In relation to a home loan, it represents 1% of the value of a loan, so that 3 points would be a fee paid to a lender equal to 3% of the amount of the loan.

Polygon [5.2] A geometric figure that has three or more straight sides that all lie in a plane so that the starting point and the ending point are the same

Polynomial [3.1] An algebraic expression that may be written as a sum (or difference) of terms. Each *term* of a polynomial contains multiplication only.

Population [10.4] The total set of items (actual or potential) defined by some characteristic of the items

Positive number [2.2] A number greater than 0

Positive sign [2.2] The symbol "+" when used in front of a number or an expression

Postulate [5.1] A statement that is accepted without proof

Pound [6.5] A unit of measurement for mass in the United States system. It is equal to 16 oz.

Power [1.5] See *Exponent.*

Precision [6.1] The accuracy of the measurement; for example, a measurement is taken to the nearest inch, nearest foot, or nearest mile. It is not to be confused with accuracy that applies to the calculation.

Premise [8.6] A previous statement or assertion that serves as the basis for an argument

Present value [7.2, 7.5] See *Compound interest.*

Previous balance method [7.4] A method of calculating credit card interest using the formula $I = Prt$ in which P is the balance owed before the current payment is subtracted

Prime factorization [1.5] The factorization of a number so that all of the factors are primes and so that their product is equal to the given number

Prime numbers [1.5] A number with exactly two factors: 1 and the number itself. That is, $P = \{2, 3, 5, 7, 11, 13, 17, 19, 23, \ldots\}$.

Principal [7.2] See *Compound interest.*

Prism [6.5] In this book, it refers to a right prism, which is also called a parallelepiped or more commonly a box.

Probability [9.1] If an experiment can result in any of n ($n \geq 1$) mutually exclusive and equally likely outcomes, and if s of these are considered favorable to event E, then $P(E) = s/n$.

Problem solving procedure [3.6] 1. *Read the problem.* Note what it is all about. Focus on processes rather than numbers. You can't work a problem you don't understand. 2. Restate the problem. Write a verbal description of the problem using operations signs and an equal sign. Look for equality. If you can't find equal quantities, you will never formulate an equation. 3. *Choose a variable.* If there is a single unknown, choose a variable. 4. *Substitute.* Replace the verbal phrases by known numbers and by the variable. 5. *Solve the equation.* This is the easy step. Be sure your answer makes sense by checking it with the original question in the problem. Use estimation to eliminate unreasonable answers. 6. *State the answer.* There were no variables defined when you started, so $x = 3$ is not an answer. Pay attention to units of measure and other details of the problem. Remember to answer the question that was asked.

Product [1.2; 2.1] The result of a multiplication

Profit formula [3.6] $P = S - C$ where P represents the profit, S represents the selling price (or revenue), and C represents the cost (or overhead).

Proof A logical argument that establishes the truth of a statement

Proper fraction [1.3] A fraction for which the numerator is less than the denominator

Proper subset [8.2] See *Subset.*

Properties of rational expressions Let P, Q, R, S, and K be any polynomials such that all values of the variable that cause division by zero are excluded from the domain.

Equality $\frac{P}{Q} = \frac{R}{S}$ if and only if $PS = QR$

Fundamental property $\frac{PK}{QK} = \frac{P}{Q}$

Addition $\frac{P}{Q} + \frac{R}{S} = \frac{PS + QR}{QS}$

Subtraction $\frac{P}{Q} - \frac{R}{S} = \frac{PS - QR}{QS}$

Multiplication $\frac{P}{Q} \cdot \frac{R}{S} = \frac{PR}{QR}$

Division $\frac{P}{Q} \div \frac{R}{S} = \frac{PS}{QR}$

Property of complements [9.2] For some event E, $P(E) + P(\overline{E}) = 1$.

Property of proportions [4.1] If the product of the means equals the product of the extremes, then the ratios form a proportion. Also, if the ratios form a proportion, then the product of the means equals the product of the extremes.

Property of zero $AB = 0$ if and only if $A = 0$ or $B = 0$ (or both).

Proportion [4.1] A statement of equality between two ratios. For example,

$$\frac{a}{b} = \frac{c}{d}$$

For this proportion, a and d are called the *extremes;* b and c are called the *means.*

Protractor [5.2] A device used to measure angles

Pyramid [6.5] A solid figure having a polygon as a base, the sides of which form the bases of triangular surfaces that meet at a common vertex

Pythagorean theorem [2.6; 5.4] If a triangle with legs a and b and hypotenuse c is a right triangle, then $a^2 + b^2 = c^2$. Also, if $a^2 + b^2 = c^2$, then the triangle is a right triangle.

Quadrant [11.1] See *Axes.*

Quadratic A second-degree polynomial

Quadratic equation An equation of the form

$$ax^2 + bx + c = 0,\ a \neq 0$$

Quadratic formula If $ax^2 + bx + c = 0$ and $a \neq 0$, then

$$x = \frac{-b \pm \sqrt{b^2 - 4ac}}{2a}$$

The number $b^2 - 4ac$ is called the *discriminant* of the quadratic.

Quadrilateral [5.2] A polygon having four sides

Quart [6.4] A measure of capacity in the United States system equal to 1/4 of a gallon

Quarterly compounding [7.5] In the compound interest formula, it is when $n = 12$.

Quartiles [10.3] Three values that divide the data into four parts

Quotient [1.2; 2.1] The result of a division

Radical [2.6] The $\sqrt{\ }$ symbol is an expression such as $\sqrt{2}$. The number 2 is called the *radicand,* and an expression involving a radical is called a radical expression.

Radius [6.2] The distance of a point on a circle from the center of the same circle

Range [10.4] In statistics, it is the difference between the largest and the smallest numbers in the data set.

Rate [4.4; 7.1; 7.3] (1) In percent problems, it is the percent. (2) In tax problems, it is the level of taxation, written as a percent. (3) In financial problems, it refers to the APR.

Ratio [4.1] The quotient of two numbers or expressions

Rational equation An equation that has at least one variable in the denominator

Rational number [1.3; 2.6] A number belonging to the set $\mathbb{Q}$ defined by

$$\mathbb{Q} = \left\{ \frac{a}{b} \middle| \ a \text{ is an integer, } b \text{ is a nonzero integer} \right\}$$

Ray [5.1] If P is a point on a line, then a ray from the point P is all points on the line on one side of P.

Real number line A line on which points are associated with real numbers in a one-to-one fashion

Real numbers [2.6] The set of all rational and irrational numbers, denoted by $\mathbb{R}$

Reciprocal [1.6; 3.4] The reciprocal of n is $\frac{1}{n}$, also called the *multiplicative inverse of n.*

Rectangle [6.2] A quadrilateral whose angles are all right angles

Rectangular coordinate system See *Cartesian coordinate system.*

Rectangular coordinates [11.1] See *Ordered pair.*

Rectangular parallelepiped [6.4] In this book, it refers to a box all of whose angles are right angles.

Reduced fraction [1.6; 2.6] A fraction in which the numerator and denominator have no common divisors (other than 1 or −1)

Reducing fractions [1.6] The process by which we make sure that there are no common factors (other than 1) for the numerator and denominator of a fraction

Reflection [5.1] Given a line L and a point P, we call the point P' the *reflection* about the line L if PP' is perpendicular to L and is also bisected by L.
Relation A set of ordered pairs
Remainder [1.3] When an integer m is divided by a positive integer n, and a quotient q is obtained for which $m = nq + r$ with $0 \le r < n$, then r is the remainder.
Repeating decimal [1.3] See *Decimal fraction.*
Revolving credit [7.3] It is the same as open-end or credit-card credit.
Right angle [5.2] An angle of 90°
Right circular cone [6.5] A cone with a circular base for which the base is perpendicular to its axis
Right circular cylinder [6.5] A cylinder with a circular base for which the base is perpendicular to its axis
Right rectangular prism [6.5] A prism whose rectangular base is perpendicular to the lateral edges, and each lateral edge is a rectangle
Root of an equation [3.4] See *Solution.*
Root of a number An nth root (n is a natural number) of a number b is a only if $a^n = b$. If $n = 2$, then the root is called a *square root;* if $n = 3$, it is called a *cube root.*
Roster method [8.1] A method of defining a set by listing the elements in the set
Rounding a number [1.4] Dropping decimals after a certain significant place. The procedure for rounding is: 1. *Locate the rounding place digit.* 2. *Determine the rounding place digit:* It stays the same if the first digit to its right is a 0, 1, 2, 3, or 4; it increases by 1 if the digit to the right is a 5, 6, 7, 8, or 9. 3. *Change digits:* All digits to the left of the rounding digit remain the same (unless there is a carry), and all digits to the right of the rounding digit are changed to zeros. 4. *Drop zeros:* If the rounding place digit is to the left of the decimal point, drop all trailing zeros or if the rounding place digit is to the right of the decimal point, drop all trailing zeros to the right of the rounding place digit.

Sale price [7.1] A reduced price usually offered to stimulate sales. It can be found by subtracting the discount from the original price, or by multiplying the original price by the complement of the markdown.
Sales tax [7.1] A tax levied by government bodies that is based on the selling price of an item
Sample [10.6] A finite part taken from a population
Sample space [9.1] The set of possible outcomes for an experiment
Satisfy [3.4; 11.3] See *Equation* or *Inequality.*
Scientific notation [1.5] Writing a number as the product of a number between 1 and 10 and a power of 10: For any real number n, $n = m \cdot 10^c$, c an integer, and $1 \le m < 10$. Calculators often switch to scientific notation to represent large or small numbers. The usual notation is 8.234 05 where the space separates the number from the power; thus 8.234 05 means 8.234×10^5.
Second component [11.1] See *Ordered pair.*
Semiannual compounding [7.5] In the compound interest formula, it is when $n = 2$.
Semicircle [6.2] Half a circle
Set [8.1] A collection of particular things, called the *members* or *elements* of the set. A set with no elements is called the *null set* or *empty set* and is denoted by the symbol $\varnothing$. All elements of a *finite set* may be listed, whereas the elements of an *infinite set* continue without end.
Set-builder notation [8.1] The notation

$$\{x \mid x \text{ has some specific property}\}$$

which is pronounced as "the set of all x such that x has some specific property."
SI system [6.1] See *Metric system.*
Signed number [2.2] An integer
Similar figures [5.4] Two geometric figures are similar if they have the same shape, but not necessarily the same size.
Similar terms [3.2] Terms that differ only in their numerical coefficients
Similar triangles [5.4] Triangles that have the same shape
Simple event [9.1] An event for which the sample space has only one element
Simple interest formula [7.2] $I = Prt$
Simplify [3.2; 3.3] (1) A *polynomial:* Combine similar terms and write terms in order of descending degree. (2) A fraction (a rational expression): Simplify numerator and denominator, factor if possible, and eliminate all common factors.

Simultaneous solution [11.4] The solution of a simultaneous system of equations

Sine [5.4] In a right triangle *ABC* with right angle *C*, sin *A* = (opposite side of *A*)/hypotenuse.

Solution [3.4; 3.7; 11.3] The values or ordered pairs of values for which an equation, a system of equations, inequality, or system of inequalities is true. Also called *roots*.

Solution set The set of all solutions to an equation

Solve [3.4; 3.5] To find the values of the variable that satisfy the equation

Solve a proportion [4.2] Find the missing term of a proportion. Procedure: First, find the product of the means or the product of the extremes, whichever does not contain the unknown term; next, divide this product by the number that is opposite the unknown term.

Sphere [6.5] The set of all points in space that are a given distance from a given point

Spreadsheet A rectangular grid used to collect and perform calculations on data. *Rows* are horizontal and labeled with numbers and *columns* are vertical and labeled with letters to designate *cells* such as A4, P604. Each cell can contain text, numbers, or formulas.

Square [6.2] A rectangle with sides that are equal

Squared [1.5] In an expression such as x^2, pronounced "*x* squared," means *xx*.

Square root [2.6] See *Root* of a number.

Square root symbol [2.6] The $\sqrt{\ }$ symbol over a number or variable

Square unit [6.3] A square with side length 1 unit by 1 unit. A *square centimeter* is a square 1 cm by 1 cm; a *square foot* is a square 1 ft by 1 ft; and a *square inch* is 1 in. by 1 in.

Standard deviation [10.4] It is a measure of the variation from a trend. In particular, it is the square root of the mean of the squares of the deviations from the mean.

Standard form of a fraction [2.6] If p and q are positive integers, then $\frac{p}{q}$ and $\frac{-p}{q}$ are called the **standard forms** of a fraction.

Statistics [10.1] Methods of obtaining and analyzing data

Sticker price [7.3] In this book, it refers to the manufacturer's total price of a new automobile as listed on the window of the car.

Straight angle [5.2] An angle whose rays point in opposite directions; an angle whose measure is 180°

Straightedge [5.1] A device used as an aid in drawing a straight line segment

Subset [8.2] A set contained within a set. There are 2^n subsets of a set with n distinct elements. A subset is *improper* if it is equivalent to the given set; otherwise, it is *proper.*

Substitution [3.6] The process of replacing one quantity or unknown by another equal quantity

Subtraction of integers [2.3] $a - b = a + (-b)$

Subtraction property of equations [3.4] The solution of an equation is unchanged by subtracting the same number from both sides of the equation.

Subtraction property of inequality See *Addition property of inequality.*

Sum [1.2; 2.1] The result of an addition

Supply [11.4] The number of items that are available at a given price

Surface [5.1] In mathematics, it is an undefined term. It is the outer face or exterior of an object; it has an extent or magnitude with length and breadth, but no thickness.

Syllogism [8.6] A logical argument that involves three propositions, usually two premises and a conclusion, the conclusion necessarily being true if the premises are true

Symbols of inclusion See *Grouping symbols.*

Symmetric property If $a = b$, then $b = a$.

System of equations [11.4] A set of equations that are to be solved *simultaneously.* A brace symbol is used to show the equations belonging to the system.

Tangent [5.4] In a right triangle *ABC* with right angle *C*, tan *A* = (opposite side of *A*)/(adjacent side of *A*).

Temperature [6.5] The degree of hotness or coldness

Term [2.1; 3.1] A number, a variable, or a product of numbers and variables. See *Polynomial.*

Test point [11.4] A point that is chosen to find the appropriate half-plane when graphing a linear inequality in two variables

Theorem [5.1; 8.6] A statement that has been proved. See *Deductive reasoning.*

Time [7.2] In a financial problem, the length of time (in years) from the present value to the future value

Ton [6.5] A measurement of mass in the United States system; it is equal to 2,000 lb.

Trailing zeros [1.3] Sometimes zeros are placed after the decimal point or after the last digit to the right of the decimal point, and if these zeros do not change the value of the number, they are called *trailing zeros.*

Transformation [5.1] A passage from one figure or expression to another, such as a reflection, translation, rotation, contraction, or dilation

Transitive law If $a = b$ and $b = c$, then $a = c$.

Translating The process of writing an English sentence in mathematical symbols

Translation [1.2] The process of changing a verbal expression into a symbolic expression

Trapezoid [6.3] A quadrilateral that has two parallel sides

Tree diagram [9.1] A device used to list all the possibilities for an experiment

Triangle [5.2, 5.3] A polygon with three sides

Trichotomy Exactly one of the following is true, for any real numbers a and b: $a < b$, $a > b$, or $a = b$.

Trigonometric ratios [5.4] The sine, cosine, and tangent ratios are known as the *trigonometric ratios.*

Trinomial [3.1] A polynomial with exactly three terms

Type I error [10.6] Rejection of a hypothesis based on sampling when, in fact, the hypothesis is true

Type II error [10.6] Acceptance of a hypothesis based on sampling when, in fact, it is false

Undefined terms [5.1; 8.1] To avoid circular definitions, it is necessary to include certain terms without specific mathematical definitions.

Union [8.3] The union of sets A and B, denoted by $A \cup B$, is the set consisting of elements in A or in B or in both A and B.

Unit circle A circle with radius 1 centered at the origin

Unit scale [2.2] The distance between the points marked 0 and 1 on a number line

United States system [6.1] The measurement system used in the United States

Universal set [8.1] The largest set under consideration for a particular discussion

Valid [8.6] In logic, refers to a correctly inferred logical argument

Variable [2.1] A symbol that represents unspecified elements of a given set. On a calculator, it refers to the name given to a location in the memory that can be assigned a value.

Variable expression [2.1] An expression that contains at least one variable

Variance [10.4] The square of the standard deviation

Venn diagram [8.2; 8.4] A diagram used to illustrate relationships among sets

Vertex [5.2; 11.5] (1) A *vertex* of a polygon is a corner point, or a point of intersection of two sides. (2) A *vertex* of a parabola is the lowest point for a parabola that opens upward, the highest point for one that opens downward, the leftmost point for one that opens to the right, and the rightmost point for one that opens to the left.

Vertical line [11.3] A line that is perpendicular to a horizontal line. Its equation has the form x = constant.

Volume [6.4] A number describing the three-dimensional content of a set. Specifically, it is the number of cubic units enclosed in a solid figure.

Weight [6.5] In everyday usage, the heaviness of an object; in scientific usage, the gravitational pull on a body

Well-defined set [8.1] A set for which there is no doubt about whether a particular element is included in the given set

Whole numbers [1.2; 1.3] The positive integers and zero; $\mathbb{W} = \{0, 1, 2, 3, \ldots\}$

***x*-axis** [11.1] The horizontal axis in a Cartesian coordinate system

***x*-intercept** The place where a graph passes through the x-axis

Yard [6.1] A linear measure in the United States system; it has the same length as 3 ft.

***y*-axis** [11.1] The vertical axis in a Cartesian coordinate system

***y*-intercept** The place where a graph passes through the y-axis. For a line $y = mx + b$, it is the point $(0, b)$.

Zero [2.2] The number that separates the positive and negative numbers; it is also called the *identity element* for addition; that is, it satisfies the property that

$$x + 0 = 0 + x = x$$

for all numbers x.

Zero multiplication theorem If a is any real number, then $a \cdot 0 = 0 \cdot a = 0$.

Zero, property of If $a \cdot b = 0$, then either $a = 0$ or $b = 0$.

APPENDIX D

TABLES

Table I *Trigonometric Ratios*

Degrees	sin x	cos x	tan x	Degrees	sin x	cos x	tan x
1	0.0175	0.9998	0.0175	46	0.7193	0.6947	1.0355
2	0.0349	0.9994	0.0349	47	0.7314	0.6820	1.0724
3	0.0523	0.9986	0.0524	48	0.7431	0.6691	1.1106
4	0.0698	0.9976	0.0699	49	0.7547	0.6561	1.1504
5	0.0872	0.9962	0.0875	50	0.7660	0.6428	1.1918
6	0.1045	0.9945	0.1051	51	0.7771	0.6293	1.2349
7	0.1219	0.9925	0.1228	52	0.7880	0.6157	1.2799
8	0.1392	0.9903	0.1405	53	0.7986	0.6018	1.3270
9	0.1564	0.9877	0.1584	54	0.8090	0.5878	1.3764
10	0.1736	0.9848	0.1763	55	0.8192	0.5736	1.4281
11	0.1908	0.9816	0.1944	56	0.8290	0.5592	1.4826
12	0.2079	0.9781	0.2126	57	0.8387	0.5446	1.5399
13	0.2250	0.9744	0.2309	58	0.8480	0.5299	1.6003
14	0.2419	0.9703	0.2493	59	0.8572	0.5150	1.6643
15	0.2588	0.9659	0.2679	60	0.8660	0.5000	1.7321
16	0.2756	0.9613	0.2867	61	0.8746	0.4848	1.8040
17	0.2924	0.9563	0.3057	62	0.8829	0.4695	1.8807
18	0.3090	0.9511	0.3249	63	0.8910	0.4540	1.9626
19	0.3256	0.9455	0.3443	64	0.8988	0.4384	2.0503
20	0.3420	0.9397	0.3640	65	0.9063	0.4226	2.1445
21	0.3584	0.9336	0.3839	66	0.9135	0.4067	2.2460
22	0.3746	0.9272	0.4040	67	0.9205	0.3907	2.3559
23	0.3907	0.9205	0.4245	68	0.9272	0.3746	2.4751
24	0.4067	0.9135	0.4452	69	0.9336	0.3584	2.6051
25	0.4226	0.9063	0.4663	70	0.9397	0.3420	2.7475
26	0.4384	0.8988	0.4877	71	0.9455	0.3256	2.9042
27	0.4540	0.8910	0.5095	72	0.9511	0.3090	3.0777
28	0.4695	0.8829	0.5317	73	0.9563	0.2924	3.2709
29	0.4848	0.8746	0.5543	74	0.9613	0.2756	3.4874
30	0.5000	0.8660	0.5774	75	0.9659	0.2588	3.7321
31	0.5150	0.8572	0.6009	76	0.9703	0.2419	4.0108
32	0.5299	0.8480	0.6249	77	0.9744	0.2250	4.3315
33	0.5446	0.8387	0.6494	78	0.9781	0.2079	4.7046
34	0.5592	0.8290	0.6745	79	0.9816	0.1908	5.1446
35	0.5736	0.8192	0.7002	80	0.9848	0.1736	5.6713
36	0.5878	0.8090	0.7265	81	0.9877	0.1564	6.3138
37	0.6018	0.7986	0.7536	82	0.9903	0.1392	7.1154
38	0.6157	0.7880	0.7813	83	0.9925	0.1219	8.1444
39	0.6293	0.7771	0.8098	84	0.9945	0.1045	9.5144
40	0.6428	0.7660	0.8391	85	0.9962	0.0872	11.4300
41	0.6561	0.7547	0.8693	86	0.9976	0.0698	14.3007
42	0.6691	0.7431	0.9004	87	0.9986	0.0523	19.0812
43	0.6820	0.7314	0.9325	88	0.9994	0.0349	28.6362
44	0.6947	0.7193	0.9657	89	0.9998	0.0175	57.2898
45	0.7071	0.7071	1.0000	90	1.0000	0.0000	undefined

Table II *Monthly Cost to Finance* $1,000

	Number of Years Financed					
Rate of Interest	**5 Years $N = 60$**	**10 Years $N = 120$**	**15 Years $N = 180$**	**20 Years $N = 240$**	**25 Years $N = 300$**	**30 Years $N = 360$**
6.0%	19.33	11.10	8.44	7.16	6.44	6.00
6.5%	19.57	11.35	8.71	7.46	6.75	6.32
7.0%	19.80	11.61	8.99	7.75	7.07	6.65
7.5%	20.04	11.87	9.27	8.06	7.39	6.99
8.0%	20.28	12.13	9.56	8.36	7.72	7.34
8.5%	20.52	12.40	9.85	8.68	8.05	7.69
9.0%	20.76	12.67	10.14	9.00	8.39	8.05
9.5%	21.00	12.94	10.44	9.32	8.74	8.41
10.0%	21.25	13.22	10.75	9.65	9.09	8.78
10.5%	21.49	13.49	11.05	9.98	9.44	9.15
11.0%	21.74	13.77	11.37	10.32	9.80	9.52
11.5%	21.99	14.06	11.68	10.66	10.16	9.90
12.0%	22.24	14.35	12.00	11.01	10.53	10.29
12.5%	22.50	14.64	12.33	11.36	10.90	10.67
13.0%	22.75	14.93	12.65	11.72	11.28	11.06
13.5%	23.01	15.23	12.98	12.07	11.66	11.45
14.0%	23.27	15.53	13.32	12.44	12.04	11.85
14.5%	23.53	15.83	13.66	12.80	12.42	12.25
15.0%	23.79	16.13	14.00	13.17	12.81	12.64
15.5%	24.05	16.44	14.34	13.54	13.20	13.05
16.0%	24.32	16.75	14.69	13.91	13.59	13.45
16.5%	24.58	17.06	15.04	14.29	13.98	13.85
17.0%	24.85	17.38	15.39	14.67	14.38	14.26
17.5%	25.12	17.70	15.75	15.05	14.78	14.66
18.0%	25.39	18.02	16.10	15.43	15.17	15.07
18.5%	25.67	18.34	16.47	15.82	15.57	15.48
19.0%	25.94	18.67	16.83	16.21	15.98	15.89
19.5%	26.22	19.00	17.19	16.60	16.38	16.30
20.0%	26.49	19.33	17.56	16.99	16.78	16.71

Table III *Mortality Table*

n = age; ℓ_n = number living at the beginning of year n (based on 100,000 births)
d_n = number dying in year n; p_n = probability of living through year n;
q_n = probability of dying in year n

n	ℓ_n	d_n	p_n	q_n	n	ℓ_n	d_n	p_n	q_n
0	100000	708	.9929	.0071	50	87624	729	.9917	.0083
1	99292	175	.9982	.0018	51	86895	792	.9909	.0091
2	99117	151	.9985	.0015	52	86103	858	.9900	.0100
3	98966	144	.9986	.0015	53	85245	928	.9891	.0109
4	98822	138	.9986	.0014	54	84317	1003	.9881	.0119
5	98684	133	.9987	.0014	55	83314	1083	.9870	.0130
6	98551	128	.9987	.0013	56	82231	1168	.9858	.0142
7	98423	124	.9987	.0013	57	81063	1260	.9845	.0156
8	98299	121	.9988	.0012	58	79803	1357	.9830	.0170
9	98178	119	.9988	.0012	59	78446	1458	.9814	.0186
10	98059	119	.9988	.0012	60	76988	1566	.9797	.0204
11	97940	120	.9988	.0012	61	75422	1677	.9778	.0222
12	97820	123	.9988	.0013	62	73745	1793	.9757	.0243
13	97697	129	.9987	.0013	63	71952	1912	.9734	.0266
14	97568	136	.9986	.0014	64	70040	2034	.9710	.0291
15	97432	142	.9986	.0015	65	68006	2159	.9683	.0318
16	97290	150	.9985	.0016	66	65847	2287	.9653	.0347
17	97140	157	.9984	.0016	67	63560	2418	.9620	.0381
18	96983	164	.9983	.0017	68	61142	2548	.9583	.0417
19	96819	168	.9983	.0017	69	58594	2672	.9544	.0456
20	96651	173	.9982	.0018	70	55922	2784	.9502	.0498
21	96478	177	.9982	.0018	71	53138	2877	.9459	.0542
22	96301	179	.9982	.0019	72	50261	2948	.9414	.0587
23	96122	182	.9981	.0019	73	47313	2993	.9368	.0633
24	95940	183	.9981	.0019	74	44320	3019	.9319	.0681
25	95757	185	.9981	.0019	75	41301	3030	.9266	.0734
26	95572	187	.9981	.0020	76	38271	3030	.9208	.0792
27	95385	190	.9980	.0020	77	35241	3020	.9143	.0857
28	95195	193	.9980	.0020	78	32221	2998	.9070	.0931
29	95002	198	.9979	.0021	79	29223	2957	.8988	.1012
30	94804	202	.9979	.0021	80	26266	2888	.8901	.1100
31	94602	207	.9978	.0022	81	23378	2790	.8807	.1194
32	94395	212	.9978	.0023	82	20588	2659	.8709	.1292
33	94183	218	.9977	.0023	83	17929	2499	.8606	.1394
34	93965	226	.9976	.0024	84	15430	2314	.8500	.1500
35	93739	235	.9975	.0025	85	13116	2113	.8389	.1611
36	93504	247	.9974	.0027	86	11003	1901	.8272	.1728
37	93257	261	.9972	.0028	87	9102	1685	.8149	.1851
38	92996	280	.9970	.0030	88	7417	1470	.8018	.1982
39	92716	301	.9968	.0033	89	5947	1263	.7876	.2124
40	92415	326	.9965	.0035	90	4684	1068	.7720	.2280
41	92089	354	.9962	.0039	91	3616	888	.7544	.2456
42	91735	383	.9958	.0042	92	2728	725	.7342	.2658
43	91352	414	.9955	.0045	93	2003	579	.7109	.2891
44	90938	447	.9951	.0049	94	1424	450	.6840	.3160
45	90491	484	.9947	.0054	95	974	341	.6499	.3501
46	90007	525	.9942	.0058	96	633	253	.6003	.3997
47	89482	569	.9937	.0064	97	380	185	.5132	.4869
48	88913	618	.9931	.0070	98	195	129	.3385	.6615
49	88295	671	.9924	.0076	99	66	66	.0000	1.0000

Courtesy of the Society of Actuaries, Chicago, Illinois

ANSWERS TO SELECTED PROBLEMS

Chapter 1

PROBLEM SET 1.1

Answers vary. The important aspect in answering these questions is that you begin to think about mathematics and your own feelings about mathematics. **17.** stop and study **19.** You need only remember the result; the derivation is optional. **21.** women's rest room **23.** restaurant **25.** kangaroo crossing **27.** F **29.** D **31.** C

PROBLEM SET 1.2

1. F; multiplication first: 52 **2.** T **3.** F; $2 \times (3 + 4) = 2 \times 3 + 2 \times 4$ is an example of the distributive property. **4.** F; first, multiply and divide (left to right), then add and subtract (left to right). **5.** F; it is a whole number. **6.** T **9.** 17; sum **11.** 12; difference **13.** 5; sum **15.** 14; sum **17.** 17; sum **19.** 7; quotient **21.** 32; sum **23.** 32; product **25.** 27; sum **27.** 11; sum **29.** 9; sum **31.** 8; difference **33.** 29; difference **35.** 8; quotient **37.** $3 \times 4 + 3 \times 8$ **39.** $12 \times 4 + 12 \times 6$ **41.** $8 \times 50 + 8 \times 5$ **43.** $4 \times 300 + 4 \times 20 + 4 \times 7$ **45.** $5 \times 800 + 5 \times 60 + 5 \times 4$ **47.** $3 + 2 \times 4$ **49.** $10(5 + 6)$ **51.** $8 \times 5 + 10$ **53.** $8(11 - 9)$ **55.** 261; difference **57.** 800; sum **59.** 1,080; sum **61.** 1,600; product **63.** 59; difference **65.** 2,700; difference **67.** 285,197; sum **69.** 2,323; sum **71.** 8,640 hours **73.** $10,600 **75.** $18,516 **77.** $37,440 **79.** 345 miles

PROBLEM SET 1.3

5. T **6.** F; can't divide by 0 **7.** T **8.** T **9.** F; $\frac{1}{3} = 0.\overline{3}$ **10.** F; $0.\overline{6} = 0.666\ldots$ **11. a.** $\boxed{1}\ \boxed{+}\ \boxed{1}\ \boxed{\div}\ \boxed{2}\ \boxed{=}$ **b.** $\boxed{6}\ \boxed{+}\ \boxed{5}\ \boxed{\div}\ \boxed{6}\ \boxed{=}$ **13.** a. $\boxed{3}\ \boxed{\div}\ \boxed{4}\ \boxed{+}\ \boxed{14}\ \boxed{=}$ **b.** $\boxed{2}\ \boxed{\div}\ \boxed{7}\ \boxed{+}\ \boxed{3}\ \boxed{=}$ **15.** a. $1\frac{1}{2}$ **b.** $1\frac{1}{3}$ **17. a.** $5\frac{1}{3}$ **b.** $6\frac{1}{4}$ **19. a.** $168\frac{1}{10}$ **b.** $149\frac{3}{10}$ **21. a.** $1\frac{11}{16}$ **b.** $17\frac{7}{10}$ **23. a.** $3\frac{5}{12}$ **b.** $12\frac{1}{5}$ **25. a.** $16\frac{3}{5}$ **b.** 14 **27. a.** $\frac{5}{2}$ **b.** $\frac{5}{3}$ **29. a.** $\frac{35}{8}$ **b.** $\frac{21}{4}$ **31. a.** $\frac{13}{2}$ **b.** $\frac{17}{5}$ **33. a.** $\frac{12}{5}$ **b.** $\frac{37}{10}$ **35. a.** $\frac{91}{8}$ **b.** $\frac{29}{10}$ **37. a.** $\frac{98}{5}$ **b.** $\frac{137}{8}$ **39. a.** 0.125 **b.** 0.375 **41. a.** 0.6 **b.** 0.875 **43. a.** $0.\overline{428571}$ **b.** $3.1\overline{6}$ **45. a.** $0.\overline{5}$ **b.** $0.\overline{7}$ **47. a.** $3.0\overline{3}$ **b.** $0.\overline{285714}$ **49. a.** $0.4\overline{6}$ **b.** $0.\overline{8}$ **51.** Answers may vary; **a.** $\frac{1}{4}$ **b.** $\frac{1}{2}$ **53.** Answers may vary; **a.** 0.50 **b.** 0.75 **55. a.** $0.\overline{09}$ **b.** $0.08\overline{3}$ **57. a.** 0.2631578947 (approx.) **b.** 0.1764705882 (approx.) **59.** $1,875 **61.** $37,950 **63.** $6,062.50

PROBLEM SET 1.4

5. F; 30.1 **6.** F; 625.98 **7.** F; 3,680,000 **8.** F; 12,456.9 **9.** T **10.** T **11.** 2.3 **13.** 6,287.45 **15.** 5.3 **17.** 6,300 **19.** 12.82 **21.** 4.818 **23.** 5 **25.** $12.99 **27.** $15.00 **29.** 690 **31.** $86,000 **33.** 0.667 **35.** 0.118 **37.** 0.137 **39.** B (Think: $4.82 \approx 5$ mi, so it is about 10 miles round trip; $5 \times 10 = 50$.) **41.** C (Think: $\$35,000 \approx \$36,000$ and $\$36,000 \div 12 = \$3,000$.) **43.** B (Think: $\$1,000 \approx \$1,200$ and $\$1,200 \div 12 = \$100$.) **45.** 0.333 **47.** 0.417 **49.** 0.318 **51.** $1,250 **53.** $12\frac{1}{8}$ or $12.13 **55.** $70.83 **57.** $112.33

PROBLEM SET 1.5

5. F; $5^2 = 5 \times 5 = 25$ **6.** F; 2^3 means $2 \times 2 \times 2$ **7.** F; $4^3 = 4 \times 4 \times 4$ (two multiplications) **8.** F; it can also be a power of ten. **9. a.** one million **b.** 10 **c.** 6 **d.** $10 \times 10 \times 10 \times 10 \times 10 \times 10$ or one million **11. a.** one-tenth **b.** 10 **c.** -1 **d.** 0.1 or one-tenth **13. a.** 3.2×10^3 **b.** 2.5×10^4 **15. a.** 4×10^{-3} **b.** 2×10^{-2} **17. a.** 5.624×10^3 **b.** 1.5824×10^4 **19. a.** 3.5×10^{10} **b.** 6.3×10^7 **21. a.** 8.61×10^{-5} **b.** 2.49×10^8 **23. a.** 6.34×10^9 **b.** 5.2019×10^{11} **25. a.** 2.02983×10^{-3} **b.** 5.209×10^{-5} **27. a.** 310 **b.** 680,000,000 **29. a.** 0.205 **b.** 0.03013 **31. a.** 5,060 **b.** 6.81 **33. a.** 64 **b.** 216 **35. a.** 64 **b.** 32 **37. a.** 1,024 **b.** 729 **39. a.** 292,140,000 **b.** 42,943,673,247,800,000,000 **41. a.** $2^2 \times 3$ **b.** $2^2 \times 5$ **43. a.** 2^8 **b.** 2×3^2 **45. a.** $2^4 \times 5^2$ **b.** $2^3 \times 5^3$ **47. a.** $7^3 \times 13$ **b.** $13^2 \times 23 \times 59$ **49. a.** $19 \times 29 \times 83$ **b.** 31^3 **51.** A **53.** C **55.** A **57.** 5.9×10^7 **59.** 5,000,000,000 **61.** 333,000 **63.** 10^{100} **65.** $150 (Estimate, do not use a calculator.) **67.** about 68 days

PROBLEM SET 1.6

7. T **8.** F; multiply 16 by $\frac{1}{2}$ **9.** F; $8 \times \frac{3}{2} = \frac{24}{2} = 12$ **10.** F; all numbers except 0 **11.** F; $\frac{2}{3} \div \frac{4}{5} = \frac{2}{3} \times \frac{5}{4} = \frac{5}{6}$ but $\frac{4}{5} \div \frac{2}{3} = \frac{4}{5} \times \frac{3}{2} = \frac{6}{5}$ **12.** T **13. a.** $\frac{1}{2}$ **b.** $\frac{1}{3}$ **c.** $\frac{1}{4}$ **d.** $\frac{1}{5}$ **15. a.** $\frac{1}{4}$ **b.** $\frac{1}{3}$ **c.** $\frac{1}{2}$ **d.** $\frac{2}{3}$ **17. a.** $\frac{24}{5}$ **b.** 3 **c.** $\frac{2}{3}$ **d.** $\frac{1}{2}$ **19. a.** $\frac{1}{3}$ **b.** $\frac{3}{20}$ **c.** $\frac{1}{4}$ **d.** $\frac{7}{8}$ **21.** Shade 15 of the 30 squares. **23.** Shade 9 of the 20 squares. **25.** Shade 3 of the 10 squares. **27. a.** $\frac{2}{3}$ **b.** $2\frac{3}{4}$ **c.** $\frac{6}{7}$ **d.** 0 **29. a.** 3 **b.** 2 **31. a.** 25 **b.** 4 **33. a.** $\frac{2}{5}$ **b.** $\frac{1}{4}$ **c.** $\frac{13}{20}$ **d.** $\frac{1}{10}$ **35. a.** $\frac{3}{2}$ **b.** $\frac{2}{3}$ **c.** $\frac{4}{3}$ **d.** $\frac{9}{8}$ **37. a.** $\frac{4}{5}$ **b.** $\frac{8}{3}$ **c.** $\frac{5}{7}$ **d.** $\frac{5}{2}$ **39. a.** $\frac{7}{5}$ **b.** $\frac{9}{7}$ **c.** $\frac{16}{27}$ **d.** 2 **41. a.** $\frac{2}{15}$ **b.** $\frac{1}{8}$ **c.** $\frac{3}{25}$ **d.** 18 **43. a.** $\frac{10}{3}$ **b.** $\frac{3}{2}$ **c.** $\frac{36}{11}$ **d.** 20 **45. a.** $\frac{65}{12}$ **b.** $\frac{19}{10}$ **c.** $\frac{36}{25}$ **d.** $\frac{55}{18}$ **47. a.** $\frac{1}{12}$ **b.** $\frac{1}{15}$ **c.** $\frac{1}{2}$ **d.** $\frac{19}{2}$ **49. a.** $\frac{7}{10}$ **b.** $\frac{9}{10}$ **c.** $\frac{4}{5}$ **51. a.** $\frac{9}{50}$ **b.** $\frac{12}{25}$ **c.** $\frac{27}{50}$ **53. a.** $\frac{101}{200}$ **b.** $\frac{3}{200}$ **c.** $\frac{1}{200}$ **55. a.** $\frac{3}{8}$ **b.** $\frac{8}{9}$ **c.** $\frac{1}{3{,}000}$ **57.** $48,000 **59.** $62,500 **61.** $2,346 **63.** $2,012.50 **65.** $13\frac{7}{8}$ **67.** $57\frac{1}{2}$ miles **69.** If you are 20, then $20 \div \frac{1}{2} = 40$; 40% off; but if you are 50, then $50 \div \frac{1}{2} = 100$ or 100% off. (It's free?)

PROBLEM SET 1.7

3. A **5.** D **7.** B **9.** A **11.** F; do not add the denominators. Answer is $\frac{7}{8}$. **12.** F; write $3\frac{5}{8} - 2\frac{7}{8} = 2\frac{13}{8} - 2\frac{7}{8} = \frac{6}{8} = \frac{3}{4}$ **13.** T **14.** F; invert the divisor **15.** T **16.** F; multiplication first: $\frac{3}{5} + \frac{1}{5} = \frac{4}{5}$ **17. a.** $\frac{3}{5}$ **b.** $\frac{8}{7}$ **c.** $\frac{8}{11}$ **19. a.** $\frac{4}{13}$ **b.** $\frac{1}{23}$ **c.** 1 **21. a.** 4 **b.** 9 **c.** 13 **23. a.** 8 **b.** 6 **c.** 10 **25. a.** 336 **b.** 630 **c.** 360 **27.** 180 **29.** 55,125 **31.** 6,300 **33.** 1,260 **35. a.** $\frac{7}{6}$ **b.** $\frac{7}{8}$ **c.** $\frac{1}{3}$ **37. a.** $\frac{1}{2}$ **b.** $\frac{35}{24}$ **c.** $\frac{7}{24}$ **39. a.** $\frac{19}{90}$ **b.** $\frac{7}{90}$ **c.** $\frac{1}{36}$ **41. a.** $7\frac{1}{4}$ **b.** $7\frac{1}{6}$ **c.** $8\frac{7}{8}$ **43. a.** $1\frac{5}{24}$ **b.** $3\frac{3}{4}$ **c.** $\frac{11}{16}$ **45.** $2\frac{23}{24}$ **47.** $10\frac{31}{40}$ **49.** $25\frac{1}{3}$ **51.** $\frac{8}{15}$ **53.** $\frac{12}{5}$ **55.** $\frac{3}{4}$ **57.** $\frac{19}{15}$ **59.** $\frac{9}{8}$ **61.** 0.8 **63.** $9.\overline{3}$ **65.** 0.4487116145 **67.** $\frac{1}{2} \times \frac{1}{4} + \frac{1}{3} \times \frac{1}{2} = \frac{7}{24}$ **69.** $\frac{1}{4} + \frac{2}{3} \times \frac{1}{2} = \frac{7}{12}$ or find the portion that is not C: $1 - \frac{5}{12} = \frac{7}{12}$ **71.** $1 - \frac{7}{24} = \frac{17}{24}$ **73.** $\frac{1}{6} \times \frac{1}{3} + \frac{1}{3} \times \frac{1}{3} = \frac{1}{6}$ **75.** $\frac{1}{2} \times \frac{1}{3} + \frac{1}{3} \times \frac{1}{3} = \frac{5}{18}$ **77.** not K: $1 - \frac{1}{6} = \frac{5}{6}$ **79.** $7\frac{5}{12}$ pounds **81.** $18\frac{1}{16}$ in. **83. a.** 100 ft **b.** 200 ft **c.** 250 ft **d.** 275 ft **e.** 287.5 ft **f.** 300 ft

CHAPTER 1 Practice Test

5. $96 \times 186{,}075{,}000 \approx 10^2 \times 1.86 \times 10^8 = 1.86 \times 10^{10}$ **7. a.** $5 \times 8 + 5 \times 2$ **b.** $2 \times 25 + 2 \times 35$ **c.** $3 \times 200 + 3 \times 50 + 3 \times 6$ **d.** $5 \times 400 + 5 \times 50 + 5 \times 9$ **9. a.** $16\frac{2}{7}$ **b.** $8\frac{1}{3}$ **c.** $16\frac{7}{10}$ **d.** $1\frac{53}{100}$ **11. a.** 0.875 **b.** $0.8\overline{3}$ **c.** $8.\overline{6}$ **d.** 2.8 **13. a.** 3.4×10^{-3} **b.** 4.0003×10^6 **c.** 1.74×10^4 **d.** 5 **15. a.** [(] [16.2] [+] [14.1] [−] [8.454] [)] [÷] [2.13] [=]; quotient **b.** [16.2] [+] [14.1] [−] [8.454] [÷] [2.13] [=]; difference **17. a.** 2×43 **b.** $2^3 \times 3^2$ **c.** 2×3^5 **d.** $2^2 \times 7^3$ **19. a.** $\frac{333}{1{,}000}$ **b.** $\frac{2}{9}$ **c.** $\frac{19}{20}$ **d.** $\frac{1}{200}$ **21. a.** 22; sum **b.** 16; quotient **c.** 17; sum **d.** 8; difference **23. a.** $\frac{5}{4}$ **b.** $\frac{2}{5}$ **c.** 2 **d.** $\frac{26}{15}$ **25. a.** $\frac{13}{24}$ **b.** $\frac{9}{20}$ **c.** $3\frac{11}{20}$ **d.** $\frac{1}{30}$ **27. a.** $\frac{8}{23}$ **b.** $\frac{31}{24}$ **29.** $\frac{1}{12}$

Chapter 2

PROBLEM SET 2.1

1. terms **7.** 5, 6, 8, 12 **9.** 1, 3, 7, 15 **11.** 100, 98, 82, 2 **13.** 36, 64, 81, 256 **15.** 0, 4, 9, 100 **17.** 100, 144, 169, 400 **18.** F; the domain is 10, 20, not x. **19.** F; it is the product of x and y. **20.** T **21.** F; $30 \div 6 \neq 6 \div 30$ **22.** T **23.** F; they are called factors. **25.** 3 **27.** 12 **29.** 18 **31.** 7 **33.** 0 **35.** 5 **37.** 10 **39.** 6 **41.** 49 **43.** 1 **45.** 15 **47.** 27 **49.** $6 + n$ **51.** $2 + n + 3$ **53.** $3n$ **55.** $10 - n$ **57.** $n - 2$ **59.** $\frac{x+3}{2}$ **61.** $1 - n$ **63.** $n - 100$ **65.** $4n$ **67.** $2 \cdot 3n$ **69.** $5 + n$ **71.** $\frac{5}{n}$

PROBLEM SET 2.2

5. minus **7.** opposite **9.** opposite *Show Problems* 10 – 25 *on number lines.* **11.** -1 **13.** 1 **15.** -6 **16.** F; the sign depends on the numbers being added. **17.** F; it is negative. **18.** T **19.** T; minus is an operation requiring two numbers. **20.** T; minus is an operation requiring two numbers. **21.** F; it is 0. **23. a.** 4 **b.** -4 **25. a.** -1 **b.** 13 **27. a.** -1 **b.** -5 **29. a.** 2 **b.** -14 **31. a.** -18 **b.** 37 **33. a.** -36 **b.** -48 **35. a.** 17 **b.** 39 **37. a.** 135 **b.** 68 **39. a.** 0 **b.** 0 **41. a.** -19 **b.** -19 **43. a.** The point N is seven units to the right of H. **b.** $+6$ **c.** -8 **45.** -2 **47.** 8 **49.** -6 **51.** 319 **53.** $-1{,}475$ **55.** -2 **57.** -799 **59.** -207 **61.** 70 **63.** -13°C **65.** 26-yard line

PROBLEM SET 2.3

3. T **4.** T **5.** F; it means $-5 + (-3)$. **6.** F; it is "the opposite of a." **7.** F; b can be positive, negative, or zero. **8.** F; it is 20. **9.** Five minus negative three **11.** The opposite of negative three **13.** x minus y **15.** Six minus the opposite of negative 1 **17. a.** $8 + (-5) = 3$ **b.** $12 + (-7) = 5$ **19. a.** $-9 + (-5) = -14$ **b.** $-9 + 5 = -4$ **21. a.** $17 + (-8) = 9$ **b.** $-17 + 8 = -9$ **23. a.** $21 + 7 = 28$ **b.** $21 + (-7) = 14$ **25. a.** $13 + 6 = 19$ **b.** $-13 + (-6) = -19$ **27. a.** $-8 + (-23) = -31$ **b.** $8 + 23 = 31$ **29. a.** $62 + 112 = 174$ **b.** $5 + 416 = 421$ **31. a.** -4 **b.** -8 **c.** -2 **33. a.** 17 **b.** 1 **c.** -22 **35.** 1 **37.** 2 **39.** 1 **41.** 3 **43.** 2 **45.** 2 **47.** 462 **49.** $-90{,}592$ **51.** -926 **53.** $-9{,}389$ **55.** 8,773 ft **57.** down 2 or -2 **59.** 3° **61.** 25°

PROBLEM SET 2.4

5. T **6.** F; the sum may be positive or negative. **7.** T **8.** T **9.** T **10.** F; it is positive. **11.** F; -36 **12.** F; -100 **13.** F; -64 **14.** T **15.** T **16.** T **17. a.** -54 **b.** -15 **19. a.** -20 **b.** -64 **21. a.** -48 **b.** 28 **23. a.** -28 **b.** -81 **25. a.** 54 **b.** -42 **27. a.** 16 **b.** 72 **29. a.** -30 **b.** -54 **31. a.** 40 **b.** 6 **33. a.** -16 **b.** 16 **35. a.** -36 **b.** 36 **37. a.** -7 **b.** -5 **39.** -6 **41.** -2 **43.** 13 **45.** 25 **47.** 25 **49.** -9 **51.** -8 **53.** 4 **55.** -41 or 41 steps to the left **57.** -47 **59.** -115 or down 115 ft **61.** moving up 360 **63.** gaining \$750

PROBLEM SET 2.5

5. F; the divisor is 20. **6.** F; it is 0. **7.** F; can't divide by 0. **8.** F; it is positive. **9.** F; an average can be positive, negative, or zero. **10.** F; it is 5 (division before addition). **11.** F; it is 5 (division before addition). **12.** F; it means $9 + \frac{12}{3}$. **13. a.** 6 **b.** 7 **15. a.** -4 **b.** -8 **17. a.** -7 **b.** 11 **19. a.** -3 **b.** -26

21. a. −46 **b.** 132 **23. a.** 0 **b.** 0 **25.** 4 **27.** 12 **29.** −5 **31. a.** 4 **b.** 10 **33.** −9 **35.** −6 **37.** 7 **39.** 4 **41.** −2 **43.** 5.5 **45.** −2 **47.** 1 **49.** 1 **51.** 2 **53.** 2 **55.** −5 **57.** 2 **59.** −2 **61.** $-\frac{7}{4}$

PROBLEM SET 2.6

3. 1, 4, 9, 16, 25, 36, 49, 64, 81, 100, 121, 144, 169, 196 **8.** F; must be in standard form: $\frac{-1}{2}$. **9.** T **10.** F; true if and only if a right triangle with sides a and b and hypotenuse c. **11.** F; $\sqrt{2}$ is an irrational number. **12.** F; $\sqrt{3}$ is an irrational number. **13.** F; $\frac{1}{3} = 0.\overline{3}$ **14.** F; $\frac{2}{3} = 0.\overline{6}$ **15.** F; $\sqrt{225} = 15$ **16.** F; $\sqrt{100} = 10$ **17.** F; repeating decimals are rational. **18.** F; $\sqrt{9} = 3$ **19.** F; $\sqrt{16} = 4$ **20.** T **21.** T **22.** F; it is between 3 and 4. **23.** T **24.** F; it is rational. **25. a.** $\frac{-7}{8}$ **b.** $\frac{-5}{8}$ **c.** $\frac{-7}{9}$ **27. a.** $\frac{x}{y}$ **b.** $\frac{y}{3}$ **c.** $\frac{-1}{x}$ **29.** 1 **31.** $\frac{-1}{2}$ **33.** $\frac{-23}{35}$ **35.** $\frac{1}{5}$ **37.** $\frac{-35}{6}$ **39.** $\frac{-8}{3}$ **41.** $\frac{37}{50}$ **43.** $\frac{-1}{6}$ **45. a.** 3 **b.** 1 **c.** 0 **d.** impossible **47. a.** 9 **b.** impossible **c.** 13 **d.** 14 **49. a.** 35 **b.** 45 **c.** 98 **d.** 78 **51. a.** 3.87 **b.** 4.12 **c.** 4.47 **53. a.** 13.78 **b.** 31.62 **c.** 44.72 **55. a.** rational; 5.0 **b.** irrational; $2 < \sqrt{5} < 3$ **c.** rational; 5.0 **d.** rational; 0.25 **57. a.** irrational; $3 < \sqrt{10} < 4$ **b.** irrational; $3 < \sqrt{15} < 4$ **c.** rational; 4.0 **d.** irrational; $4 < \sqrt{17} < 5$ **59. a.** irrational; $48 < \sqrt{2{,}400} < 49$ **b.** rational; 49.0 **c.** irrational; $49 < \sqrt{2{,}402} < 50$ **d.** irrational; $0 < \sqrt{\frac{1}{10}} < 1$ **61.** 10 ft **63.** 24 ft **65.** $\sqrt{8}$ (or $2\sqrt{2}$) in. **67.** 200 ft **69.** Length is 25 ft each; would need to purchase 100 ft. **71.** Exact length is $\sqrt{325}$; this is approximately 18 ft; would need to purchase 73 ft.

CHAPTER 2 Practice Test

3. a. −64 **b.** 26 **c.** −64 **d.** 104 **e.** 22 **f.** 24 **g.** −18 **h.** −14 **i.** 24 **j.** 5 **5. a.** 36 **b.** −20 **c.** 24 **d.** 0 **e.** −600 **f.** −150 **g.** −16 **h.** 16 **i.** −6 **j.** −18 **7. a.** 6 **b.** 30 **c.** impossible **d.** impossible **e.** 5 **f.** −17 **g.** $\frac{-1}{30}$ **h.** $\frac{-3}{4}$ **i.** $\frac{1}{21}$ **j.** $\frac{16}{5}$ **9. a.** 10 **b.** does not exist **c.** 0 **d.** 87 **11.** $n + 20$ **13.** $3n$ **15. a.** 0 **b.** −2 **c.** −3 **d.** 84 **17.** It is 200 ft shorter. **19.** $\sqrt{52}$ or $2\sqrt{13}$; 7 ft

Chapter 3

PROBLEM SET 3.1

1. A term is a number, a variable, or the product of numbers and variables. **3.** $b^m \cdot b^n = b^{m+n}$ **5. a.** degree 4; coefficient 3 **b.** degree 2; coefficient 5 **7. a.** degree 1; coefficient 1 **b.** degree 3; coefficient 1 **9. a.** second-degree binomial **b.** third-degree binomial **11. a.** zero-degree monomial **b.** third-degree binomial **13. a.** second-degree trinomial **b.** second-degree trinomial **15.** F; add the exponents: x^6. **16.** F; add the exponents: y^5. **17.** F; expression is simplified. **18.** F; expression is simplified. **19.** F; add the exponents: 2^7. **20.** F; add the exponents: 4^5. **21.** F; $2^2 \cdot 3^2 = 4 \cdot 9 = 36$ **22.** F; $3^2 \cdot 2^3 = 9 \cdot 8 = 72$ **23. a.** x^8 **b.** y^6 **25. a.** x^7 **b.** x^5y^6 **27. a.** $-6x^3y^5z^2$ **b.** $49x^3y^4$ **29. a.** $21x^2$ **b.** $21xy$ **31. a.** $6xy$ **b.** $-27x^2$ **33. a.** $-9xyz$ **b.** $-8abc$ **35. a.** $4x^2$ **b.** $9y^2$ **37. a.** $-25x^2$ **b.** $50xy^2$ **39. a.** $16x^3$ **b.** $-12x^3$ **41. a.** $20y^6$ **b.** $20y^6$ **43. a.** x^7 **b.** $-6x^6$ **45. a.** $48y^3$ **b.** $-25x^6$ **57.** $(2 + 3)^2 = 25$ **59.** $x^2 + 5$ **61.** $x^2 + 4$ **63.** $x^3 - y^3$

PROBLEM SET 3.2

1. $a(b + c) = ab + ac$ **3.** $3x$ and $10x$; $5xy$ and $7xy$; $2x^2y$ and $6x^2y$ are similar terms. **5.** F; $x + x + x = 3x$ **6.** F; $y + y + y + y = 4y$ **7.** F; $2x^2 + 3x^2 = 5x^2$ **8.** F; $6y + 2y = 8y$ **9.** F; left side is simplified. **10.** F; left side is simplified. **11.** $7x$ **13.** $11z$ **15.** y **17.** $2x$ **19.** $-2z$ **21.** $5x$ **23.** $-14x$ **25.** $-2x$ **27.** 0 **29.** $2x + 3y$ **31.** $13x$ **33.** $9x$ **35.** $3y$ **37.** $-3x + 4y$ **39.** $-b$ **41.** $-4a + 2b$ **43.** $5x$ **45.** $2y$ **47.** 8 **49.** $x + 8$ **51.** $-3x$ **53.** $-2y$ **55.** $-7x + 8y + 18$ **57.** $4x - 7y$ **59.** $4x - 4y$ **61.** $4x + 4y$ **63.** $4x^2 - 3x - 1$

PROBLEM SET 3.3

1. Answers vary; $a(b + c) = ab + ac$ **3.** F; $3x + 15$ **4.** F; $-x - 3$ **5.** T **6.** T **7.** F; 1 **8.** F; -8 **9.** F; $x^2 + 9x + 20$ **10.** F; $x^2 - 5x + 6$ **11.** $9x + 12$ **13.** $7x + 23$ **15.** $5x - 13$ **17.** $10x - 20$ **19.** $7x - 3y$ **21.** $3x - 4y$ **23.** $3x - 3$ **25.** $3x - 1$ **27.** $x - 17$ **29.** $-2x + 4$ **31.** $-3x + 9y$ **33.** $-x + 2y$ **35.** $-2x + 3y$ **37.** $x + 1$ **39.** $-x - 7$ **41.** $x - 8$ **43.** $x^2 + 3x + 2$ **45.** $x^2 + 2x - 35$ **47.** $x^2 + x - 6$ **49.** $x^2 - 5x + 4$ **51.** $x^2 - 4$ **53.** $6x^2 + 7x - 3$ **55.** $x^2 + 10x + 25$ **57.** $x^2 - 6x + 9$ **59.** $4x^2 + 4x + 1$ **61.** $3x^2y^2 + xyz - 2z^2$ **63.** $3x - 8$ **65.** $-2x^2 - 6x$ **67.** $4x^4 - x^3 + 11x^2 - 3x$ **69.** $10x^3 - 2x^2 - 9x - 28$

PROBLEM SET 3.4

5. F; should be $x = 6$; subtract 2 from both sides. **6.** F; should be $x = -1$; add 5 to both sides. **7.** F; should be $x = \frac{2}{3}$; divide both sides by 3. **8.** F; should be $x = -50$; multiply both sides by 5. **9.** F; the sum is 0. **10.** F; the product is 1. **11.** F; subtract 8 from both sides. **12.** F; add 6 to both sides. **13.** F; add 5 to both sides. **14.** F; subtract 3 from both sides. **15.** F; use the division property of equations. **16.** F; use the multiplication property of equations. **17.** 15 **19.** 2 **21.** -3 **23.** -9 **25.** -1 **27.** -88 **29.** 5 **31.** 1 **33.** -22 **35.** -4 **37.** 4 **39.** 24 **41.** 32 **43.** 21 **45.** -60 **47.** -44 **49.** 39 **51.** 40 **53.** 0 **55.** 11 **57.** 12 **59.** -14 **61.** -14 **63.** -45 **67.** Cannot multiply equals times equals; can only multiply both sides by nonzero constant.

PROBLEM SET 3.5

1. F; subtract 3 from both sides. **2.** F; add 5 to both sides. **3.** F; subtract 5 from both sides. **4.** F; add 3 to both sides. **5.** F; subtract x from both sides; solution is $x = -6$. **6.** F; subtract 5 from both sides; solution is $x = -15$. **7.** F; $x + 15 = 18$ (if you multiply both sides by 3); it is better to subtract 5 from both sides to obtain $\frac{x}{3} = 1$ or $x = 3$. **8.** F; not finished, should be $x = -7$. **9. a.** $5x$ **b.** $7y$ **11. a.** y **b.** $3x$ **13. a.** $-y$ **b.** -5 **15. a.** $4x$ **b.** $7x$ **17. a.** 4 **b.** 5 **19. a.** 5 **b.** 8 **21. a.** -54 **b.** 18 **23. a.** -1 **b.** 7 **25. a.** -19 **b.** -17 **27. a.** -4 **b.** -13 **29.** 14 **31.** 41 **33.** 0 **35.** 8 **37.** 21 **39.** 70 **41.** -14 **43.** 17 **45.** 3 **47.** 15 **49.** -1 **51.** -25 **53.** 2 **55.** 9 **57.** 13 **59.** 2 **61.** 2 **63.** -3 **65.** 5

PROBLEM SET 3.6

3. A **5.** B **7.** B **9.** B **11.** C **13.** $E = c - d$ **15.** $Y = 2x - 7$ **17.** $Y = \frac{x}{5z}$ **19.** $L = \frac{1}{2}p - w$ or $L = \frac{p - 2w}{2}$ **21.** $Y = -\frac{5}{3}x + 3$ **23.** $Y = \frac{5}{3}x - 3$ **25.** 9 **27.** 3 **29.** -1 **31.** 3 **33.** 6 **35.** 46, 48 **37.** 26 gallons **39.** 150 mi/15 gal ≈ 10 mpg; 12.2 mpg **41.** 200 mi/10 gal ≈ 20 mpg; 23.6 mpg **43.** 350 mi/10 gal ≈ 35 mpg; 41.6 mpg **45.** 1,000 mi/50 gal ≈ 20 mpg; 19.6 mpg **47.** Green Giant asparagus spears are about 0.8¢/oz less expensive. **49.** Star Kist is about 0.1¢/oz less expensive than Chicken of the Sea. **51.** The four-pack is the better buy and is about $0.06/battery less. **53.** The 1-lb, 2-oz = 18-oz box is $0.008/oz less expensive. **55.** The value of the lot is $18,700. **57.** $10.92 **59.** $91.04 **61.** $4.91 **63.** $548.42 **65.** $78.44

PROBLEM SET 3.7

1. *Verbal statements may vary.* **a.** Nineteen is greater than four. **b.** x is less than eight. **3.** *Verbal statements may vary.* **a.** Negative five is less than or equal to negative two. **b.** The opposite of x is greater than two. **c.** k is greater than or equal to negative two. **d.** m is less than negative four. **5. a.** $P > 0$ **b.** $N < 0$ **7. a.** $M \leq 4$ **b.** $Q \geq 4$

9.
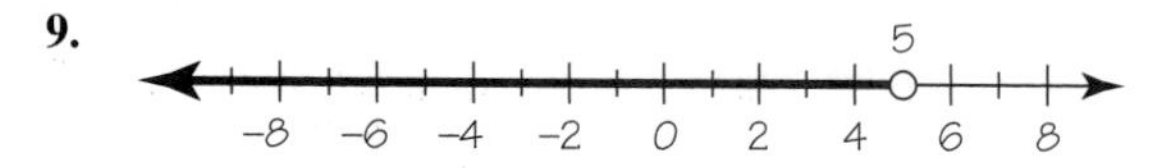

11.
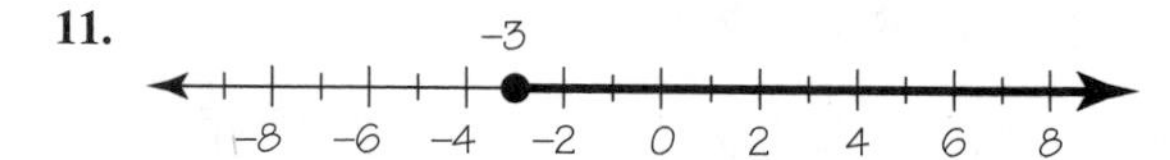

13.

15.
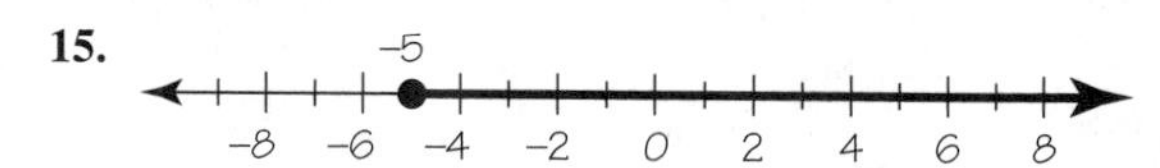

17.
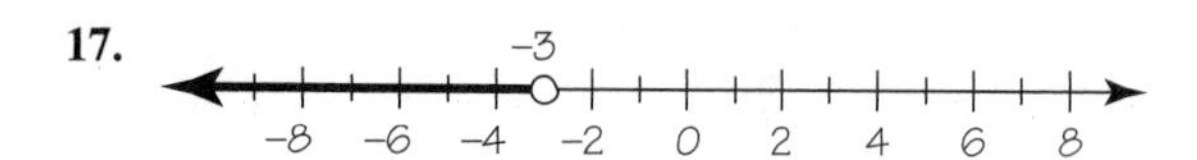

19.
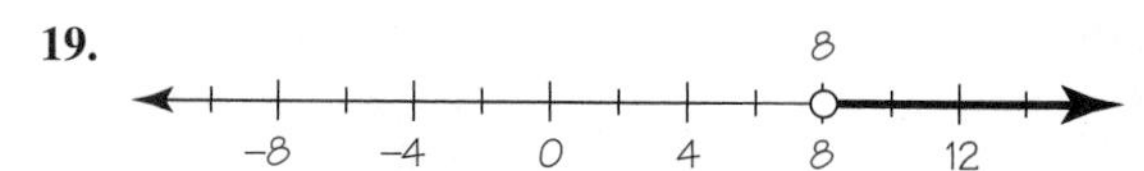

21.
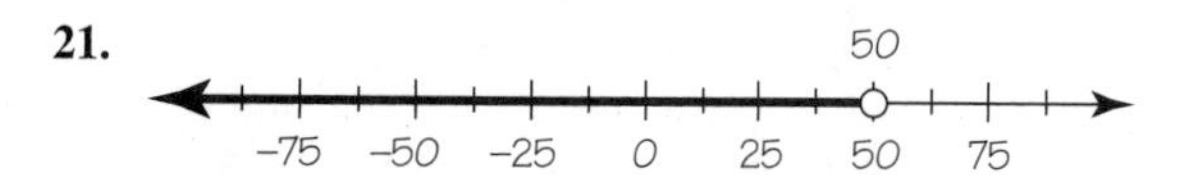

23.
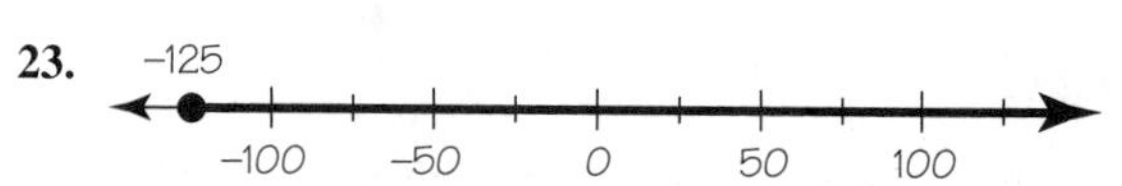

25.
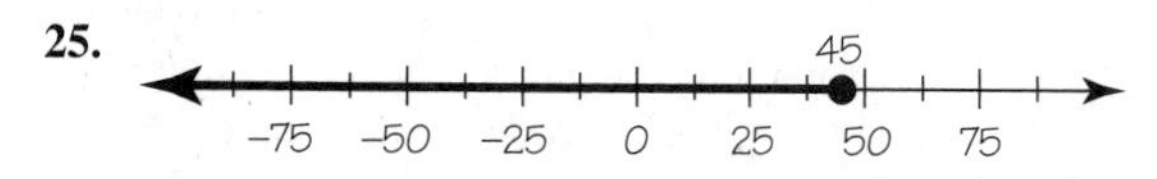

27.
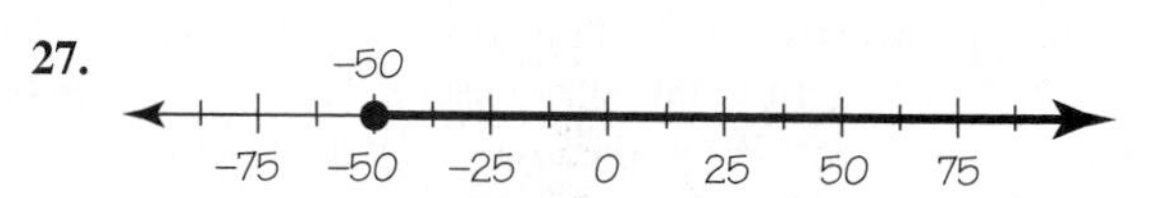

29.
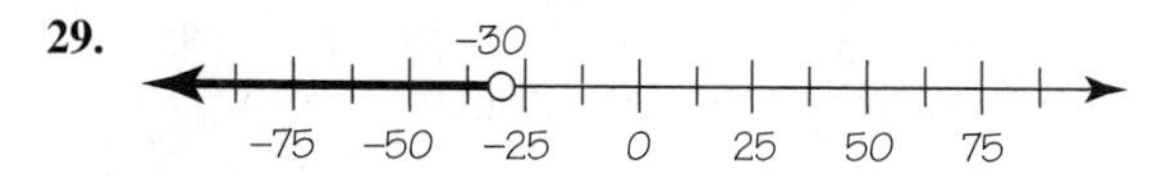

31.
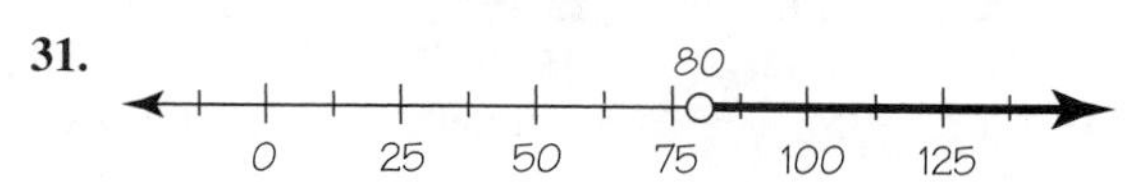

33. F; it is a solid dot. **34.** T **35.** F; reverse inequality **36.** F; reverse inequality **37.** F; reverse inequality when dividing by a negative **38.** F; reverse inequality **39.** $x \geq -4$ **41.** $x \geq -2$ **43.** $y > -6$ **45.** $s > -2$ **47.** $m < 5$ **49.** $x > -1$ **51.** $x \leq 1$ **53.** $s < -6$ **55.** $a \geq 2$ **57.** any number greater than -5 **59.** any number less than -4 **61.** any number less than 4 **63.** All values in the domain satisfy the inequality.

CHAPTER 3 Practice Test

1. a. 3; 5 **b.** 1; 2 **c.** 4; 1 **d.** 5; -3 **3. a.** x^9y^{11} **b.** $36x^2y$ **c.** $48x^2y^2$ **d.** x^3y **5. a.** $6x - 14$ **b.** $-4x - 1$ **c.** $3x - 9y$ **d.** $-2x - 1$ **7. a.** $9x^2 - 16$ **b.** $2a^2b^2 + abc - 3c^3$ **9. a.** -3 **b.** -31 **c.** 21 **d.** -10 **11. a.** 23 **b.** -5 **13. a.** 2 **b.** -20 **15. a.** $x > -3$ **b.** $x > 3$ **17.** $t = \dfrac{I}{Pr}$ **19.** $y = 2x$ **21.** The first number is 24; estimate by $75/3 = 25$. **23.** The estimated cruising range is about $10 \times 25 = 250$; by direct calculation it is 285.6 miles. **25.** The car's miles per gallon is 35.

Chapter 4

PROBLEM SET 4.1

3. $\frac{3}{2}$ **5.** $\frac{1}{3}$ **7.** $\frac{4}{7}$ **9.** $\frac{5}{1}$ **11.** $\frac{23}{1}$ **13.** $\frac{3}{1}$ **15.** $\frac{3}{1}$ **17.** $\frac{2}{5}$ **19.** $\frac{2}{5}$ **21.** $\frac{11}{25}$ **23.** $\frac{17}{7}$
25. means: 8, 35; extremes: 5, 56; 5 is to 8 as 35 is to 56. **27.** means: 2, 3; extremes: 5, x; 5 is to 3 as 2 is to x. **29.** means: 2, c; extremes: a, 3; a is to 2 as c is to 3. **31.** means: b, c; extremes: a, d; a is to b as c is to d. **33.** F; ratios involve two numbers. **34.** F; ratios involve two numbers. **35.** T **36.** T **37.** F **38.** T **39.** yes **41.** yes **43.** no **45.** no **47.** yes **49.** no **51.** no **53.** yes **55.** 20 to 1 **57.** 53 to 50 **59. a.** 3 to 4 **b.** 1 to 4 **c.** 3 to 2 **d.** 3 to 19 **61.** 14 to 1 **63.** 18 to 1 **65.** 15 to 26 **67.** 39 to 100 **69.** 8 to 25 **71.** 183 to 500

PROBLEM SET 4.2

1. F; it means to find the value of the missing or unknown quantity. **2.** T **3.** 30 **5.** 10 **7.** 8 **9.** 21 **11.** 16 **13.** $\frac{15}{2}$ or 7.5 **15.** 4 **17.** 3 **19.** 9 **21.** $\frac{1}{4}$ or 0.25 **23.** $\frac{5}{2}$ or 2.5 **25.** 22 **27.** $4.90 **29.** 10 gallons **31.** about 9 minutes **33.** 2 gallons **35.** 27 minutes **37.** 2,475 calories **39.** $8\frac{1}{3}$ in.

PROBLEM SET 4.3

9. $5\frac{1}{2}$ or $\frac{11}{2}$ **11.** $\frac{37}{1{,}000}$ **13.** $\frac{3}{50}$ **15.** 0.50; 50% **17.** $\frac{3}{4}$; 75% **19.** $\frac{2}{5}$; 0.4 **21.** 1; 1 **23.** $\frac{17}{20}$; 85% **25.** $\frac{3}{5}$; 0.6 **27.** 0.375; 37.5% **29.** $\frac{9}{20}$; 0.45 **31.** $\frac{5}{2}$; 2.5 **33.** $\frac{9}{10}$; 90% **35.** 0.2; 20% **37.** $0.\overline{6}$; $66\frac{2}{3}$% **39.** $\frac{3}{20}$; 15% **41.** $\frac{13}{20}$; 65% **43.** $\frac{7}{40}$; 17.5% **45.** $\frac{2}{25}$; 0.08 **47.** $\frac{5}{8}$; 0.625 **49.** 0.15; 15% **51.** $\frac{3}{80}$; 3.75% **53.** $0.8\overline{3}$; $83\frac{1}{3}$% **55.** $\frac{1}{400}$; 0.25% **57.** $0.13\overline{8}$; $13\frac{8}{9}$% **59.** $\frac{1}{9}$; $0.\overline{1}$ **61.** $0.\overline{7}$; $77\frac{7}{9}$% **63. a.** 70%; C **b.** 80%; B **c.** 75%; C **d.** 83%; B **e.** 95%; A **65. a.** 80%; B **b.** 75%; C **c.** 78%; C **d.** 60%; D **e.** 98%; A **67.** $8.75

PROBLEM SET 4.4

3. T **4.** T **5.** T **6.** 50% of 50% is 25%; multiply decimals — do not add. That is, "of" means multiply. *Answers to 7 – 28 may vary.* **7.** 1,000 **9.** 9,500 **11.** 5,000 **13.** 200 **15.** 750 **17.** 36 or 40 **19.** 6,000 **21.** 90 million (Think: $180 \div 2$) **23.** 45 million (Think: $180 \div 4$) **25.** 162 million (Think: $9 \times 18 = 162$; that is not an easy estimate, so you might instead use the results of Problem 24: $180 - 18 = 162$. Another possibility is $9 \times 20 = 180$.) **27.** 5.4 million (Think: $1.8 \times 3 = 5.4$ or $\approx 2 \times 3 = 6$) **29.** $\frac{15}{100} = \frac{A}{64}$; 9.6 **31.** $\frac{14}{100} = \frac{21}{W}$; 150 **33.** $\frac{P}{100} = \frac{10}{3}$; 200% **35.** $\frac{P}{100} = \frac{4}{5}$; 80% **37.** $\frac{P}{100} = \frac{9}{12}$; 75% **39.** $\frac{35}{100} = \frac{49}{W}$; 140 **41.** $\frac{120}{100} = \frac{16}{W}$; $13\frac{1}{3}$ **43.** $\dfrac{33\frac{1}{3}}{100} = \dfrac{12}{W}$; 36 **45.** $\frac{6}{100} = \frac{A}{8{,}150}$; $489 **47.** 19.8 million **49.** $10.86 **51.** 5% **53.** 59 (You can't employ half a person.) **55.** The tax withheld is $2,624. **57.** 94 points **59.** 90% **61.** The old wage was $1,250 and the new wage is $1,350. **63.** 400% **65.** 50%

CHAPTER 4 Practice Test

1. a. 20 to 1 **b.** 2 to 1 **c.** 31 to 15 **d.** 18 to 1 **3. a.** 0.6; 60% **b.** 1.5; 150% **c.** $1.\overline{3}$; $133\frac{1}{3}$% **d.** $0.1\overline{6}$; $16\frac{2}{3}$% **5. a.** $\frac{7}{20}$; 0.35 **b.** $\frac{12}{5}$; 2.4 **c.** $\frac{3}{50}$; 0.06 **d.** $\frac{3}{8}$; 0.375 **7. a.** $\dfrac{45}{100} = \dfrac{A}{120}$; 54 **b.** $\dfrac{P}{100} = \dfrac{60}{80}$; 75% **c.** $\dfrac{25}{100} = \dfrac{A}{300}$; 75 **d.** $\dfrac{120}{100} = \dfrac{90}{W}$; 75 **9.** 5 gallons **11.** 7.5 gallons **13.** 17 **15.** $62.54

Chapter 5

PROBLEM SET 5.1

1. Answers vary; an axiom is a property or principle that is assumed to be true, whereas a theorem is a property that is proved true. **3.** Answers vary; traditional (Euclidean geometry) and transformational geometry **4.** F; some words are undefined, some are assumed (axioms), and some facts are proved (theorems). **5.** F; the lines must be in the same plane. **6.** T *Constructions in Problems 7 – 15 can be verified by comparison with the art in the text.* **17.** symmetric **19.** symmetric **21.** not symmetric **23.** symmetric **25.** reflection

PROBLEM SET 5.2

1. T; if two angles are the same, then they must have the same measure. **2.** F; a right angle has a measure of 90°. **3.** T **5.** pentagon **7.** decagon **9.** hexagon **11.** dodecagon **13.** heptagon **15.** nonagon *Constructions in Problems 17 – 21 can be verified by comparison with the art in the text.* **23. a.** right **b.** acute **c.** straight **d.** acute **e.** acute **25.** $\angle AOC$ **27. a.** right **b.** acute **c.** straight **d.** acute **e.** acute **29.** $\angle AOC$ **31. a.** right **b.** acute **c.** straight **d.** acute **e.** acute **33.** $\angle AOC$

PROBLEM SET 5.3

3. $\overline{AB} \simeq \overline{ED}$; $\overline{AC} \simeq \overline{EF}$; $\overline{CB} \simeq \overline{FD}$; $\angle A \simeq \angle E$; $\angle B \simeq \angle D$; $\angle C \simeq \angle F$ **5.** $\overline{RS} \simeq \overline{TU}$; $\overline{RT} \simeq \overline{TR}$; $\overline{ST} \simeq \overline{UR}$; $\angle SRT \simeq \angle UTR$; $\angle S \simeq \angle U$; $\angle STR \simeq \angle URT$ **7.** 88° **9.** 145° **11.** 56° **13.** 75° **15.** 80° **17.** 100° *Constructions in Problems 19 – 24 can be verified by comparison with the art in the text.* **25.** 50° **27.** 120° **29.** 20° **31.** 21° **33.** 6° **35.** 30°; 60°; 90° **37.** 13°; 53°; 114° **39.** See Problem 21. **41.** See Problem 23. **43.** activity problem **45.** 540°

PROBLEM SET 5.4

5. F; $\frac{a}{c} = \frac{a'}{c'}$ **6.** F; $\overline{BC}$ and $\overline{B'C'}$ **7.** T **8.** T **9.** F; $\cos B = \frac{\text{adjacent side of } B}{\text{hypotenuse}}$ **10.** T **11.** a **13.** b **15.** c **17.** $\frac{b}{c}$ **19.** $\frac{a}{c}$ **21.** $\frac{b}{a}$ **23.** not similar **25.** similar; $\overline{GH} \simeq \overline{JK}$; $\overline{GI} \simeq \overline{JL}$; $\overline{IH} \simeq \overline{LK}$; $\angle G \simeq \angle J$; $\angle H; \simeq \angle K$; $\angle I \simeq \angle L$ **27.** similar; $\overline{TV} \simeq \overline{WY}$; $\overline{TU} \simeq \overline{WX}$; $\overline{VU} \simeq \overline{YX}$; $\angle T \simeq \angle W$; $\angle U \simeq \angle X$; $\angle V \simeq \angle Y$ **29.** $\sqrt{58}$ **31.** 9 **33.** $\frac{15}{7}$ **35.** $\frac{10}{9}$ **37.** 0.2588 **39.** 0.3090 **41.** 0.9903 **43.** 0.8290 **45.** 1.2799 **47.** 57.2900 **49.** $\sin A = \frac{12}{13}$; $\cos A = \frac{5}{13}$; $\tan A = \frac{12}{5}$ **51.** $\sin A = \frac{\sqrt{35}}{6} \approx 0.9860$; $\cos A = \frac{1}{6} \approx 0.1667$; $\tan A = \frac{\sqrt{35}}{1} \approx 5.9161$ **53.** $\sin A = \frac{1}{\sqrt{5}} \approx 0.4472$; $\cos A = \frac{2}{\sqrt{5}} \approx 0.8944$; $\tan A = \frac{1}{2} = 0.5$ **55.** $\sin A = \frac{1}{2} = 0.5$; $\cos A = \frac{\sqrt{3}}{2} \approx 0.8660$; $\tan A = \frac{1}{\sqrt{3}} \approx 0.5774$ **57.** $\sin A = \frac{\sqrt{2}}{3} \approx 0.4714$; $\cos A = \frac{\sqrt{7}}{3} \approx 0.8819$; $\tan A = \frac{\sqrt{2}}{\sqrt{7}} \approx 0.5345$ **59.** 45 ft **61.** 24 ft **63.** 13 ft **65.** 19 ft **67.** 70 ft **69.** 67,900,000 miles

CHAPTER 5 Practice Test

1. a. ——— **b.** [circle] **c.** Compare with text illustration. **d.** Compare with text illustration.

3. 85° **5. a.** straight **b.** obtuse **c.** acute **d.** acute **7.** 45°, 50°, 85° **9.** similar **11. a.** 4.3315 **b.** 0.1219 **c.** 0.9511 **d.** 0.9336 **13.** The pit is 45 ft long. **15.** The distance across the river is 45 ft.

Chapter 6

PROBLEM SET 6.1

3. A **5.** B **7.** C **9.** B **11.** A **13.** B **15.** B **17.** A
19. ——— **21.** ——— **23.** ———
25. ——— **27.** 3 cm **29.** 3.2 cm **31.** 1 in. **33.** 2 in. **35.** $1\frac{3}{8}$ in.

PROBLEM SET 6.2

3. **a.** $P = 3s$ **b.** $P = 2(\ell + w)$ **c.** $P = 4s$ **d.** $C = \pi d$ or $C = 2\pi r$ **5.** C **7.** A **9.** A **11.** C **13.** A **14.** F; it is approximately equal. **15.** F; π is an irrational number. **16.** F; it is the distance around the circle. **17.** F; the diameter is twice the radius. **18.** T **19.** T **21.** 30 ft **23.** 24 cm **25.** 960 ft **27.** 9 dm **29.** 21.99 m **31.** 157.08 ft **33.** 502.65 ft **35.** 46 ft **37.** 530 cm **39.** 20.0 cm **41.** 15.2 cm **43.** 35.7 in. **45.** 11.1 cm **47.** 65 ft **49.** The sides are 26 in., 39 in., and 52 in. **51.** 66 dm

PROBLEM SET 6.3

5. C **7.** A **9.** C **11.** A **13.** C **15.** C **17.** F; it is 9 ft^2. **18.** F; it is $\frac{9}{4}\pi$ ft^2. **19.** F; it is km^2. **20.** F; it is $0.5\ell h$. **21.** T **22.** F; two sides are parallel. **23.** 3 cm^2 **25.** 12 cm^2 **27.** 3 cm^2 **29.** 10 or 11 cm^2 **31.** 15 $in.^2$ **33.** 1,196 m^2 **35.** 100 mm^2 **37.** 7,560 ft^2 **39.** 136.5 dm^2 **41.** 5,550 cm^2 **43.** 28 $in.^2$ **45.** 314.2 $in.^2$ **47.** 78.5 $in.^2$ **49.** 19.6 dm^2 **51.** 7.6 cm^2 **53.** 216 $in.^2$ **55.** $93\frac{1}{2}$ $in.^2$ **57.** \$3,593.75 **59.** 156 ft^2 **61.** **a.** about 30 ft $\times$ 30 ft = 900 ft^2 **b.** $977\frac{1}{2}$ ft^2 **63.** 32 yd^2 are required; \$1,664 **65.** 20 pounds are necessary; \$117 **67.** 79 $in.^2$ **69.** 154 $in.^2$ **71.** about \$0.10 per square inch **73.** large **75.** Answers vary; the size is about 50 $in.^2$, and the price for a small pizza should be about \$0.11 per square inch; this is about \$5.49 (\$4.95 to \$5.95 is acceptable; calculate $50 \times 0.11 \approx 5.50$). **77.** C; lot C is \$5.59/ft^2, and lot D is \$6.14/ft^2. **79.** 196,020 ft^2

PROBLEM SET 6.4

4. T **5.** F; a liter is larger. **6.** T **7.** F; they are the same size. **8.** F; it is a cubic measure. **9.** F; 1,000 dm^3 = 1 m^3 **10.** F; since 3 ft = 1 yd, 9 ft^2 = 1 yd^2 and 27 ft^3 = 1 yd^3 **11.** 60 cm^3 **13.** 42 cm^3 **15.** 1,728 $in.^3$ **17.** 1 yd^3 **19.** 400 mm^3 **21.** 96,000 cm^3 **23.** 16 oz **25.** 13 oz **27.** 380 mL **29.** 14 oz **31.** 25 mL **33.** 75 mL **35.** C **37.** A **39.** B **41.** B **43.** B **45.** B **47.** 4.3 gal **49.** 3,000 L **51.** 1 L **53.** 500 L **55.** 24 L **57.** **a.** 45.375 ft^3 **b.** 26.375 ft^3 **59.** 2.5 yd^3 **61.** 6 yd^3 **63.** 1 yd^3 **65.** 112 kL **67.** **a.** 5.6×10^5 people/mi^2 **b.** 10,761 mi^2 **c.** 5.5 acres/person

PROBLEM SET 6.5

3. 0°C, 32°F; 100°C, 212°F; 22°C, 70°F **5.** C **7.** C **9.** B **11.** B **13.** A **15.** B **17.** C **19.** B **21.** B **23.** A **25.** centimeter **27.** meter **29.** milliliter **31.** kiloliter **33.** kilogram **35.** Celsius **37.** 30 ft^3 **39.** 19 cm^3 **41.** 600 cm^3 **43.** 47 $in.^3$

PROBLEM SET 6.6

1. **a.** $\frac{1}{12}$ **b.** $\frac{1}{36}$ **c.** $\frac{1}{63,360}$ **3.** **a.** 12 **b.** $\frac{1}{3}$ **c.** $\frac{1}{5,280}$ **5.** **a.** 36 **b.** 3 **c.** $\frac{1}{1,760}$ **7.** **a.** $\frac{1}{3}$ **b.** $\frac{1}{6}$ **c.** $\frac{1}{48}$ **9.** **a.** $\frac{1}{8}$ **b.** $\frac{1}{16}$ **c.** $\frac{1}{2}$ **11.** **a.** $\frac{1}{2}$ **b.** $\frac{1}{8}$ **c.** $\frac{1}{4}$ **13.** **a.** $\frac{1}{16}$ **b.** $\frac{1}{32,000}$ **c.** $\frac{1}{2,000}$ **15.** $\frac{9}{1,760}$ mi; 27 ft; 324 in. **17.** $\frac{1}{10,560}$ mi; $\frac{1}{6}$ yd; $\frac{1}{2}$ ft **19.** 7,040 yd; 21,120 ft; 253,440 in. **21.** $\frac{5}{2,112}$ mi; $4\frac{1}{6}$ yd; $12\frac{1}{2}$ ft **23.** $1\frac{31}{32}$ qt; $3\frac{15}{16}$ pt; $7\frac{7}{8}$ c **25.** $1\frac{3}{4}$ qt; 7 c; 56 oz **27.** 2 qt; 4 pt; 64 oz **29.** 10 pt; 20 c; 160 oz **31.** $\frac{1}{500}$ T; 64 oz **33.** $\frac{3}{1,000}$ T; 6 lb **35.** $\frac{21}{8,000}$ T; 84 oz **37.** 9,000 lb; 144,000 oz

CHAPTER 6 Practice Test

1. **a.** ______________________

b. ____

c. ______________ **d.** ____

3. a. inch and centimeter **b.** ounce and gram **c.** ounce and milliliter **d.** degree Fahrenheit and degree Celsius
5. Answers vary. **c.** 37°C; 98.6°F **7. a.** dkg, cg, mg **b.** kL, cL, mL **c.** qt, c, tsp **d.** hm, m, mm
9. a. $P = 28$ ft; $A = 33$ ft^2 **b.** $P = 60$ in.; $A = 225$ in.^{2} **c.** $P = 28$ m; $A = 32$ m^2 **d.** $P = 30$ ft; $A = 30$ ft^2
11. a. 392 in.^{3}; 2 gal **b.** 1,848 in.^{3}; 8 gal **c.** 896 dm^3; 896 L **d.** 20 dm^3; 20 L **13. a.** 34 m^3 **b.** 42 cm^3
5. 21 yd^2 **17.** 224 gal **19.**

13 12 5

Chapter 7

PROBLEM SET 7.1

5. \$22.50 **7.** \$13.75 **9.** \$3.75 **11.** \$5.00 **13.** \$1.00 **15. a.** 0.9 **b.** 0.8 **c.** 0.7 **d.** 0.6 **e.** 0.5
17. a. $\frac{2}{3}$ **b.** $\frac{1}{3}$ **c.** $\frac{1}{4}$ **d.** $\frac{2}{5}$ **e.** $\frac{5}{6}$ **19. a.** 0.17 **b.** 83% **c.** $\frac{3}{5}$ **d.** 0.81 **e.** 73% **21.** C **23.** B **25.** A
27. F; the markdown is \$10. **28.** F; the percent markdown is 20%. **29.** F; c is the complement of the percent markdown.
30. F; after solving for c, find the percent markdown by subtracting from 1. That is, find $1 - c$. **31.** F; the complement is 70%. **32.** T **33.** \$13.80 **35.** \$5.53 **37.** \$917.50 **45.** \$13.25 **47.** \$4,473.70 **49.** \$26.50
51. \$218.50 **53.** \$74.10 **55.** \$22.50 **57.** \$175.00 **59.** \$24.71 **61.** 3.9% **63.** \$525.00 **65.** 47.5%
67. b. 2 bags/min is four minutes of wages; the savings is 40¢. **c.** less than 0.5%; hardly a bundle **69.** \$89.56

PROBLEM SET 7.2

1. money paid for the use of another's money **3.** ordinary interest, divide by 360; exact interest, divide by 365
5. B; interest rate is not stated, but you should still recognize a reasonable answer. **7.** C; price is not stated, but you should still recognize a reasonable answer. **9.** C is the most reasonable. **11.** B **13.** C **15.** \$80; \$1,080
17. \$288; \$1,088 **19.** \$42.50; \$542.50 **21.** \$200; \$600 **23.** \$990; \$2,190 **25.** \$400,000; \$500,000
27. 15%; \$4,600 **29.** \$100; \$166 **31.** 9%; \$1,032 **33.** \$1,260; \$4,260 **35.** \$900; \$1,332 **37.** 13%; \$1,824
39. 18%; \$12,350 **41.** 10 yr; \$720 **43.** \$13.40; \$249.90 **45.** F; $P = \frac{I}{rt}$ **46.** F; $r = \frac{I}{Pt}$ **47.** F; use $t = \frac{315}{360}$.
48. F; use $t = \frac{5}{12}$. **49.** F; the future value is \$45.50 × 36 = \$1,638. **50.** T **51.** 17% **53.** 6 years
55. \$101,500 **57.** \$800,000 **59.** \$200,000 **61.** $122\frac{1}{2}\%$ **63.** \$400,000

PROBLEM SET 7.3

3. B **5.** C **7.** A **9.** A **11.** F; $N = 12t$ and $t = 3$, so $N = 36$. **12.** F; forgot to add the interest before dividing by 36. **13.** F; $r = 0.058$, not 5.8; $I = \$155.73$. **14.** F; $P = \$36,500$ **15.** T **16.** F; $N = 12t$ and $t = 5$, so $N = 60$. Thus, $m = \$897.29$. **17.** \$15,747 **19.** \$34,631 **21.** \$29,108 **23.** \$1,296 interest; \$136 per month **25.** \$330 interest; \$76.25 per month **27.** \$4,050 interest; \$142.50 per month **29.** \$143 interest; \$33.04 per month **31.** \$997.50 interest; \$260.31 **33.** 23% **35.** 21% **37.** 35% **39.** 21% **41.** 18%
43. \$11,430 **45.** \$23,443.20 **47.** \$8,886.48 **49.** 9.2% (0.0915266667) **51.** 6.0% (0.0600303951)

The answers for Problems 53 – 57 *recalculate the interest rate and do not use the rounded rates stated in Problems* 48 – 52.
53. 17.5% (0.1751351351) **55.** 31.0% (0.3103701913) **57.** 17.0% (0.1704881633) **59.** 5% add-on rate; APR is about 9.8%. **61.** Sears' add-on rate is 29.2% APR; bank is better.

PROBLEM SET 7.4

1. $I = Prt$ **3.** C **5.** A **7.** C **9.** A **11.** B **13.** 15% **15.** 18% **17.** 8% **19.** $4.50 **21.** $3.95 **23.** $0.75 **25.** $8.33 **27.** $7.67 **29.** $49.88 **31.** $52.50 **33.** $34.52 **34.** F; average daily balance method does not use $\frac{1}{12}$. **35.** F; adjusted balance method **36.** F; previous balance method **37.** F; use 31 days to obtain $1,045.16. **38.** F; use 28 days and (55 × 18) instead of (55 × 20) to obtain $162.14. **39.** F; $P = \$850$; $I = \$10.63$ **40.** F; $P = \$400$; $I = \$5.00$ **41.** F; $t = \frac{31}{365}$; $I = \$6.39$ **43.** Over $33 the interest would be more than $0.50.

PROBLEM SET 7.5

1. a. $n = 4$ **b.** $n = 2$ **c.** $n = 12$ **d.** $n = 360$ **3. a.** $N = nt$ **b.** $i = \dfrac{r}{n}$ **5.** $P = \dfrac{A}{(1 + i)^N}$ **7.** F; $n = 12$ **8.** F; in this book we use $n = 360$ unless otherwise stated. **9.** F; $i = 1\%$ **10.** F; $i = 0.005$ or 0.5% **11.** T **12.** F; $t = 4$; $r = 0.045$; and $n = 360$, so $N = 4(360) = 1{,}440$; $i = \dfrac{r}{n} = \dfrac{0.045}{360}$; we find, by calculator, $A \approx \$598.60$.
13. $1,400; $1,469.33; $69.33 more **15.** $2,720; $2,809.86; $89.86 more **17.** $17,000; $48,231.47; $31,231.47 more **19.** 1; 9%; 5; $1,538.62 **21.** 1; 8%; 3; $629.86 **23.** 4; 2%; 12; $634.12 **25.** 4; 4.5%; 40; $29,081.82 **27.** 4; 5%; 40; $35,199.94 **29.** 12; 2%; 60; $13,124.12 **31.** 4; 3%; 2; $954.81 **33.** 4; 4%; 5; $1,520.82 **35.** $903.46 **37.** $1,028.25 **39.** $755.59 **41.** $16,536.79 **43.** $50.73 **45.** $572,177.99 **47.** $3,019,988.95 **49. a.** $1.15 **b.** $1.92 **c.** $3.00 **d.** $2.08 **51. a.** $139.76 **b.** $201.88 **c.** $1,149.16 **d.** $1,863.51 **53.** $1,220.19 **55.** $2,181,175.28 **57.** $8,705.51 **59.** $15,415.93 **61.** 35 years (from Table II) **63.** Answers vary; pay off the loan.

PROBLEM SET 7.6

1. interest rate, origination fee, and points **3.** length of loan, down payment, and APR **7.** Use the maximum amount of a loan formula (and Table II). **9.** C **11.** D **13.** C **15.** $369 **17.** $702 **19.** $1,163 **21.** $13,980 **23.** $6,435 **25.** $33,600 **27.** $2,097 **29.** $3,217.50 **31.** $840 **33.** 7.69% **35.** 6.68% **37.** 10.91% **39.** $405.46 **41.** $583.94 **43.** $820.25 **45.** $746.30 **47.** $521.45 **49.** $1,006.88 **51.** $744.80 **52.** F; it costs $3.97 more (not less). **53.** F; it saves $27,391 (25 years not 20 years). **54.** F; forgot the 0.125 factor. The comparison rate is 6.19%. **55.** F; did not correctly change percents to decimals. It should be $0.055 + 0.125(0.0075 + 100/252{,}000) \approx 0.05599$; 5.60%. **56.** F; looked up the wrong number in Table II; it should be 7.16, so the monthly payment is $2,491.68. **57.** F; $321,354; should subtract the amount of the loan, not the price of the house. **59.** $15,200 **61.** $7,200 **63.** $102,000 **65. a.** $20,000 **b.** $80,000 **c.** $644.80 **d.** $74,752 **e.** $1,791.11 **67. a.** $45,000 **b.** $180,000 **c.** $1,857.60 **d.** $265,824 **e.** $5,160 **69.** Maximum loan is $169,527.90; down payment is $70,472.10.

CHAPTER 7 Practice Test

1. a. $6.75 **b.** $9.00 **c.** $17.99 **d.** $44.55 **3. a.** 18% **b.** 15% **c.** 21% **d.** 9% **5. a.** $4,750 **b.** $5,719.39 **c.** $6,029.29 **d.** $6,108.05 **e.** $5,918.41 **7. a.** $478,000 **b.** $49,000 **c.** $2,047,000 **d.** $329,000 **9. a.** 11.02% **b.** 7.16% **c.** 13.71% **d.** 8.51% **11.** $203.20 **13.** 40% **15.** $16.25 **17.** $12,844.33 **19.** $1,269 interest; $100.53 monthly payment; 35.0% APR **21.** 15.7% **23.** $1,216.80 **25.** $967.50 **27.** $2,627.34 **29.** $4,500

Chapter 8

PROBLEM SET 8.1

7. F; set is undefined. **8.** F; Ø = { } **9.** F; the cardinality is 1. **10.** F; the cardinality of {0} is 1, but the cardinality of Ø is 0. **11.** F; it is finite. **12.** F; the counting numbers have cardinality $\aleph_0$. **13.** well defined **15.** well defined **17.** not well defined; answers vary **19.** not well defined; answers vary **21.** {*m, a, t, h, e, i, c, s*} **23.** {1, 3, 5, 7, 9} **25.** {*p, i, e*} **27.** {6, 8, 10, 12, 14} **29.** {counting numbers less than 10} **31.** {multiples of 10 between 0 and 105} **33.** {odd numbers between 100 and 170} **35.** the set of all *x* such that *x* is an odd counting number; {1, 3, 5, 7, . . . } **37.** the set of all *x* such that *x* is a natural number not equal to 8; {1, 2, 3, 4, 5, 6, 7, 9, 10, 11, . . . } **39.** the set of all *x* such that *x* is a whole number less than 8; {0, 1, 2, 3, 4, 5, 6, 7} **43. a.** $A \longleftrightarrow C$; $B \longleftrightarrow D \longleftrightarrow E \longleftrightarrow F$ **b.** $A = C$; $D = E = F$ **47.** {1, 2, 3, . . . , *n*, . . . , 353, 354, . . . , 586, 587}

$\updownarrow$ $\updownarrow$ $\updownarrow$ $\updownarrow$ $\updownarrow$ $\updownarrow$ $\updownarrow$ $\updownarrow$

{550, 551, 552, . . . , *n* + 549, . . . , 902, 903 }

Not 1-to-1, so they do not have the same cardinality.

49. a. finite **b.** infinite **c.** finite **d.** finite

PROBLEM SET 8.2

5. F; the listed set is the improper subset of *E*. **6.** F; {1, 2, 3} and Ø are missing. **7.** Ø, {*m*}, {*y*}, {*m, y*} **9.** Ø, {*y*}, {*o*}, {*u*}, {*y, o*}, {*y, u*}, {*o, u*}, {*y, o, u*} **11.** Ø , {*m*}, {*a*}, {*t*}, {*h*}, {*m, a*}, {*m, t*}, {*m, h*}, {*a, t*}, {*a, h*}, {*t, h*}, {*m, a, t*}, {*m, a, h*}, {*m, t, h*}, {*a, t, h*}, {*m, a, t, h*} **13. a.** Ø **b.** Ø , {1} **c.** Ø, {1}, {2}, {1, 2} **d.** Ø, {1}, {2}, {3}, {1, 2}, {1, 3}, {2, 3}, {1, 2, 3} **e.** Ø, {1}, {2}, {3}, {4}, {1, 2}, {1, 3}, {1, 4}, {2, 3}, {2, 4}, {3, 4}, {1, 2, 3}, {1, 2, 4}, {1, 3, 4}, {2, 3, 4}, {1, 2, 3, 4} **f.** 32 subsets

PROBLEM SET 8.3

5. F; union is {1, 2, 3, 4, 5, 6, 7, 8, 9}. **6.** F; intersection is {4, 5}. **7.** F; $\overline{A}$ = {6, 7, 8, 9, 10, 11, 12, . . . } **8.** T **9.** {2, 6, 8, 10} **11.** {3, 4, 5} **13.** {2, 3, 5, 6, 8, 9} **15.** {1, 3, 4, 5, 6, 7, 10} **17.** {1, 2, 3, 4, 5, 6, 7, 9, 10} **19.** {1, 2, 3, 4, 5, 6} **21.** {1, 2, 3, 5, 6, 7} **23.** {3} **25.** {5, 6, 7} **27.** {1, 2, 4, 6}

29.

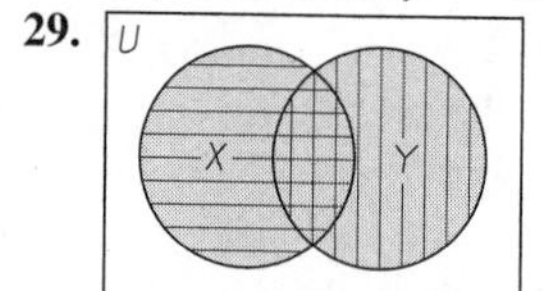

31.

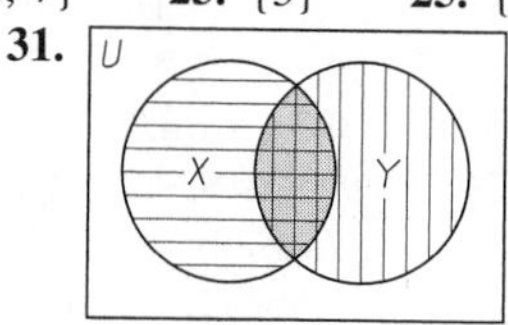

33.

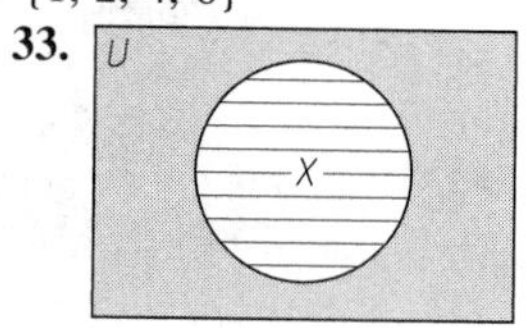

35. a. *B* **b.** *A* **c.** *U* **d.** Ø

PROBLEM SET 8.4

6. F; first find the union; then find the complement. **7.** F; complement does not distribute. **8.** F; complement does not distribute. **9.** F; use a highlighter to mark all parts shaded in either part (2) or part (3). **10.** T; there is no mistake.

11.

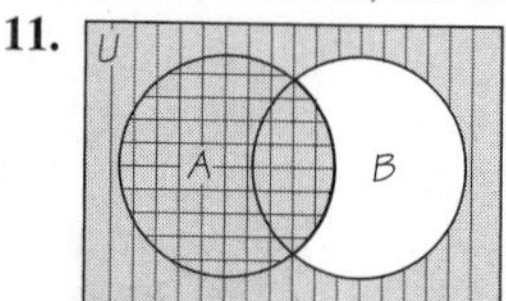

13.

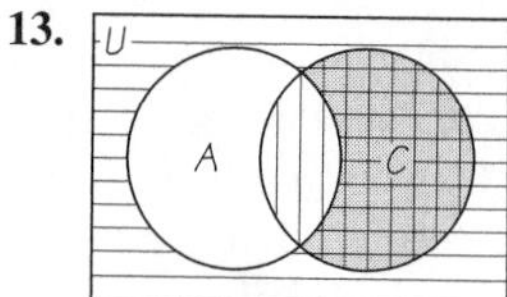

15.

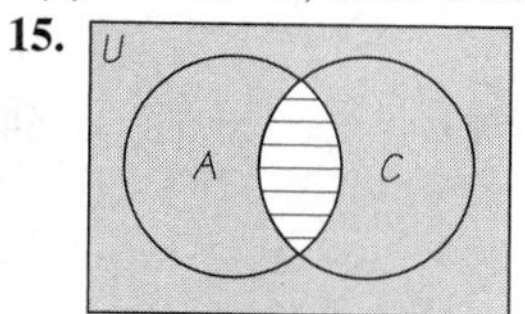

17.

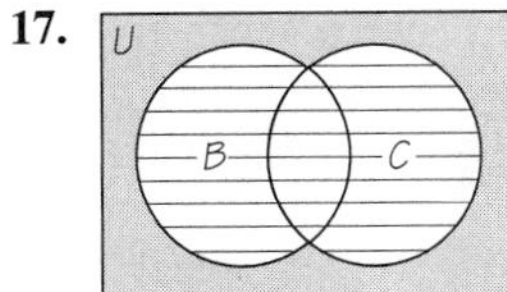

19.

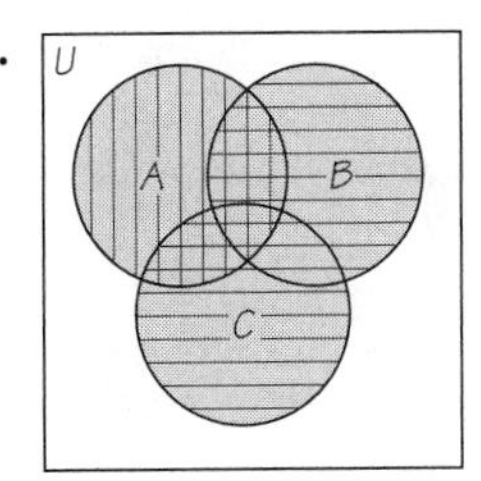

21.

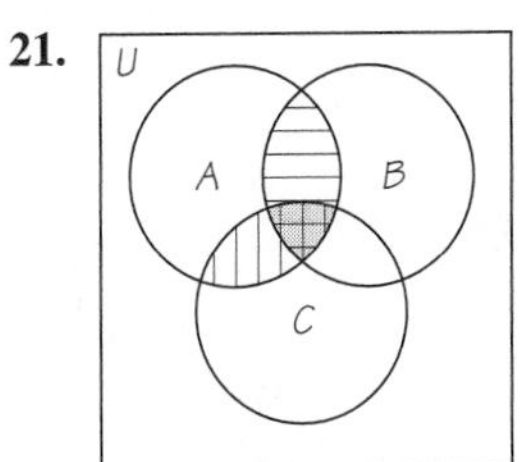

23.

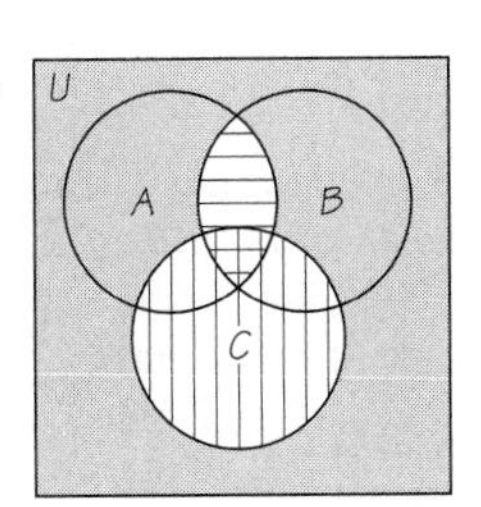

25.

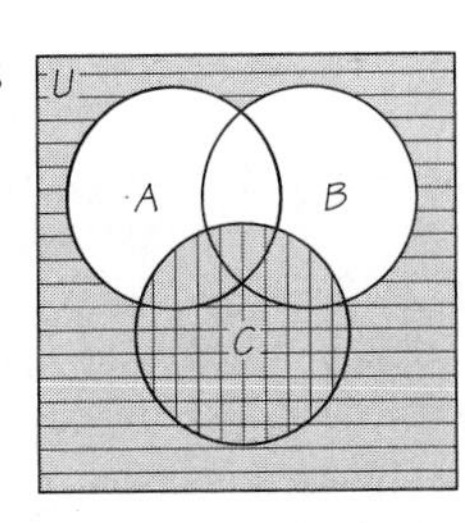

27. ∅ **29.** A or {2, 4, 6, 8}
31. ∅ **33.** {4, 6} **35.** {1, 3, 7, 10}
37. T **39.** T
41. See Example 2.

43.

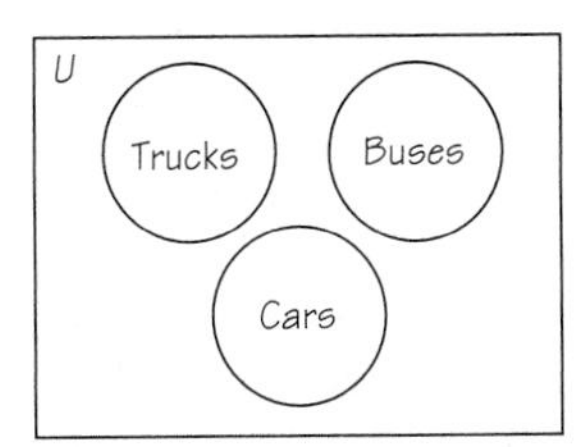

45.

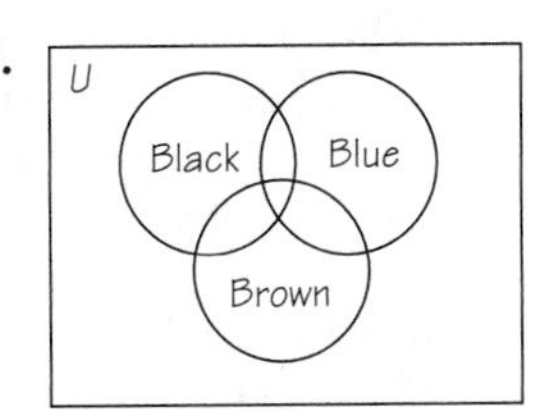

47.

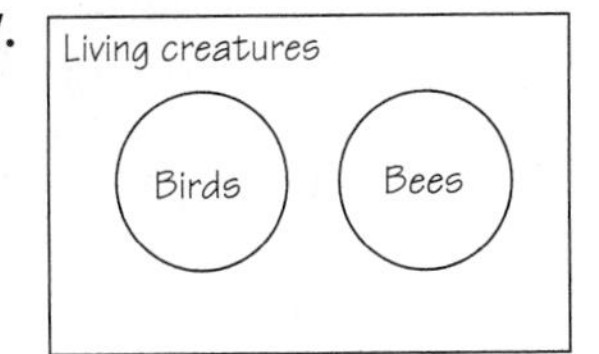

PROBLEM SET 8.5

1. 4 **3.** F; we need to know how many are in the intersection. **5.** 144

7.

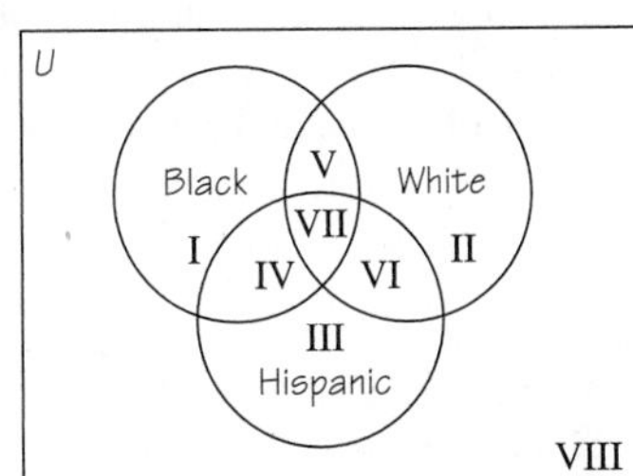

Numbers in millions;
I: 30
II: 189
III: 24
IV: 5
V: 1
VI: 3
VII: empty
VIII: 11

9. a. 50 **b.** 15
11. There are 34 people playing.

13.

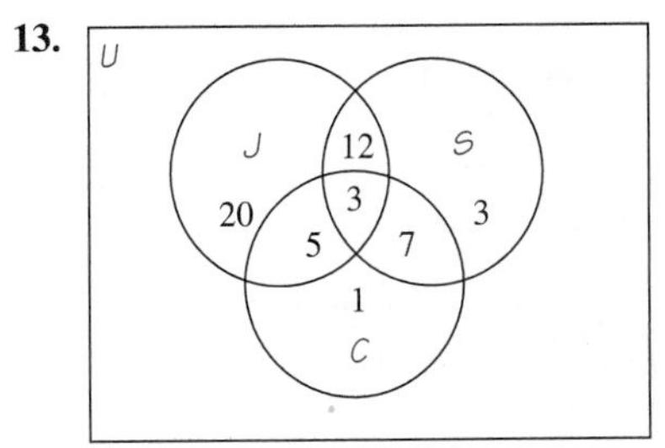

15.

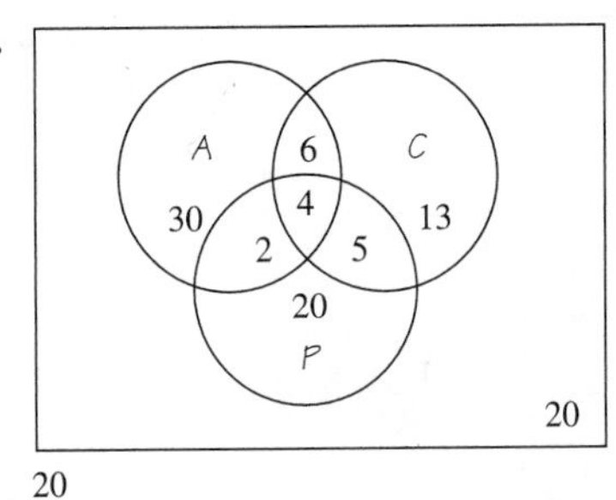

17.

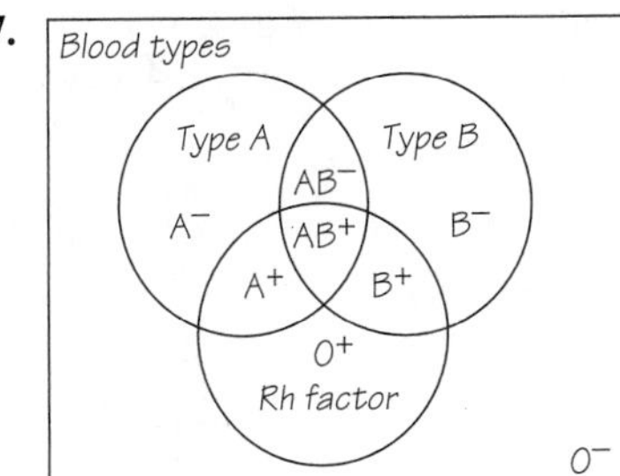

19. a. U Whites Blacks C Hispanics Others

b. Assume the total population is 263 million and roughly 80% are white; then, | white | = 210 million; 15% are black, | black | = 40 million; and 10% are Hispanic, | Hispanic | = 26 million. It is clear that
$0.27(210) \approx 57$
$0.14(40) \approx 6$
$0.13(26) \approx 3$
The sum of the smaller numbers does not even come close to the larger.

PROBLEM SET 8.6

3. inductive reasoning; explanations vary **5.** A syllogism is a type of logical reasoning that has a form consisting of two premises and a conclusion. **7.** 3, 6, 9, 3, 6, 9, . . . **9** $1{,}000^2 = 1{,}000{,}000$ **11.** $100^2 = 10{,}000$ **13. a.** 8 **b.** 188 **c.** 2,888 **d.** Answers vary. **15.** (I) figures that are not equilateral, equiangular, or quadrilaterals; (II) equilateral figures that are not equiangular and not quadrilaterals; (III) equiangular figures that are not equilateral and are also not quadrilaterals; (IV) figures that are both equilateral and equiangular, but are not quadrilaterals; (V) figures that are both equilateral and quadrilaterals, but are not equiangular; (VI) figures that are equilateral, equiangular, and quadrilaterals; (VII) figures that are both quadrilaterals and equiangular, but are not equilateral; (VIII) quadrilaterals that are not equilateral and also are not equiangular. *Counterexamples in Problems* 17–25 *may vary.* **17.** T **19.** F; 3 + 5 = 8 **21.** T **23.** F; 3 by 4 rectangle has $A = 12$ and $P = 14$. **25.** F; $S \subseteq S$ but the cardinality is not less. **27.** valid **29.** valid **31.** Set of men is the complement of the set of women; not valid **33.** not valid **35.** valid **37.** valid **39.** Phoenix to Albuquerque is about 450 miles and Albuquerque to Oklahoma City is about 560 miles. If she gets up at 6:00 A.M., and he is in Phoenix, then at about 3:00 P.M. he will be in Albuquerque if he is traveling by car. Another 12 hours and he'll be in Oklahoma, and she'll probably be sleeping. Other modes of transportation would not fit the song so well, so he is probably using a car. Inductive reasoning **41.** Answers vary; two persons: a girl or woman and the speaker (we don't know whether the speaker is male, even though it is probable). Girl or woman: (1) she works; (2) she has a door; (3) she can read; (4) she can talk; (5) she knows the speaker's name. Speaker: (1) left a note on her door; (2) is traveling east; (3) has left her before; (4) has informed her of intention of leaving at other times before this note; (5) has (had) a phone; deductive reasoning **43.** the speaker; deductive reasoning **45.** It could not have taken place on Sunday ("Talking to him . . . last Sunday"), Monday (or else would have said talked to him last night), or Saturday (or else would have said preacher is coming to dinner tomorrow). Possible days are Tuesday, Wednesday, Thursday, and Friday. Most probable day is Wednesday (next most probable is Thursday) because since Sunday night Billy Joe spent some time at the sawmill and on Choctaw Ridge with speaker. This would indicate that it probably wasn't Tuesday. The preacher would be likely to have dinner plans by Friday, so again the probable day is Wednesday; inductive reasoning **47.** yes (Every country does, even though it might not be a holiday.) **49.** white (It must be at the North Pole for all sides to have a southern exposure.) **51.** They are sisters; the beggar is a woman.

CHAPTER 8 Practice Test

Counterexamples may vary. **3.** F; 0.121121112 . . . **5. a.** not well defined **b.** well defined **c.** well defined **d.** not well defined **7.** Answers may vary. **a.** {even numbers between 9 and 20} **b.** {odd numbers between 2,000 and 3,000} **c.** {counting or natural numbers} **d.** {perfect squares less than 150} **9. a.** infinite **b.** finite **c.** finite **d.** finite **11. a.** {1, 2, 3, 4, 5, 7, 9, 10} **b.** {3, 6, 9, 10, 11, . . . , 25, 27, 30} **c.** {1, 2, 3, . . . , 10, 12, 15, 18, 21, 24, 27, 30} **d.** U **13. a.** {2, 4, 6, 8, 10, 11, 12, . . . } **b.** $\{x \mid x$ is an odd counting number$\}$ **c.** {11, 12, 13, 14, . . . } **d.** U

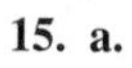

15. a.

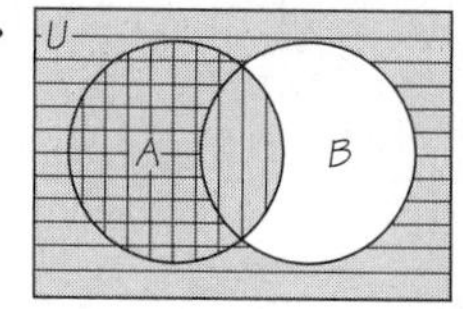

b.

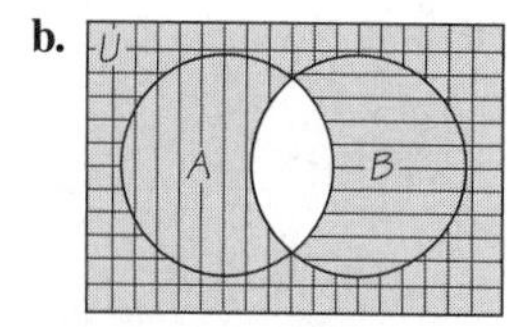

c.

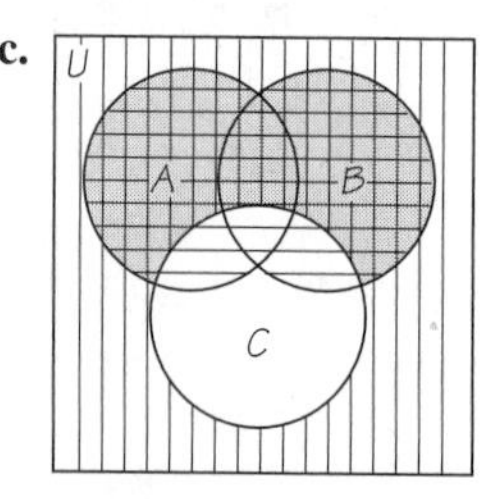

d.

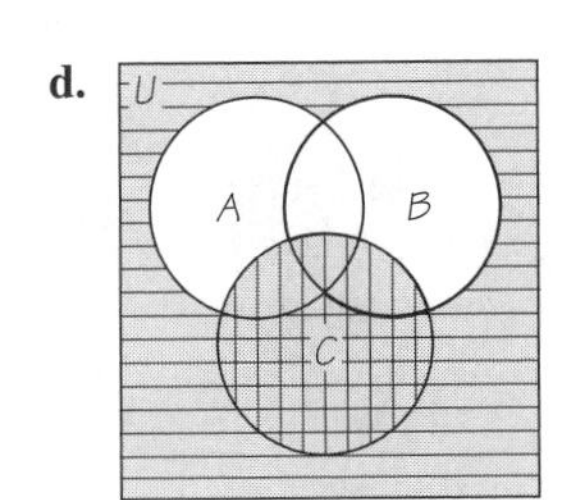

19. inductive reasoning
21. $100^2 + 1 = 10{,}001$
23. not valid
25. valid

Chapter 9

PROBLEM SET 9.1

7. T **8.** F; there are 36 events in the sample space; $P(\text{two}) = \frac{1}{36}$. **9.** F; $P(\text{two}) = \frac{4}{52} = \frac{1}{13}$ **10.** F; the outcomes must be mutually exclusive and equally likely. **11.** F; $P(A \text{ or } B) = P(A \cup B)$ **12.** F; $P(A \text{ and } B) = P(A \cap B)$ **13.** F; find the union of A and B first. **14.** F; find the intersection of A and B first. **15. a.** $\frac{1}{3}$ **b.** $\frac{1}{3}$ **c.** $\frac{1}{3}$ **17. a.** $\frac{1}{6}$ **b.** $\frac{1}{6}$ **c.** $\frac{1}{6}$ **d.** $\frac{1}{2}$ **19.** about 0.02 **21.** about 0.21 **23.** $\frac{5}{18}$ **25.** $\frac{1}{2}$ **27.** 0.00001 38516 9452 **29.** 0.00144 05762 3 **31.** 0.00392 46467 82 **33.** 0.04753 90156 1 **35.** $\frac{1}{52}$ **37.** $\frac{1}{4}$ **39.** $\frac{1}{4}$ **41.** 0 **43.** $\frac{1}{52}$ **45. a.** $\frac{1}{12}$ **b.** $\frac{7}{12}$ **c.** $\frac{1}{12}$ **47. a.** $\frac{1}{4}$ **b.** $\frac{3}{4}$ **49.** $\frac{1}{9}$ **51.** $\frac{1}{6}$ **53.** $\frac{1}{9}$ **55.** $\frac{7}{36}$ **57.** $\frac{2}{9}$ **59. a.** 0.03 **b.** 0.06 **c.** 0.08 **d.** 0.11 **e.** 0.14 **f.** 0.17 **g.** 0.14 **h.** 0.11 **i.** 0.08 **j.** 0.06 **k.** 0.03 **61.** yes **63.** $\frac{1}{9}$

65.

	1	2	3	4
1	(1,1)	(1,2)	(1,3)	(1,4)
2	(2,1)	(2,2)	(2,3)	(2,4)
3	(3,1)	(3,2)	(3,3)	(3,4)
4	(4,1)	(4,2)	(4,3)	(4,4)

67. a. $\frac{1}{4}$ **b.** $\frac{3}{16}$ **c.** $\frac{1}{8}$

Also show the sample spaces in Problems 69–73. **69.** Pick A; $P(A \text{ winning}) = \frac{3}{4}$ **71.** Pick either; fair game, $P(B) = P(C) = \frac{1}{2}$ **73.** Pick E; $P(E \text{ winning}) = \frac{2}{3}$ **75.** A beats B, B beats C, C beats D, but D beats A! Choose second; the reason this game is called WIN is that if you let your opponent choose first you can always wind up with a probability of winning of $\frac{2}{3}$. If your opponent picks A, you pick D; if your opponent picks B, you pick A; if your opponent picks C, you pick B; if your opponent picks D, you pick C. **77.** Chance, since the most likely number to roll is 7.

PROBLEM SET 9.2

3. B **5.** B **7.** They are the same. **9.** F; $P(\overline{E}) = \frac{4}{7}$ **10.** F; they are complements. **11.** F; the probability of heads on the next flip is $\frac{1}{2}$. **12.** F; the probability of black on the next spin is not changed by past spins of the wheel; the wheel does not "remember" previous spins. **13.** F; it is $\frac{13}{52} = \frac{1}{4}$ **15.** $\frac{1}{5}$ **17.** 0.995 **19.** $\frac{33}{36} = \frac{11}{12}$ **21.** 1, 2, 3, or 5; $\frac{2}{3}$ **23.** drawing a spade, club, or diamond; 0.82 **25.** your school losing; 0.17 **27.** $\frac{10}{13}$ **31.** Answers vary; your limited resources and the betting limit imposed on the game

PROBLEM SET 9.3

3. F; odds in favor are 2 to 1. **4.** F; odds against are 5 to 3. **5.** F; they are reciprocals. **7.** 1 to 3 **9.** $\frac{9}{10}$ **11.** $P(\#1 \text{ winning}) = \frac{1}{3}$; $P(\#2 \text{ winning}) = \frac{1}{16}$; $P(\#3 \text{ winning}) = \frac{2}{5}$; $P(\#4 \text{ winning}) = \frac{5}{12}$; $P(\#5 \text{ winning}) = \frac{1}{2}$. *Note:* At a horse race the odds are determined by track betting and the sum of the probabilities is not necessarily 1. **13.** $\frac{3}{7}$ **15.** 1 **17.** $\frac{1}{3}$ **19.** $\frac{1}{2}$ **21.** $\frac{4}{51}$ **23.** $\frac{4}{17}$ **25.** $\frac{26}{51}$ **27.** $P(7|2) = \frac{2}{11}$ **29.** 2 to 7 **31.** Pat is correct.

PROBLEM SET 9.4

1. EXPECTATION = (AMOUNT TO WIN)(PROBABILITY OF WINNING) **3.** A **5.** C **7.** C **9.** \$27.20 **11.** \$0.83 **13.** The expectation is about \$0.09, but the nickel is most likely to be picked. **15.** \$0.02 **17.** F; there are also green spots. $P(\text{black}) = \frac{18}{38} = \frac{9}{19}$ **18.** F; if you play the game a number of times, the *average* winnings per game should be \$5. **19.** T **20.** T **21.** F; there is a cost of playing; $E = 49{,}999\left(\frac{1}{1{,}000{,}000}\right) + (-1)\left(\frac{999{,}999}{1{,}000{,}000}\right) = -\$0.95$ **22.** F; change odds to a probability; $E = (\$49)(\frac{1}{100}) + (-\$1)(\frac{99}{100}) \approx -\$0.50$ **23.** \$12.14 **25.** \$317.47 **27.** \$43.38 **29.** $E = -\$0.50$ **31.** $E = -\$8{,}125$; they should not dig since the expectation is negative. **33.** $E = \$0$ **35.** The expectation is a loss of \$0.50 per play. It is not a good deal. **37.** $E = \$0$; fair game **39.** $E = -\$0.25$; don't play the game. **41.** $E \approx 0.19886$ or about \$0.20 **43.** $E \approx 0.05329\ 99137$ or about \$0.05 **45.** $-\$0.05$ **47.** $-\$0.05$ **49.** $-\$0.05$ **51.** $-\$0.05$ **53.** $-\$0.05$ **55.** $E = \$1(\frac{1}{2}) + \$2(\frac{1}{4}) + \$4(\frac{1}{8}) + \cdots = \$5.00$ **57.** $E = \$1(\frac{1}{2}) + \$2(\frac{1}{4}) + \$4(\frac{1}{8}) + \$8(\frac{1}{16}) + \cdots = \$0.50 + \$0.50 + \$0.50 + \cdots$ Pay *any* amount to play the game since the sum is infinite. This result is unacceptable to many mathematicians and is consequently called a paradox. See Martin Gardner's *Mathematical Puzzles and Diversions* (New York: Simon & Schuster, 1959, p. 145) for a discussion of the problem.

CHAPTER 9 Practice Test

3. $\frac{1}{5}$ **5. a.** $\frac{5}{36}$ **b.** $\frac{1}{6}$ **7. a.** $\frac{1}{13}$ **b.** $\frac{1}{4}$ **c.** $\frac{4}{13}$ **9. a.** $\frac{1}{4}$ **b.** $\frac{1}{9}$ **c.** $\frac{1}{36}$ **11.** 0.91 **13.** $\frac{15}{16}$ **15.** $\frac{20}{21}$ **17.** 50%; 1 to 1 **19.** $\frac{3}{5}$ **21.** $\frac{1}{3}$ **23.** \$96.60

Chapter 10

PROBLEM SET 10.1

5.

HEIGHT	*TALLY*	*FREQUENCY*
72	|	1
71	||	2
70	|||	3
69	||||	4
68	|||	3
67	卌	5
66	|||	3
65	|||	3
64	||||	4
63	||	2

7.

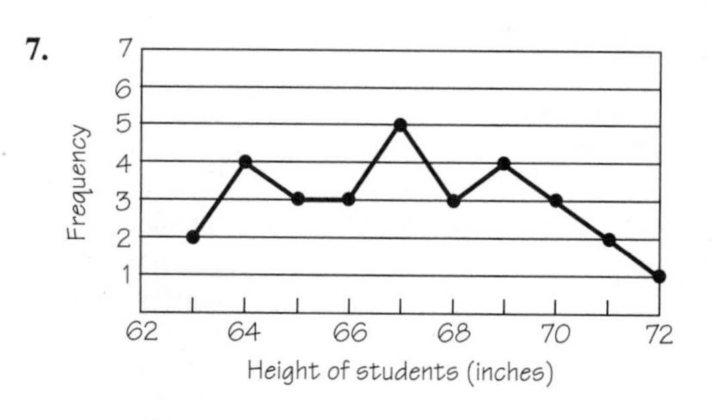

9.

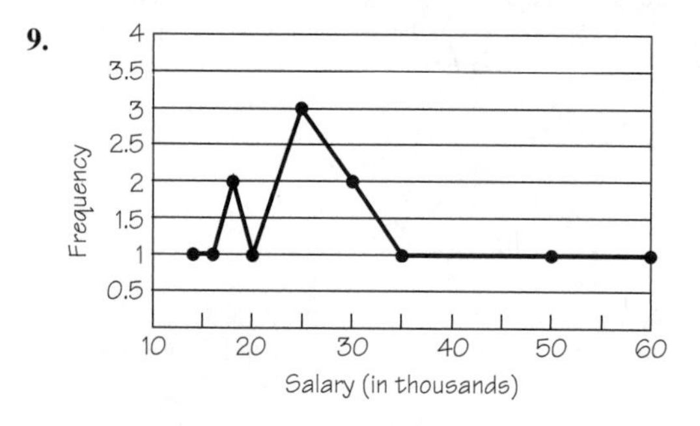

11.

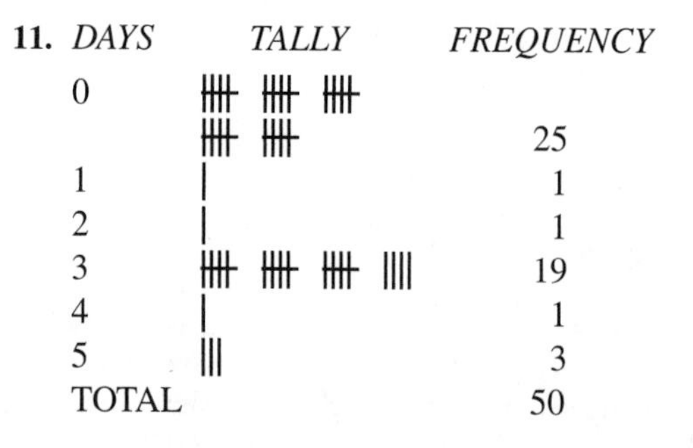

DAYS	*TALLY*	*FREQUENCY*
0	卌 卌 卌 卌 卌	25
1	|	1
2	|	1
3	卌 卌 卌 ||||	19
4	|	1
5	|||	3
TOTAL		50

13.

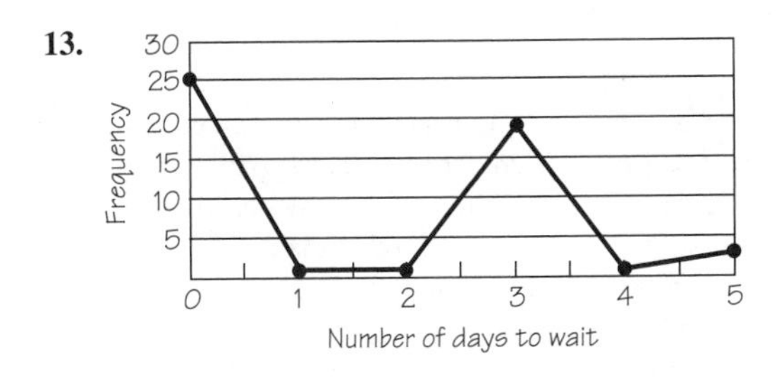

15.

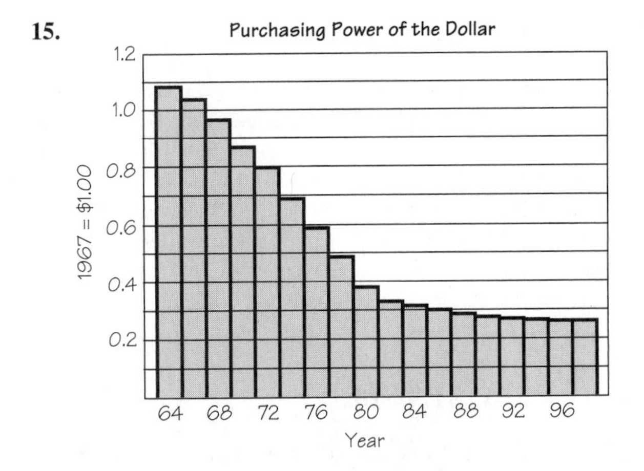

17. a. two **b.** Morton **c.** 30% **d.** 33% **e.** Sept. 1999 **f.** Sept. 2000 **19. a.** Answers vary; U.S. trade with Mexico for the years 1988–1991 **b.** more imports **c.** \$1.4 billion **d.** \$25 billion **21.** 25 times **23.** maybe illegal **25.** no **27.** 44 kwh **29.** 55 kwh **31.** appliance, 20°; automotive, 10°; clothing, 55°; grocery, 260°; nursery, 15°

33.

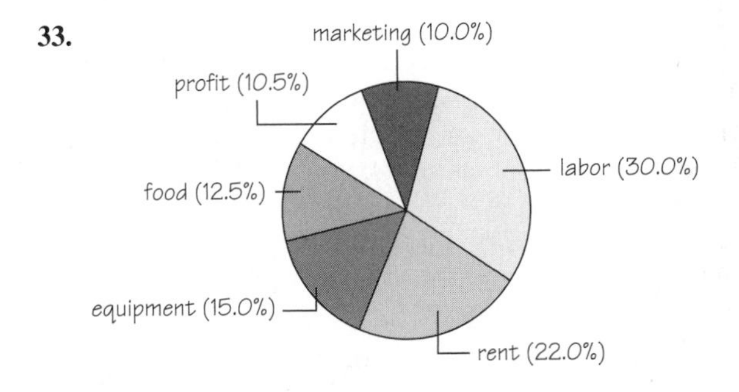

35.

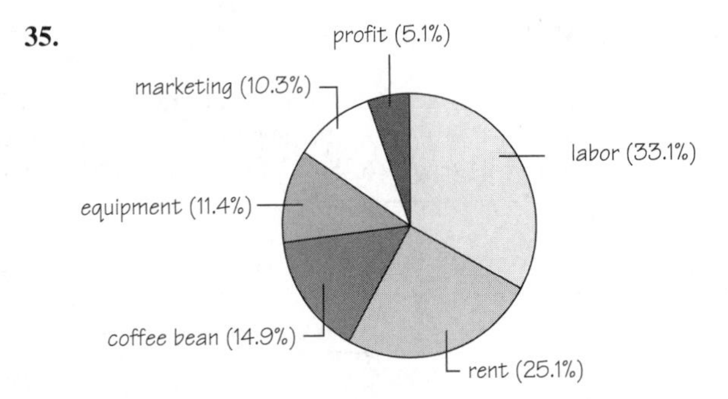

37.

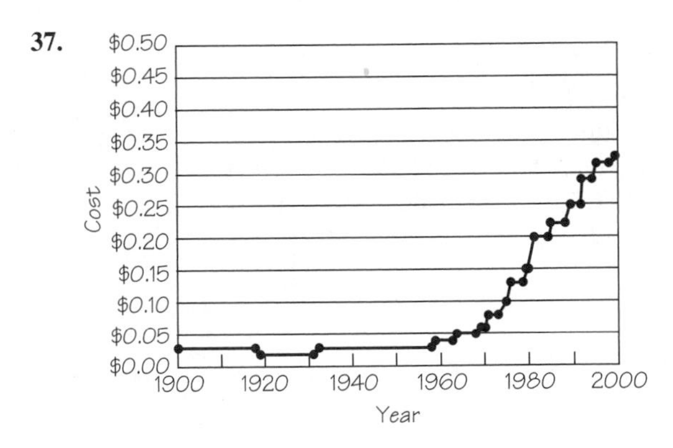

39. Answers vary; graph is meaningless without a scale. **40.** Answers vary; the graphs are based on height, but the impression created is that of area, and the area of the second tree is much more than 65% greater (even though the height is 65% greater). **41.** False; there may have been more traffic, more cars, worse weather; there may be many reasons for the increase in accidents. **42.** The graph is appropriate and accurate.

43. The advertisement says that the car is 57 in. wide on the outside, but a full 5 ft (60 in.) wide on the inside.

PROBLEM SET 10.2

5. F; the type of average was not specified. Do not assume that "average" is the mean. **6.** F; can't average averages. Combined, Dave had 136 hits for 490 at bats (0.278 avg). Combined, Andy had 253 hits for 969 at bats (0.261 avg). **7.** F; the middle number is 10. **8.** T **9.** F; the median is 10. **10.** F; the data set is bimodal with modes 10 and 15. **11.** mean = 3; median = 3; no mode **13.** mean = 105; median = 105; no mode **15.** mean = 6; median = 7; mode = 7 **17.** mean = 10; median = 8; no mode **19.** mean = 91; median = 95; mode = 95 **21.** mean = 3; median = 3; mode = 3 **23.** mean = 615 **25.** mean = \$28,900; median = \$28,000; mode = \$25,000 **27.** mean ≈ 68.1; median = 70; mode = 70 **29.** mean = 56; median = 65; mode = 70 **31.** mean = \$1,541.67; median = \$1,550.00; no mode; the median is most descriptive. **33.** mean = 2,713,956; median = 1,612,427; no mode; the mean is most descriptive.

PROBLEM SET 10.3

1. Three values that divide the data into four parts **3.** Ninety-nine values that divide the data into one hundred parts **5.** 94th percentile **7.** 87th percentile **9.** 99th percentile **11.** 6 **13.** 50 **15.** 11 **17.** 9th decile **19.** 2nd quartile **21. a.** 92nd **b.** 66 **c.** 56.5 **d.** 71 **e.** 75 **23.** 350 **25.** James **27.** 10th decile **29. a.** 3rd **b.** 12th percentile **c.** 5th decile **d.** 2nd quartile

PROBLEM SET 10.4

5. F; if the variance is between 0 and 1, the standard deviation will be larger than the variance. **6.** T **7.** range = 4; var = 2; $\sigma \approx 1.41$ **9.** range = 4; var = 2; $\sigma \approx 1.41$ **11.** range = 8; var ≈ 5.71; $\sigma \approx 2.39$ **13.** range = 18; var = 41.6; $\sigma \approx 6.45$ **15.** range = 17; var = 40.4; $\sigma \approx 6.36$ **17.** range = 4; var ≈ 1.43; $\sigma \approx 1.20$ **19.** mean = 217,851; variance = 2

21.

Q_2

0 Q_1 Q_3 100

23. $\sigma \approx 5{,}700$ **25.** $\sigma \approx 10$ **27.** $\sigma \approx 26$ **29.** mean = 19,200; range = 4,000; $\sigma = 1{,}327$

PROBLEM SET 10.5

1.

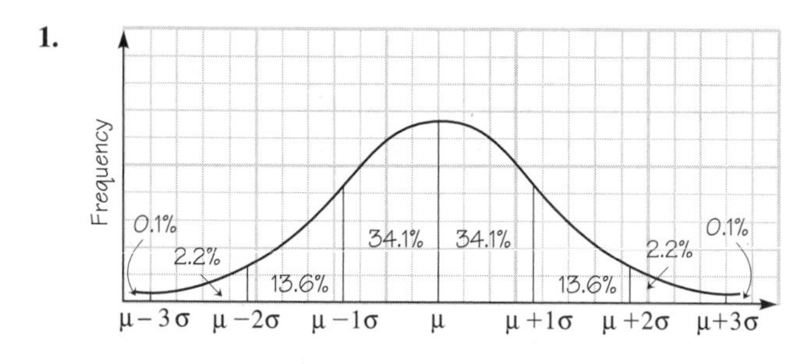

3. 49 **5.** 0.841 **7.** 31 **9.** 60 or above **11.** 25 **13.** 0.159 **15.** 0.977 **17.** 0.023 **19. a.** 50% **b.** 12.4 oz

PROBLEM SET 10.6

3. Suppose the coin lands heads 60% of the time. This means that tails comes up 40%. Heads/tails has probability $0.6 \times 0.4 = 0.24$, and tails/heads has probability $0.4 \times 0.6 = 0.24$.

CHAPTER 10 Practice Test

1.

TALLY		*FREQUENCY*
6	||	2
7	||||	4
8	~~||||~~	5
9	|||	3
10	|	1
TOTAL		15

3. 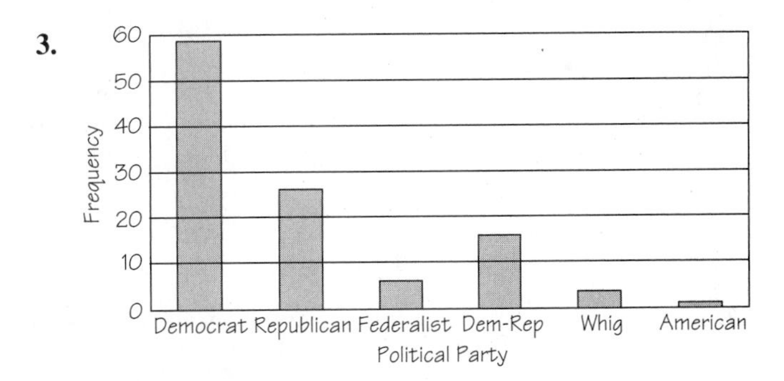

5. a. 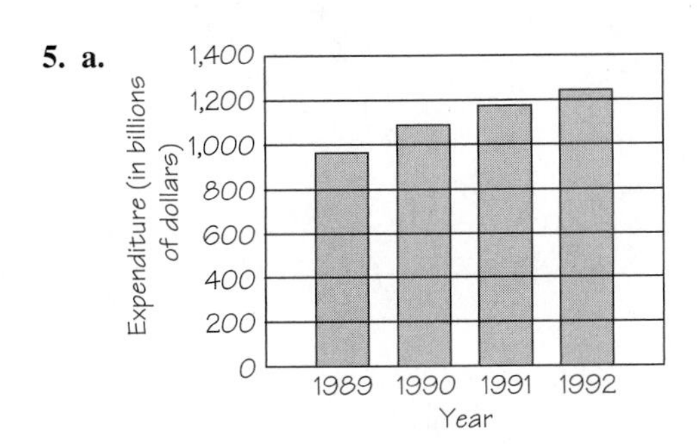

b. 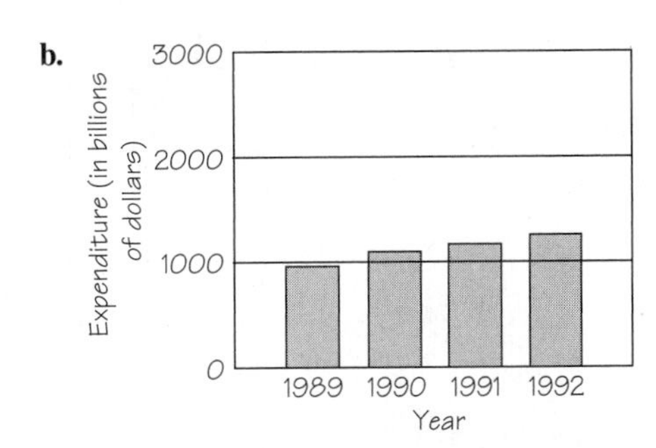

c. 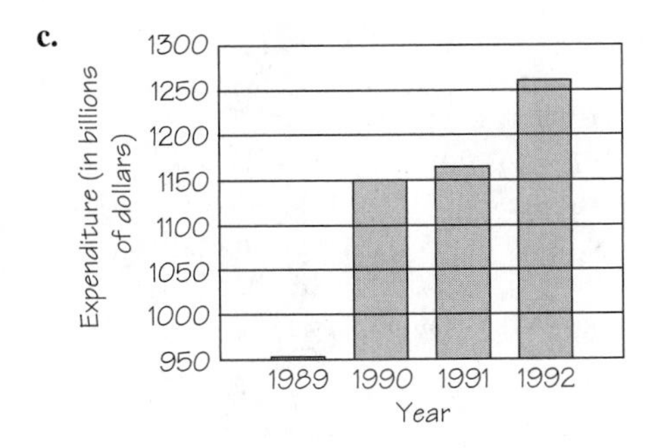

7. mean = 13.5; median = 14; mode = 16 **9.** mean = $14\frac{2}{3}$; median = 15; mode = 18 **11.** mode **13.** median **15.** range = 10; $\sigma \approx 2.82$; $\sigma^2 = 7.95$ **17.** range = 13; $\sigma \approx 4.16$; $\sigma^2 = 17\frac{1}{3}$ **19.** 230 **21.** smaller standard deviation; less variance **23.** Answers vary; accept false hypotheses. **25.** Answers vary; take a sample of representative pages.

Chapter 11

PROBLEM SET 11.1

1. Telegraph Hill **3.** Embarcadero **5.** Ferry Building **7.** Old Mint Building **9.** Coit Tower **11.** Eva Mikalson **13.** Terry Shell **15.** Laurie Pederson **17.** Otis Moorehouse **19.** Beverly Schaap

21.

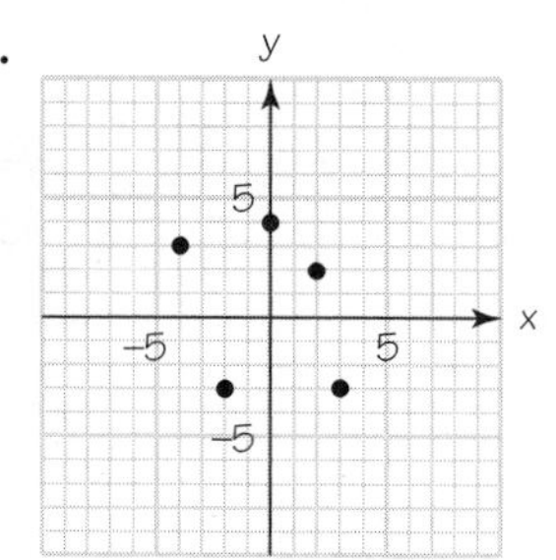

23.

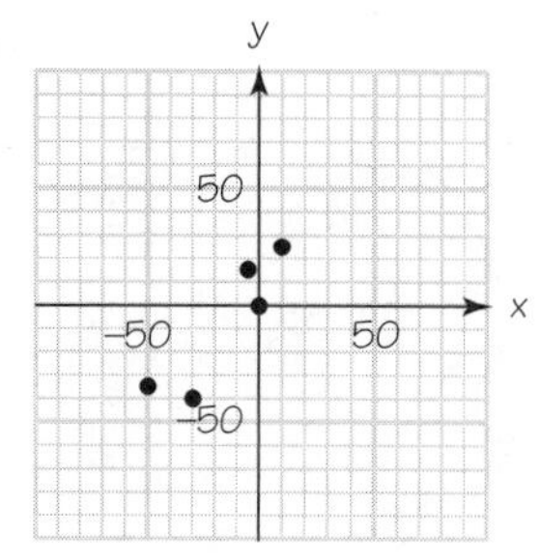

25.

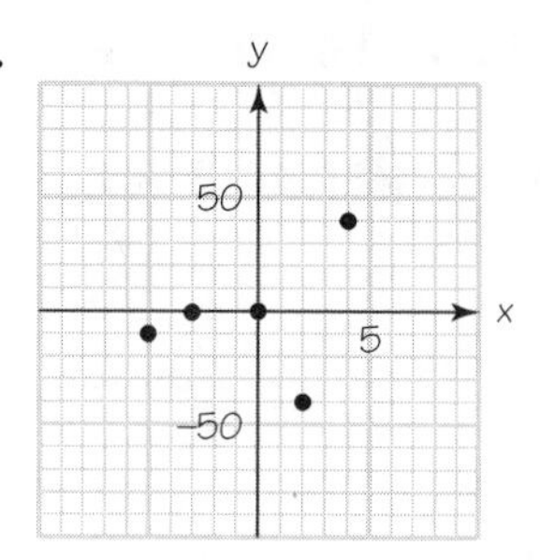

27.

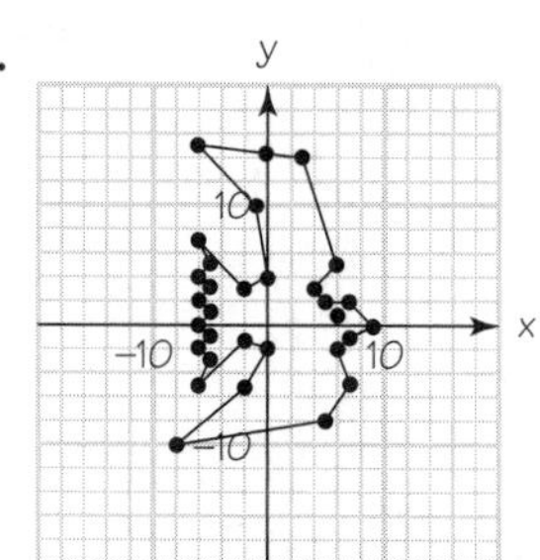

29. They all lie on the same line.

31.

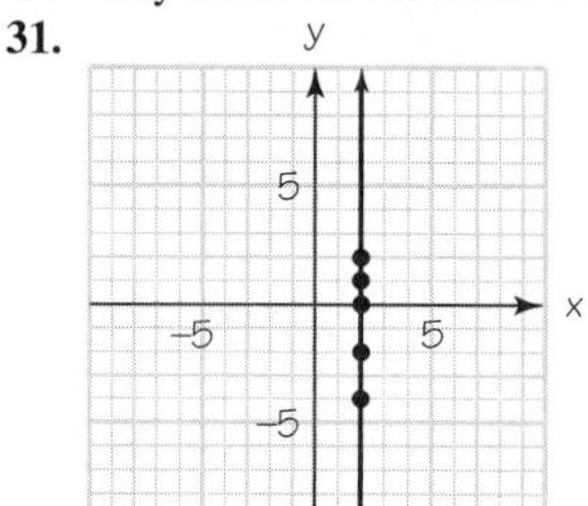

33.

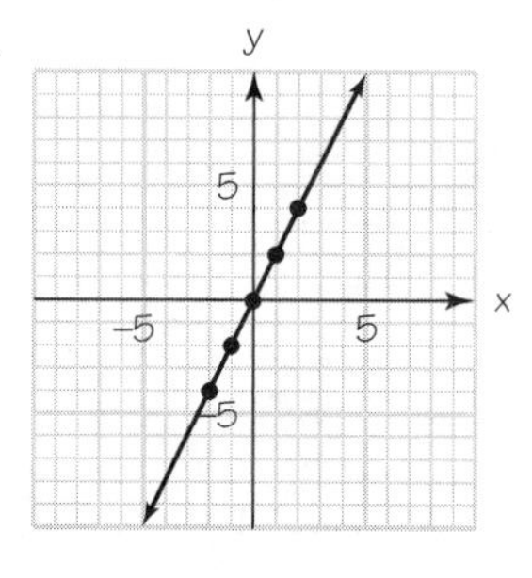

PROBLEM SET 11.2

3. F; each value 1, 2, and 3 is associated with one value, so it is a function. **4.** F; input value is 5. **5.** T **6.** F; $f(10) = 10 + 10 = 20$ **7.** F; $g(10) = 10(10) = 100$ **8.** F; $h(\pi) = 2\pi$ **9.** function **11.** function **13.** not a function **15.** function **17.** function **19.** not a function **21.** $M(4) = 17$, $M(6) = 37$, $M(-8) = 65$, $M(\frac{1}{2}) = 1\frac{1}{4}$, $M(x) = x^2 + 1$ **23.** $g(4) = 7$, $g(6) = 11$, $g(-8) = -17$, $g(\frac{1}{2}) = 0$, $g(x) = 2x - 1$ **25.** $f(4) = 12$, $f(6) = 14$, $f(-8) = 0$, $f(\frac{1}{2}) = 8\frac{1}{2}$, $f(x) = x + 8$ **27. a.** 8 **b.** −16 **c.** $p - 7$ **29. a.** −1 **b.** −31 **c.** $3a - 1$ **31. a.** 5 **b.** −2 **c.** $\frac{3}{2}$ **33. a.** 64 **b.** 96 **c.** 128 **d.** 256 **e.** 512 **35. a.** 1,430 **b.** 1,050 **c.** 670 **d.** 290 **e.** 100 **37.** not a function; $\{(a, e), (e, a), (b, e), (e, b), (b, c), (c, b), (c, e), (e, c), (e, d), (d, e)\}$ **39.** 1 mi = 5,280 ft so $5{,}280 = 16t^2$; by trail and error, it would be about 18 seconds to the bottom.

PROBLEM SET 11.3

Ordered pairs in Problems 3–13 *may vary.* **15.** F; it should be $3x - 5y = 3(5) - 5(4) = -5$. Order is important. **16.** F; $y = \frac{2}{3}x - 4$. If $x = 0$, $y = -4$; $x = 3$, $y = -2$; $x = -3$, $y = -6$. **17.** F; the graph is a horizontal line. **18.** T **19.** F; it is a vertical line. **20.** F; it is a horizontal line.

21.

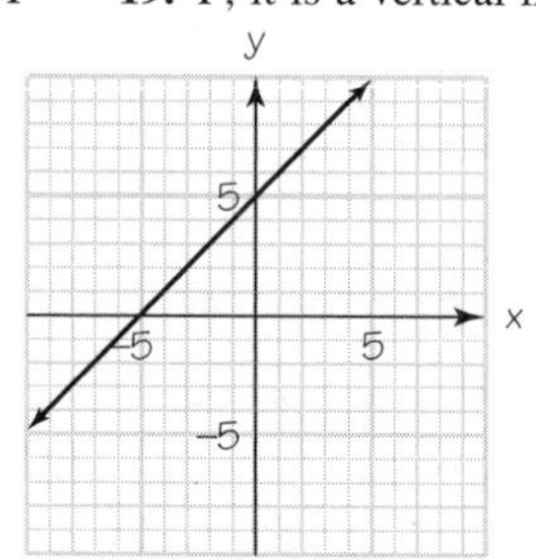

23.

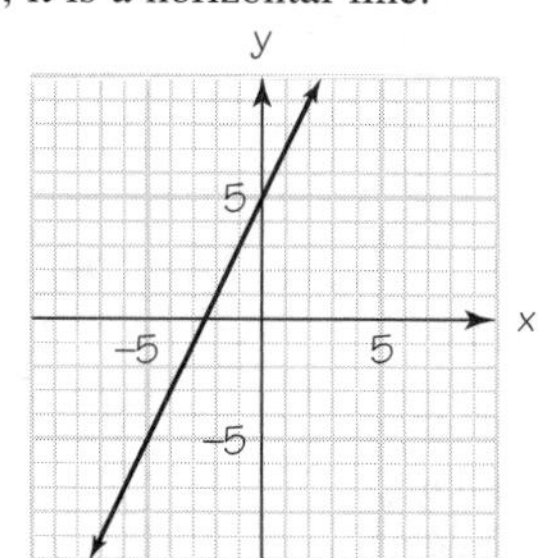

25.

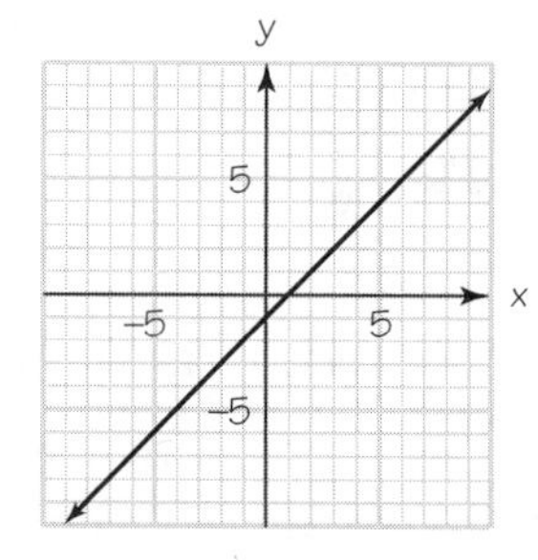

27.

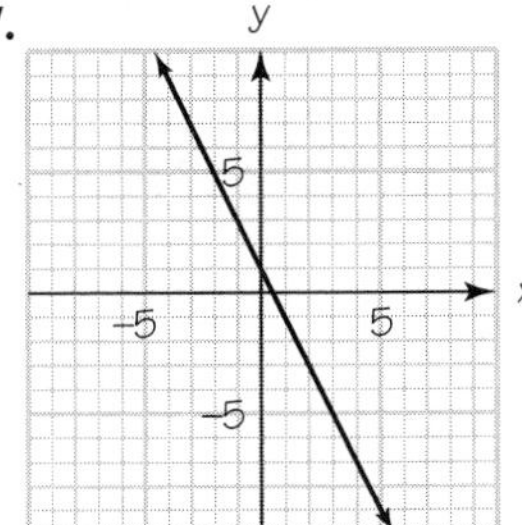

29.

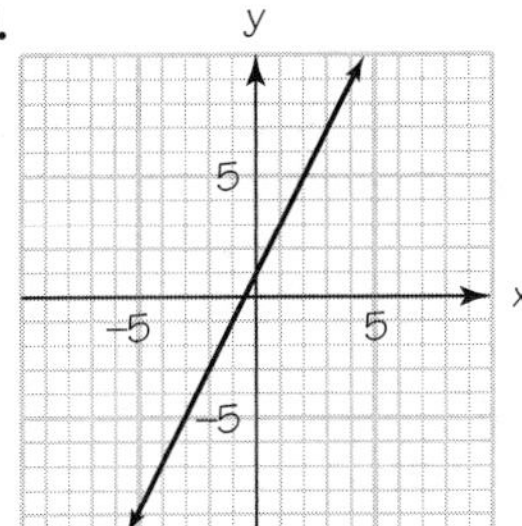

31.

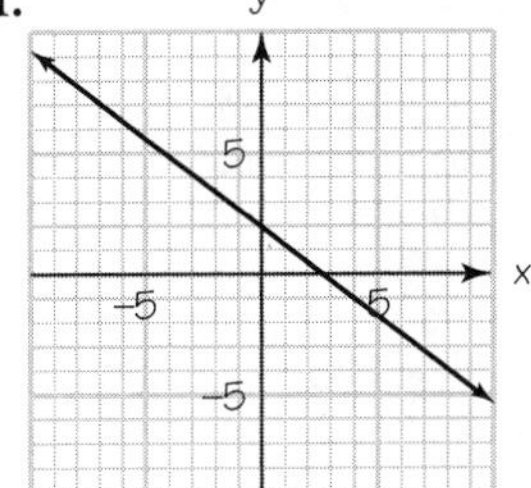

33.

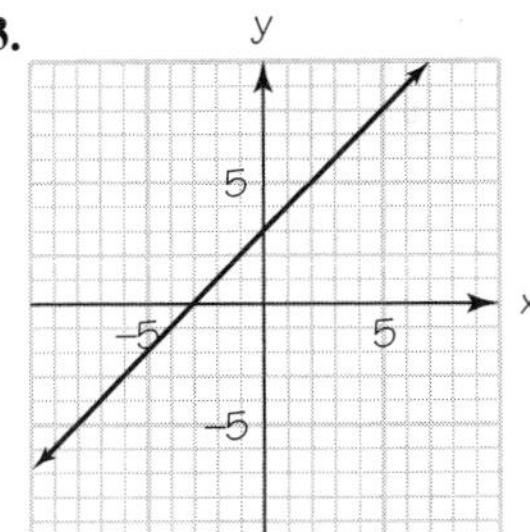

35.

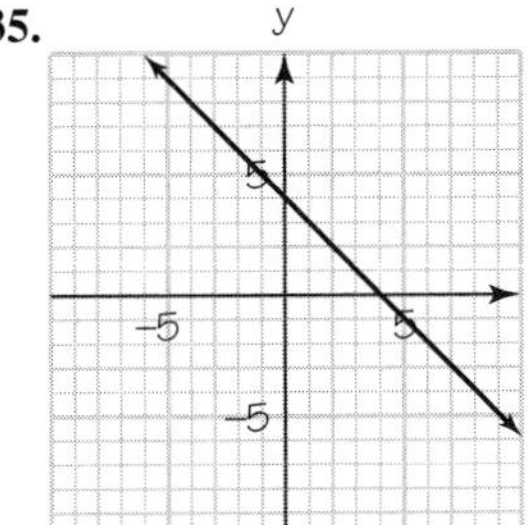

37.

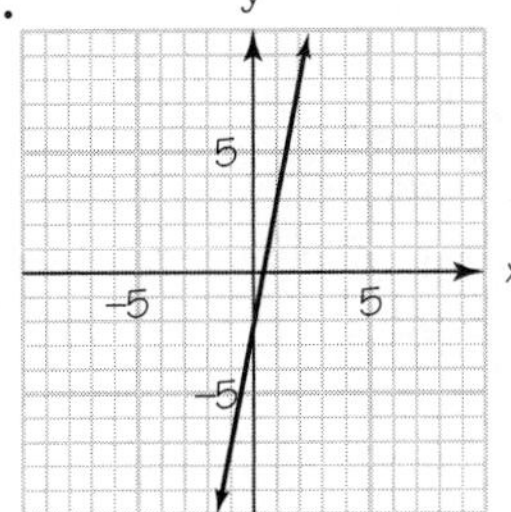

39.

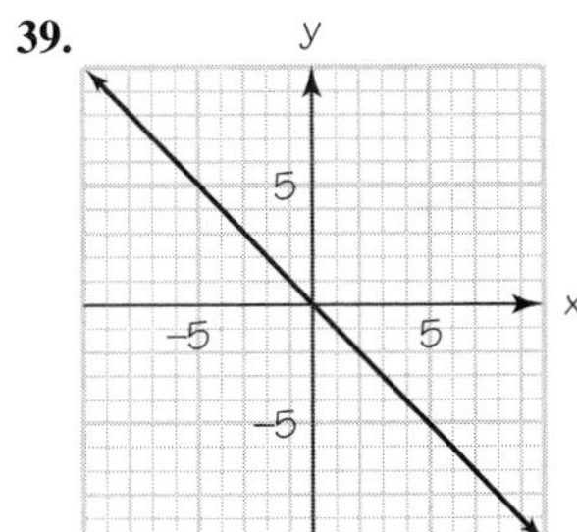

41.

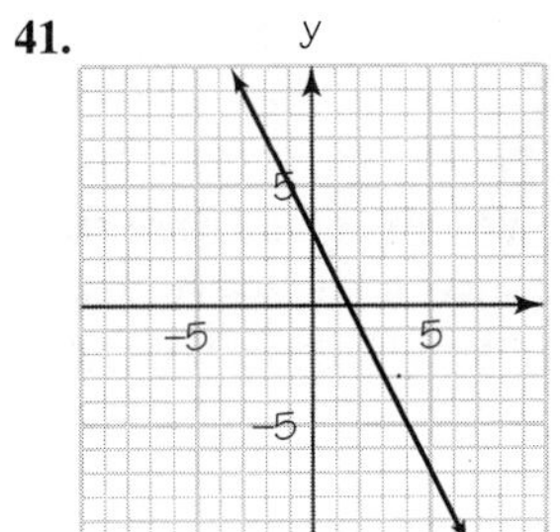

43.

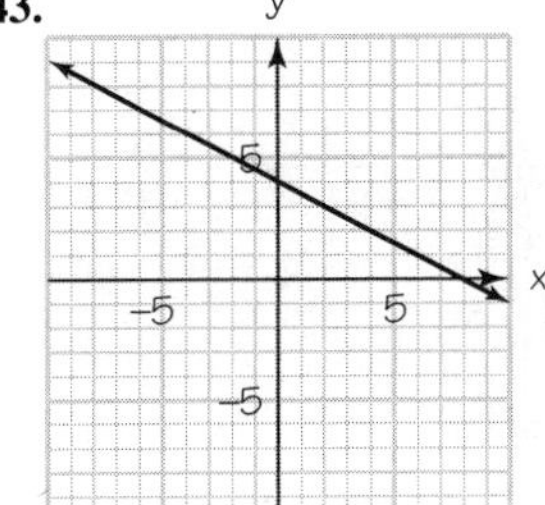

45.

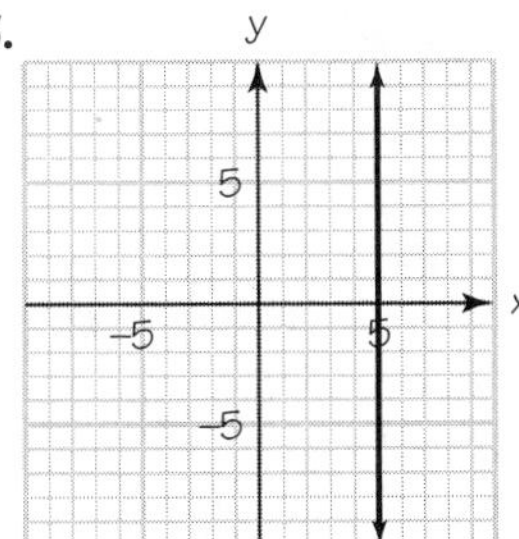

47.

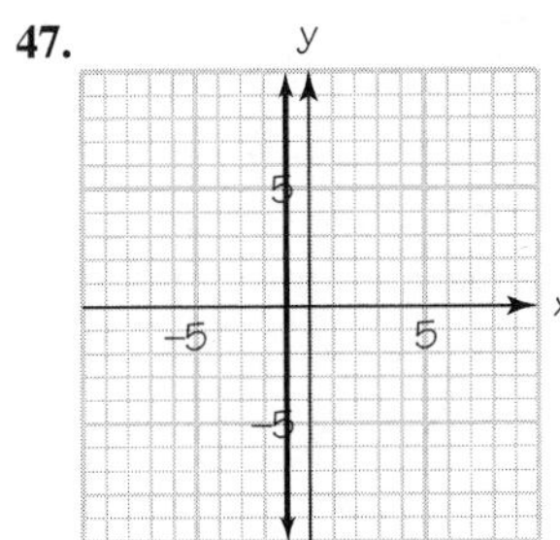

49.

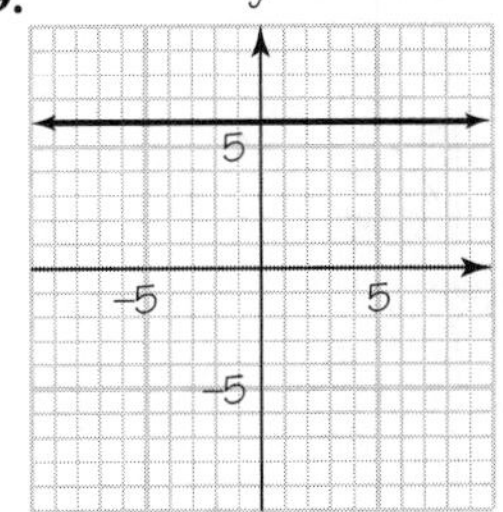

51.

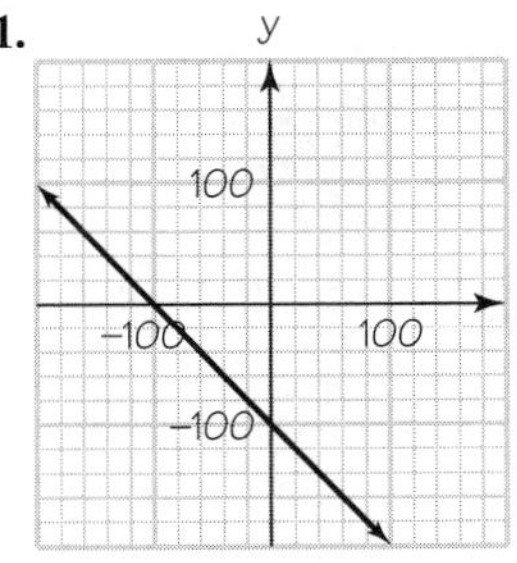

53.

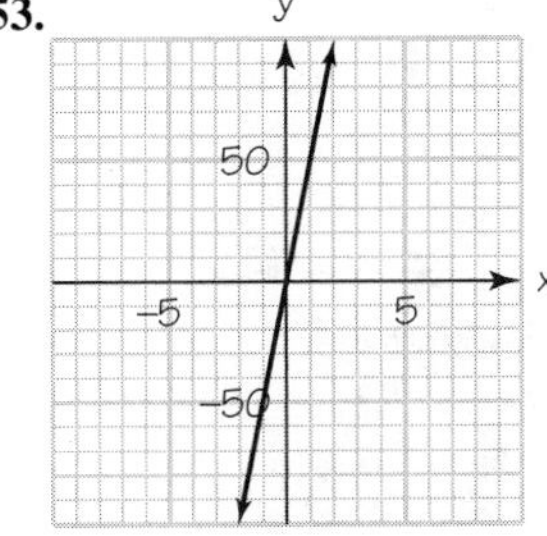

PROBLEM SET 11.4

1. F; $x = 3$ and $y = 2$, so $2(\mathbf{3}) - \mathbf{2} = 4$, not 1. **2.** T **3.** F; $\mathbf{0} = 1 - \mathbf{0}$ is not true. **4.** F; the boundary is $2x + 5y = 2$. **5.** F; can't choose a test point on the boundary line. **6.** F; graph the boundary, which is the *equation* $x - y = 2$. **7.** T **8.** T **9.** F; (0, 0) is not a test point because it lies on the boundary. **10.** T **11.** T **12.** T

13. (3, 2)

15. (2, −3)

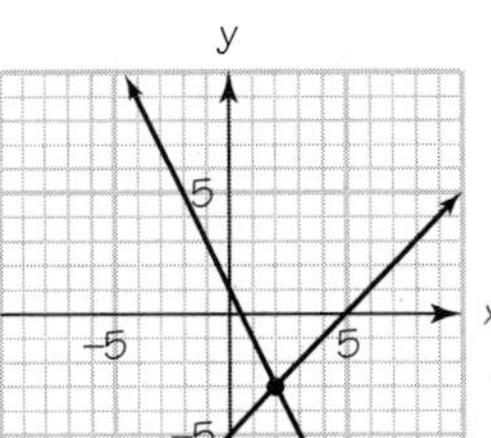

17. (2, −2)

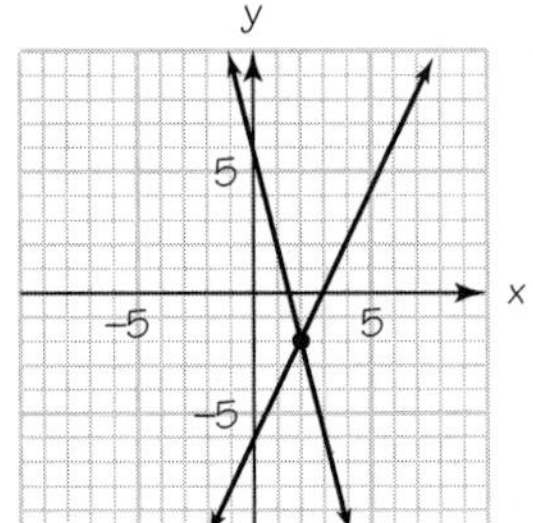

19. (−2, −3)

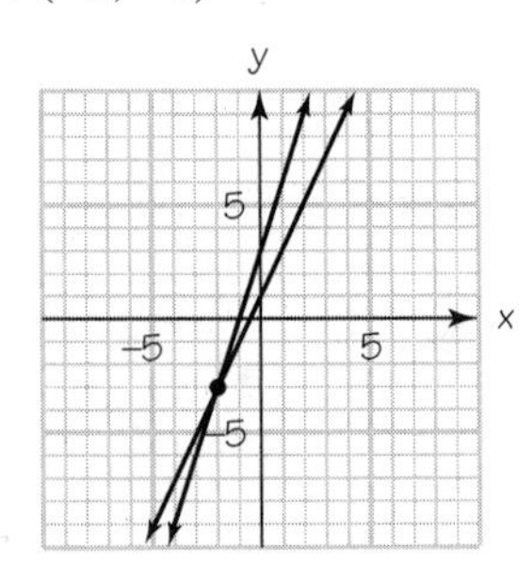

21. (−2, 5)

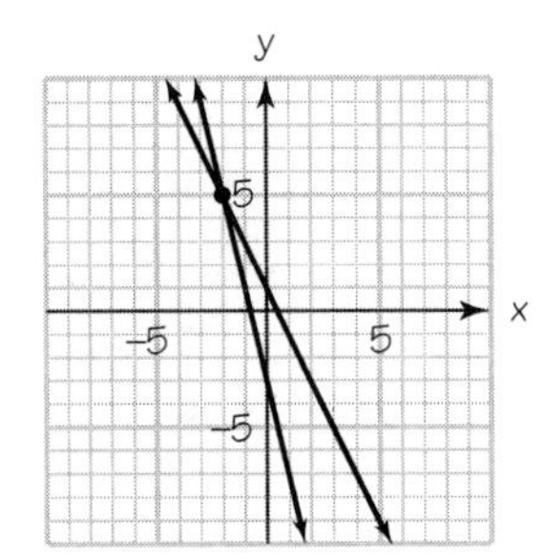

23. (−2, 2)

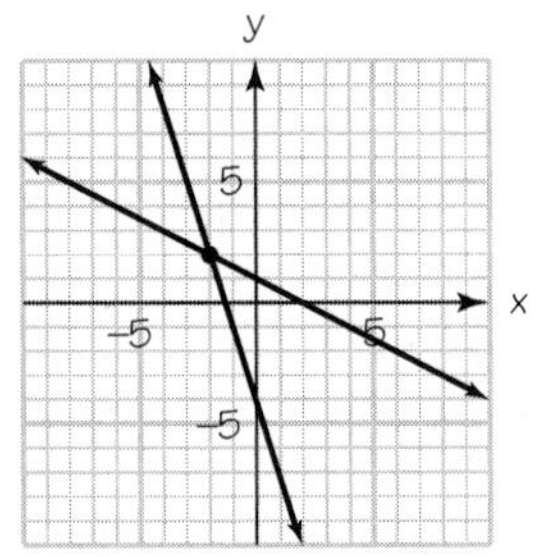

25. (6, 2)

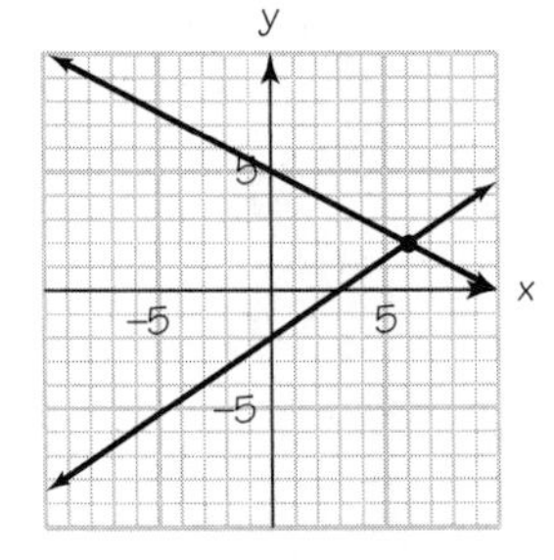

27. (−6, 7)

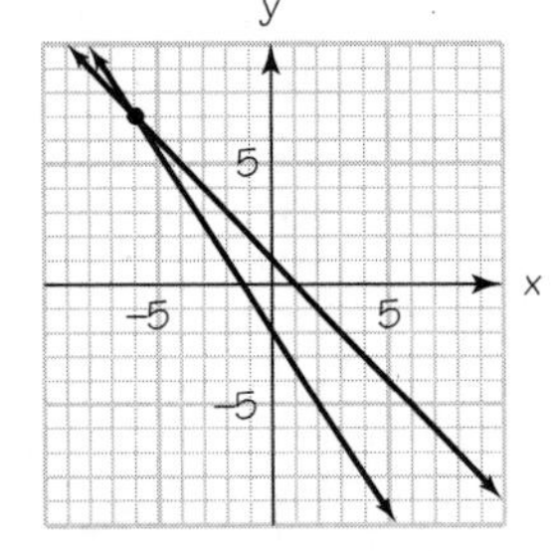

29.

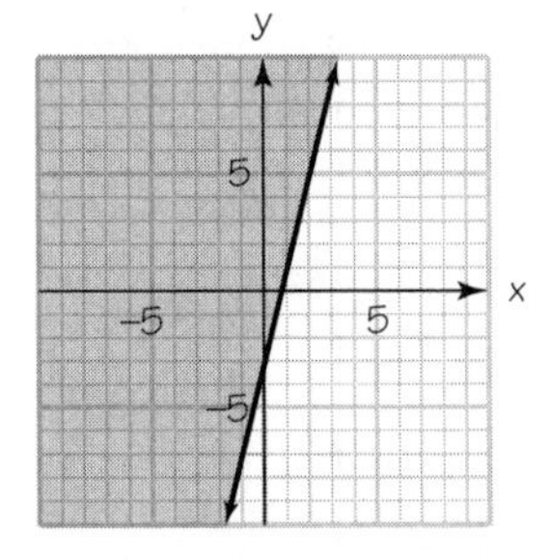

31.

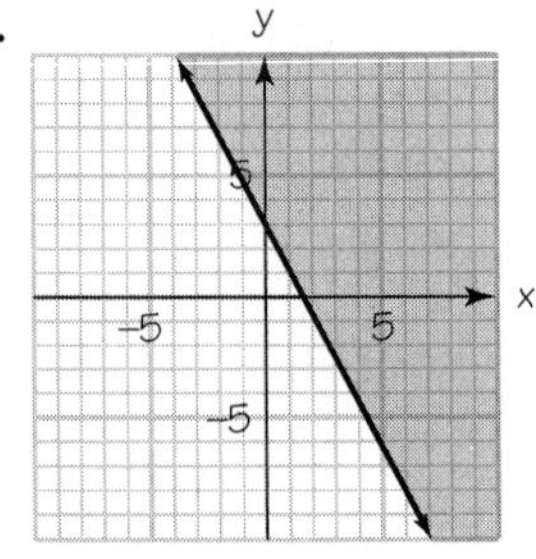

33.

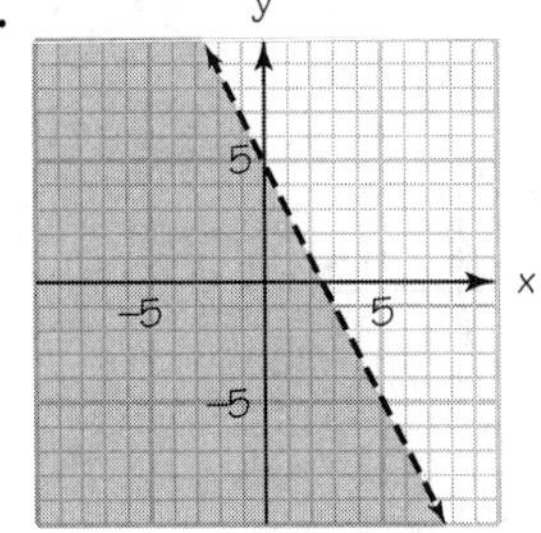

35.

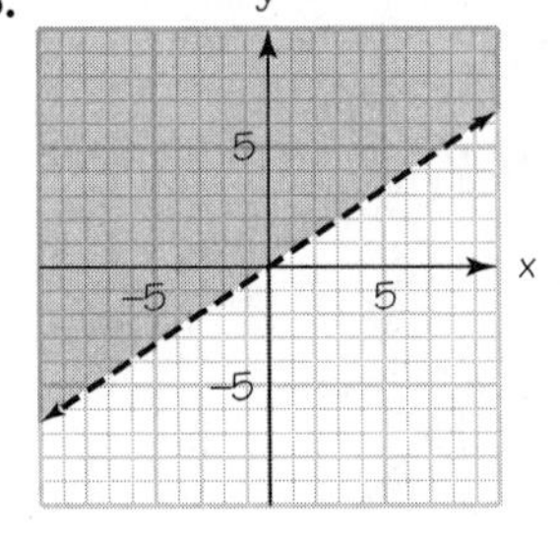

37.

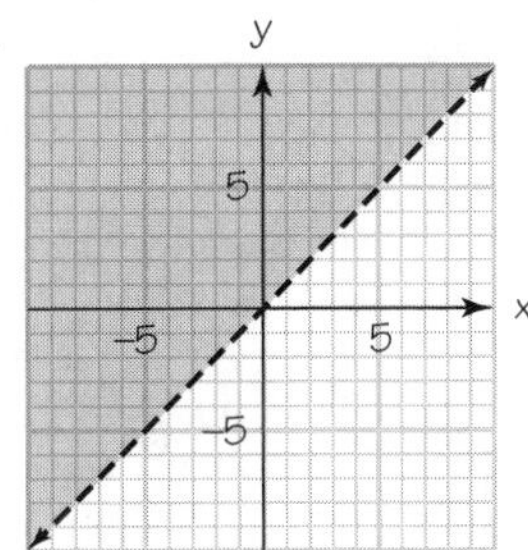

39.

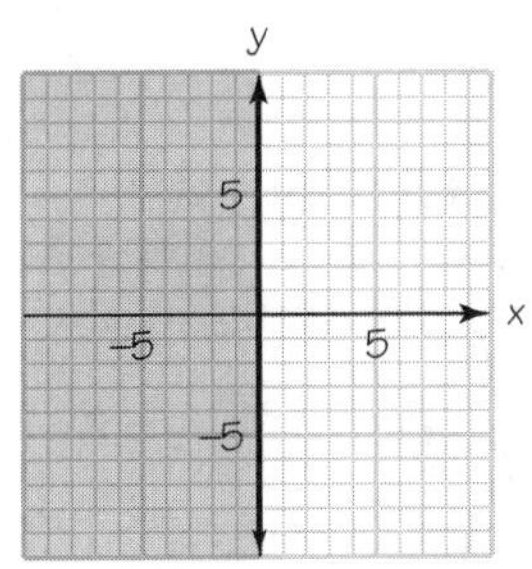

41.

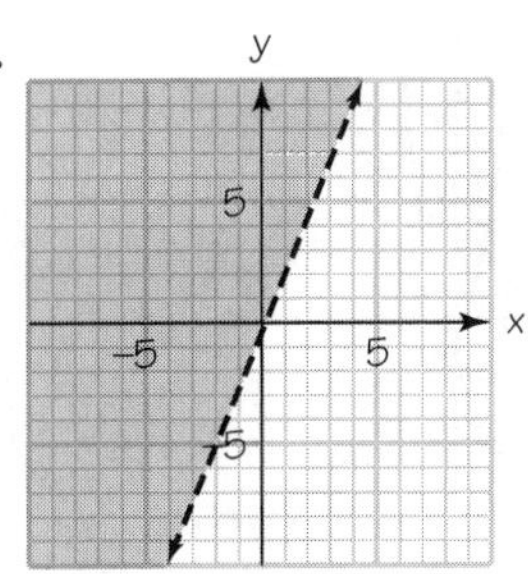

43. Option A: $c = 30 + 0.4m$; Option B: $c = 50$

45. They are the same for 50 miles.

47.

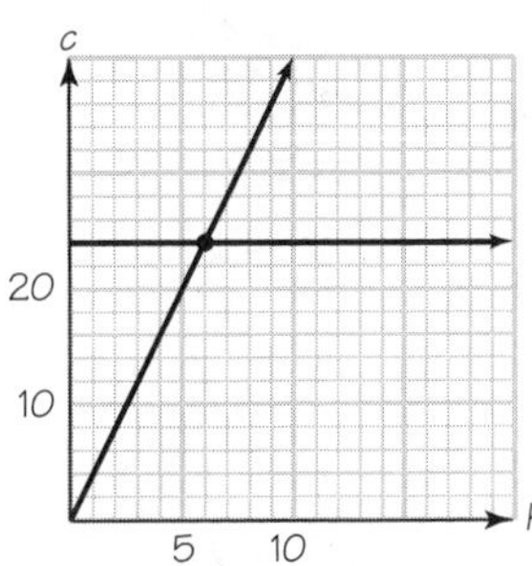

49. $4.00

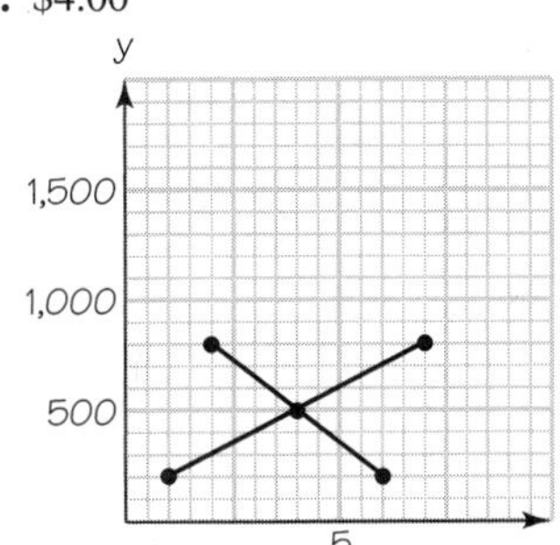

51. $4.00

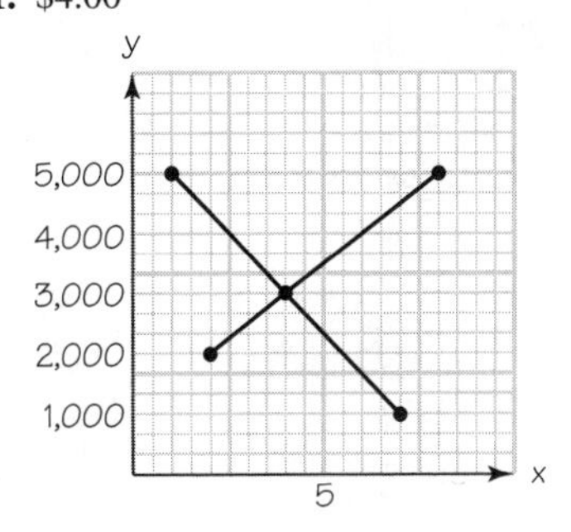

53. $25.00

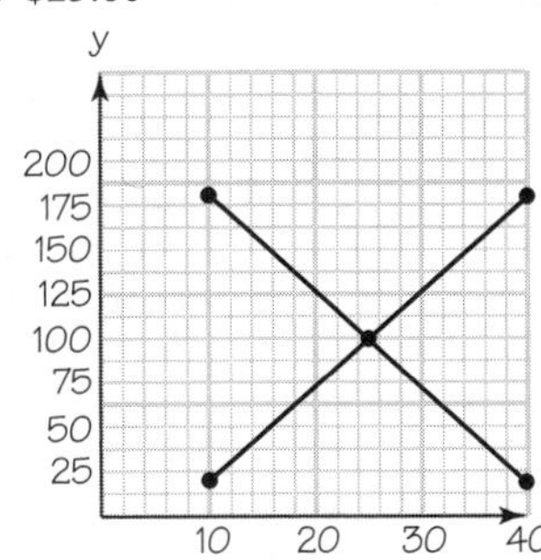

55.

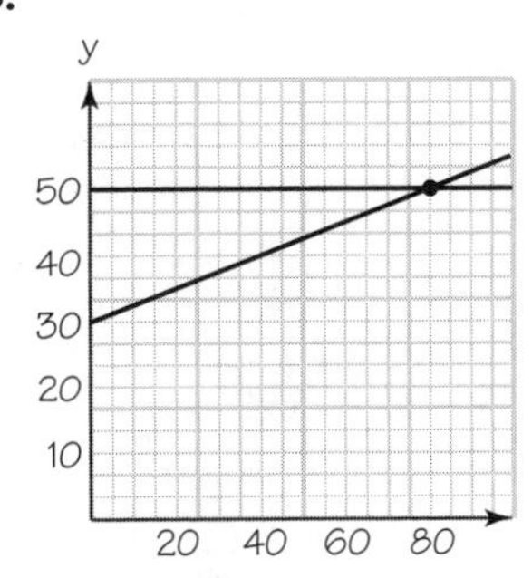

Option A, if she drives less than 80 miles.

PROBLEM SET 11.5

3. F; the points satisfying this equation do not form a line.

4. T

5. F; the correct graph is

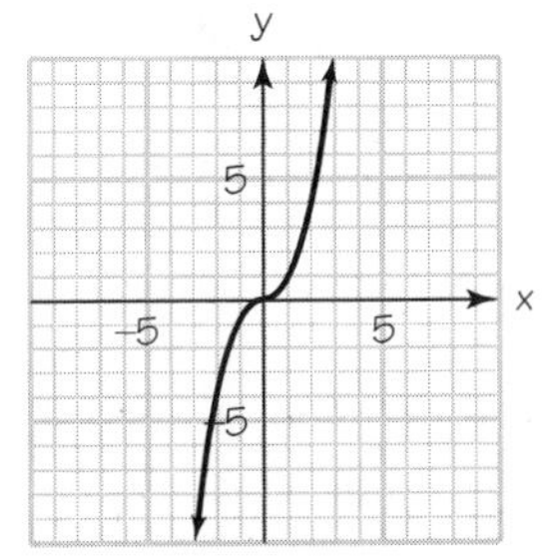

7.

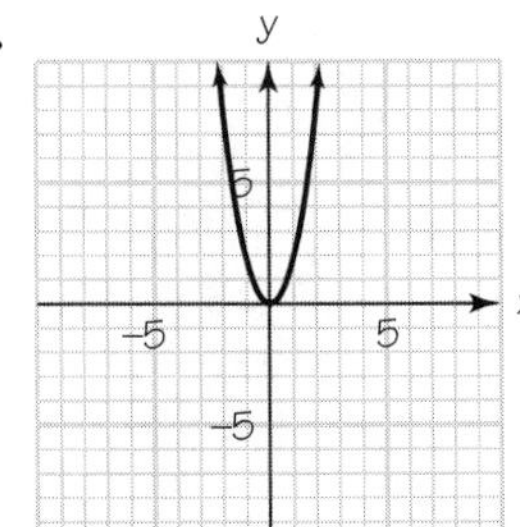

9.

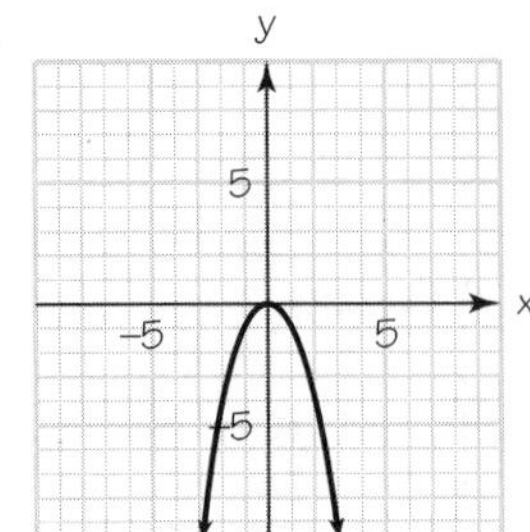

11.

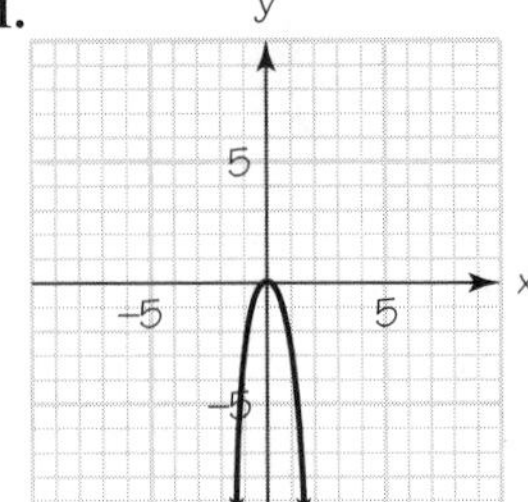

13.

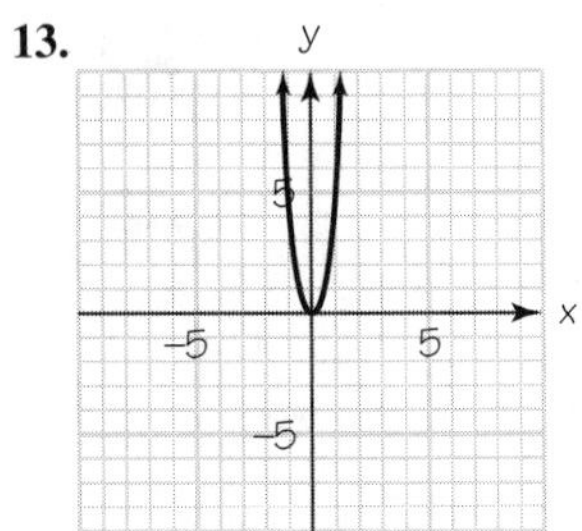

15.

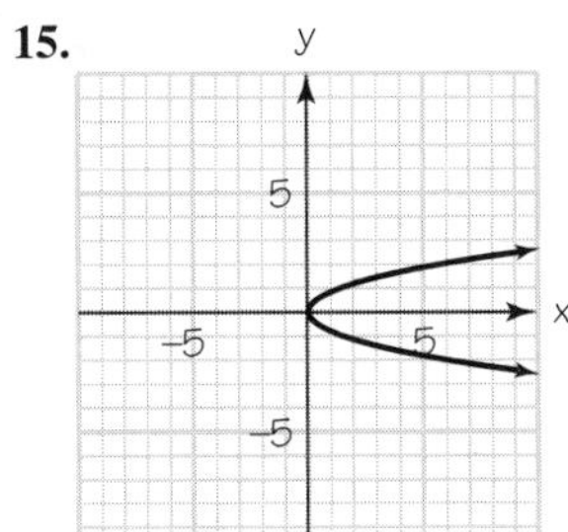

17.

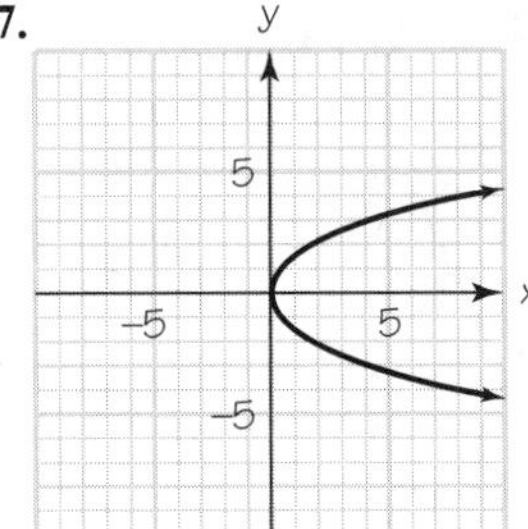

19.

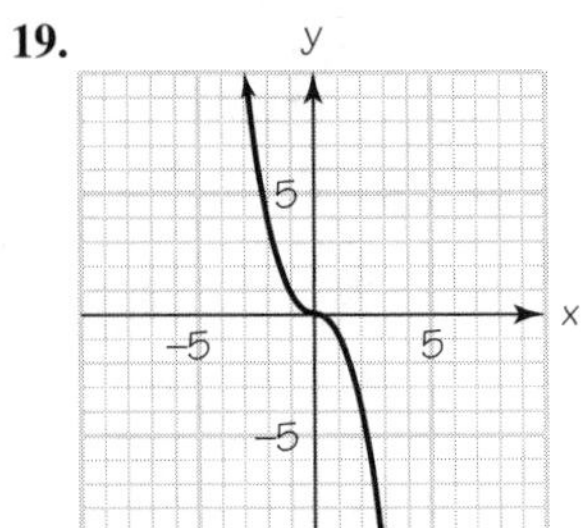

21.

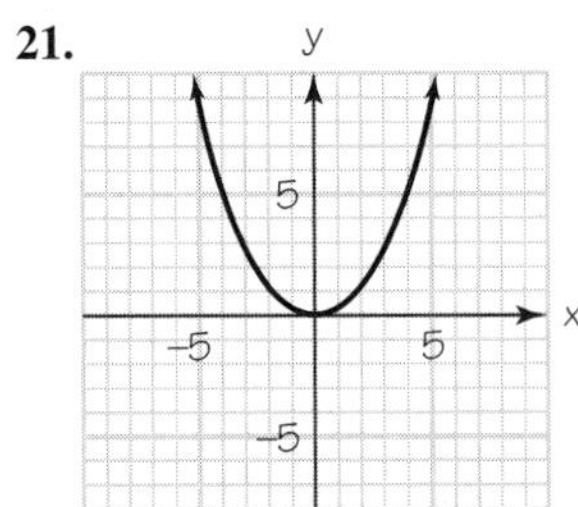

23.

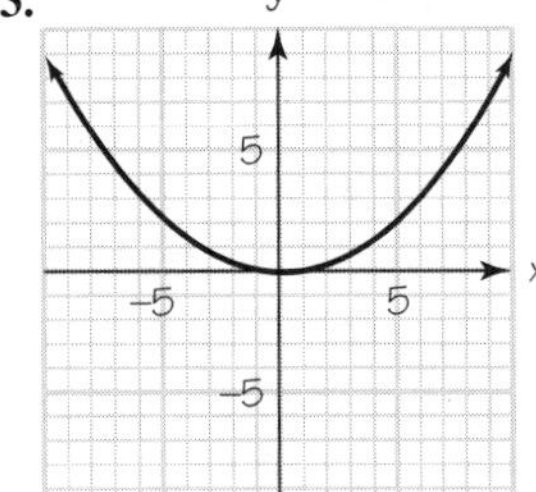

25.

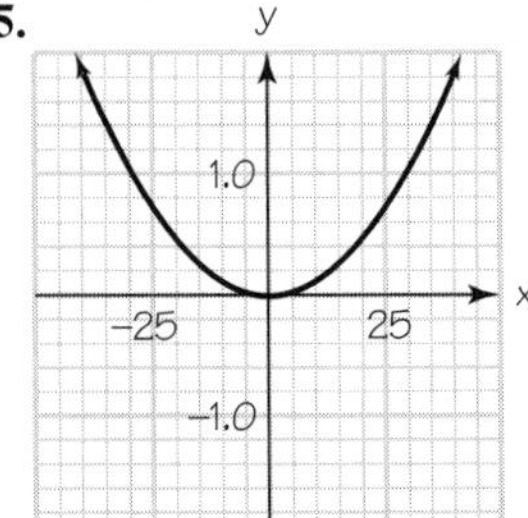

27.

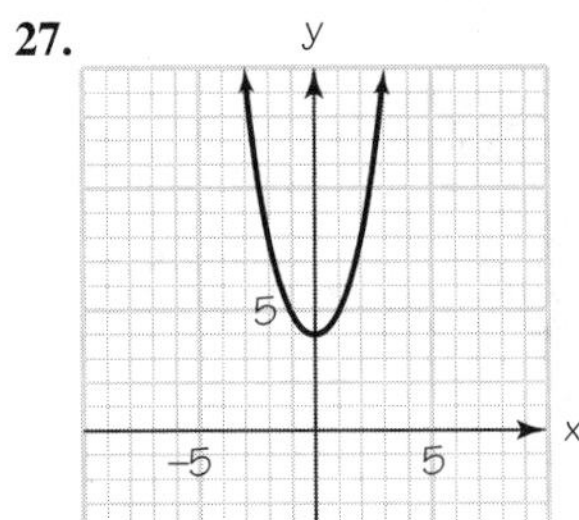

29.

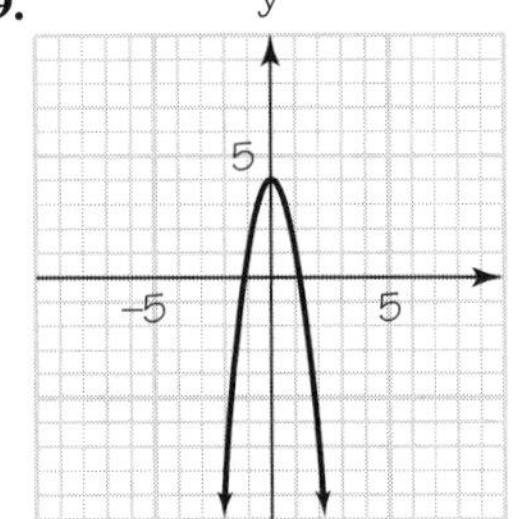

31.

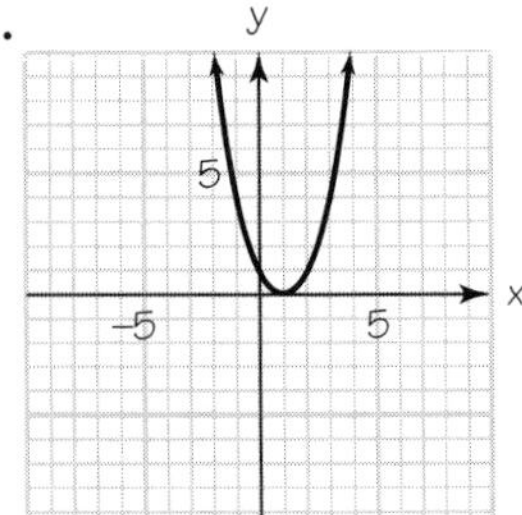

33.

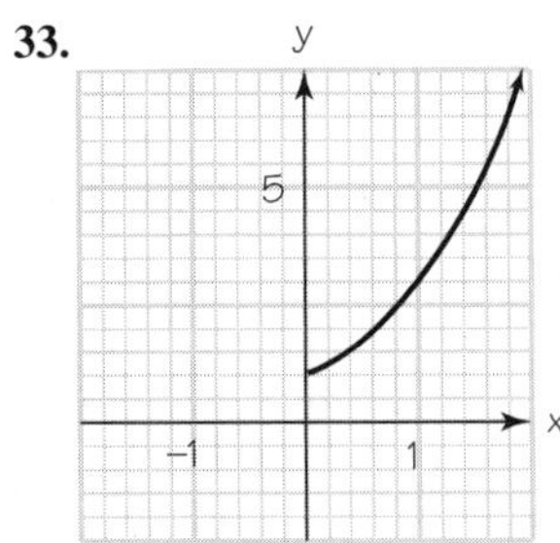

35.

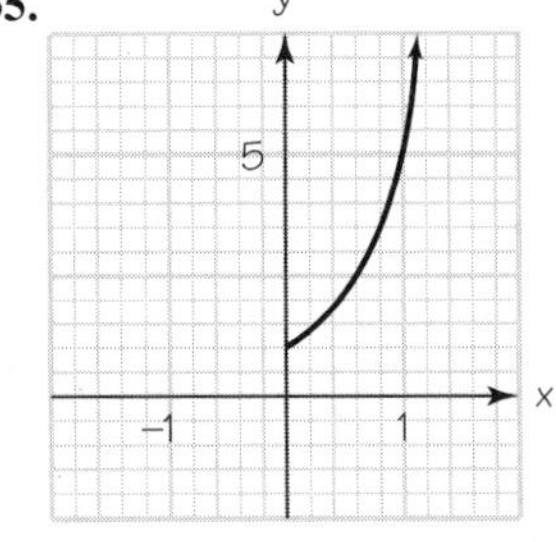

37.

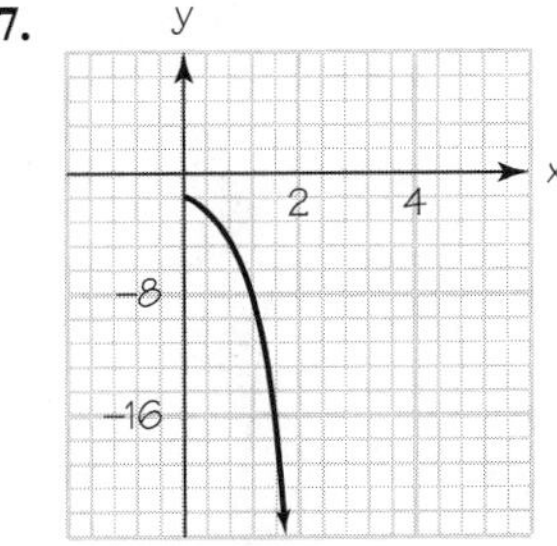

39.

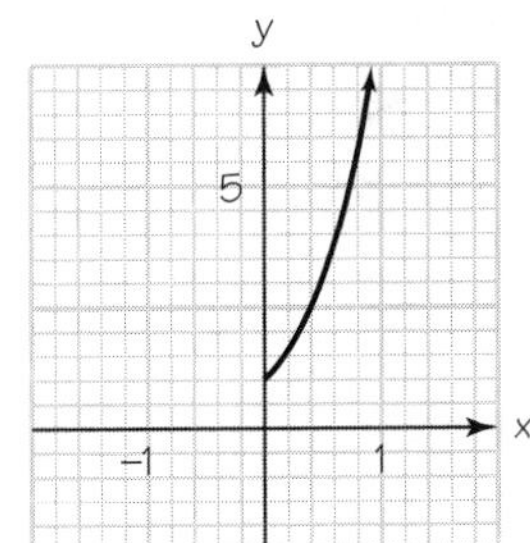

41.

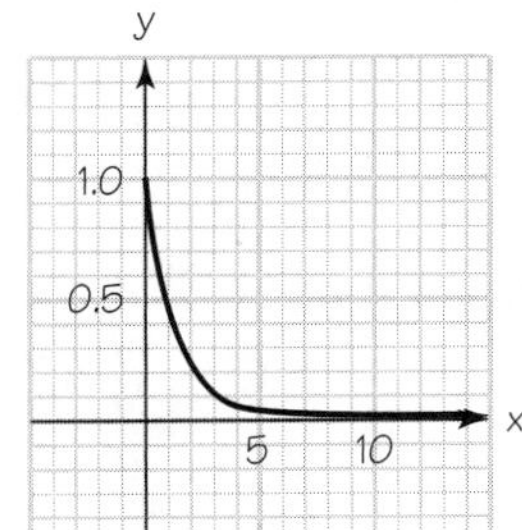

43.

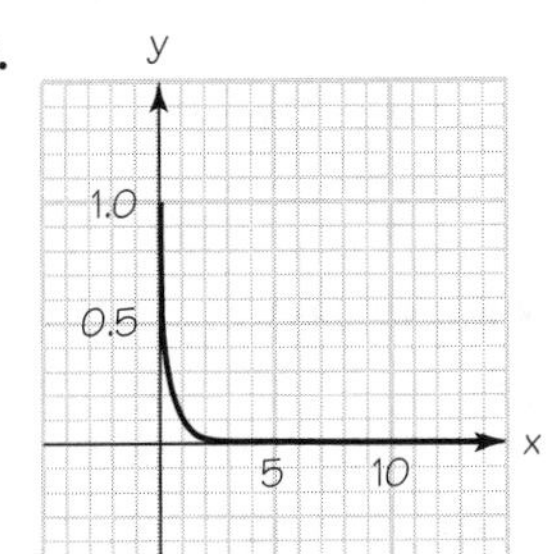

45.

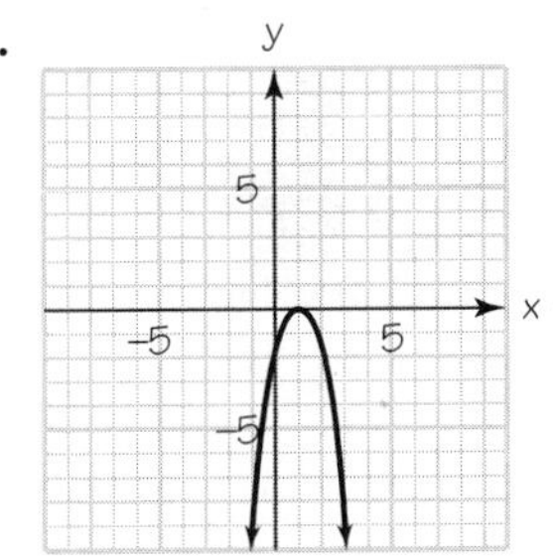

47.

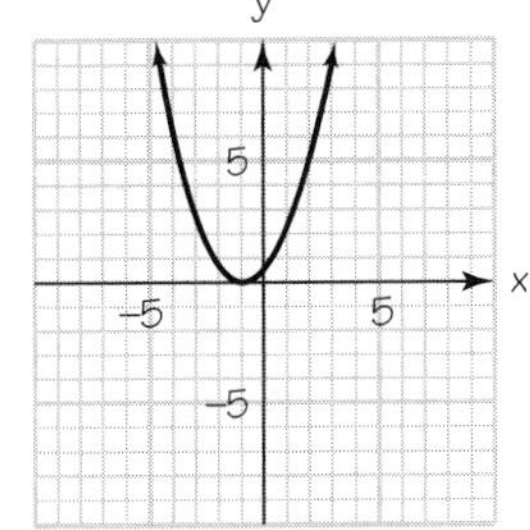

49.

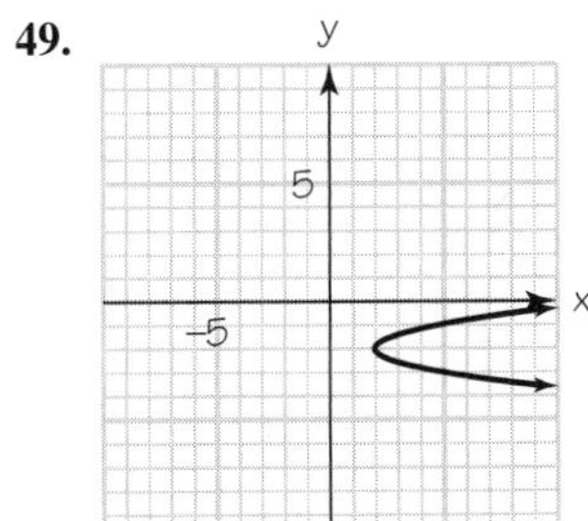

51.

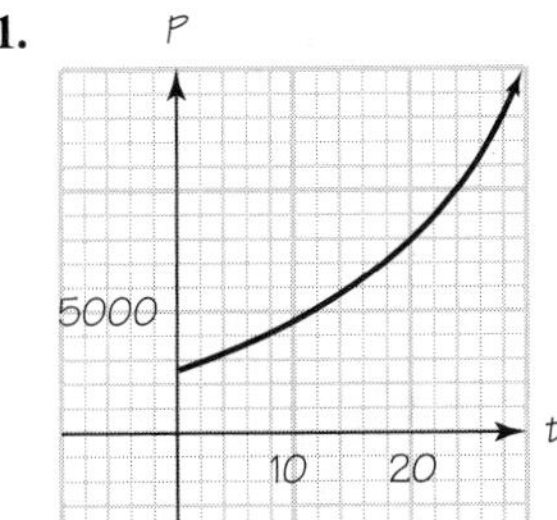

53.

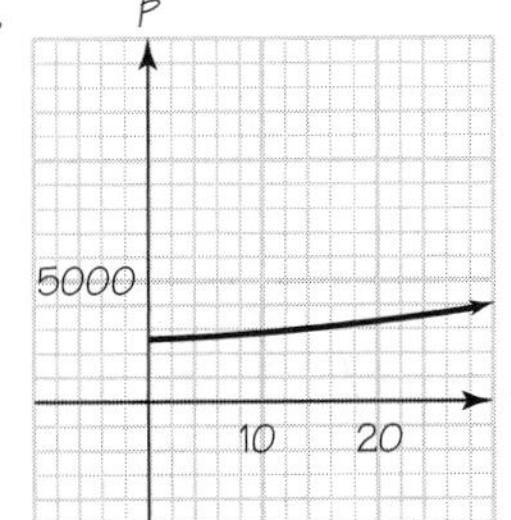

55.

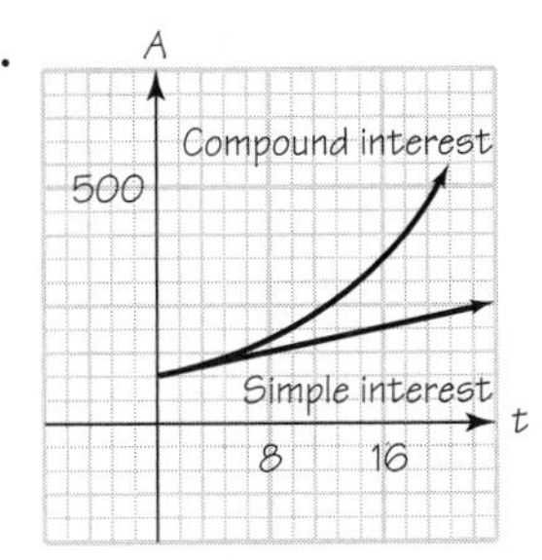

57. a.

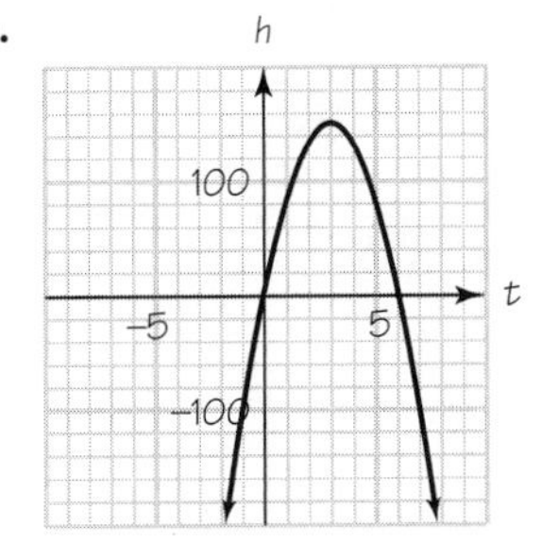

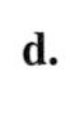

d.

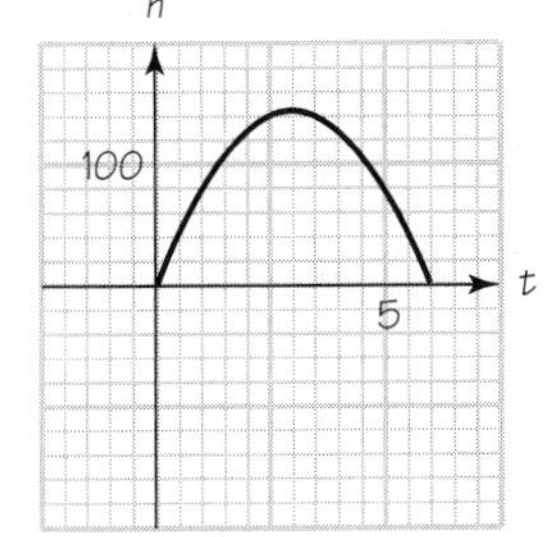

b. downward **c.** no

59.

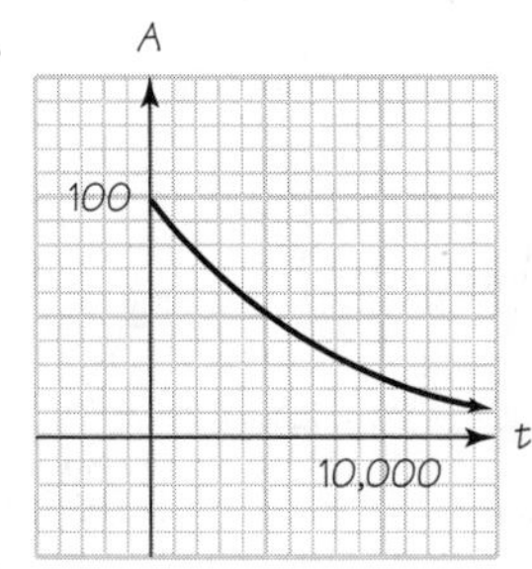

61.

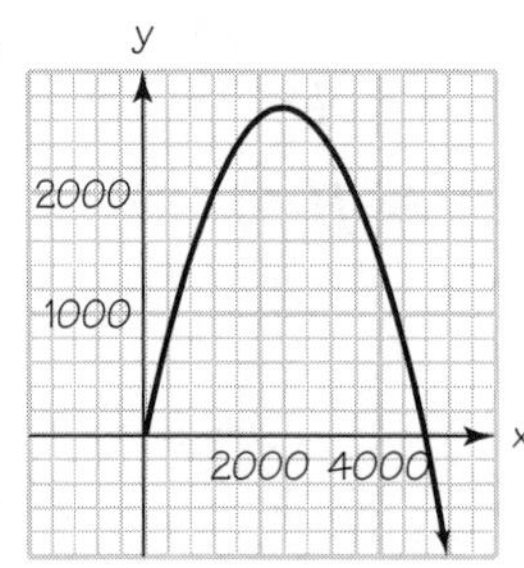

Yes, he will make it (see detail of graph):

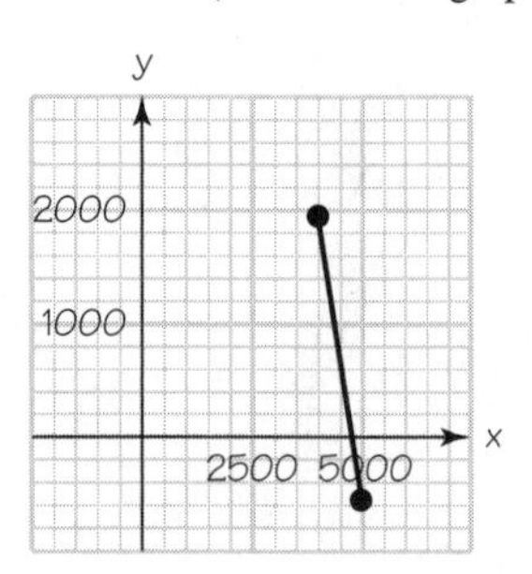

CHAPTER 11 Practice Test

1.

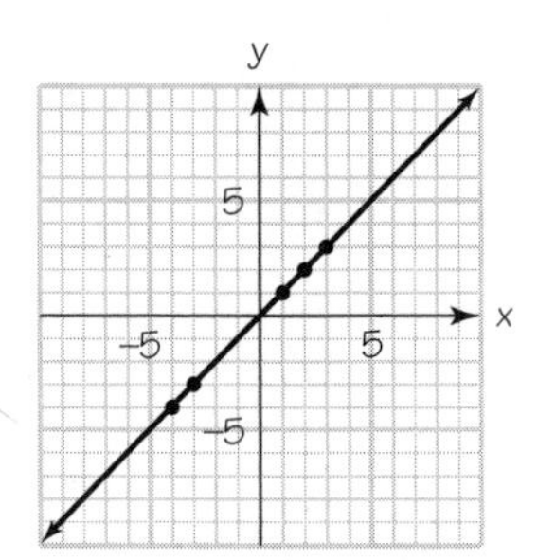

3. **a.** function **b.** not a function **c.** function **d.** not a function

5.

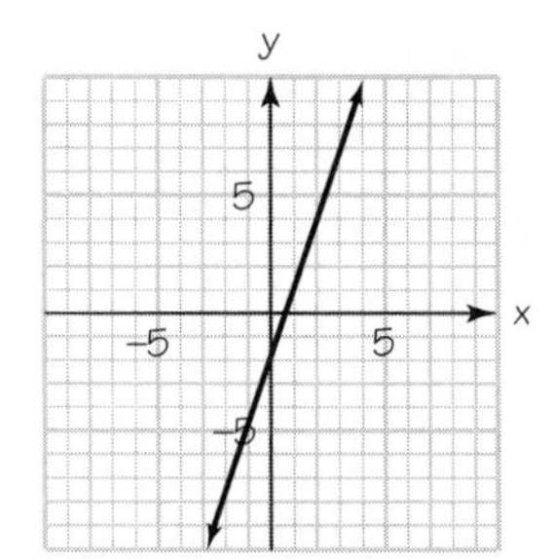

7.

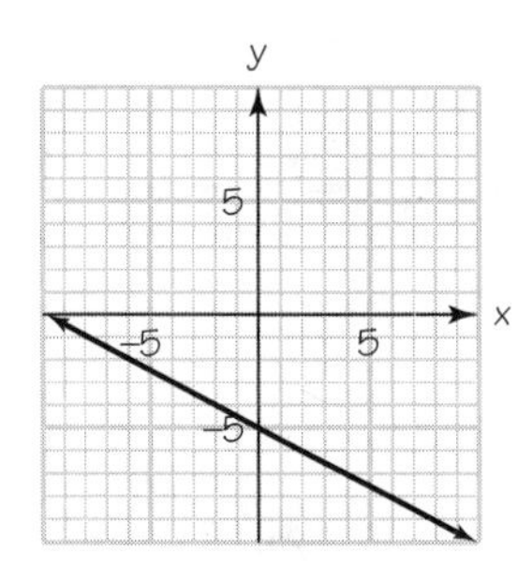

9.

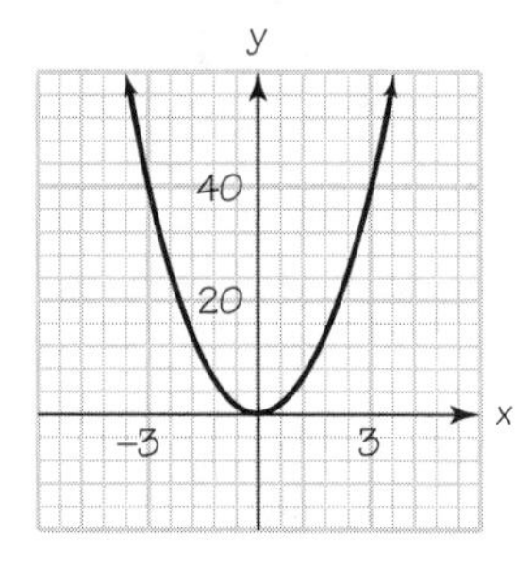

11.

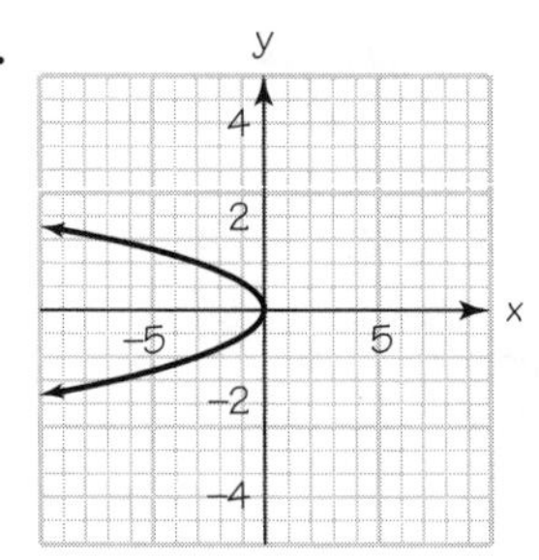

13.

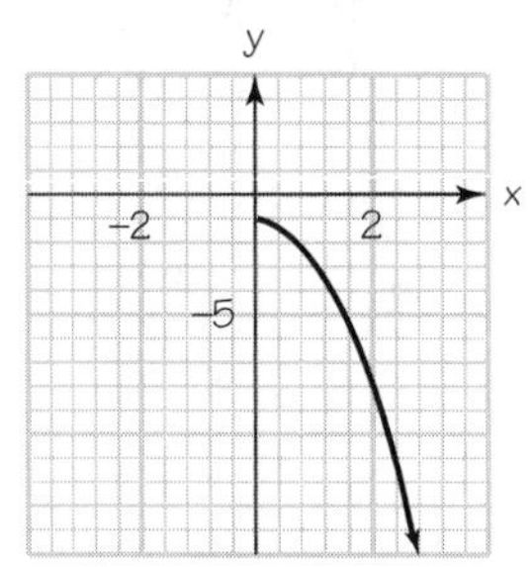

15.

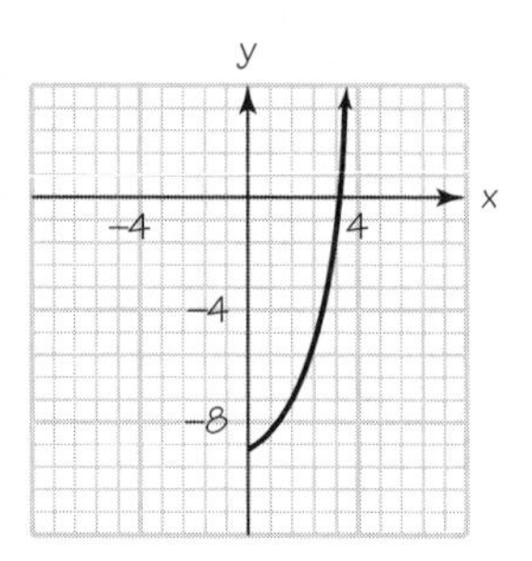

17.

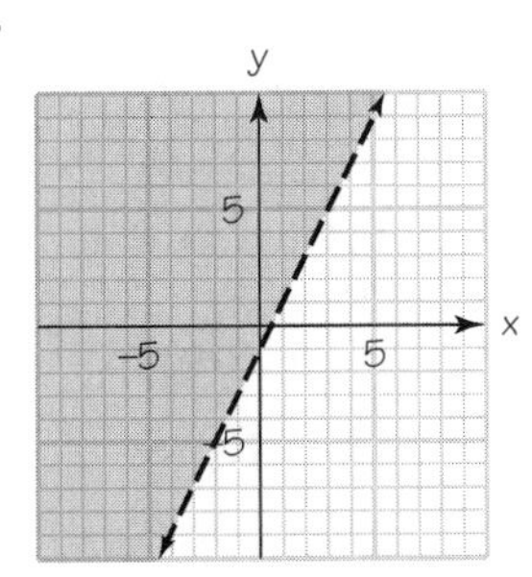

19.

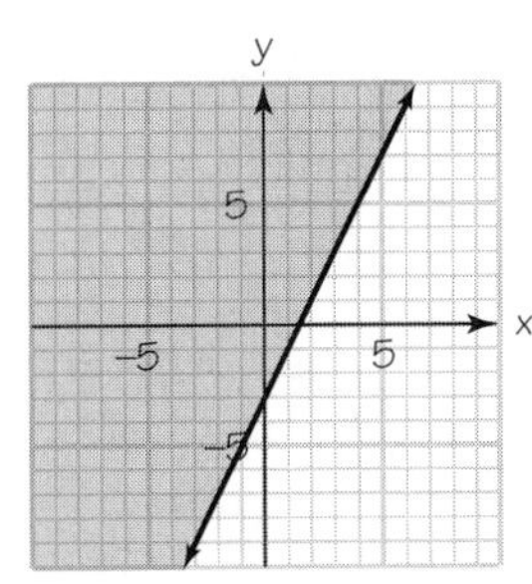

21. (2, 3)

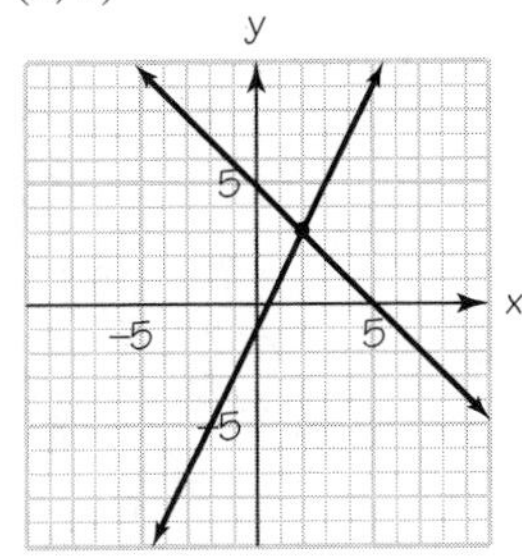

23. (2, −3)

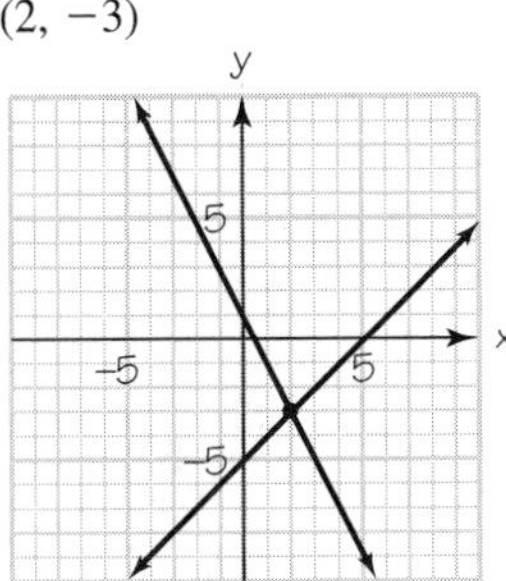

25.

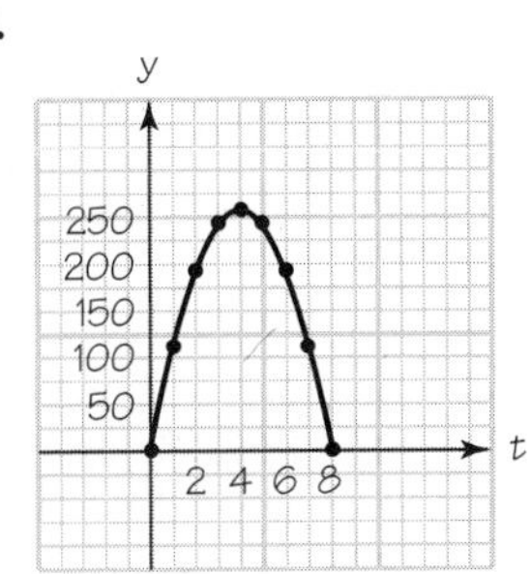

27.

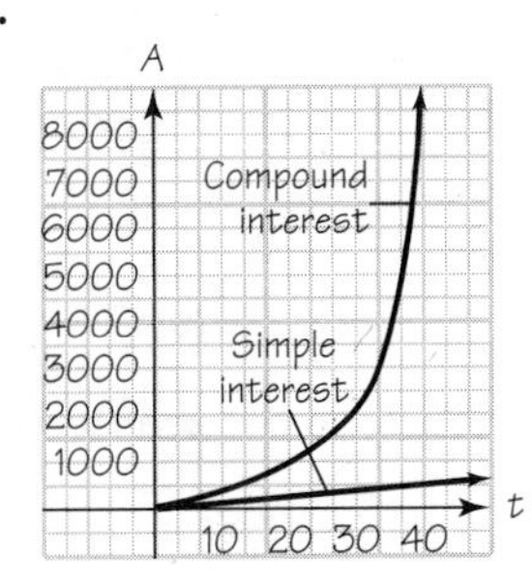

Appendices

PROBLEM SET A

5.

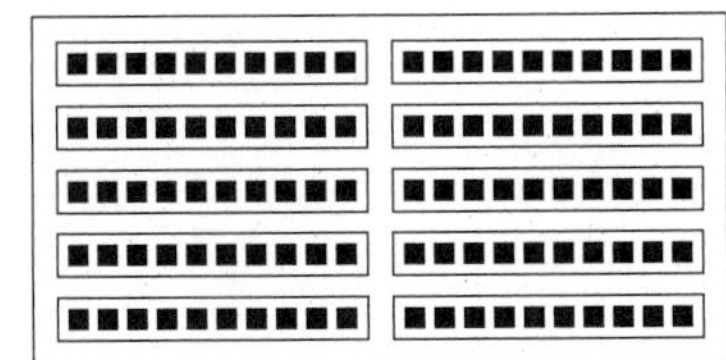

7. 5 units **9.** 5 thousandths **11. a.** 100,000 **b.** 1,000 **13. a.** 0.0001 **b.** 0.001 **15. a.** 5,000 **b.** 500 **17. a.** 0.06 **b.** 0.00009 **19. a.** 10,234 **b.** 65.089 **21. a.** 7,000,000.03 **b.** 6,000,000,000.002 **23.** 3,028.5402 **25. a.** $7 \times 10^2 + 4 \times 10 + 1$ **b.** $7 \times 10^5 + 2 \times 10^4 + 8 \times 10^3 + 4 \times 10^2 + 7$ **27. a.** $4 \times 10^1 + 7 + 2 \times 10^{-3} + 1 \times 10^{-4} + 5 \times 10^{-5}$ **b.** $5 \times 10^2 + 2 \times 10 + 1$ **29. a.** $4 \times 10^2 + 2 \times 10^1 + 8 + 3 \times 10^{-1} + 1 \times 10^{-2}$ **b.** $5 \times 10^3 + 2 \times 10^2 + 4 \times 10^1 + 5 + 5 \times 10^{-1}$ **31. a.** $8 \times 10^2 + 9 \times 10^1 + 3 + 1 \times 10^{-4}$ **b.** $8 + 5 \times 10^{-5}$

PROBLEM SET B

5. a. 13 **b.** 23 **c.** 10 **d.** 15 **e.** 1101 **7. a.** $1 \times 2^5 + 1 \times 2^4 + 1 \times 2^2 + 1 \times 2^1 + 1 \times 2^0 + 1 \times 2^{-1} + 1 \times 2^{-4}$ **b.** $5 \times 6^3 + 4 \times 6^2 + 1 \times 6^1 + 1 \times 6^0 + 1 \times 6^{-1} + 2 \times 6^{-3} + 3 \times 6^{-4}$ **9. a.** $3 \times 4^5 + 2 \times 4^4 + 3 \times 4^3 + 2 \times 4^{-1}$ **b.** $2 \times 5^5 + 3 \times 5^4 + 4 \times 5^3$ **11. a.** 343 **b.** 751 **13. a.** 116 **b.** 53 **15. a.** 807 **b.** 2,250 **17. a.** 351.125 **b.** 2,001 **19.** 10344_{five} **21.** 100000000_{two} **23.** 3122_{five} **25.** $2E79_{twelve}$ **27.** 1000000000_{two} **29.** 1221_{three} **31.** 1132_{eight} **33.** 6 days, 14 hours **35.** 2 lb, 7 oz **37.** 18 quarters, 1 nickel, 4 pennies **39.** 242_{five} **41.** 33 quarters, 1 nickel, 4 pennies **43.** $54 = 46_{twelve}$; 4 years and 6 months

TEXT CREDITS

This page constitutes an extension of the copyright page. We have made every effort to trace the ownership of all copyrighted material and to secure permission from copyright holders. In the event of any question arising as to the use of any material, we will be pleased to make the necessary corrections in future printings. Thanks are due to the following authors, publishers, and agents for permission to use the material indicated.

Page 3: *Peanuts* cartoon © 1979 United Feature Syndicate, Inc. Reprinted by permission.
Page 9: Cartoon reprinted by permission of Jerry Marcus.
Page 24: Cartoon courtesy of Patrick J. Boyle.
Page 30: *Graffiti* cartoon © 1975 Newspaper Enterprise Associates, Inc. Reprinted by permission.
Page 34: *B.C.* cartoon reprinted by permission of Johnny Hart and Creators Syndicate, Inc.
Page 43: *Peanuts* cartoon © 1963 United Feature Syndicate, Inc. Reprinted by permission.
Page 44: *Peanuts* cartoon © 1966 United Feature Syndicate, Inc. Reprinted by permission.
Page 50: Cartoon courtesy of Patrick J. Boyle.
Page 59: Cartoon courtesy of Patrick J. Boyle.
Page 79: Symbolic messages in Problems 79 and 80 by Alanna J. Mozar. Reprinted with permission from the June 1978 Reader's Digest. Copyright © 1978 by Reader's Digest Assn., Inc.
Page 84: *Funky Winkerbean* cartoon © 1973 North America Syndicate, Inc. Reprinted by permission.
Page 127: Reproduction of a Moslem manuscript.
Page 146: Quotation from *Science*, Vol. 220, No. 4596, April 29, 1983, p. 29.
Page 147: "Tips for Teens," copyright © 1979 by Newsweek, Inc. All rights reserved. Reprinted by permission.
Page 151: *Quincy* cartoon © 1970 King Features Syndicate. Reprinted by permission.
Page 152: Cartoon by Johns. Copyright © 1975 by *Industrial Research* magazine. Reprinted by permission.
Page 154: Advertisement reprinted by permission of Procter & Gamble.
Page 173: Reproduction of Euclid's *Elements* from *The VNR Concise Encyclopedia of Mathematics,* Van Nostrand Reinhold, 1975, plate 12.
Page 178: History of the percent system from *31st Yearbook of the National Council of Teachers of Mathematics,* 1969, pp. 146 – 147.
Page 197: Cartoon "The Meaning of Life," © 1971. Reprinted by permission of *Saturday Review* and Henry Martin.
Page 203: Cartoon reprinted by permission of Tribune Media Services. Copyright © Tribune Media Services, Inc. All rights reserved.
Page 239: Quotation from *A Survey of Mathematics* by Vivian Shaw Groza, Holt Rinehart & Winston, Inc., pp. 206 – 207.
Page 257: Advertisement reprinted by permission of Round Table Franchise Corporation.
Page 276: Story from "Fourth Down, 50 Centimeters to Go," by James L. Collier. Copyright © 1978 Kansas City Star Company.
Page 284: Cartoon from *The Metrics Are Coming! The Metrics Are Coming!,* by R. Cardwell. Copyright © 1975 Dorrance & Company.
Page 290: First news story, *New York Times,* November 21, 1994; second news story, *New York Times,* December 22, 1994.
Page 295: Excerpt from *Curriculum and Evaluation Standards for School Mathematics,* p. 7. Copyright © 1989 National Council of Teachers of Mathematics.
Page 296: Quotation from *A Scrap-book of Elementary Mathematics,* by William F. White, Chicago, 1908, p. 215.
Page 386: *Born Loser* cartoon © 1975 reprinted by permission of Newspaper Enterprise Association, Inc.
Page 392: *Peanuts* cartoon © 1961 United Feature Syndicate, Inc. Reprinted by permission.
Page 394: Lyrics for "By the Time I Get to Phoenix" by Jim Webb. © 1967 Charles Koppleman Music/Martin Bandier Music.
Page 395: Lyrics for "Ode to Billy Joe" by Bobbie Gentry. © 1967 Northridge Music Company. Administered by All Nations, Inc.
Page 414: *B.C.* cartoon reprinted by permission of Johnny Hart and Creators Syndicate, Inc.
Page 416: From the *Dear Abby* column. Copyright © 1974 Universal Press Syndicate. Reprinted with permission of the Universal Press Syndicate. All rights reserved.
Page 427: "Le Trente-et-un, ou la maison de pret sur nantissement," by Darcis. Reprinted by permission of Bibliotheque Nationale, Paris.
Page 438: Sweepstakes documents copyright © The Reader's Digest Association, Inc. Reprinted by permission.
Page 459: *Peanuts* cartoon © 1961 United Feature Syndicate, Inc. Reprinted by permission.
Page 478: Cartoon by David Pascal.
Page 489: Anacin advertisement courtesy of Whitehall Laboratories, Inc.
Page 501: *B.C.* cartoon reprinted by permission of Johnny Hart and Creators Syndicate, Inc.
Page 512: Avis trademark, Dollar Rent-A-Car System Trademark, and Hertz Registered Service Mark reprinted by permission.

PHOTO CREDITS

Page 7: James L. Amos/Corbis.
Page 29: Copyright 1923 by E. S. Curtis.
Page 54: Courtesy of Shirley Frye.
Page 55: Richard Hagberg.
Page 79: Deborah Davis/PhotoEdit.
Page 87: Bettmann.
Page 88: Phyllis Picardi/Stock Boston.
Page 99: Courtesy of Herman Cain.
Pages 134 and 136: Archive Photos (Elvis Presley); Archive Photos (Whoopi Goldberg); Archive Photos (Sylvester Stallone); David Maung/Impact Visuals (Bill Gates); Bettmann (Marilyn Monroe).
Page 154: Bettmann.
Page 172: Time Life.
Page 178: Terry Powell/The Photographer's Window.
Page 202: *Chambered nautilus:* Terry Powell/The Photographer's Window; *Winged Lion:* (Accession Number P60P1146) by William Abbenseth, Asian Art Museum of San Francisco, The Avery Brundage Collection; *DaVinci's Dodecahedron:* courtesy of The Metropolitan Museum of Art, Whittesley Fund; *Kaiser vase:* courtesy of Sterling Publishing Company, reprinted by permission; *Empire State Building:* Visual Dictionary of Buildings, page 52, published with permission by Dorling Kindersley, Ltd.
Page 214: VNR Concise Encyclopedia of Mathematics, Van Nostrand Reinhold, 1975, plate 12.
Page 222: Terry Powell/The Photographer's Window.
Page 253: James P. Blair/Words & Pictures/PNI.
Page 258: PNI.
Page 297: GHP Studios.
Page 305: Steve Weber/Stock Boston/PNI.
Page 313: Sears Roebuck & Co.
Page 315: Kim Sayer/Bettmann.
Page 317: Terry Powell/The Photographer's Window.
Page 322: Eric Berndt/Photo Network/PNI.
Page 328: GHP Studios.
Page 348: CocoMcCoy/Rainbow/PNI.
Page 358, top: Charles Eames, courtesy International Business Machines Corporation.
Page 358, bottom: Bruce Forster/Allstock/PNI.
Page 366: Werner Wolff/Black Star/PNI.
Page 380: GHP Studios.
Page 382: Corbis.
Page 385: G. Olson/Woodfin Camp & Associates/PNI.
Page 403: Michael Grecco/Stock Boston/PNI.
Page 422, top: Bettmann.
Page 422, bottom: Peter Turnley/Bettmann.
Page 444: GHP Studios.
Page 470: Bettmann.
Page 477: Bettmann.
Page 478: GHP Studios.
Page 482: Mike Yamashita/Woodfin Camp & Associates/PNI.
Page 483: Fred Lyon/Photo Researchers.
Page 484: Courtesy UPI/Bettmann Newsphoto.
Page 501: N. R. Rowan/Stock Boston.
Page 522: Bettmann Newsphoto.
Page 545: National Aeronautics & Space Administration.

INDEX

A

B

C

Q

R

S